AUTONOMIC NERVOUS SYSTEM

HANDBOOK OF CLINICAL NEUROLOGY

Series Editors

MICHAEL J. AMINOFF, FRANÇOIS BOLLER, AND DICK F. SWAAB

VOLUME 117

ELSEVIER

EDINBURGH LONDON NEW YORK OXFORD PHILADELPHIA
ST LOUIS SYDNEY TORONTO 2013

AUTONOMIC NERVOUS SYSTEM

Series Editors

MICHAEL J. AMINOFF, FRANÇOIS BOLLER, AND DICK F. SWAAB

Volume Editors

RUUD M. BUIJS AND DICK F. SWAAB

VOLUME 117

3rd Series

ELSEVIER

EDINBURGH LONDON NEW YORK OXFORD PHILADELPHIA
ST LOUIS SYDNEY TORONTO 2013

ELSEVIER B.V.
Radarweg 29, 1043 NX, Amsterdam, The Netherlands

ISBN: 9780444534910

British Library Cataloguing in Publication Data
A catalogue record for this book is available from the British Library

Library of Congress Cataloging in Publication Data
A catalog record for this book is available from the Library of Congress

Commissioning Editor: Thomas E. Stone
Development Editor: Michael Parkinson
Project Manager: Sujatha Thirugnana Sambandam
Designer/Design Direction: Alan Studholme

The Publisher's policy is to use **paper manufactured from sustainable forests**

Handbook of Clinical Neurology 3rd Series

Available titles

Vol. 79, The human hypothalamus: basic and clinical aspects, Part I, D.F. Swaab ISBN 9780444513571
Vol. 80, The human hypothalamus: basic and clinical aspects, Part II, D.F. Swaab ISBN 9780444514905
Vol. 81, Pain, F. Cervero and T.S. Jensen, eds. ISBN 9780444519016
Vol. 82, Motor neurone disorders and related diseases, A.A. Eisen and P.J. Shaw, eds. ISBN 9780444518941
Vol. 83, Parkinson's disease and related disorders, Part I, W.C. Koller and E. Melamed, eds. ISBN 9780444519009
Vol. 84, Parkinson's disease and related disorders, Part II, W.C. Koller and E. Melamed, eds. ISBN 9780444528933
Vol. 85, HIV/AIDS and the nervous system, P. Portegies and J. Berger, eds. ISBN 9780444520104
Vol. 86, Myopathies, F.L. Mastaglia and D. Hilton Jones, eds. ISBN 9780444518996
Vol. 87, Malformations of the nervous system, H.B. Sarnat and P. Curatolo, eds. ISBN 9780444518965
Vol. 88, Neuropsychology and behavioural neurology, G. Goldenberg and B.C. Miller, eds. ISBN 9780444518972
Vol. 89, Dementias, C. Duyckaerts and I. Litvan, eds. ISBN 9780444518989
Vol. 90, Disorders of consciousness, G.B. Young and E.F.M. Wijdicks, eds. ISBN 9780444518958
Vol. 91, Neuromuscular junction disorders, A.G. Engel, ed. ISBN 9780444520081
Vol. 92, Stroke – Part I: Basic and epidemiological aspects, M. Fisher, ed. ISBN 9780444520036
Vol. 93, Stroke – Part II: Clinical manifestations and pathogenesis, M. Fisher, ed. ISBN 9780444520043
Vol. 94, Stroke – Part III: Investigations and management, M. Fisher, ed. ISBN 9780444520050
Vol. 95, History of neurology, S. Finger, F. Boller and K.L. Tyler, eds. ISBN 9780444520081
Vol. 96, Bacterial infections of the central nervous system, K.L. Roos and A.R. Tunkel, eds. ISBN 9780444520159
Vol. 97, Headache, G. Nappi and M.A. Moskowitz, eds. ISBN 9780444521392
Vol. 98, Sleep disorders Part I, P. Montagna and S. Chokroverty, eds. ISBN 9780444520067
Vol. 99, Sleep disorders Part II, P. Montagna and S. Chokroverty, eds. ISBN 9780444520074
Vol. 100, Hyperkinetic movement disorders, W.J. Weiner and E. Tolosa, eds. ISBN 9780444520142
Vol. 101, Muscular dystrophies, A. Amato and R.C. Griggs, eds. ISBN 9780080450315
Vol. 102, Neuro-ophthalmology, C. Kennard and R.J. Leigh, eds. ISBN 9780444529039
Vol. 103, Ataxic disorders, S.H. Subramony and A. Durr, eds. ISBN 9780444518927
Vol. 104, Neuro-oncology Part I, W. Grisold and R. Sofietti, eds. ISBN 9780444521385
Vol. 105, Neuro-oncology Part II, W. Grisold and R. Sofietti, eds. ISBN 9780444535023
Vol. 106, Neurobiology of psychiatric disorders, T. Schlaepfer and C.B. Nemeroff, eds. ISBN 9780444520029
Vol. 107, Epilepsy Part I, H. Stefan and W.H. Theodore, eds. ISBN 9780444528988
Vol. 108, Epilepsy Part II, H. Stefan and W.H. Theodore, eds. ISBN 9780444528995
Vol. 109, Spinal cord injury, J. Verhaagen and J.W. McDonald III, eds. ISBN 9780444521378
Vol. 110, Neurological rehabilitation, M. Barnes and D.C. Good, eds. ISBN 9780444529015
Vol. 111, Pediatric neurology Part I, O. Dulac, M. Lassonde and H.B. Sarnat, eds. ISBN 9780444528919
Vol. 112, Pediatric neurology Part II, O. Dulac, M. Lassonde and H.B. Sarnat, eds. ISBN 9780444529107
Vol. 113, Pediatric neurology Part III, O. Dulac, M. Lassonde and H.B. Sarnat, eds. ISBN 9780444595652
Vol. 114, Neuroparasitology and tropical neurology, H.H. Garcia, H.B. Tanowitz and O.H. Del Brutto, eds. ISBN 9780444534903
Vol. 115, Peripheral nerve disorders, G. Said and C. Krarup, eds. ISBN 9780444529022
Vol. 116, Brain stimulation, A.M. Lozano and M. Hallett, eds. ISBN 9780444534972

Foreword

The long-held concept of the autonomic nervous system as a simple reflex system, of a dual system that exerts control by the balance of sympathetic and parasympathetic influences on target organs, is gradually being replaced. The autonomic nervous system is coming to be understood in very different terms today. This new volume, the latest addition to the 3rd series of the *Handbook of Clinical Neurology*, for which we take responsibility as series editors, is a very different book than others dealing with the autonomic nervous system. It stresses that the system is fully integrated with the rest of the brain and that it is multisystemic, with integration occurring at many different levels of the central nervous system. Its output is also highly diversified and involves many different neurotransmitters and modulators. One of the aims of this volume has been to facilitate an understanding of the integration between the autonomic and the rest of the central nervous system, even as the distinction between the two systems becomes blurred. It also clarifies the close interaction between the endocrine and other organ systems on the one hand and the nervous system on the other, an interaction that is essential to maintain the functional state of the organism.

The volume that has emerged is one that will appeal to biologists and medical researchers as well as to neurologists and other clinicians. The breadth of the volume is remarkable, and it contains much more than is suggested by its title. The first half of the volume is of more fundamental interest whereas the second half provides thoughtful accounts of various dysautonomias encountered in clinical practice. Readers should focus on the book as a whole, however, as it will truly enhance and even re-set their understanding of the autonomic nervous system.

As series editors, we read each of the chapters as they were submitted and made suggestions for improvements. We were greatly impressed, however, by the scholarship of the authors in producing such comprehensive accounts. We are grateful to David Goldstein of Bethesda who contributed much in an early phase of this book and to Ruud M. Buijs, of Mexico City, and Dick F. Swaab, from Amsterdam, who, as the volume editors, conceived such an exciting book and assembled an outstanding group of contributors, and we congratulate them on their achievement. We are grateful also to the team at Elsevier, and particularly to Mr. Michael Parkinson in Lochcarron, Ms Kristi Anderson in San Diego, and Mr. Thomas Stone in Boston for their assistance at all stages in the production of this book.

Michael J. Aminoff
François Boller

Preface

The autonomic nervous system (ANS) is often regarded as a distinct entity but, as becomes clear from the present volume, it is—in fact—a fully integrated part of the central nervous system (CNS). Its primary task is regulating body functions on the basis of decisions made by the brain. A myriad of sensory nerve fibers coming from the sensory and other organs provides the brain with information on the state of the body and the outside world. This information is subsequently translated into autonomic and hormonal output signals that serve to change or maintain the physiological state of the organs. This complex interaction between body and brain is, by definition, reciprocal. Disorders of the brain, such as depression or migraine, may occur in association with diseases of the body, such as hypertension or obesity, illustrating the complex autonomic interaction between brain and body. The central portions of the ANS are by no means limited to the hypothalamus. In recent years, basic and clinical research has made great strides and has shown that the ANS is composed of many different systems integrated at different levels of the CNS in such a way as to render arbitrary the distinction between the two systems. In addition, the output of the ANS is highly diversified, as it targets different tissues with different branches and neurotransmitters. The present volume therefore makes no effort to separate the ANS from the rest of CNS, nor to provide a simple working mechanism for the interaction between CNS and ANS. Rather, our approach has been to help the reader understand the integration of these two systems and the essence of their interaction with information that arises from the body and its environment.

The contributions to this volume by well-known basic and clinical scientists explain that the brain–body interaction is a feedback loop so that, particularly with disease, it is sometimes difficult to determine the origin of the disorder. This volume provides recent insights in the regulation of body temperature, cardiovascular functions, migraine, and micturition in order to provide a basis for understanding how diseases may arise from an imbalance in the ANS. It also deals with the ways in which information on body functions and dysfunctions may be transmitted to the brain via novel pathways, leading to a change in autonomic output. Moreover, the volume provides examples on how the brain, via the ANS, sensitizes endocrine organs to the effect of hormones, offering an insight into the mechanism of how hormonal balances may be affected in neurological diseases. These examples illustrate that the ANS is an important topic for clinical neurologists, not only because the central regulation of vital functions may be disturbed, but also because of the likelihood that information feeding back from the viscera to the brain by way of the spinal cord and vagus nerve may affect neurological functions. ANS dysfunction may thus be the cause of many of the symptoms that are a feature of the aging process as well as of various neurological disorders, such as Parkinson's disease, multiple system atrophy, Alzheimer's disease, pure autonomic failure, migraine, emesis, hypoglycemia and autoimmune autonomic failure.

From many chapters of this volume, the picture emerges that the ANS is not so much a simple reflex system as a sophisticated homeostasis-mediating instrument of the brain, involving a multitude of brain areas. In this way, the brain can use the ANS as a vehicle to prepare the body for action by changes in behavior, organ functions, and hormone secretion, i.e. by fine-tuning various physiological processes. A number of examples is given at the organizational level on how the brain, via its association with the ANS, synchronizes behavior, endocrine systems, and autonomic functional output to the body, and how disorders in this integrative process may lead to disease. Such mechanisms not only entail control of brain–body functions such as micturition, bone formation, or cardiovascular functions, but also the relationship between the brain and the immune system. Disturbances in neurological functions that occur in, for example, Parkinson's disease, migraine, bipolar disorder, and Cushing syndrome are therefore discussed in terms of their relation with autonomic disturbances. The volume is not limited to the way in which the ANS communicates with the body, but also provides an insight into the way in which organs of the body, on their part, relay this information back to the brain, and the content of this volume thus operates at the crossroads of neurology, psychiatry, neuroendocrinology and internal medicine.

The growing insight into the brain–body and body–brain relationship shows that a lot has happened since the publication of two volumes on the ANS in the *Handbook of Clinical Neurology* series, 15 years ago. Not only have important technical innovations been developed for the assessment of autonomic functions, such as sympathetic neuroimaging, sympathetic microneurography, and skin biopsies, but new therapeutic approaches have emerged based upon insights into ANS functions, such as carotid sinus stimulation.

We are indebted to the outstanding group of contributors to this volume, many of whom are internationally recognized as experts in the field. Their efforts have resulted in a multidisciplinary volume on the ANS that will appeal to both clinical neurologists and basic scientists. We are also grateful for the helpful Elsevier staff, and in particular to Mr. Michael Parkinson, Ms. Kristi Anderson, and Mr. Tom Stone, who provided expert assistance in the production of this volume, and for the very valuable advice of the series editors, Michael Aminoff and Francois Boller.

Ruud M. Buijs and Dick F. Swaab

Contributors

E.E. Benarroch
Department of Neurology, Mayo Clinic, Rochester, MN, USA

G. Boeckxstaens
Department of Gastroenterology, University Hospital Leuven, University of Leuven, Leuven, Belgium

R.M. Buijs
Department of Cell Biology and Physiology, Institute for Biomedical Research, Universidad Nacional Autónoma de México, Mexico City, Mexico

G. Burnstock
Autonomic Neuroscience Centre, University College Medical School, London, UK and Department of Pharmacology, University of Melbourne, Australia

B.J. Canning
Johns Hopkins Asthma and Allergy Center, Baltimore, MD, USA

M.G. Cersosimo
Parkinson's Disease and Movement Disorder Unit, Hospital de Clínicas, University of Buenos Aires, Argentina

A.D. Craig
Atkinson Research Laboratory, Barrow Neurological Institute, Phoenix, AZ, USA

H.D. Critchley
Department of Psychiatry, Brighton and Sussex Medical School and Sackler Centre for Consciousness Science, University of Sussex, Brighton, UK

P.E. Cryer
Department of Medicine, Washington University in St. Louis and Barnes-Jewish Hospital, St. Louis, Missouri, USA

I. Cygankiewicz
Department of Electrocardiology, Medical University of Lodz, Lodz, Poland

P.D. Drummond
School of Psychology and Exercise Science, Murdoch University, Perth, Western Australia

J. Eccles
Department of Psychiatry, Brighton and Sussex Medical School, Brighton, UK

C. Escobar
Department of Anatomy, Faculty of Medicine, Universidad Nacional Autónoma de México, Mexico City, Mexico

W.C. Engeland
Department of Neuroscience, University of Minnesota, Minneapolis, MN, USA

T. Erbas
Department of Endocrinology and Metabolism, Hacettepe University Medical School, Ankara, Turkey

R.D. Fealey
Department of Neurology, Mayo Clinic, Rochester, MN, USA

C.J. Fowler
Department of Uro-Neurology, National Hospital for Neurology and Neurosurgery, University College London Hospitals Foundation Trust, London, UK

Q. Fu
Institute for Exercise and Environmental Medicine, Texas Health Presbyterian Hospital Dallas, and University of Texas Southwestern Medical Center, Dallas, TX, USA

S.N. Garfinkel
Department of Psychiatry, Brighton and Sussex Medical School and Sackler Centre for Consciousness Science, University of Sussex, Brighton, UK

E.M. Garland
Autonomic Dysfunction Center, Vanderbilt University, Nashville, TN, USA

C.H. Gibbons
Center for Autonomic and Peripheral Nerve Disorders, Beth Israel Deaconess Medical Center, Boston, MA, USA

P.J. Goadsby
Department of Neurology, University of California, San Francisco, CA, USA

D.S. Goldstein
Clinical Neurocardiology Section, Clinical Neurosciences Program, Division of Intramural Research, National Institute of Neurological Disorders and Stroke, National Institutes of Health, Bethesda, MD, USA

G. Grassi
Department of Prevention and Clinical Medicine, University of Milano-Bicocca, Italy

W.L. Hasler
Division of Gastroenterology, Department of Internal Medicine, University of Michigan Health System, Ann Arbor, MI, USA

J.-Y. He
Department of Orthopaedic Surgery, Xinhua Hospital, Shanghai Jiaotong University School of Medicine, Shanghai, China

W.B. Hooper
Department of Internal Medicine, Greenwich Hospital, Greenwich, CT, USA

R. Iliescu
Cardiovascular Research Department, Gr. T. Popa Center for Biomedical Research and the University of Medicine and Pharmacology, Gr. T. Popa, Iasi, Romania

L.-S. Jiang
Department of Orthopaedic Surgery, Xinhua Hospital, Shanghai Jiaotong University School of Medicine, Shanghai, China

E.J. Johns
Department of Physiology, University College Cork, Cork, Republic of Ireland

M.J. Joyner
Department of Anesthesiology, Mayo Clinic, Rochester, MN, USA

H. Kaufmann
Dysautonomia Center, Department of Neurology, New York University School of Medicine, New York, NY, USA

F. Krismer
Division of Clinical Neurobiology, Department of Neurology, Medical University, Innsbruck, Austria

B.D. Levine
Institute for Exercise and Environmental Medicine, Texas Health Presbyterian Hospital Dallas, and University of Texas Southwestern Medical Center, Dallas, TX, USA

T.E. Lohmeier
Department of Physiology, University of Mississippi Medical Center, Jackson, MS, USA

V.G. Macefield
School of Medicine, University of Western Sydney, Sydney, and Neuroscience Research Australia, Sydney, Australia

G. Mancia
Department of Prevention, Clinica Medica, Ospedale San Gerardo, Monza (Milan), Italy

S.B. Mazzone
School of Biomedical Sciences, University of Queensland, St Lucia, Queensland, Australia

S. Muppidi
Department of Neurology and Neurotherapeutics, University of Texas Southwestern Medical Center, Dallas, TX, USA

J.N. Panicker
Department of Uro-Neurology, National Hospital for Neurology and Neurosurgery, University College London Hospitals Foundation Trust, London, UK

D. Robertson
Autonomic Dysfunction Center, Vanderbilt University, Nashville, TN, USA

J.H. Seth
Department of Uro-Neurology, National Hospital for Neurology and Neurosurgery, University College London Hospitals Foundation Trust, London, UK

D.F. Swaab
Netherlands Institute for Neuroscience, an Institute of the Royal Netherlands Academy of Arts and Sciences, Amsterdam, The Netherlands

A.Y. Tan
Beth Israel Deaconess Medical Center, Harvard Medical School, Boston, MA, USA

S. Vernino
Department of Neurology and Neurotherapeutics, University of Texas Southwestern Medical Center, Dallas, TX, USA

R.L. Verrier
Beth Israel Deaconess Medical Center, Harvard Medical School, Boston, MA, USA

A.I. Vinik
Eastern Virginia Medical School, Strelitz Diabetes Center, Division of Endocrinology and Metabolism, Eastern Virginia Medical School, Norfolk, VA, USA

N. Wang
Center for Autonomic and Peripheral Nerve Disorders, Beth Israel Deaconess Medical Center, Boston, MA, USA

E.A. Wehrwein
Department of Physiology, Michigan State University, East Lansing, MI, USA

G.K. Wenning
Division of Clinical Neurobiology,
Department of Neurology, Medical University, Innsbruck, Austria

W. Zareba
Heart Research Follow-up Program,
University of Rochester Medical Center,
Rochester, NY, USA

X.-F. Zheng
Department of Orthopaedic Surgery, Xinhua Hospital, Shanghai Jiaotong University School of Medicine, Shanghai, China

Contents

Handbook of Clinical Neurology, Vol. 117 (3rd series)
Autonomic Nervous System
R.M. Buijs and D.F. Swaab, Editors

Chapter 1

The autonomic nervous system: a balancing act

RUUD M. BUIJS*

Department of Cell Biology and Physiology, Institute for Biomedical Research, Universidad Nacional Autónoma de México, Mexico City, Mexico

INTRODUCTION AND OVERVIEW OF THE AUTONOMIC NERVOUS SYSTEM

The purpose of the present chapter is to give a brief overview of the autonomic nervous system (ANS) and to pay special attention to those parts subject to controversy or uncertainties. For well-established areas of knowledge the reader is referred to the chapter by Goldstein (Chapter 2) and to reviews such as that of Janig and Habler (2000).

Since the discovery of Otto Loewi (Loewi, 1921) that stimulation of the vagus branch slowed down the heartbeat by means of the release of a "vagusstoff" (vagus substance), and Henry Dale's discovery (1934) that the active substance in this process is acetylcholine, we accept that the parasympathetic branch of the ANS, which originates from the dorsal motor nucleus of the vagus (DMV) or from the nucleus ambiguus, uses acetylcholine (ACh) as a transmitter. Consequently, the transmitter of the parasympathetic nervous system is thought to be ACh. However, in several targets of these parasympathetic nuclei of the ANS the final neurotransmitter has not been established. This could be due to the fact that for some organs, such as the liver, no ganglion has been described that provides it with parasympathetic input; this has even led to the suggestion that parasympathetic innervation to the liver is absent. Several models of the autonomic nervous system therefore do not even indicate the innervation of the liver by the vagus, in spite of clear macroscopic evidence. Similar doubts are present for the parasympathetic innervation of the spleen, adipose tissue, and lymphoid tissue. Other aspects that demand attention are the structures in the central nervous system (CNS) that have the capacity to control or influence the output of the ANS, and thus provide input to the sympathetic and parasympathetic autonomic motor nuclei. Transneuronal tracing in the last 20 years has revealed the main brain centers in terms of the capacity to influence its autonomic outflow (Card et al., 1990; Nadelhaft et al., 1992; Buijs et al., 1999, 2003a).

Another important point is that the ANS should not only be seen as a system merely carrying out the commands of the brain; it also functions as a reflex circuit, using the sensory feedback of the organs, to change and precisely adapt the output of the ANS in order to adjust the physiological state of the body. Clearly these reflexes are also immediately modulated by input from higher neural centers located in many different areas in the brainstem, hypothalamus, and even prefrontal cortex, which contain preautonomic neurons, i.e., neurons that provide input to autonomic motor neurons. This is also evident when analyzing structures that provide input to the nucleus of the tractus solitarius (NTS), one of the main centers where sensory information from the body is integrated (Fig. 1.1) (Ruggiero et al., 1996). This sensory input leads to an immediate change in autonomic output that is immediately modulated by the input the NTS receives from many structures higher in the CNS, from the brainstem and hypothalamus up to the prefrontal cortex, but also from ascending sensory information from the spinal cord (Luiten et al., 1985; al Maskati and Zbrozyna, 1989; Powell et al., 1994; Van Eden and Buijs, 2000; Craig, 2003). Here, the general consensus is that descending CNS information is able to influence the reflex arc that exists between the NTS, and the dorsal motor nucleus of the vagus (DMV). Interestingly, since the NTS also provides this sensory information to many areas in the hind and forebrain, the descending input the NTS receives may, to a large extent, determine how that information is transmitted, i.e., the brain is modulating its own input. How spinal sensory systems may contribute to visceral information passing to higher centers in the brain is still largely uncharted territory (Craig, 2003), but the

*Correspondence to: Ruud M. Buijs, Department of Cell Biology and Physiology, Institute for Biomedical Research, Universidad Nacional Autónoma de México, Av. Universidad 3000, Mexico City, DF, 04510, Mexico. E-mail: ruudbuijs@gmail.com

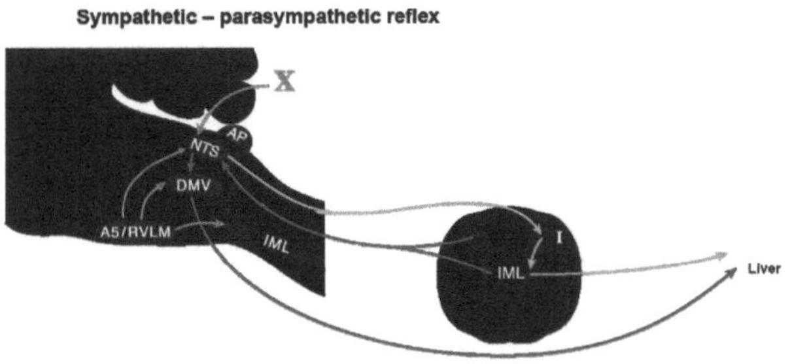

Fig. 1.1. Illustrates the sympathetic and parasympathetic reflex. The relationship between the cell groups that may influence the vagal output is indicated in green and the influence to the sympathetic output in pink. Both vagal and sympathetic outputs are influenced by each other. AP, area postrema; NTS, nucleus tractus solitarius; DMV, dorsal motor nucleus of the vagus; IML, intermediolateral column. (Modified from Buijs et al., 2001.)

conclusion that they do seems logical in view of the reflex circuitry. In the spinal cord, a similar reflex circuit is in place between the sensory neurons in the dorsal horn and the sympathetic motor neurons in the intermediolateral column (IML). Also, this circuit can be modulated by descending information. Finally, in addition to input from these sensory areas, the DMV and IML, two major output stations of the ANS, also receive direct input from higher centers. Interestingly, this gives the higher centers of the brain a powerful control, not only over the modulation of the sensory reflex but also directly, over the outgoing information from the ANS.

THE SYMPATHETIC NERVOUS SYSTEM

In principle, the central control of the outflow of sympathetic nervous information is tactically organized as a long chain of motor neurons in the intermediolateral column of the spinal cord (IML). Tactically, because due to this segmental organization, the ACh-producing motor neurons reach different and multiple ganglions along the spinal cord. Neurons in these ganglions have the capacity to produce different neurotransmitters, of which noradrenaline/norepinephrine (NA) is the main one, often together with cotransmitters such as neuropeptide Y (NPY) (Lundberg et al., 1983). Interestingly, the target structures of the ANS, although as diverse as the brain and the kidney, are very often the smooth muscles around the blood vessels within these organs. This allows the ANS, by vasoconstriction and vasodilatation, to control the blood supply, for example, not only to the kidney and the lungs but also to itself (Zhu et al., 1997).

Differentiation of function

The segmental organization of the sympathetic nervous system allows topographical signal distribution, albeit in a very general sense: i.e., the superior cervical ganglion not only serves to provide the head with sympathetic innervation, it also supplies part of the sympathetic innervation to the heart. Consequently, even with a segmental distribution, one cannot conclude that one segment or one ganglion provides only one outgoing signal. For example, a strong sympathetic signal only goes out to the pineal during the hours of darkness, when NA releases melatonin (Klein et al., 1971). At the same time, the heart, receiving input from the same ganglion, should (at least in humans) receive a sympathetic signal of lower strength. The sympathetic signal to the heart increases again early in the morning, when the autonomic signals need to support the activity of the individual (Massin et al., 2000). If this autonomic balance is disturbed, as evidenced by a higher sympathetic tone to the heart, hypertension or diabetes may ensue (Nielsen et al., 1999; Nakano et al., 2001). Interestingly, in coronary heart disease with **enhanced** sympathetic signaling to the heart (Lee et al., 2005) the melatonin secretion is lower (Brugger et al., 1995), indicating **lower** sympathetic output to the pineal and supporting the idea of a diversity of sympathetic functions.

Normally, both liver and adrenal receive a higher sympathetic signal at the beginning of the activity period, resulting in enhanced glucose production and (nor)adrenaline and cortisol secretion, respectively, just before the activity period (Buijs et al., 2003b, 2006). This differentiation of sympathetic output to liver and adrenal on the one hand and the pineal on the other is coordinated by the suprachiasmatic nucleus (SCN), the biological clock (see Chapter 15 Buijs et al.), and illustrates the capacity of higher centers in the brain to differentiate their output (Fig. 1.2A). Finally, differentiation of function within the sympathetic system is probably best illustrated by the example of sympathetic output that is essential for heat production, i.e., the sympathetic output to brown adipose tissue, compared with the sympathetic output that is essential for heat loss, i.e., the output to the sweat glands (Schotzinger and Landis, 1988). The sympathetic projections to brown adipose tissue will become active just before activity onset early in the morning in order to generate heat by burning fat (Ootsuka et al., 2009; Blessing et al., 2013). This contrasts with the sympathetic tone to the sweat glands, which is at its highest when the core temperature is at its lowest, resulting in lower thresholds for sweating during exercise in the early part of the night (Waterhouse et al., 2007). This complexity is further illustrated by studies showing that women susceptible to hot flashes experience a peak in these unwelcome events around 3 o'clock in the afternoon, in both hot and temperate environments, also indicating a different reaction of the sympathetic system under basal conditions (Freedman et al., 1995; Sievert et al., 2010).

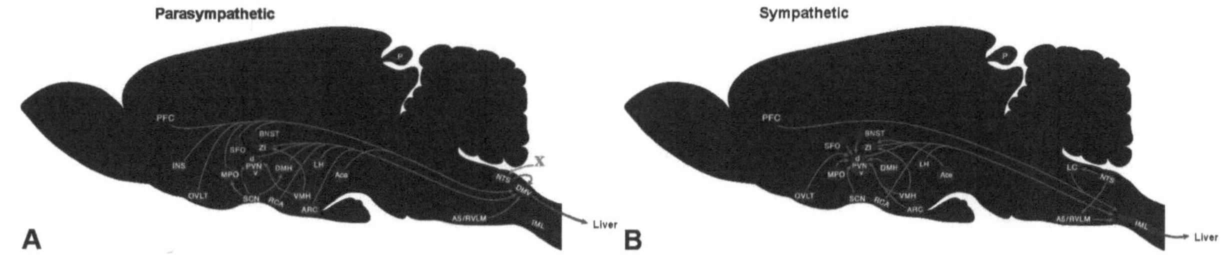

Fig. 1.2. (**A**) and (**B**) Sagittal schemes of the sites in the rat brain with autonomic neurons that are able to influence organs in the abdominal cavity with the liver as example. The direct projections to the liver are indicated in red, the second-order in blue, and third-order in yellow. Comparing the parasympathetic pattern to the sympathetic pattern illustrates that far more second-order cell groups are in control of the dorsal motor nucleus of the vagus than in control of the sympathetic motor neurons. A5/RVLM, rostroventrolateral medulla; Ace, amygdala central part; ARC, arcuate nucleus; BNST, bed nucleus of the stria terminalis; DMH, dorsomedial nucleus of the hypothalamus; DMV, dorsal motor nucleus of the vagus; IML, intermediolateral column; INS, insular cortex; LC, locus coeruleus; LH, lateral hypothalamus; MPO, medial preoptic area; NTS, nucleus tractus solitarius; OVLT, organum vasculosum of the lamina terminalis; PFC, prefrontal cortex; RCA, retrochiasmatic area; SCN, suprachiasmatic nucleus; SFO, subfornical organ; VMH, ventromedial hypothalamus; PVNdv, paraventricular nucleus dorsal ventral; ZI, zona incerta. (Modified from Buijs et al., 2001.)

Consequently, on the basis of these differences in activity and control, we cannot generalize and speak about *the* sympathetic nervous system; rather, we need to be specific and indicate which branches of the sympathetic nervous system we are examining.

Central control of different branches of the sympathetic and parasympathetic nervous system

After having concluded that a high differentiation with dissimilar activities exists between the various parts of the sympathetic nervous system, we need to consider that the activity of these diverse branches needs to be controlled somehow by separate autonomic motor neurons and separate systems in the brain. Animal studies using pseudorabies virus (PRV) as a transneuronal tracer have shed some light on this problem (Fig. 1.2A, B). PRV has the capacity to be taken up by nerve endings and is retrogradely transported to the cell bodies that innervate an organ, since the virus multiplies in the cell body and after lysis will be taken up again by synaptic terminals that innervate only this cell. PRV is then retrogradely transported again until the chain of command has been determined (Card et al., 1990). Initial studies using these viruses to trace the central preautonomic neurons showed that, for example, in the hypothalamus, the paraventricular nucleus (PVN) and the lateral hypothalamus (LH) always show PRV labeling, irrespective of whether the brain was infected via the sympathetic motor neurons or via the parasympathetic motor neurons. This suggested that the same hypothalamic neurons may provide input to the sympathetic system as well as the parasympathetic system. Investigating this possibility, it was observed that there is a complete separation between neurons in all areas of the brain that give input to the parasympathetic system or sympathetic system – this separation extends as far as the SCN (Fig. 1.3) (Buijs et al., 2003a). Moreover, the same areas in the brain also show labeled neurons irrespective of the organ into which the virus was injected (Strack et al., 1989), showing that these hypothalamic nuclei are involved not only in driving parasympathetic or sympathetic functions but also in the control of a whole variety of organs.

Different neurons in the CNS control different organs

This involvement of the same hypothalamic areas in the control of different organs provoked the question whether organs are influenced by a general drive from the hypothalamus, or whether the hypothalamus would be able to influence these organs separately, by using dissimilar neurons projecting exclusively to motor neurons that are involved only in sending signals to a particular organ. The development of genetically modified viruses that produce different markers (Kim et al., 1999), allowing the determination of the chain of command of separate organs, helped to answer this question. It will be clear that such tracer studies require that survival time and infection speed are equal for the different tracers and are very carefully controlled; if one virus infects faster or has a faster route of infection this could easily lead to populations of neurons that show the two tracers in error. Reliable analysis of transneuronal tracing is therefore only possible when the different infection stages are carefully monitored and different survival times are used;

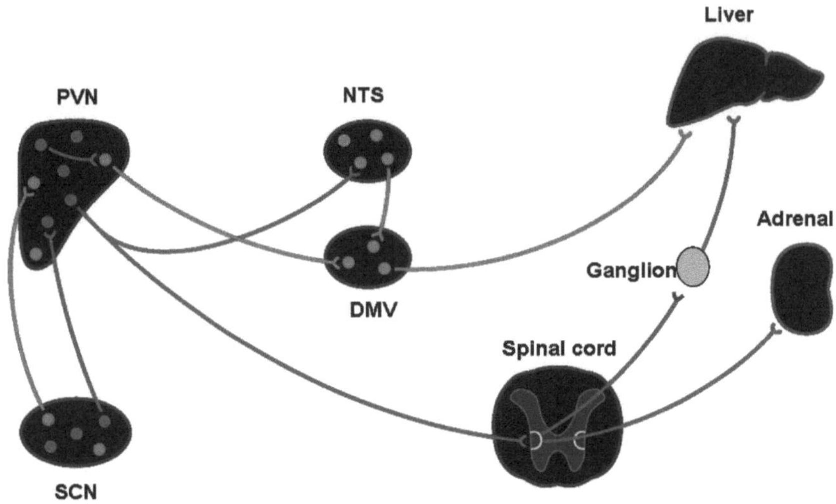

Fig. 1.3. Scheme of interaction between the hypothalamic suprachiasmatic nucleus (SCN) and the paraventricular nucleus (PVN). Separate sympathetic (red) or parasympathetic (blue) neurons of the SCN project to preautonomic neurons of the PVN, where a similar sympathetic–parasympathetic separation can be observed. Preautonomic neurons of the PVN project either to the preganglionic sympathetic neurons in the intermediolateral (IML) column of the spinal cord, or to the preganglionic neurons of the dorsal motor nucleus of the vagus (DMV). The presympathetic PVN neurons have axon collaterals to preparasympathetic neurons, either in the PVN itself, or in the nucleus tractus solitarius. (Reproduced from Buijs et al., 2003a.)

under these conditions, infections up to the third order can be analyzed. Such studies have been done for colon and bladder (Rouzade-Dominguez et al., 2003), for adrenal and ovary (Toth et al., 2008), and for subcutaneous and intra-abdominal fat (Kreier et al., 2002). The general features of these findings are that sympathetic motor neurons in the spinal cord projecting to these organs are completely separated. The results for the preautonomic neurons in brainstem and hypothalamus differ to some extent. Two studies report the presence of hypothalamic neurons containing two viruses, indicating that the neuron can interact with two organs, while also in the same sections, neurons are found that only project to one organ (Rouzade-Dominguez et al., 2003; Toth et al., 2008). In contrast, structures where only separate neurons are found, innervating either subcutaneous or intra-abdominal fat, range from the SCN to the prefrontal cortex (Kreier et al., 2002). Consequently, the various studies showed that the brain has the capacity to differentiate its output with respect to its target structure in order to separate its signal, depending on the functionality of the organ.

THE PARASYMPATHETIC NERVOUS OUTPUT

Organization of the dorsal motor nucleus of the vagus

The consensus is that the output of the parasympathetic nervous system (PNS) is mainly achieved by the release of acetylcholine (ACh) from its nerve terminals. This is true when intermediate ganglions are involved that transmit the message of the parasympathetic motor nuclei. However, as we will see, for some organs such ganglia cannot be found, and therefore the origin of the parasympathetic innervation to the organ is unclear or controversial. Liver, spleen, and adipose tissue are examples of such organs. For the transmission of information from parasympathetic ganglia, it is the case that, unlike noradrenaline, ACh exhibits a very short half-life of 1–2 milliseconds, due to the presence of acetylcholinesterase, which hydrolyzes ACh at a high rate. Close contact between cholinergic nerve terminals and cells expressing ACh receptors is therefore required for the cholinergic control of these cells. The dorsal motor nucleus of the vagus (DMV) and the nucleus ambiguus are the two nuclei in the brainstem from which the parasympathetic output to most organs arises, either directly or via an intermediate ganglion. Considering the direct input from these two parasympathetic motor nuclei, it is important to keep in mind that Ach is the neurotransmitter in just a part of these neurons, in some of these colocalized with dopamine (Jones and Beaudet, 1987; Huang et al., 1993; Tsukamoto et al., 2005). Consequently, when organs receive direct input from the DMV the parasympathetic neurotransmitter can be different from Ach. Furthermore, there are other parasympathetic nuclei in the brainstem and in the lower lumbar levels of the spinal cord that provide parasympathetic innervation to, respectively, areas of the head and the reproductive organs (Janig and Habler, 2000).

In addition, neurons of the DMV have projections to postganglionic neurons localized in the intestine (i.e., the mesenteric plexus). The role of the mesenteric plexus is to provide the innervation to the intestines and to regulate, for example, its contractions. The large intestine differs in this respect from the rest of the gastrointestinal tract: it receives a parasympathetic innervation from two distinct sources: the DMV innervates the proximal colon and the sacral parasympathetic nucleus provides vagal input to the distal colon. To some extent this agrees with the segmental somatotopic organization of the sympathetic system. However, the majority of the organs in the body cavity receive input from the DMV, and therefore the question needs to be asked, does the DMV have a somatotopic organization similar to that of the sympathetic nervous system (SNS) and do its neurons provide input to certain organs only?

Differentiation of function within the dorsal motor nucleus of the vagus

The answer to this question is revealing, showing specificity also within the DMV. A study by Hayakawa et al. (2003) analyzed the innervation of the stomach and showed that the neurons projecting to the cardia, antrum, and pylorus are located throughout the entire DMV, suggesting that the pylorus and the antrum are innervated not only by the gastric branch, but also by the hepatic branch of the vagus nerve. In addition, this study indicated that there is no distinctive topographic distribution in the DMV of gastric motor neurons that innervate different parts of the stomach, suggesting that there is no topography in the DMV. However, a careful double-labeling study of the neurons innervating different parts of the stomach showed that there is a differentiation of function in the DMV neurons, because distinct neurons in the DMV contact either the cardiac or the pyloric sphincter muscles. Additional confirmation of the target specificity of the DMV neurons can be found in the work of Cailotto et al., who, after injections of the retrograde tracer CtB into the proximal and distal part of the ileum, observed the labeling of separate motor neurons in the DMV (Cailotto et al., 2012). Consequently, these results illustrate a high differentiation of function of DMV neurons, which seems to coincide with the distinct tasks of the targeted tissues.

The question whether this means a complete separation of function has featured in a study by Kreier et al., who demonstrated that possibly not all neurons of the PNS and SNS target different organs, but that these neurons may be distinguished on the basis of their physiological function (Kreier et al., 2005). In this study it was observed that a large proportion of the DMV neurons project to the liver, and that a small number of these neurons also project to the pancreas or to intra-abdominal fat, i.e., targeting the three tissues with the same neuron. However, neurons that project to intra-abdominal fat do not project to subcutaneous fat, suggesting a divergent functionality. Not surprisingly in view of what we have seen before, within the rest of the brain a similar separation and specialization of neurons was also observed (Kreier et al., 2005). A similar division in function can be suspected in hypothalamic neurons that project to DMV neurons projecting to different parts of the stomach and different parts of the ileum. We would therefore like to propose a specialization of function within the PNS and SNS, a specialization not so much in their targets but more in the functions to be executed. If the function of the ANS is seen as "supporting and executing the physiological functions necessary to support the behavior driven by the brain," then it can be hypothesized that the organization of the PNS and SNS should follow the physiology. Consequently, it is proposed that the ANS, whose output is controlled by multiple areas within the brain, directs the body in separate compartments and different functions rather than as a whole. The reflexes will still be specifically aimed at tuning particular organs, but they will take into account the functionality of the whole compartment.

Intra-abdominal fat tissue and its relationship with disease

Considering the specialization in the projections of DMV neurons to intra-abdominal fat or to subcutaneous fat, and the association of intra-abdominal fat with diabetes and hypertension, it has also been proposed that the body, with respect to the output of the ANS, can be divided into different compartments: one that drives the region of the thorax and muscles and one that takes care of the organs in the abdomen (Kreier et al., 2003). Anatomical evidence for such compartmentalization is given above. Since parasympathetic activity is associated more with the uptake of glucose in the liver (Pocai et al., 2005) and fat tissue (Kreier et al., 2002), and with the release of insulin (Luzi et al., 1992), all neurons of the ANS involved in executing these functions can be considered as belonging to one compartment. Normally the output of the ANS to such a compartment is in balance; however, lifestyle-associated behavior and other disturbances may shift this balance and result in an enhanced parasympathetic or sympathetic tone. It is proposed that in the metabolic syndrome resulting from lifestyle changes and excessive food intake, this balance is shifted to a more sympathetic drive to the heart and muscle compartment and a more parasympathetic drive to the intra-abdominal compartment. This shift in autonomic balance consequently produces the accumulation

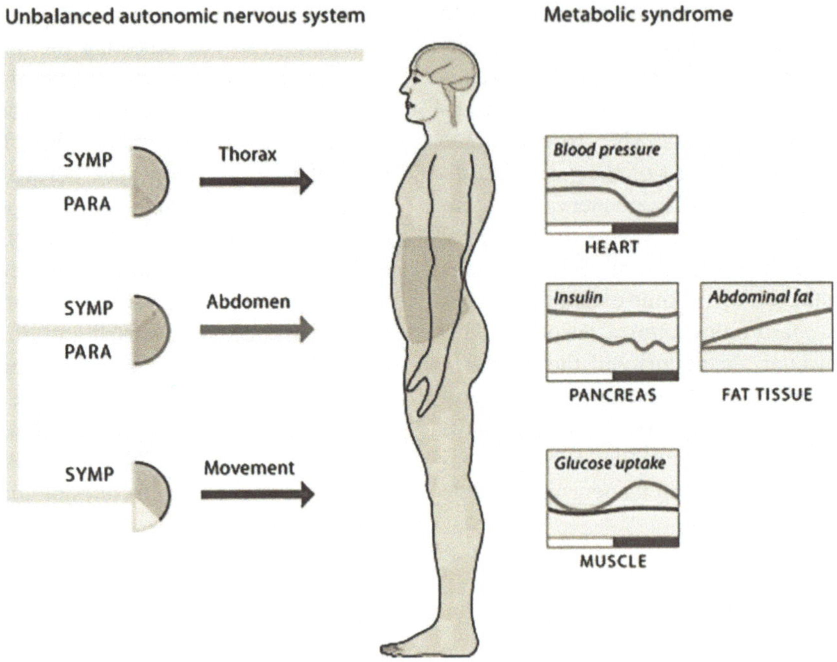

Fig. 1.4. Model of the metabolic syndrome caused by desynchronization. The disturbed output of the biological clock affects the selective balance of the ANS in different parts of the body. In the intra-abdominal compartment, the ANS is shifted in favor of the parasympathetic branch (blue), resulting in increased insulin secretion and growth of intra-abdominal fat tissue compared with normal values (gray). Contrarily, in the thorax and movement compartment the sympathetic branch (red) prevails, leading to high blood pressure and impaired glucose uptake by the muscle compared with normal values. In this model, the symptoms of the metabolic syndrome are the result and not the cause of the disease. (Reproduced from Kreier et al., 2003.)

of intra-abdominal fat, since the uptake of fat is driven by the parasympathetic output. Lastly, this shifted balance of the ANS may also result in diabetes type 2 (enhanced parasympathetic drive to abdominal compartment) or hypertension (enhanced sympathetic drive to thoracic compartment). (For more details and discussion on this topic see Kreier et al., 2003; Buijs and Kreier, 2006, and Fig. 1.4.)

THE BLANK SPOTS

Interestingly, although all organs receive input from the SNS, PNS input seems to be absent in some organs. Since it is difficult to prove the absence of something, the absence of PNS innervation is mainly based on inability to detect ACh and on the inability of some tracers to be taken up sufficiently to be detected in parasympathetic motor nuclei. Relevant organs in this respect are the spleen and white adipose tissue, while for organs such as kidney and adrenal it is generally accepted that they will not receive parasympathetic innervation, mainly due to the lack of macroscopic evidence for such input and the lack of retrograde tracing in the DMV. In spite of the fact that there is no evidence for the presence of ACh innervation of the liver (Schafer et al., 1998), tracing studies have for many years shown the presence of input arising from the DMV (e.g., Magni and Carobi, 1983; Kohno et al., 1987). The presence of retrograde labeled neurons in the DMV after tracer injection despite the absence of ACh in the liver indicates two important points, first that the DMV projects directly to the liver without the mediation of a ganglion and second, that the DMV might use neurotransmitters other than ACh to signal to the liver, for example, dopamine, just as has been shown for the stomach (Tsukamoto et al., 2005). This might be true also for the other organs where the manner of innervation by parasympathetic nerves is more controversial.

The spleen

Tracey's group has demonstrated the inhibition of inflammation by stimulation of the vagus nerve, via an action on the Ach receptor in the spleen (Borovikova et al., 2000). The confirmation of this result of vagal stimulation came later for intestinal inflammation (de Jonge et al., 2005). Following this, several studies have tried to investigate in more detail the importance of vagal input and Ach receptors in the spleen. As early as 1929, Henry Dale demonstrated the presence of acetylcholine in the spleen (Dale and Dudley, 1929). However, ACh is now suggested to be present mainly in a specific memory

T cell population producing ACh, as these cells express choline acetyltransferase (Rosas-Ballina et al., 2011). Interestingly, in spite of neuroanatomical and physiological evidence for the presence of parasympathetic input to the spleen, as shown by means of retrograde tracing and the differential effect of parasympathetic denervation or sympathetic denervation on the adaptive immune response (Buijs et al., 2008), it has been suggested that the vagal control of the spleen relies on vagal innervation of splenic postganglionic sympathetic neurons (Rosas-Ballina et al., 2008). This last suggestion was recently challenged by a study showing the absence of neuronal contacts between vagus nerve fibers and ganglion neurons projecting to the spleen (Bratton et al., 2012). Future studies will be necessary to determine how the vagus may influence the functioning of the spleen and which neurotransmitters may be used for this. With respect to the vagal inflammatory pathway, all these studies taken together suggest that, at least, the brain plays an important role in modifying the immune response by changing the autonomic outflow of the vagus. This was confirmed in a study by de Jonge et al. (2005), who used vagal stimulation and observed an amelioration of surgery-induced inflammation and postoperative ileus. The question whether the vagus may indeed serve as a reflex arc was recently answered by Cailotto et al. (2012), who demonstrated that, after manipulation of the rat intestine, which normally leads to postoperative ileus, a vagovagal reflex was indeed initiated, with a selective activation of neurons in the NTS and DMV whereby nearly 50% of the activated DMV neurons project to the zone of inflammation. Since the majority of the descending projections of the hypothalamus terminate in the NTS (Fig. 1.3), this construction allows the brain an excellent opportunity to modify this vagal reflex.

Adipose tissue

Based mainly on the absence of the presence of acetylcholine (ACh) (Bryant et al., 1983), adipose tissue was also assumed not to receive parasympathetic input, until its presence was detected in brown adipose tissue (BAT) around the heart (Schafer et al., 1998), in the form of the vesicular acetylcholine transporter. White adipose tissue (WAT) also fails to show the presence of ACh or its transporter. Even so, just as in the spleen, injections of PRV revealed the presence of retrogradely labeled neurons in the dorsal motor nucleus of the vagus. Interestingly, just as with the spleen, injections of the retrograde monosynaptic retrograde tracers CtB or fluorogold revealed retrogradely labeled neurons in the DMV, and the proof of the specificity of this tracing was that it was undetectable after parasympathetic denervation of the organ (Kreier et al., 2002; Buijs et al., 2008; Cailotto et al., 2012). Not surprisingly, these findings met with severe criticism from those claiming that such DMV labeling is caused by leakage of the tracers into the peritoneal cavity, thus causing nonspecific uptake and labeling of the DMV (Berthoud et al., 2006). However, these arguments do not explain the difference between DMV neurons specifically projecting only to intra-abdominal fat tissue next to neurons only projecting to subcutaneous tissue, and the disappearance of this labeling after vagal denervation (Kreier and Buijs, 2007). In addition, earlier studies have shown evidence of the labeling of DMV neurons after PRV injection into BAT tissue not receiving Ach innervation (Bamshad et al., 1999 – see Figures 2 and 3 in that paper). Recently, in fact, support for a vagal influence for the growth of intra-abdominal fat tissue was provided by a surgical intervention in humans. Since bodyweight loss is a common problem in patients after gastrectomy, it was hypothesized that selective vagotomy in gastrectomy would result in preferential reduction of visceral fat; indeed, sparing the vagus nerve largely prevented this weight loss (Miyato et al., 2012).

In view of the direct labeling of DMV neurons after injection of the conventional tracer CtB into the spleen, liver and adipose tissue, the most likely explanation is that these organs receive direct input from the DMV without the ganglion functioning as a go-between. This would explain both the retrograde labeling after injection of nonviral retrograde tracers that cannot pass a synapse and the absence of acetylcholine in liver, adipose tissue, and spleen nerve fibers. Research aimed at identifying the transmitters used by the DMV to transmit its message should clarify this.

Feedback

Although much is known about the organization of the output of the ANS to our organs, there is relatively little knowledge about the feedback of our organs to the brain. It is safe to assume that every organ has the capacity to reach the brain via the release of hormones and thus to provide feedback to the control center of the ANS. Examples of this are the secretion of cortisol and aldosterone by the adrenal, leptin and adiponectin by adipose tissue, renin by the kidney, angiotensinogen and fibroblast growth factors by the liver, and so on. Most of these substances also target at least the CNS, where, if they are unable to pass the blood–brain barrier, they act in areas where the blood–brain barrier is absent, such as the circumventricular organs (Smith and Ferguson, 2010). Since these circumventricular organs have extensive projections to autonomic targets within the hypothalamus or in the brainstem and spinal cord, they form an excellent feedback circuit. For example, it is known that angiotensin II acts on the circumventricular organs not only to stimulate water intake but also to increase blood pressure via the ANS in order to compensate for volume loss (Hendel and Collister, 2005). In addition, much is known about visceral feedback that reaches the brainstem in the area of the NTS via the sensory vagus nerve. This feedback innervation has a general somatotopic distribution whereby even different areas of the alimentary tract have different terminal fields in the NTS (Altschuler et al., 1989; Broussard and Altschuler, 2000). In addition, the brain receives feedback from the organs via spinal sensory terminals (Khurana and Petras, 1991; Morita et al., 1991; Qin et al., 2003). Despite the fact that the functional significance of the messages able to reach the brain via spinal afferents from the organs is largely unknown, more and more evidence has been presented that the sensory feedback via spinal sensory nerves is also associated with physiological changes perceived by the organ (de Jonge et al., 2003; Qin et al., 2003; Schliess and Haussinger, 2006; Reinehr et al., 2010; Doignon et al., 2011; Lechner et al., 2011). Most of these signals may be aimed at regulating the function of the organ in a "reflex" manner, but there is also evidence that this feedback may influence the function of other organs, or behavior, via neuronal sensory feedback (Uno et al., 2006; Warne et al., 2007).

THE BRAIN BALANCES ITS AUTONOMIC OUTPUT

As will become clear from many chapters in this book, the organs of our body need a careful balance between themselves and the brain. The autonomic nervous system is the main carrier of direct information between all these parts. If even only one organ is off balance this may, via sensory and hormonal pathways, influence the functioning of the whole system. This becomes clear when behavior is not in line with the message of the brain. For instance, when our biological clock tells us that it is time to fast and rest, and we eat or are active instead, as is the case in shift work, the result is an increased risk of hypertension, obesity, and diabetes (Gangwisch et al., 2006, 2007; Nabe-Nielsen et al., 2008; Scheer et al., 2009). Animal studies have shown that this might be due to a lack of balance not only in physiological processes but also in the molecular machinery of organs such as the liver, which could lead to liver steatosis (Salgado-Delgado et al., 2010, 2013). What emphasizes the importance of maintaining balance in the interaction between the various organs is that liver steatosis, with or without associated obesity, is often associated with kidney disease (Targher et al., 2012; Abenavoli et al., 2013). This suggests that a disbalance in the liver itself may have repercussions on the functioning of the kidney. Similarly, it has been demonstrated that inducing disbalance in the liver by adenovirus expression of peroxisome proliferator-activated receptor, which leads to liver steatosis, results in a loss of peripheral adiposity and an increase in energy expenditure. Tellingly, these effects were nullified by selective sensory denervation of the vagus nerve, which underlines the importance of sensory liver, brain signaling (Uno et al., 2006). Similarly, but in this case as a reflection of an unbalance of the autonomic output of the brain, the association between intra-abdominal fat and the metabolic syndrome might originate in a behavior that is not in line with the physiological demands of the body, but which eventually leads to symptoms of liver steatosis and/or glucose unbalance, resulting in the accumulation of intra-abdominal fat in one case and hypertension in the other (Kreier et al., 2003; Salgado-Delgado et al., 2010, 2013). Animal experiments showing a disturbance in the liver enzymes and glucose levels in the circulation after an unbalance of the autonomic interaction with the liver (Cailotto et al., 2007) also support the hypothesis that lack of balance in the autonomic output to a single organ may have effects not only on the organ itself but also on the entire physiology.

CONCLUDING REMARKS

In this chapter, the importance of the interaction between brain and body in order to maintain homeostasis has been emphasized. This is not just a matter of a top-down or reflex regulation, it is also a matter of signals from the organs influencing the functioning of the brain while the reflex regulation of blood pressure and heart is subject to modulation by descending information from several areas in hypothalamus or cortex. The output of the CNS to control its autonomic output shows an amazing differentiation; not only are there different neurons within, for example, the biological clock or prefrontal cortex, which may influence selectively the parasympathetic or sympathetic motor neurons, there are also different neurons that project to different body compartments. Visceral sensory information may reach higher centers in the CNS via vagal or spinal sensory pathways, leading to integrated responses taking into account such factors as time of day, season, reproductive status, and mood. Based on all that information, the brain sets the balance of the different parts of the ANS, changing the emphasis of the ANS output depending on the situation. If that balance is disturbed, either by behavior or by disease of the organ, this may lead to pathology that may affect the functioning of the whole individual.

References

Abenavoli L, Luigiano C, Larussa T et al. (2013). Liver steatosis in celiac disease: the open door. Minerva Gastroenterol Dietol 59: 89–95.

al Maskati HA, Zbrozyna AW (1989). Stimulation in prefrontal cortex area inhibits cardiovascular and motor components of the defence reaction in rats. J Auton Nerv Syst 28: 117–125.

Altschuler SM, Bao XM, Bieger D et al. (1989). Viscerotopic representation of the upper alimentary tract in the rat: sensory ganglia and nuclei of the solitary and spinal trigeminal tracts. J Comp Neurol 283: 248–268.

Bamshad M, Song CK, Bartness TJ (1999). CNS origins of the sympathetic nervous system outflow to brown adipose tissue. Am J Physiol 276: R1569–R1578.

Berthoud HR, Fox EA, Neuhuber WL (2006). Vagaries of adipose tissue innervation. Am J Physiol Regul Integr Comp Physiol 291: R1240–R1242.

Blessing W, Mohammed M, Ootsuka Y (2013). Brown adipose tissue thermogenesis, the basic rest-activity cycle, meal initiation, and bodily homeostasis in rats. Physiol Behav 10 doi:pii: S0031-9384(13)00104-2. 10.1016/j.physbeh.2013.03.028. [Epub ahead of print].

Borovikova LV, Ivanova S, Zhang M et al. (2000). Vagus nerve stimulation attenuates the systemic inflammatory response to endotoxin. Nature 405: 458–462.

Bratton BO, Martelli D, McKinley MJ et al. (2012). Neural regulation of inflammation: no neural connection from vagus to splenic sympathetic neurons. Exp Physiol 97: 1180–1185.

Broussard DL, Altschuler SM (2000). Brainstem viscerotopic organization of afferents and efferents involved in the control of swallowing. Am J Med 108 (Suppl 4a): 79S–86S.

Brugger P, Marktl W, Herold M (1995). Impaired nocturnal secretion of melatonin in coronary heart disease. Lancet 345: 1408.

Bryant KR, Rothwell NJ, Stock MJ et al. (1983). Parasympathethic effects on diet-induced thermogenesis. Eur J Pharmacol 95: 291–294.

Buijs RM, Kreier F (2006). The metabolic syndrome: a brain disease? J Neuroendocrinol 18: 715–716.

Buijs RM, Wortel J, Van Heerikhuize JJ et al. (1999). Anatomical and functional demonstration of a multisynaptic suprachiasmatic nucleus adrenal (cortex) pathway. Eur J Neurosci 11: 1535–1544.

Buijs RM, Chun SJ, Niijima A et al. (2001). Parasympathetic and sympathetic control of the pancreas: a role for the suprachiasmatic nucleus and other hypothalamic centers that are involved in the regulation of food intake. J Comp Neurol 431: 405–423.

Buijs RM, La Fleur SE, Wortel J et al. (2003a). The suprachiasmatic nucleus balances sympathetic and parasympathetic output to peripheral organs through separate preautonomic neurons. J Comp Neurol 464: 36–48.

Buijs RM, Van Eden CG, Goncharuk VD et al. (2003b). The biological clock tunes the organs of the body: timing by hormones and the autonomic nervous system. J Endocrinol 177: 17–26.

Buijs RM, Scheer FA, Kreier F et al. (2006). Chapter 20: Organization of circadian functions: interaction with the body. Prog Brain Res 153: 341–360.

Buijs RM, Van der Vliet J, Garidou ML et al. (2008). Spleen vagal denervation inhibits the production of antibodies to circulating antigens. PLoS One 3: e3152.

Cailotto C, Van Heijningen C, Van der Vliet J et al. (2007). Daily rhythms in metabolic liver enzymes and plasma glucose require a balance in the autonomic output to the liver. Endocrinology 149: 1914–1925.

Cailotto C, Costes LM, Van der Vliet J et al. (2012). Neuroanatomical evidence demonstrating the existence of the vagal anti-inflammatory reflex in the intestine. Neurogastroenterol Motil 24: 191–200.

Card JP, Rinaman L, Schwaber JS et al. (1990). Neurotropic properties of pseudorabies virus: uptake and transneuronal passage in the rat central nervous system. J Neurosci 10: 1974–1994.

Craig AD (2003). Interoception: the sense of the physiological condition of the body. Curr Opin Neurobiol 13: 500–505.

Dale HH, Feldberg W (1934). The chemical transmitter of vagus effects to the stomach. J Physiol 81: 320–334.

Dale HH, Dudley HW (1929). The presence of histamine and acetylcholine in the spleen of the ox and the horse. J Physiol 68: 97–123.

de Jonge WJ, van den Wijngaard RM, The FO et al. (2003). Postoperative ileus is maintained by intestinal immune infiltrates that activate inhibitory neural pathways in mice. Gastroenterology 125: 1137–1147.

de Jonge WJ, Van Der Zanden EP, The FO et al. (2005). Stimulation of the vagus nerve attenuates macrophage activation by activating the Jak2-STAT3 signaling pathway. Nat Immunol 6: 844–851.

Doignon I, Julien B, Serriere-Lanneau V et al. (2011). Immediate neuroendocrine signaling after partial hepatectomy through acute portal hyperpressure and cholestasis. J Hepatol 54: 481–488.

Freedman RR, Norton D, Woodward S et al. (1995). Core body temperature and circadian rhythm of hot flashes in menopausal women. J Clin Endocrinol Metab 80: 2354–2358.

Gangwisch JE, Boden-Albala B, Buijs RM et al. (2006). Short sleep duration as risk factor for hypertension: analysis of the NHANES I. Hypertension 47: 1–7.

Gangwisch JE, Heymsfield SB, Boden-Albala B et al. (2007). Sleep duration as a risk factor for diabetes incidence in a large U.S. sample. Sleep 30: 1667–1673.

Hayakawa T, Takanaga A, Tanaka K et al. (2003). Cells of origin of vagal motor neurons projecting to different parts of the stomach in the rat: confocal laser scanning and electron microscopic study. Anat Embryol (Berl) 207: 289–297.

Hendel MD, Collister JP (2005). Contribution of the subfornical organ to angiotensin II-induced hypertension. Am J Physiol Heart Circ Physiol 288: H680–H685.

Huang XF, Tork I, Paxinos G (1993). Dorsal motor nucleus of the vagus nerve: a cyto- and chemoarchitectonic study in the human. J Comp Neurol 330: 158–182.

Janig W, Habler HJ (2000). Specificity in the organization of the autonomic nervous system: a basis for precise neural regulation of homeostatic and protective body functions. Prog Brain Res 122: 351–367.

Jones BE, Beaudet A (1987). Distribution of acetylcholine and catecholamine neurons in the cat brainstem: a choline acetyltransferase and tyrosine hydroxylase immunohistochemical study. J Comp Neurol 261: 15–32.

Khurana RK, Petras JM (1991). Sensory innervation of the canine esophagus, stomach, and duodenum. Am J Anat 192: 293–306.

Kim JS, Enquist LW, Card JP (1999). Circuit-specific coinfection of neurons in the rat central nervous system with two pseudorabies virus recombinants. J Virol 73: 9521–9531.

Klein DC, Weller JL, Moore RY (1971). Melatonin metabolism: neural regulation of pineal serotonin: acetyl Coenzyme A N-acetyltransferase activity. Proc Natl Acad Sci U S A 68: 3107–3110.

Kohno T, Mori S, Mito M (1987). Cells of origin innervating the liver and their axonal projections with synaptic terminals into the liver parenchyma in rats. Hokkaido Igaku Zasshi 62: 933–946.

Kreier F, Buijs RM (2007). Evidence for parasympathetic innervation of white adipose tissue, clearing up some vagaries. Am J Physiol Regul Integr Comp Physiol 293: R548–R549.

Kreier F, Fliers E, Voshol PJ et al. (2002). Selective parasympathetic innervation of subcutaneous and intra-abdominal fat – functional implications. J Clin Invest 110: 1243–1250.

Kreier F, Yilmaz A, Kalsbeek A et al. (2003). Hypothesis: shifting the equilibrium from activity to food leads to autonomic unbalance and the metabolic syndrome. Diabetes 52: 2652–2656.

Kreier F, Kap YS, Mettenleiter TC et al. (2005). Tracing from fat tissue, liver and pancreas: a neuroanatomical framework for the role of the brain in type 2 diabetes. Endocrinology 147: 1140–1147.

Lechner SG, Markworth S, Poole K et al. (2011). The molecular and cellular identity of peripheral osmoreceptors. Neuron 69: 332–344.

Lee KW, Blann AD, Lip GY (2005). High pulse pressure and nondipping circadian blood pressure in patients with coronary artery disease: relationship to thrombogenesis and endothelial damage/dysfunction. Am J Hypertens 18: 104–115.

Loewi O (1921). Über humorale Übertragbarkeit der Herznervenwirkung. I. Pflugers Arch 189: 239–242.

Luiten PG, Ter Horst GJ, Karst H et al. (1985). The course of paraventricular hypothalamic efferents to autonomic structures in medulla and spinal cord. Brain Res 329: 374–378.

Lundberg JM, Terenius L, Hokfelt T et al. (1983). High levels of neuropeptide Y in peripheral noradrenergic neurons in various mammals including man. Neurosci Lett 42: 167–172.

Luzi L, Battezzati A, Perseghin G et al. (1992). Lack of feedback inhibition of insulin secretion in denervated human pancreas. Diabetes 41: 1632–1639.

Magni F, Carobi C (1983). The afferent and preganglionic parasympathetic innervation of the rat liver, demonstrated by the retrograde transport of horseradish peroxidase. J Auton Nerv Syst 8: 237–260.

Massin MM, Maeyns K, Withofs N et al. (2000). Circadian rhythm of heart rate and heart rate variability. Arch Dis Child 83: 179–182.

Miyato H, Kitayama J, Hidemura A et al. (2012). Vagus nerve preservation selectively restores visceral fat volume in patients with early gastric cancer who underwent gastrectomy. J Surg Res 173: 60–67.

Morita H, Nishida Y, Hosomi H (1991). Neural control of urinary sodium excretion during hypertonic NaCl load in conscious rabbits: role of renal and hepatic nerves and baroreceptors. J Auton Nerv Syst 34: 157–169.

Nabe-Nielsen K, Garde AH, Tuchsen F et al. (2008). Cardiovascular risk factors and primary selection into shift work. Scand J Work Environ Health 34: 206–212.

Nadelhaft I, Vera PL, Card JP et al. (1992). Central nevous system neurons labelled following the injection of pseudorabies virus into the rat urinary bladder. Neurosci Lett 143: 271–274.

Nakano Y, Oshima T, Ozono R et al. (2001). Non-dipper phenomenon in essential hypertension is related to blunted nocturnal rise and fall of sympatho-vagal nervous activity and progress in retinopathy. Auton Neurosci 88: 181–186.

Nielsen FS, Hansen HP, Jacobsen P et al. (1999). Increased sympathetic activity during sleep and nocturnal hypertension in type 2 diabetic patients with diabetic nephropathy. Diabet Med 16: 555–562.

Ootsuka Y, de Menezes RC, Zaretsky DV et al. (2009). Brown adipose tissue thermogenesis heats brain and body as part of the brain-coordinated ultradian basic rest-activity cycle. Neuroscience 164: 849–861.

Pocai A, Obici S, Schwartz GJ et al. (2005). A brain-liver circuit regulates glucose homeostasis. Cell Metab 1: 53–61.

Powell DA, Watson K, Maxwell B (1994). Involvement of subdivisions of the medial prefrontal cortex in learned cardiac adjustments in rabbits. Behav Neurosci 108: 294–307.

Qin C, Chandler MJ, Miller KE et al. (2003). Responses and afferent pathways of C(1)-C(2) spinal neurons to gastric distension in rats. Auton Neurosci 104: 128–136.

Reinehr R, Sommerfeld A, Haussinger D (2010). Insulin induces swelling-dependent activation of the epidermal growth factor receptor in rat liver. J Biol Chem 285: 25904–25912.

Rosas-Ballina M, Ochani M, Parrish WR et al. (2008). Splenic nerve is required for cholinergic antiinflammatory pathway control of TNF in endotoxemia. Proc Natl Acad Sci U. S. A 105: 11008–11013.

Rosas-Ballina M, Olofsson PS, Ochani M et al. (2011). Acetylcholine-synthesizing T cells relay neural signals in a vagus nerve circuit. Science 334: 98–101.

Rouzade-Dominguez ML, Miselis R, Valentino RJ (2003). Central representation of bladder and colon revealed by dual transsynaptic tracing in the rat: substrates for pelvic visceral coordination. Eur J Neurosci 18: 3311–3324.

Ruggiero DA, Mtui EP, Otake K et al. (1996). Central and primary visceral afferents to nucleus tractus solitarii may generate nitric oxide as a membrane-permeant neuronal messenger. J Comp Neurol 364: 51–67.

Salgado-Delgado R, Angeles-Castellanos M, Saderi N et al. (2010). Food intake during the normal activity phase prevents obesity and circadian desynchrony in a rat model of night work. Endocrinology 151: 1019–1029.

Salgado-Delgado RC, Saderi N, Basualdo MC et al. (2013). Shift work or food intake during the rest phase promotes metabolic disruption and desynchrony of liver genes in male rats. PLoS One 8: e60052.

Schafer MK, Eiden LE, Weihe E (1998). Cholinergic neurons and terminal fields revealed by immunohistochemistry for the vesicular acetylcholine transporter. II. The peripheral nervous system. Neuroscience 84: 361–376.

Scheer FA, Hilton MF, Mantzoros CS et al. (2009). Adverse metabolic and cardiovascular consequences of circadian misalignment. Proc Natl Acad Sci U S A 106: 4453–4458.

Schliess F, Haussinger D (2006). Osmosensing and signaling in the regulation of liver function. Contrib Nephrol 152: 198–209.

Schotzinger RJ, Landis SC (1988). Cholinergic phenotype developed by noradrenergic sympathetic neurons after innervation of a novel cholinergic target in vivo. Nature 335: 637–639.

Sievert LL, Reza A, Mills P et al. (2010). Diurnal rhythm and concordance between objective and subjective hot flashes: the Hilo Women's Health Study. Menopause 17: 471–479.

Smith PM, Ferguson AV (2010). Circulating signals as critical regulators of autonomic state – central roles for the subfornical organ. Am J Physiol Regul Integr Comp Physiol 299: R405–R415.

Strack AM, Sawyer WB, Hughes JH et al. (1989). A general pattern of CNS innervation of the sympathetic outflow demonstrated by transneuronal pseudorabies viral infections. Brain Res 491: 156–162.

Targher G, Pichiri I, Zoppini G et al. (2012). Increased prevalence of chronic kidney disease in patients with type 1 diabetes and non-alcoholic fatty liver. Diabet Med 29: 220–226.

Toth IE, Banczerowski P, Boldogkoi Z et al. (2008). Cerebral neurons involved in the innervation of both the adrenal gland and the ovary: a double viral tracing study. Brain Res Bull 77: 306–311.

Tsukamoto K, Hayakawa T, Maeda S et al. (2005). Projections to the alimentary canal from the dopaminergic neurons in the dorsal motor nucleus of the vagus of the rat. Auton Neurosci 123: 12–18.

Uno K, Katagiri H, Yamada T et al. (2006). Neuronal pathway from the liver modulates energy expenditure and systemic insulin sensitivity. Science 312: 1656–1659.

Van Eden CG, Buijs RM (2000). Functional neuroanatomy of the prefrontal cortex: autonomic interactions. Prog Brain Res 126: 49–62.

Warne JP, Foster MT, Horneman HF et al. (2007). Afferent signaling through the common hepatic branch of the vagus inhibits voluntary lard intake and modifies plasma metabolite levels in rats. J Physiol 583: 455–467.

Waterhouse J, Aizawa S, Nevill A et al. (2007). Rectal temperature, distal sweat rate, and forearm blood flow following mild exercise at two phases of the circadian cycle. Chronobiol Int 24: 63–85.

Zhu BS, Blessing WW, Gibbins IL (1997). Parasympathetic innervation of cephalic arteries in rabbits: comparison with sympathetic and sensory innervation. J Comp Neurol 389: 484–495.

Handbook of Clinical Neurology, Vol. 117 (3rd series)
Autonomic Nervous System
R.M. Buijs and D.F. Swaab, Editors
2013 Published by Elsevier B.V.

Chapter 2

Differential responses of components of the autonomic nervous system

DAVID S. GOLDSTEIN*
Clinical Neurocardiology Section, Clinical Neurosciences Program, Division of Intramural Research, National Institute of Neurological Disorders and Stroke, National Institutes of Health, Bethesda, MD, USA

This chapter conveys several concepts and points of view about the scientific and medical significance of differential alterations in activities of components of the autonomic nervous system in stress and disease. More detailed discussions have appeared elsewhere (Goldstein, 1995, 2001, 2006).

HISTORICAL OVERVIEW

The English physiologist John N. Langley coined the terms "autonomic nervous system" and "parasympathetic nervous system," about the turn of the 20th century – that is, relatively recently in medical history (Langley, 1921); however, the notion that automatic, unconscious, involuntary mechanisms orchestrate the functions of body organs is ancient. The 2nd-century Greek physician Galen, whose teachings dominated medical thought and practice for 14 centuries, viewed the chains of ganglia on each side of the spinal column and nerves emanating from them as tubes distributing the "animal spirit" in the body, producing consent, or "sympathy," among the organs. This nerve network came to be called the "sympathetic nervous system."

The notion that the autonomic nervous system plays a key role in regulation of the body's "inner world" is not only essentially correct but it is a founding concept of modern physiology. In 1865, in his masterpiece, *Introduction to the Study of Experimental Medicine,* Claude Bernard described his earlier observations on the role of the liver in secreting glucose formed from glycogen stores and how his studies of heat regulation led to the discovery of vascular blood flow regulation by sympathetic nerves (Bernard, 1912). By tremendously insightful induction he developed the concept that the ability of an organism to maintain a constant fluid environment bathing cells of the body – the "*milieu intérieur*" – is essential for life independent of the external environment. He postulated that the body maintains the constant internal environment by myriad, continual, compensatory reactions. These compensatory reactions tend to restore a state of equilibrium in response to any outside changes, enabling independence from the external environment. His *Lectures on the Phenomena of Life Common to Animals and Vegetables* (Bernard, 1974) contains one of the most famous passages in the history of physiology: "The constancy of the internal environment is the condition for free and independent life ... All the vital mechanisms, however varied they might be, always have one purpose, that of maintaining the integrity of the conditions of life within the internal environment." This view might seem straightforward or even simple minded today, but it was revolutionary in the history of medical ideas.

Beginning about the turn of the 20th century, the highly influential American physiologist Walter B. Cannon expanded on Bernard's theory of the *milieu intérieur*. In a series of magnificent experiments over about a quarter century, Cannon demonstrated for the first time the critical role that adrenaline, the effector compound of the sympathetic adrenergic system, plays in maintaining the constancy of the inner world of the body (Cannon, 1929a, 1939).

Cannon introduced and popularized three ideas that by now are well known and widely accepted: homeostasis, fight-or-flight responses, and the sympathoadrenal system. All three relate to the main message of this

*Correspondence to: David S. Goldstein, MD PhD, Clinical Neurocardiology Section, CNP, DIR, NINDS, NIH, Building 10, Room 5N220, 10 Center Drive, MSC 1620, 9000 Rockville Pike, Bethesda, MD 20892-1620, USA. Tel: +1-301-496-2103, Fax: +1-301-402-0180, E-mail: goldsteind@ninds.nih.gov

chapter, which is that activities of components of the autonomic nervous system are altered differentially in normal physiology and in disease processes.

"Homeostasis," a word Cannon coined (Cannon, 1929b), refers to the stability of the inner world – a direct extension from Bernard's notion of the *milieu intérieur*. According to Cannon, the brain coordinates body systems with the aim of maintaining a set of goal values for key internal variables. Internal or external disturbances threatening homeostasis arouse internal nervous and hormone systems, induce emotional and motivational states, and generate externally observable behaviors, all of which have a single, overriding goal, which is reestablishing homeostasis.

He also described "fight-or-flight" responses for the first time, asserting that not only physical emergencies such as blood loss from trauma but also psychological emergencies such as antagonistic encounters between members of the same species evoke release of adrenaline into the bloodstream. To Cannon, the body's responses in "fight" are the same as those to "flight." Adrenaline exerts several important effects in different body organs, all of which, from Cannon's point of view, maintain homeostasis in fight-or-flight situations. In the skeletal muscle of the limbs, adrenaline relaxes blood vessels, thereby increasing local blood flow – important not only to provide metabolic fuels to exercising muscle but also to remove waste products of metabolism that otherwise accumulate in skeletal muscle and interfere with performance. Adrenaline constricts cutaneous blood vessels and promotes clotting; both effects minimize blood loss from lacerations. Adrenaline releases the key metabolic fuel, glucose, from the liver into the bloodstream, via breakdown of the storage form of glucose, glycogen. (Claude Bernard discovered conversion of glycogen to glucose in the liver.) Adrenaline stimulates respiration, maximizing delivery of oxygen to the bloodstream via the lungs. Adrenaline removes the electrolyte, potassium ion, from the circulation, an effect that also promote homeostasis, because trauma destroys cells, which contain high potassium ion concentrations, building up the potassium ion content in the surrounding fluid. From a psychological point of view, adrenaline intensifies emotional experiences (Schachter and Singer, 1962) and increases what Cannon called "reservoirs of power," by antifatigue and energizing effects.

Emotion-associated behaviors such as aggressive attack, fearful flight, immobile terror, hopeless defeat, emotional fainting, and sexual activity actually differ importantly in internal physiological and biochemical patterns, just as they do in external appearances, emotional experiences, and behaviors. Conversely, an increase in the circulating level of adrenaline does not imply that the individual is having a fight-or-flight experience. For instance, adrenaline levels approximately double just by a person's standing up, and even a mild fall in the blood glucose level stimulates adrenaline release. Because Cannon used only a single dependent variable, the adrenaline response, he could not appreciate the existence of the different physiological and biochemical patterns. Hans Selye and his followers somewhat analogously emphasized responses of the hypothalamic–pituitary–adrenocortical system.

Most germane to this discussion, Cannon theorized that the sympathetic nervous system and the adrenal gland act together as a functional unit to maintain homeostasis. This probably was the first proposal for the existence of an integrated nervous-hormone, or neuroendocrine, system. Indeed, Cannon became so convinced that the sympathetic nervous system and adrenal gland functioned as a unit that in the 1930s, he formally proposed that the sympathetic nervous system uses the same chemical messenger, adrenaline, as does the adrenal gland. He recognized that stimulation of sympathetic nerves produces effects somewhat different from those produced by injected adrenaline. After he obtained evidence for either release of a substance other than adrenaline during stimulation of sympathetic nerves, or else conversion of adrenaline to a different substance in the target cells, Cannon erroneously backed the latter view. He suggested that two forms of the released chemical messenger exist, excitatory "sympathin E" and inhibitory "sympathin I." In 1939, he wrote that adrenaline was the chemical messenger of the sympathetic nerves. Differences in organ responses to adrenaline and to "sympathin" would be due to conversion of the latter to another substance in the activated target cells (Cannon and Lissak, 1939). This mistake may have cost Cannon a Nobel Prize, because in 1946, a year after Cannon's death, the Swedish physiologist U.S. von Euler correctly identified the chemical messenger of the sympathetic nerves in mammals as norepinephrine (noradrenaline), adrenaline's chemical precursor (von Euler, 1946); for this discovery, von Euler shared the Nobel Prize for Physiology or Medicine in 1970.

Cannon's notion of a unitary sympathoadrenal system persists to this day. As explained in some detail below, however, many situations entail differential regulation of the sympathetic noradrenergic and adrenomedullary hormonal systems; and at least one clinical disorder, fainting, often features a combination of shutdown of sympathetic nervous system outflows to the cardiovascular system yet marked stimulation of adrenomedullary hormonal system activity. Researchers in the area have come to question the validity of the notion of a unitary sympathoadrenal system, although clinicians often continue to lump together the two components.

COMPONENTS OF THE AUTONOMIC NERVOUS SYSTEM

The use of terms such as "the autonomic nervous system," "autonomic failure," "dysautonomia," and "autonomic dysfunction" imply the existence of a single entity. This section presents a different view, that the autonomic nervous system has functionally and neurochemically distinctive components, which are reflected in differential responses to stressors and differential involvement in pathophysiological states.

One can conceptualize the autonomic nervous system as having at least five components: the sympathetic noradrenergic system, the sympathetic cholinergic system, the parasympathetic cholinergic system, the sympathetic adrenergic system, and the enteric nervous system (Goldstein, 2006).

The sympathetic noradrenergic system

The sympathetic noradrenergic system plays a dominant role in regulation of the circulation by the brain, not only during emergencies but in continual activities of daily living such as standing up, exercise, adjustments to meal ingestion, and thermoregulation. There is substantial sympathetic noradrenergic innervation of the heart, and cardiac noradrenergic stimulation increases the force and rate of heart contraction and increases the rate of spontaneous depolarization. Sympathetic noradrenergic nerves enmesh blood vessels – especially arterioles – throughout the body. Since the caliber of the arterioles determines total peripheral resistance to blood flow, sympathetic noradrenergic innervation of the smooth muscle cells in arteriolar walls represents a focal point in neural regulation of blood pressure. In the heart, sympathetic noradrenergic nerves form lattice-like networks around myocardial cells and also supply coronary arterial vessels.

In the sheltered confines of a laboratory, with controlled temperature and *ad libitum* water, nutrients, and calories, mammals do not seem to require an intact sympathetic noradrenergic system (Cannon, 1931). It is by now clear that even under resting conditions pulse-synchronous bursts of skeletal muscle sympathetic nerve activity are detectable, and norepinephrine spills over continuously into the venous drainage of most organs. Activities of daily life, such as standing up, digesting a meal, speaking in public, and walking – i.e., not only emergencies – are associated with rapid adjustments in sympathetic noradrenergic system outflows.

Because of the close architectural association between the sympathetic nerves and myocardial and arteriolar smooth muscle cells, one might predict that lesions of the sympathetic noradrenergic system manifest clinically with dysregulation of cardiovascular performance. Indeed, orthostatic hypotension, a fall in blood pressure each time a person stands up, is a cardinal manifestation of sympathetic noradrenergic failure. By mechanisms that remain poorly understood, neurogenic orthostatic hypotension is commonly accompanied by supine hypertension (Biaggioni and Robertson, 2002), posing a difficult challenge of medical management. Failure of sympathetic noradrenergic outflows to the head produces ptosis and miosis, as in the Horner syndrome. Failure of sympathetic noradrenergic cardiovascular outflows also manifests as postprandial hypotension, inability to tolerate extremes of temperature, and exercise intolerance. The sympathetic noradrenergic system participates also in regulation of sodium balance by the kidneys and in functions of endocrine organs such as the pancreas, thyroid, and salivary glands.

The sympathetic cholinergic system

The sympathetic cholinergic system is the main component of the autonomic nervous system that mediates sweating. Failure of sympathetic cholinergic outflows results in anhidrosis, as in the facial anhidrosis that is part of the triad (along with ptosis and miosis) of the Horner syndrome. In autonomic failure from inability to synthesize norepinephrine, the patients have severe neurogenic orthostatic hypotension from sympathetic noradrenergic failure, yet normal sweating because of an intact sympathetic cholinergic system. Analogous neurotransmitter specificity occurs in Parkinson disease with orthostatic hypotension, in which sympathetic noradrenergic denervation occurs without evidence for sympathetic cholinergic denervation in most patients (Sharabi et al., 2003). Small fiber neuropathy and postural tachycardia syndrome often involve locally decreased sweating (Illigens and Gibbons, 2009; Peltier et al., 2010), generally without evidence of sympathetic noradrenergic dysfunction. In pandysautonomia, such as from autoimmune autonomic ganglionopathy, there is both orthostatic hypotension as a result of sympathetic noradrenergic failure and anhidrosis as a result of sympathetic cholinergic failure.

The parasympathetic cholinergic system

The parasympathetic cholinergic system (PCS) is commonly considered to be a "vegetative" system, involved with digestion, elimination, and calorie storage. The main nerve of the PCS, the vagus nerve, innervates the myocardium, splanchnic organs, and most of the gastrointestinal tract. Stimulation of parasympathetic cholinergic outflows increases salivation and lacrimation, evokes pupillary constriction, stimulates gastric acid secretion and gut smooth muscle contraction, augments urinary bladder tone, and contributes to penile erection in men. There is decreased heart rate and decreased atrioventricular node electrical conduction. Parasympathetic cholinergic system failure manifests as dry mouth and eyes (sicca syndrome), urinary retention, and constipation.

Because the vagus nerve innervates the enteric nervous system, which is a local nerve network in walls of the gastrointestinal tract, it is difficult if not impossible to distinguish clinically parasympathetic cholinergic denervation, enteric denervation, and disruption of reflexive modulation of autonomic outflows as determinants of symptoms such as constipation, abdominal bloating, and esophageal reflux.

The sympathetic adrenergic system

With mediation of adrenaline secretion of the bloodstream, the sympathetic adrenergic system constitutes the main hormonal component of the autonomic nervous system. Adrenaline was the first hormone to be identified and produced in a laboratory, in the early 20th century. During the decades that followed, adrenaline acquired a unique mystique and folklore, which, for reasons that were largely technical, outpaced its science. Injected as a drug, adrenaline potently produces obvious effects. The skin turns pale, the heart pounds, the blood pressure rises, and the individual feels energized. Because of this potency, however, the plasma of healthy humans at rest contains remarkably low levels of adrenaline – as little as a per picograms per milliliter (picomolar) – measurable only during the past 30 years or so. The technical problem of measuring levels of adrenaline directly in the bloodstream, as opposed to indirectly and insensitively via adrenaline's hemodynamic effects, impeded accumulation of scientific understanding about involvement of the sympathetic adrenergic system in pathophysiological states.

Adrenaline is one of the three main hormones responsible for regulation of blood glucose levels, the other two being insulin and glucagon. Sympathetic adrenergic system failure might be expected to be associated with a tendency toward hypoglycemia; however, the redundancy of effector systems results in few if any clinical manifestations, unless the other effector systems are inactivated.

SCIENTIFIC INTEGRATIVE MEDICINE

Scientific integrative medicine is not a treatment method or discipline but a way of thinking that applies systems concepts to understand normal physiology and clinical disorders, providing a framework for understanding complex and dynamic challenges to our integrity as organisms and, in turn, for developing novel treatments based on this complexity and dynamism. This section provides an overview of some concepts of scientific integrative medicine.

Negative feedback regulation

Physiological homeostatic systems entail negative feedback regulation of numerous monitored variables, including core temperature, blood pressure, serum osmolality, glucose levels, and metabolic rate. Conceptually, each system depends on a comparator – a homeostat – to compare afferent information about the monitored variables with set points or other criteria for responding.

Disruption of a negative feedback loop, by preventing afferent information from reaching the brain, inability to process the information and regulate effector functions correctly, or dysfunction or loss of effectors, increases fluctuations in levels of monitored variables. Positive feedback loops are inherently unstable, and conversion from a negative to a positive neurocirculatory feedback loop presages rapid decompensation. For instance, one can understand transitions from heat stress to heat shock and from compensated to decompensated heart failure in terms of positive feedback loops.

Multiple effectors

Multiple effector systems determine levels of most monitored variables of the body. This redundancy extends the range of control, allows at least some regulation of the monitored variable if a particular effector fails (compensatory activation), and enables elaboration of specific, adaptive effector patterns. Destruction of sympathetic noradrenergic neurons compensatorily activates the sympathetic adrenergic system (Micalizzi and Pals, 1979). Anesthetized dogs undergoing clamping of the adrenal hilar vessels, which eliminates adrenomedullary secretion, show no change in blood pressure, but chemically sympathectomized dogs undergoing the same clamping evince rapid decreases in blood pressure to shock levels (Gauthier et al., 1972). De Champlain and Van Ameringen (1972) reported similar results in rats.

Compensatory activation can for long periods be a major source of homeostasis in the face of intrinsic degeneration. Eventually, the same activation can induce positive feedback loops, resulting in instability, rapid worsening of clinical status, and death. For instance, in patients with valvular regurgitation, compensatory recruitment of cardiac sympathetic outflow can maintain myocardial contractile function for many years. Sympathetic activation augments peripheral resistance to blood flow, promotes cardiac hypertrophy, increases myocardial oxygen consumption, accelerates atherosclerosis (Manuck et al., 1988), and increases rates of spontaneous electrical depolarization. As intrinsic pump function declines, ever-increasing sympathetic outflow eventually can precipitate sudden death from myocardial infarction or a lethal arrhythmia.

Effector sharing

Different homeostatic systems can share effectors. Sharing of the sympathetic adrenergic system by the barostat and glucostat can explain hyperglycemia in any of several emergencies such as hemorrhagic shock, stroke, sepsis, and myocardial infarction. Not surprisingly, treatment of hyperglycemia by insulin infusion in these situations does not improve outcome. Analogously, sharing of the vasopressin system by the barostat and osmostat can explain hyponatremia in heart failure. From the principle of effector sharing one may predict that the most efficient means to reverse hyperglycemia and hyponatremia in these conditions is to treat the underlying cause.

Primitive specificity

Hans Selye, who popularized stress as a medical scientific idea (Selye, 1956), defined stress as (or a state resulting in) the "the nonspecific response of the body to any demand upon it" (Selye, 1974). By nonspecific he meant a shared element regardless of the nature of the demand, or stressor. Although Selye acknowledged that responses to stressors have a specific component that tends to counter effects of the stressor, he maintained that after removal of specific responses from consideration a nonspecific stress syndrome would remain. Chrousos and Gold (1992) modified Selye's doctrine of nonspecificity by proposing that above a threshold intensity, any stressor elicits the stress syndrome. More than a half century elapsed before Selye's doctrine of nonspecificity underwent experimental testing, which, as discussed below, failed to confirm it (Pacak et al., 1998). Even so, modern lay and even scientific literature continues to accept the notions of a unitary stress response, a central neural stress system, and a stress "syndrome."

In contrast with Selye's doctrine of nonspecificity, "primitive specificity" refers to patterned neuroendocrine responses that depend on the particular challenge to homeostasis as perceived by the organism. The availability of multiple effectors and obvious natural selective advantages may have fostered the evolution of such patterning. The neuroendocrine response pattern for water deprivation differs from that for salt deprivation, which differs from that for glucoprivation, etc. Even in response to different challenges to psychological homeostasis, neuroendocrine patterns have a degree of neuroendocrine specificity. The response in "fight" is not the same as in "flight," "fright," "defeat," or "faint."

Sympathetic noradrenergic system activation dominates the responses to orthostasis, moderate exercise, and exposure to cold, whereas sympathetic adrenergic system activation dominates those to glucoprivation and emotional distress. Hypothalamic–pituitary–adrenocortical system activation seems especially prominent in distressing situations sensed as novel.

Stress, allostasis, and allostatic load

According to the homeostat theory, stress is not a stereotyped, nonspecific response pattern, nor a noxious stimulus, but a condition, or a state, in which expectations, whether genetically programmed, established by prior learning, or deduced from circumstances, do not match current or anticipated perceptions of the internal or external environment, and the discrepancy elicits patterned, compensatory responses. One can conceptualize stress in terms of an error signal that reflects the difference between afferent information about conditions as sensed and a set point for responding that is determined by a regulator.

Steady state levels of monitored variable can be modified by changing the set point or other instructions for responding. Allostasis refers to this "other sameness." The flexibility comes at the cost of wear and tear – allostatic load. Allostatic load can lead eventually to a positive feedback loop and rapid system failure.

By way of analogy, suppose you went on sabbatical for a year and that when you left you forgot to close a large window. The temperature would be controlled at the programmed settings, but the air conditioner would be on more in the summer and the furnace would be on more in the winter. With these appliances being on more of the time there would be more wear and tear on them, and they would eventually become less efficient. This would aggravate the wear and tear – a positive feedback loop. In fact, when you returned, you might find that the entire heating and cooling system had failed. According to the notion of allostatic load, it is by way of prolonged activation of effectors to maintain allostasis that chronic stress can contribute to the development of chronic degenerative diseases.

From the point of view of disorders of catecholamine systems, it is important to recognize that most of the catecholamine released by exocytosis is inactivated by reuptake into the cytosol, mediated by cell membrane catecholamine transporters (Goldstein et al., 1988; Eisenhofer et al., 1996). Stress-related catecholamine release therefore results, in essence, in translocation of catecholamines from vesicular storage sites to the cytosol. The clinical significance of this translocation is discussed in the section on pathophysiological states.

DIFFERENTIAL SYMPATHETIC NORADRENERGIC, SYMPATHETIC ADRENERGIC, AND HYPOTHALAMIC–PITUITARY–ADRENOCORTICAL RESPONSES TO STRESSORS

Beginning soon after adequately sensitive assay methods of plasma levels of norepinephrine and adrenaline became available, evidence rapidly accumulated for differential noradrenergic vs. adrenergic responses in various situations (Robertson et al., 1979). These findings do not fit with Cannon's notion of a unitary sympathoadrenal system.

The largest sympathetic adrenergic system responses are seen when the organism encounters stressors that can be categorized in terms of posing a global or metabolic threat, such as hemorrhagic shock, insulin-induced hypoglycemia, or cardiac arrest (Little et al., 1985; Pacak et al., 1998). In these situations proportionate increments in adrenomedullary secretion generally exceed increments in sympathetic noradrenergic outflows.

Cold exposure

Results of studies of laboratory animals and humans exposed to cold fit with the notion of primitive specificity. In rats, cold exposure increases plasma norepinephrine levels, with little if any increase in plasma adrenaline or corticotropin levels (Fukuhara et al., 1996), consistent with sympathetic noradrenergic activation and relatively less adrenergic and adrenocortical activation. In humans, mild core hypothermia also increases antecubital venous levels of norepinephrine but not adrenaline (Frank et al., 2002). Infusion of ice-cold saline by way of a central i.v. catheter evokes relatively selective sympathetic noradrenergic activation; when the core temperature falls by about 1° centigrade, sympathetic adrenergic activation becomes evident (Frank et al., 2002).

Glucoprivation

Lack of available glucose, either because of low circulating glucose levels (e.g., after insulin injection) or because of block of intracellular glucose metabolism (e.g., after 2-deoxyglucose injection), produces marked activation of the sympathetic adrenergic system, in contrast with relatively mild sympathetic noradrenergic responses that are heterogeneous among organs (Medvedev et al., 1990; Elman et al., 2001). In humans, insulin-induced hypoglycemia produces large increases in plasma adrenaline levels with concurrent increases in forearm norepinephrine spillover (Paramore et al., 1998).

Some of the sympathetic adrenergic response to glucoprivation is the result of experienced distress; however, in humans pretreated with alprazolam for sedation, 2-deoxyglucose administration nevertheless produces massive increases in plasma adrenaline levels (Breier et al., 1992).

Associations of the sympathetic adrenergic and hypothalamic–pituitary–adrenocortical systems

Steroid-producing cells of the adrenal cortex and catecholamine-producing chromaffin cells of the adrenal medulla have until recently been regarded as independent. The cortex and medulla actually have multiple contact areas and because of the cortical-medullary direction of blood flow in the adrenal gland, adrenomedullary chromaffin cells are bathed continuously in blood containing high concentrations of adrenocortical steroids. Adrenal glucocorticoids are trophic for phenylethanolamine-N-methyltransferase, the enzyme catalyzing synthesis of adrenaline from norepinephrine (Wurtman and Axelrod, 1965). Meanwhile, ongoing catecholamine biosynthesis seems trophic for adrenocortical cells (Bornstein et al., 2000).

Corticotropin-releasing hormone plays key roles in activity of the hypothalamic–pituitary–adrenocortical system and the sympathetic adrenergic system. Rats treated with a corticotropin-releasing hormone receptor antagonist have attenuated stress-induced adrenaline responses (Brown et al., 1985), and mice with deficiency of corticotropin-releasing hormone have decreased basal and restraint-induced adrenaline levels (Jeong et al., 2000).

Adrenalectomy compensatorily activates the sympathetic noradrenergic system. Adrenalectomized monkeys have markedly decreased plasma adrenaline concentrations but accentuated plasma norepinephrine responses to subsequent cholecystectomy, and the magnitude of the norepinephrine accentuation is inversely related to the dose of cortisol administered preceding the second surgery (Udelsman et al., 1987). Norepinephrine fails to normalize completely even at the high dose of glucocorticoid replacement, calculated to be similar to the cortisol production rate during maximal stress.

Prednisone administered daily over 2 weeks decreases directly recorded skeletal muscle sympathetic nerve activity and plasma norepinephrine levels. Glucocorticoids therefore decrease sympathetic noradrenergic outflows in humans (Golczynska et al., 1995).

Meta-analysis on sympathetic and hypothalamic–pituitary–adrenocortical responses to stressors

A meta-analysis (Goldstein and Kopin, 2008) retrieved publications describing original data about plasma adrenaline, norepinephrine, and corticotropin levels measured before and during or after exposure to stressors, both in humans and in laboratory animals. A total of 15 different stressors were identified for which the available literature satisfied inclusion criteria. Mean adrenaline responses were strongly positively correlated with mean corticotropin responses and less strongly with norepinephrine responses. Plasma adrenaline responses were larger than expected for norepinephrine responses during hypoglycemia and smaller than expected for norepinephrine responses during cold exposure without hypothermia, orthostasis, and active escape/avoidance. Plasma norepinephrine responses were larger than expected for corticotropin responses during cold exposure without hypothermia and severe/exhausting exercise and smaller than expected for corticotropin responses during hypoglycemia.

Thus there seems to be at least as good a justification for the concept of coordinated adrenocortical–adrenomedullary responses as for coordinated adrenomedullary–sympathoneural responses in stress. This coordination may involve corticotropin-releasing hormone driving adre nocortical and adrenomedullary outflows (Yoshida-Hiroi et al., 2002) as well as interactions between adrenocortical and adrenomedullary chromaffin cells (Zuckerman-Levin et al., 2001).

DIFFERENTIAL INVOLVEMENT OF AUTONOMIC COMPONENTS IN PATHOPHYSIOLOGICAL STATES

Hypoglycemia-associated adrenomedullary failure in diabetes mellitus

Because of the extraordinary responsiveness of the sympathetic adrenergic system to any form of glucoprivation, insulin-induced hypoglycemia normally evokes marked increases in circulating adrenaline levels, producing a variety of symptoms and signs such as sweating, pallor, fast pulse rate, and anxiety. In hypoglycemia unawareness, the onset of neuroglycopenia, manifested by cognitive dysfunction, impaired judgment, dysphoria, depression, and fatigue, comes before or without clinical signs of sympathetic adrenergic system activation (Hoeldtke and Boden, 1994; Cryer, 2006). Permanent brain damage can result. Hypoglycemia unawareness is an important problem in the management of patients with insulin-dependent diabetes mellitus. Treatment with β-adrenoceptor blockers is unwise in such patients.

Hypoglycemia unawareness is associated with attenuated responses of the sympathetic adrenergic system. Avoidance of hypoglycemia for as little as 2–3 weeks reverses hypoglycemia unawareness and restores adrenaline responses. Real-time continuous glucose monitoring allows patients to view their blood glucose levels, and this form of biofeedback augments the adrenaline response to hypoglycemia (Bornstein et al., 2000).

Disorders of the hypothalamic–pituitary–adrenocortical system

In keeping with the view that in response to stressors, changes in hypothalamic–pituitary–adrenocortical and sympathetic adrenergic activities are more closely related than are changes in sympathetic noradrenergic and sympathetic adrenergic activities, disorders producing decreased delivery of adrenocortical steroids to the adrenal medulla are associated with decreased synthesis or release of adrenaline, without concurrent sympathetic noradrenergic inhibition.

Chromaffin cells regulate steroid hormone release by the adrenal cortex, and steroids induce catecholamine production in the medulla. Consequently, disorders of the adrenal cortex have been shown to affect chromaffin cell function, and vice versa. Mouse models of adrenal cortical dysfunction, such as from targeted disruption of the 21-hydroxylase or type 1 corticotropin-releasing hormone receptor genes, show alterations in chromaffin cell function, while disruption of tyrosine hydroxylase, a key enzyme in catecholamine synthesis, impairs adrenocortical function.

Congenital adrenal hyperplasia due to deficiency of 21-hydroxylase is a disorder of the adrenal cortex characterized by cortisol deficiency, with or without aldosterone deficiency, and androgen excess. Patients with the most severe form also have abnormalities of the adrenal medulla and epinephrine deficiency (Merke et al., 2000).

Addison disease is an autoimmune disorder manifesting with symptoms and signs of adrenocortical failure. Mean plasma adrenaline is reduced in patients with Addison disease compared to controls, and the norepinephrine: adrenaline ratio is increased (Bornstein et al., 1995). These results indicate that a physiologically high local glucocorticoid concentration is responsible for normal adrenaline production under basal conditions.

A third cause of hypoadrenalism is hypophysiotrophic hypocortisolism. Hypopituitary patients have low plasma levels or urinary excretion rates of adrenaline (Rudman et al., 1981).

Sympathoadrenal imbalance in neurocardiogenic syncope

Neurocardiogenic syncope (also called reflex syncope, vasovagal syncope, and vasodepressor syncope) is the most common cause of acute loss of consciousness in adults (Abboud, 1993). We attempted to identify neuroendocrine and hemodynamic changes before syncope that could play a pathophysiological role. Patients referred for chronic orthostatic intolerance had plasma catecholamines measured serially during tilt table testing. Patients with blood sampled before tilt-induced syncope had progressive, marked increases in plasma epinephrine levels before syncope. Simultaneously obtained norepinephrine levels increased to a much smaller extent than did epinephrine levels – hence the term "sympathoadrenal imbalance" – whereas patients who did not have syncope during tilt table testing had forearm vasoconstriction and no sympathoadrenal imbalance during tilt (Goldstein et al., 2003a). Sympathoadrenal imbalance therefore precedes tilt-evoked and spontaneous neurocardiogenic syncope. Sympathoadrenal imbalance may contribute to the hemodynamic derangements resulting in neurocardiogenic syncope. Analogous sympathoadrenal imbalance has been reported in patients with alcohol-induced syncope (Takahashi et al., 2008).

Parkinson disease: neurotransmitter-specific sympathetic lesion

Parkinson disease involves relatively selective dysfunction of the sympathetic noradrenergic system (Sharabi et al., 2003, 2008). Sympathetic adrenergic function seems to be intact (Goldstein et al., 2003b). Although some Parkinson disease patients complain of decreased sweating, others have increased sweating. Considered as a single group, Parkinson disease patients have approximately normal results of the quantitative sudomotor axon reflex test, a measure of sympathetic cholinergic innervation (Sharabi et al., 2003).

Among Parkinson disease patients there is greater loss of noradrenergic innervation in the heart than in the body as a whole (Goldstein et al., 2000). Plasma levels of both norepinephrine and of dihydroxyphenylglycol, the main neuronal metabolite of norepinephrine, are normal in Parkinson disease considered as a single group. Some clues about bases for cardioselectivity of sympathetic noradrenergic denervation in Parkinson disease may be that the myocardium contains a high tissue concentration of norepinephrine, implying relatively dense innervation; cardiac sympathetic nerves avidly remove circulating catecholamines from the coronary arterial blood; and there is greater production of dihydroxyphenylglycol, a product of oxidation of norepinephrine than other organs (Eisenhofer et al., 1992), suggesting a high rate of production of potentially toxic quinones, aldehydes, and other oxidation products.

Parkinson disease and allostatic load

Parkinson disease is well known to be a disease mainly of the elderly. Concepts of scientific integrative medicine help understand why this should be. The flexibility afforded by allostasis comes at the cost of wear and tear – allostatic load. Consider the analogy of a bank robber's getaway car. The engine is always on, idling, because when the time comes, the bank robber must get away quickly. Idling getaway cars provide obvious advantages, but at an eventual cost. If there were a design or manufacturing flaw, or the oil had the wrong formulation, or the idle were set wrong – and if enough time passed – the engine would fail. Catecholaminergic neurons are always in "idle." There have always been strong survival advantages to rapid initiation of movement and sustained performance before exhaustion. Having "leaky vesicles" enables sudden activity by "shifting gears" and stamina by "gearing down" (Eisenhofer et al., 2004). Catecholamines leak continuously from vesicles into the cytoplasm, where the catecholamines are "combusted" by enzyme-catalyzed oxidation to form catecholaldehydes, which are toxic (Panneton et al., 2010). Decreased efficiency of catecholamine storage (Goldstein et al., 2011) could predispose to aging-related deleterious positive feedback loops (Caudle et al., 2008), because catecholaldehydes promote α-synucleinopathy (Burke et al., 2008), and synucleinopathy builds up cytoplasmic catecholamines (Mosharov et al., 2006, 2009). Parkinson disease may therefore be a disease of the elderly because of allostatic load, which depends on genetic predisposition, environmental exposures, stress-related catecholamine release and reuptake into the cytosol, and time. Allostatic load links stress with degenerative diseases. All these processes, combined over a lifetime, could build up "gunk" in catecholaminergic "engines," represented by the Lewy bodies seen in the catecholaminergic neurons of patients with Parkinson disease.

References

Abboud FM (1993). Neurocardiogenic syncope. N Engl J Med 328: 1117–1120.

Bernard C (1912). Introduction à l'étude de la médecine expérimentale. Ch. Delagrave, Paris.

Bernard C (1974). Lectures on the Phenomena of Life Common to Animals and Vegetables. Charles C Thomas, Springfield, IL.

Biaggioni I, Robertson RM (2002). Hypertension in orthostatic hypotension and autonomic dysfunction. Cardiol Clin 20: 291–301, vii.

Bornstein S, Breidert M, Ehrhart-Bornstein M et al. (1995). Plasma catecholamines in patients with Addison's disease. Clin Endocrinol 42: 215–218.

Bornstein SR, Tian H, Haidan A et al. (2000). Deletion of tyrosine hydroxylase gene reveals functional interdependence of adrenocortical and chromaffin cell system in vivo. Proc Natl Acad Sci U S A 97: 14742–14747.

Breier A, Davis O, Buchanan R et al. (1992). Effects of alprazolam on pituitary-adrenal and catecholaminergic responses to metabolic stress in humans. Biol Psychiatry 32: 880–890.

Brown MR, Fisher LA, Webb V et al. (1985). Corticotropin-releasing factor: a physiologic regulator of adrenal epinephrine secretion. Brain Res 328: 355–357.

Burke WJ, Kumar VB, Pandey N et al. (2008). Aggregation of alpha-synuclein by DOPAL, the monoamine oxidase metabolite of dopamine. Acta Neuropathol 115: 193–203.

Cannon WB (1929a). Bodily Changes in Pain, Hunger, Fear and Rage. D. Appleton & Co., New York.

Cannon WB (1929b). Organization for physiological homeostasis. Physiol Rev 9: 399–431.

Cannon WB (1931). The effects of progressive sympathectomy on blood pressure. Am J Physiol 97: 592–595.

Cannon WB (1939). The Wisdom of the Body. W.W. Norton, New York.

Cannon WB, Lissak K (1939). Evidence for adrenaline in adrenergic neurones. Am J Physiol 125: 765–777.

Caudle WM, Colebrooke RE, Emson PC et al. (2008). Altered vesicular dopamine storage in Parkinson's disease: a premature demise. Trends Neurosci 31: 303–308.

Chrousos GP, Gold PW (1992). The concepts of stress and stress system disorders. Overview of physical and behavioral homeostasis. J Am Med Assoc 267: 1244–1252.

Cryer PE (2006). Hypoglycemia in diabetes: pathophysiological mechanisms and diurnal variation. Prog Brain Res 153: 361–365.

De Champlain J, Van Ameringen MR (1972). Regulation of blood pressure by sympathetic nerve fibers and adrenal medulla in normotensive and hypertensive rats. Circ Res 31: 617–628.

Eisenhofer G, Esler MD, Meredith IT et al. (1992). Sympathetic nervous function in human heart as assessed by cardiac spillovers of dihydroxyphenylglycol and norepinephrine. Circulation 85: 1775–1785.

Eisenhofer G, Friberg P, Rundqvist B et al. (1996). Cardiac sympathetic nerve function in congestive heart failure. Circulation 93: 1667–1676.

Eisenhofer G, Kopin IJ, Goldstein DS (2004). Leaky catecholamine stores: undue waste or a stress response coping mechanism? Ann N Y Acad Sci 1018: 224–230.

Elman I, Goldstein DS, Adler CM et al. (2001). Inverse relationship between plasma epinephrine and testosterone levels during acute glucoprivation in healthy men. Life Sci 68: 1889–1898.

Frank SM, Cattaneo CG, Wieneke-Brady MB et al. (2002). Threshold for adrenomedullary activation and increased cardiac work during mild core hypothermia. Clin Sci 102: 119–125.

Fukuhara K, Kvetnansky R, Cizza G et al. (1996). Interrelations between sympathoadrenal system and hypothalamo-pituitary-adrenocortical/thyroid systems in rats exposed to cold stress. J Neuroendocrinol 8: 533–541.

Gauthier P, Nadeau R, De Champlain J (1972). Acute and chronic cardiovascular effects of 6-hydroxydopamine in dogs. Circ Res 31: 207–217.

Golczynska A, Lenders JW, Goldstein DS (1995). Glucocorticoid-induced sympathoinhibition in humans. Clin Pharmacol Ther 58: 90–98.

Goldstein DS (1995). Stress, Catecholamines, and Cardiovascular Disease. Oxford University Press, New York.

Goldstein DS (2001). The Autonomic Nervous System in Health and Disease. Marcel Dekker, New York.

Goldstein DS (2006). Adrenaline and the Inner World: An Introduction to Scientific Integrative Medicine. Johns Hopkins University Press, Baltimore, MD.

Goldstein DS, Kopin IJ (2008). Adrenomedullary, adrenocortical, and sympathoneural responses to stressors: a meta-analysis. Endocr Regul 42: 111–119.

Goldstein DS, Brush JE Jr, Eisenhofer G et al. (1988). In vivo measurement of neuronal uptake of norepinephrine in the human heart. Circulation 78: 41–48.

Goldstein DS, Holmes C, Li ST et al. (2000). Cardiac sympathetic denervation in Parkinson disease. Ann Intern Med 133: 338–347.

Goldstein DS, Holmes C, Frank SM et al. (2003a). Sympathoadrenal imbalance before neurocardiogenic syncope. Am J Cardiol 91: 53–58.

Goldstein DS, Holmes C, Sharabi Y et al. (2003b). Plasma levels of catechols and metanephrines in neurogenic orthostatic hypotension. Neurology 60: 1327–1332.

Goldstein DS, Holmes C, Kopin IJ et al. (2011). Intra-neuronal vesicular uptake of catecholamines is decreased in patients with Lewy body diseases. J Clin Invest 121: 3320–3330.

Hoeldtke RD, Boden G (1994). Epinephrine secretion, hypoglycemia unawareness, and diabetic autonomic neuropathy. Ann Intern Med 120: 512–517.

Illigens BM, Gibbons CH (2009). Sweat testing to evaluate autonomic function. Clin Auton Res 19: 79–87.

Jeong KH, Jacobson L, Pacak K et al. (2000). Impaired basal and restraint-induced epinephrine secretion in corticotropin-releasing hormone-deficient mice. Endocrinology 141: 1142–1150.

Langley JN (1921). The Autonomic Nervous System. W. Heffer and Sons, Cambridge, England.

Little RA, Frayn KN, Randall PE et al. (1985). Plasma catecholamines and cardiac arrest. Lancet ii: 509–510.

Manuck SB, Kaplan JR, Adams MR et al. (1988). Effects of stress and the sympathetic nervous system on coronary artery atherosclerosis in the cynomolgus macaque. Am Heart J 116: 328–333.

Medvedev OS, Esler MD, Angus JA et al. (1990). Simultaneous determination of plasma noradrenaline and adrenaline kinetics. Responses to nitroprusside-induced hypotension and 2-deoxyglucose-induced glucopenia in the rabbit. Naunyn Schmiedebergs Arch Pharmacol 341: 192–199.

Merke DP, Chrousos GP, Eisenhofer G et al. (2000). Adrenomedullary dysplasia and hypofunction in patients with classic 21-hydroxylase deficiency. N Engl J Med 343: 1362–1368.

Micalizzi ER, Pals DT (1979). Evaluation of plasma norepinephrine as an index of sympathetic neuron function in the conscious, unrestrained rat. Life Sci 24: 2071–2076.

Mosharov EV, Staal RG, Bove J et al. (2006). Alpha-synuclein overexpression increases cytosolic catecholamine concentration. J Neurosci 26: 9304–9311.

Mosharov EV, Larsen KE, Kanter E et al. (2009). Interplay between cytosolic dopamine, calcium, and alpha-synuclein causes selective death of substantia nigra neurons. Neuron 62: 218–229.

Pacak K, Palkovits M, Yadid G et al. (1998). Heterogeneous neurochemical responses to different stressors: a test of Selye's doctrine of nonspecificity. Am J Physiol 275: R1247–R1255.

Panneton WM, Kumar VB, Gan Q et al. (2010). The neurotoxicity of DOPAL: behavioral and stereological evidence for its role in Parkinson disease pathogenesis. PLoS One 5: e15251.

Paramore DS, Fanelli CG, Shah SD et al. (1998). Forearm norepinephrine spillover during standing, hyperinsulinemia, and hypoglycemia. Am J Physiol 275: E872–E881.

Peltier AC, Garland E, Raj SR et al. (2010). Distal sudomotor findings in postural tachycardia syndrome. Clin Auton Res 20: 93–99.

Robertson DA, Johnson GA, Robertson RM et al. (1979). Comparative assessment of stimuli that release neuronal and adrenomedullary catecholamines in man. Circulation 59: 637–643.

Rudman D, Moffitt SD, Fernhoff PM et al. (1981). Epinephrine deficiency in hypocorticotropic hypopituitary children. J Clin Endocrinol Metab 53: 722–729.

Schachter S, Singer J (1962). Cognitive, social, and physiological determinants of emotional state. Psychol Rev 69: 379–399.

Selye H (1956). The Stress of Life. McGraw-Hill, New York.

Selye H (1974). Stress without Distress. New American Library, New York.

Sharabi Y, Li ST, Dendi R et al. (2003). Neurotransmitter specificity of sympathetic denervation in Parkinson's disease. Neurology 60: 1036–1039.

Sharabi Y, Imrich R, Holmes C et al. (2008). Generalized and neurotransmitter-selective noradrenergic denervation in Parkinson's disease with orthostatic hypotension. Mov Disord 23: 1725–1732.

Takahashi N, Imai S, Saito F et al. (2008). Alcohol produces imbalance of adrenal and neuronal sympathetic activity in patients with alcohol-induced neurocardiogenic syncope. Circ J 72: 979–985.

Udelsman R, Goldstein DS, Loriaux DL et al. (1987). Catecholamine-glucocorticoid interactions during surgical stress. J Surg Res 43: 539–545.

von Euler US (1946). A specific sympathomimetic ergone in adrenergic nerve fibres (sympathin) and its relations to adrenaline and nor-adrenaline. Acta Physiol Scand 12: 73–96.

Wurtman RJ, Axelrod J (1965). Adrenaline synthesis: control by the pituitary gland and adrenal glucocorticoids. Science 150: 1464–1465.

Yoshida-Hiroi M, Bradbury MJ, Eisenhofer G et al. (2002). Chromaffin cell function and structure is impaired in corticotropin-releasing hormone receptor type 1-null mice. Mol Psychiatry 7: 967–974.

Zuckerman-Levin N, Tiosano D, Eisenhofer G et al. (2001). The importance of adrenocortical glucocorticoids for adrenomedullary and physiological response to stress: a study in isolated glucocorticoid deficiency. J Clin Endocrinol Metab 86: 5920–5924.

Handbook of Clinical Neurology, Vol. 117 (3rd series)
Autonomic Nervous System
R.M. Buijs and D.F. Swaab, Editors

Chapter 3

Cotransmission in the autonomic nervous system

GEOFFREY BURNSTOCK*
Autonomic Neuroscience Centre, University College Medical School, London, UK and Department of Pharmacology, University of Melbourne, Australia

EARLY STUDIES

For many years, understanding of neurotransmission incorporated the concept that one neuron releases only a single transmitter, termed "Dale's Principle" by Eccles (1957). This idea arose from a widely adopted misinterpretation of Dale's suggestion in 1935 that the same neurotransmitter was stored in and released from all terminals of a single sensory neuron, a suggestion which did not specifically preclude the possibility that more than one transmitter may be associated with the same neuron (Dale, 1935). Early hints that nerves might release more than one transmitter began in the 1950s with evidence for the involvement of both noradrenaline/norepinephrine (NA) and acetylcholine (ACh) in sympathetic transmission. Koelle identified acetylcholinesterase in some adrenergic neurons in 1955 (Koelle, 1955), while Burn and Rand introduced the concept of a "cholinergic" link in adrenergic transmission (Burn and Rand, 1959). Another line of evidence, provided by Hillarp, concerned the coexistence of adenosine 5′-triphosphate (ATP) with catecholamines, first in adrenal chromaffin cells (Hillarp et al., 1955) and later in sympathetic nerves (Schümann, 1958). Ultrastructural studies of the enteric nervous system suggested that there were several different cotransmitters in nerves; at least nine distinguishable types of axon profiles showing different combinations of vesicle types were described in the guinea pig myenteric plexus (Cook and Burnstock, 1976). Inconsistencies in the single transmitter hypothesis provided by these and other studies, including those concerned with invertebrate neurotransmission (Brownstein et al., 1974; Cottrell, 1976), were rationalized in an article by Burnstock in 1976 with the provocative title: "Do some nerve cells release more than one transmitter?" Later, it was widely accepted that "cotransmission" is an integral feature of neurotransmission (see Cuello, 1982; Osborne, 1983; Burnstock, 1990a, 2004, 2009b; Kupfermann, 1991; Lundberg, 1996).

A role for ATP as a cotransmitter in sympathetic, parasympathetic, sensory-motor, and enteric nonadrenergic, noncholinergic (NANC) inhibitory nerves was supported by research from Burnstock and colleagues (see Burnstock 1983, 1990b, 2007, 2009b), while Hökfelt and colleagues focused on the colocalization, vesicular storage, and release of peptides from both peripheral and central nerves (see Hökfelt et al., 1977, 1986). Evidence for ATP being a cotransmitter with established neurotransmitters in the central nervous system (CNS) (including brainstem and hypothalamus, which regulate autonomic activities) as well as in the periphery has been reported (see Table 3.1).

Furness and Costa introduced the concept of "chemical coding" to describe the combination of potential neurotransmitters found in enteric nerves and this concept has since been applied to other nerve types, in both peripheral and central nervous systems (Furness et al., 1989). Colocalized substances are not necessarily cotransmitters; they can (especially peptides) act as pre- and/or postjunctional neuromodulators of the release and actions of the principal cotransmitters. The proportions of cotransmitters vary considerably between species and organs, and show plasticity of expression during development and in pathological conditions. In general, classical transmitters are contained in small synaptic vesicles, whereas peptides are stored in large granular (dense-cored) vesicles (LGVs), although small molecule transmitters are sometimes stored together with peptides in LGVs.

Pharmacological studies of pre- and postjunctional neuromodulation provide evidence, which is complementary to the concept of cotransmission. For example,

*Correspondence to: Geoffrey Burnstock, Autonomic Neuroscience Centre, University College Medical School, Royal Free Campus, Rowland Hill Street, London NW3 2PF, UK. Tel: +44-(0)20-7830-2948, Fax: +44-(0)20-7830-2949, E-mail: g.burnstock@ucl.ac.uk

Table 3.1

Cotransmitters in the peripheral and central nervous systems

	Cotransmitters
Peripheral nervous system	
Sympathetic nerves	ATP + NA + NPY
Parasympathetic nerves	ATP+ACh+VIP
Sensory-motor	ATP+CGRP+SP
NANC enteric nerves	ATP+NO+VIP
Motor nerves (in early development)	ATP+ACh
Central nervous system	
Cortex, caudate nucleus	ATP+ACh
Hypothalamus, locus ceruleus	ATP+NA
Hypothalamus, dorsal horn, retina	ATP+GABA
Mesolimbic system	ATP+DA
Hippocampus, dorsal horn	ATP+glutamate

Compiled from Burnstock, 2007.
ACh, acetylcholine; ATP, adenosine 5′-triphosphate; CGRP, calcitonin gene-related peptide; DA, dopamine; GABA, γ-amino butyric acid; NA, noradrenaline; NO, nitric oxide; NPY, neuropeptide Y; SP, substance P; VIP, vasoactive intestinal peptide.

parallel presynaptic modulation of transmitter overflow supports the concept of closely associated corelease, while postjunctional synergism between colocalized transmitters provides justification of cotransmission in terms of transmitter economy.

SYMPATHETIC NERVES

There is compelling evidence that under certain conditions *in vitro*, single sympathetic neurons may release NA, ACh, or a mixture of these two transmitter substances (Le Douarin et al., 1975; Furshpan et al., 1976; Patterson et al., 1976). It seems likely that this represents a true reflection of events that occur *in vivo* during perinatal development (Hill and Hendry, 1977). It appears that a population of sympathetic nerve cells are present at birth that have the potential to synthesize both NA and ACh. These multipotential cells require nerve growth factor (NGF) to survive and they respond to NGF with an increased production of both choline acetyltransferase and tyrosine hydroxylase, enzymes that are involved in the synthesis of ACh and NA, respectively. Under the influence of conditioning factors, most of the cells appear to differentiate into either cholinergic or adrenergic neurons soon after birth. However, it is possible that some sympathetic neurons, supplying some organs in some animals, retain the ability to produce and release both ACh and NA (see Burn and Rand, 1965; Burnstock, 1978). In late pregnancy, sympathetic innervation of guinea pig uterine blood vessels exhibits a remarkable switch from adrenergic vasoconstrictor to cholinergic vasodilator control (Bell, 1968).

There is substantial evidence to show that NA and ATP are cotransmitters in sympathetic nerves, having differentially important roles as transmitters and neuromodulators depending on the tissue, the species, and on the parameters of stimulation (see Burnstock and Sneddon, 1985; Burnstock, 1990b). The first indication that ATP might be released from sympathetic nerves was the demonstration that stimulation of sympathetic nerves led to release of tritium from taenia coli preincubated in [^{3}H]adenosine (which is taken up and converted to [^{3}H]ATP) (Su et al., 1971). Later, Langer and Pinto (1976) suggested that the substantial residual NANC response of the cat nictitating membrane observed after depletion of NA by reserpine was due to the release of ATP remaining in the sympathetic nerves. Most of the early studies of sympathetic cotransmision, involving NA and ATP, were made by Dave Westfall and colleagues on the vas deferens, a tissue with a high density of sympathetic nerves (Westfall et al., 1978; Fedan et al., 1981). Subsequently, numerous studies demonstrated that cotransmission of NA and ATP also occurs in other visceral organs and many different blood vessels in a variety of species (see Burnstock, 1990a). In retrospect, there was a good indication that the excitatory junction potentials (EJPs) recorded in the guinea pig vas deferens when the electrophysiology of sympathetic nerve-smooth muscle transmission was first described by Burnstock and Holman (1960) (Fig. 3.1A) were due to ATP released as a cotransmitter from sympathetic nerves, rather than to NA. It was puzzling at the time that adrenoceptor antagonists failed to block the EJPs, although guanethidine, a drug that prevents the release of sympathetic transmitters, was effective. It was over 20 years later that NANC EJPs were shown to be blocked by desensitization of the ATP (P2) receptors by α,β-methylene ATP (α,β-meATP) (Fig. 3.1B) and were mimicked by ATP (Sneddon and Burnstock, 1984b; Sneddon and Westfall, 1984) (Fig. 3.1C). Later, it was shown that reserpine, which depletes NA in the sympathetic nerves, did not affect the twitch contractions of the vas deferens in response to nerve stimulation or release of ATP (Kirkpatrick and Burnstock, 1987). After destruction of sympathetic nerves with 6-hydroxydopamine, purinergic nerve-mediated responses were abolished, establishing that the ATP was released from sympathetic nerves and not from separate "purinergic" nerves. ATP is costored with NA in small and large vesicles. Differential prejunctional modulation of the release of NA and ATP by various agents has been shown in the vas deferens, perhaps indicating that NA and ATP are stored in different vesicles (Ellis and Burnstock, 1990). ATP has been shown to be a cotransmitter with NA in

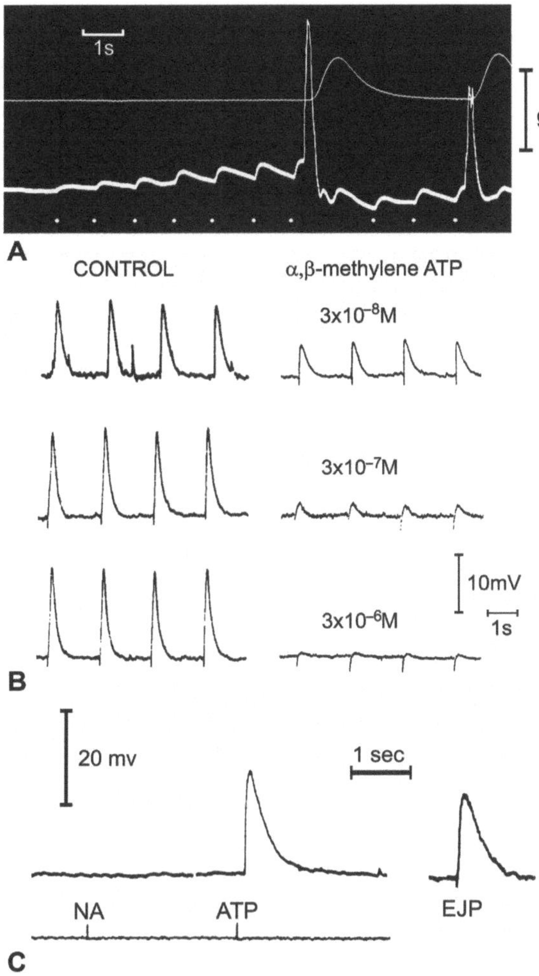

Fig. 3.1. Electrical activity in smooth muscle cells of the guinea pig vas deferens during stimulation of sympathetic nerves. (**A**) Sucrose gap recording of the activity of smooth muscle cells of the vas deferens in response to nerve stimulation (white dots). The upper trace records the tension, the lower trace electrical activity. Note both facilitation and summation of successive excitatory junction potentials (EJPs) and that at a critical depolarization threshold an action potential is initiated which results in contraction. (Reproduced from Burnstock and Costa, 1975, with permission from Springer.) (**B**) The effect of α,β-methylene (α,β-meATP) ATP on EJPs recorded from the guinea pig vas deferens (intracellular recording). The control responses to stimulation of the motor nerves at 0.5 Hz are shown on the left. After at least 10 minutes in the continuous presence of α,β-meATP, EJPs were recorded using the same stimulation parameters. The EJPs are clearly reduced in magnitude in the presence of α,β-meATP (3×10^{-8} M and 3×10^{-7} M). Notice also that in control cells several large spontaneous EJPs were seen, whereas after α,β-meATP no spontaneous EJPs were recorded. The EJPs were virtually abolished with α,β-meATP (3×10^{-6} M). (Reproduced from Sneddon and Burnstock, 1984b, with permission from Elsevier.) (**C**) The effect of NA, ATP and nerve stimulation (EJP) on guinea pig vas deferens recorded using microelectrodes (From Burnstock and Verkhratsky, 2012, with permission from Springer.)

sympathetic nerves supplying the human vas deferens (Banks et al., 2006).

Cotransmission of NA and ATP in perivascular sympathetic nerves supplying the aorta, portal vein, and saphenous, pulmonary, tail, ear, basilar, hepatic, coronary, and mesenteric arteries of different mammalian species has been described (e.g., see Sneddon and Burnstock, 1984a; Kennedy and Burnstock, 1985; Ramme et al., 1987; Warland and Burnstock, 1987; Burnstock, 1990b, 1995, 2009c). Electrophysiological studies have shown that in a number of vessels the electrical response to stimulation of perivascular sympathetic nerves is biphasic; an initial fast, transient depolarization or EJP of the vascular smooth muscle is followed by a slow, prolonged depolarization. The EJP and slow depolarization are mimicked by the effects of ATP and NA, respectively. Considerable variation exists in the proportions of NA and ATP utilized by sympathetic nerves. For example, in guinea pig submucosal arterioles both vasoconstriction and EJPs, evoked in response to electrical stimulation of sympathetic nerves, are mediated exclusively by ATP, with NA assuming the role of a neuromodulator, by acting through prejunctional α_2-adrenoceptors to depress transmitter release (Evans and Surprenant, 1992). At the other extreme, in rat renal arteries the purinergic component is relatively small. It has also been noted that the purinergic component is optimal with short bursts of low frequency stimulation, whereas longer durations of higher frequency favor adrenergic transmission.

Neuropeptide Y (NPY) has been found to be present in LGV in most sympathetic nerves. The release of NPY, as well as NA and ATP, in response to electrical stimulation of sympathetic nerve terminals is prevented by guanethidine. The major role of NPY in the vasculature, and in the vas deferens, appears to be that of a pre- and/or postjunctional modulator of sympathetic transmission, since it has little direct postjunctional action or causes contraction only at high concentrations (see Fig. 3.2A). Direct vasoconstrictor actions of NPY have, however, been demonstrated in some vessels. At the prejunctional level, NPY has potent inhibitory effects, reducing the release of NA and ATP from sympathetic nerves. Postjunctionally, NPY generally acts to enhance the actions of sympathetic nerve stimulation, NA and ATP.

Although 5-hydroxytryptamine (5-HT) immunofluorescent nerves have been localized in a number of vessels, for the most part 5-HT is not synthesized and stored in separate nerves, but is taken up, stored in, and released as a "false transmitter" from sympathetic nerves (Jackowski et al., 1989). Enkephalins have been shown to coexist with NA in cell bodies and fibers of some postganglionic sympathetic neurons (Schultzberg

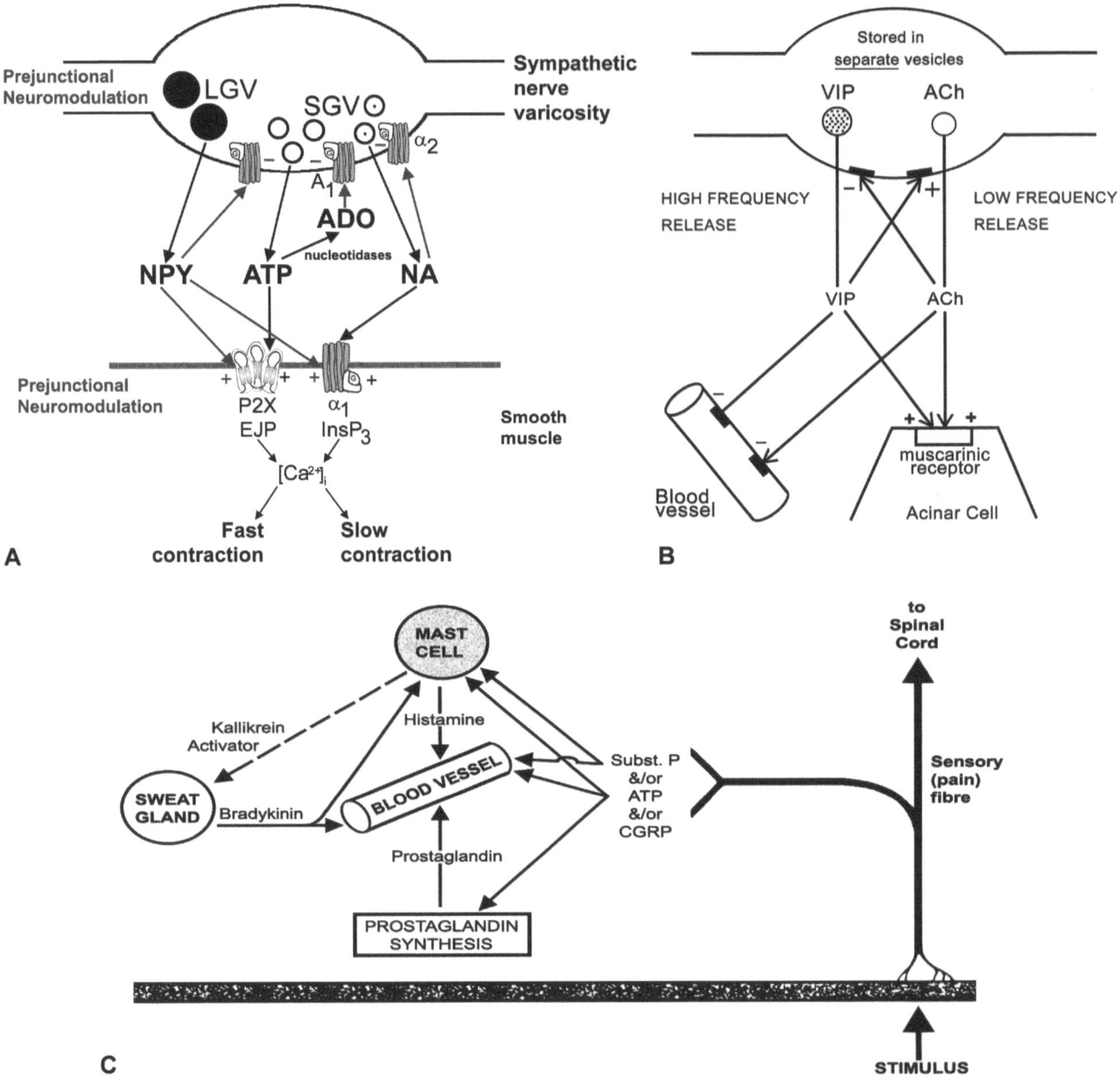

Fig. 3.2. (**A**) Schematic of sympathetic cotransmission. ATP (adenosine 5′-triphosphate) and NA (noradrenaline) released from small granular vesicles (SGV) act on P2X and α_1 receptors on smooth muscle, respectively. ATP acting on inotropic P2X receptors evokes excitatory junction potentials (EJPs), increase in intracellular calcium ($[Ca^{2+}]_i$) and fast contraction; while occupation of metabotropic α_1 adrenoceptors leads to production of inositol triphosphate (IP_3), increase in $[Ca^{2+}]_i$ and slow contraction. Neuropeptide Y (NPY) stored in large granular vesicles (LGV) acts after release both as a prejunctional inhibitory modulator of release of ATP and NA and as a postjunctional modulatory potentiator of the actions of ATP and NA. Soluble nucleotidases are released from nerve varicosities, and are also present as ectonucleotidases. (Reproduced from Burnstock, 2009e, with permission from Elsevier.) (**B**) A classic transmitter (ACh) coexists with vasoactive intestinal polypeptide (VIP) in parasympathetic nerves supplying the cat salivary gland. ACh and VIP are stored in separate vesicles; they can be released differentially at different stimulation frequencies to act on acinar cells and glandular blood vessels. ACh is released during low frequency stimulation to increase salivary secretion from acinar cells and to elicit some minor dilatation of blood vessels in the gland. At high stimulation frequencies, VIP is released to produce marked dilatation of the blood vessels in the gland and to act as a neuromodulator, postjunctionally on the acinar gland to enhance the actions of ACh, and prejunctionally on the nerve varicosities to enhance the release of ACh. ACh also has an inhibitory action on the release of VIP. (Reproduced from Burnstock, 1983, with permission from Elsevier.) (**C**). Diagram showing the basis of the "axon reflex" in the skin leading to vasodilatation and inflammation. It is suggested that calcitonin gene-related peptide (CGRP), substance P (SP), and ATP are released during antidromic activation of sensory collatorals. (Adapted from Burnstock, 1977, with permission of the Nature Publishing Group.)

et al., 1979). The functional significance of sympathetic coexistence of opioids is likely to be related to their prejunctional inhibitory effects on sympathetic transmission.

PARASYMPATHETIC NERVES

Evidence was presented for seasonal changes in release of ACh, 5-HT, histamine, and a peptide from vagus nerves supplying the frog stomach (Singh, 1964). The evidence for cotransmission of ACh and vasoactive intestinal polypeptide (VIP) in certain postganglionic parasympathetic neurons comes from pharmacological studies performed by Lundberg (1981) on cat salivary glands (see Fig. 3.2B). ACh and VIP are released from the same parasympathetic nerve terminals in response to transmural nerve stimulation. During low-frequency stimulation, ACh is released to cause an increase in salivary secretion from acinar cells and also to elicit some minor dilatation of blood vessels in the gland. VIP is preferentially released at high frequencies to cause marked vasodilatation of blood vessels and, while it has no direct effect on acinar cells, it acts as a neuromodulator to enhance both the postjunctional effect of ACh on acinar cell secretion and the release of ACh from nerve varicosities via prejunctional receptors. Vasodilator nerves to the uterine arteries in the guinea pig contain immunoreactivity to VIP, which coexists with dynorphin, NPY, and somatostatin (Morris et al., 1987). NPY-like immunoreactivity has been reported in some of the choline acetyltransferase-/VIP-containing neurons of the parasympathetic ciliary, sphenopalatine, otic, and pterygopalatine ganglia with targets including the iris and cerebral vessels (Leblanc and Landis, 1988).

Autonomic control of penile erection, involving relaxation of the smooth muscle of the corpus cavernosum as well as dilatation of other penile vascular beds, has traditionally been attributed to the vasodilator effects of ACh and VIP released from parasympathetic neurons. Nitric oxide (NO) released from nerves arising from neurons in the pelvic ganglia, have been claimed to play a role in smooth muscle relaxation leading to penile erection (Burnett et al., 1992). NO synthase (NOS)-containing fibers, shown by lesion studies to arise from parasympathetic cell bodies in the sphenopalatine ganglia, have been localized in the adventitia of cerebral arteries and many of these also contain VIP (Bredt and Snyder, 1992). A functional role for perivascular neuronal NO in cerebral arteries has been identified in studies showing that stimulation of adventitial nerve fibers causes vascular relaxation, which is attenuated by inhibitors of NOS (Toda et al., 1990).

Parasympathetic nerves supplying the urinary bladder utilize ACh and ATP as cotransmitters (Burnstock et al., 1972, 1978; Burnstock, 2001b), in variable proportions in different species, and by analogy with sympathetic nerves, ATP again acts through P2X ionotropic receptors to produce EJPs and fast contraction, while the slow component of the response is mediated by a metabotropic receptor, in this case muscarinic. There is also evidence for parasympathetic, purinergic cotransmission to resistance vessels in the heart and airways.

SENSORY-MOTOR NERVES

The neuropeptides substance P (SP) and calcitonin gene-related peptide (CGRP) are the principal transmitters of primary afferent nerves and have been shown to coexist in the same terminals (Gibbins et al., 1985; Rubino and Burnstock, 1996). Furthermore, with the use of colloidal gold particles of different sizes, they have been shown to coexist in the same large granular vesicles (Gulbenkian et al., 1986). The motor (efferent) function of sensory nerves has been demonstrated in rat mesenteric arteries where evidence exists for a role for CGRP as the mediator of vasodilatation following release from sensory motor nerves. In contrast, SP is not coreleased with CGRP by electrical stimulation and SP has little or no vasodilator action on rat mesenteric arteries. While it is possible that SP released from nerves supplying the microvasculature could produce vasodilatation via SP receptors on endothelial cells, it is most unlikely to reach the endothelium without degradation in larger blood vessels.

Other peptide and nonpeptide substances, including neurokinin A, somatostatin, VIP, and ATP, have been described in capsaicin-sensitive sensory neurons (Maggi and Meli, 1988). Unmyelinated sensory neurons containing cholecystokinin (CCK)/CGRP/dynorphin/SP have been shown to project to cutaneous arterioles in guinea pig skin (Gibbins et al., 1987). Neurons from the same ganglia which contain CCK/CGRP/SP innervate arterioles of skeletal muscle, CGRP/dynorphin/SP nerve fibers mostly supply the pelvic viscera, and CGRP/SP fibers run mainly to the heart, large arteries, and veins. There is also evidence for a sensory role for ATP and it has been proposed that ATP may coexist in sensory nerve terminals with SP and CGRP, and with glutamate in primary afferent neurons (Holton, 1959; Burnstock, 2009d) (Fig. 3.2C).

INTRINSIC ENTERIC AND CARDIAC NEURONS

Intrinsic neurons exist in most of the major organs of the body. Many of these are part of the parasympathetic nervous system, but certainly in the gut, and perhaps also in the heart and airways, some of these intrinsic neurons are derived from neural crest tissue, which differs from that which forms the sympathetic and parasympathetic systems, and appear to participate in local reflex actions independent of the CNS.

The enteric nervous system contains several hundred million neurons located in the myenteric plexuses between muscle coats and the submucous plexus. The chemical coding of these nerves has been examined in detail (see Furness, 2006). A subpopulation of these intramural enteric nerves provides NANC inhibitory innervation of the gastrointestinal smooth muscle. It seems likely that three major cotransmitters are released from these nerves. ATP produces fast inhibitory junction potentials, NO also produces inhibitory junction potentials but with a slower time course, while VIP produces slow tonic relaxations (see Burnstock, 2001a, 2008). The proportions of these three transmitters varies considerably in different regions of the gut and in different species; for example, in some sphincters the NANC inhibitory nerves utilize largely VIP, in others largely NO, while in nonsphincteric regions of the intestine ATP is more prominent. In recent papers, evidence has suggested that ACh and ATP are fast excitatory cotransmitters to myenteric neurons and that there may be colocalization of ACh, ATP, and serotonin in enteric Dogiel type I/S neurons. Detailed studies have allowed a very complete mapping of the complex neuronal markers and projections of enteric neurons. Several peptidergic substances, including NPY, VIP, enkephalin, somatostatin, peptide histidine isoleucine, galanin, SP, and CGRP, have been identified in enteric neurons, often coexisting (up to five peptides in the same neuron) with the neurotransmitters NA, ACh, 5-HT, NO, and ATP (see Table 3.2) (Furness and Costa, 1987).

Studies of intrinsic cardiac neurons in culture have shown that subpopulations of intrinsic nerves in the atrial and intra-atrial septum contain and/or release cotransmitters, including ATP, NO, NPY, ACh, and 5-HT (Hassall and Burnstock, 1986; Burnstock et al., 1987). Many of these nerves project to the coronary microvasculature and produce potent vasomotor actions (Saffrey et al., 1992). NO and ATP have been shown to be the mediators of NANC vasodilatation of the rabbit portal vein.

PHYSIOLOGICAL SIGNIFICANCE OF COTRANSMISSION

In general, autonomic cotransmission offers more diverse physiological control by mechanisms other than the all-or-none control by messages coming from the CNS that was the dominant view for many years (see Burnstock, 2004, 2009a) (Fig. 3.3).

Cotransmitters with different firing patterns

Although single presynaptic action potentials release small molecule neurotransmitters, trains of impulses are needed to release neuropeptides. For sympathetic and parasympathetic cotransmission, release of ATP is favored at low frequency stimulation, whereas NA and ACh are released at higher frequencies (Bradley et al., 2003; Ventura et al., 2003). There are instances where more than one fast cotransmitter is released (e.g., glutamate and ATP) together with one or more peptides.

Different cotransmitters act on different postjunctional cells

Neurons using multiple transmitters may project to two or more targets. For example, ACh released at low frequency stimulation from parasympathetic nerves supplying salivary glands acts on acinus cells to produce secretion and a minor dilatation of vessels, whereas, at higher frequency stimulation, its cotransmitter VIP causes powerful vasodilatation of vessels in the glands and postjunctional enhancement of ACh-induced saliva secretion (Lundberg, 1996).

Neuromodulation

A cotransmitter can feed back on presynaptic receptors that increase or decrease its own release and/or that of its cotransmitter(s) (see Vizi, 1979). For example, ATP released as a cotransmitter with glutamate from primary afferent fibers in lamina II of the spinal cord can act on prejunctional $P2X_3$ receptors to facilitate the release of its cotransmitter, glutamate, whereas adenosine resulting from ectoenzymatic breakdown of ATP acts on presynaptic P1 receptors to inhibit glutamate release. Both NA and ATP can prejunctionally modulate sympathetic transmission, NA via prejunctional α_2-adrenoceptors and ATP via P1 receptors following breakdown to adenosine or directly via P2X and P2Y receptors (see Burnstock, 2007). Modulation of cotransmitter release and presynaptic action by other local agents also occurs and might provide another level of synaptic flexibility.

Synergism

There is an increasing number of reports of the synergistic actions of cotransmitters. ATP and NA released from sympathetic nerves have synergistic actions on smooth muscle of vas deferens and blood vessels, and ATP released with ACh from motor neurons facilitates the nicotinic actions of ACh at the skeletal neuromuscular junction. The mechanisms underlying cotransmitter synergism are not well understood. However, it has been suggested that postjunctional synergism between the responses of vas deferens to NA and ATP is caused by the ability of NA to potentiate the contractile responses to ATP by sensitizing smooth muscle cells

Table 3.2

Types of neurons in the enteric nervous system† ‡

	Proportion	Chemical coding	Function/comments
Myenteric neurons			
Excitatory circular muscle motor neurons	12%	*Short*: ChAT/TK/ENK/GABA *Long*: ChAT/TK/ENK/NFP	To all regions, primary transmitter ACh, cotransmitter TK
Inhibitory circular muscle motor neurons	16%	*Short*: NOS/VIP/PACAP/ENK/NPY/GABA *Long*: NOS/VIP/PACAP/Dynorphin/BN/NFP	Several cotransmitters with varying prominence: NO, ATP, VIP, PACAP
Excitatory longitudinal muscle motor neurons	25%	ChAT/Calretinin/TK	Primary transmitter ACh, cotransmitter TK
Inhibitory longitudinal muscle motor neurons	~2%	NOS/VIP/GABA	Several cotransmitters with varying prominence: NO, ATP, VIP, PACAP
Ascending interneurons (local reflex)	5%	ChAT/Calretinin/TK	Primary transmitter ACh
Descending interneurons (local reflex)	5%	ChAT/NOS/VIP ± BN ± NPY	Primary transmitter ACh, ATP may be a cotransmitter
Descending interneurons (secretomotor reflex)	2%	ChAT/5-HT	Primary transmitter ACh, 5-HT (at NK_3 receptors)
Descending interneurons (migrating myoelectric complex)	4%	ChAT/SOM	Primary transmitter ACh
Myenteric intrinsic primary afferent (primary sensory) neurons	26%	ChAT/Calbindin/TK/NK_3 receptor	Primary transmitter TK
Intestinofugal neurons	<1%	ChAT/BN/VIP/CCK/ENK	Primary transmitter ACh
*Motor neurons to gut endocrine cells	N/A	N/A	For example, myenteric neurons innervating gastrin cells. Neurons of this type may be in submucosal ganglia
Submucosal neurons			
Noncholinergic secretomotor/vasodilator neurons	45%	VIP/GAL	Primary transmitter VIP. A small proportion of these have cell bodies in myenteric ganglia
Cholinergic secretomotor/vasodilator neurons	15%	ChAT/Calretinin/Dynorphin	Primary transmitter ACh

†This table lists the neuron types that are found in the guinea-pig small intestine, some of their defining characteristics, and percentages of occurrence in each of the ganglionated plexuses. Also listed are three types of motor neuron that are found in other parts of the tubular digestive tract, marked by an asterisk*. ACh, acetylcholine; ATP, adenosine 5′-triphosphate; BN, bombesin; CCK, cholecystokinin; CGRP, calcitonin gene-related peptide; ChAT, choline acetyltransferase; DA, dopamine; ENK, enkephalin; GABA, γ-amino butyric acid; GAL, galanin; 5-HT, 5-hydroxytryptamine; NA, noradrenaline; NFP, neurofilament protein; NK, neurokinin; NOS, nitric oxide synthase; NPY, neuropeptide Y; PACAP, pituitary adenylyl cyclase-activating peptide; SOM, somatostatin; SP, substance P; TK, tachykinin; VIP, vasoactive intestinal peptide.
‡Reproduced from Furness, 2000, with permission from Elsevier.

to Ca^{2+} via an inhibitory action on myosin light chain phosphatase, an action mediated by protein kinase C (Smith and Burnstock, 2004).

Negative cross-talk

Coapplication of nicotinic and P2X receptor agonists produces less than the additive responses predicted by independent receptor activation. Inhibitory interactions between 5-HT_3 and P2X receptors have been described in submucosal and myenteric neurons (Barajas-López et al., 2002; Boué-Grabot et al., 2003).

Trophic factors

Some costored and coreleased substances can act as long-term (trophic) factors, as well as fast neurotransmitters. For example, ATP can act on P2 receptors, or P1 (adenosine) receptors after ectoenzymatic breakdown, to promote vascular cell proliferation, motility, differentiation or death (see Burnstock, 2002; Burnstock and Verkhratsky, 2010). NPY released from sympathetic nerves has cardiovascular trophic effects in end-stage renal disease. There is growing evidence that neuropeptide trophic factors are synthesized, stored and released from nerve terminals together with fast neurotransmitters.

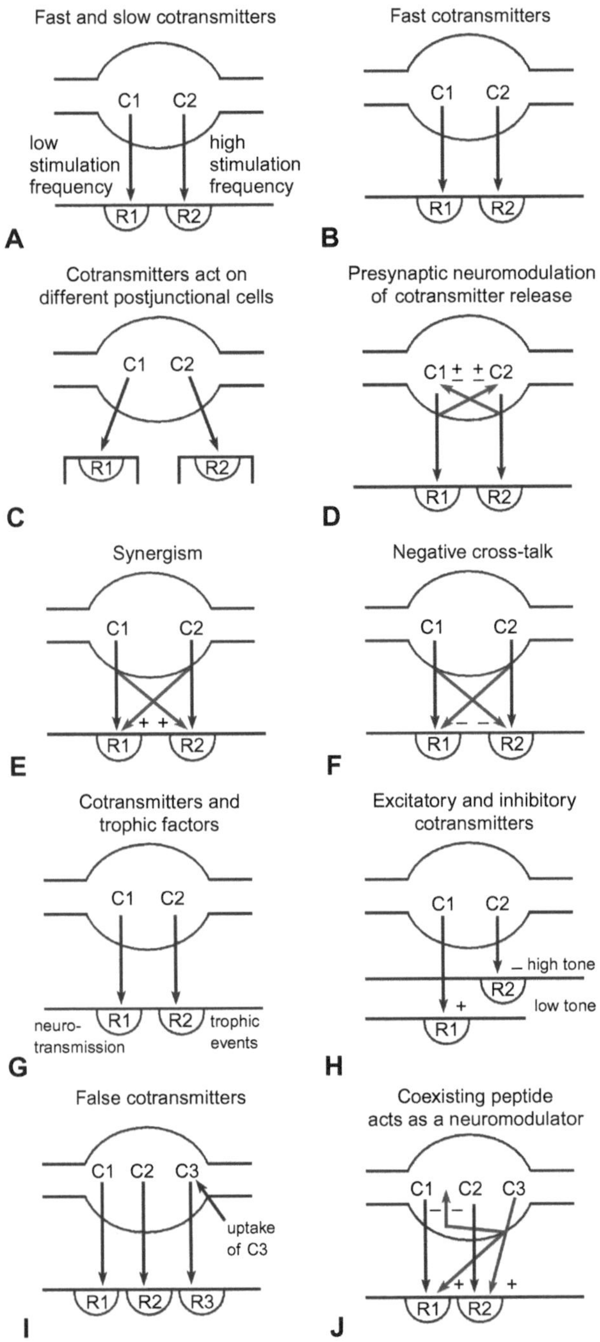

Fig. 3.3. Spectrum of signaling variations offered by cotransmission (black arrows = neurotransmission; gray arrows = pre- or postjunctional neuromodulation). (**A**) Fast transmission is usually produced by small molecules (C1) released at low frequency nerve stimulation acting on ionotropic receptors (R1), whereas slow transmission is usually produced by release of peptides (C2) or other molecules at high frequency stimulation acting on G protein-coupled receptors (R2). (**B**) Cotransmitters C1 and C2 can both be fast messengers acting via ionotropic receptors (R1 and R2). (**C**) Cotransmitters C1 and C2 act on receptors (R1 and R2) localized on different postjunctional cells. (**D**) Cotransmitters C1 and C2 not only act postjunctionally via R1 and R2 receptors but can also act as prejunctional modulators to either inhibit (−) or enhance (+) the release of C1 and/or C2. (**E**) Cotransmitters C1 and C2 act synergistically to enhance the combined responses produced via R1 and R2 receptors. (**F**) Cotransmitters C1 and C2 act to inhibit the responses evoked via R1 and/or R2 receptors. (**G**) Cotransmitter C1 evokes neurotransmission via R1 receptors, while C2 evokes long-term (trophic) responses of postjunctional cells via R2 receptors. (**H**) Cotransmitter C1 produces excitation via R1 receptors when the postjunctional smooth muscle target has low tone, with C2 having little influence; however, when the smooth muscle tone is high, the dominant response might be relaxation produced by C2 via R2 receptors. (**I**) Substance C3 is taken up by nerve terminals, rather than being synthesized and stored as is true for the cotransmitters C1 and C2. C3 can then be released on nerve stimulation to act on postjunctional R3 receptors. In these circumstances, C3 would be known as a "false transmitter." (**J**) A coexisting substance C3 (often a peptide) can be synthesized and stored in a nerve, but not act directly via a postjunctional receptor to produce changes in postjunctional cell activity. It could, however, act as a prejunctional inhibitor (−) of the release of the cotransmitters C1 and C2, or as a postjunctional enhancer (+) of the responses mediated by R1 and R2. (Reproduced from Burnstock, 2004, with permission from Elsevier.)

Excitatory and inhibitory cotransmitters

Although cotransmitters generally have similar actions on postjunctional cells, there are some examples of cotransmitters having opposite actions. For example, in the mammalian uterus, one or other cotransmitter dominates depending on the hormonal and/or tonic status of the postjunctional muscle cells. Brain-derived neurotrophic factor (BDNF) increases the release of ACh and reduces NA release from sympathetic nerves to cause a rapid shift from excitatory to inhibitory transmission (Yang et al., 2002; Landis, 2002).

False cotransmitters

For example, it has been known for some time that sympathetic nerves take up 5-HT, which can then be released as a "false transmitter," rather than a genuine "cotransmitter." A "false transmitter" is a substance actively taken up and subsequently released by a neuron that does not synthesize it.

Coexisting peptides acting as neuromodulators

In general, most neuropeptides act as neuromodulators rather than neurotransmitters (Milner and Burnstock, 1994). For example, NPY released from sympathetic nerves acts as a prejunctional inhibitor of release of ATP and NA and a postjunctional potentiator of the actions of ATP and NA (Ellis and Burnstock, 1990).

COTRANSMITTER PLASTICITY

Neurons possess the genetic potential to produce many neurotransmitters. The particular combination and quantity that results is partly preprogrammed and partly determined by "trophic" factors that trigger the expression or suppression of the appropriate genetic machinery. A number of studies have demonstrated plasticity of expression of autonomic nerves during development and aging, following trauma or surgery, after chronic exposure to drugs and in disease (Burnstock 1990c, 2006; Burnstock and Milner, 1999). NA and NPY show different expression in cerebrovascular nerves during development. Whereas there is a decrease in expression of vasoconstrictor cerebrovascular (NA and 5-HT) in aging rats, there is an increase in vasodilator neurotransmitters (VIP and CGRP) (Mione et al., 1988). BDNF was claimed to switch sympathetic neurotransmission to the heart from an adrenergic excitation to cholinergic inhibition.

In patients with bladder areflexia following lower motor spinal lesion there is increased innervation by VIP, NPY, and NA-containing nerve fibers to the striated muscle of the intrinsic external urethral sphincter, which may indicate a regulatory mechanism via the intrinsic ganglia and/or the somatic nervous system to help overcome this type of bladder dysfunction. Surgical ganglionectomy leads to increased SP levels in the iris and ciliary body, increased CGRP in pial vessels, and increased expression of NPY in parasympathetic neurons supplying cerebral vessels (Mione et al., 1990). Long-term guanethidine sympathectomy of both neonate and adult rats results in a drastic increase in the innervation of tissues by the sensory neuropeptide, CGRP, probably due to increased availability to nerve growth factor for which sensory and sympathetic neurons compete (Aberdeen et al., 1992). After extrinsic denervation of human respiratory tract by heart–lung transplantation, the intrinsic parasympathetic neurons that persist express an NA-synthesizing enzyme and NPY, substances normally found in sympathetic nerves.

During the course of experimentally induced diabetes there are marked changes in the expression of neurotransmitters/neuromodulators in nerves supplying the bowel. While there are degenerative changes in VIP and NA-containing nerves early on in the development of the disease, the expression of 5-HT, SP, and CGRP in nerve fibers changes at different times during the progression of the disease. Reduction in the expression of VIP and 5-HT, but not NPY and NA, has been demonstrated in perivascular nerves supplying the cerebral blood vessels of 8-week streptozotocin-induced diabetic rats (Lagnado et al., 1987). Changes in chemical coding of myenteric neurons in ulcerative colitis have been reported, with a shift from cholinergic to more SP-mediated cotransmission. In interstitial cystitis and outflow obstruction, there is a substantial increase in ATP as the cotransmitter with ACh in parasympathetic nerves supplying the human bladder (Palea et al., 1993; Smith and Chapple, 1994; O'Reilly et al., 2001; Andersson and Hedlund, 2002), and there is increase in ATP as a cotransmitter with NA in sympathetic nerves in spontaneous hypertensive rats (Vidal et al., 1986).

CONCLUDING COMMENTS

A compelling body of evidence is now available to support the cotransmitter hypothesis, which implies more sophisticated local control mechanisms than were envisaged in earlier times. However, a number of issues still need to be resolved, including:

1. resolution of the different roles of coexisting substances (including neurotransmitter, pre- and postjunctional neuromodulator and/or trophic roles) under different physiological conditions and patterns of discharge in the nerves
2. identification of different transmitter substances in vesicular storage sites in nerve terminals at the ultrastructural level, perhaps utilizing colloidal gold particle techniques
3. studies of the molecular mechanisms that are concerned with the regulation of transmitter and receptor expression during development (including old age) and during the compensatory changes that occur in nerves in the adult, which remain following trauma, surgery and disease.

In view of the growing number of examples of the plasticity of the nervous system in the adult in aging and disease, neuropathologists should consider *increases* in expression of transmitters and density of nerves as well as the more traditional approach of looking for degrees of damage in loss of nerves. In terms of drug development, since the expression of cotransmitters and receptors changes so markedly with age, sex, and disease states, we cannot assume that drugs tested on young healthy male volunteers will be appropriate for, say, a man of 80; thus the age, sex, and pathological history of individual patients will need to be taken into account in prescribing therapeutic strategies. Another possibility, although clearly not popular, is that the use of drug cocktails designed to match the coding of neurons projecting to a damaged area may be a more efficient strategy than using a single compound with widespread actions and then trying to find a dosage or procedure to avoid unwanted side-effects.

It has been particularly difficult to establish cotransmitter roles for the many peptides found in nerves, particularly in the enteric nervous system, where up to five neuropeptides are found in some neurons, partly because specific receptors and physiological roles have not been established for some of these and partly because of the lack of selective antagonists. It will be important to distinguish between neuromodulator, neurotransmitter and neurotrophic roles for released peptides or indeed for as yet unrecognized roles.

It has become clear that ATP is a primitive signaling molecule that has been retained as a cotransmitter in every nerve type in both the peripheral and central nervous systems, although the relative role of ATP varies considerably in different species and pathophysiological conditions. ATP appears to become a more prominent cotransmitter in stress and inflammatory conditions. Most nerves contain and release ATP as a fast cotransmitter together with classical transmitters such as ACh, NA, dopamine, glutamate, GABA, and one or more peptides. Now that cotransmission is recognized as a universal mechanism, it is recommended that the terms "adrenergic," "cholinergic," "peptidergic," "purinergic," "aminergic," and "nitrergic" should not be used when *nerves* are described, although adrenergic, cholinergic, peptidergic, purinergic, aminergic, or nitrergic *transmission* is still meaningful.

References

Aberdeen J, Milner P, Lincoln J et al. (1992). Guanethidine sympathectomy of mature rats leads to increases in calcitonin gene-related peptide and vasoactive intestinal polypeptide-containing nerves. Neuroscience 47: 453–461.

Andersson KE, Hedlund P (2002). Pharmacologic perspective on the physiology of the lower urinary tract. Urology 60: 13–20.

Banks F, Knight G, Calvert RC et al. (2006). The purinergic component of human vas deferens contraction. Fertil Steril 85: 932–939.

Barajas-López C, Montaño LM, Espinosa-Luna R (2002). Inhibitory interactions between 5-HT_3 and P2X channels in submucosal neurons. Am J Physiol Gastrointest Liver Physiol 283: G1238–G1248.

Bell C (1968). Dual vasoconstrictor and vasodilator innervation of the uterine arterial supply in the guinea pig. Circ Res 23: 269–279.

Boué-Grabot E, Barajas-Lôpez C, Chakfe Y et al. (2003). Intracellular cross talk and physical interaction between two classes of neurotransmitter-gated channels. J Neurosci 23: 1246–1253.

Bradley E, Law A, Bell D et al. (2003). Effects of varying impulse number on cotransmitter contributions to sympathetic vasoconstriction in rat tail artery. Am J Physiol Heart Circ Physiol 284: H2007–H2014.

Bredt DS, Snyder SH (1992). Nitric oxide, a novel neuronal messenger. Neuron 8: 3–11.

Brownstein MJ, Saavedra JM, Axelrod J et al. (1974). Coexistence of several putative neurotransmitters in single identified neurons of *Aplysia*. Proc Natl Acad Sci U S A 71: 4662–4665.

Burn JH, Rand MJ (1959). Sympathetic postganglionic mechanisms. Nature 184: 163–165.

Burn JH, Rand MJ (1965). Acetylcholine in adrenergic transmission. Annu Rev Pharmacol 5: 163–182.

Burnett AL, Lowenstein CJ, Bredt DS et al. (1992). Nitric oxide: a physiologic mediator of penile erection. Science 257: 401–403.

Burnstock G (1976). Do some nerve cells release more than one transmitter? Neuroscience 1: 239–248.

Burnstock G (1977). Autonomic neuroeffector junctions – reflex vasodilatation of the skin. J Invest Dermatol 69: 47–57.

Burnstock G (1978). Do some sympathetic neurones synthesize and release both noradrenaline and acetylcholine. Prog Neurobiol 11: 205–222.

Burnstock G (1983). Recent concepts of chemical communication between excitable cells. In: NN Osborne (Ed.), Dale's Principle and Communication between Neurones, Pergamon Press, Oxford, pp. 7–35.

Burnstock G (1990a). Co-transmission. The fifth Heymans memorial lecture – Ghent, February 17, 1990. Arch Int Pharmacodyn Ther 304: 7–33.

Burnstock G (1990b). Noradrenaline and ATP as cotransmitters in sympathetic nerves. Neurochem Int 17: 357–368.

Burnstock G (1990c). Overview. Purinergic mechanisms. Ann N Y Acad Sci 603: 1–17.

Burnstock G (1995). Noradrenaline and ATP: cotransmitters and neuromodulators. J Physiol Pharmacol 46: 365–384.

Burnstock G (2001a). Purinergic signalling in gut. In: Handbook of Experimental Pharmacology, MP Abbracchio, M Williams (Eds.), Purinergic and Pyrimidinergic Signalling II – Cardiovascular, Respiratory, Immune, Metabolic and Gastrointestinal Tract Function, Vol. 151/II. Springer-Verlag, Berlin, pp. 141–238.

Burnstock G (2001b). Purinergic signalling in lower urinary tract. In: Handbook of Experimental Pharmacology, MP Abbracchio, M Williams (Eds.), Purinergic and Pyrimidinergic Signalling I – Molecular, Nervous and Urinogenitary System Function, Vol. 151/I. Springer-Verlag, Berlin, pp. 423–515.

Burnstock G (2002). Purinergic signalling and vascular cell proliferation and death. Arterioscler Thromb Vasc Biol 22: 364–373.

Burnstock G (2004). Cotransmission. Curr Opin Pharmacol 4: 47–52.

Burnstock G (2006). Pathophysiology and therapeutic potential of purinergic signaling. Pharmacol Rev 58: 58–86.

Burnstock G (2007). Physiology and pathophysiology of purinergic neurotransmission. Physiol Rev 87: 659–797.

Burnstock G (2008). The journey to establish purinergic signalling in the gut. Neurogastroenterol Motil 20: 8–19.

Burnstock G (2009a). Autonomic neurotransmission: 60 years since Sir Henry Dale. Annu Rev Pharmacol Toxicol 49: 1–30.

Burnstock G (2009b). Purinergic cotransmission. Exp Physiol 94: 20–24.

Burnstock G (2009c). Purinergic regulation of vascular tone and remodelling. Auton Autacoid Pharmacol 29: 63–72.

Burnstock G (2009d). Purines and sensory nerves. Handb Exp Pharmacol 194: 333–392.

Burnstock G (2009e). Cotransmission. In: LR Squire (Ed.), Encyclopedia of Neuroscience, Academic Press, Oxford, pp. 247–254.

Burnstock G, Costa M (1975). Adrenergic Neurones: Their Organisation, Function and Development in the Peripheral Nervous System, Chapman and Hall, London, pp. 1–225.

Burnstock G, Holman ME (1960). Autonomic nerve-smooth muscle transmission. Nature 187: 951–952.

Burnstock G, Milner P (1999). Structural and chemical organisation of the autonomic nervous system with special reference to non-adrenergic, non-cholinergic transmission. In: CJ Mathias, R Bannister (Eds.), Autonomic Failure. A Textbook of Clinical Disorders of the Autonomic Nervous System, Oxford University Press, Oxford, pp. 63–71.

Burnstock G, Sneddon P (1985). Evidence for ATP and noradrenaline as cotransmitters in sympathetic nerves. Clin Sci 68: 89S–92S.

Burnstock G, Verkhratsky A (2010). Long-term (trophic) purinergic signalling: purinoceptors control cell proliferation, differentiation and death. Cell Death Dis 1: e9.

Burnstock G, Verkhratsky A (2012). Purinergic signalling and the nervous system. Springer, Heidelberg/Berlin, pp. 1–715.

Burnstock G, Dumsday B, Smythe A (1972). Atropine resistant excitation of the urinary bladder: the possibility of transmission via nerves releasing a purine nucleotide. Br J Pharmacol 44: 451–461.

Burnstock G, Cocks T, Crowe R et al. (1978). Purinergic innervation of the guinea-pig urinary bladder. Br J Pharmacol 63: 125–138.

Burnstock G, Allen TGJ, Hassall CJS et al. (1987). Properties of intramural neurones cultured from the heart and bladder. In: C Heym (Ed.), Histochemistry and Cell Biology of Autonomic Neurons and Paraganglia. Exp Brain Res Series 16, Springer Verlag, Heidelberg, pp. 323–328.

Cook RD, Burnstock G (1976). The ultrastructure of Auerbach's plexus in the guinea-pig. I. Neuronal elements. J Neurocytol 5: 171–194.

Cottrell GA (1976). Does the giant cerebral neurone of *Helix* release two transmitters: ACh and serotonin? J Physiol 259: 44P–45P.

Cuello AC (1982). Co-transmission. Proceedings Symposium 50th Anniversary Meeting British Pharmacological Society, Oxford, Macmillan Press, London, p. 264.

Dale H (1935). Pharmacology and nerve endings. Proc R Soc Med 28: 319–332.

Eccles JC (1957). The clinical significance of research work on the chemical transmitter substances of the nervous system. Med J Aust 44: 745–753.

Ellis JL, Burnstock G (1990). Neuropeptide Y neuromodulation of sympathetic co-transmission in the guinea-pig vas deferens. Br J Pharmacol 100: 457–462.

Evans RJ, Surprenant A (1992). Vasoconstriction of guinea-pig submucosal arterioles following sympathetic nerve stimulation is mediated by the release of ATP. Br J Pharmacol 106: 242–249.

Fedan JS, Hogaboom GK, O'Donnell JP et al. (1981). Contributions by purines to the neurogenic response of the vas deferens of the guinea-pig. Eur J Pharmacol 69: 41–53.

Furness JB (2000). Types of neurons in the enteric nervous system. J Auton Nerv Syst 81: 87–96.

Furness JB (2006). The enteric nervous system, Blackwell Publishing, Massachusetts.

Furness JB, Costa M (1987). The Enteric Nervous System, Churchill Livingstone, Edinburgh.

Furness JB, Morris JL, Gibbins IL et al. (1989). Chemical coding of neurons and plurichemical transmission. Annu Rev Pharmacol Toxicol 29: 289–306.

Furshpan EJ, MacLeish PR, O'Lague PH et al. (1976). Chemical transmission between rat sympathetic neurons and cardiac myocytes developing in microcultures: evidence for cholinergic, adrenergic, and dual-function neurons. Proc Natl Acad Sci U S A 73: 4225–4229.

Gibbins IL, Furness JB, Costa M et al. (1985). Co-localization of calcitonin gene-related peptide-like immunoreactivity with substance P in cutaneous, vascular and visceral sensory neurons of guinea-pigs. Neurosci Lett 57: 125–130.

Gibbins IL, Furness JB, Costa M (1987). Pathway-specific patterns of the coexistence of substance P, calcitonin gene-related peptide, cholecystokinin and dynorphin in neurons of the dorsal root ganglia of the guinea-pig. Cell Tissue Res 248: 417–437.

Gulbenkian S, Merighi A, Wharton J et al. (1986). Ultrastructural evidence for the coexistence of calcitonin gene-related peptide and substance P in secretory vesicles of peripheral nerves in the guinea pig. J Neurocytol 15: 535–542.

Hassall CJS, Burnstock G (1986). Intrinsic neurones and associated cells of the guinea-pig heart in culture. Brain Res 364: 102–113.

Hill CE, Hendry IA (1977). Development of neurons synthesizing noradrenaline and acetylcholine in the superior cervical ganglion of the rat in vivo and in vitro. Neuroscience 2: 741–749.

Hillarp NA, Nilson B, Hogberg B (1955). Adenosine triphosphate in the adrenal medulla of the cow. Nature 176: 1032–1033.

Hökfelt T, Johansson O, Kellerth JO (1977). Immunohistochemical distribution of substance P. In: US von Euler, B Pernow (Eds.), Substance P, Raven Press, New York, pp. 117–145.

Hökfelt T, Fuxe K, Pernow B (1986). Coexistence of Neuronal Messengers: A New Principle in Chemical Transmission Progress in Brain Research, Vol. 68, Elsevier, Amsterdam, pp. 3–411.

Holton P (1959). The liberation of adenosine triphosphate on antidromic stimulation of sensory nerves. J Physiol 145: 494–504.

Jackowski A, Crockard A, Burnstock G (1989). 5-Hydroxytryptamine demonstrated immunohistochemically in rat cerebrovascular nerves largely represents 5-hydroxytryptamine uptake into sympathetic nerve fibres. Neuroscience 29: 453–462.

Kennedy C, Burnstock G (1985). Evidence for two types of P_2-purinoceptor in longitudinal muscle of the rabbit portal vein. Eur J Pharmacol 111: 49–56.

Kirkpatrick K, Burnstock G (1987). Sympathetic nerve-mediated release of ATP from the guinea-pig vas deferens is unaffected by reserpine. Eur J Pharmacol 138: 207–214.

Koelle GB (1955). The histochemical identification of acetylcholinesterase in cholinergic, adrenergic and sensory neurons. J Pharmacol Exp Ther 114: 167–184.

Kupfermann I (1991). Functional studies of cotransmission. Physiol Rev 71: 683–732.

Lagnado MLJ, Crowe R, Lincoln J et al. (1987). Reduction of nerves containing vasoactive intestinal polypeptide and serotonin, but not neuropeptide Y and catecholamine, in cerebral blood vessels of the 8-week streptozotocin-induced diabetic rat. Blood Vessels 24: 169–180.

Landis SC (2002). Quick-change artist: from excitatory to inhibitory synapse in minutes. Nat Neurosci 5: 503–504.

Langer SZ, Pinto JEB (1976). Possible involvement of a transmitter different from norepinephrine in residual responses to nerve stimulation of cat nictitating membrane after pretreatment with reserpine. J Pharmacol Exp Ther 196: 697–713.

Le Douarin NM, Renad D, Teillet MA et al. (1975). Cholinergic differentiation of presumptive adrenergic neuroblasts in interspecific chimeras after heterotopic transplantation. Proc Natl Acad Sci USA 72: 728–732.

Leblanc GG, Landis SC (1988). Target specificity of neuropeptide Y-immunoreactive cranial parasympathetic neurons. J Neurosci 8: 146–155.

Lundberg JM (1981). Evidence for coexistence of vasoactive intestinal polypeptide (VIP) and acetylcholine in neurons of cat exocrine glands. Morphological, biochemical and functional studies. Acta Physiol Scand Suppl 496: 1–57.

Lundberg JM (1996). Pharmacology of cotransmission in the autonomic nervous system: integrative aspects on amines, neuropeptides, adenosine triphosphate, amino acids and nitric oxide. Pharmacol Rev 48: 113–178.

Maggi CA, Meli A (1988). The sensory-efferent function of capsaicin-sensitive sensory neurons. Gen Pharmacol 19: 1–43.

Milner P, Burnstock G (1994). Trophic factors and the control of smooth muscle development and innervation. In: D Raeburn, MA Giembycz (Eds.), Airways Smooth Muscle: Development and Regulation of Contractility. Birkhäuser Verlag, Basel, Switzerland, pp. 1–39.

Mione MC, Dhital KK, Amenta F et al. (1988). An increase in the expression of neuropeptidergic vasodilator, but not vasoconstrictor, cerebrovascular nerves in aging rats. Brain Res 460: 103–113.

Mione MC, Cavanagh JFR, Lincoln J et al. (1990). Long-term chemical sympathectomy leads to an increase of neuropeptide Y immunoreactivity in cerebrovascular nerves and iris of the developing rat. Neuroscience 34: 369–378.

Morris JL, Gibbins IL, Furness JB (1987). Increased dopamine-β-hydroxylase-like immunoreactivity in non-noradrenergic axons supplying the guinea-pig uterine artery after 6-hydroxydopamine treatment. J Auton Nerv Syst 21: 15–27.

O'Reilly BA, Kosaka AH, Chang TK et al. (2001). A quantitative analysis of purinoceptor expression in human fetal and adult bladders. J Urol 165: 1730–1734.

Osborne NN (1983). Dale's Principle and Communication between Neurones, Pergamon Press, Oxford.

Palea S, Artibani W, Ostardo E et al. (1993). Evidence for purinergic neurotransmission in human urinary bladder affected by interstitial cystitis. J Urol 150: 2007–2012.

Patterson PH, Reichardt LF, Chun LL (1976). Biochemical studies on the development of primary sympathetic neurons in cell culture. Cold Spring Harb Symp Quant Biol 40: 389–397.

Ramme D, Regenold JT, Starke K et al. (1987). Identification of the neuroeffector transmitter in jejunal branches of the rabbit mesenteric artery. Naunyn Schmiedebergs Arch Pharmacol 336: 267–273.

Rubino A, Burnstock G (1996). Capsaicin-sensitive sensory-motor neurotransmission in the peripheral control of cardiovascular function. Cardiovasc Res 31: 467–479.

Saffrey MJ, Hassall CJS, Allen TGJ et al. (1992). Ganglia within the gut, heart, urinary bladder and airways: studies in tissue culture. Int Rev Cytol 136: 93–144.

Schultzberg M, Hökfelt T, Terenius L et al. (1979). Enkephalin immunoreactive nerve fibres and cell bodies in sympathetic ganglia of the guinea-pig and rat. Neuroscience 4: 249–270.

Schümann HJ (1958). Effects of insulin and reserpine on the adrenalin and ATP content of the chromatic granules of the adrenal medulla. Naunyn Schmiedebergs Arch Exp Pathol Pharmakol 233: 237–249.

Singh I (1964). Seasonal variations in the nature of neurotransmitters in a frog vagus-stomach muscle preparation. Arch Int Physiol Biochim 72: 843–851.

Smith NCE, Burnstock G (2004). Mechanism underlying postjunctional synergism between responses of the vas deferens to noradrenaline and ATP. Eur J Pharmacol 498: 241–248.

Smith DJ, Chapple CR (1994). In vitro response of human bladder smooth muscle in unstable obstructed male bladders: a study of pathophysiological causes. Neurourol Urodyn 13: 414–415.

Sneddon P, Burnstock G (1984a). ATP as a co-transmitter in rat tail artery. Eur J Pharmacol 106: 149–152.

Sneddon P, Burnstock G (1984b). Inhibition of excitatory junction potentials in guinea-pig vas deferens by α, β-methylene-ATP: further evidence for ATP and noradrenaline as cotransmitters. Eur J Pharmacol 100: 85–90.

Sneddon P, Westfall DP (1984). Pharmacological evidence that adenosine triphosphate and noradrenaline are co-transmitters in the guinea-pig vas deferens. J Physiol 347: 561–580.

Su C, Bevan JA, Burnstock G (1971). [^{3}H]adenosine triphosphate: release during stimulation of enteric nerves. Science 173: 337–339.

Toda N, Minami Y, Okamura T (1990). Inhibitory effects of L-N^G-nitro-arginine on the synthesis of EDRF and the cerebroarterial response to vasodilator nerve stimulation. Life Sci 47: 345–351.

Ventura S, Dewalagama RK, Lau LC (2003). Adenosine 5′-triphosphate (ATP) is an excitatory cotransmitter with noradrenaline to the smooth muscle of the rat prostate gland. Br J Pharmacol 138: 1277–1284.

Vidal M, Hicks PE, Langer SZ (1986). Differential effects of α, β-methylene ATP on responses to nerve stimulation in SHR and WKY tail arteries. Naunyn Schmiedebergs Arch Pharmacol 332: 384–390.

Vizi ES (1979). Prejunctional modulation of neurochemical transmission. Prog Neurobiol 12: 181–290.

Warland JJI, Burnstock G (1987). Effects of reserpine and 6-hydroxydopamine on the adrenergic and purinergic components of sympathetic nerve responses of the rabbit saphenous artery. Br J Pharmacol 92: 871–880.

Westfall DP, Stitzel RE, Rowe JN (1978). The postjunctional effects and neural release of purine compounds in the guinea-pig vas deferens. Eur J Pharmacol 50: 27–38.

Yang B, Slonimsky JD, Birren SJ (2002). A rapid switch in sympathetic neurotransmitter release properties mediated by the p75 receptor. Nat Neurosci 5: 539–545.

Handbook of Clinical Neurology, Vol. 117 (3rd series)
Autonomic Nervous System
R.M. Buijs and D.F. Swaab, Editors

Chapter 4

Sensitization of endocrine organs to anterior pituitary hormones by the autonomic nervous system

WILLIAM C. ENGELAND*
Department of Neuroscience, University of Minnesota, Minneapolis, MN, USA

INTRODUCTION

The ability of an individual to respond to changes in the internal and external environment depends on neural and endocrine control systems designed to defend homeostasis. Signals to the brain from peripheral and central receptors are integrated and if necessary, the brain activates the appropriate response to offset the initial stimulus to prevent tissue damage. Endocrine glands are a primary effector of homeostatic control systems, secreting into the systemic circulation hormones that regulate target tissue function and feed back to modulate hormonal secretion. Brain control of endocrine organ secretion is mediated by efferent mechanisms that consist of both hormonal and neural elements. The classic example of hormonal control is the hypothalamic–pituitary system in which anterior pituitary hormones regulated by hypothalamic releasing hormones are required for peripheral endocrine gland function. Pituitary hormones regulate the function of specific endocrine glands, having both growth-promoting (trophic) and secretion-promoting (tropic) effects. Targets of hypothalamopituitary systems include the adrenal cortex, the gonads (ovary and testes), and the thyroid. In addition to hormonal control, each of these organs is innervated by autonomic nerves that provide neural signals that interact with pituitary hormones to control endocrine secretion. Instead of a tropic pituitary factor, other glands like the endocrine pancreas rely on signals provided by circulating metabolic factors that interact with autonomic neural input to control secretion (for recent reviews, see Taborsky, 2010; Thorens, 2011). In contrast to dual control by humoral and neural input, endocrine glands derived from neural tissue such as the pineal and the adrenal medulla rely primarily on autonomic nerves to regulate secretion. Although autonomic control of endocrine organs has been reviewed previously (see Unsicker, 1996), recent work has indicated that autonomic neural input affects hormone secretion in part by modulating the sensitivity of endocrine cells to humoral signals. This review will describe results that support this contention and offer some perspective on its clinical significance. Since an exhaustive review of how autonomic nerves influence each endocrine gland is beyond the scope of this report, the discussion will be focused on the adrenal cortex and the ovary, two examples in which recent findings have emerged that highlight the clinical relevance of autonomic–endocrine interactions.

EVIDENCE FOR AUTONOMIC CONTROL OF THE ADRENAL CORTEX

Stress is generally defined as a physical or psychological stimulus that threatens homeostasis, the ability of an organism to maintain internal stability in the face of a changing external environment. The secretion of glucocorticoids from the adrenal cortex is considered an essential adaptive component of the response to stress. The adrenal cortex is the final element of the hypothalamic–pituitary–adrenal (HPA) axis. The HPA response to stress is initiated by the activation of neurons in the paraventricular nucleus (PVN) of the hypothalamus. The peptides, corticotropin-releasing hormone (CRH) and vasopressin (VP), released from PVN neurons, stimulate the release of adrenocorticotropic hormone (ACTH) from the anterior pituitary (Rivier and Vale, 1983; Antoni, 1986), which in turn stimulates adrenal glucocorticoids. The stress response is imperative for survival after acute stress, since glucocorticoids mobilize glucose for metabolic needs, augment cardiovascular responses, and modulate

*Correspondence to: W.C. Engeland, Ph.D., Department of Neuroscience, 6–145 Jackson Hall, 321 Church St., Minneapolis, MN 55455, USA. Tel: +1-612-625-4976, Fax: +1-612-626-5009, E-mail: engel002@umn.edu

cognitive processes including memory (Tsigosa and Chrousos, 2002; Lupien et al., 2007). The HPA axis is also characterized by a prominent circadian rhythm in which adrenal secretion is increased at the onset of the daily activity cycle, guaranteeing that glucocorticoids are available to support the body's metabolic, cardiovascular, and neural requirements at activity onset (Dickmeis, 2009).

Although ACTH is required to support the viability and function of the adrenal cortex (Tchen et al., 1977), it is well established that changes in circulating glucocorticoids can occur without concomitant changes in ACTH (Bornstein et al., 2008). The dissociation between ACTH and glucocorticoids has been attributed to extrapituitary control of the adrenal cortex by autonomic nerves. Anatomical evidence for neural control of the adrenal cortex is clear. The adrenal cortex is innervated by sympathetic postganglionic and primary afferent nerve fibers, but there is little evidence for direct parasympathetic innervation (Kesse et al., 1988; Coupland et al., 1989; Holgert et al., 1995, 1998); in addition, cortical cells are intertwined anatomically with medullary chromaffin cells, modified postganglionic sympathetic neurons that are controlled by preganglionic sympathetic fibers (Gallo-Payet et al., 1987; Bornstein et al., 1994). Functional evidence for autonomic control of adrenal cortical function has been provided by *in vivo* and *in vitro* studies. Electrical stimulation of the thoracic splanchnic nerve in different species increases adrenal secretion of glucocorticoids in a frequency-dependent manner (Edwards and Jones, 1987a; Engeland and Gann, 1989; Bornstein et al., 1990). In addition, adrenal cortical cells express a variety of neurotransmitter receptors and adrenergic, cholinergic, and neuropeptidergic receptor agonists can alter glucocorticoid secretion *in vitro* (Ehrhart-Bornstein et al., 1998).

The importance of sympathetic input in controlling adrenocortical secretion under physiological conditions has been evaluated by assessing glucocorticoid responses in laboratory animals subjected to splanchnicectomy, severing the thoracic splanchnic nerve that provides sympathetic input to the adrenal. In conscious lambs, splanchnicectomy reduces the glucocorticoid response to norepinephrine injection without affecting the ACTH response (Edwards et al., 1986). The glucocorticoid response to dehydration in rats, but not the ACTH response, is also decreased by splanchnicectomy (Ulrich-Lai and Engeland, 2002). In both models, adrenal sensitivity to ACTH was tested and found to be reduced by splanchnicectomy (Edwards and Jones, 1987b; Ulrich-Lai and Engeland, 2002), supporting the hypothesis that sympathetic adrenal innervation augments glucocorticoid responses to stress by increasing adrenal sensitivity to ACTH.

In addition to the stress response, sympathetic neural activity has been implicated in controlling the circadian rhythm in glucocorticoid secretion in rodents. The rhythm in plasma glucocorticoids is paralleled by a rhythm in adrenal sensitivity to ACTH (Dallman et al., 1978; Kaneko et al., 1980; Ulrich-Lai et al., 2006). Splanchnicectomy disrupts the diurnal rhythm in plasma glucocorticoids in rats by reducing the peak concentration, a response that is associated with a decrease in adrenal sensitivity to ACTH (Ulrich-Lai et al., 2006). These results support the contention that rhythms in adrenal steroidogenesis are dependent both on ACTH and sympathetic neural activity, the output of a hypothalamic–spinal–adrenal circuit (Buijs et al., 1999; Engeland and Arnhold, 2005).

The specific mechanism through which sympathetic nerves alter adrenal sensitivity to ACTH has not been identified. However, a number of possibilities have been suggested. In addition to adrenal parenchyma, blood vessels are innervated by postganglionic sympathetic fibers containing vasoactive neurotransmitters (Carlsson et al., 1993) and adrenal blood flow changes differentially following stimulation (Edwards and Jones, 1987a; Engeland and Gann, 1989) or severing of the splanchnic nerve (Breslow et al., 1987; Jasper et al., 1990). The cortex and medulla receive separate arterial supplies (Coupland and Selby, 1976), and cortical and medullary flow are differentially changed by sympathetic neurotransmitters (Breslow et al., 1993; Ligier et al., 1994). Increasing cortical blood flow is thought to promote glucocorticoid secretion by delivering increased ACTH (Urquhart, 1965; L'Age et al., 1970; Jones et al., 1990). In addition to adrenal blood flow, sympathetic nerve activity controls medullary secretion of catecholamines and neuropeptides (Critchley et al., 1980; Engeland et al., 1986). Since there is physiological (Ehrhart-Bornstein et al., 1998; Bornstein et al., 2000) evidence for chromaffin cells to effect steroidogenesis, it is likely that medullary–cortical interactions participate in sympathetic control of glucocorticoid secretion. Adrenal demedullation (surgical removal of the adrenal medulla) in rodents prevents splanchnicectomy-induced reduction in glucocorticoids during stress and circadian rhythms (Ulrich-Lai and Engeland, 2002; Ulrich-Lai et al., 2006), suggesting that sympathetic drive to the medulla mediates the glucocorticoid responses. However, changes in glucocorticoid responses after demedullation must be viewed with some caution; after surgery, the cortex is revascularized and hyperinnervated (Ulrich-Lai and Engeland, 2000), changes that could offset responses to loss of medullary cells. Use of a more selective approach to remove adrenal catecholamines like genetic deletions clearly is warranted (Bornstein et al., 2000; Bao et al., 2007). Since ACTH stimulates steroidogenesis by binding to the melanocortin 2 (MC2; ACTH) receptor and generating cAMP (Schimmer, 1995; Côté et al., 1999), sympathetic input could affect adrenal sensitivity to ACTH by changing cAMP. This possibility is supported by experiments in rodents showing that splanchnicectomy reduces adrenal cAMP induced by stress or by circadian rhythms (Ulrich-Lai and Engeland, 2002; Ulrich-Lai et al., 2006). Activation of β-adrenergic receptors stimulates glucocorticoid secretion from bovine cortical cells (Walker et al., 1991), and blockade of β-adrenergic receptors in human (Bornstein et al., 1996) and rat (Hinson et al., 1992) cells prevents corticosteroid responses to neuropeptides. Since β-adrenergic receptors such as MC2 receptors are coupled to the α_s subunit of G-protein that stimulates adenylyl cyclase to produce cAMP (Levitzki, 1988), the elevation of intracellular cAMP may be a mechanism for catecholamines to enhance adrenal sensitivity to ACTH.

Recent insights into the clinical relevance of sympathetic control of the adrenal cortex

In the addition to the circadian rise in HPA activity that occurs in the morning in humans, cortisol increases in response to awakening (Spath-Schwalbe et al., 1992; Pruessner et al., 1997). The cortisol awakening response (CAR) has become a standard measure of HPA activity and a diagnostic marker of work-related stress (Schlotz et al., 2004) and other forms of chronic stress (Wust et al., 2000), including posttraumatic stress disorder (Neylan et al., 2005). Although increases in cortisol are associated with ACTH, differences in the dynamics of the pituitary–adrenal responses before and after awakening (Born et al., 1999; Wilhelm et al., 2007) suggest that extra-ACTH factors are involved (see Clow et al., 2010). Although exposure to light is not required for the CAR (Born et al., 1999; Wilhelm et al., 2007), light pulses in the morning enhance the amplitude of the CAR (Scheer and Buijs, 1999; Leproult et al., 2001). In rodents, exposure to light pulses increases glucocorticoids independently of ACTH, a response prevented by splanchnicectomy (Ishida et al., 2005). Thus, a sympathetic mechanism may also contribute to the glucocorticoid responses to awakening in humans. The physiological importance of the CAR is still being delineated. For example, a recent study has shown that blockade of the CAR reduces recall especially for emotional materials (Rimmele et al., 2010), suggesting that the glucocorticoid pulse upon awakening facilitates memory retrieval. In many studies, salivary cortisol has been shown to be a reliable method for assessing the CAR and other indices of HPA activity. However, as recently discussed by Hellhammer et al. (2009), using changes in salivary cortisol as an index of HPA activity must take into account extra-ACTH mechanisms of adrenal cortical control including sympathetic neural activity. As more clinical studies show that the CAR can be used as a prospective screen for differences in HPA responsiveness to acute and chronic stress conditions (Inslicht et al., 2011), understanding the mechanisms controlling the amplitude of the CAR has become an issue of major importance.

EVIDENCE FOR AUTONOMIC CONTROL OF THE OVARY

Ovarian secretion of steroids including estrogen and progesterone is controlled by the hypothalamic–pituitary–ovarian (HPO) axis. Hypothalamic neurons in the arcuate nucleus synthesize and release gonadotropin-releasing hormone (GnRH) to stimulate the release of the gonadotropins, luteinizing hormone (LH) and follicle-stimulating hormone (FSH), from the anterior pituitary, which then stimulate ovarian steroid secretion. Reproductive cycles are dependent on the frequency and amplitude of hypothalamic GnRH pulses that control the pattern of gonadotropin secretion that in turn controls ovarian steroid secretion and ovulation (Wildt et al., 1981).

In addition to hypothalamic–pituitary control, the ovary receives sensory, parasympathetic, and sympathetic innervation (Burden et al., 1986; Nance et al., 1988; Burden, 2005) that provides afferent and efferent input to the brain via multisynaptic pathways that integrate signals to control ovarian function (Gerendai et al., 1998). Peripheral innervation is provided by the superior ovarian nerve (SON) and the plexus nerves, supplying multiple ovarian structures including the interstitial tissue, follicles, and blood vessels; this innervation pattern is similar across multiple species, including nonhuman primates (Schultea et al., 1992). Both parasympathetic and sympathetic activity has been implicated in the control of ovarian function. Vagotomy in rats reduces compensatory increases in ovarian growth and follicular function induced by unilateral ovariectomy, suggesting that parasympathetic nerves participate in these responses (Burden et al., 1986; Morales et al., 2007). Electrical stimulation of the SON, the primary source of postganglionic sympathetic fibers, alters steroid secretion in rats (Weiss et al., 1982). In addition, sectioning the SON in rats results in decreases in estradiol and progesterone that are dependent on the phase of the estrous cycle (Aguado and Ojeda, 1984a), decreases in female reproductive behavior (Erskine and Weaver, 1988), and decreases in ovulatory response to gonadotropins (Morales et al., 1998). Since sympathetic nerves supplying nonvascular ovarian tissue release both norepinephrine and neuropeptides such as vasoactive intestinal peptide (VIP) (Aguado and Ojeda, 1984b), these neurotransmitters have been tested for their steroidogenic effects. β-adrenergic receptor agonists stimulate progesterone secretion from human ovarian tissue *in vitro* (Lara et al., 2001); in rodent tissue, priming with FSH is required for a response, suggesting that gonadotropins increase steroidogenic responsiveness to β-adrenergic activity (Adashi and Hsueh, 1981). Other work showed that progesterone responses of peripubertal ovarian tissue to β-adrenergic agonists varied with the phase of the estrous cycle in rats (Aguado et al., 1982); these results support the contention that sympathetic neural activity is important for the initiation of ovarian cycles. In contrast, VIP enhances progesterone synthesis from rat granulosa cells by synergizing with FSH, a response that is associated with both a dose- and time-dependent increase in cAMP (Davoren and Hsueh, 1985; Trzeciak et al., 1986). More recently, studies have shown that β-adrenergic agonists and VIP act to prime developing ovaries, enhancing cAMP responses to FSH in part by upregulating FSH receptors (Mayerhofer et al., 1997). Taken together, these results provide support for the contention that sympathetic nerves affect important aspects of ovarian function including follicular development and steroidogenesis by modulating responsiveness to gonadotropins via a cAMP-dependent mechanism.

Recent insights into the clinical relevance of sympathetic control of the ovary

Since sympathetic neural activity can influence ovarian function, studies have focused on a possible link between sympathetic nerves and ovarian dysfunction. Much attention has been directed toward the possible role of ovarian sympathetic nerves in polycystic ovary syndrome (PCOS). One of the most frequent endocrine problems in women of reproductive age, PCOS is a heterogeneous disorder with uncertain etiology and one of the leading causes of female infertility (Teede et al., 2010). The most common features are anovulation with irregular or loss of menstrual periods and polycystic ovarian morphology, excessive production of androgens, and insulin resistance (Azziz et al., 2004). Normal follicular development is dependent on coordinated HPO interactions that are disrupted in PCOS by hyperandrogenism, hyperinsulinemia, and altered intraovarian signaling (Goodarz et al., 2011). The possible involvement of sympathetic activity was based initially on the observation of increased density of adrenergic innervation of ovaries in PCOS (Semenova, 1969) and of the effectiveness of ovarian wedge resection to reinitiate menstrual cycles in PCOS patients (Stein, 1964). The latter observation has also been ascribed to the profound changes that occur in ovarian steroid secretion, reestablishing sufficient feedback control of the hypothalamic–pituitary axis to initiate normal follicular development (Hendriks et al., 2007).

Animal models of PCOS based on treatments to disrupt pulsatile activity of the HPO axis have been effective in producing many of the symptoms of the human disorder (Mahajan, 1988). Using estradiol-induced PCOS in rats, heightened sympathetic activity precedes the development of ovarian cysts (Lara et al., 1993) and results in exaggerated ovarian steroid responses both to gonadotropins and to β-adrenergic receptor agonists. In addition, transection of the SON not only reverses the altered responsiveness of ovarian tissue, but also attenuates the polycystic condition and restores estrous cycles (Barria et al., 1993). The increased ovarian sympathetic tone observed in estrogen-induced PCOS has been ascribed to the overproduction of nerve growth factor (NGF) and its low affinity receptor, $p75^{ntr}$, since intraovarian blockade of NGF activity reduces the follicular cysts and partially restores estrous cycles (Lara et al., 2000). More recent studies using transgenic mice have shown that overexpression of ovarian NGF results in ovarian hyperinnervation by sympathetic fibers, disrupted estrous cycles, and reduced ovulatory responses to gonadotropins (Dissen et al., 2009). Since NGF is increased in follicular fluid from PCOS patients and cultured granulosa cells from PCOS ovaries produce increased NGF (Dissen et al., 2009), it is possible that overproduction of NGF represents a contributing ovarian factor in the pathogenesis of PCOS. Ovarian NGF could play a role acting indirectly as a neurotrophic factor to promote sympathetic innervation or acting directly on ovarian cells to alter morphology and function (Dissen et al., 2000; Salas et al., 2006). Additional work is required to distinguish between these possibilities and to evaluate the importance of a derangement in sympathetic inputs to the ovary in PCOS.

CONCLUDING REMARKS

Summary of autonomic control of the adrenal cortex and the ovary

This report has highlighted work on the adrenal cortex and the ovary, demonstrating morphological and physiological evidence for autonomic innervation to influence hormonal secretion. For both the adrenal cortex and the ovary, the secretory response to their tropic anterior pituitary hormone is increased by sympathetic neural activity. Sympathetic nerves release norepinephrine and neuropeptides that act indirectly by changing organ blood flow to alter the exposure of the endocrine cell to pituitary hormones or directly by modulating endocrine cell steroidogenesis. Sympathetic-mediated changes in endocrine cell sensitivity are likely due to effects of sympathetic neurotransmitters on intracellular signal transduction pathways. For the adrenal cortex and the ovary, increases in sensitivity to tropic pituitary hormones result in part from enhanced intracellular mobilization of cAMP. Although other intracellular pathways may be involved, altering cAMP would in turn affect downstream cAMP-mediated mechanisms important for controlling adrenal cortical and ovarian function (Miller and Auchus, 2011).

The clinical importance of sympathetic control of the adrenal cortex and the ovary continues to be defined. For the adrenal cortex, changes in adrenal sensitivity to ACTH have been implicated in the CAR, an important clinical tool for assessing HPA function. As the CAR continues to be applied to evaluate HPA activity in individuals exposed to acute and chronic stress, determining the contribution of a sympathetic component in the CAR will be important. Knowing that sympathetic drive to the adrenal cortex contributes to glucocorticoid responses to ACTH could influence treatment options for adrenal dysfunction. For the ovary, increases in sympathetic tone have been implicated in PCOS, a leading cause of female infertility. The observation that increases in ovarian NGF occur in experimental animal models of PCOS and in PCOS patients has led to the intriguing possibility that NGF or its receptors may be a therapeutic target for treatment of PCOS.

Dual control of other endocrine glands

Similar to the adrenal cortex and the ovary, the testes and the thyroid receive input from both the anterior pituitary and the autonomic nervous system. Using transneuronal labeling in rats, the anatomy of central neural pathways that underlie autonomic control of the thyroid (Kalsbeek et al., 2000) and the testes (Gerendai et al., 2000; Lee et al., 2002) have been delineated. For the thyroid, early work established a functional role for autonomic innervation to affect thyroid hormone secretion (Ahrén, 1986). Although more recent animal studies also implicate increased sympathetic activity in producing glandular enlargement and hyperfunction (Young et al., 2005), defining the clinical relevance of autonomic control of the thyroid requires more extensive study. For the testes, recent studies by the Rivier laboratory have provided clear evidence for a hypothalamic–brainstem–spinal pathway that controls testosterone secretion independently of the hypothalamic–pituitary axis (Selvage and Rivier, 2003; Selvage et al., 2006). Since this circuit contributes to inhibition of testicular function induced by stress including alcohol exposure (Selvage et al., 2004; James et al., 2008), it will be important to establish its importance in affecting male infertility induced by environmental factors (Emanuele and Emanuele, 2001; Fisher et al., 2010).

ACKNOWLEDGMENT

This work was supported in part by NSF grant IOS1025199.

REFERENCES

Adashi EY, Hsueh AJ (1981). Stimulation of beta 2-adrenergic responsiveness by follicle-stimulating hormone in rat granulosa cells in vitro and in vivo. Endocrinology 108: 2170–2178.

Aguado LI, Ojeda SR (1984a). Ovarian adrenergic nerves play a role in maintaining preovulatory steroid secretion. Endocrinology 114: 1944–1946.

Aguado LI, Ojeda SR (1984b). Prepubertal ovarian function is finely regulated by direct adrenergic influences. Role of noradrenergic innervation. Endocrinology 114: 1845–1853.

Aguado LI, Petrovic SL, Ojeda SR (1982). Ovarian beta-adrenergic receptors during the onset of puberty: characterization, distribution, and coupling to steroidogenic responses. Endocrinology 110: 1124–1132.

Ahrén B (1986). Thyroid neuroendocrinology: neural regulation of thyroid hormone secretion. Endocr Rev 7: 149–155.

Antoni FA (1986). Hypothalamic control of ACTH secretion: advances since the discovery of 41-residue corticotropin-releasing factor. Endocr Rev 7: 351–378.

Azziz R, Woods KS, Reyna R et al. (2004). The prevalence and features of the polycystic ovary syndrome in an unselected population. J Clin Endocrinol Metab 89: 2745–2749.

Bao X, Lu CM, Liu F et al. (2007). Epinephrine is required for normal cardiovascular responses to stress in the phenylethanolamine N-methyltransferase knockout mouse. Circulation 116: 1024–1031.

Barria A, Leyton V, Ojeda SR et al. (1993). Ovarian steroidal response to gonadotropins and beta-adrenergic stimulation is enhanced in polycystic ovary syndrome: role of sympathetic innervation. Endocrinology 133: 2696–2703.

Born J, Hansen K, Marshall L et al. (1999). Timing the end of nocturnal sleep. Nature 397: 29–30.

Bornstein SR, Ehrhart-Bornstein M, Scherbaum WA et al. (1990). Effects of splanchnic nerve stimulation on the adrenal cortex may be mediated by chromaffin cells in a paracrine manner. Endocrinology 127: 900–906.

Bornstein SR, Gonzales-Hernandez JA, Ehrhart-Bornstein M et al. (1994). Intimate contact of chromaffin and cortical cells within the human adrenal gland forms the cellular basis for important intraadrenal interactions. J Clin Endocrinol Metab 78: 225–232.

Bornstein SR, Haidan A, Ehrhart-Bornstein M (1996). Cellular communication in the neuroadrenocortical axis: role of vasoactive polypeptide. Endocr Res 22: 819–829.

Bornstein SR, Tian H, Haidan A et al. (2000). Deletion of tyrosine hydroxylase gene reveals functional interdependence of adrenocortical and chromaffin cell system in vivo. Proc Natl Acad Sci U S A 97: 14742–14747.

Bornstein SR, Engeland WC, Ehrhart-Bornstein M et al. (2008). Dissociation of ACTH and glucocorticoids. Trends Endocrinol Metab 19: 175–180.

Breslow MJ, Jordan DA, Thellman ST et al. (1987). Neural control of adrenal medullary and cortical blood flow during hemorrhage. Am J Physiol Heart Circ Physiol 252: H521–H528.

Breslow MJ, Tobin JR, Bredt DS et al. (1993). Nitric oxide as a regulator of adrenal blood flow. Am J Physiol Heart Circ Physiol 264: H464–H469.

Buijs RM, Wortel J, van Heerikhuize JJ et al. (1999). Anatomical and functional demonstration of a multisynaptic suprachiasmatic nucleus adrenal (cortex) pathway. Eur J Neurosci 11: 1535–1544.

Burden HW (2005). Adrenergic innervation in ovaries of the rat and guinea pig. J Anat 133: 455–462.

Burden HW, Lawrence IE, Smith CP et al. (1986). The effects of vagotomy on compensatory ovarian hypertrophy and follicular activation after unilateral ovariectomy. Anat Rec 214: 61–66.

Carlsson S, Jonsdottir IH, Skarphedinsson JO et al. (1993). Evidence for an adrenergic innervation of the adrenal cortical blood vessels in rats. Acta Physiol Scand 149: 23–30.

Clow A, Hucklebridge F, Stalder T et al. (2010). The cortisol awakening response: more than a measure of HPA axis function. Neurosci Biobehav Rev 35: 97–103.

Côté M, Payet MD, Rousseau E et al. (1999). Comparative involvement of cyclic nucleotide phosphodiesterases and adenylyl cyclase on adrenocorticotropin-induced increase of cyclic adenosine monophosphate in rat and human glomerulosa cells. Endocrinology 140: 3594–3601.

Coupland RE, Selby JE (1976). The blood supply of the mammalian adrenal medulla: a comparative study. J Anat 122: 539–551.

Coupland RE, Parker TL, Kesse WK et al. (1989). The innervation of the adrenal gland. III. Vagal innervation. J Anat 163: 173–181.

Critchley JAJH, Ellis P, Ungar A (1980). The reflex release of adrenaline and noradrenaline from the adrenal glands of cats and dogs. J Physiol 298: 71–78.

Dallman MF, Engeland WC, Rose JC et al. (1978). Nycthemeral rhythm in adrenal responsiveness to ACTH. Am J Physiol Regul Integr Comp Physiol 235: R210–R218.

Davoren JB, Hsueh AJ (1985). Vasoactive intestinal peptide: a novel stimulator of steroidogenesis by cultured rat granulosa cells. Biol Reprod 33: 37–52.

Dickmeis T (2009). Glucocorticoids and the circadian clock. J Endocrinol 200: 3–22.

Dissen GA, Garcia-Rudaz C, Paredes A et al. (2009). Excessive ovarian production of nerve growth factor facilitates development of cystic ovarian morphology in mice and is a feature of polycystic ovarian syndrome in humans. Endocrinology 150: 2906–2914.

Dissen GA, Parrott JA, Skinner MK et al. (2000). Direct effects of nerve growth factor on thecal cells from antral ovarian follicles. Endocrinology 141: 4736–4750.

Edwards AV, Jones CT (1987a). The effect of splanchnic nerve stimulation on adrenocortical activity in conscious calves. J Physiol 382: 385–396.

Edwards AV, Jones CT (1987b). The effect of splanchnic nerve section on the sensitivity of the adrenal cortex to ACTH in the calf. J Physiol 390: 23–31.

Edwards AV, Jones CT, Bloom SR (1986). Reduced adrenal cortical sensitivity to ACTH in lambs with cut splanchnic nerves. J Endocrinol 110: 81–85.

Ehrhart-Bornstein M, Hinson JP, Bornstein SR et al. (1998). Intraadrenal interactions in the regulation of adrenocortical steroidogenesis. Endocr Rev 19: 101–143.

Emanuele MA, Emanuele N (2001). Alcohol and the male reproductive system. Alcohol Res Health 25: 282–287.

Engeland WC, Arnhold MM (2005). Neural circuitry in the regulation of adrenal corticosterone rhythmicity. Endocrine 28: 325–332.

Engeland WC, Gann DS (1989). Splanchnic nerve stimulation modulates steroid secretion in hypophysectomized dogs. Neuroendocrinology 50: 124–131.

Engeland WC, Bereiter DF, Gann DS (1986). Sympathetic control of adrenal secretion of enkephalins and catecholamines after hemorrhage in awake dogs. Am J Physiol Regul Integr Comp Physiol 251: R341–R348.

Erskine MS, Weaver CE (1988). The role of ovarian sympathetic innervation in the control of estrous responsiveness in the rat. Horm Behav 22: 1–11.

Fisher JR, Baker GH, Hammarberg K (2010). Long-term health, well-being, life satisfaction, and attitudes toward parenthood in men diagnosed as infertile: challenges to gender stereotypes and implications for practice. Fertil Steril 94: 574–580.

Gallo-Payet N, Pothier P, Isler H (1987). On the presence of chromaffin cells in the adrenal cortex: their possible role in adrenocortical function. Biochem Cell Biol 65: 588–592.

Gerendai I, Tóth IE, Boldogkoi Z et al. (1998). Neuronal labeling in the rat brain and spinal cord from the ovary using viral transneuronal tracing technique. Neuroendocrinology 68: 244–256.

Gerendai I, Tóth IE, Boldogkoi Z et al. (2000). Central nervous system structures labelled from the testis using the transsynaptic viral tracing technique. J Neuroendocrinol 12: 1087–1095.

Goodarz MO, Dumesic DA, Chazenbalk G et al. (2011). Polycystic ovary syndrome: etiology, pathogenesis and diagnosis. Nat Rev Endocrinol 7: 219–231.

Hellhammer DH, Wust S, Kudielka BM (2009). Salivary cortisol as a biomarker in stress research. Psychoneuroendocrinology 34: 163–171.

Hendriks ML, Ket JCF, Hompes PGA et al. (2007). Why does ovarian surgery in PCOS help? Insight into the endocrine implications of ovarian surgery for ovulation induction in polycystic ovary syndrome. Hum Reprod Update 13: 249–264.

Hinson JP, Kapas S, Orford CD et al. (1992). Vasoactive intestinal peptide stimulation of aldosterone secretion by the rat adrenal cortex may be mediated by the local release of catecholamines. J Endocrinol 133: 253–258.

Holgert H, Aman K, Cozzari C et al. (1995). The cholinergic innervation of the adrenal gland and its relation to enkephalin and nitric oxide synthase. Neuroreport 6: 2576–2580.

Holgert H, Dagerlind A, Hokfelt T (1998). Immunohistochemical characterization of the peptidergic innervation of the rat adrenal gland. Horm Metab Res 30: 315–322.

Inslicht SS, Otte C, McCaslin SE et al. (2011). Cortisol awakening response prospectively predicts peritraumatic and acute stress reactions in police officers. Biol Psychiatry 70: 1055–1061.

Ishida A, Mutoh T, Ueyama T et al. (2005). Light activates the adrenal gland: timing of gene expression and glucocorticoid release. Cell Metab 2: 297–307.

James P, Rivier C, Lee S (2008). Presence of corticotrophin-releasing factor and/or tyrosine hydroxylase in cells of a neural brain-testicular pathway that are labeled by a transganglionic tracer. J Neuroendocrinol 20: 173–181.

Jasper MS, McDermott P, Gann DS et al. (1990). Measurement of blood flow to the adrenal capsule, cortex and medulla in dogs after hemorrhage by fluorescent microspheres. J Auton Nerv Syst 30: 159–167.

Jones CT, Edwards AV, Bloom SR (1990). The effect of changes in adrenal blood flow on adrenal cortical responses to ACTH in conscious calves. J Physiol 429: 377–386.

Kalsbeek A, Fliers E, Franke AN et al. (2000). Functional connections between the suprachiasmatic nucleus and the thyroid gland as revealed by lesioning and viral tracing techniques in the rat. Endocrinology 141: 3832–3841.

Kaneko M, Hiroshige T, Shinsako J et al. (1980). Diurnal changes in amplification of hormone rhythms in the adrenocortical system. Am J Physiol Regul Integr Comp Physiol 239: R309–R316.

Kesse WK, Parker TL, Coupland RE (1988). The innervation of the adrenal gland I. The source of pre- and postganglionic nerve fibers to the rat adrenal gland. J Anat 157: 33–41.

L'Age M, Gonzalez-Luque A, Yates FE (1970). Adrenal blood flow dependence of cortisol secretion rate in unanesthetized dogs. J Physiol 219: 281–287.

Lara HE, Ferruz JL, Luza S et al. (1993). Activation of ovarian sympathetic nerves in polycystic ovary syndrome. Endocrinology 133: 2690–2695.

Lara HE, Dissen GA, Leyton V et al. (2000). An increased intraovarian synthesis of nerve growth factor and its low affinity receptor is a principal component of steroid-induced polycystic ovary in the rat. Endocrinology 141: 1059–1072.

Lara HE, Porcile A, Espinoza J et al. (2001). Release of norepinephrine from human ovary: coupling to steroidogenic response. Endocrine 15: 187–192.

Lee S, Miselis R, Rivier C (2002). Anatomical and functional evidence for a neural hypothalamic–testicular pathway that is independent of the pituitary. Endocrinology 143: 4447–4454.

Leproult R, Colecchia EF, L'Hermite-Baleriaux M et al. (2001). Transition from dim to bright light in the morning induces an immediate elevation of cortisol levels. J Clin Endocrinol Metab 86: 151–157.

Levitzki A (1988). From epinephrine to cyclic AMP. Science 241: 800–806.

Ligier B, Breslow MJ, Clarkson K et al. (1994). Adrenal blood flow and secretory effects of adrenergic receptor stimulation. Am J Physiol Heart Circ Physiol 266: H220–H227.

Lupien SJ, Maheu F, Tu M et al. (2007). The effects of stress and stress hormones on human cognition: implications for the field of brain and cognition. Brain Cogn 65: 209–237.

Mahajan DK (1988). Polycystic ovarian disease: animal models. Endocrinol Metab Clin North Am 17: 705–732.

Mayerhofer A, Dissen GA, Costa ME et al. (1997). A role for neurotransmitters in early follicular development: induction of functional follicle-stimulating hormone receptors in newly formed follicles of the rat ovary. Endocrinology 138: 3320–3329.

Miller WL, Auchus RJ (2011). The molecular biology, biochemistry, and physiology of human steroidogenesis and its disorders. Endocr Rev 32: 81–151.

Morales L, Chavez R, Ayala ME et al. (1998). Effects of unilateral or bilateral superior ovarian nerve section in prepubertal rats on the ovulatory response to gonadotrophin administration. J Endocrinol 158: 213–219.

Morales L, Ricardo B, Bolaños A et al. (2007). Ipsilateral vagotomy to unilaterally ovariectomized pre-pubertal rats modifies compensatory ovarian responses. Reprod Biol Endocrinol 5: 24.

Nance DM, Burns J, Klein CM et al. (1988). Afferent fibers in the reproductive system and pelvic viscera of female rats: anterograde tracing and immunocytochemical studies. Brain Res Bull 21: 701–709.

Neylan TC, Brunet A, Pole N et al. (2005). PTSD symptoms predict waking salivary cortisol levels in police officers. Psychoneuroendocrinology 30: 373–381.

Pruessner JC, Wolf OT, Hellhammer DH et al. (1997). Free cortisol levels after awakening: a reliable biological marker for the assessment of adrenocortical activity. Life Sci 61: 2539–2549.

Rimmele U, Meier F, Lange T et al. (2010). Suppressing the morning rise in cortisol impairs free recall. Learn Mem 17: 186–190.

Rivier C, Vale W (1983). Interaction of corticotropin-releasing factor and arginine vasopressin on adrenocorticotropin secretion in vivo. Endocrinology 113: 939–942.

Salas C, Julio-Pieper M, Valladares M et al. (2006). Nerve growth factor-dependent activation of trkA receptors in the human ovary results in synthesis of follicle-stimulating hormone receptors and estrogen secretion. J Clin Endocrinol Metab 91: 2396–2403.

Scheer F, Buijs RM (1999). Light affects morning salivary cortisol in humans. J Clin Endocrinol Metab 84: 3395–3398.

Schimmer BP (1995). The 1994 Upjohn Award Lecture. Molecular and genetic approaches to the study of signal transduction in the adrenal cortex. Can J Physiol Pharmacol 73: 1097–1107.

Schlotz W, Hellhammer J, Schulz P et al. (2004). Perceived work overload and chronic worrying predict weekend-weekday differences in the cortisol awakening response. Psychosom Med 66: 207–214.

Schultea TD, Dees WL, Ojeda SR (1992). Postnatal development of sympathetic and sensory innervation of the rhesus monkey ovary. Biol Reprod 47: 760–767.

Selvage DJ, Rivier C (2003). Importance of the paraventricular nucleus of the hypothalamus as a component of a neural pathway between the brain and the testes that modulates testosterone secretion independently of the pituitary. Endocrinology 144: 594–598.

Selvage DJ, Hales DB, Rivier CL (2004). Comparison between the influence of the systemic and central injection of alcohol on Leydig cell activity. Alcohol Clin Exp Res 28: 480–488.

Selvage DJ, Parsons L, Rivier C (2006). Role played by brainstem neurons in regulating testosterone secretion via a direct neural pathway between the hypothalamus and the testes. Endocrinology 147: 3070–3075.

Semenova II (1969). Adrenergic innervation of ovaries in Stein–Leventhal syndrome. Vestn Akad Med Nauk SSSR 24: 58–62.

Spath-Schwalbe E, Scholler T, Kern W et al. (1992). Nocturnal adrenocorticotropin and cortisol secretion depends on sleep duration and decreases in association with spontaneous awakening in the morning. J Clin Endocrinol Metab 75: 1431–1435.

Stein IF (1964). Multiple pregnancy following wedge resection in the Stein–Leventhal syndrome. Int J Fertil 9: 343–350.

Taborsky G (2010). The physiology of glucagon. J Diabetes Sci Technol 4: 1338–1344.

Tchen TT, Chan SW, Kuo TH et al. (1977). Studies on the adrenal cortex of hypophysectomized rats: a model for abnormal cellular atrophy and death. Mol Cell Biochem 15: 79–87.

Teede H, Deeks A, Moran L (2010). Polycystic ovary syndrome: a complex condition with psychological, reproductive and metabolic manifestations that impacts on health across the lifespan. BMC Med 8: 41.

Thorens B (2011). Brain glucose sensing and neural regulation of insulin and glucagon secretion. Diabetes Obes Metab 13: 82–88.

Trzeciak WH, Ahmed CE, Simpson ER et al. (1986). Vasoactive intestinal peptide induces the synthesis of the cholesterol side-chain cleavage enzyme complex in cultured rat ovarian granulosa cells. Proc Natl Acad Sci U S A 83: 7490–7494.

Tsigosa C, Chrousos GP (2002). Hypothalamic–pituitary–adrenal axis, neuroendocrine factors and stress. J Psychosom Res 53: 865–871.

Ulrich-Lai YM, Engeland WC (2000). Hyperinnervation during adrenal regeneration influences the rate of functional recovery. Neuroendocrinology 71: 107–123.

Ulrich-Lai YM, Engeland WC (2002). Adrenal splanchnic innervation modulates adrenal cortical responses to dehydration stress in rats. Neuroendocrinology 76: 79–92.

Ulrich-Lai YM, Arnhold MM, Engeland WC (2006). Adrenal splanchnic innervation contributes to the diurnal rhythm of plasma corticosterone in rats by modulating adrenal sensitivity to ACTH. Am J Physiol Regul Integr Comp Physiol 290: R1128–R1135.

Unsicker K (1996). Autononic–Endocrine Interactions. In: G Burnstock (Ed.), The Autonomic Nervous System. Vol. 10. Harwood Academic Publishers, Amsterdam.

Urquhart J (1965). Adrenal blood flow and the adrenocortical response to corticotropin. Am J Physiol 209: 1162–1168.

Walker SW, Lightly ERT, Clyne C et al. (1991). Adrenergic and cholinergic regulation of cortisol secretion from the zona fasciculata/reticularis of bovine adrenal cortex. Endocr Res 17: 237–264.

Weiss GK, Dail WG, Ratner A (1982). Evidence for direct neural control of ovarian steroidogenesis in rats. J Reprod Fertil 65: 507–511.

Wildt L, Hausler A, Marshall G et al. (1981). Frequency and amplitude of gonadotropin-releasing hormone stimulation and gonadotropin secretion in the rhesus monkey. Endocrinology 109: 376–385.

Wilhelm I, Born J, Kudielka BM et al. (2007). Is the cortisol awakening rise a response to awakening? Psychoneuroendocrinology 32: 358–366.

Wust S, Federenko I, Hellhammer DH et al. (2000). Genetic factors, perceived chronic stress, and the free cortisol response to awakening. Psychoneuroendocrinology 25: 707–720.

Young JB, Bürgi-Saville ME, Bürgi U et al. (2005). Sympathetic nervous system activity in rat thyroid: potential role in goitrogenesis. Am J Physiol Endocrinol Metab 288: E861–E867.

Handbook of Clinical Neurology, Vol. 117 (3rd series)
Autonomic Nervous System
R.M. Buijs and D.F. Swaab, Editors

Chapter 5

Central control of autonomic function and involvement in neurodegenerative disorders

MARIA G. CERSOSIMO[1] AND EDUARDO E. BENARROCH[2]*
[1]*Parkinson's Disease and Movement Disorder Unit, Hospital de Clínicas, University of Buenos Aires, Argentina*
[2]*Department of Neurology, Mayo Clinic, Rochester, MN, USA*

OVERVIEW OF THE CENTRAL AUTONOMIC NETWORK

Several interconnected areas distributed throughout the neuraxis have a critical role in the moment-to-moment control of the body's visceral functions, maintenance of homeostasis, and adaptation to internal or external challenges (Fig. 5.1). These areas not only control the sympathetic and parasympathetic outflows but also affect endocrine and motor outputs as components of integrated physiological responses. Although the functional anatomy of this central autonomic network was first characterized in experimental animals (Strack et al., 1989; McAllen et al., 1995; Hopkins and Armour, 1998; Bandler et al., 2000; Saper, 2002; Dampney and Horiuchi, 2003; Guyenet, 2006; Stornetta, 2009; Geerling et al., 2010), many of these areas are activated during autonomic responses in humans, as demonstrated in several functional neuroimaging studies (Blok et al., 1997; King et al., 1999; Bandler et al., 2000; Critchley et al., 2003, 2011; Critchley, 2005; Egan et al., 2005; McAllen et al., 2006; Fowler and Griffiths, 2010; Macefield and Henderson, 2010; Sander et al., 2010).

Forebrain components

The forebrain regions that are directly or indirectly involved in high-level control of autonomic functions include the insular cortex, anterior cingulate cortex, amygdala, and several areas of the hypothalamus.

INSULAR CORTEX

The insular cortex is the primary interoceptive cortex and integrates visceral, pain, and temperature sensations (Saper, 2002; Craig, 2003). The dorsal insula has a viscerotopic organization (Cechetto, 1987) and receives topographically organized inputs from gustatory, visceral, muscle, and skin receptors via the thalamus. The dorsal insula projects to the right anterior insula, which conveys the conscious experience of bodily sensation by integrating these interoceptive inputs with inputs from cortical areas involved in perceptive, emotional, and cognitive processing (Craig, 2003). For example, functional magnetic resonance (fMRI) studies showed increased activity of the anterior insula during tasks in which subjects attended to the timing of their heartbeats (Craig, 2005; Critchley et al., 2011). The insula is also a visceromotor area controlling both the sympathetic and parasympathetic outputs, primarily via a relay in the lateral hypothalamus (Loewy, 1982; Westerhaus and Loewy, 2001; Nagai et al., 2010).

ANTERIOR CINGULATE CORTEX

The anterior cingulate cortex is interconnected with the anterior insula and both areas show coactivation in fMRI studies (Deen et al., 2011). The anterior cingulate has been functionally subdivided into ventral (affective) and dorsal (cognitive) regions, which have different cytoarchitecture, connectivity, and functions (Vogt et al., 2005). Functional neuroimaging studies indicate that the ventral anterior cingulate is part of the brain "default mode network" that is active in resting conditions, whereas the dorsal anterior cingulate cortex is a component of the frontoparietal attention networks (Margulies et al., 2007). The ventral anterior cingulate cortex includes subcallosal and precallosal portions that are cytoarchitectonically and neurochemically distinct

*Correspondence to: Eduardo E. Benarroch, M.D., Mayo Clinic, 200 First Street SW, Rochester, MN 55905, USA. Tel: +1-507-284-3375, Fax: +1-507-266-4419, E-mail: benarroch.eduardo@mayo.edu

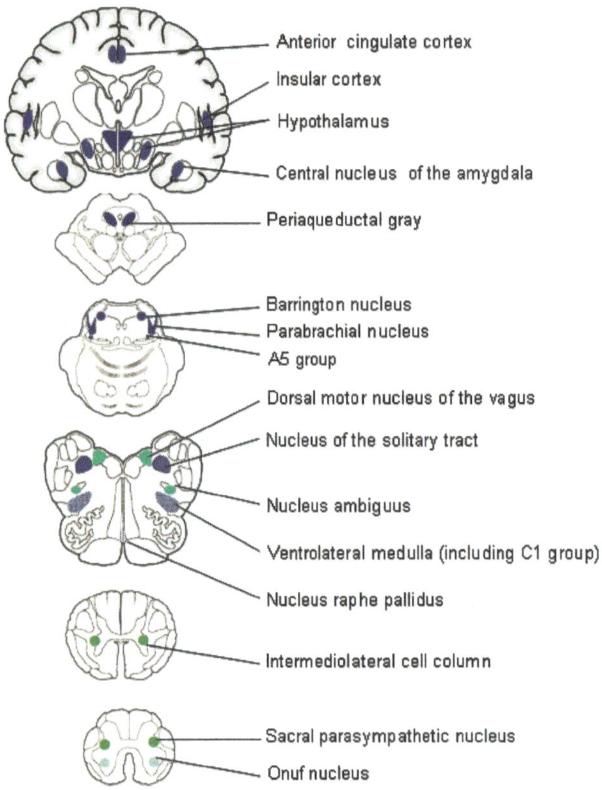

Fig. 5.1. Central autonomic control areas.

(Vogt et al., 2005). These anterior cingulate areas have extensive connections with the insula, prefrontal cortex, amygdala, hypothalamus, and brainstem. Via these projections, the anterior cingulate cortex controls sympathetic and parasympathetic functions (Terreberry and Neafsey, 1987; Hurley et al., 1991; Ter Horst et al., 1996; Verberne and Owens, 1998).

Functional MRI studies show that the patterns of activation of the anterior cingulate cortex vary according to the experimental task and the associated sympathetic or parasympathetic responses (Critchley, 2009). The dorsal anterior cingulate cortex is activated during tasks that involve awareness, attention, and behavioral engagement and are associated with an increase in sympathetic drive. In contrast, the subgenual anterior cingulate cortex and adjacent ventromedial prefrontal cortex (which are part of the "default mode network") became inactivated during these tasks (Critchley, 2009). This is consistent with evidence that the subgenual anterior cingulate projects to autonomic nuclei involved in parasympathetic control of the heart (Terreberry and Neafsey, 1987; Hurley et al., 1991; Ter Horst et al., 1996; Verberne and Owens, 1998). Functional neuroimaging studies also suggest that the anterior cingulate and associated forebrain regions may be involved with the generation of baroreflex-mediated sympathetic and cardiovascular responses (Kimmerly et al., 2005, 2007; Critchley et al., 2011).

Amygdala

The amygdala provides affective or emotional value to incoming sensory information (LeDoux, 2007). The amygdala is structurally complex and has multiple downstream targets that participate in the autonomic and neuroendocrine response to stress (Ulrich-Lai and Herman, 2009). The amygdala nuclear complex consists of two major divisions: the basolateral complex and the extended amygdala. The basolateral amygdala has an affiliation with the cerebral cortex and includes the lateral, basal, and accessory nuclei. The extended amygdala is a continuum that includes the central nucleus of the amygdala (CeA), lateral bed nucleus of the stria terminalis (BNST), and associated regions of the sublenticular substantia innominata. There is a general pattern of connectivity within the amygdala. Inputs from the cerebral cortex and thalamus reach primarily the basolateral subdivision. The basolateral complex projects back to the cerebral cortex, thalamus, and striatum to modulate sensory processing, memory, decision making, and behavioral drive. In addition, the basolateral complex sends intra-amygdaloid inputs to the extended amygdala, which projects to the hypothalamus and brainstem. The CeA is the main effector nucleus of the extended amygdala and has a major role in integration of the stress response, particularly fear responses (Davis, 1992). The CeA initiates behavioral and autonomic responses via widespread connections with the hypothalamus, particularly the paraventricular nucleus and lateral hypothalamic areas, and brainstem, particularly the periaqueductal gray, rostral ventrolateral, and ventromedial medullary reticular formation (Loewy, 1982). The CeA is differentially activated by disruption of homeostasis and systemic stressors (such as cytokines) and contributes to the integration of the autonomic responses to psychological stress (Ulrich-Lai and Herman, 2009).

Hypothalamus

The hypothalamus and the adjacent preoptic area have a central role in integrated autonomic and endocrine responses necessary for homeostasis and adaptation. These include regulation of the circadian rhythms and sleep–wake cycle, thermoregulation, glucoregulation, osmoregulation, responses to stress, and immunomodulation (Saper, 2002; Thompson and Swanson, 2003). The hypothalamus acts as a visceromotor pattern generator that initiates specific patterns of autonomic responses according to the stimulus, such as hypoglycemia, changes in blood temperature or osmolarity, or external stressors.

The preoptic–hypothalamic area is functionally subdivided into three functional zones: periventricular, medial, and lateral (Thompson and Swanson, 2003).

The periventricular zone includes the suprachiasmatic nucleus, which is the circadian pacemaker, and several areas involved in neuroendocrine control via projections to the pituitary gland. The medial zone includes several nuclei involved in homeostasis and adaptive responses, including the medial preoptic area, paraventricular nucleus (PVN), and dorsomedial nucleus (DMH), which orchestrate coordinated autonomic and endocrine outputs for thermoregulation, osmoregulation, and stress responses (Sawchenko et al., 2000; Thompson and Swanson, 2003; Dimicco and Zaretsky, 2007; Yoshida et al., 2009). The lateral zone includes nuclei that control sleep, arousal, and motivated behavior (Saper, 2002).

The main autonomic outputs of the hypothalamus originate from the PVN, DMH, and lateral hypothalamic area (Saper, 2002). The PVN is the principal integrator of stress signals. It contains different neuronal populations that are differentially activated during stress responses (Sawchenko et al., 2000; Benarroch, 2005). They include magnocellular neurons that release arginine vasopressin (AVP) to the general circulation; neurons that activate the adrenocortical axis, and different groups of neurons that project to autonomic nuclei of the brainstem and spinal cord (Sawchenko et al., 2000; Geerling et al., 2010). Via these outputs, the PVN modulates not only stress responses but also cerebral blood flow, food and sodium intake, glucose metabolism, and cardiovascular, renal, gastrointestinal, and respiratory functions. The DMH participates in stress responses (Dimicco and Zaretsky, 2007), thermoregulation (Morrison et al., 2008; Yoshida et al., 2009), and cardiovascular control (Dampney et al., 2008; Hunt et al., 2010). The posterior lateral hypothalamus contains neurons synthesizing hypocretin/orexin; these neurons contribute to control of autonomic output in the setting of arousal, feeding, and reward-driven behaviors (Saper, 2002).

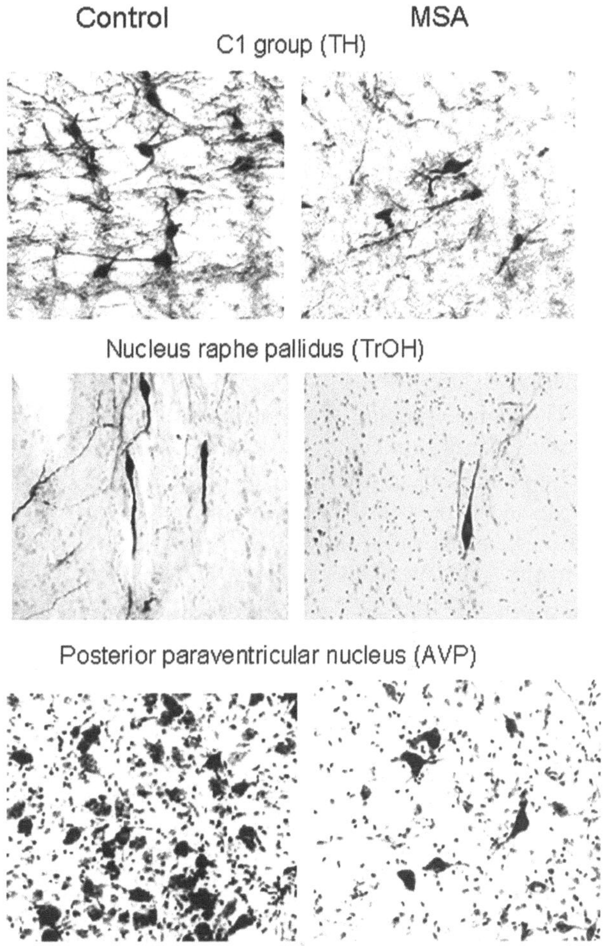

Fig. 5.2. Photomicrographs of 50 μm thick sections immunostained for tyrosine hydroxylase (TH), tryptophan hydroxylase (TrOH), or arginine vasopressin (AVP) comparing a multiple system atrophy (MSA) case with an age-matched control. In MSA, there is loss of catecholaminergic (TH) neurons in the rostral ventrolateral medulla (C1 group); serotoninergic (TrOH immunoreactive) neurons in the nucleus raphe pallidus; and AVP immunoreactive neurons in the posterior portion of the paraventricular nucleus.

Brainstem components

The brainstem areas controlling the sympathetic and parasympathetic outputs include the periaqueductal gray matter of the midbrain (PAG), the parabrachial nucleus (PBN) and adjacent areas of the pons, several medullary regions, including the nucleus of the solitary tract (NTS), ventrolateral reticular formation of the medulla, and medullary raphe (Fig. 5.2).

Periaqueductal gray

The PAG is the interface between the forebrain and the lower brainstem and has a major role in integrated autonomic and somatic responses to stress, pain modulation, and other adaptive functions. The PAG consists of different longitudinal columns; each column has different cortical and brainstem connections and coordinates different patterns of behaviors (Bandler et al., 2000). The PAG participates in cardiovascular responses associated with pain modulation (Heinricher et al., 2009); thermoregulation (de Menezes et al., 2009); coordination of the micturition reflex (Holstege, 2005); and respiratory rhythm and airway resistance (Haxhiu et al., 2002; Zhang et al., 2007; Subramanian et al., 2008). Experimental studies also indicate that the PAG may have an important role in mechanisms of arousal (Lu et al., 2006) and regulation of REM sleep (Fuller et al., 2007). These multiple functions depend on extensive PAG interconnections with the prefrontal and anterior cingulate cortex, amygdala, hypothalamus, and brainstem autonomic, motor, and pain modulatory areas.

Parabrachial complex and adjacent regions of the pons

The PBN is a major relay and coordinating center. It receives converging visceral, nociceptive, and thermoreceptive inputs from the spinal cord and conveys this information to the hypothalamus, amygdala, and thalamus (Saper, 2002). The PBN also contains separate subnuclei that participate in the control of respiratory, cardiovascular, and gastrointestinal reflexes. Some PBN neurons are necessary for maintenance of the normal pattern of respiration (Chamberlin, 2004; Smith et al., 2009).

The dorsal pontine tegmentum also contains the pontine micturition center (PMC), also referred to as Barrington's nucleus or the M- (for medial) region, which is critical for the coordination of the micturition reflex and participates in the control of the function of the lower gastrointestinal tract and sexual organs (Holstege, 2005).

Nucleus of the solitary tract

The NTS is the first relay station of taste and visceral afferent information and includes several subnuclei with a viscerotropic organization. The rostral portion of the l NTS receives taste inputs; the intermediate portion receives gastrointestinal afferents; and the caudal portion receives afferents from baroreceptors, cardiac receptors, chemoreceptors, and pulmonary receptors (Jean, 1991). The different subnuclei of the NTS relay this information, either directly or via the PBN, to the PAG, hypothalamus, amygdala and to ventral thalamic nuclei projecting to the insular cortex (Cechetto, 1987; Saper, 2002).

The second major function of the NTS is to serve as the first central relay for all afferents that trigger medullary reflexes controlling cardiovascular function, including the baroreflex and cardiac reflexes (Dampney and Horiuchi, 2003), respiration, including the carotid chemoreflex and pulmonary mechanoreflexes (Feldman et al., 2003) and gastrointestinal motility, particularly that of the esophagus and stomach (Travagli et al., 2006).

Rostral ventrolateral medulla

The rostral ventrolateral medulla (RVLM), including the C1 group of epinephrine-containing neurons, is a key area for regulation of arterial blood pressure (Dampney and Horiuchi, 2003; Guyenet, 2006). Glutamatergic neurons of the RVLM project directly to sympathetic preganglionic neurons controlling cardiac output and blood flow to the skeletal muscle and visceral organs (Morrison, 1999). The RVLM provides tonic activation to these neurons and mediates all reflexes controlling arterial blood pressure, including the baroreflex, cardiopulmonary reflexes, and chemoreflexes (Dampney and Horiuchi, 2003; Guyenet, 2006). Some of the glutamatergic RVLM neurons correspond to the C1 group and express different combinations of neuropeptides (Stornetta, 2009).

The sympathoexcitatory RVLM neurons receive and integrate a large variety of inputs from the brainstem and forebrain (Guyenet, 2006). These include inhibitory signals from baroreceptor-sensitive neurons of the NTS and mediated via inhibitory γ-aminobutyric acid (GABA)-ergic neurons of the caudal ventrolateral medulla (Dampney et al., 2003; Guyenet, 2006). Other medullary inputs to the RVLM contribute to cardiac, vestibular, exercise, and nociceptive sympathoexcitatory reflexes, as well as sympathoexcitatory responses to hypoxia (Guyenet, 2000). The RVLM also receives several inputs from the hypothalamus, including the PVN, which trigger sympathoexcitatory responses to internal or external stimuli (Dampney et al., 2008; Geerling et al., 2010). Whereas different groups of sympathoexcitatory RVLM neurons may preferentially control the skeletal muscle or splanchnic vessels, the kidney, or the heart (McAllen et al., 1995), there is high degree of convergence and divergence between the RVLM neurons and their preganglionic sympathetic targets (Guyenet, 2006).

Caudal ventrolateral medulla

The caudal ventrolateral medulla contains GABAergic neurons that maintain a tonic inhibitory control on the RVLM and relay the inhibitory inputs from barosensitive neurons of the NTS, thus mediating the sympathoinhibitory component of the arterial baroreflex (Dampney et al., 2003). However, electrical stimulation studies indicate that the caudal medulla also contains three pressor regions: the caudal pressor area, the intermediate pressor area, and the medullocervical pressor area (Goodchild and Moon, 2009). The caudal ventrolateral medulla also contains the A1 group of neurons that, together with the A2 group from the area of the NTS, provide noradrenergic innervation to the hypothalamus and other rostral autonomic areas. The A1 neurons are a component of a reflex pathway that triggers AVP release from magnocellular neurons of the hypothalamus in response to hypovolemia or hypotension (Lightman et al., 1984).

Ventromedial medulla and caudal raphe

The rostral ventromedial medulla (RVMM), including the caudal raphe nuclei, has an important role in thermoregulation (Morrison, 1999), pain modulation (Bandler et al., 2000), and control of automatic ventilation (Corcoran et al., 2009). One group of medullary raphe neurons have a thermoregulatory function and initiate sympathetic responses to cold; they provide direct input to preganglionic sympathetic neurons that activate skin vasoconstriction and nonshivering thermogenesis in the brown adipose tissue (Morrison et al., 2008; Yoshida et al., 2009). The chemical coding of bulbospinal presympathetic neurons is complex, and includes not only serotonergic, but also GABAergic and peptidergic neurons (Stornetta, 2009).

Medullary respiratory groups

The medullary reticular formation also contains the ventral respiratory column, which consists of several groups of inspiratory or expiratory neurons organized into different rostrocaudal groups and involved in respiratory rhythmogenesis (Feldman et al., 2003). The rostral medulla also contains chemoreceptive neurons involved in automatic breathing in response to hypoxia and hypercapnia (Feldman et al., 2003; Corcoran et al., 2009; Nattie and Li, 2009; Guyenet et al., 2010). These include a group of glutamatergic neurons located close to the ventral medullary surface, referred to as the retrotrapezoid nucleus in rodents (Guyenet et al., 2010) and serotonergic neurons of the medullary raphe (Corcoran et al., 2009). The medullary arcuate nucleus, located in the ventral medullary surface, has both glutamatergic and serotonergic neurons (Paterson et al., 2006) and has been proposed as a putative chemosensitive area in humans (Filiano et al., 1990).

AUTONOMIC OUTPUT OF THE CENTRAL NERVOUS SYSTEM

The central nervous system controls autonomic effectors via preganglionic sympathetic and parasympathetic neurons. All preganglionic neurons use acetylcholine as their primary neurotransmitter, which elicits fast excitation of neurons in the peripheral autonomic ganglia via ganglion-type ($\alpha3\beta4$) receptors. The autonomic ganglia contain different types of neurochemically coded neurons that innervate specific targets. A discussion of the specific anatomical, neurochemical, and physiological effects of each sympathetic and parasympathetic output is beyond the scope of this chapter. Only some salient concepts will be emphasized here.

Sympathetic preganglionic units

The sympathetic output is critical for maintenance of arterial pressure, thermoregulation, and redistribution of regional blood flow during stress and exercise.

Functional organization

The sympathetic output originates from sympathetic preganglionic neurons located in the thoracolumbar spinal cord at the T1–L2 segments, primarily in the intermediolateral cell column. These neurons are organized into functionally separate units that innervate selective subpopulations of sympathetic ganglion neurons and receive distinct segmental afferent inputs triggering segmental somato- and viscerosympathetic reflexes (Janig and Habler, 2003). These preganglionic sympathetic units include cardiomotor, muscle vasoconstrictor, visceral vasoconstrictor, skin vasoconstrictor, skin vasodilator, sudomotor, pilomotor, pupillomotor, and visceromotor units (Janig and Habler, 2003; Guyenet, 2006). The sympathetic ganglion neurons that mediate these effects are chemically coded; for example, vasoconstrictor neurons utilize norepinephrine, neuropeptide Y, and adenosine triphosphate, whereas sudomotor neurons utilize acetylcholine, vasoactive intestinal polypeptide, and nitric oxide.

Patterns of sympathetic responses

Different preganglionic sympathetic units are recruited in a coordinated fashion by premotor neurons in the brainstem and hypothalamus to initiate different patterns of responses to specific internal or external stressors, such as postural changes, exercise, hypoglycemia, dehydration, exposure to heat or cold, or stress (Saper, 2002). The main sources of premotor sympathetic innervation are the RVLM, medullary raphe, A5 noradrenergic group of the pons, PVN, and lateral hypothalamic area (Loewy, 1982; Strack et al., 1989). These descending sympathoexcitatory outputs orchestrate different patterns of responses. For example, baroreflex-sensitive neurons of the RVLM innervate sympathetic preganglionic neurons controlling blood pressure, whereas neurons of the RVMM activated by the DMH initiate sympathoexcitatory response to cold (Morrison, 1999). Preganglionic neurons receiving inputs from the PVN control sympathetic output to the kidneys or adrenal medulla and mediate responses to internal or external stressors (Guyenet, 2006).

Parasympathetic outputs

In contrast to the sympathetic system, which mediates stimulus-specific patterns of responses affecting multiple effectors, the parasympathetic system mediates reflexes activated in an organ-specific fashion. A comprehensive review of the parasympathetic outflow is beyond the scope of this chapter. Only the vagal and sacral parasympathetic outflow will be discussed, given their relevance in neurodegenerative disorders.

Vagal output

The vagus nerve has both afferent and efferent components, which are critical to trigger autonomic reflexes.

The afferent vagus provides inputs to the NTS and conveys information from cardiovascular, gastrointestinal, and respiratory receptors (Jean, 1991; Travagli et al., 2006). The vagal efferent output originates from preganglionic neurons located in the dorsal motor nucleus of the vagus (DMV) and in the ventrolateral portion of the nucleus ambiguus (NAmb) in the medulla. The DMV contains most of the vagal preganglionic parasympathetic neurons, which are organized in a viscerotropic fashion (Huang et al., 1993) and innervate the local ganglia in the respiratory tract, enteric nervous system (ENS), liver, and pancreas.

The DMV receives inputs from the NTS and mediates all vagovagal reflexes controlling gastrointestinal motility and secretion (Travagli et al., 2006) via activation of neurons of the myenteric and the submucosal plexus of the ENS. Different subgroups of DMV neurons provide parallel excitatory and inhibitory pathways controlling gastrointestinal motility (Chang et al., 2003). The excitatory vagal motor pathway innervates myenteric neurons that utilize acetylcholine and stimulate contraction of smooth muscle; the inhibitory pathway is mediated by postganglionic myenteric neurons that utilize nitric oxide, vasoactive intestinal polypeptide, or adenosine triphosphate (Grundy and Schemann, 2007). Vagal influence is most prominent in the esophagus and stomach but is less important in the small bowel and colon, where motility and secretion are primarily controlled by local ENS reflexes. (Grundy and Schemann, 2007).

Vagal preganglionic neurons located in the ventrolateral portion of the NAmb provide the primary control of the heart via the cardiac ganglia (McAllen and Spyer, 1976; Hopkins and Armour, 1998; Jordan, 2005). These cardiovagal outputs inhibit the automatism of the sinoatrial node exerting a beat-to-beat control of the heart rate. There are two main influences that control the activity of these cardiovagal neurons; they are excited by baroreceptor-sensitive NTS neurons and mediate efferent cardioinhibitory component of the baroreflex, and are inhibited by inspiratory neurons of the ventral respiratory group, which is responsible for respiratory sinus arrhythmia (McAllen and Spyer, 1976; Hopkins and Armour, 1998; Jordan, 2005). Neurons of the DMV contribute to the innervation of the cardiac ganglia and may contribute to the antagonistic effects of the vagus on sympathetic actions on the ventricles and coronary vessels (Verberne, 2004).

Sacral parasympathetic output

The sacral preganglionic output originates in neurons located in the lateral gray matter at the S2–S4 segments of the sacral spinal cord (de Groat, 2006; Fowler et al., 2008; Birder et al., 2010). These neurons are critical for normal micturition, defecation, and sexual organs; this involves their coordinated interactions with both lumbar sympathetic neurons located at T12–L2 levels and somatic motor neurons of the Onuf nucleus at the S2–S4 levels innervating the external urinary sphincter and pelvic floor. The best characterized example of this coordinated interaction is the control of urine storage and micturition. Urine storage depends on "guarding reflexes" mediated by lumbosacral sympathetic neurons (which inhibit contraction of the detrusor and promote contraction of the bladder neck) and Onuf nucleus motoneurons that elicit contraction of the external urethral sphincter (Fowler et al., 2008). Normal micturition involves a spinobulbospinal reflex that is triggered by bladder afferent inputs that reach the PAG and are then conveyed to the PMC (Holstege, 2005). During voluntary voiding, there is withdrawal of prefrontal inhibition on the PAG, which activates the PMC; descending PMC projections directly activate the sacral preganglionic neurons (promoting detrusor contraction) and, via interneurons, inhibit the sphincter motoneurons of the Onuf nucleus (allowing relaxation of the external sphincter) (Holstege, 2005; Fowler et al., 2008).

INVOLVEMENT OF THE CENTRAL AUTONOMIC NETWORK IN NEURODEGENERATIVE DISORDERS

Neurodegenerative disorders affecting the different components of the central autonomic network, the preganglionic neurons, or their target autonomic ganglia can produce autonomic failure manifested with orthostatic hypotension (OH), neurogenic bladder, erectile failure, gastrointestinal dysmotility, and impaired thermoregulation. Neurodegenerative autonomic disorders are characterized by the presence of intracellular inclusions containing α-synuclein. These synucleinopathies include two main groups, multiple system atrophy (MSA) (Gilman et al., 2008; Wenning et al., 2008) and Lewy body disorders, including Parkinson disease (PD), dementia with Lewy bodies (McKeith et al., 2005), and the so-called "pure" autonomic failure (PAF) (Kaufmann and Biaggioni, 2003). Although there are several features that may help the clinical differential diagnosis among these disorders, the clinical picture does not always predict the underlying neuropathology. Multiple system atrophy is characterized by the presence of cytoplasmic inclusion in the oligodendrocytes (glial cytoplasmic inclusion, GCIs) and to a lesser extent neurons; Lewy body disorders are characterized by the presence of α-synuclein containing intracellular Lewy bodies and Lewy neurites.

Over the past several years, the application of quantitative studies using selective immunocytochemical markers, such as tyrosine hydroxylase (to identify catecholaminergic neurons), tryptophan hydroxylase (to identify serotonergic neurons), choline acetyl transferase (to identify cholinergic neurons), nitric oxide synthase, and vesicular glutamate transporter-2 (as a marker of glutamatergic neurons), has provided insight into the involvement of putative central autonomic neurons in neurodegenerative disorders.

Multiple system atrophy

In MSA, autonomic failure is associated with abnormalities in control of automatic respiration and endocrine function. The mechanisms for these manifestations are probably multifactorial and reflect involvement of preganglionic efferent neurons, brainstem regulatory circuits, and hypothalamic regions (Benarroch, 2002; Ozawa, 2007).

Preganglionic sympathetic neurons

Loss of spinal preganglionic sympathetic neurons has been classically considered the substrate for sympathetic failure in MSA (Oppenheimer, 1980). However, there is not always a correlation between the degree of preganglionic cell depletion and severity of sympathetic failure in this disorder. It is estimated that more than 50% of cells within the intermediolateral column need to decay before symptoms become evident (Low and Dyck, 1978); pharmacological evidence indicates residual basal sympathetic activity in MSA (Diedrich et al., 2003).

Ventrolateral medulla

In MSA, OH is associated with supine hypertension, which indicates failure of the baroreflex control of sympathetic output mediated by neurons of the RVLM. There is consistent loss of C1 epinephrine-synthesizing neurons in MSA (Benarroch et al., 1998). Since these neurons may contribute glutamatergic RVLM output to preganglionic vasomotor neurons (Guyenet, 2006), their loss may contribute to OH in MSA. The involvement of the C1 group is much more consistent in MSA than in Lewy body disorders, including PD and dementia with Lewy bodies (Benarroch et al., 2005), which is consistent with evidence that OH in these disorders primarily reflects peripheral mechanisms, as discussed below. However, involvement of sympathetic nerves can also occur at a late stage of MSA (Orimo et al., 2007a), which is consistent with evidence of cardiac sympathetic denervation in some isolated cases (Nagayama et al., 2008).

The C1 neurons are not the only bulbospinal excitatory group affected in MSA. There is also involvement of medullary serotonergic neurons (Benarroch et al., 2004). Some of these bulbospinal neurons mediate sympathoexcitatory responses to cold in experimental animals (Morrison et al., 2008), but the clinical correlations of these pathological findings are yet undetermined. Involvement of these neurons may contribute to impaired vasoconstriction to cold responses in MSA compared to control subjects (Young et al., 2006).

Vagal nuclei

There is loss of cholinergic neurons in the DMV in MSA (Benarroch et al., 2006b) and this may be the primary contributor to gastrointestinal symptoms in this disorder. The involvement of vagal neurons of the NAmb controlling laryngeal function is controversial and may differ among cases. Some studies showed loss (Isozaki et al., 2000), whereas others showed relative preservation of these neurons (Benarroch et al., 2003). Thus, the contribution of loss of NAmb to laryngeal stridor in MSA is yet to be defined. Neurons on the ventrolateral portion of the NAmb are affected in MSA but relatively spared in Lewy body disorders (Benarroch et al., 2006b). Since these neurons have been implicated in cardiovagal control in experimental animals (Hopkins and Armour, 1998; Verberne, 2004), their involvement may contribute to impaired baroreflex cardioinhibition and respiratory sinus arrhythmia in MSA.

Putative medullary respiratory neurons

Respiratory manifestations, including sleep apnea, alveolar hypoventilation, respiratory dysrhythmia and laryngeal stridor are serious manifestations that may be a prominent cause of death in MSA patients (Vetrugno et al., 2004; Ghorayeb et al., 2005). Whereas the neuropathological basis of laryngeal stridor in MSA is still controversial (Isozaki et al., 2000; Benarroch et al., 2003), central hypoventilation and sleep apnea may reflect involvement of putative brainstem chemosensitive neurons. In MSA, there is loss of both glutamatergic and serotonergic neurons in the arcuate nucleus (Benarroch et al., 2007). Whereas the arcuate nucleus has been proposed as a putative chemosensitive area in humans, the relationship between these findings and impaired automatic ventilation in MSA is still to be determined. However, this possibility is supported by a recent study showing a depletion of medullary serotonergic neurons in MSA patients succumbing to sudden death (Tada et al., 2009).

Pontine micturition area

One of the typical features of MSA is the presence of prominent urological symptoms, including urinary incontinence, at early stages of the disease (Winge and Fowler, 2006; Gilman et al., 2008). Urinary incontinence is commonly associated with incomplete bladder emptying, reflecting detrusor hypocontractility and with denervation of the external urinary sphincter, as detected in sphincter electromyogram (EMG). The mechanisms of neurogenic bladder in MSA are multifactorial. There is loss of neurons in the sacral preganglionic nucleus (Kennedy and Duchen, 1985), which explains detrusor hypocontractility; and in the Onuf nucleus (Chalmers and Swash, 1987), which explains weakness of the external urinary sphincter. However, there is also loss of corticotropin-releasing factor immunoreactive neurons in the dorsolateral pons in the area corresponding to the PMC in MSA (Benarroch and Schmeichel, 2001), suggesting that impaired output to the sacral effector neurons may contribute to the urological manifestations of this disorder.

Periqueductal gray

There is involvement of the PAG in MSA. There is loss of tyrosine hydroxylase immunoreactive neurons in the ventral PAG both in MSA and cases with cortical stage Lewy body disease (Benarroch et al., 2009). These neurons may correspond to the dopaminergic neurons implicated in maintenance of wakefulness in experimental animals (Lu et al., 2006), and thus their involvement may contribute to excessive diurnal somnolence that characterizes these synucleinopathies. However, MSA also involves other neuronal groups concerned with maintenance of arousal. α-Synuclein pathology affects all columns of the PAG (Benarroch et al., 2010), which may contribute to disturbances in cardiovascular, bladder, and respiratory control in this disorder.

Hypothalamus

There is involvement of different hypothalamic cell groups in MSA. There is loss of AVP neurons in the suprachiasmatic nucleus (Ozawa et al., 1998; Benarroch et al., 2006a), which may contribute to circadian abnormalities (Ozawa, 2007). There is also loss of AVP neurons in the posterior PVN (Benarroch et al., 2006a). Experimental evidence indicates that the posterior PVN, including AVP neurons, projects to the intermediolateral cell column, and thus loss of these neurons may contribute to impaired control of sympathetic function in MSA. In contrast, there is preservation of magnocellular AVP neurons in the supraoptic and paraventricular nuclei in MSA, suggesting a selective cell vulnerability that cannot be solely explained by the neurotransmitter phenotype. However, there is loss of tyrosine-hydroxylase immunoreactive fibers innervating these magnocellular AVP neurons, reflecting in part loss of noradrenergic A1 neurons in the caudal ventrolateral medulla (Benarroch et al., 2006a). Experimental findings indicate that A1 participate in the reflex increase of AVP during baroreceptor unloading (Lightman et al., 1984). Thus, loss of noradrenergic innervation of magnocellular AVP neurons may explain, at least in part, the impairment in AVP secretion in response to orthostatic stress (Puritz et al., 1983; Niimi et al., 1999; Deguchi et al., 2004) or to clonidine (Kimber et al., 1999) in MSA.

Parkinson disease and dementia with Lewy bodies

There is extensive neuropathological evidence that in Lewy body disorders the peripheral autonomic nervous system is affected at early stages by α-synuclein neuropathology. This has been shown primarily in the ENS (Wakabayashi and Takahashi, 1997; Braak et al., 2006) and the cardiac sympathetic ganglia and sympathetic terminals (Okada et al., 2004; Orimo et al., 2007b). However, some evidence also indicates involvement of central autonomic areas in these disorders.

Medullary sympathoexcitatory groups

Although there may be involvement of C1 neurons in cases with a clinical history of PD or dementia with Lewy bodies, it is less predictable and severe than in MSA, even in cases with severe OH (Benarroch et al., 2000, 2005). This is consistent with evidence from neuropathological (Okada et al., 2004; Orimo et al., 2007b), clinical (Goldstein et al., 2005), pharmacological (Sharabi et al., 2006, 2008) and neuroimaging (Goldstein et al., 1997; Goldstein, 2007; Nakajima et al., 2008; Kobayashi et al., 2009) studies, indicating that OH may primarily reflect peripheral sympathetic involvement in Lewy body disorders (Kingsbury et al., 2010). However, there may be prominent involvement of the brainstem autonomic nuclei and intermediolateral cell column, as well as sympathetic ganglia in some cases presenting with severe autonomic failure without motor manifestations of PD (Hishikawa et al., 2000). There is also involvement of the intermediolateral cell column by α-synuclein pathology in cases of incidental Lewy body disease in the absence of autonomic symptoms.

Dorsal motor nucleus of the vagus

The DMV is involved by α-synuclein pathology at early stages of PD and in incidental cases of Lewy body disease (Braak et al., 2003). There is loss of substance P-containing preganglionic neurons in the DMV in PD (Halliday et al., 1990) and loss of cholinergic DMV neurons in cases of dementia with Lewy bodies (Benarroch et al., 2006b). Despite early involvement of the DMV, involvement of the enteric nervous system (Wakabayashi and Takahashi, 1997; Braak et al., 2006) may also have a prominent role in the gastrointestinal manifestations of PD (Cersosimo and Benarroch, 2008). Based on the early involvement of the ENS and DMV by α-synuclein pathology (Wakabayashi and Takahashi, 1997; Braak et al., 2006), Braak et al. (2003) proposed that pathogens capable of passing the mucosal barrier of the gastrointestinal tract can, via postganglionic ENS neurons, enter the central nervous system along unmyelinated preganglionic axons from neurons in the DMV. From the DMV, the neuropathology could progressively reach more rostral vulnerable susceptible regions and eventually the cerebral cortex via by way of retrograde axonal and transneuronal transport (Braak et al., 2003). This hypothesis is supported by a recent study showing a consistent involvement of the DMV in cases with diagnosis of PD and in incidental Lewy body disease (Kingsbury et al., 2010), and by experimental evidence indicating that conformationally abnormal proteins, including α-synuclein, can propagate in a prion-like fashion along neural pathways (Angot et al., 2010). However, other studies indicate that the DMV is not necessarily an early step in the development of Lewy body pathology in the central nervous system (Jellinger, 2008; Kalaitzakis et al., 2008). The relative contribution of the involvement of the DMV and the ENS and the gastrointestinal manifestations of PD is not fully understood (Natale et al., 2008).

Sacral micturition areas

The most common urodynamic abnormality in PD is detrusor overactivity. The precise mechanism is still poorly understood but is likely to be multifactorial. Whereas there is evidence of Lewy bodies in the sacral preganglionic neurons in isolated cases of PD (Oyanagi et al., 1990), there is preservation of the sacral sphincter motoneurons in the Onuf nucleus, which provides the basis for the use of sphincter electromyography to distinguish between MSA from PD at early stages of disease (Sakakibara et al., 2009).

SUMMARY

Elegant neuroanatomical and neurophysiological studies in experimental animals have provided a vast amount of information on the organization of the central nervous system areas controlling autonomic function. Recent functional neuroimaging studies, as well as insight from clinicopathological correlations in neurodegenerative diseases, indicate that the same areas identified in experimental animals have a major role in autonomic control in humans. Refinements in neuroimaging techniques, and further neuropathological studies on carefully selected cases at different stages of disease, will continue to elucidate the functional organization and neurochemistry of central autonomic control in humans, which has implications not only for understanding the pathophysiology of central autonomic disorders but also identifying potential therapeutic targets.

References

Angot E, Steiner JA, Hansen C et al. (2010). Are synucleinopathies prion-like disorders? Lancet Neurol 9: 1128–1138.

Bandler R, Keay KA, Floyd N et al. (2000). Central circuits mediating patterned autonomic activity during active vs. passive emotional coping. Brain Res Bull 53: 95–104.

Benarroch EE (2002). New findings on the neuropathology of multiple system atrophy. Auton Neurosci 96: 59–62.

Benarroch EE (2005). Paraventricular nucleus, stress response, and cardiovascular disease. Clin Auton Res 15: 254–263.

Benarroch EE, Schmeichel AM (2001). Depletion of corticotrophin-releasing factor neurons in the pontine micturition area in multiple system atrophy. Ann Neurol 50: 640–645.

Benarroch EE, Smithson IL, Low PA et al. (1998). Depletion of catecholaminergic neurons of the rostral ventrolateral medulla in multiple systems atrophy with autonomic failure. Ann Neurol 43: 156–163.

Benarroch EE, Schmeichel AM, Parisi JE (2000). Involvement of the ventrolateral medulla in parkinsonism with autonomic failure. Neurology 54: 963–968.

Benarroch EE, Schmeichel AM, Parisi JE (2003). Preservation of branchimotor neurons of the nucleus ambiguus in multiple system atrophy. Neurology 60: 115–117.

Benarroch EE, Schmeichel AM, Low PA et al. (2004). Involvement of medullary serotonergic groups in multiple system atrophy. Ann Neurol 55: 418–422.

Benarroch EE, Schmeichel AM, Low PA et al. (2005). Involvement of medullary regions controlling sympathetic output in Lewy body disease. Brain 128: 338–344.

Benarroch EE, Schmeichel AM, Sandroni P et al. (2006a). Differential involvement of hypothalamic vasopressin neurons in multiple system atrophy. Brain 129: 2688–2696.

Benarroch EE, Schmeichel AM, Sandroni P et al. (2006b). Involvement of vagal autonomic nuclei in multiple system atrophy and Lewy body disease. Neurology 66: 378–383.

Benarroch EE, Schmeichel AM, Low PA et al. (2007). Depletion of putative chemosensitive respiratory neurons in the ventral medullary surface in multiple system atrophy. Brain 130: 469–475.

Benarroch EE, Schmeichel AM, Dugger BN et al. (2009). Dopamine cell loss in the periaqueductal gray in multiple system atrophy and Lewy body dementia. Neurology 73: 106–112.

Benarroch EE, Schmeichel AM, Low PA et al. (2010). Differential involvement of the periaqueductal gray in multiple system atrophy. Auton Neurosci 158: 111–117.

Birder L, de Groat W, Mills I et al. (2010). Neural control of the lower urinary tract: peripheral and spinal mechanisms. Neurourol Urodyn 29: 128–139.

Blok BF, Willemsen AT, Holstege G (1997). A PET study on brain control of micturition in humans. Brain 120: 111–121.

Braak H, Rub U, Gai WP et al. (2003). Idiopathic Parkinson's disease: possible routes by which vulnerable neuronal types may be subject to neuroinvasion by an unknown pathogen. J Neural Transm 110: 517–536.

Braak H, Bohl JR, Muller CM et al. (2006). Stanley Fahn Lecture 2005. The staging procedure for the inclusion body pathology associated with sporadic Parkinson's disease reconsidered. Mov Disord 21: 2042–2051.

Cechetto DF (1987). Central representation of visceral function. Fed Proc 46: 17–23.

Cersosimo MG, Benarroch EE (2008). Neural control of the gastrointestinal tract: implications for Parkinson disease. Mov Disord 23: 1065–1075.

Chalmers D, Swash M (1987). Selective vulnerability of urinary Onuf motoneurons in Shy–Drager syndrome. J Neurol 234: 259–260.

Chamberlin NL (2004). Functional organization of the parabrachial complex and intertrigeminal region in the control of breathing. Respir Physiol Neurobiol 143: 115–125.

Chang HY, Mashimo H, Goyal RK (2003). Musings on the wanderer: what's new in our understanding of vago-vagal reflex? IV. Current concepts of vagal efferent projections to the gut. Am J Physiol Gastrointest Liver Physiol 284: G357–G366.

Corcoran AE, Hodges MR, Wu Y et al. (2009). Medullary serotonin neurons and central CO2 chemoreception. Respir Physiol Neurobiol 168: 49–58.

Craig AD (2003). Interoception: the sense of the physiological condition of the body. Curr Opin Neurobiol 13: 500–505.

Craig AD (2005). Forebrain emotional asymmetry: a neuroanatomical basis? Trends Cogn Sci 9: 566–571.

Critchley HD (2005). Neural mechanisms of autonomic, affective, and cognitive integration. J Comp Neurol 493: 154–166.

Critchley HD (2009). Psychophysiology of neural, cognitive and affective integration: fMRI and autonomic indicants. Int J Psychophysiol 73: 88–94.

Critchley HD, Mathias CJ, Josephs O et al. (2003). Human cingulate cortex and autonomic control: converging neuroimaging and clinical evidence. Brain 126: 2139–2152.

Critchley HD, Nagai Y, Gray MA et al. (2011). Dissecting axes of autonomic control in humans: Insights from neuroimaging. Auton Neurosci 161: 34–42.

Dampney RA, Horiuchi J (2003). Functional organisation of central cardiovascular pathways: studies using c-fos gene expression. Prog Neurobiol 71: 359–384.

Dampney RA, Polson JW, Potts PD et al. (2003). Functional organization of brain pathways subserving the baroreceptor reflex: studies in conscious animals using immediate early gene expression. Cell Mol Neurobiol 23: 597–616.

Dampney RA, Horiuchi J, McDowall LM (2008). Hypothalamic mechanisms coordinating cardiorespiratory function during exercise and defensive behaviour. Auton Neurosci 142: 3–10.

Davis M (1992). The role of the amygdala in fear and anxiety. Annu Rev Neurosci 15: 353–375.

de Groat WC (2006). Integrative control of the lower urinary tract: preclinical perspective. Br J Pharmacol 147 (Suppl 2): S25–S40.

de Menezes RC, Zaretsky DV, Fontes MA et al. (2009). Cardiovascular and thermal responses evoked from the periaqueductal grey require neuronal activity in the hypothalamus. J Physiol 587: 1201–1215.

Deen B, Pitskel NB, Pelphrey KA (2011). Three systems of insular functional connectivity identified with cluster analysis. Cereb Cortex 21: 1498–1506.

Deguchi K, Sasaki I, Touge T et al. (2004). Abnormal baroreceptor-mediated vasopressin release as possible marker in early diagnosis of multiple system atrophy. J Neurol Neurosurg Psychiatry 75: 110–115.

Diedrich A, Jordan J, Tank J et al. (2003). The sympathetic nervous system in hypertension: assessment by blood pressure variability and ganglionic blockade. J Hypertens 21: 1677–1686.

Dimicco JA, Zaretsky DV (2007). The dorsomedial hypothalamus: a new player in thermoregulation. Am J Physiol Regul Integr Comp Physiol 292: R47–R63.

Egan GF, Johnson J, Farrell M et al. (2005). Cortical, thalamic, and hypothalamic responses to cooling and warming the skin in awake humans: a positron-emission tomography study. Proc Natl Acad Sci U S A 102: 5262–5267.

Feldman JL, Mitchell GS, Nattie EE (2003). Breathing: rhythmicity, plasticity, chemosensitivity. Annu Rev Neurosci 26: 239–266.

Filiano JJ, Choi JC, Kinney HC (1990). Candidate cell populations for respiratory chemosensitive fields in the human infant medulla. J Comp Neurol 293: 448–465.

Fowler CJ, Griffiths D, de Groat WC (2008). The neural control of micturition. Nat Rev Neurosci 9: 453–466.

Fowler CJ, Griffiths DJ (2010). A decade of functional brain imaging applied to bladder control. Neurourol Urodyn 29: 49–55.

Fuller PM, Saper CB, Lu J (2007). The pontine REM switch: past and present. J Physiol 584: 735–741.

Geerling JC, Shin JW, Chimenti PC et al. (2010). Paraventricular hypothalamic nucleus: axonal projections to the brainstem. J Comp Neurol 518: 1460–1499.

Ghorayeb I, Bioulac B, Tison F (2005). Sleep disorders in multiple system atrophy. J Neural Transm 112: 1669–1675.

Gilman S, Wenning GK, Low PA et al. (2008). Second consensus statement on the diagnosis of multiple system atrophy. Neurology 71: 670–676.

Goldstein DS (2007). Cardiac denervation in patients with Parkinson disease. Cleve Clin J Med 74 (Suppl 1): S91–S94.

Goldstein DS, Holmes C, Cannon RO 3rd et al. (1997). Sympathetic cardioneuropathy in dysautonomias. N Engl J Med 336: 696–702.

Goldstein DS, Eldadah BA, Holmes C et al. (2005). Neurocirculatory abnormalities in Parkinson disease with orthostatic hypotension: independence from levodopa treatment. Hypertension 46: 1333–1339.

Goodchild AK, Moon EA (2009). Maps of cardiovascular and respiratory regions of rat ventral medulla: focus on the caudal medulla. J Chem Neuroanat 38: 209–221.

Grundy D, Schemann M (2007). Enteric nervous system. Curr Opin Gastroenterol 23: 121–126.

Guyenet PG (2000). Neural structures that mediate sympathoexcitation during hypoxia. Respir Physiol 121: 147–162.

Guyenet PG (2006). The sympathetic control of blood pressure. Nat Rev Neurosci 7: 335–346.

Guyenet PG, Stornetta RL, Bayliss DA (2010). Central respiratory chemoreception. J Comp Neurol 518: 3883–3906.

Halliday GM, Blumbergs PC, Cotton RG et al. (1990). Loss of brainstem serotonin- and substance P-containing neurons in Parkinson's disease. Brain Res 510: 104–107.

Haxhiu MA, Yamamoto BK, Dreshaj IA et al. (2002). Activation of the midbrain periaqueductal gray induces airway smooth muscle relaxation. J Appl Physiol 93: 440–449.

Heinricher MM, Tavares I, Leith JL et al. (2009). Descending control of nociception: specificity, recruitment and plasticity. Brain Res Rev 60: 214–225.

Hishikawa N, Hashizume Y, Hirayama M et al. (2000). Brainstem-type Lewy body disease presenting with progressive autonomic failure and lethargy. Clin Auton Res 10: 139–143.

Holstege G (2005). Micturition and the soul. J Comp Neurol 493: 15–20.

Hopkins DA, Armour JA (1998). Brainstem cells of origin of physiologically identified cardiopulmonary nerves in the rhesus monkey (*Macaca mulatta*). J Auton Nerv Syst 68: 21–32.

Huang XF, Tork I, Paxinos G (1993). Dorsal motor nucleus of the vagus nerve: a cyto- and chemoarchitectonic study in the human. J Comp Neurol 330: 158–182.

Hunt JL, Zaretsky DV, Sarkar S et al. (2010). Dorsomedial hypothalamus mediates autonomic, neuroendocrine, and locomotor responses evoked from the medial preoptic area. Am J Physiol Regul Integr Comp Physiol 298: R130–R140.

Hurley KM, Herbert H, Moga MM et al. (1991). Efferent projections of the infralimbic cortex of the rat. J Comp Neurol 308: 249–276.

Isozaki E, Matsubara S, Hayashida T et al. (2000). Morphometric study of nucleus ambiguus in multiple system atrophy presenting with vocal cord abductor paralysis. Clin Neuropathol 19: 213–220.

Janig W, Habler HJ (2003). Neurophysiological analysis of target-related sympathetic pathways – from animal to human: similarities and differences. Acta Physiol Scand 177: 255–274.

Jean A (1991). The nucleus tractus solitarius: neuroanatomic, neurochemical and functional aspects. Arch Int Physiol Biochim Biophys 99: A3–A52.

Jellinger KA (2008). A critical reappraisal of current staging of Lewy-related pathology in human brain. Acta Neuropathol 116: 1–16.

Jordan D (2005). Vagal control of the heart: central serotonergic (5-HT) mechanisms. Exp Physiol 90: 175–181.

Kalaitzakis ME, Graeber MB, Gentleman SM et al. (2008). The dorsal motor nucleus of the vagus is not an obligatory trigger site of Parkinson's disease: a critical analysis of alpha-synuclein staging. Neuropathol Appl Neurobiol 34: 284–295.

Kaufmann H, Biaggioni I (2003). Autonomic failure in neurodegenerative disorders. Semin Neurol 23: 351–363.

Kennedy PG, Duchen LW (1985). A quantitative study of intermediolateral column cells in motor neuron disease and the Shy–Drager syndrome. J Neurol Neurosurg Psychiatry 48: 1103–1106.

Kimber J, Watson L, Mathias CJ (1999). Abnormal suppression of arginine-vasopressin by clonidine in multiple system atrophy. Clin Auton Res 9: 271–274.

Kimmerly DS, O'Leary DD, Menon RS et al. (2005). Cortical regions associated with autonomic cardiovascular regulation during lower body negative pressure in humans. J Physiol 569: 331–345.

Kimmerly DS, Wong SW, Salzer D et al. (2007). Forebrain regions associated with postexercise differences in autonomic and cardiovascular function during baroreceptor unloading. Am J Physiol Heart Circ Physiol 293: H299–H306.

King AB, Menon RS, Hachinski V et al. (1999). Human forebrain activation by visceral stimuli. J Comp Neurol 413: 572–582.

Kingsbury AE, Bandopadhyay R, Silveira-Moriyama L et al. (2010). Brain stem pathology in Parkinson's disease: an evaluation of the Braak staging model. Mov Disord 25: 2508–2515.

Kobayashi S, Tateno M, Morii H et al. (2009). Decreased cardiac MIBG uptake, its correlation with clinical symptoms in dementia with Lewy bodies. Psychiatry Res 174: 76–80.

LeDoux J (2007). The amygdala. Curr Biol 17: R868–R874.

Lightman SL, Todd K, Everitt BJ (1984). Ascending noradrenergic projections from the brainstem: evidence for a major role in the regulation of blood pressure and vasopressin secretion. Exp Brain Res 55: 145–151.

Loewy AD (1982). Descending pathways to the sympathetic preganglionic neurons. Prog Brain Res 57: 267–277.

Low PA, Dyck PJ (1978). Splanchnic preganglionic neurons in man. III. Morphometry of myelinated fibers of rami communicantes. J Neuropathol Exp Neurol 37: 734–740.

Lu J, Jhou TC, Saper CB (2006). Identification of wake-active dopaminergic neurons in the ventral periaqueductal gray matter. J Neurosci 26: 193–202.

McAllen RM, Spyer KM (1976). The location of cardiac vagal preganglionic motoneurones in the medulla of the cat. J Physiol 258: 187–204.

McAllen RM, May CN, Shafton AD (1995). Functional anatomy of sympathetic premotor cell groups in the medulla. Clin Exp Hypertens 17: 209–221.

McAllen RM, Farrell M, Johnson JM et al. (2006). Human medullary responses to cooling and rewarming the skin:

a functional MRI study. Proc Natl Acad Sci U S A 103: 809–813.

Macefield VG, Henderson LA (2010). Real-time imaging of the medullary circuitry involved in the generation of spontaneous muscle sympathetic nerve activity in awake subjects. Hum Brain Mapp 31: 539–549.

McKeith IG, Dickson DW, Lowe J et al. (2005). Diagnosis and management of dementia with Lewy bodies: third report of the DLB Consortium. Neurology 65: 1863–1872.

Margulies DS, Kelly AM, Uddin LQ et al. (2007). Mapping the functional connectivity of anterior cingulate cortex. Neuroimage 37: 579–588.

Morrison SF (1999). RVLM and raphe differentially regulate sympathetic outflows to splanchnic and brown adipose tissue. Am J Physiol 276: R962–R973.

Morrison SF, Nakamura K, Madden CJ (2008). Central control of thermogenesis in mammals. Exp Physiol 93: 773–797.

Nagai M, Hoshide S, Kario K (2010). The insular cortex and cardiovascular system: a new insight into the brain–heart axis. J Am Soc Hypertens 4: 174–182.

Nagayama H, Yamazaki M, Ueda M et al. (2008). Low myocardial MIBG uptake in multiple system atrophy with incidental Lewy body pathology: an autopsy case report. Mov Disord 23: 1055–1057.

Nakajima K, Yoshita M, Matsuo S et al. (2008). Iodine-123-MIBG sympathetic imaging in Lewy-body diseases and related movement disorders. Q J Nucl Med Mol Imaging 52: 378–387.

Natale G, Pasquali L, Ruggieri S et al. (2008). Parkinson's disease and the gut: a well known clinical association in need of an effective cure and explanation. Neurogastroenterol Motil 20: 741–749.

Nattie E, Li A (2009). Central chemoreception is a complex system function that involves multiple brain stem sites. J Appl Physiol 106: 1464–1466.

Niimi Y, Ieda T, Hirayama M et al. (1999). Clinical and physiological characteristics of autonomic failure with Parkinson's disease. Clin Auton Res 9: 139–144.

Okada Y, Ito Y, Aida J et al. (2004). Lewy bodies in the sinoatrial nodal ganglion: clinicopathological studies. Pathol Int 54: 682–687.

Oppenheimer DR (1980). Lateral horn cells in progressive autonomic failure. J Neurol Sci 46: 393–404.

Orimo S, Kanazawa T, Nakamura A et al. (2007a). Degeneration of cardiac sympathetic nerve can occur in multiple system atrophy. Acta Neuropathol 113: 81–86.

Orimo S, Takahashi A, Uchihara T et al. (2007b). Degeneration of cardiac sympathetic nerve begins in the early disease process of Parkinson's disease. Brain Pathol 17: 24–30.

Oyanagi K, Wakabayashi K, Ohama E et al. (1990). Lewy bodies in the lower sacral parasympathetic neurons of a patient with Parkinson's disease. Acta Neuropathol 80: 558–559.

Ozawa T (2007). Morphological substrate of autonomic failure and neurohormonal dysfunction in multiple system atrophy: impact on determining phenotype spectrum. Acta Neuropathol 114: 201–211.

Ozawa T, Oyanagi K, Tanaka H et al. (1998). Suprachiasmatic nucleus in a patient with multiple system atrophy with abnormal circadian rhythm of arginine-vasopressin secretion into plasma. J Neurol Sci 154: 116–121.

Paterson DS, Thompson EG, Kinney HC (2006). Serotonergic and glutamatergic neurons at the ventral medullary surface of the human infant: observations relevant to central chemosensitivity in early human life. Auton Neurosci 124: 112–124.

Puritz R, Lightman SL, Wilcox CS et al. (1983). Blood pressure and vasopressin in progressive autonomic failure. Response to postural stimulation, L-dopa and naloxone. Brain 106: 503–511.

Sakakibara R, Uchiyama T, Yamanishi T et al. (2009). Sphincter EMG as a diagnostic tool in autonomic disorders. Clin Auton Res 19: 20–31.

Sander M, Macefield VG, Henderson LA (2010). Cortical and brain stem changes in neural activity during static handgrip and postexercise ischemia in humans. J Appl Physiol 108: 1691–1700.

Saper CB (2002). The central autonomic nervous system: conscious visceral perception and autonomic pattern generation. Annu Rev Neurosci 25: 433–469.

Sawchenko PE, Li HY, Ericsson A (2000). Circuits and mechanisms governing hypothalamic responses to stress: a tale of two paradigms. Prog Brain Res 122: 61–78.

Sharabi Y, Eldadah B, Li ST et al. (2006). Neuropharmacologic distinction of neurogenic orthostatic hypotension syndromes. Clin Neuropharmacol 29: 97–105.

Sharabi Y, Imrich R, Holmes C et al. (2008). Generalized and neurotransmitter-selective noradrenergic denervation in Parkinson's disease with orthostatic hypotension. Mov Disord 23: 1725–1732.

Smith JC, Abdala AP, Rybak IA et al. (2009). Structural and functional architecture of respiratory networks in the mammalian brainstem. Philos Trans R Soc Lond B Biol Sci 364: 2577–2587.

Stornetta RL (2009). Neurochemistry of bulbospinal presympathetic neurons of the medulla oblongata. J Chem Neuroanat 38: 222–230.

Strack AM, Sawyer WB, Hughes JH et al. (1989). A general pattern of CNS innervation of the sympathetic outflow demonstrated by transneuronal pseudorabies viral infections. Brain Res 491: 156–162.

Subramanian HH, Balnave RJ, Holstege G (2008). The midbrain periaqueductal gray control of respiration. J Neurosci 28: 12274–12283.

Tada M, Kakita A, Toyoshima Y et al. (2009). Depletion of medullary serotonergic neurons in patients with multiple system atrophy who succumbed to sudden death. Brain 132: 1810–1819.

Ter Horst GJ, Hautvast RW, De Jongste MJ et al. (1996). Neuroanatomy of cardiac activity-regulating circuitry: a transneuronal retrograde viral labelling study in the rat. Eur J Neurosci 8: 2029–2041.

Terreberry RR, Neafsey EJ (1987). The rat medial frontal cortex projects directly to autonomic regions of the brainstem. Brain Res Bull 19: 639–649.

Thompson RH, Swanson LW (2003). Structural characterization of a hypothalamic visceromotor pattern generator network. Brain Res Brain Res Rev 41: 153–202.

Travagli RA, Hermann GE, Browning KN et al. (2006). Brainstem circuits regulating gastric function. Annu Rev Physiol 68: 279–305.

Ulrich-Lai YM, Herman JP (2009). Neural regulation of endocrine and autonomic stress responses. Nat Rev Neurosci 10: 397–409.

Verberne AJ (2004). Differential cardiac parasympathetic innervation – what is the functional significance? Am J Physiol Regul Integr Comp Physiol 287: R485–R486.

Verberne AJ, Owens NC (1998). Cortical modulation of the cardiovascular system. Prog Neurobiol 54: 149–168.

Vetrugno R, Provini F, Cortelli P et al. (2004). Sleep disorders in multiple system atrophy: a correlative video-polysomnographic study. Sleep Med 5: 21–30.

Vogt BA, Vogt L, Farber NB et al. (2005). Architecture and neurocytology of monkey cingulate gyrus. J Comp Neurol 485: 218–239.

Wakabayashi K, Takahashi H (1997). Neuropathology of autonomic nervous system in Parkinson's disease. Eur Neurol 38 (Suppl 2): 2–7.

Wenning GK, Stefanova N, Jellinger KA et al. (2008). Multiple system atrophy: a primary oligodendrogliopathy. Ann Neurol 64: 239–246.

Westerhaus MJ, Loewy AD (2001). Central representation of the sympathetic nervous system in the cerebral cortex. Brain Res 903: 117–127.

Winge K, Fowler CJ (2006). Bladder dysfunction in Parkinsonism: mechanisms, prevalence, symptoms, and management. Mov Disord 21: 737–745.

Yoshida K, Li X, Cano G et al. (2009). Parallel preoptic pathways for thermoregulation. J Neurosci 29: 11954–11964.

Young TM, Asahina M, Nicotra A et al. (2006). Skin vasomotor reflex responses in two contrasting groups of autonomic failure: multiple system atrophy and pure autonomic failure. J Neurol 253: 846–850.

Zhang W, Hayward LF, Davenport PW (2007). Respiratory responses elicited by rostral versus caudal dorsal periaqueductal gray stimulation in rats. Auton Neurosci 134: 45–54.

Handbook of Clinical Neurology, Vol. 117 (3rd series)
Autonomic Nervous System
R.M. Buijs and D.F. Swaab, Editors

Chapter 6

Interaction between cognition, emotion, and the autonomic nervous system

HUGO D. CRITCHLEY[1,2]*, JESSICA ECCLES[1], AND SARAH N. GARFINKEL[1,2]
[1]*Department of Psychiatry, Brighton and Sussex Medical School, Brighton, UK*
[2]*Sackler Centre for Consciousness Science, University of Sussex, UK*

INTRODUCTION

Autonomic control is integrated with cognition and emotion. The expression of advanced cognitive capabilities in humans, involving consciousness and language alongside a rich repertoire of emotions within complex social environments, ultimately arises through evolutionary imperatives determined through selective pressure to survive and reproduce. Homeostatic control, to which the autonomic nervous system is crucial, maintains the basics for physical survival. Social and emotional behavior is thus shaped by environmental challenges in the context of homeostatic needs. Superordinate cognitive processes are also subject to these influences. Brain and body are intrinsically and dynamically coupled; perceptions, emotions, and cognitions respond to, and change, the state of the body. These interactions form much of the content of psychophysiological research identifying bodily signatures of mental activity.

The concept of arousal is interesting in this regard, referring to mental sharpening that enhances attention to, and the processing of, potentially important information (salience is a term for this potential motivational importance). Arousal can improve the performance of effortful tasks. Arousal is also used to refer to physical states of action-readiness often identified through autonomic signatures such as changes in sympathetic skin response. In many instances, these psychological and peripheral physiological concepts of arousal overlap, as in coma and wakefulness or the attentional orienting and sympathetic reaction ("fight or flight response") elicited by an unexpected loud sound. However, the general notion of arousal, when applied to autonomic function, underplays the precision of autonomic control expressed through detailed organ specificity and patterning of visceral responses. Moreover, there are relatively common occasions when physiological preparedness (or the related concept, "activation") and psychological alertness dissociate: examples include disinhibiting effects of sedative drugs, exaggerated startle responses in delirium, or conversely, the physiological calm necessary to undertake demanding mental tasks (e.g., as demonstrated by the detrimental effects of stress-related arousal on working memory) (Schoofs et al., 2008).

Neuroimaging in humans can provide mechanistic insight into interactions between cognition, emotion and the autonomic nervous system. Functional imaging techniques such as positron emission tomography (PET) and functional magnetic resonance imaging (fMRI) allow regional changes in brain function (via proxies in local cerebral hemodynamics) to be related to perception, thinking, and feeling. Once concerns are managed regarding the potentially confounding nature of autonomically mediated changes in physiological state on regional hemodynamic signals of brain neural activity, the combination of functional neuroimaging with autonomic monitoring provides novel and clinically relevant data. This chapter draws together evidence in humans for the integration of autonomic nervous control with cognition and emotion, highlighting clinical implications. It is an attempt to illustrate principles rather than comprehensively summarize the field.

*Correspondence to: Professor Hugo D. Critchley, Department of Psychiatry, Room 05, Brighton and Sussex Medical School, BN1 9RR, UK. Tel: +44-1273-873818, E-mail: H.Critchley@bsms.ac.uk

AUTONOMIC INTERACTION WITH COGNITION AND EMOTION

Perceptions, cognitions and emotions interact with autonomic nervous control in specific ways and at different levels of the neuraxis:

1. Mental processes influence autonomic responses to alter the physical state of the body. This can occur: (i) directly and automatically through autonomic nerves; (ii) indirectly though changes in respiration, skeletomotor activity or posture under partial or full volitional control; or (iii), as intentional or incidental consequences of enacting behavioral decisions.
2. The internal physiological state of the body can influence mental processes. Thus: (i) sensory representation of the internal bodily state can grab attention, compete for representational or cognitive resources, and interrupt ongoing thoughts and feelings; (ii) information arising from the body may also be fully integrated with perceptions and cognitions, potentially adding to impact, for example, arousal will enhance encoding into memory; (iii) autonomic/visceral state acts as a variable context (e.g., "occasion setter") (Bouton, 1993; Bouton et al., 2001) for emotional and cognitive processes, for example, information learned in low arousal is best recalled in low arousal.
3. The central interaction of processes supporting the generation and the representation of autonomically mediated changes in visceral state may be the critical mediator of autonomic influences on cognition and emotion. Central viscerosensory and visceromotor representations are exchanged as afference and efference copies to allow error signaling. Where there is mismatch between intended and actual autonomic state, corrective efferent reactions are accompanied by interpretative processes. The unconscious operation of the autonomic nervous system can be interrupted by deviations from expected state, i.e., we become aware of our autonomic bodily state when we experience changes in internal state that are "unpredicted" by control centers.

AUTONOMIC INTEGRATION AND INTERACTION WITH COGNITIVE PROCESSES

Autonomic interaction with engagement, attention, cognitive and mental effort

Engagement with stimuli in the environment engenders a distinct state of body and of mind, characterized by psychological and physiological arousal and reactivity. These changes are a focus of much psychophysiological research (including autonomic, electroencephalographic and neuroimaging studies) that attempts to understand this basic level of mind–body integration. Here we focus on neuroimaging data.

Almost from the outset, functional imaging studies of human brain revealed increased activity in dorsal anterior cingulate cortex (dACC) when people are engaged in cognitively demanding tasks. A typical "cognitive activation" neuroimaging experiment (notably those using $H_2{}^{15}O$ PET), would contrast performance of a task containing the cognitive process of interest with a control task that was similar in terms of sensorimotor performance but lacked the extra cognitive element. For example, performing a mathematical puzzle would be contrasted with counting. This would control for aspects of inner speech and mental number representation but lack the extra mathematical cognitive element. The subtraction of the counting part of the task in principle allows the specific cognitive aspect of mathematics to be looked at in isolation. Nevertheless, general processes relating to effort may remain: In PET, the regional increase in brain activity is inferred from task-coupled regional increase radio-labeled blood flow or metabolism. In a review of 107 PET studies, the same region of dorsal anterior cingulate cortex was shown to be activated across a range of different experimental tasks (Paus et al., 1998), suggesting that the region was involved nonspecifically in behavioral or cognitive effort (arising from the extra elements in the active versus control task). Similar observations were made in relation to deactivation of ventromedial prefrontal, subgenual and posterior cingulate cortices when people engage in demanding tasks (Raichle et al., 2001), leading to the proposal of "default mode" or resting network. These observations link performance of effortful, attentionally demanding tasks to the predictable activation and deactivation of regions of cingulate cortex. By extension, these studies suggested that the changes in autonomic bodily state that accompany and presumably facilitate effortful behavior and cognitive performance are linked to the same neural processes.

The simultaneous measurement of an autonomic parameter during functional brain imaging of cognitive processing provides a means of dissecting neural substrates, supporting cognition along different psychophysiological dimensions (examples given in Fig. 6.1). Studies have been undertaken, testing for brain correlates of experimentally provoked changes in autonomic physiology including heart rate (King et al., 1999; Wager et al., 2009a, b), sympathetic skin response (Fredrikson et al., 1998; Critchley et al., 2000b; Patterson et al., 2002), blood pressure (e.g., Critchley et al., 2000a; Harper et al., 2000; Gianaros et al., 2005) and baroreflex suppression (e.g., Gianaros et al., 2012). Following the

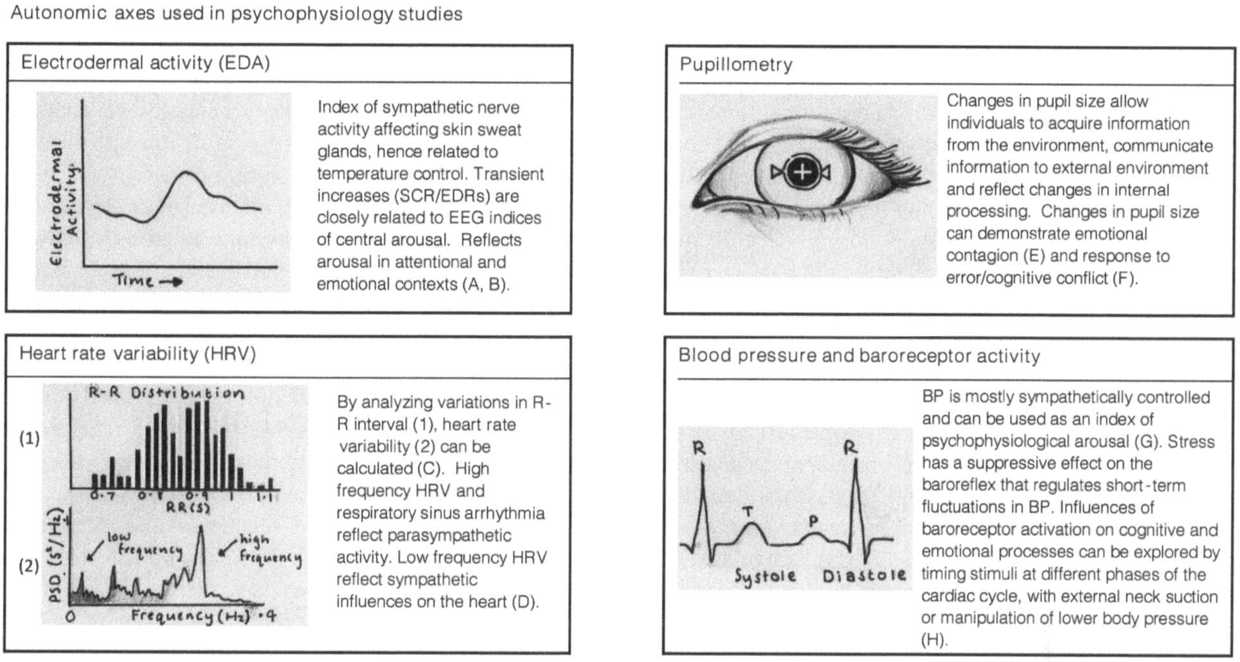

Fig. 6.1. This figure illustrates the role of psychophysiology in investigating integration of autonomic nervous system with cognition and emotion. The above measures provide axes of autonomic activity that can be monitored during neuroimaging and laboratory tasks, allowing integrative analyses. (A) SCL negatively correlates with vmPFC activity. (B) SSR positively correlates with dACC and insula activity (Nagai et al., 2004b). (C) High-frequency HRV correlates with PAG and left insula activity. (D) Low-frequency HRV relates to changes in dACC and bilateral insula. Reduced high frequency HRV may relate to endophenotypic marker of vulnerability to affective disorder (Gray et al., 2009). (E) Observing pupil changes in another activitates Edinger–Westphal nucleus in observer (Harrison et al., 2006); discordance between observer and observed papillary changes activates bilateral insula, left amygdala and ACC (Harrison et al., 2009). (F) Stroop errors increase pupil changes, correlated with activity in dACC and dMPF (Critchley et al., 2005c). (G) Increased dCC reflects increased blood pressure after stress (mental or physical exercise) (Critchley et al., 2000a). (H) Using fMRI, stimuli presented before and during systole elicited differential changes in amygdala, anterior insula and pons activity, engendering different effects on BP (Gray et al., 2009).

observation of anterior cingulate changes described (Paus et al., 1998), the integration of autonomic changes with task engagement was explicitly explored in a set of PET and fMRI studies. When healthy participants performed easy and difficult versions of mental arithmetic (serial subtractions) and a motor exercise (isometric handgrip) tasks during PET scanning, dorsal cingulate activity was enhanced and importantly this activation correlated with task increases in blood pressure (Critchley et al., 2000a). These findings were replicated initially in healthy older subjects and found to be perturbed in patients who lacked peripheral autonomic response (Critchley et al., 2001c). An adapted version of this task was also undertaken in fMRI with simultaneous electrocardiography (ECG): frequency components of heart rate variability were measured during isometric exercise and high- and low-demand working memory tasks (Critchley et al., 2003). Measures of heart rate variability power at high and low frequency bands were derived using rectified band-pass filtering and regressed against measured changes in regional brain activity, controlling for the stimulation and movement parameters of the experimental task. The low frequency component, encompassing sympathetic neural influences on heart rate variability, positively correlated with changes in neural activity within dorsal cingulate (mid, extending anteriorly toward the callosal genu) and bilateral insula cortex (along with regions of thalamus and dorsal brainstem). This observation also complemented findings implicating anterior cingulate in behaviorally integrated sympathetic drive. Correspondingly, abnormalities in sympathetic cardiovascular response were observed in patients with dorsal cingulate lesions (Critchley et al., 2003).

Studies have also targeted electrodermal activity as an autonomic index of orientation, task engagement, cognitive possessing or stimulus potency (Fredrikson et al., 1998; Critchley et al., 2000b; Williams et al., 2000; Patterson et al., 2002). As with cardiac studies, activity within medial frontal and cingulate cortices typically correlates with sympathetic neural changes in the skin, but there appears to be more variability, suggesting characteristics of the experimental task (perceptual, cognitive, motivational) influence more this relationship between cortical and electrodermal autonomic activity (Critchley, 2009). One autonomic biofeedback study required participants to volitionally increase and decrease skin conductance during neuroimaging and examined the brain correlates of short-term phasic fluctuations in sympathetic activity (corresponding to sympathetic skin responses, SSRs) as distinct from tonic drifts in skin conductance level (SCL) (Nagai et al., 2004b). In keeping with earlier observations (e.g., Fredrikson et al., 1998), SSRs were coupled to dorsal cingulate activity increases (alongside insular and subcortical activation), while the SCL measure of sympathetic tone was negatively correlated with a region of ventromedial prefrontal cortex, extending to subgenual cingulate. This study made an explicit link between regions of the proposed "default mode network" and the control of bodily arousal, suggesting close coupling between resting brain activity with antisympathetic effects (Nagai et al., 2004b).

These and related observations led to a heuristic model of the functional topography of medial frontal cortices in relation to the integration of autonomic control with cognition, emotion, and volitional behavior. Supragenual regions of anterior and mid cingulate were linked to efferent sympathetic influences while ventromedial prefrontal and subgenual cingulate cortices were linked to antisympathetic and parasympathetic (Gianaros et al., 2004; Matthews et al., 2004) influences on autonomic bodily state (Critchley, 2004). Recently, findings endorsed the "two channel" heuristic model: using a social stress task to provoke changes in cardiovascular state, increased activity within dorsal/midcingulate and decreased activity within ventromedial prefrontal cortex both independently predicted increases in evoked heart rate (via thalamus and periaqueductal gray matter) (Wager et al., 2009a). In such experiments, neuroimaging (with fMRI) is not typically very sensitive to proximate, brainstem (and hypothalamic) correlates of autonomic control. This is largely for technical reasons but also perhaps because the housekeeping/homeostatic role of medullary autonomic centers is more continuous, showing less step-change in regional neural activity than the dorsal pons and periaqueductal gray matter (Critchley et al., 2001b; Gray et al., 2010a). Medullary centers are hence potentially less "imageable," requiring more specialized, nonstandard imaging approaches.

Autonomic expression of cognitive conflict and error detection

Psychophysiology highlights the link between changes in bodily arousal state and the behavioral significance of stimuli and actions. In the case of threat, a change in bodily arousal is part of a protective response, i.e., an evolutionarily selected, survival-related response repertoire where the evoked shift in bodily state facilitates an adaptive behavior, e.g., escape motor action. The same type of shifts in body state also occur during types of "cold" cognitive processing, for example, when behavioral strategies need to be changed. One view, promoted especially by Damasio and colleagues, is that fluctuations in bodily arousal contribute to cognitive processes themselves, and feed back to influence and bias thoughts, judgments, and behaviors (Damasio et al., 1991a). This may be particularly useful in correcting suboptimal behaviors.

Certain cognitive tasks, such as the Stroop interference task, evoke demands on attentional processes through cognitive conflicts that are measurable behaviorally from response times, autonomic reactions, and behavioral error. In the "standard" color-word Stroop task, a participant reads from a list of words written in different ink colors. The words are all names of colors. The participant then goes through the list again, naming the ink color of the words while ignoring what is actually written. It is harder (i.e., participants are slower and/or make more errors) to name the ink color (e.g., red) of a word for a different color (e.g., blue). Such cognitive conflict (interference) is generated where a response is required that goes against a prepotent or "more natural" response to the stimuli. Stroop task performance is particularly associated with the engagement of dorsal anterior cingulate cortex. One influential formulation attributes cingulate activation to more than attentional demand, suggesting its central role in rapid detection and signaling of cognitive conflict and behavioral error (e.g., Bechtereva et al., 1990; Dehaene et al., 1994; Bush et al., 2000). An imaging study set out to unpick how the role of dorsal anterior cingulate in effort-related autonomic control might relate to cognitive conflict and error detection: The study measured task-evoked changes in brain activity and autonomic response (here sympathetic influences on pupil size) during performance of a "numerical" Stroop task. It was observed that errors in Stroop task performance elicited greatest effects on pupil diameter, always on trials where there was a conflict (Critchley et al., 2005c). Activity within regions of dorsal anterior cingulate and dorsomedial frontal cortex reflected fluctuations in pupil size and the presence of cognitive conflict in the stimuli. One area in particular was engaged when errors were made that evoked specific high-magnitude pupillary arousal responses. Findings from other studies suggest that this type of autonomic response to error occurs when there is conscious awareness of having made the error (Nieuwenhuis et al., 2001; Hajcak et al., 2003). The imaging study therefore provided insight into a brain area (within dorsal anterior cingulate) integrating adaptive changes in bodily state with conscious self-monitoring. This physiological signal of a cognitive process may represent what Damasio might term a "somatic marker," shifting the physiological context and feeding back to bias cognitive behavior to avoid further mistakes over ensuing trials (see Fig. 6.2).

Autonomic interaction with decision making

Evidence for interaction between the autonomic nervous system and cognitive processes is perhaps most evident in the context of decision making. Damasio's "somatic marker hypothesis" provided an influential model for this: motivationally important (salient) events trigger automatic changes in bodily state. These changes are mainly envisaged as being autonomic and as reflecting prior experience of that event, usually a negative consequence: Bodily responses mark the occurrence of salient stimuli through feedback, i.e., a parallel somatic/visceral representation of that bodily response. Thus, using information provided by brainstem nuclei, somatosensory context, insula cortex, and amygdala, and mediated through medial frontal cortices (lesion data particularly implicate the ventromedial frontal cortex), visceral/autonomic information is integrated with perceptual information to enhance the representation of important stimuli while shaping both the cognitive and behavioral response. In early accounts of the somatic marker hypothesis, emphasis was given to the unconscious influence of the bodily response on behavior. Experimentally, such effects were illustrated using performance on a gambling task as the "behavioral probe."

Neuroimaging studies following on from the work of Damasio's group have examined the brain activity supporting interaction between autonomic responses and motivational decision making; i.e., gambling task performance. For example, one study, using a card game, revealed activity within dorsal anterior cingulate cortex in anticipation of the outcome of risky decisions that reflected both the degree of risk in the gamble and the state of sympathetic electrodermal arousal when anticipating the outcome of the gamble (Critchley et al., 2001a). A similar observation using "wheels of fortune," indicated the importance of agency in the decision-making process: heart rate increased if a participant actively selected which gamble they wanted to play, despite outcomes having the same monetary value. This physiological shift occurred in anticipation of the outcome and in response to the feedback given at outcome. Notably the genual and dorsal anterior cingulate cortex was engaged at outcome, reflecting both the agency and the autonomic state accompanying that agency (Coricelli et al., 2005). These studies highlight the integrative contribution of autonomic response to brain activity during decision making.

The attention drawn by the somatic marker hypothesis to decision making and the influence of autonomic/somatic bodily state coincided with broadening of interest in choice selection in motivational behavior accompanying advances in economic theory. The result is the growth of a discipline now known as neuroeconomics that attempts to understand the basis of rational and irrational choices. Economic theory used to emphasize the notion that optimal decision-making is rational, maximizing gains, minimizing loss with some cost to risk (i.e., Bayesian maximization of expected utility). Such models assume that deciders possess infinite knowledge and information-processing power to inform their decisions (Naqvi et al., 2006). This view also implicitly argues that emotions should either be excluded from the study of rational decision making, or studied as a detrimental influence. Emotional neuroscience and psychophysiological research now challenges the view that rational choice and emotional processing are unrelated or opposed, with evidence for potential beneficial effects of emotion responses on decision making (Bechara and Damasio, 2005). The autonomic nervous system is an important mediator of these effects: autonomic responses reflect learning of the behavioral value of stimuli, and the central feedback of autonomic bodily changes can influence behavioral judgments. Experimentally this influence can be shown to act implicitly or explicitly, to guide behaviors that maximize reward and avoid punishment (Damasio et al., 1991b; Bechara et al., 1997) (Fig. 6.2).

Changes in peripheral physiology prepare the body for behavior modification; thus it follows that brain areas sensitive to peripheral physiological fluctuations may also modify decision making. The ventromedial and orbitofrontal cortices are implicated in the computation of reward and punishment contingencies (Roberts, 2006). The anterior cingulate is activated during a variety of decision-making processes such as conflict resolution (see above and Pochon et al., 2008). Anterior cingulate with the anterior insula is implicated in the production of subjective feeling states arising out of visceral bodily sensations and the coordination of appropriate responses to internal and external events (Medford and Critchley, 2010). In the formulation of the somatic marker hypothesis it was noted that individuals with vmPFC damage are impaired in their ability to modulate behavioral choice in a gambling task following prior losses (Bechara et al., 1997), an effect coupled to attenuated autonomic reactivity. It is worth noting that the bodily response can be specific: during a stochastic learning task, error-related feedback elicited a deceleration in heart rate (correlating with a specific electroencephalogram potential), while feedback indicating success induced a corresponding increase in heart rate (Groen et al., 2007).

Emotions through central and peripheral states can be detrimental to particular types of decisions. Patients with deficits in the expression of emotion can show enhanced, rational decision making, reflecting this removal of emotional bias (Shiv et al., 2005a, b; Di Martino et al., 2009). Changing moods influence the cognitive weighting of decision parameters, with low mood (and implicitly low autonomic reactivity) linked to a preference for low-risk, low-reward outcomes (Smith and Ellsworth, 1985), while anxiety (typically engendering heightened sympathetic tone) leads to a heightened intolerance of uncertainty (Stern et al., 2009). Together these studies illustrate the integration of autonomic control with cognitive processes underlying decision making.

Autonomic effects on memory

Memories are influenced by attention, arousal, and emotion: processes coupled to each other and to autonomic state. Increased central arousal, typically induced by emotional events, enhance memory encoding and facilitate subsequent memory recall (Cahill and McGaugh, 1998). Moreover, memory retention can be predicted by autonomic indicants such as heart rate response and electrodermal activity. Pharmacological agents that enhance peripheral and central sympathetic action can enhance memory (Cahill and McGaugh, 1998); in contrast, β-adrenergic antagonists that attenuate sympathetic activity can impair memory, particularly of emotional material (Cahill and McGaugh, 1998; van Stegeren et al., 1998; Kroes et al., 2010). Interestingly, memories founded upon familiarity (in the absence of recollection) are also marked by increased autonomic activity, as indexed by electrodermal (Morris et al., 2008) or pupillometric (Heaver and Hutton, 2011) change, raising the intriguing possibility that the representation of these bodily changes may mediate the feeling of "knowing" ascribed to such memories (Morris et al., 2008). Correspondingly, subjective memory impairment is a common symptom in people with autonomic failure (Heims et al., 2006).

Sensitivity to autonomically mediated changes in one's body, i.e., "interoceptive awareness," plays a role in mediating the influence of autonomic state on memory (Fig. 6.2). The notion of interoceptive awareness will be

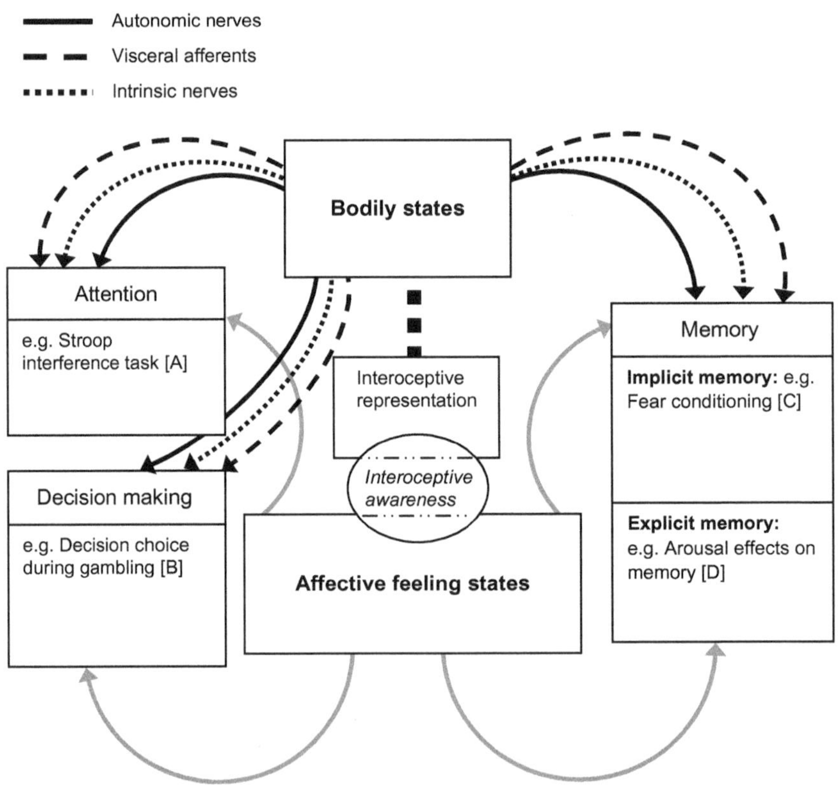

Fig. 6.2. Fluctuations in bodily arousal and affective feeling states can influence cognitive processes. Errors in a stroop interference task are associated with increased pupillary responses ([A] Critchley et al., 2005c). Bodily states and affective feeling states can also guide behaviors, such as decision making during gambling (e.g. [B] (e.g., Bechara et al., 1997; Critchley et al., 2000a)). Emotional events can influence both affective feeling states and increase central arousal, leading to enhanced memory encoding and facilitated memory ([D] Cahill and McGaugh, 1998). Bodily states, synonymous with "gut feelings," can also guide implicit memory processes ([C] Katkin et al., 2001). The propensity for individuals to utilize information conferred in bodily states is mediated by intersubject variability in interoceptive awareness, with some subjects better able to detect fluctuations in bodily processes and thus use the information inherent within those fluctuations to guide cognitions and behaviors (e.g. [C] Katkin et al., 2001).

revisited in this chapter. It is predicated on the concept that some individuals are more influenced by their internal bodily states (of arousal) and as a consequence have better conscious awareness of autonomic processes. Differences in the degree of interoceptive awareness can be tested using behavioral tasks, for example, heartbeat detection, which attempt to measure the accuracy of conscious awareness of individual heartbeats at rest. One version of this task asks the participant to judge if audible notes are in time or delayed relative to individual heart beats (e.g., Whitehead et al., 1977; Wiens et al., 2000); another version asks the participant to count the number of heartbeats occurring over fixed time intervals (Schandry, 1981). Using such approaches participants can be categorized into groups of high and low interoceptive awareness.

Individuals with high interoceptive awareness perform better on specific memory tests, perhaps because the pattern of autonomic change during stimulus processing, or evoked by recollection, can provide additional information to support memory endorsements. In a seminal paper investigating the relationship between interoceptive awareness and memory for emotional information, a fear conditioning paradigm was used in conjunction with the subliminal presentation of emotional stimuli (Katkin et al., 2001). During fear acquisition, snakes and spiders were subliminally presented, and shocks followed one class of stimuli (e.g., snakes; CS+) but not the other (e.g. spiders; CS−). Participants rated the likelihood that these stimuli would be followed by a shock and it was found that interoceptive awareness predicted participants' shock prediction accuracy. The interpretation was that participants with high interoceptive awareness use the perception of visceral cues generated in the nonconscious fear paradigm to facilitate the subsequent recognition of emotional stimuli. Participants with high interoceptive awareness also show enhanced later recognition memory for emotional pictures (Pollatos and Schandry, 2008) and words (Werner et al., 2010). These observations are interesting since they extend a theoretical prediction that those individuals with either greater autonomic response or enhanced access to bodily information have extra cues to facilitate memory retrieval. Thus one way through which autonomic responses influence memory is mediated by interoceptive sensitivity (Katkin et al., 2001) (Fig. 6.2). It is noteworthy the cognitive impact of autonomic arousal states and their central representation is most evident in the processing of emotional information.

AUTONOMIC INTEGRATION WITH EMOTION

Definition and autonomic expression of emotions

There exist different conceptualizations and seemingly contradictory theories of emotion in the literature. A prevalent view sees emotion as a transient perturbation in ongoing behavioral functioning evoked by an external or internal triggering stimulus (defined as emotive by this capacity). A working definition is that emotions are stereotyped response sequences and biases to subsequent behavior, which include these experiential, physiological and behavioral components. Emotions enrich human experience and have communicative social impact; emotive stimuli are assigned heightened value, bias attention and memory, and affect motivational drives and reorient ongoing behavior (Scherer, 2000; Dolan, 2002). Evolution appears to have selected for a set of "basic" emotions, fear, happiness, anger, disgust, surprise, and sadness (Darwin, 1872; Ekman et al., 1969; Ekman and Friesen, 1971), that are universal; they are present across all human cultures.

The obligatory role of physiological response, notably autonomically mediated bodily changes to emotion, is one area of controversy (Harrison et al., 2011). There is some empirical evidence for the coupling of individual basic emotions to discrete patterns of autonomic response, e.g., changes in heart rate, skin conductance, and skin temperature (Levenson, 1992), or distinguishable patterns of cardiorespiratory activity (Rainville et al., 2006). No single autonomic dimension (e.g., sympathetic arousal) appears to be sufficient to characterize different emotions (Rainville et al., 2006). Characteristic (modal) and diverse alterations in physiology with specific basic emotions continue to be mapped out (see Kreibig, 2010, and Harrison et al., 2011, for recent reviews). In sum, there are degrees of emotion-specific patterning that variably overlap across emotion categories.

Stimuli within basic emotion categories elicit differential physiological responses depending upon factors such as psychological state of the viewer (e.g., viewing anger stimuli can elicit a fear response, or lead to the mirroring of an anger response) and motivational direction (e.g., withdrawal-orientated anger can elicit a decrease in heart rate, self-directed anger an increase in heart rate (Stemmler et al., 2007)). Within disgust, class of stimuli can also influence physiological responding: core contamination disgust (e.g., unpalatable food, foul smells, dirty toilets) causes increases in both sympathetic and parasympathetic indices, manifest as increased/unchanged heart rate, increased heart rate variability (HRV), decreased stroke volume, enhanced peripheral resistance, and an increase in respiration rate and effects on the stomach (tachygastria). Body boundary violation disgust (e.g., injections, mutilation scenes, bloody injuries) are characterized by reduced heart rate, increased electrodermal activity, increased respiration rate, unchanged stroke volume, and total peripheral resistance (Harrison et al., 2010).

The heterogeneity of emotion-specific physiological responses found in meta-analytical studies (Cacioppo et al., 2000) has led to dismissal of any predictive association, e.g., "it is not possible to confidently claim that there are kinds of emotion with unique and invariant autonomic signatures" (Barrett, 2006). Instead, patterns of physiological response are proposed to follow more general dimensions of threat and challenge, and positive versus negative affect with autonomic activity is "mobilized in response to the metabolic demands associated with actual behavior or expected behavior." Since behaviors are neither emotion-specific nor context-invariant (Lang et al., 1990), *a priori* emotion-specific autonomic patterns are therefore rather improbable (Barrett, 2006). A different view asks why autonomic responses would not convey emotion-specific activation patterns, given their specific functions of individual emotions for human adaptation (Stemmler, 2004). Since emotions have distinct goals, they ought to require differentiated autonomic activity for body protection and behavior preparation. Nevertheless, the emphasis is on integrated patterns of response rather than single isolated changes (Hilton, 1975; Stemmler, 2004; Mauss and Robinson, 2009).

Meta-analyses of physiological responding in emotion indicate an intermediate position with context-specific emotional effects of autonomic activity (Cacioppo et al., 2000). Valence-specific (good or bad) patterning is more generally consistent than emotion-specific patterning. However, it has been argued that the greater autonomic responses to negative emotions may be an artifactual consequence of the tendency to compare one positive emotion (happiness) with many more negative emotions (e.g., fear, sadness, anger, disgust) (Kreibig, 2010). Inclusion of multiple positive emotions (affection, amusement, contentment, happiness, joy, pleasure, pride, and surprise) can improve the observed degree of emotion-specific physiological patterning (Kreibig, 2010).

Central generators of emotional autonomic change

Anterior cingulate activity is associated with generation of sympathetic autonomic responses in the context of emotional processing; again there appears to be some relation to the nature of the task and with the precise autonomic response measured. The effects of emotional processing on cardiac responses have been studied using forewarned reaction time tasks: such tasks induce predictable changes in both brain activity and autonomic bodily responses. A warning stimulus (cue) signals that the participant should respond to the next imperative stimulus with a reaction time response. In the period between cue and imperative stimuli, there is an orienting response to the cue stimulus followed by cardiac deceleration prior to the imperative stimulus and response. An electrocortical potential (the contingent negative variation, CNV) also occurs during this anticipatory interval, indicative of thalamocortical excitability and ascribable in part to dorsal cingulate activation (e.g., Nagai et al., 2004a). An emotional version of the forewarned reaction time task replaced the imperative stimuli with faces depicting different emotional expressions. The reaction time response was a choice judgment of the emotion portrayed. By measuring cardiac responses to the face stimuli it was possible to examine brain activity associated with differential cardiac responses (orienting acceleration and deceleration) to different emotional expressions. In fact the accelerative and decelerative cardiac responses were closely correlated. Processing of happy and disgust faces attenuated, while processing sad and angry faces enhanced heart rate responses. Brain activity reflecting emotional effects on heart rate within and between the different categories of emotional processing was enhanced in regions including anterior cingulate cortex, amygdala, and temporal lobe (lingual and fusiform) cortices (Critchley et al., 2005a). Similar brain areas are activated in sexual arousal, reflecting activity in autonomic, cognitive, and emotional areas, including hypothalamus activity and claustrum (Georgiadis et al., 2009; Kuhn and Gallinat, 2011).

The emotional effects observed on heart rate were consistent with those observed in other contexts, notably Ekman and coworkers' studies on differential autonomic responses to emotion (Ekman et al., 1983). The basic principle, greater parasympathetic and perhaps less sympathetic activation associated with happiness and disgust and a shift toward sympathetic dominance and parasympathetic activity with sadness and anger, is replicated in other contexts and to other stimulus sets (e.g., Umeda et al., unpublished observations). These studies highlight a degree of emotion-specificity of autonomic response that is nevertheless crude compared to the organ-specific control visible as changes in posture and skin perfusion that characterize the expression of different emotions for which autonomic patterning is the rule.

The central nucleus of the amygdala is implicated as a generator of emotional autonomic activity, particularly in the context of fear processing. The amygdala is involved in learning fear, such that patients with temporal lobe lesions affecting the amygdala demonstrate impaired fear conditioning through absent autonomic (electrodermal) reactivity to threat (LaBar et al., 1995). In fact, the degree to which healthy people activate amygdala when learning about threat correlates with the degree of autonomic response to the threat (LaBar et al., 1998). The role of the amygdala in signaling behaviorally and autonomically the salience of emotive stimuli has led to its inclusion, alongside anterior cingulate and insula cortex, within the "salience" network (Menon and

Uddin, 2010). However, it is worth noting that neuroimaging studies using emotional paradigms are variable in the extent to which autonomic responses are found to correlate with amygdala activity (e.g., Wager et al., 2009a). Moreover, there is evidence to link the amygdala also to afferent representation of internal physiological state (Harrison et al., 2009).

Autonomic contributions to emotional feeling states

Historically, theories of emotion have given different weights to the necessity of autonomic responses for emotional feeling states. Peripheral theories of emotion argue that the origins of emotional feelings stem from bodily responses. Based on this premise, the seminal James and Lange theory of emotion argued that emotions depend entirely on (feedback from) the bodily response:

> *We owe all the emotional side of our mental life, our joys and sorrows, our happy and unhappy hours to our vasomotor system. If the impressions which fall upon our senses did not possess the power of stimulating it, we would wander through life unsympathetic and passionless, all impressions of the outer word would only enrich our experience, increase our knowledge, but would arouse neither joy nor anger, would give us neither care nor fear.*
>
> *(Lange and James, 1967)*

However, Cannon argued that bodily arousal responses are too undifferentiated to account for the variety of distinct emotional feeling states (Cannon, 1927). Subsequent attempts have been made to reconcile these distinct views by incorporating the role of cognitive appraisal. The two-stage model of Schachter and Singer argued changes in peripheral arousal can elicit subjective emotionality "of corresponding intensities," yet, the nature of the emotion category is determined by cognitive context (Schachter and Singer, 1962). Thus, emotional states are states of arousal, whose type (e.g., fear, happiness, sadness) is determined by the cognitive interpretation of what evoked that state. The somatic marker hypothesis, discussed above (Autonomic interaction with decision making), conceptualizes emotions and emotional feeling states along similar lines, with perhaps a more negative bias.

Studies in clinical populations with altered peripheral physiological responses and altered affect provide insight into the relationship between autonomic patterning and subjective emotional feeling states. It is recognized that emotion behaviors (e.g., pathological laughter) can deviate from peripheral physiological state but such instances are the exception (Parvizi et al., 2001). The absence of peripheral autonomic response, for example, in patients with pure autonomic failure (PAF), an acquired failure of peripheral autonomic regulation, may lead to subtle blunting of emotional feeling (appearing less anxious relative to comparably debilitated Parkinson's patients without autonomic failure (Mathias, 2003; Critchley, unpublished observations)), but these effects are subtle relative to other emotional changes (see below). The uncoupling of brain and body that occurs in part following high spinal cord transection has been suggested to lead to altered emotions. However, while changes in brain response to emotive stimuli are demonstrable (Nicotra et al., 2006), there is little evidence for attenuation of emotional feelings in patients with high spinal cord transaction or for change in emotion reactivity that cannot be explained by adjustment to disability or chronic pain (Deady et al., 2010).

Interoceptive sensitivity and insula cortex

Interoceptive awareness is mentioned above (Autonomic effects on memory) and much of the motivation for defining individual differences in interoceptive sensitivity comes from peripheral theories of emotion, where the prediction is that people more attuned to (autonomic) bodily responses experience emotions with heightened intensity. As noted, the emphasis has been on heartbeat detection, both because heartbeats are distinct internal events that can be easily measured and because people's ability to detect heartbeats correlates with their ability to detect changes in other autonomically innervated organs (Whitehead and Drescher, 1980). Cranial nerves (mainly vagus) and spinal cord pathways (particularly the lamina I spinothalamocortical pathway) (Craig, 2002) proved the major routes of visceral functional and autonomic feedback to the brain. Insula cortex appears to be particularly important in the cortical representation of internal state, with right anterior insula subserving interoceptive awareness and its expression as emotional feeling states (see below) (Critchley et al., 2004) and the interaction of interoceptive information with the conscious appraisal of other information (Critchley, 2004; Singer et al., 2009).

A neuroimaging study used a heartbeat detection task to reveal engagement of insula, anterior cingulate and somatosensory cortices when people focus on their internal bodily processes. Awareness and sensitivity to the occurrence of individual heartbeats was revealed by how well people distinguished the presence of a heart-note delay in auditory tones triggered by individual heartbeats. Across participants, activity and even gray matter volume within one brain region, right anterior insular cortex, predicted differences in interoceptive sensitivity (Critchley et al., 2004). The link with emotional process was revealed by showing that interoceptive sensitivity and activity/volume within right insula cortex also reflected day-to-day experiences of anxiety (and to a lesser extent other negative emotions). This study reinforced proposals put forward by Craig that right anterior insula cortex in primates is the terminus of afferent viscerosensory information. Since information mapped by this system was essentially of motivational significance, Craig proposed that insula cortex represents motivational state and through rerepresentations becomes the substrate for conscious feeling states (Craig, 2002, 2003).

The role of insula cortex in the integrative representation of emotional feeling states that originate in bodily responses was particularly highlighted by a study of disgust: combining electrocardiography, electrogastrography, and functional magnetic resonance imaging, the distinct physiological effects of core (nauseating) disgust and body boundary violation disgust (evoked by video sequences) were mapped into activity within discrete subregions of insula cortex that predicted the magnitude of subjective disgust reported by the observer (Harrison et al., 2010). This study suggested again the origin of emotional feelings in the cortical representation of bodily states.

Perturbation of visceral afferent information: relationship to neural correlates and affect

Visceral stimulation modulates activity within brain areas implicated in affective processing. Investigating changes in internal bodily state via direct gastrointestinal stimulation of esophagus or large bowel reveals enhanced activity within cingulate and insula cortex. This occurs even in the absence of pain, though potentially still reflects affective processing of rectal sensation (Hobday et al., 2001). Anorectal stimulation provides segregation of somatosensory (anal) versus visceral (rectal) afferents within the nervous system (Eickhoff et al., 2006). Bilateral insula and dorsal anterior cingulate cortex are activated by rectal (but not anal) stimulation, reflecting heightened propensity for visceral sensations to be displayed in those areas involved in affective processing (Hobday et al., 2001). The ventromedial prefrontal cortex is another key area; observations in patients with spinal cord lesions and from a single patient treated with vagus nerve stimulation for depression highlight its role in the integration of afferent visceral sensory information perturbed by interference with spinal and vagus nerve routes (Nicotra et al., 2006; Critchley et al., 2007). Surprisingly, there does not seem to be an obvious functional hierarchy in these effects, something that other data also suggest. The conclusion here is that anterior insula, dorsal anterior cingulate, and ventromedial prefrontal cortex contribute to the integration of visceral afferent information, generated by salient stimuli with stimulus processing itself.

Beat-to-beat effects: relationship to neural correlates and affect

General perturbations in interoceptive state modulate activity within similar brain areas to those resulting from visceral stimulation (i.e., areas involved in affective processing). Afferent information from the viscera influences stimulus processing even at the level of individual heartbeats (Lacey and Lacey, 1978, 1980). At systole the pressure wave of cardiac ejection activates aortic and carotid baroreceptors. Baroreceptor effects influence the processing of painful and strong unexpected somatosensory stimuli to modify cortical reflexive and autonomic response (Donadio et al., 2002; Edwards et al., 2003; Wallin, 2007). Efferent sympathetic responses to stimuli occurring at systole are modulated and notably there is inhibition of muscle sympathetic nerve traffic while sympathetic electrodermal responses are unaltered. These effects are amplified in patients with blood phobia and syncope, suggesting that these mechanisms are central to traits in emotional behavior and reactivity (Donadio et al., 2007). The phasic signature of this channel, reflecting short-term fluctuations in visceral state within each cardiac cycle, can be exploited in the study of autonomic interactions with emotion to reveal corresponding neuronal correlates. Our own neuroimaging observations (Gray et al., 2010b) reveal that the interaction between stimuli and baroreceptor activation within the cardiac cycle is mediated by differential engagement of a discreet set of brainstem, cortical, and subcortical centers to elicit further characteristic changes in autonomic state. Significantly, these effects can influence processing of emotional stimuli (Gray et al., 2012). Gray et al. (2012) investigated the interaction between emotional appraisal and cardiovascular activity by presenting emotional stimuli at distinct points within the cardiac cycle. Combining a forewarned reaction time task with discrete emotional facial expressions, face stimuli were presented at either the end of diastole (indexed by the ECG R-wave) or systole (approximately at the ECG T-wave) when baroreceptor impulses are processed centrally. Using a behavioral index of emotional intensity, ratings of disgust increased when stimuli were processed at systole. This was in contrast to other emotions (happiness, anger, sadness), where intensity ratings were not modified by cardiac cycle. PAG appeared to underlie the moderating effect of cardiac timing on emotional appraisal, whereby the negative association between PAG activity and subjective intensity was significantly enhanced when emotional stimulus presentation was synchronous with R-wave relative to T-wave. Rebound heart rate responses were attenuated following disgust and happy stimuli, replicating prior research (Critchley et al., 2005a), and localized orbital frontal activity predicted heart rate response to emotional stimuli. This finding is in line with the integrative role of orbitofrontal cortex in combining "cognitive and perceptual cross-modal representations with motivationally-salient physiological information and is a cortical source of descending influences on internal bodily state" (Gray et al., 2012).

Autonomic communication and empathy

The neural integration of cognitive, affective and autonomic response is proposed to be a guide to adaptive social behavior (Damasio et al., 1991a; Damasio, 1994, 1999). The mirroring of changes in visceral state during emotion may be mirrored in the responses of others, permitting a corresponding representation in the observer. The degree to which individuals are susceptible to this "contagion" predicts individual differences in questionnaire ratings of empathy. The notion that affective feelings are coupled to autonomically mediated visceral responses implies that the sharing of emotional feelings empathetically embodies a sharing of visceral autonomic response across individuals. A dependence of emotional interchange on aspects of autonomic reactivity is suggested from observations in patients with PAF, who manifest a blunting of empathetic responses on a questionnaire measure of emotional empathy (Chauhan et al., 2008). With many autonomic responses serving as visual signals of emotion, reciprocation is visibly evident in the contagion of fear (including facial pallor) and anger responses (facial flushing), but in many cases the exchange is subtle and the signals are covert.

Pupillary signals influence emotional processing (Harrison et al., 2006, 2007). Student volunteers were asked in a behavioral task to make visual analog ratings of positive/negative, intensity and attractiveness attributes of pictures of emotional faces. For each identity/expression combination, the size of the stimuli pupils was digitally manipulated to cover a range of (biologically plausible) sizes. Participants were not informed of this image manipulation and none were aware of this at debriefing after the task was performed. Contrary to the initial prediction that enlarged pupils, reflecting sympathetic activity, would produce greater intensity ratings of all the emotions, no significant (linear) effect of pupil size was observed for any of the emotions except sadness. Here the smaller the pupils the more negative and more emotionally intense sad faces were rated. Interestingly, the degree to which participants were affected by this covert manipulation of perceived pupil size correlated with individual differences in questionnaire measures of emotional empathy. A parallel neuroimaging study was conducted with the same pupil-manipulated face stimuli. Participants made age judgments of the faces, processing both emotion and the manipulated pupil size incidentally. During scanning with fMRI, it was also possible to use pupillometry to record the pupillary response of the participant to the stimuli. Activity changes in amygdala, insula, superior temporal sulcus, and brainstem (in the region of the Edinger–Westphal nucleus) reflected the interaction between perceived pupil size and sad versus non-sad emotion. When the pupil responses of the participants were examined, it was found that there was coherence in the observed and observer's pupils only in the context of sadness (i.e., seeing a sad face with small pupils caused the participants' own pupils to constrict more). The extent to which this occurred was again predicted by activity within the Edinger–Westphal nucleus (the autonomic nucleus responsible for pupillary control).

These observations show a contagion of autonomic responses in emotional processing, highlighting automatic mirroring only in the context of sadness perception that influences the judgment of sadness in a manner related to individual differences in empathy for others. Moreover, the findings highlight the presence of organ specificity in emotional autonomic responses that are integrated with cognitive and emotional aspects of empathy within a discrete neural system.

"Contagion" of emotion can occur when changes in visceral state during emotion may be mirrored in an observer, inducing a corresponding representation (Critchley, 2009). It thus follows that those individuals with enhanced sensitivity to visceral perception and monitoring have a consequent increased sensitivity to empathetic feeling. Indeed, empirical research has linked enhanced proficiency for interoceptive awareness with a greater capacity for empathy (Fukushima et al., 2011). The amplitude of heartbeat-evoked potentials (HEPs, for affective vs. physical judgments), thought to reflect cortical representation of cardiac signal, correlated with self-rated empathy scores (Fukushima et al., 2011).

INTEGRATIVE MECHANISMS OF STRESS AND RELEVANCE TO HEALTH

Cardiovascular risk

Cardiac risk is enhanced by psychological conditions, including grief and depression, and is associated with differences in personality and socioeconomic circumstances. Low vagus parasympathetic tone and exaggerated evoked sympathetic (blood pressure and heart rate) response represent putative physiological mediators of cardiac risk (e.g., Carroll et al., 2001; Thayer and Lane, 2007). Studies of at-risk recently bereaved individuals highlight abnormalities in regions, including subgenual cingulate, implicated in parasympathetic influences on the heart that correspond to withdrawal of this cardioprotective "break" (O'Connor et al., 2007). Abnormalities within similar regions of subgenual cingulate and ventromedial prefrontal cortex are commonly reported in depressed patients (who also show diminished parasympathetic tone) (O'Connor et al., 2007). Interestingly, the structural morphology of these and related brain regions in healthy individuals relate to social predictors of cardiac risk, including perceived life stress and social standing (Gianaros et al., 2007a, b). Exaggerated cardiovascular responses to stress also represent a psychophysiological predictor of cardiac morbidity and mortality: across individuals, the response within posterior cingulate cortex predicts individual differences in cardiovascular responsivity, hence, by inference, cardiac risk (Gianaros et al., 2004, 2005). Salient environmental stimuli may motivate alterations in cardiovascular function, over and above physiological motivators such as physical stress, exercise, faint, or hemorrhage. The range of these specific environmental (or experimental) situations that can impact on cardiovascular state is extremely large, reflecting both primary and learned challenges that ultimately evoke mental and physiological stress. Associations are apparent at the population level between environmental stressors and adverse cardiac outcomes such as acute coronary syndromes or sudden cardiac death. For example, the major earthquakes in Los Angeles (Leor et al., 1996) and the Kobe region of Japan (Suzuki et al., 1995) were both followed by significantly increased hospital admissions for acute myocardial infarction (Bhattacharyya and Steptoe, 2007). Similar research suggests that war, terrorist acts, industrial disasters, and even sporting events are associated with higher rates of infarction, disruption of normal cardiac rhythm, and sudden cardiac death. More predictably, increased hospitalization of cardiac patients is observed during Christmas (Reedman et al., 2008), New Year's Eve (Kloner et al., 1999; Phillips et al., 2004), and on Mondays (Barnett and Dobson, 2005), further highlighting associations between times of potential psychosocial stress and cardiovascular dysfunction. Specific emotional states are particularly associated with increased risk of clinically significant cardiac events (Bhattacharyya and Steptoe, 2007). Acute episodes of anger (Mittleman et al., 1995, 1997; Strike et al., 2006), stress (Moller et al., 2005; Vlastelica, 2008), or depression (Steptoe et al., 2006) are significantly over-reported in cardiac patients prior to admission, suggesting negative emotions may trigger adverse cardiac events. Likewise, longer-term exposure to stress may also increase cardiac risk (Willich et al., 1993; Deljanin et al., 2007; Bhattacharyya et al., 2010).

Brain imaging of autonomic risk factors

Blood pressure reactivity during emotional or cognitive stress is a recognized indicator of risk for hypertension, ventricular hypertrophy, and premature atherosclerosis (Krantz and Manuck, 1984; Treiber et al., 2003). Functional brain imaging studies also frequently identify amygdala activity during tasks which induce alterations in cardiovascular function. During incongruent conditions of a Stroop color-word interference task, mean arterial pressure alterations were positively associated with change in bilateral amygdala activity, and volumetric analysis of participants' structural MRI scans revealed gray matter volume within the amygdala negatively correlated with blood pressure reactivity. This same pattern was also observed within genual anterior cingulate cortex and pontine nuclei (Gianaros et al., 2008). Carotid artery intimal media thickness (IMT), measured with ultrasonography, is an indicator of atherosclerosis and a risk factor for a range of cardiovascular diseases including angina, myocardial infarction, and stroke (Lorenz et al., 2007). IMT positively correlates with activity in anterior cingulate cortex and dorsal amygdala during a facial expression matching task (Gianaros et al., 2009). These results add further support for the notion of a cortical network including the amygdala, (insula) and anterior cingulate cortex through which cognitive emotional stressors are translated into changes in cardiovascular function. Moreover, they suggest a pattern whereby exaggerated amygdala responses to emotional stressors are linked to cardiovascular damage as a direct consequence of excessive reactivity to stress (Gianaros et al., 2009).

Arrhythmogenesis

Brain responses to emotional challenges may trigger potentially fatal arrhythmic cardiac events. The presence of pre-existing heart disease greatly increases this risk. Lateralization of efferent autonomic drive from the brain may also be a mediating factor (Lane and Schwartz, 1987): hemispheric dominance of brain responses to powerful emotional challenges may engender a lateralization of sympathetic outflow to the heart, affecting myocardial repolarization. If the electrical initiation of contraction reaches regions of heart muscle before they are fully repolarized, the sequential coordination of contraction is disrupted, a process already destabilized in cardiac disease. Proarrhythmic changes in myocardial repolarization can be quantified from the morphology of electrocardiographic T-waves across chest leads. Mental stress shifts the heart toward proarrhythmic state, even in healthy people (Taggart et al., 2005). In a PET brain imaging study of cardiac patients, the lateralization of brain responses during stress was tested to identify regions predictive of proarrhythmic changes (Critchley et al., 2005b). Interestingly, right-sided midbrain activity (in the region of the parabrachial nucleus) predicted proarrhythmic cardiac changes and which of the patients were at greatest risk of arrhythmia. A further study using electroencephalography suggests this may be a consequence, in patients with pre-existing heart disease, of lateral dominance of a cardiocerebrocardiac loop governing cardiac contractility (Gray et al., 2007b).

INTEGRATION IN SELF-REPRESENTATION

The concepts

There is a longstanding notion that the mental representation of self is ultimately grounded on the representation of the body, and the consistency of the representation of the internal body ultimately provides the primary reference for the limbs and special senses as the body interacts with the environment. Homeostatic maintenance of the *milieu intérieur* is highly dynamic and controlled through interacting and competing mechanisms. While there is a tendency to see the arrangement of these mechanisms as hierarchical, with conscious experience and volitional behavior somewhere near the top and emotions more proximate to the basic workings of the body-machine, this model is at best heuristic. The concept of allostasis touches on this instability, recognizing that there is inherently no internal stability and that it is more, perhaps, the adaptive process of control of internal state that can be viewed as a consistent factor and the basis for self-reference in its broad sense. Here, by adaptive control, we refer to the integration of the autonomic or behavioral adjustments to change internal states with the afferent feedback signaling the direction of change in internal state and motivating further adjustment. Within the brain, this integration is achieved through prediction errors and efference copies.

Taking as a model the notion that the psychological representation of self is grounded on the control of internal bodily state, there are a number of predictions. First, that the notion of agency is important. Uncoupling autonomic efferent control from the visceral feedback might, if transient, provoke a sense of unease that would require behavioral or psychological adjustment, yet which if persistent might destabilize cognitive and emotional processes established over the life course that are grounded on intact self-representation. Second, processes related to control and representation of internal bodily state will determine the strength of self-representation.

Uncoupling of internal control in clinical groups

Patients with acquired (typically traumatic) high spinal cord transection experience a major uncoupling of their internal and external bodily control. In patients with pure autonomic failure there is a gradual compromise in autonomic effector function. Yet in neither patient group is there any compelling evidence that self-representation is compromised. Clearly, following spinal cord transection, individuals need to adjust psychologically to paralysis and loss of independence with even basic functions. They are vulnerable to depression and chronic pain. Nevertheless, spinal cord transection appears not to limit the range of emotions an individual can experience nor, in itself, compromise emotional responsivity. Psychological adjustment reactions, depression, pain, and loss of physical function often enhance self-reference and bolster subjective sense of self. Similar observations can be made of patients with pure autonomic failure whose apparent pragmatism is often quite noteworthy and in whom it is difficult to identify emotional deficits.

Signatures of self-disturbance

Mismatch and misattribution of interoceptive cues appears key in disorders of the self. In a "comparator model" of schizophrenia, it is proposed that disturbances of self (e.g., delusions of control) arise as a consequence of problems in predictive coding, reflecting confusion between evaluation of changes in sensations caused by the self and changes associated with external causes: Sensory effects of self are attributed instead to external forces (Frith, 2012). A similar prediction error is implicated as a central mechanism in generation of anxiety (a core symptom of many affective disorders including obsessive compulsive disorder and autistic spectrum conditions), with heightened anxiety a consequence of augmented detection of difference between observed and expected bodily states. Insula cortex is proposed to mediate anxiety via (mis)match of interoceptive signals; subjective anxiety is associated with enhanced interoceptive prediction error signals (Paulus and Stein, 2006). Highly anxious individuals show increased anterior insula cortex activity during emotion processing. Inducing mismatch between predicted and actual interoceptive signals via false physiological feedback further demonstrates role of insula cortex (Gray et al., 2007a). Depersonalization and derealization disorders, where the concept of self is also disturbed, show similar mismatch of interoceptive cues, pain signals may be dulled and their origin misattributed. For example, in an earlier account of the disorder, a patient of Mayer-Gross (1935) reported "I feel pains in my chest, but they seem to belong to someone else, not to me" (p. 114). Patients with depersonalization disorder and schizophrenia show corresponding autonomic disturbances. Depersonalization is associated with blunted autonomic responses, e.g., skin conductance abnormalities (Lader, 1975; Griffin et al., 1997). In bipolar disorder, negative correlations are reported between level of dissociation and autonomic parameters (Latalova et al., 2010). Low heart rate variability and decreased baroreflex sensitivity is found in schizophrenia (Koponen et al., 2008). Interestingly, a signal mismatch is suspected: Kring and Moran report that subjects with schizophrenia exhibit greater skin conductance reactivity in response to emotionally evocative film clips, despite displaying few observable facial expressions (Kring and Moran, 2008). This paradox of increased emotionality (autonomic reactivity and subjective arousal) in the face of deficits in perception and expression of emotion has been attributed to specific intra-amygdala abnormalities (Aleman and Kahn, 2005).

Susceptibility to the "rubber hand illusion," in which synchronous tactile and visual stimulation may lead individuals to experience a false hand as their own, is a useful experimental probe into mechanisms of self-representation and coherence within concepts of agency. Those with disordered self-representation (e.g., schizophrenia patients) are more prone to the illusion (Peled et al., 2000). This can be equally induced by dissociative agents such as ketamine (Morgan et al., 2011) and enhanced by administration of dexamphetamine (Albrecht et al., 2011), which results in elevated dopamine, a mediator of salience in psychosis (Kapur, 2003). Interestingly, a recent study by Tsakaris and colleagues finds that more interoceptively aware individuals are less likely to perceive the rubber hand as their own (Tsakiris et al., 2011). This raises important questions of insight (and its modulation) in disordered self-representation and is a target for further integrated research.

CONCLUSIONS: A FUNCTIONALLY INTEGRATED SYSTEM

We have outlined the motivational integration of cognitive emotional and cardiovascular function, providing examples from recent neuroimaging research which highlight neural systems important in mediating these relationships. At a general level, this research provides insight into brain–body interactions promoting broader understandings of health and disease, consistent with known risk factors for adverse cardiac events including depression, anxiety, and psychosocial triggers. More specifically, we have endeavored to highlight the importance of cortical activity in cognitive emotional and cardiovascular integration. Anterior cingulate cortical activity is recurrently observed during cognitive and emotional experimental paradigms, and also has a demonstrated role in cardiovascular regulation accompanying cognitive and emotional processing, consistent with a role as "visceromotor" cortex. Likewise, insula cortex activity is important in both the regulation and representation of cardiovascular function and may be conceptualized as a "viscerosensory" cortical region (see Craig, 2005). Activity here may assist the integration of visceral and emotional processing, facilitating the coloring of emotional experience by concomitant cardiovascular and visceral activity. Further, the amygdala demonstrates a functional correspondence with stress and emotion effects in the heart and vasculature in addition to its demonstrated contributions to processing emotional and salient stimuli. Activity within these regions may regulate hypothalamic, pons and medullary circuits responsible for coordinated sympathetic regulation of the heart. Recent methodological advances in neuroimaging (i.e., moderation analysis) allow for the construction of statistical models which directly assess the degree to which neural change during stressor tasks mediate cardiovascular changes also accompanying stressor tasks (Wager et al., 2009a). The continuing development of functional neuroimaging techniques that allow the simultaneous investigation of neural and psychophysiological activity *in vivo* (Gray et al., 2009) promise further advances in delineating the integration of cardiac function with cognitive and emotional processing.

References

Albrecht MA, Martin-Iverson MT, Price G et al. (2011). Dexamphetamine effects on separate constructs in the rubber hand illusion test. Psychopharmacology (Berl) 217: 39–50.

Aleman A, Kahn RS (2005). Strange feelings: do amygdala abnormalities dysregulate the emotional brain in schizophrenia? Prog Neurobiol 77: 283–298.

Barnett AG, Dobson AJ (2005). Excess in cardiovascular events on Mondays: a meta-analysis and prospective study. J Epidemiol Community Health 59: 109–114.

Barrett LF (2006). Are emotions natural kinds? Perspect Psychol Sci 1: 28.

Bechara A, Damasio AR (2005). The somatic marker hypothesis: a neural theory of economic decision. Games Econ Behav 52: 336–372.

Bechara A, Damasio H, Tranel D et al. (1997). Deciding advantageously before knowing the advantageous strategy. Science 275: 1293–1295.

Bechtereva NP, Kropotov JD, Ponomarev VA et al. (1990). In search of cerebral error detectors. Int J Psychophysiol 8: 261–273.

Bhattacharyya MR, Steptoe A (2007). Emotional triggers of acute coronary syndromes: strength of evidence biological processes and clinical implications. Prog Cardiovasc Dis 49: 353–365.

Bhattacharyya MR, Perkins-Porras L, Wikman A et al. (2010). The long-term effects of acute triggers of acute coronary syndromes on adaptation and quality of life. Int J Cardiol 138: 246–252.

Bouton ME (1993). Context time and memory retrieval in the interference paradigms of Pavlovian learning. Psychol Bull 114: 80–99.

Bouton ME, Mineka S, Barlow DH (2001). A modern learning theory perspective on the etiology of panic disorder. Psychol Rev 108: 4–32.

Bush G, Luu P, Posner MI (2000). Cognitive and emotional influences in anterior cingulate cortex. Trends Cogn Sci 4: 215–222.

Cacioppo JT, Berntson GG, Larsen JT et al. (2000). The psychophysiology of emotion. In: MJ Lewis, JM Haviland-Jones (Eds.), Handbook of Emotions, 2nd edn. Guilford Press New York.

Cahill L, McGaugh JL (1998). Mechanisms of emotional arousal and lasting declarative memory. Trends Neurosci 21: 294–299.

Cannon WB (1927). Bodily changes in pain hunger fear and rage : an account of recent researches into the function of emotional excitement. D Appleton, New York.

Carroll D, Smith GD, Shipley MJ et al. (2001). Blood pressure reactions to acute psychological stress and future blood pressure status: a 10-year follow-up of men in the Whitehall II Study. Psychosom Med 63: 737–743.

Chauhan B, Mathias CJ, Critchley HD (2008). Autonomic contributions to empathy: evidence from patients with primary autonomic failure. Auton Neurosci 140: 96–100.

Coricelli G, Critchley HD, Joffily M et al. (2005). Regret and its avoidance: a neuroimaging study of choice behavior. Nat Neurosci 8: 1255–1262.

Craig AD (2002). How do you feel? Interoception: the sense of the physiological condition of the body. Nat Rev Neurosci 3: 655–666.

Craig AD (2003). Interoception: the sense of the physiological condition of the body. Curr Opin Neurobiol 13: 500–505.

Craig AD (2005). Forebrain emotional asymmetry: a neuroanatomical basis? Trends Cogn Sci 9: 566–571.

Critchley HD (2004). The human cortex responds to an interoceptive challenge. Proc Natl Acad Sci U S A 101: 6333.

Critchley HD (2009). Psychophysiology of neural cognitive and affective integration: fMRI and autonomic indicants. Int J Psychophysiol 73: 88–94.

Critchley H, Corfield D, Chandler M et al. (2000a). Cerebral correlates of autonomic cardiovascular arousal: a functional neuroimaging investigation in humans. J Physiol 523: 259–270.

Critchley HD, Elliott R, Mathias C et al. (2000b). Neural activity relating to generation and representation of galvanic skin conductance responses: a functional magnetic resonance imaging study. J Neurosci 20: 3033.

Critchley HD, Mathias CJ, Dolan RJ (2001a). Neural activity in the human brain relating to uncertainty and arousal during anticipation. Neuron 29: 537–545.

Critchley HD, Mathias CJ, Dolan RJ (2001b). Neuroanatomical basis for first- and second-order representations of bodily states. Nat Neurosci 4: 207–212.

Critchley HD, Melmed RN, Featherstone E et al. (2001c). Brain activity during biofeedback relaxation: a functional neuroimaging investigation. Brain 124: 1003–1012.

Critchley HD, Mathias CJ, Josephs O et al. (2003). Human cingulate cortex and autonomic control: converging neuroimaging and clinical evidence. Brain 126: 2139.

Critchley HD, Wiens S, Rotshtein P et al. (2004). Neural systems supporting interoceptive awareness. Nat Neurosci 7: 189–195.

Critchley HD, Rotshtein P, Nagai Y et al. (2005a). Activity in the human brain predicting differential heart rate responses to emotional facial expressions. Neuroimage 24: 751–762.

Critchley HD, Taggart P, Sutton PM et al. (2005b). Mental stress and sudden cardiac death: asymmetric midbrain activity as a linking mechanism. Brain 128: 75–85.

Critchley HD, Tang J, Glaser D et al. (2005c). Anterior cingulate activity during error and autonomic response. Neuroimage 27: 885–895.

Critchley HD, Lewis PA, Orth M et al. (2007). Vagus nerve stimulation for treatment-resistant depression: behavioral and neural effects on encoding negative material. Psychosom Med 69: 17–22.

Damasio AR (1994). Descartes' error: emotion reason and the human brain. GP Putnam, New York.

Damasio AR (1999). The feeling of what happens: body and emotion in the making of consciousness. Harcourt Brace, New York.

Damasio AR, Tranel D, Damasio HC (1991a). Somatic markers and the guidance of behavior: theory and preliminary testing. In: HS Levin, HM Eisenberg, AL Benton (Eds.), Frontal Lobe Function and Dysfunction. Oxford University Press, New York.

Damasio H, Kuljis RO, Yuh W et al. (1991b). Magnetic resonance imaging of human intracortical structure in vivo. Cereb Cortex 1: 374–379.

Darwin C (1872). The Expression of the Emotions in Man and Animals [SI]. Murray, London.

Deady DK, North NT, Allan D et al. (2010). Examining the effect of spinal cord injury on emotional awareness expressivity and memory for emotional material. Psychol Health Med 15: 406–419.

Dehaene S, Posner MI, Tucker DM (1994). Localization of a neural system for error detection and compensation. Psychol Sci 5: 303–305.

Deljanin Z, Rancic N, Tiodorovic B et al. (2007). Association of stressful life events with acute myocardial infarction in population in the city of Nis within the period from 1998–2000. Vojnosanit Pregl 64: 463–468.

Di Martino A, Ross K, Uddin LQ et al. (2009). Functional brain correlates of social and nonsocial processes in autism spectrum disorders: an activation likelihood estimation meta-analysis. Biol Psychiatry 65: 63–74.

Dolan RJ (2002). Emotion cognition and behavior. Science 298: 1191–1194.

Donadio V, Kallio M, Karlsson T et al. (2002). Inhibition of human muscle sympathetic activity by sensory stimulation. J Physiol 544: 285–292.

Donadio V, Liguori R, Elam M et al. (2007). Arousal elicits exaggerated inhibition of sympathetic nerve activity in phobic syncope patients. Brain 130: 1653–1662.

Edwards L, McIintyre D, Carroll D et al. (2003). Effects of artificial and natural baroreceptor stimulation on nociceptive responding and pain. Psychophysiology 40: 762–769.

Eickhoff SB, Lotze M, Wietek B et al. (2006). Segregation of visceral and somatosensory afferents: an fMRI and cytoarchitectonic mapping study. Neuroimage 31: 1004–1014.

Ekman P, Friesen WV (1971). Constants across cultures in the face and emotion. J Pers Soc Psychol 17: 124–129.

Ekman P, Sorenson ER, Friesen WV (1969). Pan-cultural elements in facial displays of emotion. Science 1644: 86–88.

Ekman P, Levenson RW, Friesen WV (1983). Autonomic nervous system activity distinguishes among emotions. Science 2211: 1208–1210.

Fredrikson M, Furmark T, Olsson MT et al. (1998). Functional neuroanatomical correlates of electrodermal activity: a positron emission tomographic study. Psychophysiology 355: 179–185.

Frith C (2012). Explaining delusions of control: the comparator model 20 years on. Conscious Cogn 21: 52–54.

Fukushima H, Terasawa Y, Umeda S (2011). Association between interoception and empathy: evidence from heartbeat-evoked brain potential. Int J Psychophysiol 799: 259–265.

Georgiadis JR, Reinders AA, Paans AM et al. (2009). Men versus women on sexual brain function: prominent differences during tactile genital stimulation but not during orgasm. Hum Brain Mapp 300: 3089–3101.

Gianaros PJ, Van Der Veen FM, Jennings JR (2004). Regional cerebral blood flow correlates with heart period and high-frequency heart period variability during working-memory tasks: implications for the cortical and subcortical regulation of cardiac autonomic activity. Psychophysiology 411: 521–530.

Gianaros PJ, Derbyshire SW, May JC et al. (2005). Anterior cingulate activity correlates with blood pressure during stress. Psychophysiology 422: 627–635.

Gianaros PJ, Horenstein JA, Cohen S et al. (2007a). Perigenual anterior cingulate morphology covaries with perceived social standing. Soc Cogn Affect Neurosci 22: 161–173.

Gianaros PJ, Jennings JR, Sheu LK et al. (2007b). Heightened functional neural activation to psychological stress covaries with exaggerated blood pressure reactivity. Hypertension 499: 134–140.

Gianaros PJ, Sheu LK, Matthews KA et al. (2008). Individual differences in stressor-evoked blood pressure reactivity vary with activation volume and functional connectivity of the amygdala. J Neurosci 288: 990–999.

Gianaros PJ, Hariri AR, Sheu LK et al. (2009). Preclinical atherosclerosis covaries with individual differences in reactivity and functional connectivity of the amygdala. Biol Psychiatry 655: 943–950.

Gianaros PJ, Onyewuenyi IC, Sheu LK et al. (2012). Brain systems for baroreflex suppression during stress in humans. Hum Brain Mapp 33: 1700–1716.

Gray MA, Harrison NA, Wiens S et al. (2007a). Modulation of emotional appraisal by false physiological feedback during fMRI. PLoS One 22: E546.

Gray MA, Taggart P, Sutton PM et al. (2007b). A cortical potential reflecting cardiac function. Proc Natl Acad Sci U S A 1044: 6818–6823.

Gray MA, Minati L, Harrison NA et al. (2009). Physiological recordings: basic concepts and implementation during functional magnetic resonance imaging. Neuroimage 477: 1105–1115.

Gray M, Nagai Y, Critchley HD (Eds.), (2010a). Brain Imaging of Stress and Cardiovascular Responses. Springer-Verlag, London Limited.

Gray MA, Minati L, Paoletti G et al. (2010b). Baroreceptor activation attenuates attentional effects on pain-evoked potentials. Pain 1511: 853–861.

Gray MA, Beacher FD, Minati L et al. (2012). Emotional appraisal is influenced by cardiac afferent information. Emotion 12: 180–191.

Griffin MG, Resick PA, Mechanic MB (1997). Objective assessment of peritraumatic dissociation: psychophysiological indicators. Am J Psychiatry 1544: 1081–1088.

Groen Y, Wijers AA, Mulder LJ et al. (2007). Physiological correlates of learning by performance feedback in children: a study of EEG event-related potentials and evoked heart rate. Biol Psychol 766: 174–187.

Hajcak G, Mcdonald N, Simons RF (2003). To err is autonomic: error-related brain potentials and activity and post-error compensatory behavior. Psychophysiology 400: 895–903.

Harper RM, Bandler R, Spriggs D et al. (2000). Lateralized and widespread brain activation during transient blood pressure elevation revealed by magnetic resonance imaging. J Comp Neurol 4177: 195–204.

Harrison NA, Singer T, Rotshtein P et al. (2006). Pupillary contagion: central mechanisms engaged in sadness processing. Soc Cogn Affect Neurosci 11: 5–17.

Harrison NA, Wilson CE, Critchley HD (2007). Processing of observed pupil size modulates perception of sadness and predicts empathy. Emotion 77: 724–729.

Harrison NA, Brydon L, Walker C et al. (2009). Neural origins of human sickness in interoceptive responses to inflammation. Biol Psychiatry 666: 415–422.

Harrison NA, Gray MA, Gianaros PJ et al. (2010). The embodiment of emotional feelings in the brain. J Neurosci 300: 12878–12884.

Harrison NA, Kreibig S, Critchley HD (2011). Physiology of Emotion Responses. Handbook of Human Affective Neuroscience, Cambridge University Press, Cambridge.

Heaver B, Hutton SB (2011). Keeping an eye on the truth? Pupil size changes associated with recognition memory. Memory 199: 398–405.

Heims HC, Critchley HD, Martin NH et al. (2006). Cognitive functioning in orthostatic hypotension due to pure autonomic failure. Clin Auton Res 166: 113–120.

Hilton SM (1975). Ways of viewing the central nervous control of the circulation – old and new. Brain Res 877: 213–219.

Hobday DI, Aziz Q, Thacker N et al. (2001). A study of the cortical processing of ano-rectal sensation using functional MRI. Brain 1244: 361–368.

Kapur S (2003). Psychosis as a state of aberrant salience: a framework linking biology, phenomenology, and pharmacology in schizophrenia. Am J Psychiatry 160: 13–23.

Katkin ES, Wiens S, Ohman A (2001). Nonconscious fear conditioning visceral perception and the development of gut feelings. Psychol Sci 122: 366–370.

King AB, Menon RS, Hachinski V et al. (1999). Human forebrain activation by visceral stimuli. J Comp Neurol 4133: 572–582.

Kloner RA, Poole WK, Perritt RL (1999). When throughout the year is coronary death most likely to occur? A 12-year population-based analysis of more than 220000 cases. Circulation 1000: 1630–1634.

Koponen H, Alaraisanen A, Saari K et al. (2008). Schizophrenia and sudden cardiac death: a review. Nord J Psychiatry 622: 342–345.

Krantz DS, Manuck SB (1984). Acute psychophysiologic reactivity and risk of cardiovascular disease: a review and methodologic critique. Psychol Bull 966: 435–464.

Kreibig SD (2010). Autonomic nervous system activity in emotion: a review. Biol Psychol 844: 394–421.

Kring AM, Moran EK (2008). Emotional response deficits in schizophrenia: insights from affective science. Schizophr Bull 344: 819–834.

Kroes MC, Strange BA, Dolan RJ (2010). Beta-adrenergic blockade during memory retrieval in humans evokes a sustained reduction of declarative emotional memory enhancement. J Neurosci 300: 3959–3963.

Kuhn S, Gallinat J (2011). A quantitative meta-analysis on cue-induced male sexual arousal. J Sex Med 88: 2269–2275.

Labar KS, Ledoux JE, Spencer DD et al. (1995). Impaired fear conditioning following unilateral temporal lobectomy in humans. J Neurosci 155: 6846–6855.

Labar KS, Gatenby JC, Gore JC et al. (1998). Human amygdala activation during conditioned fear acquisition and extinction: a mixed-trial fMRI study. Neuron 200: 937–945.

Lacey BC, Lacey JI (1978). Two-way communication between the heart and the brain. Significance of time within the cardiac cycle. Am Psychol 333: 99–113.

Lacey BC, Lacey JI (1980). Presidential Address 1979. Cognitive modulation of time-dependent primary bradycardia. Psychophysiology 177: 209–221.

Lader M (1975). The Psychophysiology of Mental Illness [SL], Routledge Kegan Paul, London.

Lane RD, Schwartz GE (1987). Induction of lateralized sympathetic input to the heart by the CNS during emotional arousal: a possible neurophysiologic trigger of sudden cardiac death. Psychosom Med 499: 274–284.

Lang PJ, Bradley MM, Cuthbert BN (1990). Emotion attention and the startle reflex. Psychol Rev 977: 377–395.

Lange CG, James W (1967). The Emotions. (Ed. Knight Dunlap), Hafner Publishing, New York.

Latalova K, Prasko J, Diveky T et al. (2010). Autonomic nervous system in euthymic patients with bipolar affective disorder. Neuro Endocrinol Lett 311: 829–836.

Leor J, Poole WK, Kloner RA (1996). Sudden cardiac death triggered by an earthquake. N Engl J Med 3344: 413–419.

Levenson RW (1992). Autonomic nervous system differences among emotions. Psychol Sci 33: 23.

Lorenz MW, Markus HS, Bots ML et al. (2007). Prediction of clinical cardiovascular events with carotid intima-media thickness: a systematic review and meta-analysis. Circulation 1155: 459–467.

Mathias CJ (2003). Autonomic diseases: clinical features and laboratory evaluation. J Neurol Neurosurg Psychiatry 7 (Suppl 3): iii31–iii41.

Matthews SC, Paulus MP, Simmons AN et al. (2004). Functional subdivisions within anterior cingulate cortex and their relationship to autonomic nervous system function. Neuroimage 222: 1151–1156.

Mauss IB, Robinson MD (2009). Measures of emotion: a review. Cognit Emot 233: 209–237.

Mayer-Gross W (1935). On depersonalization. British Journal of Medical Psychology 15: 103–122.

Medford N, Critchley HD (2010). Conjoint activity of anterior insular and anterior cingulate cortex: awareness and response. Brain Struct Funct 2144: 535–549.

Menon V, Uddin LQ (2010). Saliency switching attention and control: a network model of insula function. Brain Struct Funct 2144: 655–667.

Mittleman MA, Maclure M, Sherwood JB et al. (1995). Triggering of acute myocardial infarction onset by episodes of anger. Determinants of Myocardial Infarction Onset Study Investigators. Circulation 922: 1720–1725.

Mittleman MA, Maclure M, Nachnani M et al. (1997). Educational attainment anger and the risk of triggering myocardial infarction onset. The Determinants of Myocardial Infarction Onset Study Investigators. Arch Intern Med 1577: 769–775.

Moller J, Theorell T, De Faire U et al. (2005). Work related stressful life events and the risk of myocardial infarction case-control and case-crossover analyses within the Stockholm Heart Epidemiology Programme (SHEEP). J Epidemiol Community Health 599: 23–30.

Morgan HL, Turner DC, Corlett PR et al. (2011). Exploring the impact of ketamine on the experience of illusory body ownership. Biol Psychiatry 699: 35–41.

Morris AL, Cleary AM, Still ML (2008). The role of autonomic arousal in feelings of familiarity. Conscious Cogn 177: 1378–1385.

Nagai Y, Critchley HD, Featherstone E et al. (2004a). Brain activity relating to the contingent negative variation: an fMRI investigation. Neuroimage 211: 1232–1241.

Nagai Y, Critchley HD, Featherstone E et al. (2004b). Activity in ventromedial prefrontal cortex covaries with sympathetic skin conductance level: a physiological account of a "default mode" of brain function. Neuroimage 222: 243–251.

Naqvi N, Shiv B, Bechara A (2006). The role of emotion in decision making. Curr Dir Psychol Sci 155: 260.

Nicotra A, Critchley HD, Mathias CJ et al. (2006). Emotional and autonomic consequences of spinal cord injury explored using functional brain imaging. Brain 1299: 718–728.

Nieuwenhuis S, Ridderinkhof KR, Blom J et al. (2001). Error-related brain potentials are differentially related to awareness of response errors: evidence from an antisaccade task. Psychophysiology 388: 752–760.

O'Connor MF, Gundel H, Mcrae K et al. (2007). Baseline vagal tone predicts bold response during elicitation of grief. Neuropsychopharmacology 322: 2184–2189.

Parvizi J, Anderson SW, Martin CO et al. (2001). Pathological laughter and crying: a link to the cerebellum. Brain 1244: 1708–1719.

Patterson JC 2nd, Ungerleider LG, Bandettini PA (2002). Task-independent functional brain activity correlation with skin conductance changes: an fMRI study. Neuroimage 177: 1797–1806.

Paulus MP, Stein MB (2006). An insular view of anxiety. Biol Psychiatry 600: 383–387.

Paus T, Koski L, Caramanos Z et al. (1998). Regional differences in the effects of task difficulty and motor output on blood flow response in the human anterior cingulate cortex: a review of 107 PET activation studies. Neuroreport 99: R37–R47.

Peled A, Ritsner M, Hirschmann S et al. (2000). Touch feel illusion in schizophrenic patients. Biol Psychiatry 488: 1105–1108.

Phillips DP, Jarvinen JR, Abramson IS et al. (2004). Cardiac mortality is higher around Christmas and New Year's than at any other time: the holidays as a risk factor for death. Circulation 1100: 3781–3788.

Pochon JB, Riis J, Sanfey AG et al. (2008). Functional imaging of decision conflict. J Neurosci 288: 3468–3473.

Pollatos O, Schandry S (2008). Emotional processing and emotional memory are modulated by interoceptive awareness. Cognit Emot 22: 272–287.

Raichle ME, Macleod AM, Snyder AZ et al. (2001). A default mode of brain function. Proc Natl Acad Sci U S A 988: 676–682.

Rainville P, Bechara A, Naqvi N et al. (2006). Basic emotions are associated with distinct patterns of cardiorespiratory activity. Int J Psychophysiol 611: 5–18.

Reedman LA, Allegra JR, Cochrane DG (2008). Increases in heart failure visits after Christmas and New Year's day. Congest Heart Fail 144: 307–309.

Roberts AC (2006). Primate orbitofrontal cortex and adaptive behaviour. Trends Cogn Sci 100: 83–90.

Schachter S, Singer JE (1962). Cognitive social and physiological determinants of emotional state. Psychol Rev 699: 379–399.

Schandry R (1981). Heart beat perception and emotional experience. Psychophysiology 188: 483–488.

Scherer KR (2000). Psychological models of emotion. In: JC Borod (Ed.), The Neuropsychology of Emotion. Oxford University Press, Oxford.

Schoofs D, Preuss D, Wolf OT (2008). Psychosocial stress induces working memory impairments in an n-back paradigm. Psychoneuroendocrinology 333: 643–653.

Shiv B, Loewenstein G, Bechara A (2005a). The dark side of emotion in decision-making: when individuals with decreased emotional reactions make more advantageous decisions. Brain Res Cogn Brain Res 233: 85–92.

Shiv B, Loewenstein G, Bechara A et al. (2005b). Investment behavior and the negative side of emotion. Psychol Sci 166: 435–439.

Singer T, Critchley HD, Preuschoff K (2009). A common role of insula in feelings empathy and uncertainty. Trends Cogn Sci 133: 334–340.

Smith CA, Ellsworth PC (1985). Patterns of cognitive appraisal in emotion. J Pers Soc Psychol 488: 813–838.

Stemmler G (Ed.), (2004). Physiological Processes During Emotion. Lawrence Erlbaum, Mahwah, NJ.

Stemmler G, Aue T, Wacker J (2007). Anger and fear: separable effects of emotion and motivational direction on somatovisceral responses. Int J Psychophysiol 666: 141–153.

Steptoe A, Strike PC, Perkins-Porras L et al. (2006). Acute depressed mood as a trigger of acute coronary syndromes. Biol Psychiatry 600: 837–842.

Stern ER, Gonzalez R, Welsh RC et al. (2009). Neural correlates of uncertainty during decision making in obsessive compulsive disorder. Biol Psychiatry 65: 125S.

Strike PC, Perkins-Porras L, Whitehead DL et al. (2006). Triggering of acute coronary syndromes by physical exertion and anger: clinical and sociodemographic characteristics. Heart 922: 1035–1040.

Suzuki S, Sakamoto S, Miki T et al. (1995). Hanshin-Awaji earthquake and acute myocardial infarction. Lancet 3455: 981.

Taggart P, Sutton P, Redfern C et al. (2005). The effect of mental stress on the non-dipolar components of the T wave: modulation by hypnosis. Psychosom Med 677: 376–383.

Thayer JF, Lane RD (2007). The role of vagal function in the risk for cardiovascular disease and mortality. Biol Psychol 744: 224–242.

Treiber FA, Kamarck T, Schneiderman N et al. (2003). Cardiovascular reactivity and development of preclinical and clinical disease states. Psychosom Med 655: 46–62.

Tsakiris M, Tajadura-Jimenez A, Costantini M (2011). Just a heartbeat away from one's body: interoceptive sensitivity predicts malleability of body-representations. Proc Biol Sci 2788: 2470–2476.

Van Stegeren AH, Everaerd W, Cahill L et al. (1998). Memory for emotional events: differential effects of centrally versus peripherally acting beta-blocking agents. Psychopharmacology (Berl) 1388: 305–310.

Vlastelica M (2008). Emotional stress as a trigger in sudden cardiac death. Psychiatr Danub 200: 411–414.

Wager TD, Van Ast VA, Hughes BL et al. (2009a). Brain mediators of cardiovascular responses to social threat. Part II. Prefrontal-subcortical pathways and relationship with anxiety. Neuroimage 477: 836–851.

Wager TD, Waugh CE, Lindquist M et al. (2009b). Brain mediators of cardiovascular responses to social threat. Part I. Reciprocal dorsal and ventral sub-regions of the medial prefrontal cortex and heart-rate reactivity. Neuroimage 477: 821–835.

Wallin BG (2007). Interindividual differences in muscle sympathetic nerve activity: a key to new insight into cardiovascular regulation? Acta Physiol (Oxf) 1900: 265–275.

Werner NS, Peres I, Duschek S et al. (2010). Implicit memory for emotional words is modulated by cardiac perception. Biol Psychol 855: 370–376.

Whitehead WE, Drescher VM (1980). Perception of gastric contractions and self-control of gastric motility. Psychophysiology 177: 552–558.

Whitehead WE, Drescher VM, Heiman P et al. (1977). Relation of heart-rate control to heartbeat perception. Biofeedback Self Regul 22: 371–392.

Wiens S, Mezzacappa ES, Katkin ES (2000). Heartbeat detection and the experience of emotions. Cognit Emot 144: 417–427.

Williams LM, Brammer MJ, Skerrett D et al. (2000). The neural correlates of orienting: an integration of fMRI and skin conductance orienting. Neuroreport 111: 3011–3015.

Willich SN, Maclure M, Mittleman M et al. (1993). Sudden cardiac death. Support for a role of triggering in causation. Circulation 877: 1442–1450.

Handbook of Clinical Neurology, Vol. 117 (3rd series)
Autonomic Nervous System
R.M. Buijs and D.F. Swaab, Editors

Chapter 7

Interoception and autonomic nervous system reflexes thermoregulation

ROBERT D. FEALEY*
Department of Neurology, Mayo Clinic, Rochester, MN, USA

ANATOMICAL AND PHYSIOLOGICAL ASPECTS OF HUMAN THERMOREGULATION

Afferent and efferent pathways

Critical human thermoregulatory effector mechanisms include both behavioral and autonomic aspects. The latter include shivering and nonshivering thermogenesis for heat production and sweating and cutaneous vasomotor responses for heat dissipation (Aronoff and Romanovsky, 2007; Flouris, 2011). Important conceptual changes concerning human thermoregulation have occurred in the last decade.

Previously proposed mechanisms of a central integrative center in the hypothalamus which attempts to maintain body temperature at a set-point have been replaced by the concept of thermoregulation by multiple, independent thermoeffector loops collectively producing a balance-point core temperature that is closely maintained. Cutaneous, core, and ambient temperature level will influence the activity and contribution of the independent thermoeffectors to maintenance of body temperature (Aronoff and Romanovsky, 2007). Behavioral adjustments to skin temperature changes have been recently emphasized as the neuroanatomical substrates of behavioral thermoregulation have been investigated (Flouris, 2011) (Fig. 7.1).

The importance of temperature-sensitive neurons in the preoptic anterior hypothalamus remains. Warm-sensitive neurons exhibit spontaneous membrane depolarization, while incoming signals influence the rate of this depolarization between action potentials. The neurons are viewed as pacemakers in the new model. Warm-sensitive neurons comprise 20–40% of the neuronal population and cold about 5–10%. Intracellular recordings reveal the synaptic and intrinsic mechanisms responsible for neuronal thermosensitivity. Many temperature-sensitive and temperature-insensitive neurons display a depolarizing prepotential that precedes action potentials. In warm-sensitive neurons, the prepotential's depolarization is elevated by warming, and this increases the firing rate. Intracellular cAMP can increase neuronal thermosensitivity by enhancing the thermal response of the prepotential, via thermosensitive ionic conductances. Warm-sensitive neurons also receive inhibitory synaptic input (IPSPs) from temperature-insensitive neurons, enhancing the thermosensitivity (Boulant, 2006). Animal studies suggest warm-sensitive neurons in the preoptic-anterior hypothalamus (POAH) exert a tonic inhibitory influence on cold-defense-related thermogenic and heat conservation mechanisms. Human observations also suggest POAH thermosensitive neurons inhibit the posterior hypothalamus (where shivering-cold defense responses are organized). Patients undergoing the Wada test for pharmacoresistant epilepsy often shiver when there is no filling of the posterior circulation upon carotid injections. In these patients, shivering may result from a transient but selective functional lesion of the anterior hypothalamus produced by the effects of sodium amobarbital, which may result in disinhibition of the posterior hypothalamus and other brainstem thermoregulatory centers (Shah et al., 2010).

Independent thermoeffector loops

Thermoeffector pathways conveying temperature from large areas of the shell are sent to the brain via a spinoreticulohypothalamic pathway. Interaction with POAH neurons is via the dendritic tree oriented mediolateral catching signals from medial ventricular structures and laterally via the median forebrain bundle.

*Correspondence to: Robert D. Fealey, M.D, Department of Neurology, Mayo Clinic, 200 First Street SW, Rochester, MN 55905, USA. E-mail: rfealey@mayo.edu

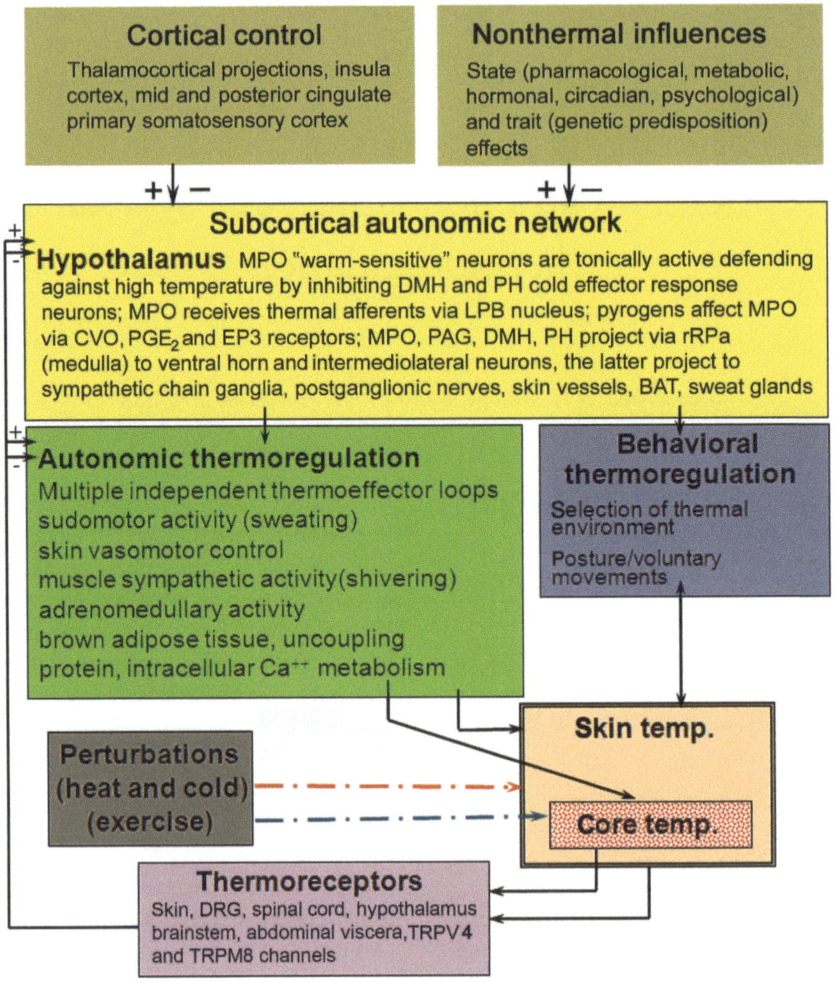

Fig. 7.1. Anatomical and physiological schema of human thermoregulatory sweating. MPO, medial preoptic; DMH, dorsomedial hypothalamus; PH, posterior hypothalamus; LPB, lateral parabrachial; CVO, circumventricular organs; PAG, periaqueductal gray; rRPa rostral raphe pallidus; TRPV, transient receptor potential vanilloid; TRPM, transient receptor potential melastatin. (Modified from Fealey, 2000.)

Local spinal–brainstem thermoeffector loops may exist (Craig, 2002). Skin thermoeffector loops operate independently, triggered by a unique combination of shell and core temperature. For example, the control of the sweat rate by skin temperature at stable core temperatures near the threshold temperature of sweating can be quite exquisite (Fig. 7.2).

Thermal receptors and afferent pathways have been targeted in more recent research. Skin has cold sensors just below epidermis with Aδ myelinated fibers; warm sensors are deeper, with an unmyelinated, C-fiber connection. They show a powerful dynamic (phasic) component but quickly adapt to a stable temperature.

Thermo-channels

TRPM8 (transient receptor potential melastatin-8) channels exist in the primary afferent nerve terminal (Knowlton and McKemy, 2011). Blockade of TRPM8 channels can affect the lateral parabrachial nucleus (LPB), thus indicating a site of action within the cutaneous afferent pathway that includes the dorsal root ganglion and laminar 1 of the dorsal horn neurons and the latter's central projections. TRPM8 is considered a likely universal cold receptor. TRPM8 channels also exhibit diversity in sensory signaling, providing perception of both pleasantly cool and painfully cold stimuli.

It has been shown in humans that skin temperature, not core temperature, plays a signaling role in the decision to behaviorally thermoregulate. However, this behavior does not occur in the complete absence of autonomic thermoregulatory responses (Brown et al., 2013).

Another transient receptor potential cation channel, subfamily V (TRPV), is expressed in the epidermis and TRPV4 is considered to be a sensor of extrinsic stimuli such as temperature and other physical or chemical

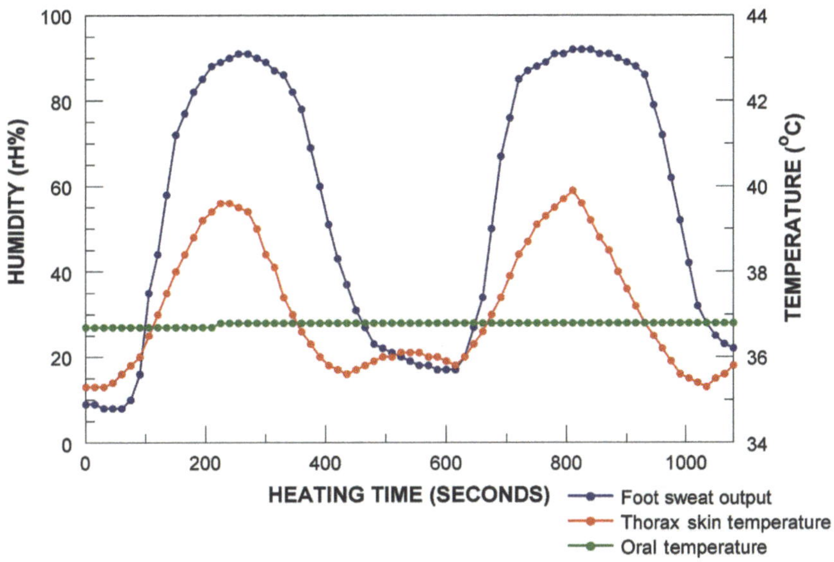

Fig. 7.2. Skin temperature reflex regulation of sweating. Note the exquisitely close regulation of sweat rate (on foot) during skin heating (of trunk) while core temperature (oral) remains constant. Subject was at the threshold temperature of generalized thermoregulatory sweating. (Modified from Fealey, 2000.)

factors (Denda et al., 2007; Akazawa et al., 2012; Almeida et al., 2012; Premkumar and Abooj, 2013).

Central thermoregulation observations in humans

Recent work in humans using controlled skin temperature suits and functional MRI with a region of interest containing the ventral midline of the rostral medulla concluded that human rostral medullary raphe neurons are selectively activated in response to a thermoregulatory challenge and point to the location of thermoregulatory neurons homologous to those of the raphe pallidus nucleus (rRPa) in rodents and cats (Horiuchi et al., 2004; Asahina et al., 2007).

Cortical elements involved in both autonomic and behavioral thermoregulation have been investigated recently. Studies utilizing PET and functional MRI brain activity imaging techniques show focal activation or suppression of activity in response to manipulation of core and skin temperature that produce autonomic and behavioral thermoregulatory responses. Brain areas involved include the insular cortex, the anterior cingulate, the amygdala, somatosensory cortex, ventromedial prefrontal and frontal opercular cortices. Activity in the anterior hypothalamus was shown to correlate with heat dissipation (Egan et al., 2005; Flouris, 2011).

Mental activity can increase sweat activity. The latter tends to affect the palms and plantar areas. Thermal sweating due to exercise and heat exposure is generalized and generally not affecting the palms and soles (Sato et al., 1989) except in patients with primary focal hyperhidrosis (Atkinson et al., 2011).

Genetic predisposition (trait) and clinical (pharmacological, anesthetic, circadian, metabolic, hormonal) state of the individual may have profound effects on thermoregulation (Gurrera, 1999).

Hypothalamic temperature-sensitive neurons are the target of multiple biochemical influences. Substances that elevate balance-point and core temperature when administered into this region include acetylcholine, histamine, prostaglandins, serotonin, TRH, and endorphins; substances that decrease temperature include dopamine, norepinephrine, neurotensin, and AVP. μ-Opioid agonists inhibit activity and decrease temperature sensitivity of warm-sensitive neurons, raising the balance-point temperature at which heat-loss mechanisms (sweating, vasodilation) commence. Progesterone decreases firing rate of warm-sensitive neurons and increases that of cold-sensitive neurons, producing an upward shift of the thermoregulatory balance-point and elevating core temperature, whereas estrogen has opposite effects. This explains the variations of core temperature that occur during the menstrual cycle (Benarroch, 2006; Charkoudian, 2010).

Details of central thermoregulatory pathways in primates

Morrison has summarized more recent studies of primate autonomic pathways involved in thermogenesis, recognizing the importance of the afferent projections of the lateral parabrachial nucleus to the POAH and the efferent projections of the dorsomedial hypothalamus to the raphe pallidus nucleus, and thence to the spinal cord skeletal muscle and sympathetic preganglionic motor neurons. Details of the specific nuclei and transmitters involved are beautifully shown in Figure 1 of his publication (Morrison, 2011). As most studies involve nonsweating animals, the central neural pathways discussed focus on those for cold defense. The pathways for mammalian thermolysis (sweating and vasodilation) are less well confirmed but are surmised to utilize the same afferent and efferent connections.

Efferent pathways for cutaneous vasoconstriction and vasodilation, for shivering and brown adipose tissue (BAT) thermogenesis, and probably sweating travel mainly ipsilaterally in the brainstem although there is evidence for crossing of fibers at lower brainstem and spinal segmental levels. The ventral medial medulla facilitory sudomotor and vasomotor pathways begin with neurons in the rostral raphe pallidus (rRpa), tend to be diffusely distributed in the medial portion of the lateral funiculus, and terminate on ventral horn α motor neurons (shivering), the intermediolateral (IML) and intermediomedial (IMM) cell columns of spinal segments T1 to L2 or L3 (sweating and vasoconstriction and BAT), with several neurotransmitters identified including excitatory (glutamate, serotonin and orexin A and B) and inhibitory (dopamine, GABA, and adenosine) (Antunes et al., 2001; Llewellyn-Smith et al., 2002; Brooke et al., 2004). Preganglionic sympathetic output (via the ventral root and white rami) innervate the paravertebral sympathetic chain ganglia via nicotinic acetylcholine synapses. Postganglionic neurons of the sympathetic ganglia send axons via the gray rami and sympathetic trunk to innervate sweat glands (via muscarinic (M3) cholinergic synapses) and blood vessels (via norepinephrine and neuropeptide Y synapses) (Schulze et al., 1997). The adult human sweat gland innervation, however, is not only cholinergic but coexpresses all of the proteins required for full noradrenergic function as well, including tyrosine hydroxylase, aromatic amino acid decarboxylase, dopamine β-hydroxylase, and the vesicular monoamine transporter VMAT2. Cholinergic transmission predominates in adults, however. Conversion from noradrenergic to cholinergic transmission occurs in the first decade. Sympathetic neurons innervating specifically the cutaneous arteriovenous anastomoses in humans also possess a full cholinergic/noradrenergic cophenotype. Cholinergic/noradrenergic coexpression is absent from other portions of the human sympathetic nervous system but is extended in the parasympathetic nervous system to intrinsic neurons innervating the heart (Weihe et al., 2005).

Mechanisms of heat loss and heat conservation

Charkoudian (2010) has summarized the importance of human skin blood flow responses to body heating and cooling. Increases in skin blood flow provide the necessary augmentation of convective heat loss during environmental heat stress and/or exercise. Cutaneous vasoconstriction is key to preventing excessive heat dissipation during cold exposure. Human sympathetic innervation of the cutaneous circulation includes a sympathetic-noradrenergic vasoconstrictor system, and a non-noradrenergic active vasodilator system. In normothermic environments noradrenergic vasoconstrictor nerves are tonically active and increase their activity during cold exposure, releasing both norepinephrine and cotransmitter neuropeptide Y. The active vasodilator system in human skin is activated during increases in body temperature, brought about by heat exposure or exercise. Active cutaneous vasodilation occurs via cholinergic nerve cotransmission and has been shown to be mediated by nitric oxide, vasoactive intestinal peptide, prostaglandins, and substance P (and/or neurokinin-1 receptors) (Charkoudian, 2010).

Shivering involves the contraction/relaxation cycle of skeletal muscle. This involves calcium release from the sarcoplasmic reticulum, binding to contractile proteins (to troponin), inhibition of further calcium release and the activation of calcium pumps moving sarcoplasmic calcium back to the sarcoplasmic reticulum against a steep gradient, relaxation of muscle and the oxidation reactions using creatine phosphokinase, steps all producing energy release and heat. Sarcoplasm free calcium regulates contraction–relaxation and agents that sustain elevation of this calcium can induce persistent contraction in the absence of muscle action potentials.

In many primates and the human neonate, nonshivering thermogenesis (NST) occurs in brown adipose tissue (BAT), an autonomic response initiated in the POAH and depending on β_1 and β_3 adrenergic innervations (Mattsson et al., 2011; Ueta et al., 2012). α_1 Adrenergic receptors and possibly thyroid hormone are also involved in this form of thermogenesis. A mitochondrial protein called uncoupling protein (UCP1) is important in creating a pathway that allows dissipation of the proton electrochemical gradient permitting respiratory chain oxidation to liberate energy in the form of heat (Klingenberg, 1999). This pathway has been implicated in the regulation of body temperature, body composition and glucose metabolism (Fleury et al., 1997). In the adult human brown adipose tissue is scarce, and evidence suggests there are other factors involved in thermogenesis (Hesselink et al., 2003). There are other UPC proteins now identified (2–5); however, their role in thermoregulation is not well defined (Zaninovich, 2005).

Other neurotransmitters/hormones identified as having a role in thermoregulation include: serotonin (in the POAH and rRPa), thyrotropin-releasing hormone (TRH), arginine vasopressin, adrenocorticotropin hormone (ACTH), and bombesin. Vasopressin, ACTH, and TRH are endogenous substances that are thermolytic in humans and may function to protect the brain from hyperthermia during fever. Bombesin (Jensen et al., 2008) and neurotensin (Katz et al., 2004) appear to suppress neurons in the POAH, producing a state of relative poikilothermia and regulated hypothermia.

The anatomical vascular arrangements of the head and trunk suggest important roles for pial arterial ramifications, the dura mater vessels, the spinal and skin venous plexuses and their interface with the spinal fluid, in convective cooling of the brain and spinal cord (Zenker and Kubik, 1996) Such countercurrent cooling mechanisms may be significant when ambient temperature does not exceed skin temperature or when therapeutic hypothermia is utilized (Neimark et al., 2007; Harris et al., 2008).

Nonthermal factors affecting thermoregulation

A comprehensive review of the literature on thermal and nonthermal factors influencing human thermoregulation has been published (Shibasaki and Crandall, 2010). In this review the authors conclude that while the primary controller of sweating is the integration between internal and skin temperatures, a number of nonthermal factors, including exercise, baroreceptor loading state, and body fluid status, modulate the sweating response. The concept of mean body temperature (MBT), which represents the sum of a fraction of internal and skin temperatures (e.g., $0.9 \times$ internal temperature $+ 0.1 \times$ mean skin temperature), is summarized. MBT is now frequently used when expressing sweating responses during exercise and during exposure to elevated ambient air temperatures.

Studies of the nonthermal factors of exercise in control of sweating have highlighted the contribution of muscle metaboreflex, mechanoreceptor and central command inputs (Shibasaki et al., 2003). The same group has studied cutaneous vasoconstriction (CVC) during exercise in heat stressed individuals, finding that neural and non-neural mechanisms contribute to the reduction in forearm CVC during isometric hand grip exercise in heat stressed humans (Shibasaki et al., 2009). However, studies relating to exercise have yielded conflicting results on the effect of nonthermal factors. For example, in those studies in which sweating is already engaged or is very close to the threshold for being engaged, stimulating the aforementioned nonthermal factors associated with exercise more consistently increases sweat rate. This is in contrast to those studies in which exercise began with the subject in a normothermic state and thus nonsweating condition, where sweating is reported to be less affected by nonthermal factors associated with exercise. The differences may be due to much larger contributions of cutaneous and other afferent stimuli when near the threshold temperature for generalized sweating. Indeed from personal observations, integrated mean skin temperature has a strong impact on sweat rate at core temperatures near the threshold of sweating but at higher temperatures it has much less influence on sweat rate (autonomic thermoregulation). However, skin temperature maintains a strong influence on behavioral thermoregulation (Ogawa and Asayama, 1982; Brown et al., 2013).

Older studies by Cheng have shown an 18–20% contribution of mean skin temperature in determining the shivering and peripheral vasoconstriction response thresholds in normal men and women (Cheng et al., 1995). However, there is good evidence for independent control of skin blood flow (and hence mean skin temperature) and thermoregulatory sweating when competing stimuli are encountered, such as when exercising in a warm environment in the upright position; in such a situation posture appears to have no effect on core temperature–sweating relationship but being upright elevates the threshold core temperature for vasodilation (Johnson and Park, 1981). Many studies done on sweat rate during differing baroreflex states have not been definitive and although findings remain controversial, acute unloading of baroreceptors is unlikely to modulate sweat rate.

Body fatness significantly affects core temperature only in the cool exercise condition, where the insulative effect of fat is pronounced. In warmer environments, high skin blood flows offset the heat retaining properties of peripheral adipose tissue (Havenith, 2001). In hot working conditions, high sweat rates with excessive loss of body fluids may result in dehydration and electrolyte imbalance. It is well established that dehydration and/or electrolyte loss, if severe, can pose a serious risk for hyperthermia complications (Cheung and McLellan, 1998). About 500 cc/hour oral fluid intake is needed during exercise in most environments. Electrolyte supplements are required for exercise periods of > 3 hours according to some studies.

Acclimatization to exertion in hot environments results in a significant decrease in sweat sodium and increased sweat rate, lower relative heart rate, and increased plasma volume. This advantageous physiological adaptation requires around 10 hours of heat exposure to initiate (Bates et al., 1996). Studies have noted that aerobic conditioning with high VO_2 max does not assure heat adaptation/acclimation, especially if exerting in a hot environment (Sawka and Montain, 2000) where evaporative cooling capacity is limited.

With profuse sweating in a hot and humid environment, constant wetting of the epidermis causes swelling of the horny layer and keratin ring of the sweat duct, resulting in narrowing and occlusion of the duct and decreased sweat drippage, a phenomenon known as hidromeiosis (Candas et al., 1983). Sweat gland secretory capacity can be augmented by repeated heat/exercise activity and conditioning and gland size increase occurs (Candas et al., 1980; Ogawa et al., 1982).

More recently, the effect of nonthermal conditions on sweat rate in humans has been investigated. Takamata and colleagues (Takamata et al., 2001) assessed sweat rate in heat-stressed subjects who received an infusion of 0.9% or 3% saline. They found that the threshold for sweating in the hyperosmotic condition (i.e., 3% saline infusion) was greatly shifted to a higher internal temperature relative to the iso-osmotic condition. Past studies (Fortney et al., 1981) found that iso-osmotic hypovolemia reduced the slope of the relationship between the change in sweating relative to the change in internal temperature, without altering the internal temperature threshold for the onset of sweating.

There are differences in thermoregulatory responses related to sex and age. For example, during exposure to extreme dry heat, female subjects showed a significantly higher rise in mean skin temperature than the male subjects. Similarly, a significantly higher rise in skin temperature was observed in elderly male subjects as compared to the youngest male group ($p < 0.05$). The differences mainly resulted from differences in mean skin blood flow (Shoenfeld et al., 1978). In another study, no relation was found between core temperature increase and the dynamics of sweating in women. The dynamics of sweating played a decisive role in limiting core temperature rise in men under dry heat exposure. The later onset of sweating in women did not influence the rectal temperature increase significantly; rather it is probably limited by a complex interaction of sweating, skin blood flow increase, and metabolic rate decrease (Grucza et al., 1985). Women may have a higher threshold temperature to activation and lower sweat gland secretory capacity than men. Threshold temperature is especially higher in the luteal phase of the menstrual cycle (Carpenter and Nunneley, 1988; Kolka and Stephenson, 1997). More recent studies, however, suggest females have a lower thermosensitivity of the whole-body sudomotor response compared to males during passive heating. Evidence for a peripheral modulation of sudomotor activity but not vasomotor activity in females was found (Gagnon et al., 2013).

Thermoregulation and fever

The activity of these independent thermoeffector mechanisms and circulating thermoactive substances determine a balance-point core temperature. This balance-point can shift, for example during fever (thought to be a regulated, elevated balance-point), when exogenous bacterial pyrogens stimulate phagocytic leukocytes to produce endogenous pyrogen (interleukins 1 beta and 6, tumor necrosis factor-alpha).

Consensus has been reached that within the BBB endothelial and perivascular cells, signaling molecules (e.g., proinflammatory cytokines) activate the release of the mediator prostaglandin (PGE2), which causes the activity of warm-sensitive neurons to decrease and that of cold-sensitive neurons to increase, raising the balance-point temperature. Of the neuronal PGE2 receptors on warm-sensitive neurons, EP3 is likely the primary fever receptor. The efferent pathways of fever start from these EP3-bearing POAH neurons which project to midbrain reticulospinal neurons for shivering and premotor neurons in the medulla oblongata (raphe pallidus) for skin vasomotor control (Nagashima et al., 2000; Romanovsky et al., 2005; Morrison, 2011). Initially, an individual feels chilled, becomes vasoconstricted and dry, and may shiver as heat-producing and heat conservation changes are occurring so as to raise core temperature to the new balance-point.

CLINICAL TESTING AND SELECTED EXAMPLES OF ABNORMAL THERMOREGULATION

Thermoregulatory sweat test

The thermoregulatory sweat test (TST) (Fealey, 2009) consists of giving a controlled passive heat stimulus to produce a maximal thermoregulatory sweat response. Skin temperature is maintained at 39.0 °C and the supine subject is enclosed in a cabinet where environmental conditions are an air temperature between 45 and 50 °C and relative humidity of 35–40%. An indicator powder placed on the skin before heating provides high color contrast between sweating (purple) and nonsweating (yellow) skin areas. The test endpoint is whole-body sweating or, if not fully sweating, raising core temperature to 38.0 °C or a 1.0 °C increase above baseline (whichever yields the higher temperature).

When the TST is combined with a test of axon reflex sweating such as the quantitative sudomotor axon reflex test (QSART) (Low et al., 1983), it becomes possible to test the entire efferent sudomotor pathway and determine whether an abnormality of sweating is due to central (preganglionic) or peripheral (postganglionic) impairment.

In many disorders, both neurological and general medical and pharmacological substances can impair sweating (Fealey, 2007), but this chapter will discuss only a few examples that especially illustrate components of thermoregulation.

Examples of abnormal human thermoregulation

Clinical examples of the effect of fever, a μ-opioid agonist, and B-cell lymphoma (cytokine) inhibition of the action of warm-sensitive neurons are presented in Figure 7.3.

All patients exhibited little sweating to a thermoregulatory challenge in spite of a marked elevation of core temperature. The febrile patient in (A) did not sweat at the usual endpoint temperature of 38.0 °C. Testing weeks later, when fever had subsided, produced sweating at much lower core temperature, revealing the underlying distribution of lepromatous neuropathy. In (B) a dose of morphine just prior to sweat testing elevated the balance point temperature for sweat onset. When retested the following day without the μ-opioid medication, sweating developed at a much lower core temperature, revealing the patient's mononeuritis multiplex distribution of anhidrosis. This action of a μ-opioid agonist in a human has been confirmed in rats to be a suppression of the activity of warm-sensitive neurons in the POAH (Yakimova et al., 1996). The effect of fever and cytokines has been shown in animal

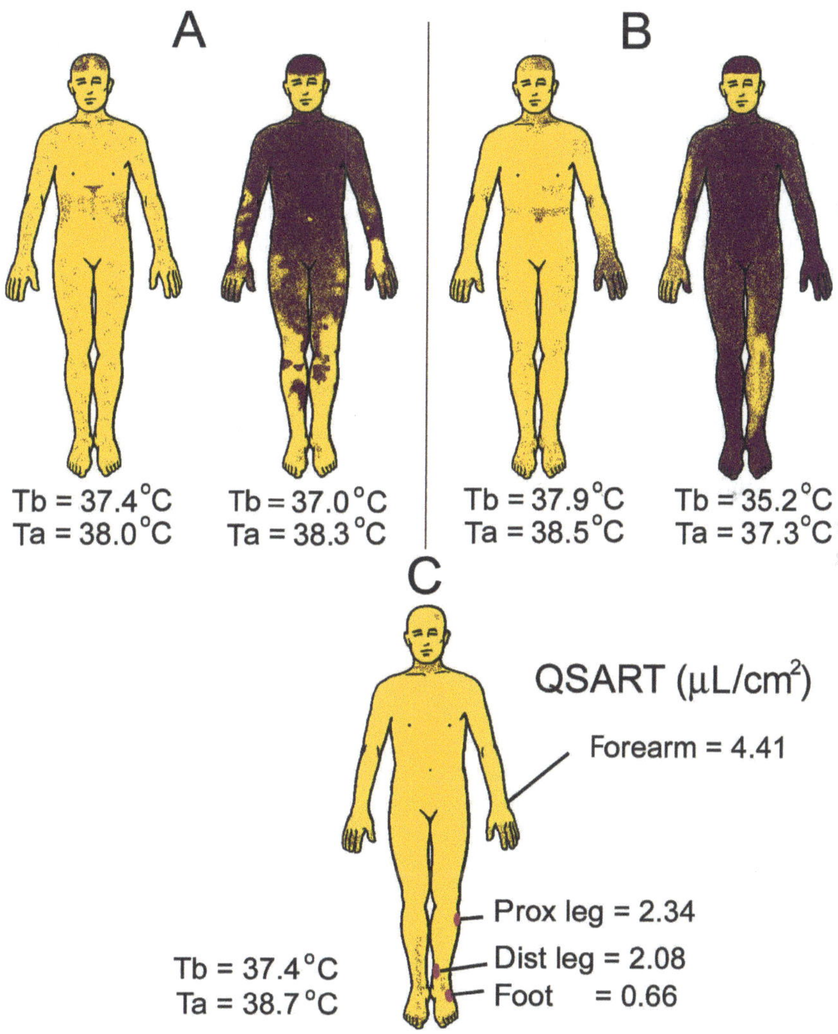

Fig. 7.3. Central thermoregulatory abnormalities as revealed by unexpected thermoregulatory sweat test results. Patient (**A**) (left), was mildly febrile when initially tested and didn't sweat at the usual endpoint temperature of 38.0 °C. Retesting (right) when afebrile revealed a sweat test pattern compatible with the patient's known lepromatous leprosy. Patient (**B**) had just ingested a mu-opioid agonist and did not sweat (left); when tested the next morning without the drug's effect, sweating did occur revealing the pattern of multiple plexopathy affecting the right arm and left leg. Both 'before' (Tb) and 'after' (Ta) TST core temperatures are shown below the TST figures. Patient (**C**) had cytokines associated with B-cell lymphoma and did not sweat even when heated to a high core temperature. Preserved QSART responses with absent TST sweating indicates a central lesion likely due to cytokine inhibition of warm sensitive neurons. Both 'before' (Tb) and 'after' (Ta) TST core temperatures are shown to the left of the TST figure (sweating in purple shade, anhidrosis in yellow shade). Reproduced with permission from Fealey, 2008.

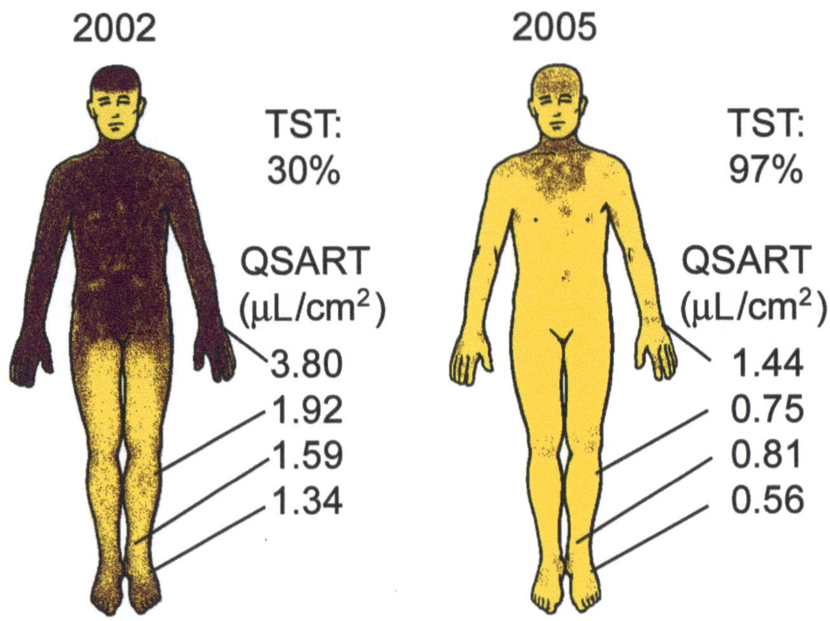

Fig. 7.4. Progressive central thermoregulatory failure due to multiple system atrophy (MSA). This patient with autopsy confirmed MSA and symptoms of parkinsonism and severe autonomic failure underwent the thermoregulatory sweat test (TST) and quantitative sudomotor autonomic reflex testing (QSART) in 2002 (left) and 2005 (right). The abnormal TST and normal QSART in the lower extremities suggested a central or preganglionic lesion. Progressive sweat loss by 2005 was mainly preganglionic although peripheral (QSART) sweating was starting to decline.

studies to involve the action of PGE2 on the EP3 receptor on hypothalamic warm-sensitive neurons (Sanchez-Alavez et al., 2006).

Disorders such as multiple system atrophy (MSA) cause loss of thermoregulatory neurons in the intermediolateral cell columns (Low, 1977) as well as the rostral raphe pallidus (Benarroch et al., 2005). These preganglionic lesions are likely responsible for the abnormal sweating patterns most commonly seen in MSA (Iodice et al., 2012). As shown in Figure 7.4, widespread loss of thermoregulatory sweating with preserved postganglionic axon reflex sweating is a typical clinical finding in early MSA. The subsequent reduction of postganglionic sweating may be due to sweat gland atrophy or loss of a trophic factor from the CNS or progression of MSA to include loss of postganglionic neurons.

CONCLUSION

The elucidation of thermo-TRP channels and their role in interoceptive autonomic and behavioral thermoregulation and the determination of central thermoregulatory pathways for cold-defense during the past decade has provided a far greater understanding of how human homeotherms can survive thermal challenges. I anticipate this knowledge will translate to new treatments of thermoregulatory disorders in the next decade.

References

Akazawa Y, Yuki T, Yoshida H et al. (2012). Activation of TRPV4 strengthens the tight-junction barrier in human epidermal keratinocytes. Skin Pharmacol Physiol 26: 15–21.

Almeida MC, Hew-Butler T, Soriano RN et al. (2012). Pharmacological blockade of the cold receptor TRPM8 attenuates autonomic and behavioral cold defenses and decreases deep body temperature. J Neurosci 32: 2086–2099.

Antunes VR, Brailoiu GC, Kwok EH et al. (2001). Orexins/hypocretins excite rat sympathetic preganglionic neurons in vivo and in vitro. Am J Physiol Regul Integr Comp Physiol 281: R1801–R1807.

Aronoff DM, Romanovsky AA (2007). Eicosanoids in non-febrile thermoregulation. Prog Brain Res 162: 15–25.

Asahina M, Sakakibara R, Liu Z et al. (2007). The raphe magnus/pallidus regulates sweat secretion and skin vasodilation of the cat forepaw pad: a preliminary electrical stimulation study. Neurosci Lett 415: 283–287.

Atkinson JL, Fode-Thomas NC, Fealey RD et al. (2011). Endoscopic transthoracic limited sympathotomy for palmar-plantar hyperhidrosis: outcomes and complications during a 10-year period. Mayo Clin Proc 86: 721–729.

Bates G, Gazey C, Cena K (1996). Factors affecting heat illness when working in conditions of thermal stress. J Hum Ergol (Tokyo) 25: 13–20.

Benarroch EE (2006). Basic neurosciences with clinical applications. Butterworth Heinemann Elsevier, Philadelphia.

Benarroch EE, Schmeichel AM, Low PA et al. (2005). Involvement of medullary regions controlling sympathetic output in Lewy body disease. Brain 128: 338–344.

Boulant JA (2006). Neuronal basis of Hammel's model for set-point thermoregulation. J Appl Physiol 100: 1347–1354.

Brooke RE, Deuchars J, Deuchars SA et al. (2004). Input-specific modulation of neurotransmitter release in the lateral horn of the spinal cord via adenosine receptors. J Neurosci 24: 127–137.

Brown SJ, Raman A, Schlader Z et al. (2013). Ventilatory efficiency in juvenile elite cyclists. J Sci Med Sport 16: 266–270.

Candas V, Libert JP, Vogt J (1980). Effect of hidromeiosis on sweat drippage during acclimation to humid heat. Eur J Appl Physiol 44: 123–133.

Candas V, Libert JP, Vogt JJ (1983). Sweating and sweat decline of resting men in hot humid environments. Eur J Appl Physiol 50: 223–234.

Carpenter AJ, Nunneley SA (1988). Endogenous hormones subtly alter women's response to heat stress. J Appl Physiol 65: 2313–2317.

Charkoudian N (2010). Mechanisms and modifiers of reflex induced cutaneous vasodilation and vasoconstriction in humans. J Appl Physiol 109: 1221–1228.

Cheng C, Matsukawa T, Sessler DI et al. (1995). Increasing mean skin temperature linearly reduces the core-temperature thresholds for vasoconstriction and shivering in humans. Anesthesiology 82: 1160–1168.

Cheung SS, McLellan TM (1998). Influence of short-term aerobic training and hydration status on tolerance during uncompensable heat stress. Eur J Appl Physiol 78: 50–58.

Craig AD (2002). How do you feel? Interoception: the sense of the physiological condition of the body. Nat Rev Neurosci 3: 655–666.

Denda M, Sokabe T, Fukumi-Tominaga T et al. (2007). Effects of skin surface temperature on epidermal permeability barrier homeostasis. J Invest Dermatol 127: 654–659.

Egan GF, Johnson J, Farrell M et al. (2005). Cortical, thalamic, and hypothalamic responses to cooling and warming the skin in awake humans: a positron-emission tomography study. Proc Natl Acad Sci U S A 102: 5262–5267.

Fealey RD (2000). Thermoregulatory failure. In: O Appenzeller (Ed.), The Autonomic Nervous System. Part II. Dysfunctions. Handbook of Clinical Neurology. Vol. 75. Elsevier, Amsterdam, pp. 54–57.

Fealey RD (2007). Thermoregulation and its disorders. In: S Gilman (Ed.), Neurobiology of Disease. Elsevier Academic Press, Burlington, MA, pp. 987–996.

Fealey RD (2008). Thermoregulatory sweat test. In: PA Low, EE Benarroch (Eds.), Clinical Autonomic Disorders. Lippincott Williams and Wilkins, Philadelphia, p. 258.

Fealey RD (2009). Thermoregulatory sweat test. In: JR Daube, DI Rubin (Eds.), Clinical Neurophysiology. 3rd edn. Oxford University Press, New York, pp. 645–660.

Fleury C, Neverova M, Collins S et al. (1997). Uncoupling protein-2: a novel gene linked to obesity and hyperinsulinemia. [see comments]. Nat Genet 15: 269–272.

Flouris AD (2011). Functional architecture of behavioural thermoregulation. Eur J Appl Physiol 111: 1–8.

Fortney SM, Nadel ER, Wenger CB et al. (1981). Effect of blood volume on sweating rate and body fluids in exercising humans. J Appl Physiol 51: 1594–1600.

Gagnon D, Crandall CG, Kenny GP (2013). Sex differences in postsynaptic sweating and cutaneous vasodilation. J Appl Physiol 114: 394–401.

Grucza R, Lecroart JL, Hauser JJ (1985). Dynamics of sweating in men and women during passive heating. Eur J Appl Physiol 54: 309–314.

Gurrera RJ (1999). Sympathoadrenal hyperactivity and the etiology of neuroleptic malignant syndrome. Am J Psychiatry 156: 169–180.

Harris BA, Andrews PJ, Marshall I et al. (2008). Forced convective head cooling device reduces human cross-sectional brain temperature measured by magnetic resonance: a non-randomized healthy volunteer pilot study. Br J Anaesth 100: 365–372.

Havenith G (2001). Individualized model of human thermoregulation for the simulation of heat stress response. J Appl Physiol 90: 1943–1954.

Hesselink MK, Mensink M, Schrauwen P (2003). Human uncoupling protein-3 and obesity: an update. Obes Res 11: 1429–1443.

Horiuchi J, McAllen RM, Allen AM et al. (2004). Descending vasomotor pathways from the dorsomedial hypothalamic nucleus: role of medullary raphe and RVLM. Am J Physiol Regul Integr Comp Physiol 287: R824–R832.

Iodice V, Lipp A, Ahlskog JE et al. (2012). Autopsy confirmed multiple system atrophy cases: Mayo experience and role of autonomic function tests. J Neurol Neurosurg Psychiatry 83: 453–459.

Jensen RT, Battey JF, Spindel ER et al. (2008). International Union of Pharmacology. LXVIII. Mammalian bombesin receptors: nomenclature, distribution, pharmacology, signaling, and functions in normal and disease states. Pharmacol Rev 60: 1–42.

Johnson JM, Park MK (1981). Effect of upright exercise on threshold for cutaneous vasodilation and sweating. J Appl Physiol 50: 814–818.

Katz LM, Young AS, Frank JE et al. (2004). Regulated hypothermia reduces brain oxidative stress after hypoxic-ischemia. Brain Res 1017: 85–91.

Klingenberg M (1999). Uncoupling protein – a useful energy dissipator. J Bioenerg Biomembr 31: 419–430.

Knowlton WM, McKemy DD (2011). TRPM8: from cold to cancer, peppermint to pain. Curr Pharm Biotechnol 12: 68–77.

Kolka MA, Stephenson LA (1997). Effect of luteal phase elevation in core temperature on forearm blood flow during exercise. J Appl Physiol 82: 1079–1083.

Llewellyn-Smith IJ, Martin CL, Minson JB (2002). Glutamate and GABA content of calbindin-immunoreactive nerve terminals in the rat intermediolateral cell column. Auton Neurosci 98: 7–11.

Low P (1977). Splanchnic preganglionic neurons in man. I. Morphometry of preganglionic cytons. Acta Neuropathol 40: 55–61.

Low PA, Caskey PE, Tuck RR et al. (1983). Quantitative sudomotor axon reflex test in normal and neuropathic subjects. Ann Neurol 14: 573–580.

Mattsson CL, Csikasz RI, Chemogubova E et al. (2011). beta (1)-Adrenergic receptors increase UCP1 in human MADS brown adipocytes and rescue cold-acclimated beta(3)-adrenergic receptor-knockout mice via nonshivering thermogenesis. Am J Physiol Endocrinol Metab 301: E1108–E1118.

Morrison SF (2011). 2010 Carl Ludwig Distinguished Lectureship of the APS Neural Control and Autonomic Regulation Section. Central neural pathways for thermoregulatory cold defense. J Appl Physiol 110: 1137–1149.

Nagashima K, Nakai S, Tanaka M et al. (2000). Neuronal circuitries involved in thermoregulation. Auton Neurosci 85: 18–25.

Neimark MA, Konstas AA, Laine AF et al. (2007). Integration of jugular venous return and circle of Willis in a theoretical human model of selective brain cooling. J Appl Physiol 103: 1837–1847.

Ogawa T, Asayama M, Miyagawa T (1982). Effects of sweat gland training by repeated local heating. Jpn J Physiol 32: 971–981.

Premkumar LS, Abooj M (2013). TRP channels and analgesia. Life Sci 92: 415–424.

Romanovsky AA, Almeida MC, Aronoff DM et al. (2005). Fever and hypothermia in systemic inflammation: recent discoveries and revisions. Front Biosci 10: 2193–2216.

Sanchez-Alavez M, Tabarean IV, Behrens MM et al. (2006). Ceramide mediates the rapid phase of febrile response to IL-1beta. Proc Natl Acad Sci U S A 103: 2904–2908.

Sato K, Kang WH, Saga K et al. (1989). Biology of sweat glands and their disorders. II. Disorders of sweat gland function [see comments]. J Am Acad Dermatol 20: 713–726.

Sawka MN, Montain SJ (2000). Fluid and electrolyte supplementation for exercise heat stress. Am J Clin Nutr 72 (2 Suppl): 564S–572S.

Schulze E, Witt M, Fink T et al. (1997). Immunohistochemical detection of human skin nerve fibers. Acta Histochem 99: 301–309.

Shah AK, Atkinson MD, Gupta P et al. (2010). Transient shivering during Wada test provides insight into human thermoregulation. Epilepsia 51: 745–751.

Shibasaki M, Crandall CG (2010). Mechanisms and controllers of eccrine sweating in humans. Front Biosci (Schol Ed) 2: 685–696.

Shibasaki M, Kondo N, Crandall CG et al. (2003). Nonthermoregulatory modulation of sweating in humans. Exerc Sport Sci Rev 31: 34–39.

Shibasaki M, Rasmussen P, Secher NH et al. (2009). Neural and non-neural control of skin blood flow during isometric handgrip exercise in the heat stressed human. J Physiol 587: 2101–2107.

Shoenfeld Y, Udassin R, Shapiro Y et al. (1978). Age and sex difference in response to short exposure to extreme dry heat. J Appl Physiol 44: 1–4.

Takamata A, Yoshida T, Nishida N et al. (2001). Relationship of osmotic inhibition in thermoregulatory responses and sweat sodium concentration in humans. Am J Physiol Regul Integr Comp Physiol 280: R623–R629.

Ueta CB, Fernandes GW, Capelo LP et al. (2012). β(1) Adrenergic receptor is key to cold- and diet-induced thermogenesis in mice. J Endocrinol 214: 359–365.

Weihe E, Schutz B, Hartschuh W et al. (2005). Coexpression of cholinergic and noradrenergic phenotypes in human and nonhuman autonomic nervous system. J Comp Neurol 492: 370–379.

Yakimova KS, Sann H, Pierau FK (1996). Neuronal basis for the hyperthermic effect of mu-opioid agonists in rats: decrease in temperature sensitivity of warm-sensitive hypothalamic neurons. Neurosci Lett 218: 115–118.

Zaninovich AA (2005). Role of uncoupling proteins UCP1, UCP2 and UCP3 in energy balance, type 2 diabetes and obesity. Synergism with the thyroid. Medicina (B Aires) 65: 163–169.

Zenker W, Kubik S (1996). Brain cooling in humans – anatomical considerations. Anat Embryol (Berl) 193: 1–13.

Handbook of Clinical Neurology, Vol. 117 (3rd series)
Autonomic Nervous System
R.M. Buijs and D.F. Swaab, Editors

Chapter 8

Regulation of blood pressure by the arterial baroreflex and autonomic nervous system

ERICA A. WEHRWEIN[1]* AND MICHAEL J. JOYNER[2]
[1]*Department of Physiology, Michigan State University, East Lansing, MI, USA*
[2]*Department of Anesthesiology, Mayo Clinic, Rochester, MN, USA*

INTRODUCTION

Mean arterial pressure (MAP) is a critical hemodynamic factor. The absence of proper regulation of MAP can have important pathophysiological consequences. Low MAP can cause inadequate blood flow to organs, syncope, and shock. On the other hand, elevated MAP contributes to increased oxygen demand by the heart, ventricular remodeling, vascular injury, end organ damage, and stroke.

MAP is tightly regulated in order to maintain appropriate perfusion of vital organs. There are a variety of mechanisms in place to sense and regulate MAP; the arterial baroreflex mechanism is explored in this chapter. As is typical with any reflex, the arterial baroreflex is composed of neural sensors, afferent pathways, central integrating centers, efferent pathways, and effector organs. Briefly, pressure sensors located primarily in the arterial walls of carotid artery sinus and aortic arch sense changes in MAP and drive afferent neural feedback centrally in proportion to changes in MAP. After central processing, the arterial baroreflex regulates MAP by altering efferent parasympathetic and sympathetic neural pathways to the heart and blood vessels (Fig. 8.1).

MAP is the product of cardiac output and total peripheral vascular resistance. Importantly, both of these parameters are under the control of the arterial baroreflex and autonomic nervous system. Since cardiac output is the product of heart rate and stroke volume, changes in either of these parameters also influence MAP. In humans, heart rate at rest is largely under the control of the parasympathetic vagus nerve, while vascular tone is sympathetically mediated.

OVERVIEW OF THE BAROREFLEX AND TERMINOLOGY

Integrative nature of the arterial baroreflex

The arterial baroreflex system is complex. It can be considered in its entirety as an integrative physiological system or in terms of its regulated component parts. Those component parts include MAP, mechanosensory transduction, afferent pathways, central neural circuits, efferent pathways, receptor pharmacology, integration with other key homeostatic inputs, molecular biology, and/or other elements. As such, the baroreflex has been extensively "dissected" experimentally and a chapter could be dedicated to each of these components. This chapter provides an overview of each of these individual components but stresses the importance of the integrative nature of this reflex.

The baroreflex system with its component parts is not unlike the famed Indian legend of the blind men and the elephant. In that tale, six blind men each feel a different part of an elephant and thus each describes what they are feeling differently. For example, one man feels the leg and describes it as a pillar while another feels the tail and describes it as a rope. None of the men realize that the different parts comprise a complete animal. The baroreflex system is the "elephant"; however, there are many regulated component parts in this system that are vital in its integrative function.

Anatomical and functional overview of the arterial baroreflex

As first described by Corneille Heymans, the sensors in the arterial baroreflex are the mechanosensitive afferent nerve endings called baroreceptors. These pressure or

*Correspondence to: Erica A. Wehrwein, PhD, Michigan State University, Department of Physiology, 567 Wilson Road, Room 2195, East Lansing, MI 48912, USA. Tel: +1-517-884-5043, E-mail: ewehrwein@cns.msu.edu

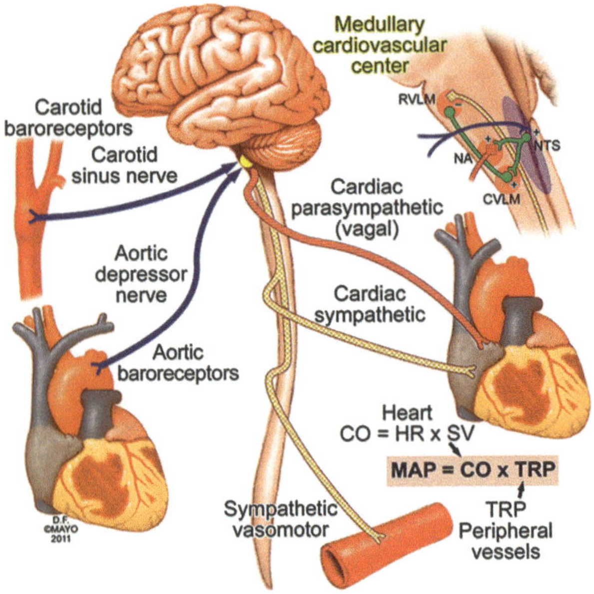

Fig. 8.1. Overview of the arterial baroreflex and central integration of baroreceptor inputs. (Main figure: Overview; Inset: Central integration.) Mean arterial pressure (MAP) is the product of cardiac output (CO) and total peripheral vascular resistance (TPR). CO is the product of heart rate (HR) and stroke volume (SV); changes in either of these parameters also influence MAP. The arterial baroreflex is a key regulator of MAP. The sensors in the baroreflex are baroreceptors located primarily in the carotid artery sinuses and aortic arch. The baroreceptor sensors transduce stretch of the vessel wall into an electrical signal in the afferent nerves that is relayed to an integrating center in the brain. Baroreceptor afferents from the carotid sinus nerve and aortic depressor nerve synapse in the nucleus of the solitary tract (NTS). The parasympathetic (vagus nerve) baroreflex is processed through the nucleus ambiguous (NA) *en route* to the sinoatrial node of the heart. The sympathetic baroreflex pathway continues from the NTS through the caudal ventrolateral medulla (CLVM) and then to the rostral ventrolateral medulla (RVLM). The sympathetic outflow then goes to the interomediolateral cell column in the spinal cord and onto its effector organs (heart and blood vessels). The parasympathetic and sympathetic neural pathways are responsible for regulation of MAP via their effects on heart rate, cardiac output, and total peripheral vascular resistance.

"baro" sensors are located primarily in the carotid artery sinuses and aortic arch where they provide instantaneous feedback on arterial stretch proportional to changes in pressure in the arterial lumen. When the artery is distended due to elevated pressure, there is activation of the embedded barosensitive afferent nerve endings (i.e., as pressure increases there is increased vessel wall stretch and increased afferent firing). The baroreceptor sensors transduce stretch of the vessel wall into an electrical signal in the afferent nerves that is relayed to an integrating center in the brain. The signal is processed centrally and then relayed via both sympathetic and parasympathetic efferent pathways to effector organs. The sympathetic efferent pathways are targeted to both the blood vessels and heart, while the parasympathetic efferent pathways are targeted to the pacemaker cells in the sinoatrial node of the heart (Fig. 8.1).

What is baroreflex sensitivity?

Baroreflex sensitivity (BRS) (also known as baroreflex gain) is used as a measure of autonomic control of the cardiovascular system. Generally, it is a measure of the autonomic effector response to a given change in arterial pressure. There are a variety of methods used to determine BRS that are further explored below (see Assessment of baroreflex function for detailed methodology).

As an example, when MAP is altered pharmacologically using the vasoconstrictor phenylephrine and vasodilator sodium nitroprusside, a large range of pressures can be observed. The changes in pressure, both high and low, are associated with the baroreflex counterregulation in heart rate. The heart rate responses over the wide range of pressures generated with this type of intervention are plotted as in Figure 8.2. A sigmoidal relationship exists and the linear portion of the curve is analyzed

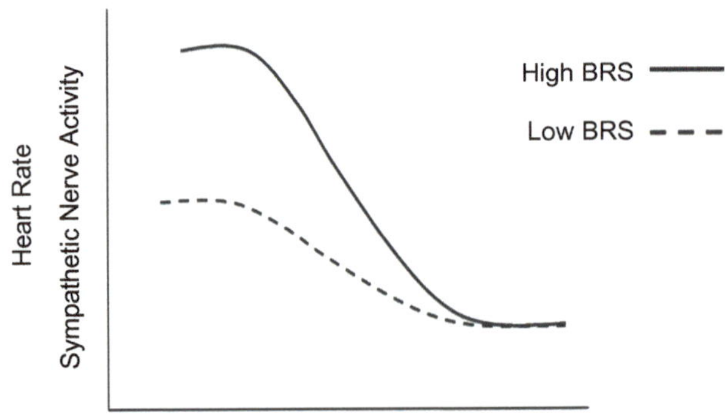

Fig. 8.2. Baroreflex sensitivity (BRS) (also called baroreflex gain). BRS is used as a measure of autonomic control of the cardiovascular system. BRS is determined from the slope of the relationship of mean arterial pressure (MAP) (systolic and diastolic arterial pressures or carotid sinus pressures can also be used) and heart rate (or sympathetic nerve activity). The heart rate (or sympathetic nerve activity) over a range of blood pressures is plotted and analyzed by linear regression. The slope of this regression line obtained from the linear part of the curve is the BRS. The BRS is high when the slope is steep (solid line) such that there is a large heart rate response to a given change in MAP. The BRS is low when the slope is flat (dashed line) such that there is a minimal heart rate response to a change in MAP.

by regression to determine slope and BRS. A greater fall in heart rate to a given pressure elevation is indicative of a higher BRS. The BRS is low when the slope is flat such that there is a minimal heart rate response to a change in MAP (Fig. 8.2). In other words, for a given change in MAP sensed by the baroreceptors there is a baroreflex-driven change in heart rate and the nature of this relationship determines the BRS.

Clinical relevance of BRS

Reduced BRS (i.e., flatter slope) is observed in a number of diseases and carries a poor prognostic outcome (Lanfranchi and Somers, 2002; Benarroch, 2008; La Rovere et al., 2008). Reductions in BRS have been noted in aging, diabetes, obesity, coronary artery disease, hypertension, and other conditions (Skrapari et al., 2006). After a myocardial infarction, patients with a reduced BRS have a greater likelihood of cardiac mortality and fatal arrhythmias (La Rovere et al., 2008). In this case, the blunting of BRS is associated with the cardiac efferent pathways, specifically parasympathetic vagal tone to the sinoatrial node. In congestive heart failure, patients who present with reduced BRS are similarly at high risk of sudden cardiac death (La Rovere et al., 2008). Importantly, exercise can improve BRS. Therefore, exercise is protective against age-related decline in baroreflex function and important as an intervention in a number of diseases (Joyner, 2006; Joyner and Green, 2009). Exercise has a "dose-dependent" effect on baroreflex function such that those engaged in regular endurance programs have the most benefit. The detailed mechanisms that determine BRS have been recently reviewed by Chapleau et al. (2001) and are beyond the scope of this chapter.

Terminology relating to the baroreflex

The terms "cardiac baroreflex" (also known as cardiac vagal baroreflex) and "sympathetic baroreflex" are often used to indicate that different efferent arms of the arterial baroreflex were assessed in a particular study. Cardiac baroreflex is determined by assessing the relationship of heart rate (or RR interval) to a given change in MAP and is technically easier to assess than the more invasive sympathetic baroreflex measurements. The sympathetic baroreflex represents the relationship between diastolic arterial pressure and vasoconstrictor sympathetic nerve activity. In humans, the sympathetic baroreflex is determined based on recordings of muscle sympathetic nerve activity (MSNA) commonly measured in the superficial peroneal nerve (Sundlof and Wallin, 1978; Kienbaum et al., 2001). In animals, the sympathetic baroreflex can be assessed using any sympathetic neural outflow such as renal sympathetic nerve activity.

Readers must be alert to whether cardiac or sympathetic baroreflex is being discussed because, while both approaches both fall under the umbrella of a "baroreflex," they are different in terms of their analysis and implications. Additionally, the cardiac and sympathetic baroreflexes are not comparable in humans since each system has a different sensitivity (Dutoit et al., 2010) and each efferent arm can be regulated differently under certain circumstances. For example, there is a clear distinction of cardiac and sympathetic baroreflex control in panic disorder patients (Lambert et al., 2002).

Cardiopulmonary baroreflex

Baroreceptive nerve endings also exist as "cardiopulmonary baroreceptors" in the intrathoracic vessels and heart. Cardiopulmonary baroreceptors exist in the atria, ventricles, and coronary vessels, as well as throughout the thoracic cavity including in the superior and inferior vena cava, and pulmonary arteries. This population of baroreceptors is quite heterogeneous both in location and function. They are difficult to study in humans because of their anatomical position. Cardiopulmonary baroreceptor afferent fibers track into the same central sites and are integrated with afferent inputs from the arterial baroreceptors. Therefore, a discussion of arterial baroreflex is incomplete without considering the influence of the cardiopulmonary baroreceptors.

Cardiopulmonary baroreceptors are responsive to changes in central blood pressure and volumes such as that occurring with orthostatic stress generated by mild tilting or exposure to lower body negative pressure (LBNP) (Zoller et al., 1972). Mild LBNP results in a slight reduction in central venous pressure and an increase in efferent sympathetic outflow; however, there is no change in MAP or heart rate (Zoller et al., 1972). This observation has been interpreted to indicate that the cardiopulmonary baroreceptors are not essential for regulating MAP but do work in coordination with the arterial baroreceptors to control MAP by altering sensitivity (i.e., gain) of the baroreflex. An alternate explanation is that the subtle changes in the arterial wave form are responsible for increased sympathetic responses (Taylor et al., 1995).

THE REFLEX ARC: FUNCTIONAL ANATOMY AND PHYSIOLOGY

Baroreceptors and sensory transduction

Baroreceptors are mechanosensitive afferent nerve endings that are interspersed in the arterial elastic layers. Baroreceptors detect mechanical deformation of the vessel wall (i.e., vascular wall stretch due to changes in intraluminal pressure). There are two components of vessel stretch that are important in this context: the absolute stretch and the rate at which the stretch occurs. That is to say that it is not only the absolute pressure in the artery and the change in pressure that occurs with each pulse (pulse pressure), but the rate at which the pressure changes. A wide range of arterial pressures can be accurately detected because baroreceptors produce a graded electrical response in associated afferent fibers based on changes in arterial pressure.

There are two primary sites that contain clusters of arterial baroreceptors: the internal carotid artery at the carotid sinus (carotid sinus baroreceptors) and the aortic arch (aortic baroreceptors). These strategic sites allow sensing of both the pressure of blood entering the brain and the pressure of blood leaving the heart. A unique feature of these baroreceptive regions is that the arterial wall has limited collagen and is highly compliant to allow distention of the wall with each pulse.

Recently, the molecular mechanisms of mechanosensitivity in baroreceptor endings have been explored. The degererin/epithelial Na^+ channel (DEG/ENaC) family is expressed in barosensitive afferent terminals (Lu et al., 2009). Of particular interest is the acid sensing ion channel (ASIC2). ASIC2 knockout mice display impaired baroreflex function including a reduction in BRS. The pathophysiology associated with the loss of ASIC2 in mice mirrors the clinical observations in human cardiovascular diseases including manifestation of hypertension and heart failure. In addition, the transient receptor potential vanilloid 1 (TRPV1) receptor has been localized in baroreceptive nerve endings. When TRPV1-containing afferent fibers are chemically ablated or pharmacologically blocked, there is a reduction in BRS (Sun et al., 2009).

Afferent pathways

From the baroreceptive endings, information on arterial pressure is carried centrally to the medulla via afferent fibers. Information from the carotid baroreceptors travels via afferent fibers in the carotid sinus nerve (a branch of cranial nerve IX; glossopharyngeal) and the associated sensory cell bodies exist in the petrosal ganglia. Information from the aortic arch baroreceptors travels in the aortic depressor nerve (a branch of cranial nerve X; vagus) with sensory cell bodies located primarily in the nodose ganglia.

There are multiple classifications of baroreflex-associated afferent fibers that track to the medulla oblongata from the baroreceptors. Both myelinated A fibers of large and small caliber and small unmyelinated C fibers are involved in afferent signaling. Arterial baroreceptor afferent fibers have been further classified functionally into type I and type II (Seagard et al., 1990). Type I are generally large caliber A fibers that display a high pressure threshold, discharge abruptly to a rise in pressure, and are generally inactive at baseline pressures. In contrast, type II fibers are generally smaller caliber A fibers and unmyelinated C fibers that are spontaneously active at baseline pressures and show continuous firing to an increase in pressure in a sigmoidal manner (Seagard et al., 1990; Chapleau, 1991). The nature of the afferent input to central sites, whether tonic (continuous) or phasic (discontinuous), is important. This has been reviewed by Chapleau (1991). Briefly, the central integration and determination of sympathetic outflow depends upon afferent activity: a phasic firing pattern results in sympathetic inhibition and a nonphasic pattern results in a disinhibition of sympathetic outflow.

Central integration

The medulla oblongata, specifically the medullary cardiovascular center (Fig. 8.1, inset), is the primary site of cardiovascular and baroreflex integration. In this region, the nucleus of the solitary tract (NTS) serves as the primary site for the first synapse of the baroreceptor afferents and is the key integrating site for all baroreceptor input including the cardiopulmonary baroreceptors. The NTS is a bilateral structure that receives monosynaptic inputs from afferents using glutamate as the primary neurotransmitter. The NTS integrates and relays baroreceptor afferent information via a polysynaptic pathway to other important medullary centers to control parasympathetic and sympathetic pathways to the heart and blood vessels.

The baroreflex-mediated parasympathetic pathways to the heart are regulated by a simple autonomic reflex arc. Excitatory pathways from the NTS project onto the cardioinhibitory area containing the nucleus ambiguus and dorsal motor nucleus of the vagus. The cardiac component of the baroreflex is largely mediated in these regions where activation of vagal preganglionic efferent fibers results in vagal stimulation to the heart and associated bradycardia (Fig. 8.1, inset).

The sympathetic pathways are more complex, with important control points in the NTS, caudal ventrolateral medulla (CVLM), and rostral ventrolateral medulla (RVLM). The NTS sends projections to inhibitory interneurons in the CVLM and also directly to the RVLM. Activation of GABAergic fibers in the CVLM results in inhibition of sympathoexcitatory neurons in the RVLM resulting in a reduction in MAP and sympathetic tone. The RVLM contains presympathetic neurons that project to the thoracic spinal cord and control sympathetic preganglionic fibers. Neurons in the rostral RVLM display spontaneous firing with each cardiac cycle and exhibit a variable firing frequency with each pulse of arterial pressure (Granata, 2003). Several reviews cover this material in more detail (Guyenet et al., 1990, 2010; Madden and Sved, 2003).

Much of the work on central integration has come from animal studies; however, a recent study has addressed this issue in humans. Macefield and Henderson showed that these same brain regions in resting humans are activated coincident to pulsatile changes in blood pressure and to spontaneous sympathetic nerve activity (Macefield and Henderson, 2010). Key medullary brain regions (including NTS, CVLM, and RVLM) were studied using functional magnetic resonance imaging (fMRI) during resting conditions with a simultaneous recording of MSNA. MSNA fluctuates in a phasic pattern linked to beat-to-beat blood pressure changes. As would be predicted from studies on animals, spontaneous increases in MSNA were associated with increases in neuronal activity in the presympathetic RVLM and decreases in activity in both the NTS and CVLM regions.

There is substantial central integration of inputs to the medullary cardiovascular centers including input from the arterial baroreceptors, cardiopulmonary baroreceptors, the arterial and central chemoreceptors, the cerebral cortex, the hypothalamus, and possibly others. The integration of these components determines efferent outflow of both the sympathetic and parasympathetic systems which serve to modulate MAP as part of the baroreflex circuitry.

Efferent pathways

The efferent arm of the arterial baroreflex is composed of parasympathetic and sympathetic neural pathways that influence the heart and blood vessels. In the parasympathetic division, information from the nucleus ambiguus and dorsal motor nucleus of the vagus control preganglionic fibers that exit the spinal column and synapse in the parasympathetic ganglia near the heart to activate postganglionic fibers. The ganglionic synapse and the cardiac postganglionic neuroeffector junction both rely on acetylcholine as the neurotransmitter acting on α 4 β 2 nicotinic cholinergic receptors (Li et al., 2010). Of particular importance relating to the baroreflex is the parasympathetic vagus nerve which has strong influence on the heart. In the sympathetic division, information from the RVLM is passed to the interomediolateral cell column, sympathetic preganglionic fibers, and then into sympathetic ganglia. Acetylcholine acting on α 7 nicotinic cholinergic receptors is the neurotransmitter in sympathetic ganglionic synapses (Li et al., 2010). Sympathetic postganglionic fibers exit the ganglia and travel to the key baroreflex-associated sites of the heart and blood vessels. Norepinephrine is the key neurotransmitter released that acts on both α and β subtypes of adrenergic receptors on these target organs.

Further consideration of the baroreflex reflex arc as a feedback loop

For simplicity, most textbooks refer to the arterial baroreflex as a "negative feedback loop." While that is technically correct in terms of the pressure input and output of the system, there is a caveat in terms of efferent neural mechanisms. The sympathetic efferent pathway fits this definition of a classic negative feedback loop in which *increased* MAP sensed by the baroreceptors results in *decreased* sympathetic vasomotor tone and vasodilation. Similarly, *decreased* MAP results in an *increase* in sympathetic vasomotor tone and vasoconstriction. There are similar directional changes with sympathetic efferent traffic to heart. An important caveat to the "negative feedback loop" is that the changes in parasympathetic vagus nerve traffic evoked by either a rise or fall in blood pressure are directionally the opposite of the changes in sympathetic traffic. The parasympathetic vagal efferent outflow to the heart follows a "positive" neural feedback loop in that *stimulation* of the baroreceptors with increased MAP results in an *activation* of the parasympathetic vagal efferent traffic to the heart. However, the net effect of parasympathetic activation with an elevation in MAP is still physiologically "negative" because acetylcholine released from parasympathetic fibers on the heart has a powerful inhibitory affect on the sinoatrial node and therefore slows the heart rate.

The baroreflex responses occur in stages that include both positive and negative neural feedback systems. For example, when MAP drops there is a baroreflex-mediated increase in both heart rate and total peripheral resistance that is mediated by a series of neural changes. The increase in heart rate in humans is initially mediated by a withdrawal of parasympathetic vagal tone alone. If needed to correct for a large drop in MAP, there will be a delayed elevation in sympathetic tone to the heart to further raise heart rate. Both vagal withdrawal and sympathetic activation act to increase cardiac output and ultimately MAP. Lastly, there is an elevation of sympathetic outflow to the vasculature to raise total peripheral resistance. This stereotypical series of events does not hold true during exercise or extreme orthostatic stress. When heart rate is high under these conditions, sympathetic traffic to the heart increases immediately to enhance the corrective responses of the baroreflex on MAP. There is little vagal tone to withdraw, so heart rate can only be increased by augmented cardiac sympathetic nerve activity.

BAROREFLEX CONTROL OF ARTERIAL PRESSURE: SHORT VS. LONG TERM

There is no debate as to the importance of the arterial baroreflex in the acute regulation of MAP. However, the notion that the arterial baroreflex is involved in the long-term control of blood pressure was all but dismissed for many years, and to date, many textbooks state definitively that the baroreflex does not control blood pressure in the long term. This concept is based on a series of studies in the sinoaortic denervation model (SAD), starting in the 1970s, that demonstrated maintenance of 24 hour MAP at normal levels when the carotid sinus and aortic baroreceptors were denervated (Cowley et al., 1973). The SAD model was considered the "gold standard" for chronic baroreflex studies. These studies are reviewed below, followed by a number of important recent studies that challenge this notion.

Early papers using the SAD model were convincing and represented a paradigm shift in the thinking about long-term buffering of blood pressure by the arterial baroreflex. The conclusions were that the loss of baroreceptor afferent input did not influence 24 hour MAP, but the range and distribution of arterial pressure became more variable (Cowley et al., 1973) (Fig. 8.3). This evidence has traditionally been used to support the idea that arterial baroreceptors are primarily involved in the short-term regulation of blood pressure by ensuring that cardiac output and peripheral resistance were appropriately matched to maintain MAP.

Data from the SAD model have led to a more renal-centric view of long-term blood pressure control. This view has supported an important role for the kidney in long-term blood pressure control. For example, MAP is related to glomerular filtration rate, influenced by changes in body fluid volume, and adjusted so that a balance is reached between fluid intake and fluid output. While this line of study has been fruitful and has led to a deep understanding of the role of the kidney and volume regulation in blood pressure control, it has been at the expense of understanding the importance of long-term neural control of blood pressure.

The emerging view, which has received strong support recently, is that the baroreflex does in fact play a role in long-term blood pressure regulation. Briefly, the data challenging the conclusions of the classic SAD model are discussed. First, Osborn and Hornfeldt (1998) used the SAD model in combination with salt loading and demonstrated that the arterial baroreceptors play a role in the long-term regulation of MAP. Second, the chronic baroreceptor unloading (CBU) model, a modified approach to the classic SAD model, has revealed further supporting evidence for long-term baroreflex control of arterial pressure (Thrasher, 2004, 2005, 2006). Although aortic baroreceptor denervation is used in the CBU model, it is coupled with a unilateral carotid baroreceptor denervation in order to maintain an anatomical connection of the baroreceptor afferents to the NTS (Thrasher, 2006).

The SAD model completely eliminates arterial baroreceptor afferent input to the NTS (and therefore results in reduced inhibitory input from the CVLM to presympathetic neurons in the RVLM). The CBU model maintains unilateral baroreceptor input to the NTS. There is substantial central remodeling in the medullary circuit, including the NTS, with the SAD model that does not occur with CBU. In the SAD model, the NTS no longer provides inhibitory inputs to the presympathetic RVLM (Schreihofer et al., 2005; Thrasher, 2006) and

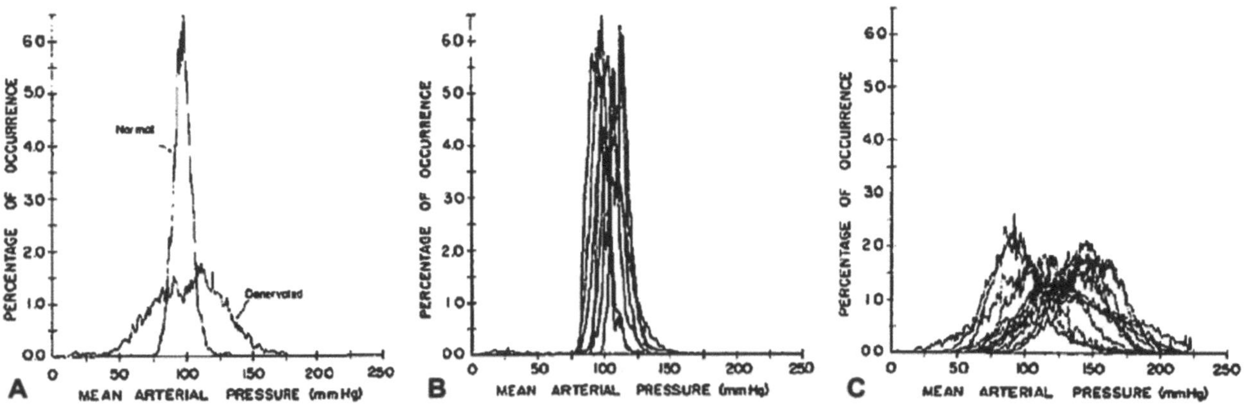

Fig. 8.3. Sinoaortic denervation (SAD) and chronic baroreflex control of mean arterial pressure. Frequency distribution of 24 hour mean arterial pressure in dogs before and after SAD. In a dog model of SAD, there is no difference in mean arterial pressure with denervation but there is a large change in variability of 24 hour blood pressure. (**A**) Individual dog before and after denervation; (**B**) overlay of normal animals ($n = 10$); (**C**) overlay of denervated animals ($n = 12$). Figure from Cowley (Cowley et al., 1973).

there is a rearrangement of γ-aminobutyric acid (GABA) inhibition to the NTS (Ito and Sved, 1997). Both of these observations would explain the suppression of sympathetic nerve activity and MAP back to baseline, even though the loss of baroreceptor input due to denervation should result in sustained elevation of both. Thus, the central remodeling is partially responsible for the observation that there is no change to MAP in the SAD model. The understanding of the central remodeling in the SAD model challenges the original conclusion that there is no role of the arterial baroreflex in long-term MAP regulation.

Data from several models of chronic baroreceptor stimulation also support a role for the baroreflex in long-term control of MAP. After 3 weeks of carotid sinus stimulation in a dog model (i.e., simulation of an increased pressure and stretch in the sinus resulting in sympathoinhibition and reduced MAP), there was no central or peripheral compensation and blood pressure remained lowered until the stimulation was stopped (Lohmeier et al., 2010). As shown in Figure 8.4, there is a sustained suppression of MAP in this model compared to the SAD model in which MAP returns to baseline. Heusser et al. (2010) demonstrated that acute (1 week) electrical field stimulation of the carotid sinus in humans with resistant hypertension resulted in a reduction in blood pressure and sympathetic nerve activity (Heusser et al., 2010). Importantly, this did not alter the physiological function of the baroreflex and there was no evidence of baroreflex failure. In fact, there was a shift of baroreflex control to a lower set point (i.e., a leftward shift of the baroreflex curve) and a reduction of central sympathetic nerve drive with attenuated sympathetic activity to the vasculature, kidney, and heart. Taken together, these studies are highly supportive of a previously underappreciated role of the arterial baroreflex in long-term blood pressure control.

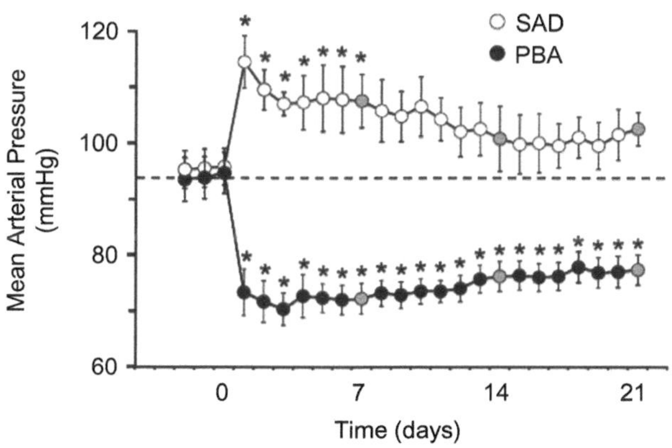

Fig. 8.4. Comparison of chronic changes in mean arterial pressure in sinoaortic denervation (SAD) versus prolonged baroreflex activation (PBA). There is a chronic reduction of mean arterial pressure (MAP) with PBA that is maintained long term, suggesting a role for the baroreflex in chronic MAP regulation. The SAD model shows the expected return of pressure to baseline. (Reproduced from Lohmeier et al., 2010.)

BAROREFLEX RESETTING

Overview of baroreflex resetting

Increases in MAP result in activation of baroreceptors as vessel wall stretch increases; however, baroreceptor activation declines rapidly within seconds of an increase in pressure resulting in less afferent activity to the same pressure. When MAP is acutely or chronically elevated for more than several seconds, there is peripheral baroreceptor "resetting." Resetting, in this context, refers to a shift in the threshold for baroreceptor activation leading to a decline in afferent activation observed to a given elevation in pressure (Fig. 8.5).

As shown in Figure 8.5, as pressure increases from point 1 to point 2 there is an initial rise in afferent activity. This is followed by a reduction in the afferent

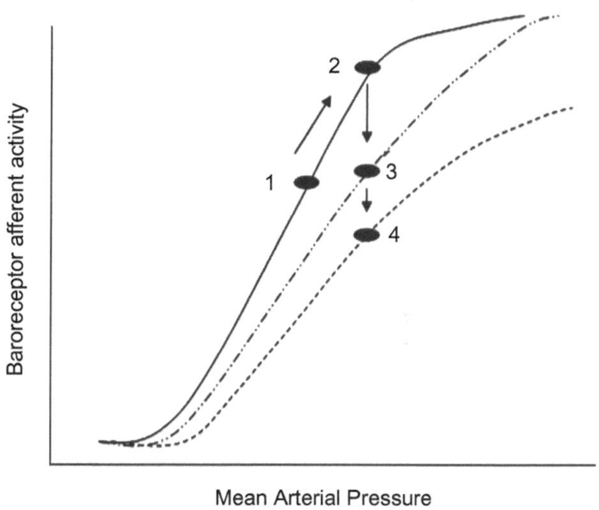

Fig. 8.5. Acute and chronic resetting of the baroreflex. Resetting is demonstrated using a change in baroreceptor afferent activity to a change in mean arterial pressure in this example. Solid line (left curve): normal response; dashed-dotted line (middle curve): acute resetting; dashed line (right curve): chronic resetting. In the left curve, a rise in arterial pressure from point 1 to 2 results in an increased stretch on the baroreceptors and a rise in afferent traffic. This occurs on the steep part of the curve so that the initial rise in pressure has a robust response; however, with pressures at point 2, there is a flattening of the curve and thus a limit to further responsiveness to a further increase in pressure. In acute resetting (middle curve, shift from point 2 to 3), the sensitivity (gain) of the response is restored by a shift in the curve. The afferent activity to a given pressure is reduced from point 2 to 3. In chronic resetting, afferent activity to a given pressure is further reduced from point 3 to 4 and there is a flattening of the slope. (Modified from Chapleau et al., 1989).

traffic (point 3) that is generated in response to the same elevated pressure. This blunting of afferent response is the consequence of resetting. Notably, this new level of afferent activity after resetting is still higher than at baseline, suggesting that there is incomplete resetting during this acute rise in pressure. At point 2 on the graph, the afferent response is nearing the flatter portion of the sigmoidal curve and this would limit further afferent activation to increasing pressures. The resetting response allows the baroreflex to continue to function over a dynamic range on the linear portion of the curve rather than to have the prevailing pressure shifted to the flatter part of the sigmoidal relationship. Additionally, this shift allows for the higher pressures to be maintained rather than suppressed by baroreflex feedback.

Baroreceptor resetting can occur at many points along the baroreflex arc including changes in mechanotransduction, signaling, afferent activity, and/or central modifications. For example, rapid resetting to acute changes in pressure occurs in myelinated fibers, whereas chronic resetting to more sustained increases in pressure is related to unmyelinated afferents (Jones and Thoren, 1977; Seagard et al., 1992; Kougias et al., 2010). There are a host of other mechanisms that are responsible for this baroreceptor resetting that have been extensively reviewed and are beyond the scope of this chapter (Chapleau et al., 1988, 1989; Kougias et al., 2010).

Acute baroreflex resetting during exercise

With acute exercise of almost any type, heart rate and blood pressure rise. There has been an evolution in thought about the role of the baroreflex in these responses during exercise. The initial thought was that the arterial baroreflex was "turned off" during exercise, allowing the heart rate and blood pressure to rise without being corrected by the baroreflex (Joyner, 2006). It is now understood that the baroreflex is intact and functional during exercise but there is resetting to the higher prevailing arterial pressure.

This notion was demonstrated in the 1960s in the work of Bevegard and Shepherd, who used the neck chamber technique (described below) to assess baroreflex function during exercise (Bevegard and Shepherd, 1966). Briefly, neck suction was used to stimulate the baroreceptors in the carotid sinus by increasing the transmural pressure in the carotid sinus, resulting in bradycardia and a fall in MAP. This response to neck suction is preserved during supine exercise in humans on a par with resting responses, further supporting the notion that the baroreflex loop is indeed intact during exercise. In addition, the work of Donald and colleagues in a canine model is supportive

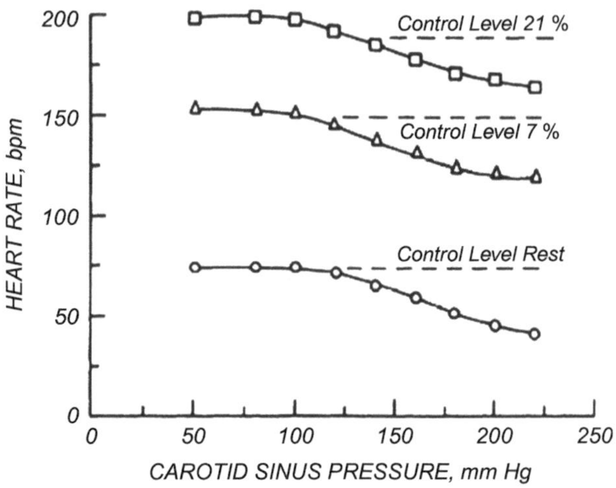

Fig. 8.6. Acute resetting of the baroreflex with exercise. Baroreflex relationship of heart rate to changes in carotid sinus pressure in dogs during rest, mild exercise (7% grade on treadmill), and moderate exercise (21% grade on treadmill). There is a shift in the baroreflex curve upward with each level of exercise but there is no change in the curve characteristics or gain. (Reproduced from Melcher and Donald, 1981.)

of arterial baroreceptors playing a major role during exercise (Melcher and Donald, 1981) (Fig. 8.6).

Given that the baroreflex is still functional during the higher pressures seen with exercise, one must consider how this occurs without a baroreflex-induced suppression of sympathetic outflow and reduction in arterial pressures during exercise. Normally, baroreceptors would sense a higher pressure, such as those seen in exercise, and would be maximally activated, resulting in suppression of sympathetic outflow. Yet, higher pressures achieved during exercise are actually maintained rather than suppressed. In addition, there is still a dynamic baroreflex response range during exercise, indicating that the entire stimulus–response curve for the baroreflex must be shifted to a higher operating point. The baroreceptor resetting during exercise means that the operating point for blood pressure is shifted to match the higher prevailing pressures during exercise while the BRS remains the same (i.e., for a given stretch on the carotid sinus there is a similar change in heart rate and total peripheral resistance).

Acute resetting is reversible and parallels changes in GABA signaling in the NTS (Potts, 2006), supporting a strong role of central neural mechanisms in the process. While the mechanisms of acute resetting are not completely understood, it is also thought that so-called central command (an index of volitional central motor drive critical for voluntary exercise) provides descending input to the medulla and can interact with key regulatory sites in the cardiovascular control centers (Raven, 2008; Williamson, 2010).

Chronic baroreceptor resetting with hypertension

Chronic resetting of the baroreflex occurs in hypertension. Similar to acute resetting with exercise, there is a shift in baroreflex threshold toward the prevailing higher pressure; however, the resetting is not reversible and there is a concurrent reduction in BRS (Sleight et al., 1977; Munch et al., 1983; Kougias et al., 2010). Thus, the issues here are twofold: the system is adjusted to favor chronic maintenance of higher pressure and BRS is lowered with negative physiological and prognostic outcomes. Figure 8.5 shows the further reduction in afferent activity to a given pressure as well as the flattening of the slope representation of a lower BRS.

A number of mechanisms have been shown to be involved in chronic baroreflex resetting (Kougias et al., 2010). The primary changes occur at the level of mechanosensory transduction. This includes direct damage to mechanosensitive baroreceptors, a loss of coupling of vascular wall stretch to the baroreceptors, and an increased stiffness of blood vessels leading to reduced distensibility. The latter is most relevant to hypertension resetting as there is strong evidence for vessel stiffening as well as associated issues such as atherosclerosis. Second, there are a number of studies suggesting that there is central reorganization in hypertension. For example, alterations in $GABA_B$ receptor function in the NTS play a role in altered baroreflex function in the renal-wrap animal model (Zhang and Mifflin, 2011).

ASSESSMENT OF BAROREFLEX FUNCTION: NONINVASIVE AND INVASIVE METHODOLOGY

The baroreceptors are tonically active at baseline arterial pressures and provide a continuous yet variable input centrally with each pulse. Given that they are mechanosensitive receptors, any manipulation that alters stretch of the carotid sinus wall is sensed by the baroreceptive endings. These baroreceptor features allow a number of experimental manipulations that either indirectly or directly alter vessel wall stretch, to be used in the measurement of baroreflex function.

Noninvasive methods

Spontaneous baroreflex analysis

Spontaneous baroreflex analysis takes advantage of the subtle variations in blood pressure and heart rate that occur in the absence of a perturbation. For example, arterial pressure varies with each heartbeat and MAP varies slightly over the course of minutes or hours. Measurement of spontaneous variation in blood pressure and heart rate can be done noninvasively in humans using beat-to-beat finger arterial pressure monitoring equipment and simple electrocardiogram. Therefore, the data collection involved in this type of analysis is relatively inexpensive and easy.

Analysis of spontaneous baroreflex can be done using computer-assisted processing of either sequence or spectral methodology, and either time- or frequency-domain measurements, respectively (La Rovere et al., 2008). In general, the sequence method (time domain) described by Parati et al. (1988) identifies a sequence of three or more consecutive heartbeats that are associated with either an increase or a decrease in arterial pressure. With an increase in arterial pressure, the heart rate is reduced so that there is a lengthening of RR interval. The sequences of beats are selected automatically when arterial pressure increases (or decreases) more than 1 mmHg over a series of heartbeats. The BRS is determined as the slope of the regression of RR interval on systolic pressure and the data from individual sequences of all lengths (numbers of heartbeats) are averaged to determine BRS.

The spectral analysis method for determining BRS relies on the notion that each spontaneous oscillation in arterial pressure results in an opposite oscillation in RR interval due to the baroreflex. These oscillations in arterial pressure and heart rate occur at the same frequencies, around 0.1 Hz for low frequency and in the range of 0.15–0.35 Hz for high frequency (Robbe et al., 1987; La Rovere et al., 2008). The spectral method uses longer periods of data for analysis than does the sequence method. There are several methods for computing the spectral indexes of BRS from this data, as reviewed by LaRovere et al. (2008), including the fast Fourier transformation or autoregressive analysis. Data are reported as gain of the transfer function in the case of transformation or square root of the RR interval and systolic pressure spectral components in the case of autoregressive analysis.

The utility of this spontaneous baroreflex method is subject to a number of concerns and it does not correlate well with other more invasive means of BRS assessment (Lipman et al., 2003; Barrett and Malpas, 2005). It should also be noted that under many physiological conditions, such as exercise or mental stress, heart rate and blood pressure both rise. This is likely due in part to acute baroreflex resetting, as discussed above. These factors limit the utility of the sequence and spectral techniques to the resting condition and in part explain why only a modest fraction of heart rate and blood pressure sequences meet the criteria for analysis.

Neck chamber

Neck suction/pressure is a method by which the external pressure on the neck, and thereby the pressure on deeper neck structures such as the carotid sinus, is altered by means of an airtight chamber or collar placed around the neck. Negative pressure works to increase transmural pressure at the carotid artery which increases baroreceptor activation as the wall is stretched. Therefore, negative neck pressure is the functional equivalent of increased arterial pressure in the carotid artery. Positive pressure has the opposite effect as it will deactivate the receptors similar to a reduction in arterial pressure.

The advantage of this method is that it provides a way to stimulate the carotid baroreceptors directly without inducing systemic changes in pressure that would also impact the aortic baroreceptors. The disadvantage is that the assessment of MSNA/sympathetic baroreflex response with the neck pressure method is problematic because there is buffering of the sympathetic response due to input from the aortic baroreceptors. The aortic baroreceptors provide an opposite input than the carotid baroreceptors.

Valsalva maneuver

Techniques that alter cardiac output can be used to assess the arterial baroreflex by causing changes in stretch of the carotid vessel wall. If venous return to the right heart is reduced experimentally there will be an associated reduction in cardiac output with subsequent stimulation of baroreceptors and vice versa. In this context, there are several common methods by which blood return to the heart can be altered, the most simple of which are the orthostatic challenge of standing or the execution of a Valsalva maneuver. The Valsalva maneuver is used to alter blood pressure to test autonomic reflex responses, can easily be performed over 15–30 seconds, and is a common means by which to assess autonomic function. The Valsalva maneuver is useful for a qualitative bedside evaluation, but is more difficult to use as a quantitative research tool.

Subjects voluntarily exhale forcefully against resistance or a closed glottis, resulting in an abrupt and transient increase in pressure in the thorax and abdomen. There are four phases of blood pressure responses that are stereotypical of the Valsalva maneuver. Phases 2 and 4 are of importance to the baroreflex. Specifically, during phase 2, the baroreceptors sense a fall in pressure and therefore central afferent traffic is reduced, resulting in an elevated heart rate and vasoconstriction (La Rovere et al., 2008). In phase 4, there is an "overshoot" of pressure induced by the rapid increase in cardiac output and this is sensed immediately by the baroreceptors leading to the bradycardia observed in phase 4. BRS is determined from the Valsalva maneuver by analyzing the heart rate response to the changes in systolic pressure during phase 4 or selectively in the blood pressure "overshoot" of phase 4. The Valsalva maneuver is a nonspecific stimulus that is more amenable to assessment of the integrity of the autonomic nervous system than as a specific test of BRS, but it is simple and can be done noninvasively.

Lower body negative pressure

There are a number of other techniques for assessing baroreflex function that require specialized testing equipment, including application of LBNP via a chamber around the lower body, tilt table testing whereby the subject is passively raised towards an upright position, or spinning the subject in a human centrifuge. All of these change central venous pressures and/or volumes. The advantage of using techniques that work on the premise of shifting central blood volumes is that they are less invasive than pharmacological procedures. However, these can also be used in combination with other more invasive methods for a more extreme challenge such as the use of LBNP combined with either tilt table testing, neck suction, and/or infusion of vasoconstrictors; thus, there may be the need for invasive measurements of central venous volumes and infusion catheters.

LBNP can be used to assess both cardiopulmonary baroreceptors and arterial baroreceptors (Convertino, 2001). In the case of cardiopulmonary baroreceptors, a low level of LBNP is applied to reduce central venous pressure without reducing MAP. This requires an invasive measure of central venous pressure along with a measure of forearm blood flow and MAP. MAP divided by forearm blood flow gives forearm vascular resistance. A plot of central venous pressure against forearm vascular resistance is generated and analyzed by linear regression to reveal BRS. In the case of arterial baroreflex, LBNP is combined with an infusion of vasoconstrictor and neck suction to determine sensitivities of both carotid and aortic baroreceptors or the aortic baroreceptors alone (Convertino, 2001). Briefly, supine baseline values for MAP, heart rate, and venous pressures are obtained and compared to those observed during LBNP combined with an infusion of phenylephrine and LBNP plus neck pressure. The changes in heart rate and MAP are calculated as a ratio (Δ heart rate/Δ MAP) used to determine BRS (Shi et al., 1993; Crandall et al., 1994).

Invasive methods

Pharmacological changes in blood pressure

One of the most common means to asses baroreflex function is to alter blood pressure experimentally using pharmacological agents: the Oxford and Modified Oxford techniques. Vasoactive drugs that are known to act on peripheral vessels, with minimal direct action on the heart, are administered to transiently raise or lower arterial pressure. There are a variety of techniques in this regard, including bolus or continuous infusions of vasoactive substances alone or in combination. Typically, the vasoconstrictors used are phenylephrine or angiotensin while the vasodilator substance is nitroprusside. The advantage of the pharmacological method is that all baroreceptors are stimulated by the systemic change in blood pressure so that, unlike methods that attempt to isolate the carotid sinus baroreceptors like neck suction, there is no concern of any buffering of the response by the aortic baroreceptors. Also, assessments can be made with either steady state infusion or bolus injections of drug permitting the rate of change of pressure to be controlled. The disadvantages are that this method is invasive requiring intravenous catheters for drug administration and possibly an arterial catheter for blood pressure assessment. There is also concern that the drugs can alter the mechanical properties of the vessel walls and thus alter the relationships between pressure, vessel deformation, and afferent firing. Finally, it should be noted that well defined differences exist in baroreflex responses depending upon whether pressure is rising or falling. This pattern is referred to as baroreflex hysteresis (Rudas et al., 1999; Studinger et al., 2007). For example, when using the modified Oxford technique there are periods of rising and falling pressure that result in a different BRS.

The sensitivity of the baroreflex is determined by evaluation of heart rate or pulse interval (cardiac baroreflex), or sympathetic nerve activity (sympathetic baroreflex) in response to a pharmacologically induced change in arterial blood pressure. If the desire is to assess cardiac baroreflex, the subject would be instrumented for electrocardiography, and if the sympathetic baroreflex was of interest then MSNA would also be used.

Baroreflex threshold analysis of the sympathetic baroreflex using MSNA

Measurement of the sympathetic baroreflex in humans requires a direct recording of MSNA and a beat-to-beat analysis of blood pressure using a finger blood pressure or an arterial catheter. At rest, the spontaneous variations in diastolic pressure that occur with each heartbeat determine the likelihood of sympathetic burst occurrence, such that lower diastolic pressures are associated with more frequent bursting and vice versa, creating an inverse relationship (Kienbaum et al., 2001; Hart et al., 2010; Wehrwein et al., 2010).

Kienbaum et al. (2001) used spontaneous fluctuations in diastolic blood pressure and MSNA to calculate a "baroreflex threshold analysis" curve. Subsequently, Hart et al. demonstrated that this threshold analysis can serve as a surrogate to pharmacological assessments of sympathetic BRS determined by using the modified Oxford technique in combination with MSNA. Briefly, a baseline recording of MSNA and blood pressure is obtained. All diastolic pressures over the baseline period for each heartbeat are analyzed then are "binned" into groups of 1 mmHg. For each bin of blood pressure, the percentage of heartbeats in the bin associated with a burst of MSNA is determined. A computer-generated "threshold slope" is calculated as the sympathetic BRS and is comparable to BRS determined by modified Oxford (Hart et al., 2010). The advantage is that this technique eliminates the need for intravenous catheters for drug infusions. The disadvantage is that it requires specialized equipment and expertise for the recording and analysis of MSNA.

BAROREFLEX IN DISEASE

Postural tachycardia syndrome

Postural tachycardia syndrome (POTS) is a condition of orthostatic intolerance characterized by an excessive elevation in heart rate during orthostatic stress. This feature of excessive heart rate elevation is also observed in POTS patients during exercise. Given that heart rate is under strong baroreflex control, it is tempting to speculate that there is an abnormal baroreflex underpinning the condition. Masuki and colleagues addressed this issue directly in humans by performing BRS testing using a modified Oxford technique at rest and during varying exercise workloads in POTS patients compared to controls (Masuki et al., 2007). Although heart rate was higher in POTS patients compared to controls during exercise, the BRS during upright and supine exercise was similar between groups with the exception of the highest workload of upright exercise (Masuki et al., 2007). Therefore, it is not abnormal baroreflex function that is responsible for excessive tachycardia in POTS patients.

Baroreflex failure

There are a number of reasons for patients to experience a failure in the baroreflex. As reviewed and summarized by Benarroch, there are three classifications of baroreflex failure: carotid baroreceptor, baroreceptor afferents, and central lesions (Benarroch, 2008). The resulting symptoms of baroreflex failure are that of an acute hypertensive crisis and a highly variable arterial pressure.

Reflex syncope

The term reflex syncope is synonymous with neurally mediated syncope. It can include vasovagal, situational, carotid sinus, and mechanical syncope (van Dijk et al., 2009). The baroreflex is powerful in its control of cardiac vagal tone and therefore if activated inappropriately the baroreflex is a potent inhibitor of the heart leading to a drastic slowing of the heart rate or even asystole. Sudden and inappropriate baroreflex activation can lead to syncope due to severe cardioinhibition. As reviewed by Bennaroch, there are two common causes of baroreflex activation related to syncope: carotid sinus hypersensitivity and glossopharyngeal neuralgia (Cicogna et al., 1993; Benarroch, 2008). There is evidence for carotid sinus hypersensitivity with aging (Humm and Mathias, 2006) that could underlie unexplained syncope in aging patients.

Neurogenic orthostatic hypotension

Neurogenic orthostatic hypotension is defined as a fall in systolic blood pressure of 20 mmHg of more and/or a fall in diastolic blood pressure of 10 mmHg or more (van Dijk et al., 2009). It can occur due to a lack of proper baroreflex control. In this case, there is blunted activation of the sympathetic vasoconstrictor nerves to the peripheral muscle and mesenteric vessels that can result in syncope if the resulting fall in arterial pressure is substantial.

CONCLUSION

In conclusion, there are two key recent concepts related to the arterial baroreflex to reiterate: (1) the importance of reduced BRS in diseases and the implications of BRS to clinical prognostic outcomes, and (2) the emerging paradigm shift on the role played by the arterial baroreflex in the long-term regulation of MAP. First, it is now well established that BRS is an important clinical measure with strong prognostic value. Second, while the tight regulation of MAP in the short term has long been recognized to be associated with the arterial baroreflex, the role of the arterial baroreflex in long-term control of blood pressure was dismissed for many years and is still neglected in many textbooks. Importantly, this chapter reviews the growing body of evidence in support of the critical role played by the arterial baroreflex in long-term control of MAP.

References

Barrett CJ, Malpas SC (2005). Problems, possibilities, and pitfalls in studying the arterial baroreflexes' influence over long-term control of blood pressure. Am J Physiol Regul Integr Comp Physiol 288: R837–R845.

Benarroch EE (2008). The arterial baroreflex: functional organization and involvement in neurologic disease. Neurology 71: 1733–1738.

Bevegard BS, Shepherd JT (1966). Circulatory effects of stimulating the carotid arterial stretch receptors in man at rest and during exercise. J Clin Invest 45: 132–142.

Chapleau MW (1991). Types of baroreceptor afferent neurons. Circ Res 68: 619–620.

Chapleau MW, Hajduczok G, Abboud FM (1988). Mechanisms of resetting of arterial baroreceptors: an overview. Am J Med Sci 295: 327–334.

Chapleau MW, Hajduczok G, Abboud FM (1989). Peripheral and central mechanisms of baroreflex resetting. Clin Exp Pharmacol Physiol Suppl 15: 31–43.

Chapleau MW, Li Z, Meyrelles SS et al. (2001). Mechanisms determining sensitivity of baroreceptor afferents in health and disease. Ann N Y Acad Sci 940: 1–19.

Cicogna R, Bonomi FG, Curnis A et al. (1993). Parapharyngeal space lesions syncope-syndrome. A newly proposed reflexogenic cardiovascular syndrome. Eur Heart J 14: 1476–1483.

Convertino VA (2001). Lower body negative pressure as a tool for research in aerospace physiology and military medicine. J Gravit Physiol 8: 1–14.

Cowley AW Jr, Liard JF, Guyton AC (1973). Role of baroreceptor reflex in daily control of arterial blood pressure and other variables in dogs. Circ Res 32: 564–576.

Crandall CG, Engelke KA, Convertino VA et al. (1994). Aortic baroreflex control of heart rate after 15 days of simulated microgravity exposure. J Appl Physiol 77: 2134–2139.

Dutoit AP, Hart EC, Charkoudian N et al. (2010). Cardiac baroreflex sensitivity is not correlated to sympathetic baroreflex sensitivity within healthy, young humans. Hypertension 56: 1118–1123.

Granata AR (2003). Modulatory inputs on sympathetic neurons in the rostral ventrolateral medulla in the rat. Cell Mol Neurobiol 23: 665–680.

Guyenet PG, Darnall RA, Riley TA (1990). Rostral ventrolateral medulla and sympathorespiratory integration in rats. Am J Physiol 259: R1063–R1074.

Guyenet PG, Stornetta RL, Abbott SB et al. (2010). Central CO2 chemoreception and integrated neural mechanisms of cardiovascular and respiratory control. J Appl Physiol 108: 995–1002.

Hart EC, Joyner MJ, Wallin BG et al. (2010). Baroreflex control of muscle sympathetic nerve activity: a nonpharmacological measure of baroreflex sensitivity. Am J Physiol Heart Circ Physiol 298: H816–H822.

Heusser K, Tank J, Engeli S et al. (2010). Carotid baroreceptor stimulation, sympathetic activity, baroreflex function, and blood pressure in hypertensive patients. Hypertension 55: 619–626.

Humm AM, Mathias CJ (2006). Unexplained syncope – is screening for carotid sinus hypersensitivity indicated in all patients aged >40 years? J Neurol Neurosurg Psychiatry 77: 1267–1270.

Ito S, Sved AF (1997). Influence of GABA in the nucleus of the solitary tract on blood pressure in baroreceptor-denervated rats. Am J Physiol 273: R1657–R1662.

Jones JV, Thoren PN (1977). Characteristics of aortic baroreceptors with non-medullated afferents arising from the aortic arch of rabbits with chronic renovascular hypertension. Acta Physiol Scand 101: 286–293.

Joyner MJ (2006). Baroreceptor function during exercise: resetting the record. Exp Physiol 91: 27–36.

Joyner MJ, Green DJ (2009). Exercise protects the cardiovascular system: effects beyond traditional risk factors. J Physiol 587: 5551–5558.

Kienbaum P, Karlssonn T, Sverrisdottir YB et al. (2001). Two sites for modulation of human sympathetic activity by arterial baroreceptors? J Physiol 531: 861–869.

Kougias P, Weakley SM, Yao Q et al. (2010). Arterial baroreceptors in the management of systemic hypertension. Med Sci Monit 16: RA1–RA8.

La Rovere MT, Pinna GD, Raczak G (2008). Baroreflex sensitivity: measurement and clinical implications. Ann Noninvasive Electrocardiol 13: 191–207.

Lambert EA, Thompson J, Schlaich M et al. (2002). Sympathetic and cardiac baroreflex function in panic disorder. J Hypertens 20: 2445–2451.

Lanfranchi PA, Somers VK (2002). Arterial baroreflex function and cardiovascular variability: interactions and implications. Am J Physiol Regul Integr Comp Physiol 283: R815–R826.

Li YF, Lacroix C, Freeling J (2010). Cytisine induces autonomic cardiovascular responses via activations of different nicotinic receptors. Auton Neurosci 154: 14–19.

Lipman RD, Salisbury JK, Taylor JA (2003). Spontaneous indices are inconsistent with arterial baroreflex gain. Hypertension 42: 481–487.

Lohmeier TE, Iliescu R, Dwyer TM et al. (2010). Sustained suppression of sympathetic activity and arterial pressure during chronic activation of the carotid baroreflex. Am J Physiol Heart Circ Physiol 299: H402–H409.

Lu Y, Ma X, Sabharwal R et al. (2009). The ion channel ASIC2 is required for baroreceptor and autonomic control of the circulation. Neuron 64: 885–897.

Macefield VG, Henderson LA (2010). Real-time imaging of the medullary circuitry involved in the generation of spontaneous muscle sympathetic nerve activity in awake subjects. Hum Brain Mapp 31: 539–549.

Madden CJ, Sved AF (2003). Rostral ventrolateral medulla C1 neurons and cardiovascular regulation. Cell Mol Neurobiol 23: 739–749.

Masuki S, Eisenach JH, Schrage WG et al. (2007). Arterial baroreflex control of heart rate during exercise in postural tachycardia syndrome. J Appl Physiol 103: 1136–1142.

Melcher A, Donald DE (1981). Maintained ability of carotid baroreflex to regulate arterial pressure during exercise. Am J Physiol 241: H838–H849.

Munch PA, Andresen MC, Brown AM (1983). Rapid resetting of aortic baroreceptors in vitro. Am J Physiol 244: H672–H680.

Osborn JW, Hornfeldt BJ (1998). Arterial baroreceptor denervation impairs long-term regulation of arterial pressure during dietary salt loading. Am J Physiol 275: H1558–H1566.

Parati G, Di Rienzo M, Bertinieri G et al. (1988). Evaluation of the baroreceptor-heart rate reflex by 24-hour intra-arterial blood pressure monitoring in humans. Hypertension 12: 214–222.

Potts JT (2006). Inhibitory neurotransmission in the nucleus tractus solitarii: implications for baroreflex resetting during exercise. Exp Physiol 91: 59–72.

Raven PB (2008). Recent advances in baroreflex control of blood pressure during exercise in humans: an overview. Med Sci Sports Exerc 40: 2033–2036.

Robbe HW, Mulder LJ, Ruddel H et al. (1987). Assessment of baroreceptor reflex sensitivity by means of spectral analysis. Hypertension 10: 538–543.

Rudas L, Crossman AA, Morillo CA et al. (1999). Human sympathetic and vagal baroreflex responses to sequential nitroprusside and phenylephrine. Am J Physiol 276: H1691–H1698.

Schreihofer AM, Ito S, Sved AF (2005). Brain stem control of arterial pressure in chronic arterial baroreceptor-denervated rats. Am J Physiol Regul Integr Comp Physiol 289: R1746–R1755.

Seagard JL, van Brederode JF, Dean C et al. (1990). Firing characteristics of single-fiber carotid sinus baroreceptors. Circ Res 66: 1499–1509.

Seagard JL, Gallenberg LA, Hopp FA et al. (1992). Acute resetting in two functionally different types of carotid baroreceptors. Circ Res 70: 559–565.

Shi X, Andresen JM, Potts JT et al. (1993). Aortic baroreflex control of heart rate during hypertensive stimuli: effect of fitness. J Appl Physiol 74: 1555–1562.

Skrapari I, Tentolouris N, Katsilambros N (2006). Baroreflex function: determinants in healthy subjects and disturbances in diabetes, obesity and metabolic syndrome. Curr Diabetes Rev 2: 329–338.

Sleight P, Robinson JL, Brooks DE et al. (1977). Characteristics of single carotid sinus baroreceptor fibers and whole nerve activity in the normotensive and the renal hypertensive dog. Circ Res 41: 750–758.

Studinger P, Goldstein R, Taylor JA (2007). Mechanical and neural contributions to hysteresis in the cardiac vagal limb of the arterial baroreflex. J Physiol 583: 1041–1048.

Sun H, Li DP, Chen SR et al. (2009). Sensing of blood pressure increase by transient receptor potential vanilloid 1 receptors on baroreceptors. J Pharmacol Exp Ther 331: 851–859.

Sundlof G, Wallin BG (1978). Human muscle nerve sympathetic activity at rest. Relationship to blood pressure and age. J Physiol 274: 621–637.

Taylor JA, Halliwill JR, Brown TE et al. (1995). "Non-hypotensive" hypovolaemia reduces ascending aortic dimensions in humans. J Physiol 483: 289–298.

Thrasher TN (2004). Baroreceptors and the long-term control of blood pressure. Exp Physiol 89: 331–335.

Thrasher TN (2005). Baroreceptors, baroreceptor unloading, and the long-term control of blood pressure. Am J Physiol Regul Integr Comp Physiol 288: R819–R827.

Thrasher TN (2006). Arterial baroreceptor input contributes to long-term control of blood pressure. Curr Hypertens Rep 8: 249–254.

van Dijk JG, Thijs RD, Benditt DG et al. (2009). A guide to disorders causing transient loss of consciousness: focus on syncope. Nat Rev Neurol 5: 438–448.

Wehrwein EA, Joyner MJ, Hart EC et al. (2010). Blood pressure regulation in humans: calculation of an "error signal" in control of sympathetic nerve activity. Hypertension 55: 264–269.
Williamson JW (2010). The relevance of central command for the neural cardiovascular control of exercise. Exp Physiol 95: 1043–1048.
Zhang W, Mifflin S (2011). Chronic hypertension enhances presynaptic inhibition by baclofen in the nucleus of the solitary tract. Hypertension 55: 481–486.
Zoller RP, Mark AL, Abboud FM et al. (1972). The role of low pressure baroreceptors in reflex vasoconstrictor responses in man. J Clin Invest 51: 2967–2972.

Handbook of Clinical Neurology, Vol. 117 (3rd series)
Autonomic Nervous System
R.M. Buijs and D.F. Swaab, Editors

Chapter 9

Cooling, pain, and other feelings from the body in relation to the autonomic nervous system

A.D. CRAIG*
Atkinson Research Laboratory, Barrow Neurological Institute, Phoenix, AZ, USA

INTRODUCTION

This chapter reviews information on the long missing sensory complement of the efferent autonomic nervous system (ANS). When Langley (1903) first described the ANS, he noted explicitly that he had omitted a description of the sensory inputs that are required for the ANS to function efficiently; he did this for the simple reason that he could not unequivocally identify such elements anatomically. As Cannon (1932) later strongly emphasized, the neural processes (autonomic, neuroendocrine, and behavioral) that maintain an energy-efficient, optimal physiological balance across conditions in the body, i.e., homeostasis, must receive modality-selective afferent inputs that report the condition of the tissues of the body. Prechtl and Powley (1990) were the first authors to suggest that the category of homeostatic afferents could be identified as the small, dark B cells in the dorsal root ganglia, but their proposal was summarily rejected by nearly all of the discussants of their publication. Unfortunately, they were not aware that the central continuation of the pathway they had envisioned provides unmistakable confirmation of their proposal.

The sensory pathway that is described in the following sections provides not only the basis for organotopic homeostatic control of the body's condition, but also the pathway for human feelings from the body, such as cool, warm, pricking pain, burning pain, itch, sensual touch, muscle ache, bowel distension, urge to urinate, vascular flush, hunger, taste, thirst and "air hunger." These feelings, which are all related to the condition of the body and underlie mood and emotional state, are all associated with strong affective motivations that are the correlates of behavioral responses needed to maintain the health of the body. Thus, these feelings can all be viewed as homeostatic emotions, a concept which emphasizes the essential autonomic role of these sensations. The anatomical pathways described below underpin the homeostatic nature of these feelings, and conversely, reveal a fundamental relationship with the ANS that explains why the affectively charged feelings from the body all have strong autonomic sequelae. So, pain is accompanied by autonomic changes because it is the perceptual correlate of a behavioral motivation generated in response to a condition which the homeostatic system cannot rectify automatically.

In the conventional view, the well-discriminated feelings of temperature, itch, and pain are associated with a somatosensory system that maps the sense of touch to a recognized map of the body (homunculus) in Rolandic cortex. In contrast, the less distinct visceral feelings of vasomotor activity, hunger, thirst, and internal sensations are said to be associated with a separate visceral system in more archaic regions. That categorization, however, obscures fundamental discrepancies, such as the fact that stimulation of Rolandic somatosensory cortices almost never produces feelings of pain or temperature, or that lesions of Rolandic cortex have no effect on temperature or pain sensations. Further, it ignores the inherent emotional (affective/motivational) qualities and reflexive autonomic effects that all feelings from the body share, which distinguish them from tactile mechanoreception and from the sense of limb position (proprioception).

The findings described below compel a conceptual shift that resolves these issues by showing that all feelings from the body are represented in a phylogenetically new system in primates that evolved from the afferent limb of the evolutionarily ancient, hierarchical

*Correspondence to: Dr. A.D. Craig, Atkinson Research Laboratory, Barrow Neurological Institute, 350 West Thomas Road, Phoenix, AZ 85013, USA. Tel: +1-602-406-3385, Fax: +1-602-406-4121, E-mail: bcraig@chw.edu

homeostatic system that maintains the integrity of the body, that is, the sensory complement of the ANS. These feelings thus represent a sense of the physiological condition of the entire body, providing a broad redefinition of the term "interoception." More importantly, from the clinical perspective, these findings reveal that feelings from the body, such as pain, are inherently linked with autonomic conditions, such as plasma extravasation or cardiac rhythmicity, because they are, respectively, sensory and motor aspects of the same homeostatic system. In humans, a metarepresentation of primary interoceptive activity is engendered in the anterior insula, which seems to provide the basis for the subjective image of the material self as a feeling (sentient) entity, that is, emotional awareness. More detailed reviews of the evidence for these views are available elsewhere (Craig, 2002, 2003a, 2009).

THE ASCENDING PATHWAY AND ITS PRINCIPLES OF ORGANIZATION

Spinal cord

The small-diameter Aδ and C primary afferent fibers that emerge late in development from small dorsal root ganglion cells (B cells) innervate every tissue of the body, and they project to the superficial dorsal horn of the spinal cord or to the medullary nucleus of the solitary tract (NTS). Such peripheral sensory fibers course with somatic and sympathetic nerves if they project to the superficial dorsal horn or with parasympathetic nerves (vagal, glossopharyngeal) if they project to NTS. They conduct information regarding all manner of physiological conditions, including the mechanical, thermal, chemical, metabolic, hormonal, and immunological status of skin, muscle, joints, teeth, bone, and viscera. For example, C fibers from deep tissues, such as muscle, can respond to warmth, hypoxia, hypoglycemia, hypo-osmolarity, acidity, ATP, and the presence of muscle metabolic products (e.g., lactate), and they are highly responsive to combinations of these stimuli (e.g., Adreani and Kaufman, 1998; Light et al., 2008). Accordingly, neurons in lamina I (the output cells of the superficial dorsal horn of the spinal cord) comprise several modality-selective, morphologically distinct classes that receive input from specific subsets of small-diameter primary afferent fibers (Craig, 2003b). These classes can be differentiated on the basis of afferent responses, electrophysiological properties, axonal projections, descending modulation, morphology, and pharmacological properties, and they correspond psychophysically with distinct feelings from the body. They include cells selectively responsive to Aδ nociceptors (first, sharp pain), C-fiber polymodal nociceptors (second, burning pain), Aδ cooling-specific thermoreceptors (cool), C-fiber warming-specific receptors (warmth), ultraslow histamine-selective C fibers (itch), tactile C fibers (sensual touch, slow brushing), and Aδ and C mechanoreceptors and metaboreceptors (so-called ergoreceptors) in muscles and joints.

The output neurons from these first synaptic regions (i.e., lamina I and the NTS) project to the homeostatic integration sites and preautonomic motor regions in the brainstem. Lamina I neurons first project heavily to the spinal autonomic nuclei, where sympathetic preganglionic neurons are found, thus forming a spinospinal loop for somatoautonomic reflexes. Altogether, these substrates provide the sensory and motor components of the hierarchically organized homeostatic (autonomic) nervous system; this conclusion is underscored by the observation that these sites are the exclusive targets of descending projections from the hypothalamus (which can be viewed as the diencephalic autonomic control center). The modality-selective responses of spinal preganglionic sympathetic neurons, e.g., the specific responses of cutaneous vasoconstrictors to skin warming and cooling (Janig, 2006), can best be explained by this afferent pathway.

Indeed, in myelin-stained transverse sections of the human spinal cord, the superficial dorsal horn that is part of the homeostatic afferent system is easily distinguishable from the deep dorsal horn (also called nucleus proprius), where large neurons receive input from large-diameter primary afferents innervating cutaneous mechanoreceptors and muscle/joint proprioceptors (which emerge early in development from large dorsal root ganglion cells, or A cells) and project to motor neurons in the ventral horn and to motor control sites centrally. Thus, the terms interoception and exteroception can be redefined (Craig, 2002) in order to differentiate these two systems (i.e., one that controls smooth muscle, as distinct from one that receives large-diameter inputs and controls striate muscle). An important principle of the hierarchical homeostatic system is that it has identifiable sensory and motor components at each level that are tightly interconnected.

It is important to note at this point that particular C fibers are exquisitely sensitive to light (sensual) touch (Bjornsdotter et al., 2009). They can be differentiated from large-diameter mechanoreceptors quantitatively by their distinct response to slow brushing, and in two polyneuropathy patients who have only C fibers remaining, they seem to produce a vaguely pleasant feeling that can be topographically localized but does not convey a sense of discriminative directionality. These afferents activate insular cortex (see below) in the same way that temperature and pain stimuli do, and with a similar posterior-to-anterior topography. This pathway is interpreted as signaling emotional safety amongst conspecifics – for example, think of monkeys grooming each other at the zoo or the emotional value of a hug. The existence of this pathway strongly distinguishes the interoceptive pathway from the exteroceptive pathway, and it demonstrates that interoceptive afferent activity participates in homeostasis beyond emergency conditions. Furthermore, it provides definitive evidence of evolutionary selection at the species level.

Brainstem

Lamina I and NTS neurons project densely and selectively to preautonomic sites in the brainstem, thus extending the afferent limb to the next rungs of the homeostatic hierarchy (Fig. 9.1) and generating spinobulbospinal loops for somatoautonomic reflexes (Sato and Schmidt, 1973; Craig, 1995). These sites include the catecholamine cell groups; the strong input to the A1 group is particularly noteworthy, because its projections to the hypothalamus are crucial for neuroendocrine responses to changes in tissue conditions (Pan et al., 1999). The major target of lamina I and the NTS in the upper brainstem is the parabrachial nucleus (PB) bilaterally; the PB is the main integration site for all homeostatic afferent activity, and thus, it is essential for the maintenance of cardiovascular, respiratory, energy (feeding and glucose), and fluid (electrolyte and water) balances (Saper, 2002). The integrative role of lamina I, NTS, and PB in the homeostatic afferent pathway is clearly consistent with the dense projections of PB to the periaqueductal gray (PAG; the mesencephalic homeostatic motor center) and to the hypothalamus (the diencephalic homeostatic motor center), which guide goal-directed autonomic, neuroendocrine and behavioral activity (Swanson, 2000).

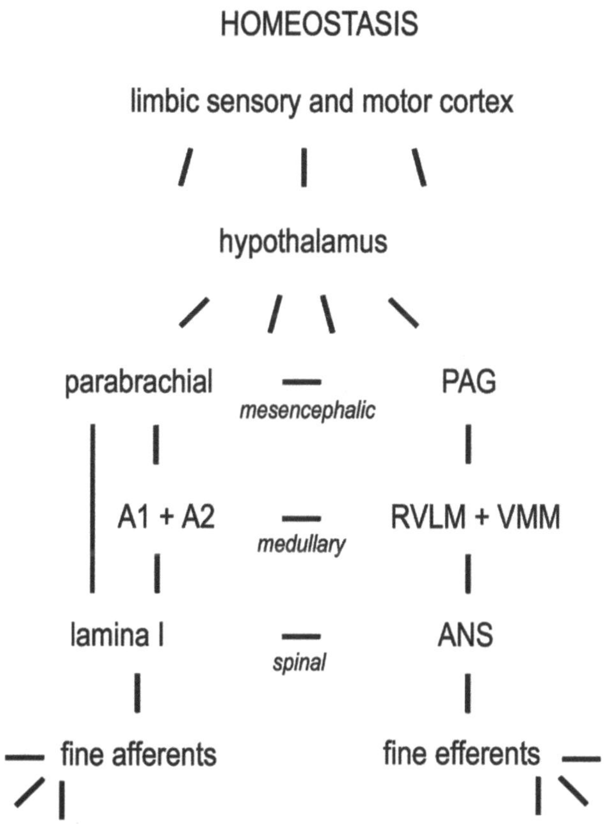

Fig. 9.1. A schematic drawing of the hierarchical central homeostatic network. PAG, periaqueductal gray; RVLM, rostral ventrolateral medulla; VMM, ventromedial medulla; ANS, autonomic nervous system.

The modality-selective somatoautonomic adjustments that are continually being made by homeostatic processing in the brainstem require input from the distinct classes of lamina I neurons that convey the physiological condition of the body. For example, innocuous thermoreceptive (cool) activity, which is represented only in lamina I neurons, linearly modulates respiratory parameters (Diesel et al., 1990), consistent with the primordial role of thermoreception in homeostasis (i.e., thermoregulation). Similarly, small-diameter afferent activity from muscle, which occurs with every movement, modulates cardiovascular and respiratory activity on an ongoing basis (the so-called exercise pressor reflex). Lamina I activity can be directly related to human pain sensation and also to essential homeostatic processes. In particular, the distinct feeling of second (burning) pain is uniquely encoded by the C-fiber-selective polymodal nociceptive (multipolar) lamina I cells, yet their activity primarily provides a homeostatic afferent signal. First, their ongoing activity is directly related to the strength of their C-fiber input (Craig, 2003b), consistent with the hypothesis that C-fiber activity signals metabolic status on an ongoing basis. Second, they are sensitive to noxious heat, to pinch, and to noxious cold, yet their static responses to cold begin at around 24 °C (about 75 °F, a comfortable ambient temperature), consistent with the increasing thermal discomfort that humans feel below that temperature. Third, the burning pain generated by their activity depends on integration in the forebrain with cooling-specific lamina I activity (as demonstrated by the thermal grill illusion – for an explanation of the thermal grill illusion see Craig (2003b)), and with core temperature. These facts directly imply that such thermal distress is a homeostatic behavioral motivation.

In all mammals, the highest level of the homeostatic hierarchy in the upper brainstem consists of the PB and the PAG. These sites can also be viewed as the lowest level of the so-called limbic system that controls emotional behavior (Heimer and Van Hoesen, 2006), because together they organize whole-body behaviors that serve life-supporting functions (cardiorespiratory control, ingestion, elimination, reproduction, etc.). Combined physiological and behavioral experiments revealed approach/avoidance columns in the PAG that have correlative, opposing cardiorespiratory actions; this organization supplies the fundamental underlying pattern for the combined behavioral and autonomic opponent organization that many authors have envisaged in the forebrain of all vertebrates (Craig, 2009; MacNeilage et al., 2009). Opponent organization is found throughout physiology, e.g., in color vision, antagonist muscles, and cardiac function, probably because it is the most energy-efficient way to provide precise control. Thus, the PB and PAG can be viewed as identifiable, complementary sensory and motor regions, respectively, that support homeostasis with a coordinate behavioral/autonomic opponent organization.

Forebrain

In mammals other than primates, the ascending pathway from lamina I and the NTS generates widely scattered projections, and the main homeostatic afferent pathway to their forebrains conveys integrated activity from PB to hypothalamus, amygdala, and, by way of the thalamus, to striatum and insular and cingulate cortices (Saper, 2002). Accordingly, multimodal context-dependent responses have been recorded in these regions in the rat. The insular cortex can be viewed as limbic sensory cortex, because it provides descending control of brainstem homeostatic integration in PB, and the cingulate cortex can be viewed as limbic motor cortex, because it projects densely to the behavioral/autonomic columns of the PAG. Lesions at these sites disrupt homeostatic behavior in rodents (Johansen et al., 2001; Saper, 2002). The emotional behavior of nonprimate mammals suggests the anthropomorphic inference that they experience feelings from the body in the same way that humans do. However, the neuroanatomical evidence indicates that they cannot, because the phylogenetically new pathway that underlies feelings from the body in humans (described below) is either rudimentary or absent in nonprimates (Craig, 2002).

In primates, lamina I neurons and NTS neurons extend such pathways by projecting topographically to a pair of relay nuclei in the posterolateral thalamus, the posterior ventral medial nucleus (VMpo) and the adjoined basal ventral medial nucleus (VMb). The lamina I axons ascend in the lateral spinothalamic tract, precisely where lesions can selectively interrupt the feelings from the body in human patients (Craig et al., 2002). The VMpo/VMb is organized anteroposteriorly, orthogonal to the mediolateral topography of the main somatosensory ventral posterior (VP) nuclei, to which it is connected at the representation of the mouth. The VMb receives direct input from NTS in addition to the integrated input that it receives from PB in all mammals (Beckstead et al., 1980). The VMpo is small in macaque monkeys, but in the human thalamus it is almost half as large as the VP (Blomqvist et al., 2000).

The VMpo and VMb project topographically to interoceptive cortex in the dorsal margin of the insula (a cortical "island" buried within the lateral sulcus that has intimate connections with the ACC, amygdala, hypothalamus, and orbitofrontal cortex). The projections of VMpo and VMb extend over the entire posterior-to-anterior extent of the insula in the macaque monkey (Ito and Craig, 2008), approximately 6–8 mm. However, in humans, the insula extends approximately 50–60 mm anteroposteriorly, and functional imaging studies indicate that lamina I input (e.g., pain, temperature, or itch stimuli) activates the most posterior 10–15 mm, while vagal and gustatory input (e.g., gastric distension, salty intensity) activates the next 10 mm or so (Craig, 2010). In other words, primary interoceptive cortex occupies the entire dorsal insula in monkeys, but only the posterior third in humans. This pathway contains modality-selective components that each generate a distinct "feeling" from the body in humans, including first (pricking) pain, second (burning) pain, cool, warm, itch, muscle ache, gastric distension, vasomotor flush, sweet, salty, and so on. The interoceptive cortex in the monkey is distinctly demarcated by *in situ* labeling for receptors of corticotropin-releasing factors (Sanchez et al., 1999), consistent with a major role in homeostasis as limbic sensory cortex. Lesion, stimulation, and evoked potential studies in humans confirm the role of this primary cortical region in pain and temperature sensation and in autonomic function (e.g., Greenspan et al., 1999).

The pattern of activation observed in many functional imaging studies suggests a posterior-to-anterior processing gradient in the human insular cortex, and it has been proposed that subjective feelings are based directly on rerepresentations of homeostatic sensory integration in the anterior insula, consistent with the James–Lange theory of emotion and Damasio's "somatic marker" hypothesis (Bjornsdotter et al., 2009; Craig, 2010). From an evolutionary perspective, it makes sense that an integrated representation of the activity in all brain networks was needed in order to improve behavior from the perspective of homeostatic efficiency. The goal of homeostatic efficiency (with respect to both the individual and the species) provides a plausible explanation of the progressive posterior-to-anterior integration in the insula.

A corollary VMpo projection to area 3a in sensorimotor cortex may relate cutaneous pain to (exteroceptive) somatic motor activity and to cortical control of spinal viscerosomatic reflex activity. In addition, a direct lamina I pathway to the ACC is also present in primates by way of a topographic projection to the ventral caudal portion of the medial dorsal nucleus (MDvc).

In nonprimates, in contrast, the ACC (i.e., limbic motor cortex) receives integrated homeostatic information from the PB by way of the medial thalamus, and lamina I activity is relayed instead to ventrolateral orbitofrontal cortex through the submedial nucleus. Physiological and behavioral studies validate the primordial role of the ACC in homeostatic behavior in rats (e.g., Johansen et al., 2001) and in the affective/motivational component of human pain by way of the direct lamina I path to MDvc.

HOMEOSTATIC EMOTIONS

The direct activation of both the interoceptive cortex and the ACC by the distinct homeostatic modalities corresponds with the simultaneous generation of both a sensation and a motivation. In humans, an emotion can be defined as a feeling and a motivation; thus, these feelings from the body can be viewed as homeostatic emotions that reflect the survival needs of the body. Pain, temperature, and itch are homeostatic emotions that drive behavior, just as hunger and thirst do (Craig, 2002, 2003a, 2009). Convergent functional imaging data indicate that the anterior insula is associated with subjective feelings from the body and with virtually all human emotions, and so it seems to provide an image of the physical self as a feeling (sentient) entity. The association of the anterior insula with the subjective perception of pain, the anticipation of pain, the subjective reduction of pain (placebo analgesia), and the subjective generation of pain (hypnotic psychogenic pain) underscores the importance of the metarepresentation of interoceptive state in the anterior insula for clinical understanding of the effects of emotion and belief on health. Furthermore, the recognition that sensual touch is incorporated in the interoceptive system emphasizes the need to incorporate the neurobiological basis of conspecific human contact in therapies for emotional and physical health (Bjornsdotter et al., 2009).

THE CONCEPT OF A "PAIN PATHWAY"

Pain has traditionally been classified as an aspect of somatic sensation, for the obvious reason that pain is, first of all, a physical sensation we feel from our bodies. Accordingly, it was thought for many years that pain is represented within the same somatosensory system that represents the exteroceptive sense of touch and the proprioceptive sense of limb movement (the dorsal column – medial lemniscal system). The highly organized somatotopic map in the somatosensory cortex was regarded as necessary in order to explain the fact that pain and temperature, like touch, are discriminative sensations that can be localized quite well on the skin surface. However, pain is certainly distinct from touch, since it can originate from any tissue of the body and since it has a strong emotional quality – a motivational affect – as well as other mystifying qualities (sensitization, radiation, "wind-up," persistence, and so on). Furthermore, stimulation of the somatosensory cortex, or a lesion of somatosensory cortex, which will definitely alter mechanoreception and proprioception, rarely has any affect on pain, temperature, itch, and other feelings from the body.

The evidence described above indicates that the feeling of pain is generated by activity in a phylogenetically distinct extension of an ancient, hierarchical system that conveys ascending afferent activity associated with homeostasis – the maintenance of the physiological condition of the body. In primates, this system provides the basis for a high-resolution, modality-specific sense of the condition of the body (interoception redefined) in dorsal posterior insular cortex. In humans, higher-order rerepresentations of this system in progressively more anterior portions of the insula provide a substantive basis for the subjective awareness of the "material me," that is, "how you feel." These rerepresentations are involved in all subjective feelings and emotions, consistent with the James–Lange theory of emotion and Damasio's "somatic marker" hypothesis of consciousness.

Thus, the feelings from the body (warm, hot, cold, pricking pain, burning pain, itch, muscle burn, ache and cramp, visceral movement and cramp, visceromotor "flush," sensual touch, and so on) all relate the physiological condition of the body, as well as the physiological condition of the homeostatic control system. In other words, when patients describe these feelings, they are providing not merely a symptomotology, but rather important clues to the condition of the homeostatic and immune systems which need to be treated. For instance, a patient with multiple sclerosis who complains of burning cold feet may be providing crucial information about the central modulation of the homeostatic afferent pathway, because the lesion that is producing these feelings is most likely central, perhaps in the spinal cord itself, rather than in the feet or in the forebrain.

Finally, the concept that human pain sensation is underpinned by a distinct and separate pathway for nociceptive-specific lamina I neurons to the primary interoceptive cortex in the dorsal posterior insula contradicts the conventional view that spinal lamina V ("wide dynamic range") neurons are "necessary and sufficient" for pain (Price et al., 2003). Evidence to the contrary has accumulated for years (Craig, 2003b), and recently the conventional view was definitively refuted by evidence that almost all such lamina V neurons convey group II muscle afferent activity, respond tonically to limb position, and project directly onto ventral horn motor neurons and other motor-related sites. That is, such lamina V neurons are an integral component of the skeletal motor system (Craig, 2008). Recent functional imaging results on so-called "windup" (see Staud et al., 2008) in fact corroborates strongly the fundamental role of lamina I spinothalamocortical projections in pain. Nevertheless, in light of the fundamental role of the lamina I pathway in homeostasis, it would be more appropriate to call it a homeostatic afferent pathway than simply a "pain pathway" (Craig, 2006).

CONCLUSION

It is important for clinicians to recognize that feelings from the body, such as pain or temperature, are neurologically distinct from tactile mechanoreception and proprioception at all levels, contrary to the conventional textbook view. Such feelings from the body represent homeostatic afferent activity, that is, the sensory input representing the physiological condition of the body which drives the central network that controls the ANS. The spinal and brainstem levels of the hierarchical homeostatic network are present in all mammals, but in humans there is a high-resolution representation of the condition of the body in primary interoceptive cortex in the posterior insula, which generates an energy-efficient map of all brain activity in the anterior insula that underpins the subjective "material me." Thus, there is a fundamental relationship between the physiological condition of the body and subjective feelings of all kinds, just as William James hypothesized. These facts strongly support a comprehensive perspective on the physical and emotional health of each patient, and they provide a rational basis for complementary medical approaches, for example, therapeutic touch (Bjornsdotter et al., 2009). Similarly, the anterior insula is crucial for both the subjective sense of time and emotional appreciation of music (see Craig, 2009); thus, these findings provide strong support for music therapy, which by affecting the subjective sense of self could have a direct affect on the homeostatic condition of the body.

ABBREVIATIONS

ACC	anterior cingulate cortex
ANS	autonomic nervous system
ATP	adenosine triphosphate
MDvc	ventral caudal part of the medial dorsal nucleus (of thalamus)
NTS	nucleus of the solitary tract
PAG	periaqueductal gray
PB	parabrachial nucleus
VMpo	posterior part of the ventral medial nucleus (of thalamus)
VMb	basal part of the ventral medial nucleus (of thalamus)
VP	ventral posterior nuclei (of thalamus).

ACKNOWLEDGMENT

Grant support: the James S. McDonnell and Barrow Neurological Foundations.

References

Adreani CM, Kaufman MP (1998). Effect of arterial occlusion on responses of group III and IV afferents to dynamic exercise. J Appl Physiol 84: 1827–1833.

Beckstead RM, Morse JR, Norgren R (1980). The nucleus of the solitary tract in the monkey: projections to the thalamus and brain stem nuclei. J Comp Neurol 190: 259–282.

Bjornsdotter M, Loken L, Olausson H et al. (2009). Somatotopic organization of gentle touch processing in the posterior insular cortex. J Neurosci 29: 9314–9320.

Blomqvist A, Zhang ET, Craig AD (2000). Cytoarchitectonic and immunohistochemical characterization of a specific pain and temperature relay, the posterior portion of the ventral medial nucleus, in the human thalamus. Brain 123: 601–619.

Cannon WB (1932). The Wisdom of the Body. W. W. Norton, NY.

Craig AD (1995). Distribution of brainstem projections from spinal lamina I neurons in the cat and the monkey. J Comp Neurol 361: 225–248.

Craig AD (2002). How do you feel? Interoception: the sense of the physiological condition of the body. Nat Rev Neurosci 3: 655–666.

Craig AD (2003a). Interoception: the sense of the physiological condition of the body. Curr Opin Neurobiol 13: 500–505.

Craig AD (2003b). Pain mechanisms: labeled lines versus convergence in central processing. Annu Rev Neurosci 26: 1–30.

Craig AD (2006). Evolution of pain pathways. In: JH Kaas (Ed.), Evolution of Nervous Systems. Vol. 3. Academic Press, Oxford, pp. 227–236.

Craig AD (2008). Retrograde analyses of spinothalamic projections in the macaque monkey: input to the ventral lateral nucleus. J Comp Neurol 508: 315–328.

Craig AD (2009). How do you feel – now? The anterior insula and human awareness. Nat Rev Neurosci 10: 59–70.

Craig AD (2010). The sentient self. Brain Struct Funct 214: 563–577.

Craig AD, Zhang ET, Blomqvist A (2002). Association of spinothalamic lamina I neurons and their ascending axons with calbindin-immunoreactivity in monkey and human. Pain 97: 105–115.

Diesel DA, Tucker A, Robertshaw D (1990). Cold-induced changes in breathing pattern as a strategy to reduce respiratory heat loss. J Appl Physiol 69: 1946–1952.

Greenspan JD, Lee RR, Lenz FA (1999). Pain sensitivity alterations as a function of lesion location in the parasylvian cortex. Pain 81: 273–282.

Heimer L, Van Hoesen GW (2006). The limbic lobe and its output channels: implications for emotional functions and adaptive behavior. Neurosci Biobehav Rev 30: 126–147.

Ito SI, Craig AD (2008). Thalamocortical projections of the vagus-responsive region of the basal part of the ventral medial nucleus in monkeys. Soc Neurosci Abstr, Online 364.10.

Janig W (2006). The integrative action of the autonomic nervous system. Cambridge University Press, Cambridge, UK.

Johansen JP, Fields HL, Manning BH (2001). The affective component of pain in rodents: direct evidence for a contribution of the anterior cingulate cortex. Proc Natl Acad Sci USA 98: 8077–8082.

Langley JN (1903). The autonomic nervous system. Brain 26: 1–26.

Light AR, Hughen RW, Zhang J et al. (2008). Dorsal root ganglion neurons innervating skeletal muscle respond to physiological combinations of protons, ATP, and lactate mediated by ASIC, P2X, and TRPV1. J Neurophysiol 100: 1184–1201. http://dx.doi.org/10.1152/jn.01344.2007. Epub 2008 May 28. PMID: 18509077.

MacNeilage PF, Rogers LJ, Vallortigara G (2009). Origins of the left and right brain. Sci Am 301: 60–67.

Pan B, Castro-Lopes JM, Coimbra A (1999). Central afferent pathways conveying nociceptive input to the hypothalamic paraventricular nucleus as revealed by a combination of retrograde labeling and c-fos activation. J Comp Neurol 413: 129–145.

Prechtl JC, Powley TL (1990). B-afferents: a fundamental division of the nervous system mediating homeostasis. Behav Brain Sci 13: 289–332.

Price DD, Greenspan JD, Dubner R (2003). Neurons involved in the exteroceptive function of pain. Pain 106: 215–219.

Sanchez MM, Young LJ, Plotsky PM et al. (1999). Autoradiographic and in situ hybridization localization of corticotropin-releasing factor 1 and 2 receptors in nonhuman primate brain. J Comp Neurol 408: 365–377.

Saper CB (2002). The central autonomic nervous system: conscious visceral perception and autonomic pattern generation. Annu Rev Neurosci 25: 433–469.

Sato A, Schmidt RF (1973). Somatosympathetic reflexes: afferent fibers, central pathways, discharge characteristics. Physiol Rev 53: 916–947.

Staud R, Craggs JG, Perlstein WM et al. (2008). Brain activity associated with slow temporal summation of C-fiber evoked pain in fibromyalgia patients and healthy controls. Eur J Pain 12: 1078–1089.

Swanson LW (2000). Cerebral hemisphere regulation of motivated behavior (1). Brain Res 886: 113–164.

Handbook of Clinical Neurology, Vol. 117 (3rd series)
Autonomic Nervous System
R.M. Buijs and D.F. Swaab, Editors

Chapter 10

The neurological organization of micturition

JAI H. SETH*, JALESH N. PANICKER, AND CLARE J. FOWLER
Department of Uro-Neurology, National Hospital for Neurology and Neurosurgery, University College London Hospitals Foundation Trust, London, UK

INTRODUCTION

The two main functions of the lower urinary tract (LUT) are the storage and periodic elimination of urine. This is achieved by coordinated activity of smooth and skeletal muscles which are regulated by complex neural pathways distributed in the central nervous system (CNS), peripheral nerves and ganglia. There are substantial differences between the neurological control of the lower urinary tract and other visceral organs; unlike other viscera, the control of micturition is organized as a switch-like activity, coordinated by voluntary and involuntary neural input, which is in marked contrast to the tonic neural control characteristic of most visceral organs.

The LUT controlling innervation receives a major input from the central nervous system (CNS). Indeed, micturition is very much a voluntary act and is influenced by learned behavior that develops during maturation of the nervous system. The decision to void represents a convergence of human behavior and physiology. The degree of bladder fullness and sensations is matched with an individual's social circumstances and emotional state, allowing for micturition to be postponed, or even hastened. The degree of dependence on central neural control becomes evident when disease processes affect the CNS and result in aberrant reflexes.

The urinary tract consists of the upper urinary tract (UUT), the kidneys and ureters, and the lower urinary tract (LUT), the bladder and urethra. In health, the kidneys continuously produce urine, which then requires convenient, low-pressure drainage from the UUT to the LUT for low-pressure storage. The elimination of urine from the LUT is performed at timely intervals. Numerous mechanisms serve to prevent retrograde flow of urine to the UUT and avoid involuntary release of urine at a socially inappropriate time.

Important properties of the bladder include a high compliance for bladder filling at low pressure, and then the ability to generate an adequate pressure to release urine to completion, and co-ordinated activity of the smooth and striated muscles of the lower urinary tract (de Groat, 1998).

In this review, we will be discussing the organization of the neurological control of micturition.

THE INNERVATION OF THE LOWER URINARY TRACT

The neurological control of micturition involves complex interactions between the autonomic and somatic nervous systems that control the muscles of the LUT. The principal muscle of the bladder, the detrusor muscle, is a smooth muscle with fibers arranged in spiral, longitudinal, and circular bundles around the hollow viscus, with the urethral sphincters located at the bladder neck. The external urethral sphincter is a striated muscle at the level of the membranous urethra and is under voluntary control. The internal urethral sphincter, on the other hand, is considered as an extension of the smooth muscle fibers of the detrusor muscle and is not under voluntary control. These muscles are controlled by three sets of peripheral nerves: parasympathetic, sympathetic, and somatic nerves (Fowler et al., 2008).

Parasympathetic pathways

The sacral parasympathetic outflow is through the pelvic nerves and provides the major excitatory input to the urinary bladder (see Fig. 10.1). Preganglionic neurons are located in the intermediolateral region of the S2, S3, and S4 sacral spinal cord segments, and axons pass through the pelvic nerves to synapse at ganglia in the pelvic plexus and wall of the bladder. Although nicotinic cholinergic

*Correspondence to: Jai H. Seth, Department of Uro-Neurology, National Hospital for Neurology and Neurosurgery, University College London Hospitals Foundation Trust, London, UK. E-mail: seth.jai@gmail.com

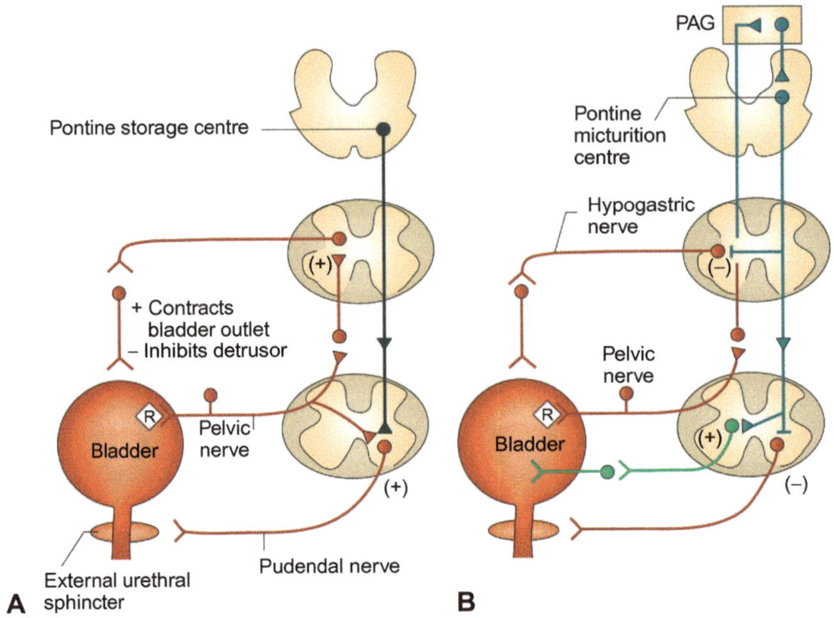

Fig. 10.1. Neural circuits that control continence and micturition. (**A**) Urine storage reflexes. During the storage of urine, distention of the bladder produces low-level vesical afferent firing. This in turn stimulates the sympathetic outflow in the hypogastric nerve to the bladder outlet (the bladder base and the urethra) and the pudendal outflow to the external urethral sphincter. These responses occur by spinal reflex pathways and represent guarding reflexes, which promote continence. Sympathetic firing also inhibits contraction of the detrusor muscle and modulates neurotransmission in bladder ganglia. A region in the rostral pons (the pontine storage center) might increase striated urethral sphincter activity. (**B**) Voiding reflexes. During the elimination of urine, intense bladder-afferent firing in the pelvic nerve activates spinobulbospinal reflex pathways (shown in blue) that pass through the pontine micturition center. This stimulates the parasympathetic outflow to the bladder and to the urethral smooth muscle (shown in green) and inhibits the sympathetic and pudendal outflow to the urethral outlet (shown in red). Ascending afferent input from the spinal cord might pass through relay neurons in the periaqueductal gray (PAG) before reaching the pontine micturition center. Note that these diagrams do not address the generation of conscious bladder sensations, nor the mechanisms that underlie the switch from storage to voiding, both of which presumably involve cerebral circuits above the PAG. R represents receptors on afferent nerve terminals. (Reproduced with permission from Fowler et al., 2008.)

receptors are involved in transmission at the ganglia, these can be modulated by activation of various receptors including muscarinic, adrenergic, purinergic, and enkephalinergic (Kawatani et al., 1983; de Groat, 2006). The postganglionic fibers release acetylcholine which excites detrusor smooth muscle, mediated by muscarinic receptors. Although both M2 and M3 receptor subtypes are expressed in the detrusor muscle, it is the M3 subtype that is the principal receptor involved in excitatory transmission (Matsui et al., 2002). Nonadrenergic-noncholinergic transmission is mediated through adenosine triphosphate (ATP), acting on P2X purinergic receptors (Burnstock, 2012). Nitric oxide (NO) released by parasympathetic nerves has an inhibitory input to the urethral smooth muscles (Andersson and Arner, 2004).

Sympathetic pathways

The thoracolumbar sympathetic outflow originates from the T12 to L2 thoracolumbar spinal segments, which travel to paravertebral ganglia. Postganglionic fibers originating from the lumbosacral sympathetic chain ganglia and inferior mesenteric ganglia travel through the hypogastric and pelvic nerves (see Fig. 10.1) (de Groat, 2006). Noradrenaline release from the postganglionic fibers inhibits the detrusor muscle through β-adrenergic receptors (Fowler et al., 2008) whereas its release in the bladder neck and urethra has an excitatory effect causing sphincteric contraction and promoting continence.

Somatic pathways

Somatic motor nerves which innervate the striated external urethral sphincter originate from a circumscribed region of the lateral ventral horn in the S2–S4 segments of the sacral spinal cord, described by neuropathologist Onufrowicz more than a century ago, and so-called "Onuf's nucleus" (Onufrowicz, 1899). Axons from these motor neurons pass through the pudendal nerve (see Fig. 10.1) and excite the external urethral sphincter muscle via cholinergic receptors causing sphincteric contraction and promoting continence (Fowler et al., 2008).

Afferent pathways

Three sets of nerves carry afferent axons innervating the urinary tract. Whereas sensations of bladder fullness are conveyed to the spinal cord by the pelvic and hypogastric nerves (Janig and Morrison, 1986), sensory input from the bladder neck and the urethra is carried in the pudendal and hypogastric nerves.

The afferent axons in these nerves consist of Aδ (thinly myelinated) and C (unmyelinated) fibers. In the bladder, the Aδ fibers tend to be located within the detrusor muscle and respond to passive distention, hence known as "tension receptors," and active contractions (Fowler et al., 2008). They conduct action potentials more rapidly than C fibers (3–10 m/s versus <2 m/s for C fibers). C fibers comprise up to two thirds of bladder afferent nerves and tend to be dispersed more widely in the detrusor and suburothelium. These fibers are insensitive to physiological filling, and hence are also termed "silent" fibers. However, they respond to noxious stimuli, provide nociception, and are sensitive to changes in temperature and pH. It is these nerves that are thought to be important in the pathogenesis of urgency and the Overactive Bladder (OAB) (Michel and Chapple, 2009).

The cell bodies of these nerves are found in the dorsal root ganglia (DRG) of the S2–S4 and T11–L2 dorsal roots. Their axons enter the dorsal horn, where they may synapse with interneurons and either transmit impulses to higher brain centers or synapse with neurons involved in spinal reflexes (Fowler et al., 2008). Afferent fibers from the bladder are thought to lie in the lateral part of the lateral column (Nathan and Smith, 1958).

In rat studies, interneurons within the dorsal commissure of the lumbosacral cord have been shown to either make local connections to form segmental spinal reflexes, or send long projections to the brain (Fowler et al., 2008). The overlap of sensory afferent and somatic nerves leaving the bladder, urethral sphincter, and urethra at the lateral dorsal horn and dorsal commissure suggests this region of viscerosomatic integration may be of importance in coordinating bladder and sphincter control (Fowler et al., 2008).

THE UROTHELIUM

It is now well recognized that the urothelium has a major role communicating with the nerves within the different layers of the bladder wall. The urothelium, comprising three main layers, forms a dynamic surface between the urine and the underlying nerves, connective tissue, cells, vessels, and muscle. The most superficial of these is the apical layer, which is composed of large hexagonal "umbrella" cells. These cells rest on an intermediate layer of columnar cells which lies on the deeper basement membrane or basal lamina (Birder, 2011).

Unmyelinated nerve fibers form a dense nexus in the suburothelial space, and occasional fibers can be seen to penetrate the basal lamina (Wiseman et al., 2002). A variety of neural receptors expressed by structures in the superficial bladder layers have been identified, including the capsaicin receptor TRPV1, purinergic receptor P2X3, and TrKA (Apostolidis et al., 2006). It is thought that the stimulation of urothelium by stretch, inflammation, acidity, and temperature change causes local release of a variety of neuropeptides, including nerve growth factor (NGF), ATP, substance P (SP), neurokinin A (NKA), and calcitonin gene-related peptide (CGRP). Some of these locally released transmitters have paracrine actions by targeting the locally expressed receptors and further modulating the expression of receptors, thus modulating afferent activity (Apostolidis et al., 2006). After a neurological or inflammatory insult, it is thought that the C fibers become the major afferent input from the bladder to the spinal tract to initiate the micturition reflex (Yoshimura et al., 2008). Increased amounts of neurotransmitter substance being released by the urothelium result in increased bladder excitability, hypersensitivity, and ultimately Overactive Bladder (OAB) symptoms (Yoshimura et al., 2008). Non-neuronal release of neurotransmitters such as ACh from the urothelium may also have a direct stimulatory effect on C fibers, local nicotinic and muscarinic receptors, implicating the urothelium in generating the sensation of urgency (Tyagi et al., 2006; Yoshida et al., 2008). *In vitro* experiments in rats have shown that after application of muscarinic agonists to the bladder, spontaneous, nonvoiding bladder contractions are initiated (Kanai et al., 2007). This observation is supported by *in vivo* studies, where administration of anticholinesterases or muscarinic agonists in animals has facilitated reflex bladder activity (Hawthorn et al., 2000). This may explain the efficacy of antimuscarinic agents, used to treat OAB, which provide an increased capacity and reduction in the sensation of urgency (Fowler et al., 2008).

There is growing evidence that the urothelium releases ATP in response to stretch, which in turn activates suburothelial purinergic receptors on afferent neurons, and so conveys information to the CNS to activate micturition (Burnstock, 2011). The purinergic P2X3 receptor appears to be important in normal bladder function, as mice lacking this receptor exhibit poor voiding and reduced bladder activity (Cockayne et al., 2000). Urothelial release of evoked ATP was increased in animal models of detrusor overactivity (DO) (spinal cord injury and cyclophosphamide induced cystitis) and botulinum toxin type A (BTX-A) application significantly reduced this evoked release when compared to controls (Smith et al., 2005). The detrusor injection of BTX-A in the bladder has been used with great success clinically to treat OAB and DO. The precise mechanism of action is unclear, but it is thought to be a dual action (Dmochowski et al., 2010). Application of ATP in rat bladders *in vivo* precipitates DO, and this effect is also reduced by BTX-A (Atiemo et al., 2005). Also, by reducing urothelial ACh release, BTX-A may block both excitatory afferent signaling and detrusor parasympathetic transmission during urine storage.

There has been much recent interest in a population of cells found to exist within the suburothelial layers of the bladder. These have been called "myofibroblasts" or "interstitial cells," and it is thought that they may play a role in the sensory response to bladder wall stimuli. The cells have been shown to be linked by gap junctions and have close contact with nerves, and it is thought that the syncytial structure so formed may act to amplify bladder stretch responses (Wiseman et al., 2003). Interstitial cells have been probed by patch clamp recordings which demonstrated an intrinsic excitability through spontaneous electrical activity after external application of ATP or carbachol, a muscarinic agonist (McCloskey, 2010). ATP-sensitive P2Y receptors have been demonstrated on the interstitial cells using immunohistochemical techniques which may imply an ability to respond to urothelial ATP release (Sui et al., 2006).

An accumulation of the evidence thus far suggests that the urothelial cells, along with these interstitial cells, nerves, various receptors and neurotransmitter substances, form a mechanosensory unit, which underlies physiological bladder sensation and function, and if disturbed can lead to lower urinary tract dysfunction.

THE NEUROLOGICAL CONTROL OF THE BLADDER

Micturition is under voluntary control, and is a learned process which develops with the maturing nervous system. To regulate micturition, a complex neural control system maintains the bladder and sphincter in the storage mode for approximately 99.8% of life, switching to the voiding mode when socially appropriate. Over the last two decades, functional brain imaging has emerged as a powerful tool to study the neurological control of the bladder.

Bladder filling

The urinary bladder functions as a reservoir, storing urine and providing continence (Panicker and Fowler, 2010). This is achieved by inhibition of bladder contraction whilst there is ongoing conscious appreciation of the extent of bladder filling and an assessment of the appropriateness of when to void. During storage, Aδ afferent nerves which originate in the bladder synapse with spinal tracts in the lumbosacral spinal cord and stimulate reflex firing in the sympathetic and somatic efferents to the smooth and striated elements of the bladder outlet, maintaining continence (de Groat, 1998). These reflexes are known collectively as the "guarding reflex" (de Groat, 1995) (Fig. 10.1). Activation of pudendal urethral efferents contracts the external urethral sphincter and prevents stress urinary incontinence.

Bladder afferent activity also ascends via pelvic afferent nerves and the spinobulbospinal pathway (de Groat, 1998), reaching the periaqueductal gray (PAG) and from thence travelling to higher centers. It is thought that through the connections between the PAG and the pontine micturition center (PMC), also known as Barrington's nucleus, switching from storage to voiding mode is enabled. While the PAG has numerous connections, the PMC has a restricted number of afferent inputs and input from the PAG determines its efferent activity, which initiates micturition (de Groat, 1998).

Functional brain imaging studies have done much to improve our understanding of the role of higher centers in the control of voluntary voiding and initiation of micturition (Griffiths and Tadic, 2008). Figure 10.2 shows

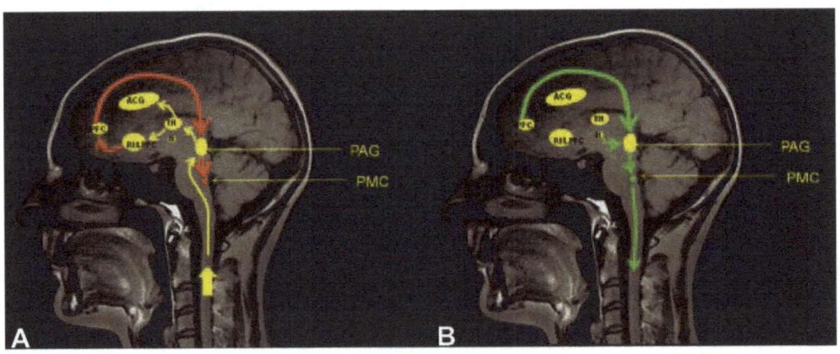

Fig. 10.2. A working model of lower urinary tract control by higher brain centers. (**A**) During storage. (**B**) Voiding. See text for a description of the sequence of events. MPFC, medial prefrontal cortex; PAG, periaqueduct gray; H, hypothalamus; PMC, pontine micturition center. (Reproduced with permission from Fowler and Griffiths, 2010.)

afferents (yellow) ascending to the midbrain PAG. From there, activity is relayed via the hypothalamus (H) and thalamus (TH) to the dorsal anterior cingulate cortex (ACC) and to the right insula (RI) and the lateral prefrontal cortex (LPFC). During storage, afferent signals pass to the medial prefrontal cortex (MPFC, red arrow) where the decision to void or not to void may be made. In the storage phase the decision is "not to void," and this situation is maintained by chronic inhibition of the PAG via a long pathway (red arrows) from the MPFC; consequently PMC activity is also suppressed, and voiding does not occur.

Voiding

Reflexes exist to promote micturition known as the "bladder-bladder reflex." The bladder afferent nerves connect with sacral spinal cord interneurons (Fig. 10.1). These interneurons synapse with preganglionic efferent parasympathetic nerves to complete this reflex pathway. The bladder-bladder reflex is an excitatory one, which becomes activated by the sensing of a full bladder but is inhibited until it is socially appropriate to void. Interneurons activated by bladder afferent fibers also synapse with urethral parasympathetic efferent nerves to form a "bladder-urethral reflex" (Leng and Chancellor, 2005). This is an inhibitory reflex, which relaxes the smooth muscle of the proximal urethra and bladder outlet to open immediately before the onset of bladder contraction.

In humans, the majority of descending fibers concerned with micturition lie in the lateral column, on an equatorial plane passing through the central canal. This location remains the same in the cervical, thoracic, lumbar, and sacral segments (Nathan and Smith, 1958).

Efferents from the PMC project to the parasympathetic motor neurons of the detrusor in the sacral cord, whilst inhibitory interneuronal activity results in suppression of activity of motor neurons in Onuf's nucleus (Blok et al., 1997), thus achieving the reciprocal synergistic contraction of the detrusor muscle and relaxation of the sphincter. The detrusor contraction and sphincter relaxation result in the elimination of urine from the bladder. Recordings from cats have shown that several neuronal populations within the PMC exhibit firing at different times during the micturition cycle. Some of these are termed *direct neurons* and fire prior to and during reflex bladder contraction; another group are called *inverse neurons* and are active during the periods between bladder contractions; there are also "on–off" neurons which fire at the beginning and end of the bladder contractions (de Groat et al., 1998). Only a few of the inverse neurons project to the lumbosacral cord, and so it is thought that they may have a local inhibitory function in the PMC.

When the bladder reaches more than 400 mL volume, a strong sensation of the urge to void is generated. It seems probable that activity in the MPFC is involved in assessment of social appropriateness of the decision to void. Figure 10.2 shows a tentative hypothesis of the series of activations in higher control centers which have been demonstrated by human functional imaging experiments. The MPFC relaxes its inhibition of the PAG (green arrow) and the hypothalamus provides a "safe" signal; consequently the PAG excites the PMC which in turn sends descending motor output (green arrow) to the sacral spinal cord that ultimately relaxes the urethral sphincter and contracts the detrusor so that coordinated voiding occurs. Voiding is continued to completion by continuing afferent input, probably to the PAG (Fowler and Griffiths, 2010).

Numerous neurotransmitter systems are thought to be involved in the control of micturition. These include excitatory substances such as including glutamic acid, tachykinins, nitrous oxide, and ATP. Glutamic acid acts on *N*-methyl-D-aspartate (NMDA) and non-NMDA receptors, and appears to be an essential transmitter in spinal and supraspinal reflex pathways (Yoshiyama and de Groat, 2005). Inhibitory amino acids glycine and γ-aminobutyric acid (GABA), and opioid peptides (enkephalins) exhibit an inhibitory control in the PMC. Other receptors can behave in an excitatory or inhibitory fashion, depending on the type and location of the receptor in the CNS. These include dopamine, serotonin, 5-hydroxytryptamine (5-HT), noradrenaline (norepinephrine), and ACh.

DISRUPTION OF CONTROL MECHANISMS

Storage phase dysfunction most commonly results from lesions of the spinal or suprapontine pathways controlling micturition. This results in involuntary spontaneous or induced contractions of the detrusor muscle (detrusor overactivity). This overactivity is commonly associated with a preceding sensory urgency which is attributed to the usually quiescent C fibers in health, which become hyperactive in pathological conditions (Michel and Chapple, 2009). Voiding phase dysfunction usually results from lesions of the spinal or infrasacral pathways.

In myelopathy, this is due to simultaneous contraction of the external urethral sphincter and detrusor muscle – detrusor sphincter dyssynergia – which can result in both incomplete bladder emptying and abnormally high bladder pressures. Impaired bladder emptying may also result from reduced parasympathetic drive from descending bulbospinal pathways causing impaired detrusor contractions. In lesions infrasacrally, such as the cauda equina (or peripheral neuropathy),

Table 10.1

Locations of different common neurological conditions which can cause lower urinary tract dysfunction

Suprapontine	Stroke
	Trauma
	Degeneration: Parkinson's disease, Alzheimer's disease, dementia with Lewy bodies
	Hydrocephalus, normal pressure hydrocephalus
	Cerebral palsy
	Neoplasm
Suprasacral/ infrapontine	Demyelination: multiple sclerosis, transverse myelitis
	Trauma
	Vascular: arteriovenous malformations, spinal cord infarction
	Neoplasm: metastasis, primary
	Hereditary: hereditary spastic paraparesis
	Infections: tropical spastic paraparesis (HTLV-I)
	Cervical spondylosis
Infrasacral	Spinal dysraphism
	Arachnoiditis
	Intervertebral disc prolapse
	Diabetes mellitus
	Iatrogenic: pelvic or retroperitoneal surgery
	Peripheral neuopathy

voiding dysfunction results from poorly sustained detrusor contractions and possibly nonrelaxing sphincters (Panicker and Fowler, 2010).

The causes for neurogenic lower urinary tract dysfunction (LUTD) can be classified according to the level of the location of pathology. These have been traditionally divided into three levels, suprapontine, suprsacral, and infrasacral, and this is illustrated in Table 10.1.

CONCLUSION

The neurological control of micturition is a complex system which involves a number of neural circuits working closely, through gated channels. The initial primitive control of filling and emptying in infancy, possible through simple reflexes, changes with age to a system dominated by input from higher centers. Although not fully uncovered, the depth of knowledge of these complex neural networks has grown significantly over the last two decades. This has given us a better insight into the control of bladder storage and voiding in health and disease. The development of this knowledge is vital to discover more effective treatments to tackle different types of neurological lower urinary tract dysfunction, such as involuntary detrusor overactivity and incomplete bladder emptying, which can cause symptoms such as urinary incontinence and urinary retention respectively. These symptoms, which are associated with significant morbidity, can be better addressed as our understanding continues, providing improvements to patient quality of life.

ACKNOWLEDGMENT

J. Seth, J. Panicker, and C. Fowler would like to acknowledge the help of University College London Hospitals/ University College London Institute of Neurology and received a proportion of funding from the UK Department of Health's National Institute for Health Research Biomedical Research Centres funding scheme.

References

Andersson KE, Arner A (2004). Urinary bladder contraction and relaxation: physiology and pathophysiology. Physiol Rev 84: 935–986.

Apostolidis A, Dasgupta P, Fowler CJ (2006). Proposed mechanism for the efficacy of injected botulinum toxin in the treatment of human detrusor overactivity. Eur Urol 49: 644–650.

Atiemo H, Wynes J, Chuo J et al. (2005). Effect of botulinum toxin on detrusor overactivity induced by intravesical adenosine triphosphate and capsaicin in a rat model. Urology 65: 622–626.

Birder LA (2011). Urothelial signaling. Handb Exp Pharmacol 202: 207–231.

Blok BF, Weerd H, Holstege G (1997). The pontine micturition center projects to sacral cord GABA immunoreactive neurons in the cat. Neurosci Lett 233: 109–112.

Burnstock G (2011). Therapeutic potential of purinergic signalling for diseases of the urinary tract. BJU Int 107: 192–204.

Burnstock G (2012). Purinergic signalling: its unpopular beginning, its acceptance and its exciting future. Bioessays 34: 218–225.

Cockayne DA, Hamilton SG, Zhu QM et al. (2000). Urinary bladder hyporeflexia and reduced pain-related behaviour in P2X3-deficient mice. Nature 407: 1011–1015.

de Groat WC (1995). Mechanisms underlying the recovery of lower urinary tract function following spinal cord injury. Paraplegia 33: 493–505.

de Groat WC (1998). Anatomy of the central neural pathways controlling the lower urinary tract. Eur Urol 34 (Suppl 1): 2–5.

de Groat WC (2006). Integrative control of the lower urinary tract: preclinical perspective. Br J Pharmacol 147 (Suppl 2): S25–S40.

de Groat WC, Araki I, Vizzard MA et al. (1998). Developmental and injury induced plasticity in the micturition reflex pathway. Behav Brain Res 92: 127–140.

Dmochowski R, Chapple C, Nitti VW et al. (2010). Efficacy and safety of onabotulinumtoxinA for idiopathic overactive bladder: a double-blind, placebo controlled, randomized, dose ranging trial. J Urol 184: 2416–2422.

Fowler CJ, Griffiths DJ (2010). A decade of functional brain imaging applied to bladder control. Neurourol Urodyn 29: 49–55.

Fowler CJ, Griffiths D, de Groat WC (2008). The neural control of micturition. Nat Rev Neurosci 9: 453–466.

Griffiths D, Tadic SD (2008). Bladder control, urgency, and urge incontinence: evidence from functional brain imaging. Neurourol Urodyn 27: 466–474.

Hawthorn MH, Chapple CR, Cock M et al. (2000). Urothelium-derived inhibitory factor(s) influences on detrusor muscle contractility in vitro. Br J Pharmacol 129: 416–419.

Janig W, Morrison JF (1986). Functional properties of spinal visceral afferents supplying abdominal and pelvic organs, with special emphasis on visceral nociception. Prog Brain Res 67: 87–114.

Kanai A, Roppolo J, Ikeda Y et al. (2007). Origin of spontaneous activity in neonatal and adult rat bladders and its enhancement by stretch and muscarinic agonists. Am J Physiol Renal Physiol 292: F1065–F1072.

Kawatani M, Lowe IP, Booth AM et al. (1983). The presence of leucine-enkephalin in the sacral preganglionic pathway to the urinary bladder of the cat. Neurosci Lett 39: 143–148.

Leng WW, Chancellor MB (2005). How sacral nerve stimulation neuromodulation works. Urol Clin North Am 32: 11–18.

Matsui M, Motomura D, Fujikawa T et al. (2002). Mice lacking M2 and M3 muscarinic acetylcholine receptors are devoid of cholinergic smooth muscle contractions but still viable. J Neurosci 22: 10627–10632.

McCloskey KD (2010). Interstitial cells in the urinary bladder – localization and function. Neurourol Urodyn 29: 82–87.

Michel MC, Chapple CR (2009). Basic mechanisms of urgency: roles and benefits of pharmacotherapy. World J Urol 27: 705–709.

Nathan PW, Smith MC (1958). The centrifugal pathway for micturition within the spinal cord. J Neurol Neurosurg Psychiatry 21: 177–189.

Onufrowicz B (1899). Notes on the arrangement and function of the cell groups in the sacral region of the spinal cord. J Nerv Ment Dis 26: 498–504.

Panicker JN, Fowler CJ (2010). The bare essentials: uro-neurology. Pract Neurol 10: 178–185.

Smith CP, Vemulakonda VM, Kiss S et al. (2005). Enhanced ATP release from rat bladder urothelium during chronic bladder inflammation: effect of botulinum toxin A. Neurochem Int 47: 291–297.

Sui GP, Wu C, Fry CH (2006). Characterization of the purinergic receptor subtype on guinea-pig suburothelial myofibroblasts. BJU Int 97: 1327–1331.

Tyagi S, Tyagi P, Van-le S et al. (2006). Qualitative and quantitative expression profile of muscarinic receptors in human urothelium and detrusor. J Urol 176: 1673–1678.

Wiseman OJ, Brady CM, Hussain IF et al. (2002). The ultrastructure of bladder lamina propria nerves in healthy subjects and patients with detrusor hyperreflexia. J Urol 168: 2040–2045.

Wiseman OJ, Fowler CJ, Landon DN (2003). The role of the human bladder lamina propria myofibroblast. BJU Int 91: 89–93.

Yoshida M, Masunaga K, Satoji Y et al. (2008). Basic and clinical aspects of non-neuronal acetylcholine: expression of non-neuronal acetylcholine in urothelium and its clinical significance. J Pharmacol Sci 106: 193–198.

Yoshimura N, Kaiho Y, Miyazato M et al. (2008). Therapeutic receptor targets for lower urinary tract dysfunction. Naunyn Schmiedebergs Arch Pharmacol 377: 437–448.

Yoshiyama M, de Groat WC (2005). Supraspinal and spinal alpha-amino-3-hydroxy-5-methylisoxazole-4-propionic acid and N-methyl-D-aspartate glutamatergic control of the micturition reflex in the urethane-anesthetized rat. Neuroscience 132: 1017–1026.

Handbook of Clinical Neurology, Vol. 117 (3rd series)
Autonomic Nervous System
R.M. Buijs and D.F. Swaab, Editors

Chapter 11

The clinical importance of the anti-inflammatory vagovagal reflex

G. BOECKXSTAENS*
Department of Gastroenterology, University Hospital Leuven, University of Leuven, Leuven, Belgium

INTRODUCTION

One of the most important and difficult tasks of the immune system is to protect the organism against foreign invaders. This implies that the immune system must defend the host against pathogens and entry of excessive microbes, for example, at the skin or intestine, but at the same time has to tolerate microbes resident in the external and internal milieu (Abraham and Medzhitov, 2011; Minnicozzi et al., 2011; Schenten and Medzhitov, 2011). The first line of defense is provided by the innate immune system, which continuously monitors the resident microbiota. By recognition of pathogen-associated molecular patterns (PAMPs), dendritic cells and phagocytic cells such as macrophages and monocytes detect the presence of pathogens initiating the appropriate cellular and molecular defense. Upon activation, phagocytes secrete cytokines and chemokines and increase antimicrobial killing, and initiate the adaptive T- and B-cell responses. This reaction has to be tightly regulated to prevent spread of inflammation and collateral tissue damage. Diseases such as rheumatoid arthritis, inflammatory bowel disease, and sepsis are examples of an unbalanced inflammatory response, more harmful than beneficial to the host. To prevent collateral damage, counter-regulatory mechanisms are triggered to dampen inflammation and restore homeostasis (Sternberg, 2006). Anti-inflammatory cytokines are secreted by immune cells to locally control cytokine production. On a systemic level, neuroendocrine mechanisms reduce inflammation by the hypothalamic–pituitary–adrenal (HPA) axis through the anti-inflammatory effect of glucocorticoids, by the hypothalamic–pituitary–gonadal axis through sex hormones, and by the hypothalamic–pituitary–thyroid hormone axis through thyroid hormones. Although modulation of the immune system by the nervous system, in particular the adrenergic nervous system, first attracted attention decades ago (Bellinger et al., 2008), interest in the role of the autonomic nervous system as an important player in immune homeostasis has recently increased exponentially. In 2000, Tracey and coworkers demonstrated that vagus nerve stimulation potently suppresses cytokine production in a rodent model of sepsis (Borovikova et al., 2000a). This discovery has led to the introduction of the concept of the "inflammatory reflex" (Tracey, 2002), a hard-wired connection between the immune and nervous system closely interacting to regulate inflammation. In the present chapter, the current knowledge and the clinical importance of this anti-inflammatory system will be discussed.

THE CHOLINERGIC ANTI-INFLAMMATORY PATHWAY

Whilst studying the anti-inflammatory effect of CNI-1493, an inhibitor of p38 MAP kinase, it became clear that this compound suppressed carrageenan-induced paw edema at doses at least 6 logs lower when injected intracerebroventricularly than required for a systemic effect (Borovikova et al., 2000b). This action was abrogated after bilateral vagotomy, while recording of the efferent vagus nerve electrical activity revealed an increase in discharge rate after infusion of CNI-1493. Similarly, electrical stimulation of the transected peripheral vagus nerve for 20 minutes also prevented the development of an acute inflammation in response to carrageenan injected in the paw. This anti-inflammatory potential of the vagus nerve was further evaluated in a model of sepsis.

*Correspondence to: Prof. Dr. G. Boeckxstaens, Department of Gastroenterology, Translational Research Center for Gastrointestinal Disorders (TARGID), University Hospital Leuven, Catholic University Leuven, Herestraat 49, 3000 Leuven, Belgium. Tel: +32-16-345750/ +32-16-330237, E-mail: Guy.Boeckxstaens@med.kuleuven.be

Injection of endotoxin (or lipopolysaccharide (LPS)), produced by all Gram negative bacteria, activates macrophages to release cytokines leading to life-threatening hypotension or septic shock. Borovikova et al. (2000a) demonstrated that vagus nerve stimulation increased survival following endotoxin by reducing cytokine (tumor necrosis factor (TNF)) production in the liver, but especially in the spleen, the major source of TNF production. This anti-inflammatory effect could be reproduced *in vitro* using isolated human macrophage cultures; the release of TNF, interleukin (IL)-1β, IL-6 and IL-18 in response to endotoxin was significantly reduced by acetylcholine and nicotine. In a search to mimic pharmacologically the effect of vagus nerve stimulation, Wang et al. (2003) identified the α7 subtype of the nicotinic acetylcholine receptor (α7 nAChR) as the main receptor by which splenic macrophages are modulated. The anti-inflammatory effect of vagus nerve stimulation is lost in α7 nAChR knockout mice, can be blocked by specific antagonists of α7 nAChR, and is mimicked both *in vivo* and *in vitro* by α7 nAChR agonists (Wang et al., 2003). Based on these findings, the "cholinergic anti-inflammatory pathway" was introduced whereby the vagus nerve modulates the immune response in the spleen, providing an additional protective mechanism to the host (Fig. 11.1). This mechanism protects against the lethal effects of cytokines by restraining the magnitude of a potentially fatal peripheral immune response (Borovikova et al., 2000a; Tracey, 2002, 2009).

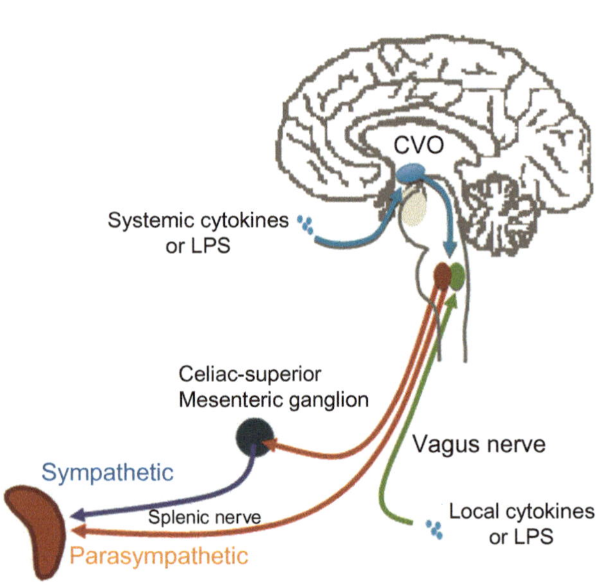

Fig. 11.1. Schematic representation of the "inflammatory reflex." Inflammation is sensed by the central nervous system via the circumventricular organs (CVO) and vagal afferents. After integration, this sensory input will trigger the efferent arm either directly or indirectly (via the celiac superior mesenteric ganglion) modulating the immune response in the spleen.

THE INFLAMMATORY REFLEX

In view of its potent anti-inflammatory properties, the cholinergic anti-inflammatory pathway has been forwarded as an additional endogenous mechanism to regulate the immune system. In comparison with the HPA axis or the local production of anti-inflammatory cytokines, cholinergic control seems to have several properties favoring a central role in immune homeostasis. Considering the speed of neural conductance, it is capable of providing an instantaneous modulatory input to the region of inflammation. Moreover, as the nervous system can adapt its output based on information obtained from various parts of the host, the modulatory influence of the cholinergic anti-inflammatory pathway is not only fast, but integrated with respect to the general wellbeing of the host. The latter, however, implies that the central nervous system receives "sensory" input on the immune status of the periphery, similar to the afferent limb of neural reflex, in order to fulfill its modulatory role.

The afferent limb of the inflammatory reflex

In case of *systemic inflammation*, cytokines and/or endotoxin are present in detectable amounts in the circulation. Animal studies have clearly shown that their presence is sensed by the brain at the level of the circumventricular organs, i.e., the area postrema, the organum vasculosum of the lamina terminalis, the subfornical organ, and at the level of the brainstem the area postrema (Fig. 11.1). Importantly, these areas are characterized by a leaky blood–brain barrier where information of the blood can be signaled to specialized neurons. Intravenous injection of endotoxin indeed induces signs of neural activation revealing neurons containing c-Fos in the circumventricular organs (Buijs et al., 2008). As these areas project to the autonomic motor neurons in the brainstem connected to the spleen, one can postulate that systemic inflammation with circulating cytokines is detected by the brain subsequently activating the cholinergic input to the spleen, thereby modulating the immune response.

The physiological role of the neuromodulation of the spleen (immune system) has been compared with the autonomic regulation of homeostasis as demonstrated in other organs, e.g., the heart (Tracey, 2009). It can be postulated that the neural input to the spleen determines the set point for the magnitude of the innate immune response towards pathogens or tissue damage. Decreased or absent neural input will increase the set point with exaggerated proinflammatory responses and tissue damage, whereas increased input results in inadequate immune defense. As part of this concept, the "tone" of the nervous input to the immune system may thus have a major impact on the host defense and development of disease.

In case of *localized peripheral inflammation*, cytokine levels are absent or too low to activate the circumventricular organs. Yet, the brain is informed about the presence of inflammation through cytokine (IL-1) receptors on vagal afferents and glomus cells adjacent to the vagus nerve (Watkins et al., 1995; Goehler et al., 1997; Hansen et al., 2001; Marquette et al., 2003) (Fig. 11.1). Brains of mice infected with *Campylobacter jejuni*, for example, revealed activation of visceral sensory nuclei in the brainstem (nucleus of the solitary tract and the lateral parabrachial nucleus), even though cytokines (TNF, IL-6, IL-1β) could not be detected in the serum (Marvel et al., 2004). Similarly, we found activation of the nucleus of the solitary tract in response to subtle intestinal inflammation evoked by manipulation of the intestine. Interestingly, motor neurons of the dorsal nucleus of the vagus nerve connected to the inflamed area were also activated compatible with the existence of a hard-wired inflammatory reflex (Cailotto et al., 2012).

The efferent limb of the inflammatory reflex

As the spleen is the major source of TNF and HMGB1, two key cytokines in the pathophysiology of sepsis (Huston et al., 2006, 2008; Qin et al., 2006), it represents an ideal target for neural modulation of the immune response. Indeed, initial experiments indicated the spleen as target organ of the vagus nerve for controlling TNF production during sepsis (Huston et al., 2006). The anatomy of the innervation, in particular the cholinergic or vagal component, is, however, still a matter of debate. Close interposition with nerve terminals could be identified, but these fibers were adrenergic in nature. Recent studies in fact suggest that vagus nerve stimulation indirectly affects splenic function by activating α7 nAChRs on adrenergic neurons located in the celiac ganglion (Fig. 11.1). Although these data indicate adrenergic modulation of splenic macrophages, *in vitro* data strongly suggest cholinergic inhibition of splenic macrophages via α7 nAChRs. One explanation could be that acetylcholine is synthesized by splenocytes when activated, providing a negative feedback to regulate its cytokine synthesis. Alternatively, the spleen is innervated by the vagus nerve but the immunohistochemical techniques used so far are not sensitive enough to depict cholinergic nerve fibers. As pointed out by Buijs et al. (2008), the lack of immunohistochemical evidence may not exclude vagal innervation of the spleen. Retrograde tracing studies using pseudorabies virus clearly illustrated parasympathetic input to the spleen via motor neurons located in the brainstem. Moreover, selective parasympathetic denervation of the spleen blocked the modulatory effect of the vagus nerve on antibody production, whereas sympathetic denervation had no effect. Clearly, additional research is required to exactly determine the neuroimmune interaction within the spleen.

The situation may be entirely different in the case of more localized or subtle inflammation. If the concept of an inflammatory reflex implies that vagal efferents dampen the immune cells at the site of inflammation, vagal fibers should innervate almost each part of the organism. Skin and joints, for example, are most likely not densely innervated by the vagus nerve, yet abundant data are available indicating an important role of the cholinergic inflammatory pathway in acute inflammation of the skin and in models of arthritis (Borovikova et al., 2000b; Saeed et al., 2005; van Maanen et al., 2009b). Potential explanations may be that vagal innervation of lymph nodes or blood vessels supplying the inflamed area may mediate the anti-inflammatory control. In the skin, for example, acute inflammation can be induced by injection of the irritant carrageenan in an air pouch created in the subcutis, a process that is reduced by vagus nerve stimulation and nicotine treatment (Saeed et al., 2005). Both *in vivo* and *in vitro* data reveal that this effect results from reduction of the endothelial expression of adhesion molecules (VCAM-1, E-selectin) and cytokine production (Saeed et al., 2005), suggesting that cholinergic innervation of blood vessels may in fact be a potent mechanism to interfere with leukocyte migration. Interestingly however, in the same model, Huston et al. (2009) provided evidence that the anti-inflammatory effect of vagal nerve stimulation resulted from altered expression of CD11b, a β2-integrin involved in cell adhesion and leukocyte chemotaxis, on the surface of neutrophils. Only in mice with an intact and innervated spleen was vagus nerve stimulation capable of exerting its anti-inflammatory effect. These findings would indicate also that peripheral inflammation is mainly controlled in the spleen.

THE GASTROINTESTINAL IMMUNE SYSTEM AND THE CHOLINERGIC ANTI-INFLAMMATORY PATHWAY

The gastrointestinal tract is of major interest as it receives the majority of the vagal efferent nerve fibers and harbors the most immune cells in the body. The mucosal immune system is constantly challenged by the intestinal microbiota, requiring a perfectly balanced equilibrium between tolerance and defense against foreign antigens (Abraham and Medzhitov, 2011). Deceased tolerance to microbiota is proposed as the main pathogenetic mechanism in inflammatory bowel disease (Crohn's disease and ulcerative colitis) characterized by severe inflammation of the intestinal mucosa. On the other hand, food allergy is another well-known example of loss of tolerance in which the immune system reacts to food antigens potentially leading to severe anaphylactic shock and even death (Bischoff, 2007, 2010). In both cases, the immune system overreacts to innocent antigens leading to severe tissue damage, morbidity, and mortality. It is becoming increasingly clear that the microenvironment in the mucosa and submucosa determines the immune response to initial exposure to foreign antigens. Key players in this process are the antigen-presenting cells, in particular the dendritic cells (DCs) (Rescigno, 2010). DCs are specialized cells that orchestrate both innate and adaptive immune responses. After uptake of antigens, they migrate from the mucosa to the mesenteric lymph nodes to instruct T cells to become effector cells (Th1, Th2, or Th17) important for immunity or induce tolerance via the generation of regulatory T cells. How the DC instructs the naïve T cells in the lymph node is strongly determined by the microenvironment where the DC encounters the respective antigen. The intestinal epithelium, for example, releases factors such as TGF-β, thymic stromal lymphopoietin (TSLP), and retinoic acid which have been demonstrated to be important for induction of tolerogenic DCs (Rescigno, 2010). Similarly, mucosal macrophages are more specialized in phagocytosis, although they also contribute to antigen presentation, and play a crucial role in the innate response to pathogens. Given the potent anti-inflammatory effect of the cholinergic innervation, one might assume that the cholinergic tone in the submucosal compartment may have an important impact on mucosal immune homeostasis. Studies in vagotomized mice indeed revealed increased susceptibility to development of colitis after exposure to the mucosal irritant dextran sulfate sodium (DSS) (Ghia et al., 2006, 2008; O'Mahony et al., 2009). Notably, reduced mucosal levels of acetylcholine in a murine model of depression were also associated with a more severe colitis in response to DSS (Ghia et al., 2008, 2009). As adoptive transfer of macrophages from depressive mice induced a similar increased susceptibility to the development of DSS colitis (Ghia et al., 2011), macrophages were proposed to be (one of) the target cells modulated by the cholinergic tone in the submucosal microenvironment. Obviously, other immune cells (T cells, DCs, mast cells) residing in the mucosa/submucosa carry nicotinic receptors and may be affected by the anti-inflammatory pathway as well.

Another important but less studied population of intestinal immune cells are the macrophages located between the longitudinal and circular muscle layer at the level of the myenteric plexus. These resident macrophages play a role in diabetes-induced gastroparesis (Choi et al., 2008, 2010), postoperative ileus (Kalff et al., 1998; Wehner et al., 2007; Boeckxstaens and de Jonge, 2009), and LPS-induced septic ileus (Eskandari et al., 1997, 1999; Turler et al., 2002; de Winter et al., 2005), and seem to represent the gatekeepers of the enteric nervous system, or the "little brain of the gut." Vagus nerve stimulation prevents muscular inflammation and ileus following intestinal manipulation (de Jonge et al., 2005), suggesting that this subpopulation of macrophages is also under cholinergic control.

Given the anatomy of the vagal innervation of the intestinal tract, it is unlikely that vagal efferents directly interact with macrophages or DCs (Fig. 11.2). In general, vagal efferents synapse with postganglionic neurons close to or within the end organ. In the gut, using an anterograde tracing technique, we confirmed that the vagus nerve mainly synapses with the myenteric plexus and submucosal plexus, but maintains no direct contact with the mucosal/submucosal or myenteric/muscular network of immune cells. Although cholinergic nerve fibers are abundantly detected in the vicinity of immune cells, these fibers are not vagal but rather of enteric origin. If the modulation of the gastrointestinal immune system occurs within the gut wall, it most likely occurs via interaction with the enteric nervous system.

By determining the cholinergic tone in the enteric nervous system, the vagal innervation may modulate the intestinal microenvironment (Fig. 11.3). Hence, the set point (balance) of the gut immune system will be affected, influencing not only macrophages or DCs, but theoretically any immune cell carrying cholinergic receptors, i.e., T and B lymphocytes, monocytes, and mast cells (de Jonge and Ulloa, 2007). Alternatively, mesenteric lymph nodes or the spleen may be the site of neuroimmune interaction, although splenic denervation did not abolish the anti-inflammatory effect in our model of postoperative ileus (Matteoli et al. (in press)).

THE α7 NICOTINIC RECEPTOR AS PHARMACOLOGICAL TARGET

As noted earlier, Wang et al. identified the α7 nAChR as the cholinergic receptor responsible for the reduction in cytokine secretion and increased survival in a model of sepsis following vagus nerve stimulation (Wang et al., 2003). This receptor is a well-characterized member of the ligand-gated ion channel superfamily (Sharma and Vijayaraghavan, 2008), and until the discovery of the cholinergic anti-inflammatory pathway, considered as a prototype nicotinic receptor involved in synaptic neurotransmission. In neurons, nicotinic receptors are pentameric complexes composed of α and β subunits assembled to form a central ion channel that is gated

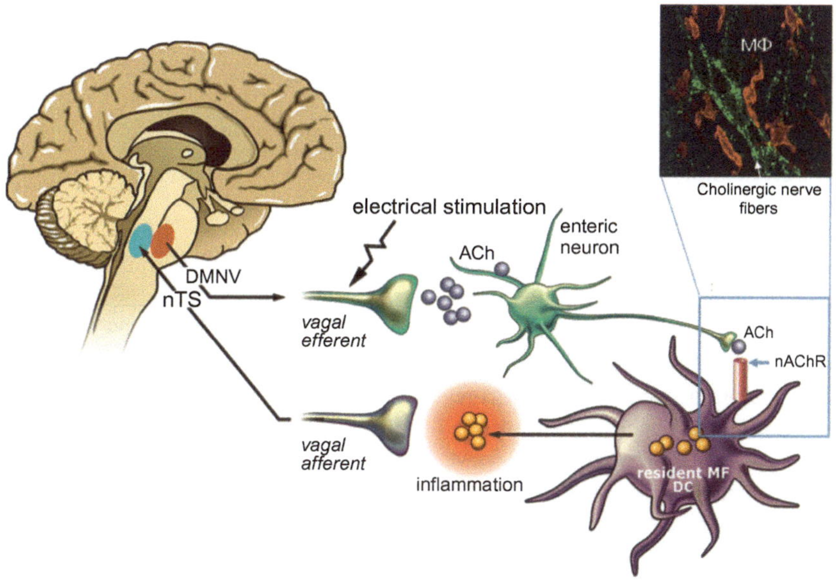

Fig. 11.2. Schematic representation of the neuroimmune interaction within the gastrointestinal wall. The function of immune cells in the mucosal/submucosal and the muscular compartment are modulated by the cholinergic tone generated by cholinergic neurons under vagal control. (Modified from Van der Zanden et al., 2009.)

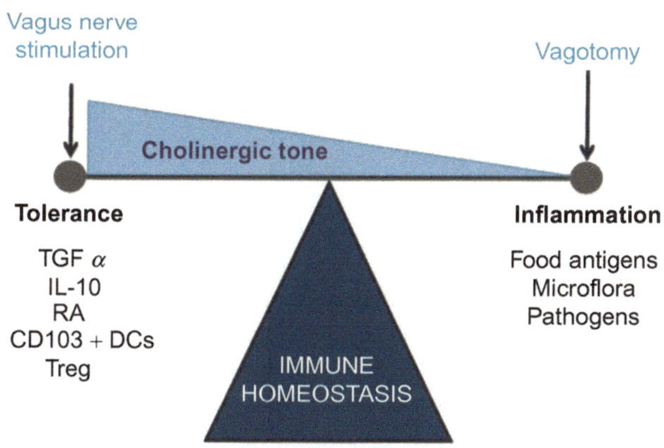

Fig. 11.3. The cholinergic tone determines immune homeostasis, either shifting the balance towards tolerance (normal to enhanced tone) or inflammation (decreased tone).

by binding of acetylcholine or other nicotinic ligands. The α7 nAChR is composed of five α7 subunits, each with a molecular weight of approximately 56 kDa. Each subunit is encoded by 10 exons, with four transmembrane domains, three putative glycosylation sites located extracellularly and three competitive agonist-binding sites (Gault et al., 1998). Human, rat, and mouse α7 nAChR are more than 90% homologous (Gallowitsch-Puerta and Tracey, 2005). The assembled subunits form a central pore with ligand binding at subunit junctions responsible for changes in the state of the receptor. The process of opening of the pore in response to agonist binding is very complex and certainly not a subject of the current chapter. It should be emphasized, though, that functional coupling of binding domains and pore domains strongly depends on the complex interaction of protein loops (Bouzat et al., 2004). Altered folding and interaction of protein loops induced by changes in amino acid sequences results in alterations of agonist binding (including α-bungarotoxin) and failure of pore opening in response to ACh binding (Bouzat et al., 2004). The high Ca^{2+} permeability of the α7 nAChR suggests involvement of Ca^{2+}-dependent events upon receptor activation in neurons (Berg and Conroy, 2002). Finally, it is important to mention that the snake toxin α-bungarotoxin is a specific and high-affinity antagonist for the α7 nAChR (Garcia-Guzman et al., 1995), functionally differentiating this receptor subtype from other neuronal nAChRs. The binding domain for this toxin is contained in the extracellular N terminal of α7 nAChR, encoded by exons 1–5 (McLane et al., 1990).

The human α7 nAChR gene (*CHRNA7*) has been mapped to chromosome 15q14. Interestingly, six mRNA splice variants in addition to the wild type α7 gene have been described (Gault et al., 1998). To date, however, it remains uncertain if any of the splice variants is expressed in cytokine-producing cells. In addition to splice variants, a partial duplication of the human α7 nAChR gene has been described (Gault et al., 1998). Exons 5–10 of the wild-type gene have been duplicated and combined with four novel exons (A–D) to form a new gene, the "duplicated α7" or the "cholinergic receptor family with sequence similarity 7A" (*CHRNA7-FAM7A*), which also maps on chromosome 15, about 2 Mb from *CHRNA7*. To date, there is evidence that this gene is transcribed but evidence of translation is lacking, and its possible function is unknown (Garcia-Guzman et al., 1995).

Immune cells also express other nicotinic receptors than the α7 nAChR. For example, rat alveolar macrophages do not express α7 nAChR, but instead are equipped with α9, α10, β1 and β2 subunits. Nicotine, however, did reduce the excitatory response of the alveolar macrophages to ATP, but without inducing changes in intracellular Ca^{2+} or membrane current (Mikulski et al., 2010). Notably, expression of different subunits can be up- or downregulated by macrophages *in vitro* in response to stimulatory triggers such as endotoxin (Chernyavsky et al., 2010). Moreover, vagus nerve stimulation not only dampens cytokine production, but also increases phagocytic capacity of intestinal macrophages (van der Zanden et al., 2009), an effect mediated by α4β2 nAChRs. On the other hand, α7 nAChR-deficient mice show an increased severity of experimental colitis, responsive to nicotine, suggesting a role for this receptor subtype in modulation of inflammation (Orr-Urtreger et al., 2005). So clearly other receptors than the α7 nAChR may be involved.

Nevertheless, as the α7 nAChR is abundantly expressed and has been identified to play a crucial role in the cholinergic modulation of the immune response, this receptor is now considered an important therapeutic target for several immune-mediated inflammatory disorders (see below) (Fig. 11.4). Selective nicotinic agonists were originally designed to mimic the cognitive effects of nicotine in patients with neurological disorders while avoiding the toxicity of nicotine. The most characterized specific α7 nAChR agonists are GTS21, 4OHGTS, ARR17779, CAP55, Exo2, and PNU-282987 (Ulloa, 2005). GTS21 has been evaluated in patients with schizophrenia (Tregellas et al., 2010, 2011) and healthy subjects (Kox et al., 2011) and was well tolerated, suggesting that α7 nAChR agonists may represent an interesting group of compounds for human use to be evaluated as new class of anti-inflammatory agents.

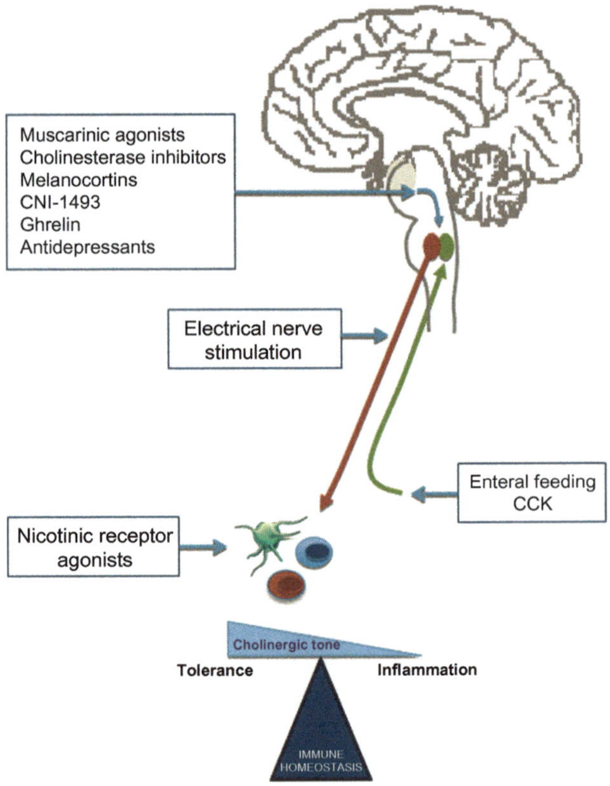

Fig. 11.4. The cholinergic anti-inflammatory pathway as therapeutic tool to modulate the immune system.

ALTERNATIVE APPROACHES TO MIMIC/ STIMULATE THE CHOLINERGIC ANTI-INFLAMMATORY PATHWAY

As electrical stimulation of the cervical vagus nerve requires a surgical procedure to implant the electrode and a pacemaker, this approach to treating patients with immune-mediated inflammatory disorders may seem to belong to science fiction (Fig. 11.4). Notwithstanding, it is currently safely used in patients with intractable epilepsy, depression, and migraine (Groves and Brown, 2005; Beekwilder and Beems, 2010). Although the exact mechanism of action is unknown, activation of vagal afferents is believed to alter neurotransmission in the brain, i.e., in the thalamus and thalamocortical projection pathways, and the limbic system (Beekwilder and Beems, 2010). Preliminary data on its effect on the immune system are somewhat conflicting, showing no effect or even an increase in both pro- and anti-inflammatory circulating cytokines (Corcoran et al., 2005; Barone et al., 2007) in patients with refractory epilepsy. In contrast, De Herdt et al. demonstrated a significant decrease in IL-8 induction by LPS-stimulated peripheral blood mononuclear cells after 6 months of vagus nerve stimulation in comparison to the prestimulation state in patients with refractory epilepsy (De Herdt et al., 2009). Alternatively, the vagus nerve can be stimulated transcutaneously by application of electrical stimuli to the sensory auricular branch of the vagus nerve (Fallgatter et al., 2003, 2005). However, to what extent vagal efferents are activated using this approach remains unclear.

Interestingly, several compounds have been identified with anti-inflammatory properties due to central activation of the cholinergic anti-inflammatory pathway (Fig. 11.4). The cholinesterase inhibitors galantamine, physostigmine, neostigmine, and the highly selective centrally-acting huperzine A have been shown to improve survival and decrease proinflammatory cytokines in experimental sepsis (Hofer et al., 2008; Pavlov et al., 2009; Peter et al., 2010), most likely by activation of the cholinergic anti-inflammatory pathway via central muscarinic receptors (Pavlov et al., 2006). A similar mechanism of action has been ascribed to CNI-1493, a p38 MAP kinase inhibitor with muscarinic agonistic properties shown to be effective in endotoxin-induced shock, necrotizing enterocolitis, acute carrageenan-induced paw edema and postoperative ileus (Borovikova et al., 2000b; Bernik et al., 2002; Zamora et al., 2005; The et al., 2011). Ghrelin (Wu et al., 2007, 2008; Cheyuo et al., 2011) and melanocortins (via the melanocortin 4 receptor) (Guarini et al., 2004; Ottani et al., 2009; Bitto et al., 2011; Minutoli et al., 2011) also improve outcome and reduce the inflammatory response in experimental models of sepsis, ischemic stroke, myocardial ischemia/reperfusion, multiple organ dysfunction, pancreatitis, hemorrhagic shock, and intestinal ischemia-reperfusion, an effect blocked by vagotomy, illustrating activation of the cholinergic anti-inflammatory pathway.

Finally, physiological stimulation of vagal afferents has been elegantly shown to suppress inflammation and restore organ function. Enteral feeding, especially with lipid-enriched enteral nutrition, leads to chylomicron formation, subsequently triggering cholecystokinin release and activation of CCK-1 receptors on the afferent limb of the vagal anti-inflammatory pathway (Fig. 11.4). Reduction of proinflammatory cytokines, improvement of intestinal barrier function, and improved outcome have been demonstrated in models of hemorrhagic shock, endotoxemia, Gram negative sepsis, and postoperative ileus (Lubbers et al., 2009, 2010a, b, 2011). These data indicate that treatment of some life-threatening conditions may be much easier than anticipated.

CLINICAL SIGNIFICANCE OF THE CHOLINERGIC ANTI-INFLAMMATORY PATHWAY

Preclinical evidence

In general, the significance of new discoveries in medicine is largely determined by its impact on clinical management. Accepting the general principle that the brain, through the vagus nerve, fulfills an important role in modulating the immune system, the next important question clearly is how this knowledge can be translated into improved clinical management of immune-mediated inflammatory disease. Preclinical models have provided a plethora of evidence supporting a beneficial effect of activation or mimicry of the cholinergic anti-inflammatory pathway in a wide variety of disorders. Most studies reported to date focus on acute innate responses triggered by ischemia, tissue damage, or bacterial products (endotoxin), i.e., conditions involving sterile inflammation. Before discussing these in more detail, it is of utmost importance to stress that dampening the immune response may have detrimental effects in case of infectious diseases. Although vagus nerve stimulation and nicotine increase intestinal macrophage phagocytosis (van der Zanden et al., 2009), reduced cytokine and chemokine production leads to impaired attraction and activation of neutrophils and T lymphocytes, impairing bacterial clearance. In a model of bacterial peritonitis, α7 nAChR knockout mice efficiently cleared infection with *Escherichia coli* from their peritoneal cavity and had sterile blood cultures, whereas wild-type animals had high bacterial loads at the primary site of infection and were bacteremic (Giebelen et al., 2008). The knockout animals displayed a more robust influx of neutrophils early after the infection, resulting in improved bacterial clearance, as evidenced by the much lower bacterial burden in liver, spleen, kidney, and lungs. Similarly, impaired survival was also observed after nicotine treatment in septic peritonitis induced by intraperitoneal injection of *E. coli* or fecal contamination (van Westerloo et al., 2005; Boland et al., 2011). Also in a model of colitis, selective α7 nAChR agonists were reported to worsen disease (Snoek et al., 2010), again illustrating that dampening cytokine production is not always beneficial to the host, especially in infectious conditions.

Finally, although most studies so far have focused on the innate immune system, modulation of the adaptive immune response is to be anticipated, especially the DCs and macrophages as key players initiating this process. One study elegantly demonstrated that antibody production against trinitrophenyl-ovalbumin was dependent on vagal innervation, suggesting that the adaptive immune system is under cholinergic control (Buijs et al., 2008). If future studies confirm this, the therapeutic and clinical impact may be even larger than assumed so far.

Preclinical evidence supporting the potential therapeutic effect of the cholinergic anti-inflammatory pathway has been recently reviewed in great detail (Pavlov, 2008; Bencherif et al., 2011). Here, I will summarize data relevant to major immune-mediated inflammatory diseases and clinical practice.

Sepsis

Most studies on the significance of the cholinergic anti-inflammatory pathway focus on sepsis, especially as this life-threatening condition remains a daily challenge to the treating physicians. Even the introduction of anti-TNF antibodies, a major recent medical advance, to the therapeutic arsenal has failed to satisfy the high expectancies of physicians, i.e., mortality rates are not significantly affected by anti-TNF treatment (Panacek et al., 2004).

Most likely, the failure of anti-TNF treatment results from the fact that many other cytokines are involved in the pathogenesis of sepsis, and most importantly, survival is determined by HMHB1 (high mobilbity group box 1 protein) levels, the late mediator of lethal systemic inflammation in sepsis (Wang et al., 1999), rather than by TNF.

In the majority of preclinical studies, sepsis is evoked by injection of endotoxin. In these studies, activation of the vagus nerve (either electrically or via enteral feeding) and nicotinic receptor activation have been abundantly proven to be very effective in improving survival, blood pressure, and intestinal barrier function and reducing inflammation, cytokine release, and organ damage (Borovikova et al., 2000a; Wang et al., 2003, 2004; Lubbers et al., 2010a). As previously indicated, outcome following therapeutic interventions, either electrical stimulation or nicotinic receptor activation, is, however, strongly dependent on the presence of a bacterial infection. Improved survival in the cecal ligation and puncture model, for example, is only achieved if antibiotics are started soon after contamination of the peritoneal cavity with fecal contents (Borovikova et al., 2000a), to prevent bacterial dissemination. Obviously, this resembles the clinical situation where antibiotics compose a major part of treatment. In clinical practice, however, patients will not be treated before sepsis is established, as in most experimental protocols. Patients are referred to the intensive care unit mainly as the clinical picture deteriorates and the inflammatory cascade leading to sepsis has reached a critical phase. Wang et al. nicely demonstrated that even when treatment with nicotine was started after the onset of sepsis, HMGB1 levels were dramatically reduced, leading to improved survival (Wang et al., 2004). This effect was absent in α7 nAChR knockout mice and mimicked by the α7 nAChR agonist choline (Parrish et al., 2008). The major progress here is the fact that the release of several cytokines, including HMGB1, by immune cells, including, most importantly, macrophages, is reduced. Based on these considerations, activation of the cholinergic anti-inflammatory pathway, either by α7 nAChR agonists or via one of the other potential approaches discussed above, could represent an important therapeutic step forward in the treatment of sepsis.

Hemorrhagic shock and ischemic (reperfusion) tissue damage

Ischemia due to severe blood loss (hemorrhagic or hypovolemic shock), cardiac arrest or arterial occlusion results in severe cell damage and organ injury. Not only tissue damage but also endotoxemia resulting from disruption of the intestinal barrier due to ischemia are potent triggers for the innate immune system, explaining the systemic inflammatory response. Tissue macrophages, or microglia in the brain, are activated and secrete large amounts of TNF and other cytokines, impairing the function of life-supporting organs such as lung, liver, and intestine. In addition, recent data indicate that brain function may be significantly affected as well. Even if patients, for example, survive sudden cardiac arrest as a result of successful cardiopulmonary resuscitation, brain function is often irreversibly hampered, with long-term neurological, psychological, and physical consequences of global cerebral ischemia (Bunch et al., 2003). Neuronal damage or cell death and disruption of blood–brain barrier induced by inflammatory cytokines, metalloproteases, nitric oxide, and reactive oxygen species released by activated microglia and macrophages are believed to underlie this syndrome (Danton and Dietrich, 2003).

Activation of the cholinergic anti-inflammatory pathway has been reported to be effective in reducing neuroinflammation and neuronal damage following cardiac arrest (Norman et al., 2011). Global cerebral ischemia activated microglia, increased mRNA expression of the proinflammatory cytokines TNF, IL-1, and IL-6, and decreased the number of choline acetyltransferase + neurons and the level of acetylcholinesterase in the hippocampus and forebrain of mice. Interestingly, treatment with the selective α7 nAChR agonist GST-21 administered 24 hours *after* the cardiac arrest and resuscitation significantly decreased neuroinflammation and neural damage. These data indicate that pharmacological activation of α7 nAChRs provides significant protection against ischemia-related cell death and inflammation, even if applied after the ischemic event. Similar results were reported for neuroinflammation in the brain following intracerebroventricular injection of endotoxin in rats (Tyagi et al., 2010a, b).

Also ischemia-reperfusion damage in intestine, kidney, heart, and liver is reduced by the activation of the cholinergic anti-inflammatory pathway (Luyer et al., 2005; Crockett et al., 2006; Sadis et al., 2007; de Haan et al., 2008, 2010; Wu et al., 2008; Yeboah et al., 2008; Ottani et al., 2010). Administration of nicotinic agonists, ghrelin, melanocortins, and even enteral feeding introduced after the episode of ischemia reduces the inflammatory response and restores organ function in this variety of preclinical models. Electrical stimulation of the vagus nerve during acute myocardial infarction reduced TNF production and tissue damage in a rat model (Kong et al., 2011), further illustrating the large potential of this approach.

Inflammatory bowel disease

Inflammatory bowel disease (IBD) is a debilitating and chronic inflammatory disease affecting the gastrointestinal tract (Rutgeerts et al., 2009). Based on clinical presentation, endoscopic appearance and histology, two major subtypes of the disease have been identified, namely Crohn's disease (typically located in the terminal ileum but any region of the gut can be affected) with transmural inflammation and ulcerative colitis (typically affecting the large intestine) which involves mostly superficial inflammation confined to the mucosa. In the past decade, treatment of IBD has improved significantly with the introduction of biologicals such as anti-TNF antibodies (Danese et al., 2011; Van Assche et al., 2011). Nevertheless, there is much room for improvement as a significant proportion of patients still require intestinal resection for intractable disease or complications such as abcesses, stricture formation, or fistulae.

Preclinical models provide convincing evidence that the cholinergic innervation of the gut has a major impact on the intestinal immune system (see above). Vagotomized animals develop more severe colitis (both microscopic and macroscopic), with increased levels of NF-κB and cytokines such as IL-1β, IL-6 and TNF-α following DSS administration (Ghia et al., 2006, 2007; O'Mahony et al., 2009), an effect that was reversed by pretreatment with nicotine. In the same line, the acetylcholinesterase inhibitors neostigmine or physostigmine attenuated colitis evoked by intrarectal dinitrobenzene sulfonic acid (Miceli and Jacobson, 2003). As DSS colitis was more severe in α7 nAChR knockout mice and the α7 nAChR agonist choline chloride improved macroscopic colitis, myeloperoxidase (MPO), IL-6 and IL-1β levels (Ghia et al., 2009), α7 nAChR receptors were identified as cholinergic receptors meditating the anti-inflammatory effect.

In clinical practice, exacerbations of colitis, especially of ulcerative colitis, may coincide with episodes of depression or psychological stress. Interestingly, increased susceptibility to developing colitis was also observed in an animal model of depression (maternal separation), a phenomenon that was associated with a reduced level of acetylcholine in the intestine and could be reversed by treatment with an antidepressant (Ghia et al., 2008, 2009). The beneficial effect of the antidepressant was eliminated by vagotomy and most interestingly normalized the intestinal acetylcholine levels. The exact underlying mechanism is not completely understood, but increased susceptibility to colitis can be transferred to a recipient mouse by transfusing macrophages isolated from mice with depression-like behavior. These data would suggest that similar to sepsis, macrophages (via α7 nAChR) are the target cell of the cholinergic anti-inflammatory pathway. Another study, however, provided evidence that $CD4^+$ T lymphocytes may be involved. Transfer of $CD4^+$ $CD25^-$ (activation marker) T lymphocytes isolated from vagotomized animals resulted in increased susceptibility to the development of DSS colitis. Clearly, more work is needed. Nevertheless, the risk of developing colitis seems to depend on the intestinal cholinergic tone due to its impact on the intestinal immune system. Of significance to the human situation, depression and other psychological conditions such as stress and anxiety may trigger exacerbations by reducing this cholinergic tone.

Based on this knowledge, nicotinic agonists have been evaluated as a potential new pharmacological approach. In addition to nicotine, α7 nAChR agonists such as choline and anabaseine have been evaluated in models of colitis. As discussed above, choline effectively improved DSS colitis and reduced cytokine production. In a model of trinitrobenzene sulphonic acid (TNBS) colitis, mice treated with the α7 nAChR agonist anabaseine developed less weight loss and less severe colitis and lower levels of MPO, NF-κB and TNF (Bai et al., 2007). Taken together, these data would favor α7 nAChR agonists as treatment for IBD; however, we obtained opposite results using the specific α7 nAChR agonists AR-R17779 and GSK1345038A. Although both agonists reduced NF-κB transcriptional activity, IL-6 and TNF release by peritoneal macrophages stimulated with LPS or heat-killed *E. coli*, clinical parameters (body weight, colon weight, and length), IL-6 and IL-17 levels were not augmented *in vivo* in two different colitis models, i.e., DSS- and TNBS-induced colitis (Snoek et al., 2010). Most likely, the protective effect of α7 nAChR activation against excessive inflammation and collateral tissue damage compromises the host defense against the increased exposure to intraluminal microbiota following mucosal damage, suggesting that care should be taken using these compounds in patients with IBD.

Arthritis

Somewhat related to IBD, rheumatoid arthritis (RA) is characterized by chronic inflammation of the joints, involving cytokines such as IL-1, IL-6, and TNF that play an important role in tissue damage, i.e., cartilage and bone destruction in the rheumatoid joint. The introduction of anti-TNF antibodies to the therapeutic armamentarium has greatly improved clinical management of patients with RA. Nevertheless, there is still a great need for more effective drugs.

Although there is no explicit evidence that joints are innervated by the vagus nerve, clinical arthritis scores and synovial inflammation are markedly increased in α7 nAChR-deficient mice in both the acute and chronic phase of the disease (van Maanen et al., 2010). Moreover, treatment with the α7 nAChR agonist AR-R17779 and vagus nerve stimulation ameliorated collagen-induced arthritis whereas unilateral vagotomy exacerbated the disease (Zhang et al., 2008; van Maanen et al., 2009a). Of note, AR-R17779 was more potent than nicotine and was associated with delayed onset of the disease and protection against joint destruction. Especially as AR-R17779 poorly crosses the blood–brain barrier, its effect is attributable to peripheral interaction reducing the risk of central side-effects. The target cells in the joint expressing α7 nAChR are located in synovial tissue on RA fibroblast-like synoviocytes, and most likely monocytes and macrophages. Taken together, these data strongly support the concept that stimulation or mimicry of the cholinergic anti-inflammatory pathway may be an innovative treatment of RA (van Maanen et al., 2009b; Koopman et al., 2011); however, future clinical studies in humans will have to confirm the relevance of these findings for clinical practice.

Postoperative and endotoxin-induced ileus

Each abdominal surgical intervention leads to impaired motility of the entire gastrointestinal tract lasting several days, with symptoms such as nausea, vomiting, intolerance to food, and absence of defecation, referred to as postoperative ileus (POI). Although some would argue that this represents a physiological response to the surgical insult and should not be regarded as a clinical problem, this iatrogenic condition is a major source of patient morbidity and a significant economic burden to health care.

For over a decade, a subtle microscopic inflammation of the intestinal muscularis, triggered by activation of resident macrophages residing between the longitudinal and circular muscle layer of the intestinal wall, has been identified as a key process in the pathophysiology of POI and endotoxin-induced ileus (Eskandari et al., 1997; Kalff et al., 1998, 1999). Activation of these phagocytes results in cytokine and chemokine release, followed by influx of mainly leukocytes and monocytes starting approximately 3–4 hours after surgery. As this inflammatory response has a major impact on neuromuscular function, compounds or interventions hampering this process may be instrumental in reducing POI (Boeckxstaens and de Jonge, 2009).

Several lines of preclinical evidence indicate that activation of the cholinergic anti-inflammatory pathway may be an efficient strategy to prevent POI. Vagus nerve stimulation via electrical current (de Jonge et al., 2005), enteral feeding (Lubbers et al., 2009) or central application of semapimod (The et al., 2011), as well as α7 nAChR activation via AR-R17779 (The et al., 2007), all resulted in reduction of influx of immune cells into the muscularis, dampened cytokine production, and improved gastrointestinal transit. In line with the findings reported by Wang et al. (2003), the anti-inflammatory effect of electrical vagus nerve stimulation is lost in α7 nAChR-deficient mice (Matteoli et al. (in press)), confirming the involvement of α7 nAChR. In contrast to the other preclinical models, though, the vagal modulation of the resident macrophages results from a direct input to the gut, as selective denervation of the spleen left the modulatory effect of the vagus nerve untouched (Matteoli et al. (in press)). Moreover, local intestinal blockade of nicotine receptors via incubation of the intestine in a solution containing hexamethonium abolished the effect of vagus nerve stimulation (de Jonge et al., 2005), further arguing against the spleen as site of neuromodulation, at least in this model of subtle intestinal inflammation.

Miscellaneous diseases

Evidence suggesting that pancreatitis, diabetes, asthma, atherosclerosis, and cystic fibrosis may potentially benefit from activation of the cholinergic anti-inflammatory system has been reported as well. The reader is referred to excellent reviews summarizing this literature (see Pavlov, 2008; Bencherif et al., 2011; Huston and Tracey, 2011) .

Human evidence

By now it is widely accepted that immune cells carry neurotransmitter receptors allowing communication between the immune and nervous system. Human peripheral blood mononuclear cells, B and T cells, macrophages, dendritic cells, and synovial fibroblasts express a variety of cholinergic muscarinic and nicotinic receptors (Guinet et al., 2004; Wessler and Kirkpatrick, 2008; De Rosa et al., 2009; Westman et al., 2009; Takahashi et al., 2010). From the initial *in vitro* work by Tracey and coworkers (Wang et al., 2003), we know that cytokine release by human macrophages is reduced by nicotine via the α7 nAChR. Not only cytokine production, but also T-cell skewing by DCs and B-cell antibody production are affected by cholinergic input. The critical question clearly is to what extent modulation of the immune system via these receptors affects human physiology and disease.

Accepting that the vagus nerve has a major impact on the immune system through the spleen, one would anticipate increased incidence of inflammatory conditions in patients who underwent vagotomy, for example, as treatment for peptic ulcer disease (before introduction of *Helicobacter pylori* eradication) or during esophagectomy. To date, no data are available supporting this. In the gastrointestinal tract, cholinergic tone, however, may be restored by compensatory increase within the enteric nervous system, possibly explaining the normalization of the susceptibility to DSS colitis after several weeks in vagotomized mice (Ghia et al., 2007). To what extent similar compensatory mechanisms are activated in other regions of the body remains to be studied. Splenectomy, on the other hand, is associated with an increased risk for potentially lethal bacterial infections, but this seems to be more related to the clearance of encapsulated bacteria at the spleen.

There is, however, increasing indirect evidence supporting an immune-modulatory role for the cholinergic anti-inflammatory pathway in humans. By using heart rate variability, several studies have evaluated the relationship between vagal tone and the activation status of the immune system (reviewed in Huston and Tracey, 2011). Heart rate variability is the time difference between successive heartbeats and determined by the balance between cholinergic (parasympathetic) and sympathetic input to the heart. In healthy subjects, reduced heart rate variability indices (=low vagal tone) are independently associated with increased CRP and IL-6 levels (Sloan et al., 2007; von Känel et al., 2008; Haarala et al., 2011). In the same line, low vagal tone was reported to be associated with impaired stress recovery of TNF levels, although these changes were very subtle (Weber et al., 2010). Similarly, TNF and IL-6 production by whole blood stimulated with endotoxin in healthy subjects was associated with vagal activity assessed by heart rate variability (Marsland et al., 2007).

Also, in immune-mediated diseases like rheumatoid arthritis, vagus nerve activity was reduced compared to healthy controls and associated with increased levels of serum HMGB1 (Goldstein et al., 2007). Similarly, the majority of studies in patients with cardiovascular disease report that parasympathetic tone as inferred from heart rate variability is inversely related to inflammatory markers (IL-6 and CRP) (Lanza et al., 2006; Haensel et al., 2008). Finally, increased morbidity and mortality following cardiac surgery, myocardial infarction, sepsis, RA, IBD, lupus erythematosus, and sarcoidosis has been reported to be associated with decreased vagus nerve activity (Huston and Tracey, 2011). Whereas these data suggest that reduced vagal tone increases the set point of the immune response with a subtle rise in proinflammatory cytokines and increased disease risk, increased vagal tone resulting from elevated intracranial pressure has been proposed as the mechanism underlying the observed immune paralysis in patients with traumatic brain injury (Kox et al., 2008). Taken together, these data provide indirect evidence that basal vagal tone may represent an important determinant of the risk of developing or having progression of disease. The finding that exercise or dietary interventions such as fish oil supplementation enhance vagal tone may partly explain their beneficial effect on general health (Goldhammer et al., 2005; Holguin et al., 2005).

The final evidence that the vagal anti-inflammatory pathway is indeed a breakthrough for the clinical management of immune-mediated inflammatory disease will ultimately have to come from clinical trials evaluating the different strategies discussed above. Selective α7 nAChR agonists are currently evaluated as potential therapeutic drugs for the treatment of cognitive impairments in schizophrenia and Alzheimer's disease (Olincy et al., 2006; Thomsen et al., 2010; Toyohara and Hashimoto, 2010; Tregellas et al., 2011; Wallace et al., 2011). To date, however, only a very limited number of studies have been performed to evaluate their anti-inflammatory potential in humans. The effect of nicotine (rectal enemas or transdermal application) has been extensively evaluated in patients with ulcerative colitis, though with controversial results (McGrath et al., 2004; Ingram et al., 2005; Nikfar et al., 2010). Most likely side-effects due to interaction with all nAChR subtypes prevent effective dosing of nicotine. Interestingly, these studies were based on the observation that ulcerative colitis patients relapse when stopping, and improve when resuming smoking, though why the opposite is observed in patients with Crohn's disease remains a mystery. Based on the observation that the selective α7 nAChR agonist GTS-21 attenuated cytokine production by stimulated whole blood and human monocytes more potently than nicotine (Kox et al., 2009; Rosas-Ballina et al., 2009), its effect was tested in a human endotoxin model. However, no significant reduction in cytokine response to endotoxin injection in healthy subjects was observed between GTS-21 and placebo-treated subjects (Kox et al., 2011). Clinical studies evaluating the effect of specific α7 nAChR are therefore awaited with great interest.

Finally, electrical stimulation of the vagus nerve, either in the neck or the ear, may be considered as treatment for chronic inflammatory diseases such as IBD and RA. A clinical trial in patients with RA will start soon, hopefully providing the first proof-of-concept study that activation of the cholinergic anti-inflammatory pathway is indeed a powerful new therapeutic tool.

CONCLUSION

Communication between the autonomic nervous system, in particular of the parasympathetic part, is increasingly considered to have a major impact on the immune system. The cross-talk between nerves and immune cells may be of great importance for the maintenance of immune homeostasis, thereby controlling the inflammatory response and preventing collateral damage or disseminated disease. Although the mechanisms involved are gradually being unraveled, the exact anatomy, for example, how cholinergic modulation in the spleen is obtained, remains a matter of debate. Furthermore, a proof-of-concept study in man is still lacking, although α7 nAChR agonists are available for human use. Nevertheless, the concept of the "inflammatory reflex" and the discovery of the anti-inflammatory potential of the vagus nerve is to be considered as a major breakthrough with an enormous potential for treatment. Future studies will hopefully confirm this and fulfill the high expectations so in the end, our patients will benefit with decreased morbidity and increased quality of life.

References

Abraham C, Medzhitov R (2011). Interactions between the host innate immune system and microbes in inflammatory bowel disease. Gastroenterology 140: 1729–1737.

Bai A, Guo Y, Lu N (2007). The effect of the cholinergic anti-inflammatory pathway on experimental colitis. Scand J Immunol 66: 538–545.

Barone L, Colicchio G, Policicchio D et al. (2007). Effect of vagal nerve stimulation on systemic inflammation and cardiac autonomic function in patients with refractory epilepsy. Neuroimmunomodulation 14: 331–336.

Beekwilder JP, Beems T (2010). Overview of the clinical applications of vagus nerve stimulation. J Clin Neurophysiol 27: 130–138.

Bellinger DL, Millar BA, Perez S et al. (2008). Sympathetic modulation of immunity: relevance to disease. Cell Immunol 252: 27–56.

Bencherif M, Lippiello PM, Lucas R et al. (2011). Alpha7 nicotinic receptors as novel therapeutic targets for inflammation-based diseases. Cell Mol Life Sci 68: 931–949.

Berg DK, Conroy WG (2002). Nicotinic alpha 7 receptors: synaptic options and downstream signaling in neurons. J Neurobiol 53: 512–523.

Bernik TR, Friedman SG, Ochani M et al. (2002). Pharmacological stimulation of the cholinergic antiinflammatory pathway. J Exp Med 195: 781–788.

Bischoff SC (2007). Role of mast cells in allergic and non-allergic immune responses: comparison of human and murine data. Nat Rev Immunol 7: 93–104.

Bischoff SC (2010). Food allergy and eosinophilic gastroenteritis and colitis. Curr Opin Allergy Clin Immunol 10: 238–245.

Bitto A, Polito F, Altavilla D et al. (2011). Melanocortins protect against multiple organ dysfunction syndrome in mice. Br J Pharmacol 162: 917–928.

Boeckxstaens GE, de Jonge WJ (2009). Neuroimmune mechanisms in postoperative ileus. Gut 58: 1300–1311.

Boland C, Collet V, Laterre E et al. (2011). Electrical vagus nerve stimulation and nicotine effects in peritonitis-induced acute lung injury in rats. Inflammation 34: 29–35.

Borovikova LV, Ivanova S, Zhang M et al. (2000a). Vagus nerve stimulation attenuates the systemic inflammatory response to endotoxin. Nature 405: 458–462.

Borovikova LV, Ivanova S, Nardi D et al. (2000b). Role of vagus nerve signaling in CNI-1493-mediated suppression of acute inflammation. Auton Neurosci 85: 141–147.

Bouzat C, Gumilar F, Spitzmaul G et al. (2004). Coupling of agonist binding to channel gating in an ACh-binding protein linked to an ion channel. Nature 430: 896–900.

Buijs RM, van der Vliet J, Garidou ML et al. (2008). Spleen vagal denervation inhibits the production of antibodies to circulating antigens. PLoS One 3: e3152.

Bunch TJ, White RD R, Gersh BJ et al. (2003). Long-term outcomes of out-of-hospital cardiac arrest after successful early defibrillation. N Engl J Med 348: 2626–2633.

Cailotto C, Costes LM, van der Vliet J et al. (2012). Neuroanatomical evidence demonstrating the existence of the vagal anti-inflammatory reflex in the intestine. Neurogastroenterol Motil. 24: 191–200.

Chernyavsky AI, Arredondo J, Galitovskiy V et al. (2010). Upregulation of nuclear factor-kappaB expression by SLURP-1 is mediated by alpha7-nicotinic acetylcholine receptor and involves both ionic events and activation of protein kinases. Am J Physiol Cell Physiol 299: C903–C911.

Cheyuo C, Wu R, Zhou M et al. (2011). Ghrelin suppresses inflammation and neuronal nitric oxide synthase in focal cerebral ischemia via the vagus nerve. Shock 35: 258–265.

Choi KM, Gibbons SJ, Nguyen TV et al. (2008). Heme oxygenase-1 protects interstitial cells of Cajal from oxidative stress and reverses diabetic gastroparesis. Gastroenterology 135: 2055–2064 2064 e1–e2.

Choi KM, Kashyap PV, Dutta N et al. (2010). CD206-positive M2 macrophages that express heme oxygenase-1 protect against diabetic gastroparesis in mice. Gastroenterology 138: 2399–2409 2409 e1.

Corcoran C, Connor TJ, O'Keane V et al. (2005). The effects of vagus nerve stimulation on pro- and anti-inflammatory cytokines in humans: a preliminary report. Neuroimmunomodulation 12: 307–309.

Crockett ET, Galligan JJ, Uhal BD et al. (2006). Protection of early phase hepatic ischemia-reperfusion injury by cholinergic agonists. BMC Clin Pathol 6, 3.

Danese S, Colombel JF, Reinisch W et al. (2011). Review article: infliximab for Crohn's disease treatment – shifting therapeutic strategies after 10 years of clinical experience. Aliment Pharmacol Ther 33: 857–869.

Danton GH, Dietrich WD (2003). Inflammatory mechanisms after ischemia and stroke. J Neuropathol Exp Neurol 62: 127–136.

de Haan JJ, Lubbers T, Hadfoune M et al. (2008). Postshock intervention with high-lipid enteral nutrition reduces inflammation and tissue damage. Ann Surg 248: 842–848.

de Haan JJ, Thuijls G, Lubbers T et al. (2010). Protection against early intestinal compromise by lipid-rich enteral nutrition through cholecystokinin receptors. Crit Care Med 38: 1592–1597.

De Herdt V, Bogaert S, Bracke KR et al. (2009). Effects of vagus nerve stimulation on pro- and anti-inflammatory cytokine induction in patients with refractory epilepsy. J Neuroimmunol 214: 104–108.

de Jonge WJ, Ulloa L (2007). The alpha7 nicotinic acetylcholine receptor as a pharmacological target for inflammation. Br J Pharmacol 151: 915–929.

de Jonge WJ, van der Zanden EP, The FO et al. (2005). Stimulation of the vagus nerve attenuates macrophage activation by activating the Jak2-STAT3 signaling pathway. Nat Immunol 6: 844–851.

De Rosa MJ, Dionisio L, Agriello E et al. (2009). Alpha 7 nicotinic acetylcholine receptor modulates lymphocyte activation. Life Sci 85: 444–449.

de Winter BY, van Nassauw L, de Man JG et al. (2005). Role of oxidative stress in the pathogenesis of septic ileus in mice. Neurogastroenterol Motil 17: 251–261.

Eskandari MK, Kalff JC, Billiar TR et al. (1997). Lipopolysaccharide activates the muscularis macrophage network and suppresses circular smooth muscle activity. Am J Physiol 273: G727–G734.

Eskandari MK, Kalff JC, Billiar TR et al. (1999). LPS-induced muscularis macrophage nitric oxide suppresses rat jejunal circular muscle activity. Am J Physiol 277: G478–G486.

Fallgatter AJ, Neuhauser B, Herrmann MJ et al. (2003). Far field potentials from the brain stem after transcutaneous vagus nerve stimulation. J Neural Transm 110: 1437–1443.

Fallgatter AJ, Ehlis AC, Ringel TM et al. (2005). Age effect on far field potentials from the brain stem after transcutaneous vagus nerve stimulation. Int J Psychophysiol 56: 37–43.

Gallowitsch-Puerta M, Tracey KJ (2005). Immunologic role of the cholinergic anti-inflammatory pathway and the nicotinic acetylcholine alpha 7 receptor. Ann N Y Acad Sci 1062: 209–219.

Garcia-Guzman M, Sala F, Sala S et al. (1995). alpha-Bungarotoxin-sensitive nicotinic receptors on bovine chromaffin cells: molecular cloning, functional expression and alternative splicing of the alpha 7 subunit. Eur J Neurosci 7: 647–655.

Gault J, Robinson M, Berger R et al. (1998). Genomic organization and partial duplication of the human alpha7 neuronal nicotinic acetylcholine receptor gene (CHRNA7). Genomics 52: 173–185.

Ghia JE, Blennerhassett P, Kumar-Ondiveeran H et al. (2006). The vagus nerve: a tonic inhibitory influence associated with inflammatory bowel disease in a murine model. Gastroenterology 131: 1122–1130.

Ghia JE, Blennerhassett P, El-Sharkawy RT et al. (2007). The protective effect of the vagus nerve in a murine model of chronic relapsing colitis. Am J Physiol Gastrointest Liver Physiol 293: G711–G718.

Ghia JE, Blennerhassett P, Collins SM (2008). Impaired parasympathetic function increases susceptibility to inflammatory bowel disease in a mouse model of depression. J Clin Invest 118: 2209–2218.

Ghia JE, Blennerhassett P, Deng Y et al. (2009). Reactivation of inflammatory bowel disease in a mouse model of depression. Gastroenterology 136: 2280–2288, e1–4.

Ghia JE, Park AJ, Blennerhassett P et al. (2011). Adoptive transfer of macrophage from mice with depression-like behavior enhances susceptibility to colitis. Inflamm Bowel Dis 17: 1474–1489.

Giebelen IA, Le Moine A, van den Pangaart PS et al. (2008). Deficiency of alpha7 cholinergic receptors facilitates bacterial clearance in *Escherichia coli* peritonitis. J Infect Dis 198: 750–757.

Goehler LE, Relton JK, Dripps D et al. (1997). Vagal paraganglia bind biotinylated interleukin-1 receptor antagonist: a possible mechanism for immune-to-brain communication. Brain Res Bull 43: 357–364.

Goldhammer E, Tanchilevitch A, Maor I et al. (2005). Exercise training modulates cytokines activity in coronary heart disease patients. Int J Cardiol 100: 93–99.

Goldstein RS, Bruchfeld A, Yang L et al. (2007). Cholinergic anti-inflammatory pathway activity and high mobility group box-1 (HMGB1) serum levels in patients with rheumatoid arthritis. Mol Med 13: 210–215.

Groves DA, Brown VJ (2005). Vagal nerve stimulation: a review of its applications and potential mechanisms that mediate its clinical effects. Neurosci Biobehav Rev 29: 493–500.

Guarini S, Cainazzo MM, Giuliani D et al. (2004). Adrenocorticotropin reverses hemorrhagic shock in anesthetized rats through the rapid activation of a vagal anti-inflammatory pathway. Cardiovasc Res 63: 357–365.

Guinet E, Yoshida K, Nouri-Shirazi M (2004). Nicotinic environment affects the differentiation and functional maturation of monocytes derived dendritic cells (DCs). Immunol Lett 95: 45–55.

Haarala A, Kähönen M, Eklund C et al. (2011). Heart rate variability is independently associated with C-reactive protein but not with Serum amyloid A. The Cardiovascular Risk in Young Finns Study. Eur J Clin Invest 41: 951–957.

Haensel A, Mills PJ, Nelesen RA et al. (2008). The relationship between heart rate variability and inflammatory markers in cardiovascular diseases. Psychoneuroendocrinology 33: 1305–1312.

Hansen MK, O'Connor KA, Goehler LE et al. (2001). The contribution of the vagus nerve in interleukin-1beta-induced fever is dependent on dose. Am J Physiol Regul Integr Comp Physiol 280: R929–R934.

Hofer S, Eisenbach C, Lukic IK et al. (2008). Pharmacologic cholinesterase inhibition improves survival in experimental sepsis. Crit Care Med 36: 404–408.

Holguin F, Tellez-Rojo MM, Lazo M et al. (2005). Cardiac autonomic changes associated with fish oil vs soy oil supplementation in the elderly. Chest 127: 1102–1107.

Huston JM, Tracey KJ (2011). The pulse of inflammation: heart rate variability, the cholinergic anti-inflammatory pathway and implications for therapy. J Intern Med 269: 45–53.

Huston JM, Ochani M, Rosas-Ballina M et al. (2006). Splenectomy inactivates the cholinergic antiinflammatory pathway during lethal endotoxemia and polymicrobial sepsis. J Exp Med 203: 1623–1628.

Huston JM, Wang H, Ochani M et al. (2008). Splenectomy protects against sepsis lethality and reduces serum HMGB1 levels. J Immunol 181: 3535–3539.

Huston JM, Rosas-Ballina M, Xue X et al. (2009). Cholinergic neural signals to the spleen down-regulate leukocyte trafficking via CD11b. J Immunol 183: 552–559.

Ingram JR, Thomas GA, Rhodes J et al. (2005). A randomized trial of nicotine enemas for active ulcerative colitis. Clin Gastroenterol Hepatol 3: 1107–1114.

Kalff JC, Schraut WH, Simmons RL et al. (1998). Surgical manipulation of the gut elicits an intestinal muscularis inflammatory response resulting in postsurgical ileus. Ann Surg 228: 652–663.

Kalff JC, Buchholz BM, Eskandari MK et al. (1999). Biphasic response to gut manipulation and temporal correlation of cellular infiltrates and muscle dysfunction in rat. Surgery 126: 498–509.

Kong SS, Liu JJ, Hwang TC et al. (2011). Tumour necrosis factor-alpha and its receptors in the beneficial effects of vagal stimulation after myocardial infarction in rats. Clin Exp Pharmacol Physiol 38: 300–306.

Koopman FA, Stoof SP, Straub RH et al. (2011). Restoring the balance of the autonomic nervous system as an innovative approach to the treatment of rheumatoid arthritis. Mol Med 17: 937–948.

Kox M, Pompe JC, Pickkers P et al. (2008). Increased vagal tone accounts for the observed immune paralysis in patients with traumatic brain injury. Neurology 70: 480–485.

Kox M, van Velzen JF, Pompe JC et al. (2009). GTS-21 inhibits pro-inflammatory cytokine release independent of the Toll-like receptor stimulated via a transcriptional mechanism involving JAK2 activation. Biochem Pharmacol 78: 863–872.

Kox M, Pompe JC, Gordinou de Gouberville MC et al. (2011). Effects of the alpha7 nicotinic acetylcholine receptor agonist Gts-21 on the innate immune response in humans. Shock 36: 5–11.

Lanza GA, Sgueglia GA, Cianflone D et al. (2006). Relation of heart rate variability to serum levels of C-reactive protein in patients with unstable angina pectoris. Am J Cardiol 97: 1702–1706.

Lubbers T, Luyer MD, de Haan JJ et al. (2009). Lipid-rich enteral nutrition reduces postoperative ileus in rats via activation of cholecystokinin-receptors. Ann Surg 249: 481–487.

Lubbers T, de Haan JJ, Hadfoune M et al. (2010a). Lipid-enriched enteral nutrition controls the inflammatory response in murine Gram-negative sepsis. Crit Care Med 38: 1996–2002.

Lubbers T, de Haan JJ, Luyer MD et al. (2010b). Cholecystokinin/cholecystokinin-1 receptor-mediated peripheral activation of the afferent vagus by enteral nutrients attenuates inflammation in rats. Ann Surg 252: 376–382.

Lubbers T, de Haan JJ, Hadfoune M et al. (2011). Chylomicron formation and glucagon-like peptide 1 receptor are involved in activation of the nutritional anti-inflammatory pathway. J Nutr Biochem 22: 1105–1111.

Luyer MD, Greve JW, Hadfoune M et al. (2005). Nutritional stimulation of cholecystokinin receptors inhibits inflammation via the vagus nerve. J Exp Med 202: 1023–1029.

Marquette C, Linard C, Galonnier M et al. (2003). IL-1beta, TNFalpha and IL-6 induction in the rat brain after partial-body irradiation: role of vagal afferents. Int J Radiat Biol 79: 777–785.

Marsland AL, Gianaros PJ, Prather AA et al. (2007). Stimulated production of proinflammatory cytokines covaries inversely with heart rate variability. Psychosom Med 69: 709–716.

Marvel FA, Chen CC, Badr N et al. (2004). Reversible inactivation of the dorsal vagal complex blocks lipopolysaccharide-induced social withdrawal and c-Fos expression in central autonomic nuclei. Brain Behav Immun 18: 123–134.

Matteoli G, Gomez-Pinilla PJ, Nemethova A et al. A distinct vagal anti-inflammatory pathway modulates intestinal muscularis resident macrophages independent of the spleen. Gut. (in press).

McGrath J, McDonald JW, Macdonald JK (2004). Transdermal nicotine for induction of remission in ulcerative colitis. Cochrane Database Syst Rev, CD004722.

McLane KE, Wu XD, Conti-Tronconi BM (1990). Identification of a brain acetylcholine receptor alpha subunit able to bind alpha-bungarotoxin. J Biol Chem 265: 9816–9824.

Miceli PC, Jacobson K (2003). Cholinergic pathways modulate experimental dinitrobenzene sulfonic acid colitis in rats. Auton Neurosci 105: 16–24.

Mikulski Z, Hartmann P, Jositsch G et al. (2010). Nicotinic receptors on rat alveolar macrophages dampen ATP-induced increase in cytosolic calcium concentration. Respir Res 11: 133.

Minnicozzi M, Sawyer RT, Fenton MJ (2011). Innate immunity in allergic disease. Immunol Rev 242: 106–127.

Minutoli L, Squadrito F, Nicotina PA et al. (2011). Melanocortin 4 receptor stimulation decreases pancreatitis severity in rats by activation of the cholinergic anti-inflammatory pathway. Crit Care Med 39: 1089–1096.

Nikfar S, Ehteshami-Ashar S, Rahimi R et al. (2010). Systematic review and meta-analysis of the efficacy and tolerability of nicotine preparations in active ulcerative colitis. Clin Ther 32: 2304–2315.

Norman GJ, Morris JS, Karelina K et al. (2011). Cardiopulmonary arrest and resuscitation disrupts cholinergic anti-inflammatory processes: a role for cholinergic alpha7 nicotinic receptors. J Neurosci 31: 3446–3452.

O'Mahony C, van der Kleij H, Bienenstock J et al. (2009). Loss of vagal anti-inflammatory effect: in vivo visualization and adoptive transfer. Am J Physiol Regul Integr Comp Physiol 297: R1118–R1126.

Olincy A, Harris JG, Johnson LL et al. (2006). Proof-of-concept trial of an alpha7 nicotinic agonist in schizophrenia. Arch Gen Psychiatry 63: 630–638.

Orr-Urtreger A, Kedmi M, Rosner S et al. (2005). Increased severity of experimental colitis in alpha 5 nicotinic acetylcholine receptor subunit-deficient mice. Neuroreport 16: 1123–1127.

Ottani A, Giuliani D, Mioni C et al. (2009). Vagus nerve mediates the protective effects of melanocortins against cerebral and systemic damage after ischemic stroke. J Cereb Blood Flow Metab 29: 512–523.

Ottani A, Giuliani D, Galantucci M et al. (2010). Melanocortins counteract inflammatory and apoptotic responses to prolonged myocardial ischemia/reperfusion through a vagus nerve-mediated mechanism. Eur J Pharmacol 637: 124–130.

Panacek EA, Marshall JC, Albertson TE et al. (2004). Efficacy and safety of the monoclonal anti-tumor necrosis factor antibody F(ab')2 fragment afelimomab in patients with severe sepsis and elevated interleukin-6 levels. Crit Care Med 32: 2173–2182.

Parrish WR, Rosas-Ballina M, Gallowitsch-Puerta M et al. (2008). Modulation of TNF release by choline requires alpha7 subunit nicotinic acetylcholine receptor-mediated signaling. Mol Med 14: 567–574.

Pavlov VA (2008). Cholinergic modulation of inflammation. Int J Clin Exp Med 1: 203–212.

Pavlov VA, Ochani M, Gallowitsch-Puerta M et al. (2006). Central muscarinic cholinergic regulation of the systemic inflammatory response during endotoxemia. Proc Natl Acad Sci U S A 103: 5219–5223.

Pavlov VA, Parrish WR, Rosas-Ballina M et al. (2009). Brain acetylcholinesterase activity controls systemic cytokine levels through the cholinergic anti-inflammatory pathway. Brain Behav Immun 23: 41–45.

Peter C, Schmidt K, Hofer S et al. (2010). Effects of physostigmine on microcirculatory alterations during experimental endotoxemia. Shock 33: 405–411.

Qin S, Wang H, Yuan R et al. (2006). Role of HMGB1 in apoptosis-mediated sepsis lethality. J Exp Med 203: 1637–1642.

Rescigno M (2010). Intestinal dendritic cells. Adv Immunol 107: 109–138.

Rosas-Ballina M, Goldstein RS, Gallowitsch-Puerta M et al. (2009). The selective alpha7 agonist GTS-21 attenuates cytokine production in human whole blood and human monocytes activated by ligands for TLR2, TLR3, TLR4, TLR9, and RAGE. Mol Med 15: 195–202.

Rutgeerts P, Vermeire S, Van Assche G (2009). Biological therapies for inflammatory bowel diseases. Gastroenterology 136: 1182–1197.

Sadis C, Teske G, Stokman G et al. (2007). Nicotine protects kidney from renal ischemia/reperfusion injury through the cholinergic anti-inflammatory pathway. PLoS One 2: e469.

Saeed RW, Varma S, Peng-Nemeroff T et al. (2005). Cholinergic stimulation blocks endothelial cell activation and leukocyte recruitment during inflammation. J Exp Med 201: 1113–1123.

Schenten D, Medzhitov R (2011). The control of adaptive immune responses by the innate immune system. Adv Immunol 109: 87–124.

Sharma G, Vijayaraghavan S (2008). Nicotinic receptors containing the alpha7 subunit: a model for rational drug design. Curr Med Chem 15: 2921–2932.

Sloan RP, McCreath H, Tracey KJ et al. (2007). RR interval variability is inversely related to inflammatory markers: the CARDIA study. Mol Med 13: 178–184.

Snoek SA, Verstege MI, van der Zanden EP et al. (2010). Selective alpha7 nicotinic acetylcholine receptor agonists worsen disease in experimental colitis. Br J Pharmacol 160: 322–333.

Sternberg EM (2006). Neural regulation of innate immunity: a coordinated nonspecific host response to pathogens. Nat Rev Immunol 6: 318–328.

Takahashi HK, Liu K, Wake H et al. (2010). Effect of nicotine on advanced glycation end product-induced immune response in human monocytes. J Pharmacol Exp Ther 332: 1013–1021.

The FO, Boeckxstaens GE, Snoek SA et al. (2007). Activation of the cholinergic anti-inflammatory pathway ameliorates postoperative ileus in mice. Gastroenterology 133: 1219–1228.

The F, Cailotto C, van der Vliet J et al. (2011). Central activation of the cholinergic anti-inflammatory pathway reduces surgical inflammation in experimental post-operative ileus. Br J Pharmacol 163: 1007–1016.

Thomsen MS, Hansen HH, Timmerman DB et al. (2010). Cognitive improvement by activation of alpha7 nicotinic acetylcholine receptors: from animal models to human pathophysiology. Curr Pharm Des 16: 323–343.

Toyohara J, Hashimoto K (2010). alpha7 Nicotinic receptor agonists: potential therapeutic drugs for treatment of cognitive impairments in schizophrenia and Alzheimer's disease. Open Med Chem J. 4: 37–56.

Tracey KJ (2002). The inflammatory reflex. Nature 420: 853–859.

Tracey KJ (2009). Reflex control of immunity. Nat Rev Immunol 9: 418–428.

Tregellas JR, Olincy A, Johnson L et al. (2010). Functional magnetic resonance imaging of effects of a nicotinic agonist in schizophrenia. Neuropsychopharmacology 35: 938–942.

Tregellas JR, Tanabe J, Rojas DC et al. (2011). Effects of an alpha 7-nicotinic agonist on default network activity in schizophrenia. Biol Psychiatry 69: 7–11.

Turler A, Schwarz NT, Turler E et al. (2002). MCP-1 causes leukocyte recruitment and subsequently endotoxemic ileus in rat. Am J Physiol Gastrointest Liver Physiol 282: G145–G155.

Tyagi E, Agrawal R, Nath C et al. (2010a). Cholinergic protection via alpha7 nicotinic acetylcholine receptors and PI3K-Akt pathway in LPS-induced neuroinflammation. Neurochem Int 56: 135–142.

Tyagi E, Agrawal R, Nath C et al. (2010b). Inhibitory role of cholinergic system mediated via alpha7 nicotinic acetylcholine receptor in LPS-induced neuro-inflammation. Innate Immun 16: 3–13.

Ulloa L (2005 May). The vagus nerve and the nicotinic anti-inflammatory pathway. Nat Rev Drug Discov 4: 673–684.

Van Assche G, Vermeire S, Rutgeerts P (2011). Management of acute severe ulcerative colitis. Gut 60: 130–133.

van der Zanden EP, Snoek SA, Heinsbroek SE et al. (2009). Vagus nerve activity augments intestinal macrophage phagocytosis via nicotinic acetylcholine receptor alpha4-beta2. Gastroenterology 137: 1029–1039, 1039 e1–4.

van Maanen MA, Lebre MC, van der Poll T et al. (2009a). Stimulation of nicotinic acetylcholine receptors attenuates collagen-induced arthritis in mice. Arthritis Rheum 60: 114–122.

van Maanen MA, Vervoordeldonk MJ, Tak PP (2009b). The cholinergic anti-inflammatory pathway: towards innovative treatment of rheumatoid arthritis. Nat Rev Rheumatol 5: 229–232.

van Maanen MA, Stoof SP, Larosa GJ et al. (2010). Role of the cholinergic nervous system in rheumatoid arthritis: aggravation of arthritis in nicotinic acetylcholine receptor alpha7 subunit gene knockout mice. Ann Rheum Dis 69: 1717–1723.

van Westerloo DJ, Giebelen IA, Florquin S et al. (2005). The cholinergic anti-inflammatory pathway regulates the host response during septic peritonitis. J Infect Dis 191: 2138–2148.

von Känel R, Nelesen RA, Mills PJ et al. (2008). Relationship between heart rate variability, interleukin-6, and soluble tissue factor in healthy subjects. Brain Behav Immun 22: 461–468.

Wallace TL, Ballard TM, Pouzet B et al. (2011). Drug targets for cognitive enhancement in neuropsychiatric disorders. Pharmacol Biochem Behav 99: 130–145.

Wang H, Bloom O, Zhang M et al. (1999). HMG-1 as a late mediator of endotoxin lethality in mice. Science 285: 248–251.

Wang H, Yu M, Ochani M et al. (2003). Nicotinic acetylcholine receptor alpha7 subunit is an essential regulator of inflammation. Nature 421: 384–388.

Wang H, Liao H, Ochani M et al. (2004). Cholinergic agonists inhibit HMGB1 release and improve survival in experimental sepsis. Nat Med 10: 1216–1221.

Watkins LR, Goehler LE, Relton JK et al. (1995). Blockade of interleukin-1 induced hyperthermia by subdiaphragmatic vagotomy: evidence for vagal mediation of immune-brain communication. Neurosci Lett 183: 27–31.

Weber CS, Thayer JF, Rudat M et al. (2010). Low vagal tone is associated with impaired post stress recovery of cardiovascular, endocrine, and immune markers. Eur J Appl Physiol 109: 201–211.

Wehner S, Behrendt FF, Lyutenski BN et al. (2007). Inhibition of macrophage function prevents intestinal inflammation and postoperative ileus in rodents. Gut 56: 176–185.

Wessler I, Kirkpatrick CJ (2008). Acetylcholine beyond neurons: the non-neuronal cholinergic system in humans. Br J Pharmacol 154: 1558–1571.

Westman M, Engstrom M, Catrina AI et al. (2009). Cell specific synovial expression of nicotinic alpha 7 acetylcholine receptor in rheumatoid arthritis and psoriatic arthritis. Scand J Immunol 70: 136–140.

Wu R, Dong W, Cui X et al. (2007). Ghrelin down-regulates proinflammatory cytokines in sepsis through activation of the vagus nerve. Ann Surg 245: 480–486.

Wu R, Dong W, Ji Y et al. (2008). Orexigenic hormone ghrelin attenuates local and remote organ injury after intestinal ischemia-reperfusion. PLoS One 3: e2026.

Yeboah MM, Xue X, Duan B et al. (2008). Cholinergic agonists attenuate renal ischemia-reperfusion injury in rats. Kidney Int 74: 62–69.

Zamora R, Grishin A, Wong C et al. (2005). High-mobility group box 1 protein is an inflammatory mediator in necrotizing enterocolitis: protective effect of the macrophage deactivator semapimod. Am J Physiol Gastrointest Liver Physiol 289: G643–G652.

Zhang P, Han D, Tang T et al. (2008). Inhibition of the development of collagen-induced arthritis in Wistar rats through vagus nerve suspension: a 3-month observation. Inflamm Res 57: 322–328.

Handbook of Clinical Neurology, Vol. 117 (3rd series)
Autonomic Nervous System
R.M. Buijs and D.F. Swaab, Editors

Chapter 12

The role of the autonomic nervous system in cardiac arrhythmias

ALEX Y. TAN AND RICHARD L. VERRIER*
Beth Israel Deaconess Medical Center, Harvard Medical School, Boston, MA, USA

INTRODUCTION

The influences of the autonomic nervous system on the stability of heart rhythm are potent and diverse. This complexity relates to the elaborate neurocircuitry, with multifold connections within the central nervous system (CNS), periphery, extrinsic cardiac nervous system, and intrinsic cardiac innervation. While the normal heart can withstand the daily barrage of autonomic activity associated with mental and physical stress, the diseased or aging myocardium can respond with significant rhythm abnormality. This inherent vulnerability characterizes both the atrial and ventricular myocardium, with significant differences in their responses to autonomic stimulation. The response of sinus node pacemakers and atrial tissue ranges from rapid rhythm to arrhythmias including atrial flutter and fibrillation. The latter abnormality, which afflicts 2.2 million U.S. citizens, is significant clinically, as it is the leading cause of hospital admissions due to arrhythmia in patients over 50 years of age (Calkins et al., 2007; Benjamin et al., 2009).

Derangements in ventricular rhythm due to intense autonomic activity can be the most severe. Notably, in patients with ischemic heart disease, there is evidence that neural activity, particularly in the form of enhanced sympathetic nerve drive, can trigger sudden cardiac death due to ventricular fibrillation (Verrier and Antzelevitch, 2004; Zhou et al., 2008a). Neural influences may also be highly conducive to tachyarrhythmias such as torsades de pointes in individuals with channelopathies including the long QT and Brugada syndromes and sudden unexplained death in patients with epilepsy (Verrier and Schachter, 2010).

INTEGRATION OF AUTONOMIC CONTROL OF HEART RHYTHM

Integrated regulation of cardiac neural activity is achieved at multiple levels (Fig. 12.1) (Lathrop and Spooner, 2001), including elaborate pathways within the hypothalamus and medullary cardiovascular regulatory sites. Baroreceptor mechanisms are integral to autonomic control, as evidenced by heart rate variability and baroreceptor sensitivity testing, while intrinsic cardiac nerves and fat pads provide local neural regulation independent of higher brain centers. The circuits operate at the level of the myocardial cell and autonomic receptors to influence G proteins and control ionic channels, pumps, and exchangers.

Regulation of the sinoatrial node is accomplished through both intrinsic and extrinsic mechanisms. Intrinsic regulation is achieved through an influence on the I_f pacemaker current, which determines the slope of spontaneous diastolic depolarization. Sinoatrial node depolarization is under control of a "voltage clock" regulated by voltage-sensitive membrane currents, namely, the hyperpolarization-activated pacemaker current, I_f, which is regulated by cyclic adenosine monophosphate (cAMP). Recent evidence supports combined roles for membrane voltage and Ca^{2+} clocks in regulating the sinoatrial node (Joung et al., 2009). Spontaneous rhythmic Ca^{2+} release from sarcoplasmic reticulum in sinoatrial nodal cells results in intracellular calcium elevation, which in turn activates membrane sodium calcium exchanger current and leads to membrane depolarization. This process is highly regulated by cAMP and the autonomic nervous system. The intrinsic heart rate of healthy individuals is

*Correspondence to: Richard L. Verrier, Ph.D., F.A.C.C., Associate Professor of Medicine, Harvard Medical School, Beth Israel Deaconess Medical Center, Harvard-Thorndike Electrophysiology Institute, 99 Brookline Avenue, RN-301, Boston, MA 02215-3908, USA. Tel: +1-617-667-0733, Fax: +1-617-975-5270, E-mail: rverrier@bidmc.harvard.edu

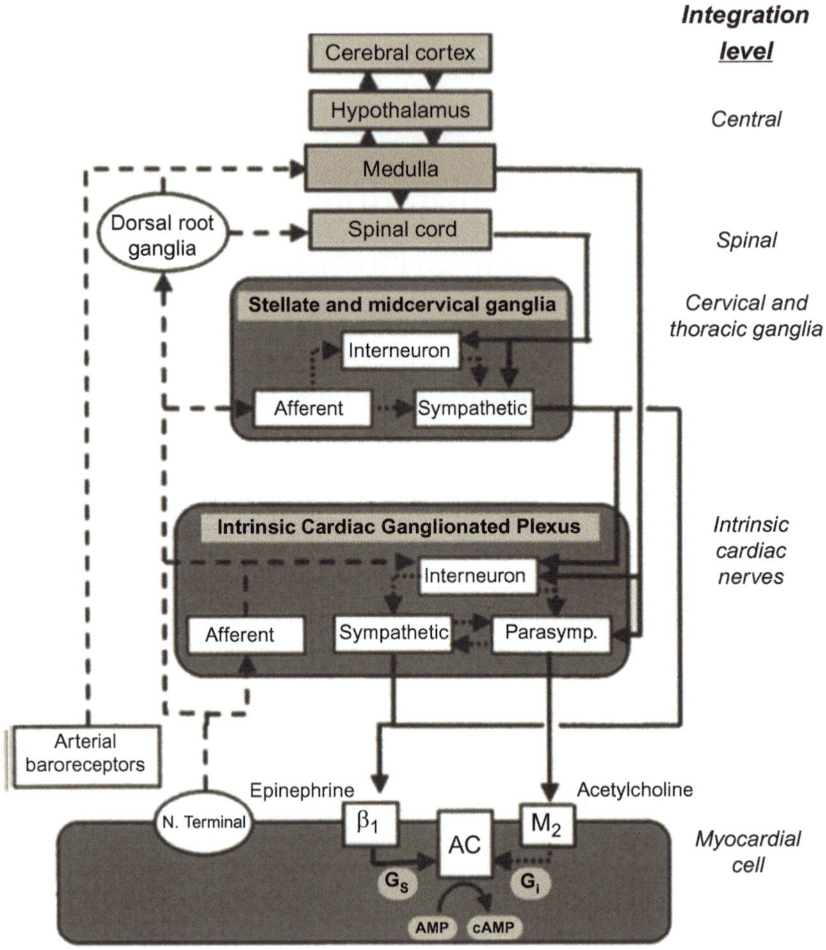

Fig. 12.1. Traditional concepts of neural control of cardiovascular function focused on afferent tracts (*dashed lines*) arising from myocardial nerve terminals and reflex receptors (e.g., baroreceptors) that are integrated centrally within hypothalamic and medullary cardiostimulatory and cardioinhibitory brain centers and on central modulation of sympathetic and parasympathetic outflow (*solid lines*) with little intermediary processing at the level of the spinal cord and within cervical and thoracic ganglia. More recent views incorporate additional levels of intricate processing within the extraspinal cervical and thoracic ganglia and within the cardiac ganglionic plexus, where interneurons are envisioned to provide new levels of noncentral integration. Release of neurotransmitters from postganglionic sympathetic neurons is believed to enhance excitation in the sinoatrial node and myocardial cells through norepinephrine binding to β_1 receptors, which enhances adenyl cyclase (AC) activity through intermediary stimulatory G proteins (Gs). Increased parasympathectomy outflow enhances postganglionic release and binding of acetylcholine to muscarinic (M2) receptors and, through coupled inhibitory G proteins (Gi), inhibits cyclic adenosine monophosphate production. The latter alters electrogenesis and pacemaking activity by affecting the activity of specific membrane Na, K, and Ca channels. (Reproduced from Lathrop and Spooner, 2001, with permission from Blackwell Publishing.)

~100 beats/minute (Katona et al., 1982) and decreases with advancing age (Opthof, 2000).

The sinoatrial node is regulated in response to physical and mental activity and sleep states (Verrier and Josephson, 2009) through the autonomic nervous system, circulating hormones, and reflex regulation associated with cardiorespiratory and baroreceptor inputs. Autonomic function increases heart rate by affecting both voltage- and calcium-dependent mechanisms. Release of norepinephrine and epinephrine steepens the slope of spontaneous diastolic depolarization and depolarizes the resting potential. Sympathetic stimulation also accelerates heart rate by protein kinase A-dependent phosphorylation of proteins that regulate intracellular calcium balance and spontaneous sarcoplasmic reticulum calcium cycling. Vagus nerve activity results in slowing heart rate with a decrease in the slope of diastolic depolarization through an effect on I_f and through hyperpolarization due to increased potassium permeability. The sinus node is also responsive to other influences including hypoxia, exercise, and temperature.

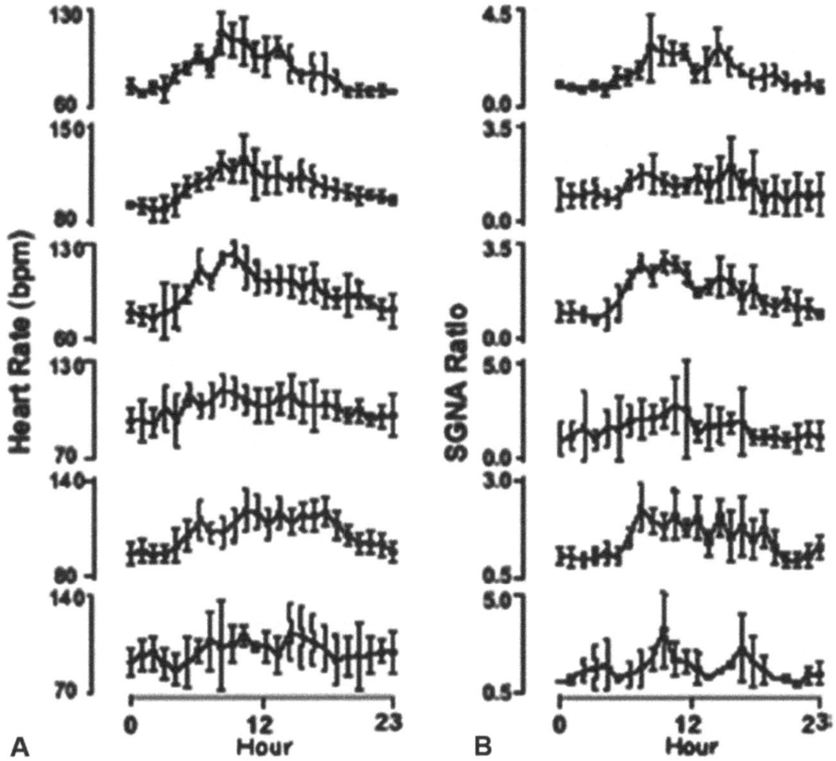

Fig. 12.2. Circadian variation of heart rate and normalized stellate ganglion nerve activity (SGNA) over a 24 hour period. *Vertical bars* at each data point represent standard deviations. Panel **B** shows left SGNA in dogs 1–3 and right SGNA in dogs 4–6. SGNA was normalized to the midnight (0–1 a.m.) value. (Reproduced from Jung et al., 2006.)

Heart rate exhibits a circadian pattern with a rise in the early morning that corresponds to the surge in sympathetic nerve activity (Fig. 12.2) (Jung et al., 2006). During nighttime, there is vagal dominance in the control of heart rate, particularly during nonrapid eye movement sleep, with periodic interruptions by rapid eye movement sleep, when heart rate increases as vagus nerve tone is diminished and sympathetic nerve activity attains levels higher than during waking (Verrier and Josephson, 2009).

Respiratory sinus arrhythmia constitutes an important measure of cardiovascular health. This rhythmic fluctuation in heart rate during breathing is mediated largely through the Hering–Breuer reflex. During inspiration, cardiac efferent vagal tone is suppressed and sympathetic efferent tone is augmented, leading to heart rate accelerations. During expiration, opposite changes in autonomic balance prevail, slowing heart rate (Fig. 12.3) (Opie, 2004). Respiratory sinus arrhythmia appears to provide a "measure of biologic cardiac age" (Hrushesky et al., 1984), as it decreases with advancing age, reflecting decreases in cardiac and vascular elasticity and compliance or in the capacity of the pacemaker to be activated (Opthof, 2000).

ADRENERGIC INFLUENCES AND SUSCEPTIBILITY TO ARRHYTHMIAS

It is well recognized that adrenergic inputs represent major neural triggers for both atrial and ventricular arrhythmias. Activation of the sympathetic nerve structures, including the posterior hypothalamus or stellate ganglia, increases susceptibility to ventricular fibrillation. As illustrated in Table 12.1 (Olshansky, 2005), a wide variety of adrenergic stimuli can lead to enhanced sympathetic nerve activity and may trigger atrial fibrillation in individuals with predisposing myocardial risk factors. The mechanisms by which autonomic nerve activation is proarrhythmic are complex and different for ventricular versus atrial arrhythmias.

In the atria, simultaneous sympathovagal activation is proarrhythmic and facilitates the onset of paroxysmal atrial fibrillation (Patterson et al., 2007; Tan et al., 2008) by a mechanism termed "calcium transient triggering," whereas in the ventricle, sympathetic activation is proarrhythmic and vagal activation is antiarrhythmic. Sympathetic nerve activation prolongs intracellular calcium transients and vagal activation shortens cardiac

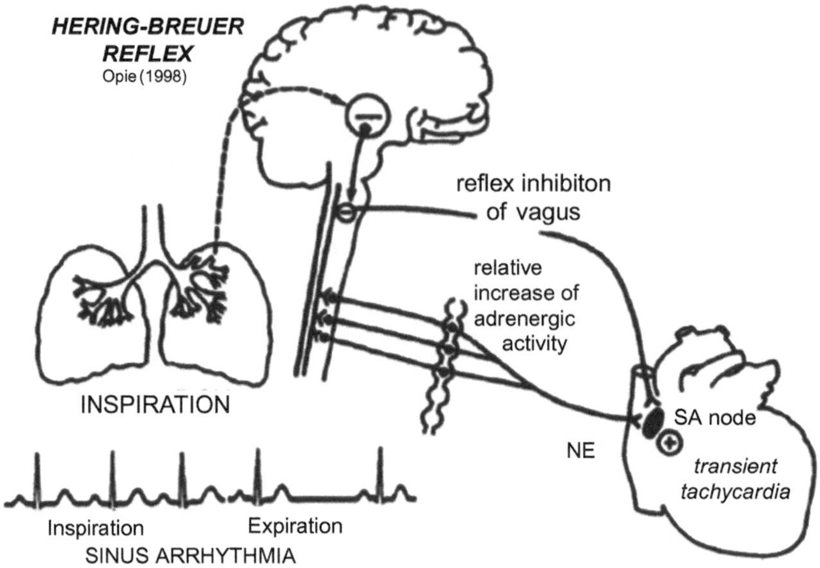

Fig. 12.3. Respiratory sinus arrhythmia. During inspiration, the Hering–Breuer reflex is stimulated to inhibit the medullary cardiovascular regulatory centers. The latter normally depresses the sinus heart rate, and its inhibition results in a relative increase in adrenergic activity. Consequently, there is a transient sinus tachycardia during inspiration and a subsequent slowing during expiration. NE, norepinephrine; SA, sinoatrial node. (Reproduced from Opie, 2004, p. 142, with permission from Dr. Opie.)

Table 12.1

Differentiation of autonomic atrial fibrillation

Adrenergically driven atrial fibrillation	Cholinergically driven atrial fibrillation
Tachycardia	Predominantly in men
Postoperative atrial fibrillation	Age at onset 40–50
Alcohol	Bradycardia
Exercise	Vomiting
Emotional stress	Severe constipation
Polyuria	Postprandial after large meal
Improvement with β-adrenergic blocker therapy	Cold carbonated beverages
More common in patients with heart disease	Rest
	Coughing
	Diving into cold water
	Valsalva maneuver
	Exacerbated by digoxin, β-adrenergic blocker therapy
	More common in patients without heart disease

(Reproduced from Olshansky, 2005.)

action potentials. The discrepancy between action potential duration and intracellular calcium transients, which are normally tightly coupled, leads to increased forward Na/Ca exchanger current, which contributes to the generation of triggered activity.

In the ventricle, the main recurring theme is that intense sympathetic nerve activity elicits arrhythmias by inducing triggered activity due to catecholamine-mediated increases in intracellular calcium and to increases in heterogeneity of repolarization. In the ventricle, the main precondition for life-threatening arrhythmias is the presence of ischemic heart disease. During acute myocardial ischemia, a striking surge in sympathetic nerve activity occurs within a few minutes of experimental left anterior descending coronary artery occlusion, as documented by direct nerve recording. Zhou et al. (2008a) performed direct recording of cardiac sympathetic nerves in ambulatory dogs. They demonstrated that ventricular fibrillation and sudden death are immediately preceded by spontaneous sympathetic nerve discharge, indicating that sympathetic nerve discharge may serve as a trigger for the arrhythmia. Such an enhancement in sympathetic nerve activity has also been shown by others to be associated with a marked increase in susceptibility to ventricular fibrillation, as evidenced by a fall in ventricular fibrillation threshold (Fig. 12.4) (Lombardi et al., 1983), as well as by spontaneous occurrence of the arrhythmia, and correlated increase in T-wave alternans (TWA) level, a repeating ABAB pattern in the T wave of the electrocardiogram indicating cardiac electrical instability (Nearing et al., 1991, 1994; Shusterman et al., 2006).

The mechanisms whereby enhanced sympathetic nerve activity increases cardiac vulnerability in the normal and ischemic heart are complex. The major indirect effects include impairment of oxygen supply–demand ratio due to increased cardiac metabolic activity, α-adrenergically

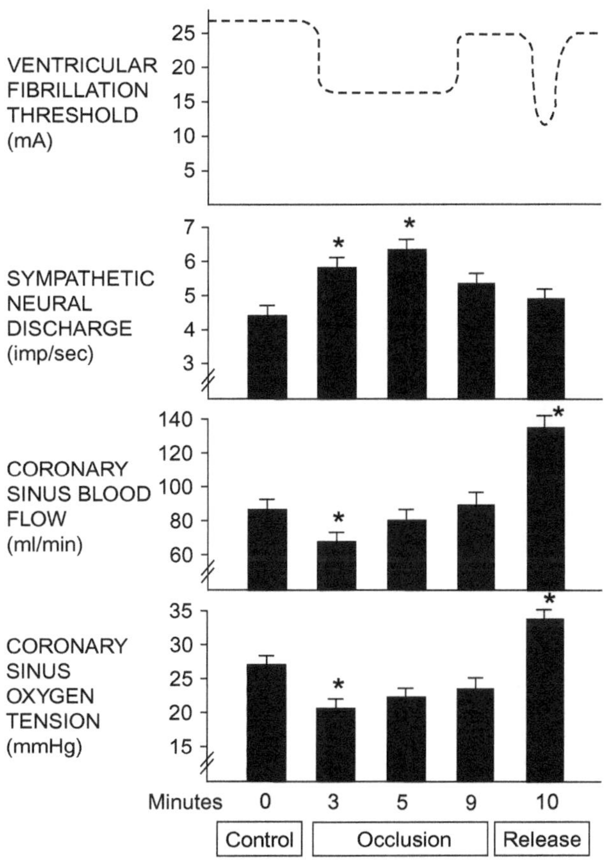

Fig. 12.4. Effects of a 10 minute period of left anterior descending (LAD) coronary artery occlusion and release on neural sympathetic activity, coronary sinus blood flow, and oxygen tension. A schematic representation of the time course of changes in ventricular fibrillation threshold is also displayed. The LAD coronary artery occlusion results in a consistent activation of sympathetic preganglionic fibers, which corresponds to the period of maximal increase in vulnerability to ventricular fibrillation (*$p < 0.05$ compared to control period). The concomitant changes in coronary sinus blood flow and reperfusion are also displayed. (Reproduced from Lombardi et al., 1983, with permission from Mosby, Inc.)

mediated coronary vasoconstriction, especially in vessels with damaged endothelium, and changes in preload and afterload. The direct arrhythmogenic effects on cardiac electrophysiological function, which are primarily mediated through β_1-adrenergic receptors, are multifold. They include derangements in impulse formation, conduction, repolarization alternans, and heterogeneity of repolarization, with the potential for culmination in ventricular tachycardia and fibrillation.

Cardiac β_1-adrenergic receptor blockade is capable of negating the profibrillatory effect of direct sympathetic nerve stimulation (Verrier et al., 1974b) by an action at the neurocardiac effector junction. β_1-adrenergic receptor blockade has also been shown to suppress T-wave alternans, consistent with the capacity of this class of antiarrhythmic agents to reduce the incidence of sudden cardiac death (Olsson et al., 1992; Verrier and Nieminen, 2010). But, cardiac β_2-adrenergic receptors do not appear to play a significant role in modulating ventricular excitable properties. The role of β_3-adrenergic receptors in cardiac arrhythmias remains unclear. β_3 receptors are present in human cardiac myocytes in smaller quantities and activate at higher concentrations of catecholamines than β_1 and β_2 receptors. In canines following myocardial infarction, infusion of β_3 agonist was shown to reduce the occurrence of spontaneous ventricular tachycardia (Zhou et al., 2008b). However, the mechanisms by which β_3 agonism is antiarrhythmic are the subject of further investigation.

In the normal heart, α-adrenergic receptor stimulation or blockade does not appear to affect ventricular electrical stability, as evidenced by the fact that administration of α-adrenergic agonists such as phenylephrine or methoxamine does not influence excitable properties (Verrier et al., 1974a; Kowey et al., 1983). During myocardial ischemia, α-adrenergic blockade may play an indirect role in alleviating coronary vasoconstriction and reducing platelet aggregability.

PARASYMPATHETIC ACTIVITY AND SUSCEPTIBILITY TO CARDIAC ARRHYTHMIAS

Vagal influences on the heart are complex, particularly because of the interconnections with sympathetic nerve fibers. As shown in Figure 12.5 (Levy, 1988), the vagus nerve exerts a prominent modulatory role on sympathetic effects. Levy and Blattberg (1976) observed that the effects of the vagus nerve are contingent upon the prevailing level of adrenergic tone, a phenomenon that they designated "accentuated antagonism." The mechanism of the opposition of adrenergic effects is presynaptic inhibition of norepinephrine release from nerve endings and a muscarinically mediated action at the second messenger level, attenuating the response to catecholamines at receptor sites. In general, vagus nerve activation in the normal heart results in significant heart rate slowing due to an influence on the pacemaker current, I_f. More intense activation can lead to varying degrees of heart block, from prolongation of the PR interval (first degree AV delay), to second degree AV block with 2:1 conduction or even to complete heart block. A remarkable degree of block can be observed in highly trained athletes during sleep or in individuals with SA node dysfunction (Katona et al., 1982). Simultaneous activation of the sympathetic and parasympathetic nervous system is thought to be highly conducive to atrial arrhythmias, due to synergistic effects on action potential characteristics, refractoriness, and propensity for

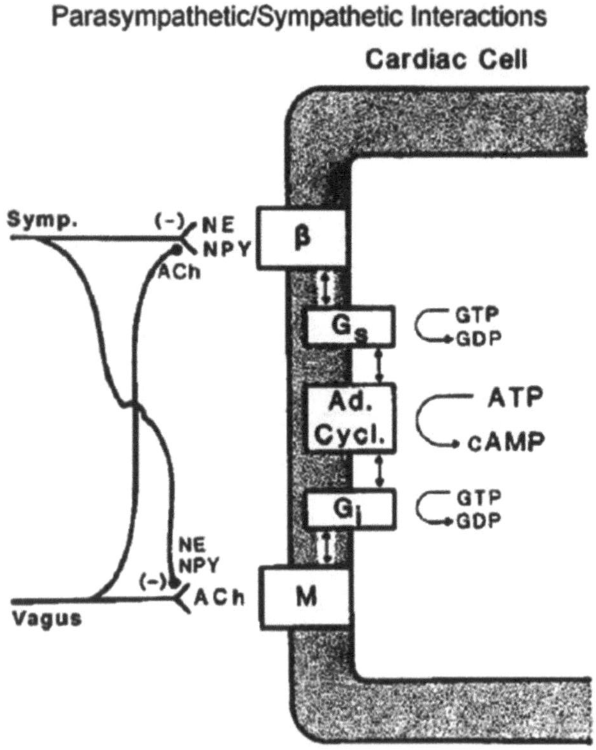

Fig. 12.5. Cardiac cell membrane indicating specific sympathetic and parasympathetic interactions that occur presynaptically. NE, norepinephrine; NPY, neuropeptide Y; ACh, acetylcholine; Symp, sympathetic. (Reproduced from Levy, 1988, with permission from Futura.)

triggered activity (Patterson et al., 2007; Tan et al., 2008). It has been estimated that 10–25% of atrial arrhythmias are facilitated by vagus nerve influences (Bettoni and Zimmerman, 2002). The types of stimuli and conditions predisposing to cholinergic atrial fibrillation are shown in Table 12.1 (Olshansky, 2005).

With respect to the ventricles, vagus nerve activity is generally antiarrhythmic, as it inhibits the profibrillatory effects of sympathetic nerve activation. Enhanced vagus nerve activity also provides indirect protection against ventricular fibrillation by reducing excess heart rate (Kolman et al., 1975), which can otherwise compromise diastolic perfusion time during acute myocardial ischemia to increase ischemic insult. Vagus nerve stimulation has been shown in experimental studies to protect against ventricular arrhythmias during myocardial ischemia and postinfarction heart failure (Lopshire et al., 2009) but its protection during reperfusion arrhythmias is attributable to decreased heart rate (Zuanetti et al., 1987). However, beneficial effects of vagus nerve activity may be annulled if profound bradycardia and hypotension ensue. For example, sinus bradycardia or AV block is often observed following acute inferior myocardial infarction due to enhanced vagal tone (Feigl et al., 1984).

BAROREFLEXES AND ARRHYTHMIAS

The protective effect of the baroreceptor mechanism has been linked primarily to the antifibrillatory influence of vagus nerve activity (Billman et al., 1982). The latter effect improves diastolic coronary perfusion, minimizing the ischemic insult from coronary artery occlusion.

The importance of baroreceptor sensitivity (BRS) was documented in human subjects in whom baroreceptor function was evaluated with the pressor agent phenylephrine. La Rovere and colleagues (1998) demonstrated that patients who experienced a myocardial infarction were less likely to experience sudden cardiac death if their baroreceptor function was not depressed. During a 10-year follow-up, cardiac mortality among the trained patients who had an exercise-induced increase in BRS$>$ 3 ms/mmHg was strikingly lower compared to that of the trained patients without such a baroreflex response and to that of the untrained patients (La Rovere et al., 2002).

In the last few years, BRS testing has been pursued by noninvasive monitoring of heart rate turbulence. This phenomenon refers to fluctuations of sinus rhythm cycle length after a single ventricular premature beat and appears to be mechanistically linked with baroreceptor sensitivity (Bauer et al., 2008). The basic principle, introduced by Schmidt and coworkers (1999), is that the reaction of the cardiovascular system to the ventricular premature beat and the subsequent decrease in arterial blood pressure is a direct function of baroreceptor responsiveness, since reflex activation of the vagus nerve controls the pattern of sinus rhythm.

INTRINSIC CARDIAC INNERVATION

An elaborate intrinsic neural network within the heart provides local, independent heart rhythm control (Randall and Ardell, 1985; Chiou et al., 1997; Arora et al., 2001; Verrier and Zhao, 2002). Myocardial ischemia can damage cardiac intrinsic neurons residing in the fat pad and thus has the potential to increase electrical inhomogeneity and susceptibility to arrhythmias (Armour, 2004). Intrinsic innervation is also vulnerable to diabetic neuropathy, which accordingly could exacerbate vulnerability to arrhythmias (Stevens et al., 1998). Surgical incisions through the atrial walls and radiofrequency ablation may isolate sinoatrial node pacemaker cells and damage the fat pads, resulting in arrhythmias due to iatrogenically induced autonomic imbalance (Randall et al., 1991).

NERVE GROWTH AND DEGENERATION

The importance of restructuring of cardiac innervation also requires attention (Olgin et al., 1998; Jayachandran et al., 2000; Chen et al., 2001). Chen and coworkers (2001) found that nerve sprouting can affect ventricles, increasing potential risk of sudden cardiac death. These investigators demonstrated a significant correlation between increased sympathetic nerve density as reflected in immunocytochemical markers and history of ischemia in native hearts of human transplant recipients. Induction of nerve sprouting with nerve growth factor resulted in increased incidence of ventricular tachycardias and sudden death, with concomitant T-wave alternans, a noninvasive marker of risk for ventricular arrhythmias (Tsai et al., 2002). Myocardial infarction led to increased local release of nerve growth factor in the peri-infarct tissue. Nerve growth factor is transported retrogradely into the left stellate ganglion resulting in stimulation of nerve sprouting in the heart (Zhou et al., 2004). The pathways involved are illustrated in Figure 12.6 (Verrier and Kwaku, 2004). In addition, sustained atrial fibrillation has been correlated with increased sympathetic nerve sprouting within the atria (Chang et al., 2001), suggesting a role for increased sympathetic hyperinnervation as a substrate for atrial fibrillation.

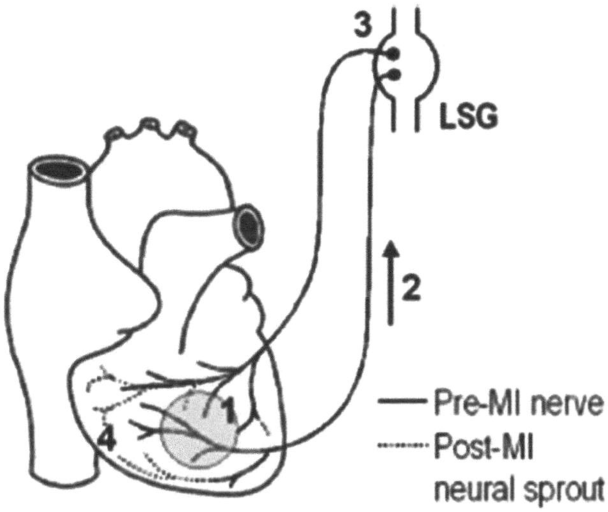

Fig. 12.6. Signaling of neural remodeling after myocardial infarction. Myocardial injury (shaded area) results in early local nerve growth factor (NGF) release, presumably from damaged cells, followed by upregulated NGF and growth-associated protein 43 (GAP43) expression, especially in the infarct area (1). These signal proteins are then retrogradely transported (2) to the nerve cell bodies in the ganglia (3) where they stimulate the sprouting of new cardiac nerve endings in the heart (4), predominantly in noninfarcted regions, leading to heterogeneous hyperinnervation. (Reproduced from Verrier and Kwaku, 2004, with permission from Lippincott Williams and Wilkins.)

BEHAVIORAL STRESS AND ARRHYTHMIAS

The view that behavioral factors may predispose to malignant arrhythmias has gained strong support. In patients with implantable cardioverter-defibrillators (ICDs), Lampert and colleagues (2002) systematically examined the potential of emotional stressors in provoking spontaneous ventricular arrhythmias. In the 15 minute period preceding shocks, there was a significant incidence of high levels of anger, with odds ratios of 1.83. Anger-induced T-wave alternans, which provides an indirect measure of autonomic function and electrical instability of the myocardial substrate (Verrier et al., 2009), has been found to predict appropriate ICD shock, suggesting that emotion-induced repolarization instability may constitute a mechanism linking stress to sudden death (Lampert et al., 2009).

Correlative findings were reported by Fries et al. (2002), who found ninefold increased risk of ICD shock with acute mental stress. These observations are consistent with the demonstration in ICD patients that mental stress as well as exercise is capable of significantly increasing TWA independent of effects on heart rate (Kop et al., 2004). By contrast, normal matched control subjects did not experience significant exercise- or mental stress-induced increases in TWA. Such surges in TWA indicate sympathetically mediated cardiac electrical instability, as they presage the onset of ventricular tachycardia and fibrillation (Nearing and Verrier, 2002a, b, 2003; Shusterman et al., 2006). The dynamic influence of mental and physical activity on cardiac electrical function is further supported by results of ambulatory electrocardiogram-based TWA analysis in postmyocardial infarction patients (Verrier et al., 2003) and those with ischemic or nonischemic cardiomyopathy (Sakaki et al., 2009).

SLEEP AS AN AUTONOMIC STRESS TEST FOR THE HEART

In healthy individuals, sleep is beneficial. However, during sleep in patients with cardiorespiratory disease, the brain can trigger breathing disorders and cardiovascular events. An estimated 250 000 nocturnal myocardial infarctions and 46 500 nocturnal sudden deaths occur annually in the U.S. population, as 20% of myocardial infarctions and 15% of sudden deaths occur during the period from midnight to 6:00 a.m. (Lavery et al., 1997). Thus, sleep is not an entirely protected state. The distribution of deaths and myocardial infarctions during nighttime is nonuniform, a pattern consistent

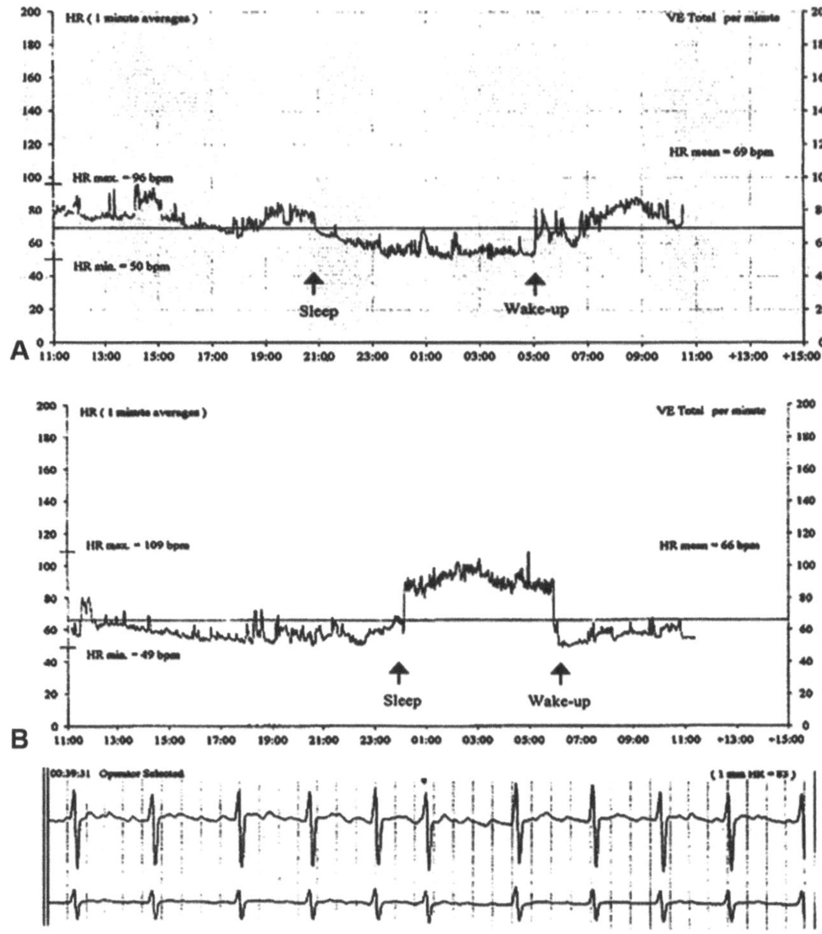

Fig. 12.7. Heart rate trend from an ambulatory ECG (AECG) recording showing a normal circadian rhythm with a sleep-induced decrease in heart rate (**A**), as compared with a nocturnal increase in heart rate caused by paroxysmal atrial fibrillation at the onset of sleep and a drop in heart rate after awakening due to spontaneous conversion to sinus rhythm (**B**). The ECG (below) documents atrial fibrillation during the sleep period. (Reproduced from Singh et al., 2004, with permission from Lippincott Williams and Wilkins.)

with provocation by pathophysiological triggers. In certain individuals, there is a nocturnal propensity for atrial fibrillation, which is thought to be attributable to the unstable autonomic changes associated with sleep-state transitions (Fig. 12.7) (Verrier and Josephson, 2009). High-risk populations for nocturnal cardiorespiratory events include a number of sizeable patient groups (Table 12.2) (Verrier and Mittleman, 2010).

The two main factors that have been implicated in nocturnal cardiac events are sleep-state-dependent surges in autonomic nervous system activity and depression of respiratory control mechanisms, which impact on a vulnerable cardiac substrate. The brain, in subserving its needs for periodic re-excitation during rapid eye movement sleep and dreaming, imposes significant demands on the heart by inducing bursts in sympathetic nerve activity, which reaches levels higher than during wakefulness. In susceptible individuals, this degree of sympathetic nerve activity may reduce coronary artery blood flow, as metabolic demand outstrips supply, and may trigger sympathetically mediated life-threatening arrhythmias. Obstructive sleep apnea, which impairs ventilation during sleep and can generate reductions in arterial oxygen saturation, afflicts 5–10 million Americans, or 2–4% of the population (Young et al., 1993). This condition has been strongly implicated, when severe, in the etiology of hypertension, ischemia, arrhythmias, myocardial infarction, and sudden death in individuals with coexisting ischemic heart disease. The surge in sympathetic nerve activity associated with apnea has been linked to cardiac electrical instability, as indicated by quantifiable levels of T-wave alternans (Takasugi et al., 2009). Excess parasympathetic activity or respiratory disturbances during sleep may trigger atrial fibrillation in certain patient populations. A challenge is also presented by nonrapid eye movement sleep, when impaired perfusion of the heart and brain may result from hypotension with

Table 12.2

Patient groups at potentially increased risk for nocturnal cardiac events

Condition (U.S. patients/year)	Possible mechanism
Angina, MI, arrhythmias, ischemia, or cardiac arrest at night (20% of MIs (~250 000 cases/year) and 15% of sudden deaths (~46 500 cases/year) occur between midnight and 6:00 a.m.)	The nocturnal pattern suggests a sleep-state-dependent autonomic trigger or respiratory distress
Unstable angina	Nondemand ischemia and angina peak between midnight and 6:00 a.m.
Acute MI (1.5 million)	Disturbances in sleep, respiration, and autonomic balance may be factors in nocturnal arrhythmogenesis. Nocturnal onset of MI is more frequent in older and sicker patients and carries a higher risk of congestive heart failure
Heart failure (5.3 million)	Sleep-related breathing disorders are pronounced in the setting of heart failure and may contribute to its progression and to mortality risk
Spousal or family report of highly irregular breathing, excessive snoring, or apnea in patients with coronary disease (15 million U.S. patients with apnea)	Patients with hypertension or atrial or ventricular arrhythmias should be screened for the presence of sleep apnea
Long QT syndrome	The profound cycle-length changes associated with sleep may trigger pause-dependent torsades de pointes in these patients
Near miss or siblings of victims of SIDS (2500 total SIDS deaths annually in the U.S., or 1 death per 2000 live births)	SIDS commonly occurs during sleep with characteristic cardiorespiratory symptoms
Brugada syndrome in Western populations; Asians with warning signs of SUNDS	SUNDS is a sleep-related phenomenon in which night terrors may play a role. It is genetically related to the Brugada syndrome
Atrial fibrillation (2.2 million)	Twenty-nine percent of episodes occur between midnight and 6:00 a.m. Respiratory and autonomic mechanisms are suspected

MI, myocardial infarction; SIDS, sudden infant death syndrome; SUNDS, sudden unexplained nocturnal death syndrome. (Adapted from Verrier and Mittleman, 2010, p. 1354.)

decreased blood flow in stenosed vessels. These conditions may be confounded by medications that cross the blood–brain barrier, alter sleep structure and/or provoke nightmares with severe cardiac autonomic discharge.

CONCLUSIONS

The autonomic nervous system plays an important role in the generation of cardiac arrhythmias. From the basic science perspective, there appears to be great potential benefit to improved understanding of the organization and function of the intrinsic nervous system. Such advances can aid diagnoses and lead to novel therapeutic approaches based on pharmacological and electrical targeted neuromodulation (Issa et al., 2005; Schwartz et al., 2008; Tan et al., 2008; Ogawa et al., 2009) to decrease cardiac sympathetic nerve activity while measures augmenting vagus nerve tone are being pursued.

References

Armour JA (2004). Cardiac neuronal hierarchy in health and disease. Am J Physiol Regul Integr Comp Physiol 287: R262–R271.

Arora RC, Hirsch GM, Johnson Hirsch K et al. (2001). Function of human intrinsic cardiac neurons in situ. Am J Physiol Regul Integr Comp Physiol 280: R1736–R1740.

Bauer A, Malik M, Schmidt G et al. (2008). Heart rate turbulence: standards of measurement, physiological interpretation, and clinical use. International Society for Holter and Noninvasive Electrocardiology Consensus. J Am Coll Cardiol 52: 1353–1365.

Benjamin EJ, Chen PS, Bild DE et al. (2009). Prevention of atrial fibrillation: report from a National Heart, Lung, and Blood Institute workshop. Circulation 119: 606–618.

Bettoni M, Zimmerman M (2002). Autonomic tone variations before the onset of paroxysmal atrial fibrillation. Circulation 105: 2753–2759.

Billman GE, Schwartz PJ, Stone HL (1982). Baroreceptor reflex control of heart rate: a predictor of sudden cardiac death. Circulation 66: 874–880.

Calkins H, Brugada J, Packer DL et al. (2007). HRS/EHRA/ECAS expert consensus statement on catheter and surgical ablation of atrial fibrillation: recommendations for personnel, policy, procedures and follow-up. A report of the Heart Rhythm Society Task Force on catheter and surgical ablation of atrial fibrillation. Heart Rhythm 4: 816–861.

Chang CM, Wu TJ, Zhou S et al. (2001). Nerve sprouting and sympathetic hyperinnervation in a canine model of atrial fibrillation produced by prolonged right atrial pacing. Circulation 103: 22–25.

Chen PS, Chen LS, Cao JM et al. (2001). Sympathetic nerve sprouting, electrical remodeling and the mechanisms of sudden cardiac death. Cardiovasc Res 50: 409–416.

Chiou CW, Eble JN, Zipes DP (1997). Efferent vagal innervation of the canine atria and sinus and atrioventricular nodes. The third fat pad. Circulation 95: 2573–2584.

Feigl D, Ashkenazy J, Kishon Y (1984). Early and late atrioventricular block in acute inferior myocardial infarction. J Am Coll Cardiol 4: 35–38.

Fries R, Konig J, Schafers HJ et al. (2002). Triggering effect of physical and mental stress on spontaneous ventricular tachyarrhythmias in patients with implantable cardioverter defibrillators. Clin Cardiol 25: 474–478.

Hrushesky WJM, Fader D, Schmitt O et al. (1984). The respiratory sinus arrhythmia: a measure of cardiac age. Science 224: 1001–1004.

Issa ZF, Zhou X, Ujhelyi MR et al. (2005). Thoracic spinal cord stimulation reduces the risk of ischemic ventricular arrhythmias in a postinfarction heart failure canine model. Circulation 111: 3217–3220.

Jayachandran JV, Sih HJ, Winkle W et al. (2000). Atrial fibrillation produced by prolonged rapid atrial pacing is associated with heterogeneous changes in atrial sympathetic innervation. Circulation 101: 1185–1191.

Joung B, Tang L, Maruyama M et al. (2009). Intracellular calcium dynamics and acceleration of sinus rhythm by beta-adrenergic stimulation. Circulation 119: 788–796.

Jung B-C, Dave AS, Tan AY et al. (2006). Circadian variations of stellate ganglion nerve activity in ambulatory dogs. Heart Rhythm 3: 78–85.

Katona PG, McLean M, Dighton DH et al. (1982). Sympathetic and parasympathetic cardiac control in athletes and nonathletes at rest. J Appl Physiol 52: 1652–1657.

Kolman BS, Verrier RL, Lown B (1975). The effect of vagus nerve stimulation upon vulnerability of the canine ventricle. Role of sympathetic-parasympathetic interactions. Circulation 52: 578–585.

Kop WJ, Krantz DS, Nearing BD et al. (2004). Effects of acute mental and exercise stress on T wave alternans in patients with implantable cardioverter defibrillators and controls. Circulation 109: 1864–1869.

Kowey PR, Verrier RL, Lown B (1983). Effect of alpha-adrenergic receptor stimulation on ventricular electrical properties in the normal canine heart. Am Heart J 105: 366–371.

La Rovere MT, Bigger JT Jr, Marcus FI et al. (1998). Baroreflex sensitivity and heart-rate variability in prediction of total cardiac mortality after myocardial infarction. ATRAMI (Autonomic Tone and Reflexes after Myocardial Infarction) Investigators. Lancet 351: 478–484.

La Rovere MT, Bersano C, Gnemmi M et al. (2002). Exercise-induced increase in baroreflex sensitivity predicts improved prognosis after myocardial infarction. Circulation 106: 945–949.

Lampert R, Joska T, Burg MM et al. (2002). Emotional and physical precipitants of ventricular arrhythmia. Circulation 106: 1800–1805.

Lampert R, Shusterman V, Burg M et al. (2009). Anger-induced T-wave alternans predicts future ventricular arrhythmias in patients with implantable cardioverter-defibrillators. J Am Coll Cardiol 53: 774–778.

Lathrop DA, Spooner PM (2001). On the neural connection. J Cardiovasc Electrophysiol 12: 841–844.

Lavery CE, Mittleman MA, Cohen MC et al. (1997). Nonuniform nighttime distribution of acute cardiac events: a possible effect of sleep states. Circulation 96: 3321–3327.

Levy MN (1988). Cardiac sympathetic-parasympathetic interactions. In: H Kulbertus, G Franck (Eds.), Neurocardiology. Futura, New York, pp. 85–98.

Levy MN, Blattberg B (1976). Effect of vagal stimulation on the overflow of norepinephrine into the coronary sinus during cardiac sympathetic nerve stimulation in the dog. Circ Res 38: 81–84.

Lombardi F, Verrier RL, Lown B (1983). Relationship between sympathetic neural activity, coronary dynamics, and vulnerability to ventricular fibrillation during myocardial ischemia and reperfusion. Am Heart J 105: 958–965.

Lopshire JC, Zhou X, Dusa C et al. (2009). Spinal cord stimulation improves ventricular function and reduces ventricular arrhythmias in a canine postinfarction heart failure model. Circulation 120: 286–294.

Nearing BD, Verrier RL (2002a). Modified moving average method for T-wave alternans analysis with high accuracy to predict ventricular fibrillation. J Appl Physiol 92: 541–549.

Nearing BD, Verrier RL (2002b). Progressive increases in complexity of T-wave oscillations herald ischemia induced VF. Circ Res 91: 727–732.

Nearing BD, Verrier RL (2003). Tracking heightened cardiac electrical instability by computing interlead heterogeneity of T-wave morphology. J Appl Physiol 95: 2265–2272.

Nearing BD, Huang AH, Verrier RL (1991). Dynamic tracking of cardiac vulnerability by complex demodulation of the T-wave. Science 252: 437–440.

Nearing BD, Oesterle SN, Verrier RL (1994). Quantification of ischaemia-induced vulnerability by precordial T-wave alternans analysis in dog and human. Cardiovasc Res 28: 1440–1449.

Ogawa M, Tan AY, Song J et al. (2009). Cryoablation of stellate ganglia and atrial arrhythmia in ambulatory dogs with pacing-induced heart failure. Heart Rhythm 6: 1772–1779.

Olgin JE, Sih HJ, Hanish S et al. (1998). Heterogeneous atrial denervation creates substrate for sustained atrial fibrillation. Circulation 98: 2608–2614.

Olshansky B (2005). Interrelationships between the autonomic nervous system and atrial fibrillation. Prog Cardiovasc Dis 48: 57–78.

Olsson G, Wikstrand J, Warnold I et al. (1992). Metoprolol-induced reduction in postinfarction mortality: pooled results from five double-blind randomized trials. Eur Heart J 13: 28–32.

Opie LH (2004). Heart Physiology: From Cell to Circulation. 4th edn. Lippincott Williams and Wilkins, Philadelphia, p. 142.

Opthof T (2000). The normal range and determinants of the intrinsic heart rate in man. Cardiovasc Res 45: 173–176.

Patterson E, Jackman WM, Beckman KJ et al. (2007). Spontaneous pulmonary vein firing in man: relationship to tachycardia-pause early after depolarizations and triggered arrhythmia in canine pulmonary veins in vitro. J Cardiovasc Electrophysiol 18: 1067–1075.

Randall WC, Ardell JL (1985). Selective parasympathectomy of automatic and conductile tissues of the canine heart. Am J Physiol 248: H61–H68.

Randall WC, Wurster RD, Duff M et al. (1991). Surgical interruption of postganglionic innervation of the sinoatrial nodal region. J Thorac Cardiovasc Surg 101: 66–74.

Sakaki K, Ikeda T, Miwa Y et al. (2009). Time-domain T-wave alternans measured from Holter electrocardiograms predicts cardiac mortality in patients with left ventricular dysfunction: a prospective study. Heart Rhythm 6: 332–337.

Schmidt G, Malik M, Barthel P et al. (1999). Heart-rate turbulence after ventricular premature beats as a predictor of mortality after acute myocardial infarction. Lancet 353: 1390–1396.

Schwartz PJ, De Ferrari GM, Sanzo A et al. (2008). Long term vagal stimulation in patients with advanced heart failure: first experience in man. Eur J Heart Fail 10: 884–891.

Shusterman V, Goldberg A, London B (2006). Upsurge in T-wave alternans and nonalternating repolarization instability precedes spontaneous initiation of ventricular tachyarrhythmias in humans. Circulation 113: 2880–2887.

Singh J, Mela T, Ruskin J (2004). Images in cardiovascular medicine: sleep (vagal)-induced atrial fibrillation. Circulation 110: e32–e33.

Stevens MJ, Raffel DM, Allman KC et al. (1998). Cardiac sympathetic dysinnervation in diabetes: implications for enhanced cardiovascular risk. Circulation 98: 961–968.

Takasugi N, Nishigaki K, Kubota T et al. (2009). Sleep apnoea induces cardiac electrical instability assessed by T-wave alternans in patients with congestive heart failure. Eur J Heart Fail 11: 1063–1070.

Tan AY, Zhou S, Ogawa M et al. (2008). Neural mechanisms of paroxysmal atrial fibrillation and paroxysmal atrial tachycardia in ambulatory canines. Circulation 118: 916–925.

Tsai J, Cao JM, Zhou S et al. (2002). T wave alternans as a predictor of spontaneous ventricular tachycardia in a canine model of sudden cardiac death. J Cardiovasc Electrophysiol 13: 51–55.

Verrier RL, Antzelevitch CA (2004). Autonomic aspects of arrhythmogenesis: the enduring and the new. Curr Opin Cardiol 19: 2–11.

Verrier RL, Josephson ME (2009). Impact of sleep on arrhythmogenesis. Circ Arrhythmia Electrophysiol 2: 450–459.

Verrier RL, Kwaku KF (2004). Frayed nerves in myocardial infarction: the importance of rewiring. Circ Res 95: 5–6.

Verrier RL, Mittleman MA (2010). Sleep-related cardiac risk. In: MH Kryger, T Roth, WC Dement (Eds.), Principles and Practice of Sleep Medicine. 5th edn. WB Saunders, Philadelphia, pp. 1353–1362.

Verrier RL, Nieminen T (2010). T-wave alternans as a therapeutic marker for antiarrhythmic agents. J Cardiovasc Pharmacol 55: 544–554.

Verrier RL, Schachter SC (2010). Neurocardiac interactions in sudden unexpected death in epilepsy: can ambulatory ECG-based assessment of autonomic function and T-wave alternans help to evaluate risk? In: CM Lathers, PL Schraeder, MW Bungo et al. (Eds.), Sudden Death in Epilepsy: Forensic and Clinical Issues. Taylor and Francis, Boca Raton, FL, pp. 693–709.

Verrier RL, Zhao SX (2002). The enigmatic cardiac fat pads: critical but underappreciated neural regulatory sites. J Cardiovasc Electrophysiol 13: 902–903.

Verrier RL, Calvert A, Lown B et al. (1974a). Effect of acute blood pressure elevation on the ventricular fibrillation threshold. Am J Physiol 226: 893–897.

Verrier RL, Thompson PL, Lown B (1974b). Ventricular vulnerability during sympathetic stimulation: role of heart rate and blood pressure. Cardiovasc Res 8: 602–610.

Verrier RL, Nearing BD, LaRovere MT et al. for the ATRAMI Investigators (2003). Ambulatory ECG-based tracking of T-wave alternans in post-myocardial infarction patients to assess risk of cardiac arrest or arrhythmic death. J Cardiovasc Electrophysiol 14: 705–711.

Verrier RL, Kumar K, Nearing BD (2009). Basis for sudden cardiac death prediction by T-wave alternans from an integrative physiology perspective. Heart Rhythm 6: 416–422.

Young T, Palta M, Dempsey J et al. (1993). The occurrence of sleep-disordered breathing among middle-aged adults. N Engl J Med 328: 1230–1235.

Zhou S, Chen LS, Miyauchi Y et al. (2004). Mechanisms of cardiac nerve sprouting after myocardial infarction in dogs. Circ Res 95: 76–83.

Zhou S, Jung BC, Tan AY et al. (2008a). Spontaneous stellate ganglion nerve activity and ventricular arrhythmia in a canine model of sudden death. Heart Rhythm 5: 131–139.

Zhou S, Tan AY, Paz O et al. (2008b). Antiarrhythmic effects of beta$_3$-adrenergic receptor stimulation in a canine model of ventricular tachycardia. Heart Rhythm 5: 289–297.

Zuanetti G, De Ferrari GM, Priori SG et al. (1987). Protective effect of vagal stimulation on reperfusion arrhythmias in cats. Circ Res 61: 429–435.

Handbook of Clinical Neurology, Vol. 117 (3rd series)
Autonomic Nervous System
R.M. Buijs and D.F. Swaab, Editors

Chapter 13

Exercise and the autonomic nervous system

QI FU[1,2] AND BENJAMIN D. LEVINE[1,2]*
[1]*University of Texas Southwestern Medical Center, Dallas, TX, USA*
[2]*Institute for Exercise and Environmental Medicine, Texas Health Presbyterian Hospital Dallas, TX, USA*

INTRODUCTION

Physical activity (dynamic exercise) is a key component of a healthy lifestyle. Increased physical activity or exercise training is not only protective against cardiovascular disease, hypertension, colon and breast cancer, type II diabetes, and obesity, but also effective in improving functional capacity in patients with these medical conditions, as well as in patients with autonomic disorders. Exercise training also improves mental health, helps to prevent depression, and promotes or maintains positive self-esteem. Endurance or "dynamic" exercise, such as running, jogging, cycling, swimming, rowing, or walking, and strength training, such as weight-lifting, are two major forms of exercise. In this chapter, we predominately focus on dynamic exercise.

There is now convincing evidence that some of the protective and therapeutic effects of exercise training are related, in a substantive fashion, to effects on the autonomic nervous system (Joyner and Green, 2009). In addition, training-induced improvement in vascular endothelial function, blood volume expansion, cardiac remodeling, insulin resistance, and renal-adrenal function may also contribute to the protection and treatment of cardiovascular, metabolic, and autonomic disorders.

ACUTE EXERCISE

Oxygen uptake

During dynamic exercise, the uptake and transport of oxygen is required for oxidative phosphorylation and the efficient production of adenosine triphosphate (ATP) to support the metabolic demands of the body (Levine, 2001). The primary unit of oxygen uptake is liters of oxygen per minute (L/min), and, in absolute terms, is a direct function of body size. However, as a measure of work capacity, it is usually normalized to body mass, and in its most familiar form is expressed as milliliters per kilogram per minute (mL/kg/min). One of the most inviolate relationships in all of exercise physiology is that between oxygen uptake and cardiac output. Regardless of age, sex, or the presence of various disease states, in general, about 5–6 liters of cardiac output are required for every liter of oxygen uptake above rest (Fig. 13.1) (Rowell, 1993b; Levine, 2001). When this relationship is depressed, as in patients with congestive heart failure (Chomsky et al., 1996), it may be a sign of severe underlying disease with impending decompensation (Martin et al., 1989). Conversely, when it is exaggerated, as in patients with metabolic myopathies (Haller et al., 1991; Taivassalo et al., 2001, 2003), it gives strong clues to the processes regulating cardiac output.

The Fick principle

Oxygen uptake is a function of the triple-product of heart rate and stroke volume (i.e., cardiac output) and arterial-mixed venous oxygen difference or total system arteriovenous oxygen difference (Rowell, 1993b) – the Fick principle:

$$\begin{aligned}\dot{V}O_2 &= \dot{Q}c \times a\text{–}vO_2\ \textit{difference} \\ &= HR \times SV \times a\text{–}vO_2\ \textit{difference}\end{aligned}$$

where $\dot{V}O_2$ is oxygen uptake, $\dot{Q}c$ is cardiac output, *HR*, is heart rate, *SV* is stroke volume, and *a-vO₂ difference* is total system arteriovenous oxygen difference. The degree to which each of the variables can increase determines the upper limit for whole-body oxygen consumption, and this limit is called the maximal oxygen uptake ($\dot{V}O_2$max) (Levine, 2008). Maximal heart rate and maximal arteriovenous difference are usually similar in athletes and nonathletes alike; the factor most commonly

*Correspondence to: B.D. Levine, Institute for Exercise and Environmental Medicine, Texas Health Presbyterian Hospital Dallas, 7232 Greenville Avenue, Suite 435, Dallas, TX 75231, USA. E-mail: benjaminlevine@TexasHealth.org

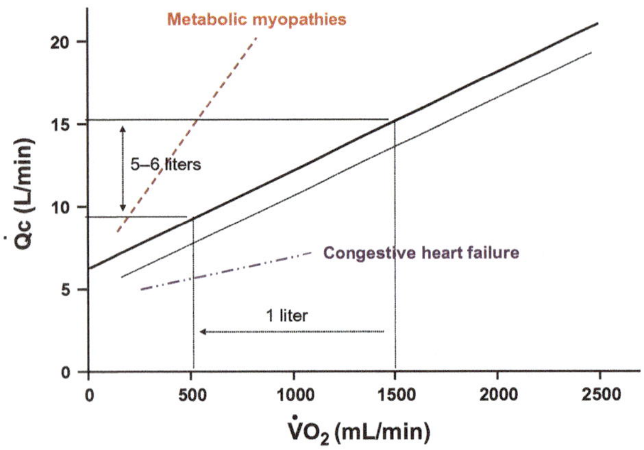

Fig. 13.1. Relationship between increases in oxygen uptake ($\dot{V}O_2$) and the corresponding increases in cardiac output ($\dot{Q}c$) during exercise in humans, which in most cases is 5–6/1. When this relationship is depressed, as in patients with congestive heart failure, it may be a sign of severe underlying disease with impending decompensation. Conversely, when it is exaggerated, as in patients with metabolic myopathies, it gives strong clues to the processes regulating $\dot{Q}c$. (Adapted with permission from Levine, 2001.)

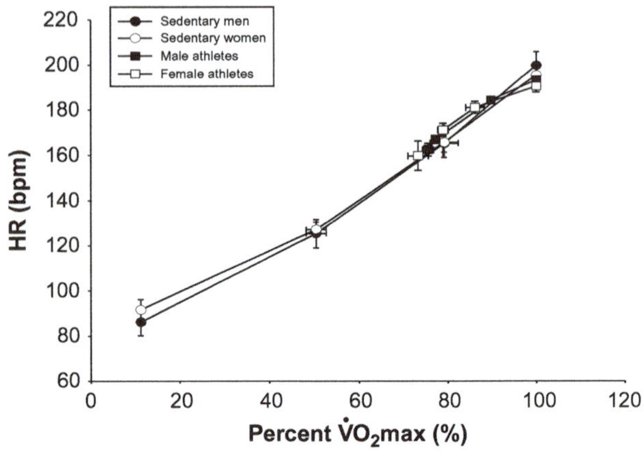

Fig. 13.2. Comparisons of heart rate (HR) responses during submaximal and maximal exercise in sedentary men and women and during submaximal exercise in highly trained male and female athletes at the same relative work rate, expressed as the percentage of maximal oxygen uptake ($\dot{V}O_2$max), which represents a matter of cardiovascular control during exercise in normal humans. (Adapted with permission from Fu and Levine, 2005.)

accounting for the different values of $\dot{V}O_2$max in different individuals of different fitness levels is stroke volume (Rowell, 1993b).

At low levels of exercise, heart rate increases almost exclusively via vagal withdrawal, with little evidence for systematic increases in sympathetic activity until the intensity of exercise is at or above the maximal steady state (Ray et al., 1992; Rowell, 1993b). The key determinant of the magnitude of the heart rate response to exercise is the relative intensity as well as the absolute amount of muscle mass engaged, while central command plays an essential role in the increase in heart rate during exercise (Levine, 2001). If heart rate is normalized or scaled by expressing it in relation to the relative oxygen uptake, for example, percent of $\dot{V}O_2$max, then diverse individuals appear to respond in virtually identical fashion (Fig. 13.2) (Rowell, 1993b; Fu and Levine, 2005), indicating that the cardiovascular response is more closely related to the relative than absolute metabolic demands.

Gravity plays a critical role in determining the distribution of blood within the cardiovascular system, and body posture markedly affects the relative importance of changes in stroke volume (Levine, 2001). In the upright position, stroke volume is only about one-half its value in the supine position due to blood pooling in the legs and a reduction in left ventricular end-diastolic volume. At the onset of exercise, the pumping action of skeletal muscle acts to augment venous return substantially, and stroke volume normally increases > 50% via the Starling mechanism (Levine, 2001). At least for most nonathletic individuals, maximal stroke volume is achieved for the most part at relatively low levels of exercise intensity (approximately 50% of maximal oxygen uptake), as pericardial constraint serves to limit left ventricular end-diastolic volume. At rest, arteriovenous oxygen difference is normally 4.5 mL/100 mL/min (approximately 23% extraction), and at $\dot{V}O_2$max this difference is commonly close to 16 mL/100 mL/min (about 80–85% extraction) (Rowell, 1993b).

Neural mechanisms for the exercise response

The cardiovascular response during exercise is initiated by a feed-forward mechanism, termed "*central command*," which involves higher brain centers such as the motor cortex, hypothalamic and mesencephalic locomotor regions that activate parallel circuits controlling locomotor, cardiovascular, and ventilatory functions (Krogh and Lindhard, 1913; Mitchell, 1990). As exercise continues, both mechanical and metabolic signals from active skeletal muscle provide feedback to cardiovascular centers in the brain through group III and IV muscle afferents, the so-called "*exercise pressor reflex*," to precisely match systemic oxygen delivery with metabolic demand (Alam and Smirk, 1937; Mitchell, 1990; Michelini and Stern, 2009).

Contracting skeletal muscle produces vasodilatory metabolites (i.e., adenosine, prostaglandins, nitric oxide) which attenuate sympathetically mediated vasoconstriction, and this phenomenon is known as "functional sympatholysis" (Remensnyder et al., 1962; Dinenno and Joyner, 2003; Keller et al., 2004). However, it has been found that nitric oxide is not obligatory for functional sympatholysis in contracting skeletal muscles of healthy humans (Dinenno and Joyner, 2003). Enhanced endothelium-derived hyperpolarizing factor activity in conditions of nitric oxide deficiency may contribute to functional sympatholysis (Ozkor et al., 2011). Vascular resistance in active skeletal muscle decreases to facilitate increases in muscle perfusion, while in inactive skeletal muscle vascular resistance increases so that systemic arterial blood pressure can be supported (Joyner and Thomas, 2003; Keller et al., 2004). If sympathetically medicated vasoconstriction is impaired, for instance, in patients with autonomic failure, hypotension will occur during exercise, and at times this hypotension may be profound. Additionally, cardiac output increases proportionate with oxygen uptake, thus allowing the maintenance or, in most cases, an increase in mean arterial pressure. An overview of the neural mechanisms responsible for the cardiovascular response during exercise is shown in Figure 13.3.

There is a parallel increase in arterial pressure and heart rate at the beginning and during exercise, indicating a resetting of the arterial baroreflex. Indeed, many human and animal studies have shown that the baroreflex control of heart rate and arterial pressure resets with little change in gain or sensitivity during exercise (Bevegard and Shepherd, 1966; Joyner, 2006; Raven, 2008). Sensitivity refers to the change in arterial pressure caused by a given change in pressure at the arterial baroreceptors (most commonly the isolated carotid sinus) (Rowell, 1993a). It also refers to the slope of the relationship between heart rate or arterial pressure and carotid sinus pressure, as shown in Figure 13.4. Upward resetting of the operating point of the arterial baroreflex appears to be the major factor responsible for the sympathy-excitatory response and increasing blood pressure during exercise (DiCarlo and Bishop, 1992). It has been proposed that central command resets the arterial baroreflex and in turn raises cardiac output by sudden withdrawal of tonic outflow to the heart (Rowell, 1993a). Conversely, activation of the exercise pressor reflex may also contribute to the resetting of the arterial baroreflex (Rowell and O'Leary, 1990; Ichinose et al., 2008).

Heart rate recovery from exercise

Early recovery of the heart rate after acute bouts of exercise appears to be dominated by vagal reactivation with sympathetic withdrawal becoming more important later in recovery (Imai et al., 1994; Pierpont et al., 2000). This notion is supported by the findings of Goldberger et al. showing that administration of atropine, a muscarinic receptor antagonist, completely abolishes heart rate recovery as well as time domain measures of heart rate variability during recovery from exercise (Goldberger et al., 2006). It is further supported by the results of Dewland et al. showing that pyridostigmine, an acetylcholinesterase inhibitor that selectively augments the parasympathetic (vagal) efferent signals, enhances heart rate recovery (Dewland et al., 2007). The rate of heart rate return to baseline after exercise has been proposed

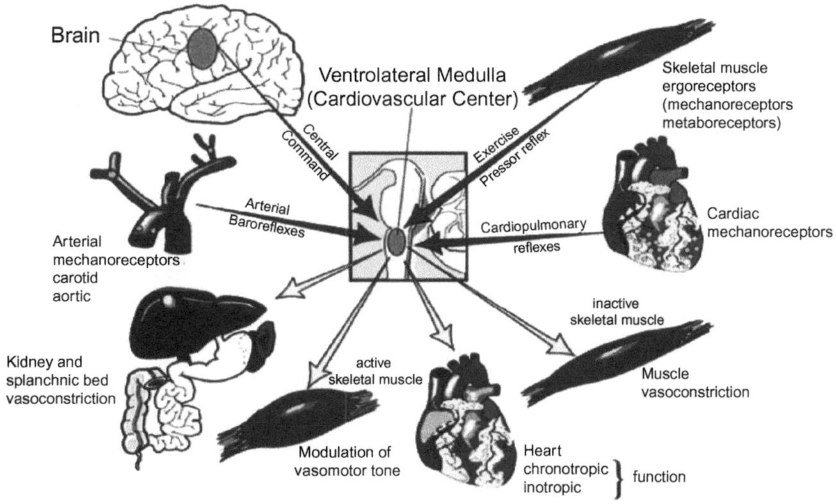

Fig. 13.3. Autonomic neural control of the cardiovascular response during exercise. Central command initiates the exercise pressor response, which is maintained and augmented via feedback from baroreceptors, as well as by stimulation of skeletal muscle mechanically and metabolically sensitive receptors. After integration in the brain, efferent responses via the parasympathetic (vagal) and sympathetic nervous systems result in increased heart rate and contractility, vasoconstriction in inactive muscle, and vasodilation in active muscle beds mediated by release of local vasodilating substances ("functional sympatholysis"). (Adapted with permission from Levine, 2001.)

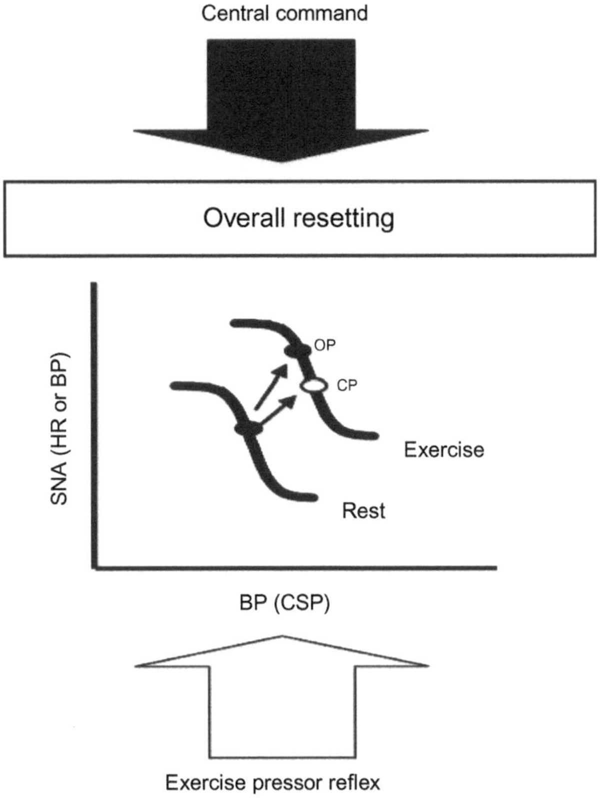

Fig. 13.4. Hypothetical stimulus-response curves for arterial baroreflex at rest and during exercise, expressed as relationship between sympathetic nerve activity (SNA) and blood pressure (BP) (or often in experiments as relationship between BP or HR and isolated carotid sinus pressure (CSP)). Resetting of the baroreflex occurs from rest to exercise. Central command plays a determinant role in the baroreflex resetting, while the exercise pressor reflex is also involved in this resetting. OP, operating point; CP, centering point. (Adapted with permission from Rowell and O'Leary, 1990; Raven et al., 2006.)

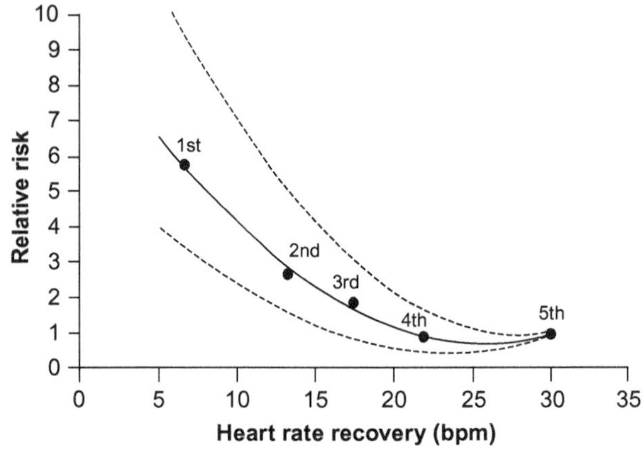

Fig. 13.5. Estimates of the relative risk of death within 6 years according to heart rate recovery 1 minute after cessation of exercise. Circles represent the relative risk of death for each of the quintiles as compared with the quintile with the greatest reduction (5th). Dashed lines represents the 95% confidence interval. The abbreviation bpm denotes beats per minute. (Adapted with permission from Cole et al., 1999.)

to be associated with fitness and good health (Shetler et al., 2001), and is usually calculated from peak exercise to minute 1 or 2 of recovery (Cole et al., 1999, 2000; Nishime et al., 2000; Rosenwinkel et al., 2001). A reduced rate of heart rate recovery, which may be a reflection of decreased vagal activity, is a powerful predictor of overall mortality, independent of workload, the presence or absence of myocardial perfusion defects, and changes in heart rate during exercise (Fig. 13.5) (Cole et al., 1999).

Cole et al. proposed that an abnormal heart rate recovery was defined as a reduction of 12 beats per minute or less from the heart rate at peak exercise to 1 minute after the cessation of exercise in a cool-down period (Cole et al., 1999). Subsequently, Watanabe et al. confirmed that even in the absence of a cool-down period and after accounting for left ventricular systolic function, heart rate recovery was still a powerful and independent predictor of death (Watanabe et al., 2001). However, the rate of heart rate recovery is substantially greater between standing and supine positions, and could lead to misclassification of some patients (Levine et al., unpublished results). Central command is restored more quickly when supine compared with standing. Thus, for accurate measurements of heart rate recovery, values for distinguishing high from low risk patients should be standardized for body position during recovery. Also, patients who achieve a low peak heart rate will have a correspondingly low heart rate recovery compared to those who achieve a higher peak heart rate, even though control of R-R interval is the same, based on the reciprocal relationship between heart rate and R-R interval. Thus, a low heart rate recovery may reflect a low peak heart rate rather than an intrinsic defect in autonomic control.

Exercise response in autonomic disorders

In general, patients with autonomic disorders have low levels of $\dot{V}O_2max$, indicating reduced physical fitness and exercise capacity. Moreover, the vast majority of the patients have blunted or abnormal cardiovascular response to exercise, especially during maximal exercise.

Parkinson's disease

The effect of Parkinson's disease on exercise is not well characterized. A recent report on Parkinson's disease showed that during submaximal exercise, cardiac contractility was the lowest in patients with reduced

^{123}I-metaiodobenzylguanidine (MIBG) uptake, an indication of cardiac sympathetic denervation, when compared with innerved patients and healthy controls; however, arterial pressure and heart rate responses were not significantly different between the three groups (Nakamura et al., 2010). Although the cardiovascular responses at submaximal or lower levels of exercise appear to be similar between patients with Parkinson's disease and healthy individuals, it has been found that half of the patients exhibit blunted responses at peak or maximal exercise (Werner et al., 2006).

Spinal cord injury

Different injury levels in patients with spinal cord injury may have different exercise responses as a result of the interruption of pathways in the spinal cord to the peripheral sympathetic nervous system in addition to the motor paralysis. For example, it was found that tetraplegics had low epinephrine and norepinephrine at rest and only slightly increased during exercise, while their impaired cardiac sympathetic innervations caused restricted cardioacceleration and markedly reduced $\dot{V}O_2max$ (Schmid et al., 1998). High-lesion paraplegics were found to have comparable norepinephrine but lower epinephrine levels as a result of partial innervation of the noradrenergic system and denervation of the adrenal medulla (Schmid et al., 1998). Conversely, spinal cord injury below T_5 showed an augmented basal and exercise-induced upper spinal thoracic sympathetic activity compared with healthy controls (Schmid et al., 1998). Claydon et al. also found that abnormal cardiovascular responses to exercise and transient postexercise hypotension were common in cervical but not thoracic spinal cord injury, which may be partly related to loss of descending sympathetic nervous control of the heart and vasculature following high spinal cord injury (Claydon et al., 2006). Additionally, it was recently reported that metaboreflex-induced blood pressure rise associated with arm cycling exercise was blunted in patients with spinal cord injury compared with healthy controls (Crisafulli et al., 2009).

Multiple system atrophy

The cardiovascular response to exercise differs between two clinical subgroups of multiple system atrophy (Shy–Drager syndrome), with a larger fall in arterial pressure during supine leg exercise but not during standing after exercise in the cerebellar form than in the Parkinsonian form (Smith et al., 1995; Smith and Mathias, 1996). Heart rate increased similarly in both groups, systemic vascular resistance fell similarly, but cardiac output rose more in patients with the Parkinsonian form compared with the cerebellar form (Smith and Mathias, 1996). Resting plasma norepinephrine concentrations were subnormal in both forms and did not increase with exercise, while postural hypotension was enhanced after exercise to the same extent in both forms (Smith and Mathias, 1996). The greater cardiovascular abnormalities in response to exercise in the cerebellar form suggest that cerebellar or brainstem autonomic pathways are impaired to a larger extent in the cerebellar form than in the Parkinsonian form of multiple system atrophy (Smith and Mathias, 1996).

Pure autonomic failure

It has been shown that patients with pure autonomic failure have an abnormal fall in blood pressure and an attenuated increase in heart rate during exercise (Smith et al., 1995, 1996a; Smith and Mathias, 1996; Humm et al., 2008), which is presumably attributable to reduced sympathetic nerve activity to the heart and vasculature and blunted sympathetic α-adrenergic vasoconstriction in inactive (nonexercising) skeletal muscle (Smith et al., 1996b; Schrage et al., 2004). Humm et al. reported that water drinking did not change the abnormal cardiovascular responses to supine cycling exercise but improved orthostatic tolerance postexercise in patients with pure autonomic failure (Humm et al., 2008). However, in autonomic failure due to familial amyloid polyneuropathy and in selective sympathetic failure due to dopamine β-hydroxylase deficiency, blood pressure usually remains unchanged with exercise (Smith et al., 1995; Smith and Mathias, 1996).

Baroreflex failure

In contrast to patients with autonomic failure who usually have severe orthostatic hypotension and reduced sympathetic activation, patients with baroreflex failure have little orthostatic hypotension and their illness is dominated by volatile hypertension (Robertson et al., 1993). It was found that baroreflex failure patients had an excessive increase in blood pressure during exercise (Ziegler et al., 1993), which is probably due to impaired baroreflex buffering. This notion seems to be supported by the results of previous studies showing that blood pressure is more variable in patients who have no carotid baroreflex, either as a result of bilateral glomectomy for glomus tumors or because of neck irradiation (Smit et al., 2002; Timmers et al., 2003a, b).

Neuromuscular myopathies

Neuromuscular diseases (myopathies) produce two major patterns of exercise intolerance; in muscular dystrophies, patients have reduced $\dot{V}O_2max$ due to the loss of functional muscle mass; in disorders of muscle energy metabolism, an imbalance in muscle energy production and utilization in exercise results in exertional muscle pain, cramping, weakness, or fatigue (Haller and Lewis, 1984). Patients with mitochondrial myopathies have significantly lower $\dot{V}O_2max$, impaired muscle oxygen extraction indicated by subnormal system arteriovenous oxygen difference at maximal exercise despite normal oxygen carrying capacity, but markedly higher cardiac output when compared with sedentary healthy individuals (Haller et al., 1991; Taivassalo et al., 2001, 2003).

Postural orthostatic tachycardia syndrome

Patients with the postural orthostatic tachycardia syndrome (POTS, also called chronic orthostatic intolerance) usually have reduced exercise tolerance and low $\dot{V}O_2max$. Recent investigations have demonstrated that POTS patients have greater heart rate and smaller stroke volume for each level of absolute workload during submaximal and maximal exercise when compared with healthy but sedentary individuals (Masuki et al., 2007a, b; Shibata et al., 2012; Shook et al., 2007). However, when expressed at the percent of $\dot{V}O_2max$ (relative workload), there is no difference in the heart rate responses between POTS patients and healthy controls (Shook et al., 2007). These results may suggest that POTS patients have no intrinsic abnormality of heart rate regulation during exercise. The tachycardia during exercise in POTS patients is not due to abnormal baroreflex control of heart rate (Masuki et al., 2007a), but rather due to a reduced stroke volume.

CHRONIC EXERCISE TRAINING

Both central command and the exercise pressor reflex are important in determining the cardiovascular response during exercise, while dynamic interactions between these feed-forward and feedback circuits are associated with beneficial long-term adjustments in the sympathetic and parasympathetic nervous systems. The beneficial effects can be observed soon after the initiation of exercise and are sustained as long as activity is continued. Figure 13.6 depicts the major beneficial effects of exercise training in humans.

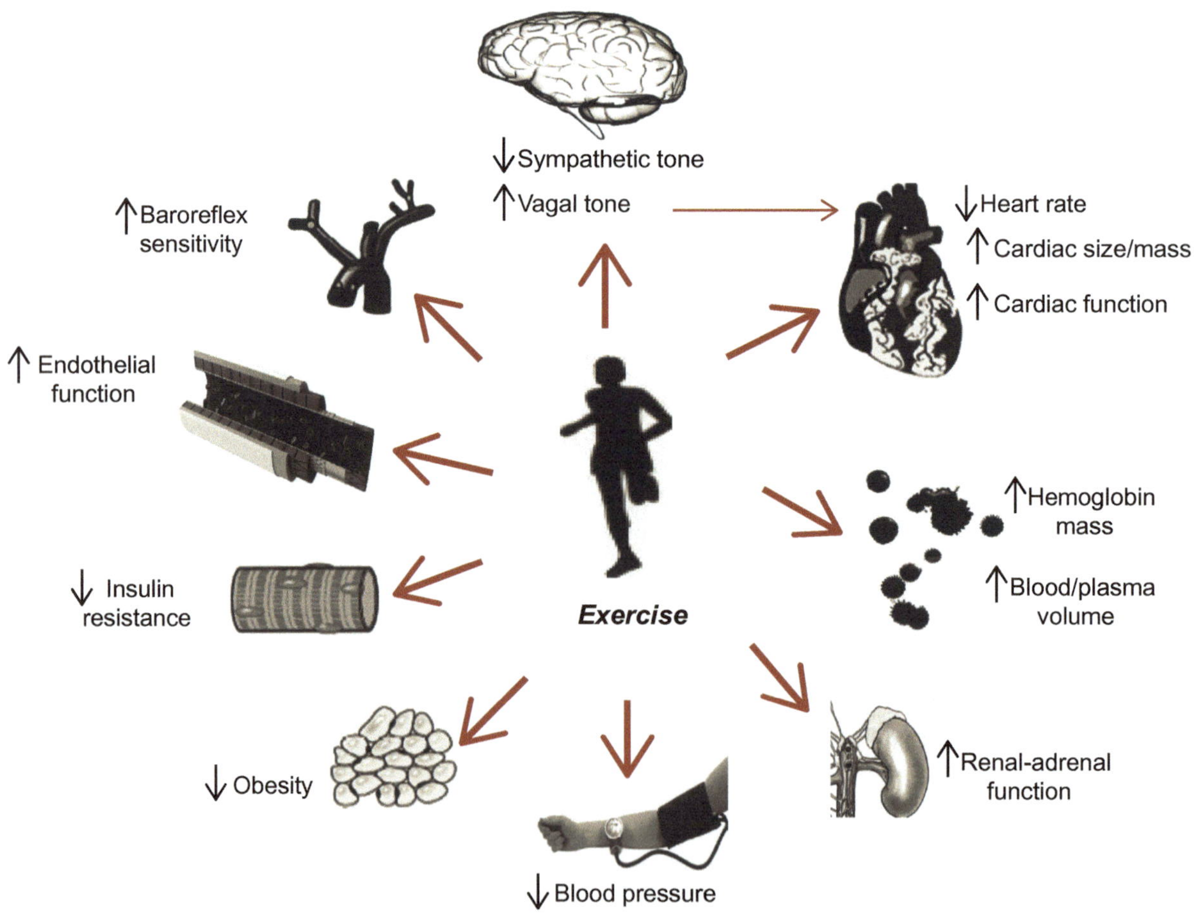

Fig. 13.6. Effects of exercise training on the autonomic nervous system, cardiovascular system, and renal-adrenal system in humans.

Exercise training and vascular function

Exercise training can increase the release of nitric oxide through shear stress during exercise, while chronic increases in nitric oxide may lead to functional and histological alterations of vascular endothelium, causing enhanced vascular structure and function (Niebauer and Cooke, 1996; Green et al., 2004). Many studies have shown that exercise training augments endothelial, nitric oxide-dependent vasodilation in both large and small vessels, while the extent of the improvement in humans depends upon the muscle mass subjected to training; with forearm exercise, changes are restricted to the forearm vessels, while lower body training can induce more generalized benefit (Green et al., 2004). It has been found that regular exercise or exercise training can prevent the age-associated loss in endothelium-dependent vasodilatation and restore levels in previously sedentary middle-aged and older healthy individuals, which may contribute to the reduced risk of cardiovascular disease in this population (DeSouza et al., 2000). However, the impact of exercise training on vascular endothelial function may depend on the balance between reactive oxygen species, antioxidant defenses, and their impact on nitric oxide bioavailability (Green et al., 2004).

Exercise training was reported to improve arterial compliance in healthy individuals (Tanaka et al., 2000) and induce structural enlargement of conduit vessels (Lloyd et al., 2001; Prior et al., 2003). Vascular conductance was also reported to be improved with training, which is not endothelial mediated and probably reflects structural adaptations required to accommodate a high muscle blood flow (Martin et al., 1987; Snell et al., 1987). We recently found that in people over the age of 65 years lifelong daily exercise training may minimize arterial stiffening, but the training effect is limited particularly when started later in life (Shibata et al., 2008). In contrast to its ability to favorably modulate the stiffness of large elastic arteries, exercise training does not consistently modulate the changes in left ventricular structure and diastolic function that occur with physiological aging in humans (Gates et al., 2003; Shibata et al., 2008; Fujimoto et al., 2010). It does, however, increase stroke volume and the Starling mechanism, presumably by improved ventricular-arterial coupling via endothelial mechanisms (Fujimoto et al., 2010).

Adaptation of the autonomic nervous system

Recent research has indicated that exercise training remodels cardiorespiratory centers, and thereby reduces sympathetic and enhances parasympathetic (vagal) outflow (Green et al., 2004; Nelson et al., 2005; Billman and Kukielka, 2007). Since the sympathetic nervous system is activated during each bout of exercise, repeated activation of this system may result in an attenuation of sympathetic activity (Grassi et al., 2000; Fraga et al., 2007; Fu et al., 2010). Animal studies have suggested that nitric oxide decreases overall sympathetic excitability within the brainstem and possibly through actions in higher brain regions (i.e., hypothalamus) (Goodson et al., 1994; Patel et al., 1996). Because of the inevitable experimental restrictions, it is unclear whether the increased release of nitric oxide with exercise training has a central sympathoinhibitory effect in humans. On the other hand, exercise is protective against weight gain and visceral obesity, which may also contribute to the reduction in sympathetic activity (Joyner and Green, 2009). Training-induced muscle adaptations appear to be important in attenuating insulin mediated sympathetic activation (Julius et al., 1991; Baron et al., 1993; Kohno et al., 2000; Henriksen, 2002), and may be especially enhanced by strength training which increases overall muscle mass (Kraus and Levine, 2007; Church et al., 2010).

Exercise training decreases resting heart rate as a result of an increase in vagal tone and a decline in the intrinsic heart rate, while a reduction in sympathetic activity likely has, at most, minimal impact (Rosenwinkel et al., 2001). Trained individuals recover from an acute bout of exercise more rapidly than untrained individuals (Hagberg et al., 1979, 1980; Imai et al., 1994; Rosenwinkel et al., 2001). The more rapid heart rate recovery in trained individuals seems to be independent of a sympathetic response, because an early study showed that the clearance of plasma norepinephrine during recovery did not change significantly after 2 months of detraining (Hagberg et al., 1979). Rather, a vagally mediated mechanism is more likely to be responsible for the rapid heart rate recovery in trained people (Rosenwinkel et al., 2001). Numerous studies have found that exercise training improves cardiac autonomic balance (i.e., increasing parasympathetic while decreasing sympathetic regulation of the heart) and increases heart rate variability (Levy et al., 1998; Iwasaki et al., 2003; Rennie et al., 2003; Okazaki et al., 2005; Galbreath et al., 2011). Additionally, exercise training can result in cardiac remodeling, increase cardiac size and mass, and improve cardiac function (Levine et al., 1991a, b; Levine, 1993; Fu et al., 2010).

Exercise training improves baroreflex function in different populations (Iwasaki et al., 2003; Okazaki et al., 2005; Galbreath et al., 2011), and is protective against age-related reductions in baroreflex sensitivity (Monahan et al., 2000; Joyner and Green, 2009). The improvement in baroreflex function with exercise training could be a result of both greater blood vessel distensibility and better signal transduction in barosensitive areas of the carotid sinus and aortic arch (Monahan et al., 2000; Tanaka et al., 2000), or it could also represent improved or maintained central integration in the brainstem cardiovascular centers (Joyner and Green, 2009).

Numerous studies have shown that exercise training lowers blood pressure. Training-induced reduction in blood pressure is likely to be attributable to an increase in nitric oxide release resulting from increased vascular shear stress during exercise (Kingwell, 2000). Conversely, training has been found to improve aerobic capacity and vascular conductance, and lower body fat, each of which could also contribute to the reduction in blood pressure (Blair et al., 1984; Duncan et al., 1985; Snell et al., 1987; Hall et al., 2001). However, many investigators believe that the blood pressure reduction with exercise training is primarily mediated by a neural mechanism, because vasomotor sympathetic nerve activity decreases after training (Grassi et al., 1992; Brown et al., 2002; Laterza et al., 2007; Fu et al., 2010).

Improvement of renal-adrenal function

Exercise training may improve renal-adrenal function and decrease circulating levels of angiotensin II or angiotensin II-induced vasoconstriction (Rush and Aultman, 2008). Zucker et al. observed in animals with heart failure that exercise training reduces the expression of the angiotensin II type I (AT1) receptors in the paraventricular nucleus of the hypothalamus and in the rostral ventrolateral medulla and nucleus tractus solitarius (Zucker et al., 2004). A previous human study showed that exercise training may suppress the renin–angiotensin–aldosterone system (Hespel et al., 1988). Using a radiotracer technique, Meredith et al. found in initially sedentary healthy men that resting renal but not cardiac norepinephrine spillover decreased after 1 month of exercise training (Meredith et al., 1991), indicating that the reduction in resting sympathetic activity with training is largely confined to the kidney. Research in patients with chronic heart failure showed that the reduction in sympathetic nerve activity and the improvement in baroreflex function after exercise training are due to the concomitant reduction in angiotensin II, as well as angiotensin receptors in the central nervous system (Mousa et al., 2008). A recent study from our laboratory found that exercise training improved renal-adrenal function in patients with POTS (Fu et al., 2011). Additionally, exercise training can increase total hemoglobin mass, red blood cell volume, and plasma/blood volume (Saltin et al., 1968; Harris et al., 2003; El-Sayed et al., 2005; Goodman et al., 2005; Fu et al., 2010).

EXERCISE THERAPY

There is abundant evidence showing that regular exercise or exercise training is not only protective against cardiovascular disease (Mora et al., 2007; Blair and Morris, 2009), hypertension (Fu et al., 2008), type II diabetes (Tuomilehto et al., 2001; Knowler et al., 2002; Lindstrom et al., 2006), breast and colon cancer (Hardman, 2001; Khan et al., 2010), and obesity (Wolf and Woodworth, 2009; Cheriyath et al., 2010), but also effective in improving functional capacity in patients with heart failure, myocardial infarction, or after coronary artery bypass surgery by increasing vagal modulation and decreasing sympathetic tone (Routledge et al., 2010). Exercise training also seems to be effective in the prevention of sudden cardiac death by augmenting baroreflex sensitivity and heart rate variability (La Rovere et al., 2001).

Clinical studies have proven that exercise training improves functional capacity in patients with autonomic disorders, such as Parkinson's disease (Baatile et al., 2000; de Goede et al., 2001; Yousefi et al., 2009), stroke (Harrington et al., 2010), multiple sclerosis (Dalgas et al., 2009; Dettmers et al., 2009; Snook and Motl, 2009), spinal cord injury (Hicks et al., 2003; Valent et al., 2007, 2009; Millar et al., 2009), Guillain–Barré syndrome (Pitetti et al., 1993; Garssen et al., 2004; Bussmann et al., 2007), muscular dystrophy (Olsen et al., 2005; Sveen et al., 2008; Voet et al., 2010), or metabolic myopathies (Taivassalo et al., 2001; Wenz et al., 2009). Exercise training also improves mental health (Raglin, 1990), helps to prevent depression (Craft et al., 2008), and promotes or maintains positive self-esteem (Piko and Keresztes, 2006). Recent human research has demonstrated that training improves older people's cognitive function, and is neuroprotective in many neurodegenerative and neuromuscular diseases.

Although the effects of exercise training on orthostatic tolerance in healthy individuals are controversial, increased orthostatic tolerance after mild to moderate training has been found in patients with unexplained syncope or orthostatic hypotension (Mtinangi and Hainsworth, 1998). Increased baroreflex sensitivity and decreased frequency of syncopal episodes after training were observed in patients with neurally mediated syncope (Gardenghi et al., 2007). Additionally, it was observed that exercise training has a beneficial effect on physiological and subjective parameters in patients with chronic fatigue syndrome (Joosen et al., 2008), and also leads to an improvement of symptoms in the majority of patients with orthostatic intolerance (Winker et al., 2005).

We have recently found that a "personalized" short-term (3 months) exercise training program improves or even cures POTS in most patients; more importantly, patient quality of life, as assessed by the 36-Item Short-Form Health Survey, is significantly improved after training (Fu et al., 2010, 2011). This carefully controlled exercise training program was initially developed for spaceflight and bed rest countermeasures (Iwasaki et al., 2003; Shibata et al., 2010). Numerous studies have shown that microgravity exposure can elicit a "POTS-like" syndrome in astronauts and fit individuals, while "cardiovascular deconditioning" (i.e., cardiac atrophy and hypovolemia) is one primary underlying mechanism. Results from our laboratory and others have suggested that POTS *per se* is a consequence or signature of "cardiovascular deconditioning" (Masuki et al., 2007a, b; Joyner and Masuki, 2008; Fu et al., 2010, 2011; Galbreath et al., 2011), and, therefore, increases in physical fitness (i.e., "reconditioning") with exercise training can improve or cure this syndrome. It has been well demonstrated that exercise training expands plasma and blood volume (Saltin et al., 1968), increases cardiac size and mass (Dorfman et al., 2007), prevents cardiac atrophy and increases orthostatic tolerance in healthy individuals after prolonged period of bed rest (Dorfman et al., 2007).

The training program implemented in our laboratory is shown in Table 13.1 (Iwasaki et al., 2003; Okazaki et al., 2005; Dorfman et al., 2007; Fu et al., 2010; Shibata et al., 2010). Table 13.2 depicts the calculations of heart rate zones for the exercise training program. The heart rate at the "maximal steady state" is set at approximately 75% of the maximal heart rate. The workout zones are structured around the maximal steady state value. A template of workouts prescribed over 3 months of training program is displayed in Table 13.3. The majority of the training sessions, particularly during the early phases, are prescribed as "base pace training" with target heart rate equivalent to ~75–85% of maximal. Initially, patients train 3–4 times per week for 20–40 minutes per session. As the patients become relatively fit, the duration of the base pace training is prolonged and subsequently sessions of increased intensity (i.e., maximal steady state) are added first once per 2 weeks and then once per week, and are always followed by recovery sessions. In addition to the endurance training, resistance training using weight lifting is also undertaken. Weight lifting starts from once a week, 15–20 minutes per session, and gradually increases to twice a week, 30–40 minutes per session.

Given the beneficial effects of exercise training, the American Heart Association, the US Surgeon General, the Centers for Disease Control and Prevention, and the American College of Sports Medicine recommend at least 30 minutes per day of at least moderate-intensity exercise, including brisk walking, jogging, cycling, swimming, or running on most, and preferably all, days of the week (Marcus et al., 2006). Physicians' advice to increase physical activity can be a strong motivator to patients, and advice conveyed as a written prescription may enhance success. Supervised exercise training is preferable to maximize functional capacity. Heart rate can be used as an easily measured estimate of relative exercise intensity, and the target training heart rate is

Table 13.1

Training zones

Training zone	Heart rate
Intervals	> Race pace
Race pace	10–15 bpm > MSS
Maximal steady state (MSS)	(Take 220 − age) ± 5 bpm
Base pace	20 bpm < MSS
Recovery	<Base pace

Table 13.2

Calculations of heart rate zones for the exercise training program

1. Determine approximate maximal heart rate (HR): take 220 − age = maximal HR
2. Determine the heart rate reserve (HRR): take maximal HR − supine resting HR = HRR
3. Take HRR × 0.75 = 75% of HRR
4. Determine the mid maximal steady state (MSS) HR: add 75% of HRR + supine resting HR = mid MSS HR
5. Determine MSS zone: 10 beats around the mid MSS HR
6. Determine base pace zone: 20 beats < lowest MSS zone HR
7. Determine recovery: <lowest base pace HR
8. Determine race pace: 10–15 beats > highest MSS HR
9. Determine interval training zone: >highest race pace but not higher than maximal HR (at least 5 beats lower than maximal HR)

Table 13.3

A template of workouts prescribed over a 3 month training program

	Training		
Month	Recovery	Base pace	Max steady state
1	2@40 minutes	10@30 minutes	2@20 minutes
2	3@40 minutes	10@40 minutes	3@25 minutes
3	3@40 minutes	10@50 minutes	3@35 minutes

usually set at approximately 75% of the maximal heart rate [(take 220 − age) ± 5 beats/min] (Table 13.2). However, it is important to emphasize that these are only guidelines that in some patients, such as those who are taking β-blockers or other medications or with underlying autonomic disorders, may affect the heart rate response to exercise and may not accurately reflect exercise intensity.

For patients with autonomic disorders, heat and body temperature during exercise may compromise blood pressure further (by further increases in vascular conductance); cooling the skin or semirecumbent exercise is recommended. For patients with POTS or orthostatic intolerance, exercise training should be initiated by using a recumbent bike, rowing, or swimming (Fu et al., 2010). The use of only semirecumbent exercise at the beginning is a critical strategy, allowing patients to exercise while avoiding the upright posture that elicits their symptoms. As the patients become relatively fit, the duration and intensity of exercise should be progressively increased, and upright exercise (e.g., upright bike, walking on the treadmill, or jogging) can be gradually added as tolerated (Fu et al., 2010).

In summary, a sedentary lifestyle is considered to be one of the most important modifiable risk factors for morbidity and mortality in humans. Physical activity or exercise training is necessary to maintain overall health and functional capacity, and it plays a crucial role in the prevention of cardiovascular disease, sudden cardiac death, hypertension, type II diabetes, colon cancer, breast cancer, and obesity. Exercise training improves functional capacity in patients with autonomic disorders, and can be therapeutic for patients with orthostatic intolerance, syncope, or POTS. In addition, exercise training improves mental health, helps to prevent depression, and promotes or maintains positive self-esteem. Adaptations involving the autonomic nervous system play a large role in the protective and therapeutic effects of exercise training. Moderate-intensity exercise at least 30 minutes per day and at least 5 days per week is recommended for the vast majority of people. Supervised exercise training is preferable to maximize functional capacity, and may be particularly important for patients with autonomic disorders.

References

Alam M, Smirk FH (1937). Observations in man upon a blood pressure raising reflex arising from the voluntary muscles. J Physiol 89: 372–383.

Baatile J, Langbein WE, Weaver F et al. (2000). Effect of exercise on perceived quality of life of individuals with Parkinson's disease. J Rehabil Res Dev 37: 529–534.

Baron AD, Brechtel-Hook G, Johnson A et al. (1993). Skeletal muscle blood flow. A possible link between insulin resistance and blood pressure. Hypertension 21: 129–135.

Bevegard BS, Shepherd JT (1966). Circulatory effects of stimulating the carotid arterial stretch receptors in man at rest and during exercise. J Clin Invest 45: 132–142.

Billman GE, Kukielka M (2007). Effect of endurance exercise training on heart rate onset and heart rate recovery responses to submaximal exercise in animals susceptible to ventricular fibrillation. J Appl Physiol 102: 231–240.

Blair SN, Morris JN (2009). Healthy hearts – and the universal benefits of being physically active: physical activity and health. Ann Epidemiol 19: 253–256.

Blair SN, Goodyear NN, Gibbons LW et al. (1984). Physical fitness and incidence of hypertension in healthy normotensive men and women. JAMA 252: 487–490.

Brown MD, Dengel DR, Hogikyan RV et al. (2002). Sympathetic activity and the heterogenous blood pressure response to exercise training in hypertensives. J Appl Physiol 92: 1434–1442.

Bussmann JB, Garssen MP, van Doorn PA et al. (2007). Analysing the favourable effects of physical exercise: relationships between physical fitness, fatigue and functioning in Guillain–Barre syndrome and chronic inflammatory demyelinating polyneuropathy. J Rehabil Med 39: 121–125.

Cheriyath P, Duan Y, Qian Z et al. (2010). Obesity, physical activity and the development of metabolic syndrome: the Atherosclerosis Risk in Communities Study. Eur J Cardiovasc Prev Rehabil 17: 309–313.

Chomsky DB, Lang CC, Rayos GH et al. (1996). Hemodynamic exercise testing. A valuable tool in the selection of cardiac transplantation candidates. Circulation 94: 3176–3183.

Church TS, Blair SN, Cocreham S et al. (2010). Effects of aerobic and resistance training on hemoglobin A1c levels in patients with type 2 diabetes: a randomized controlled trial. JAMA 304: 2253–2262.

Claydon VE, Hol AT, Eng JJ et al. (2006). Cardiovascular responses and postexercise hypotension after arm cycling exercise in subjects with spinal cord injury. Arch Phys Med Rehabil 87: 1106–1114.

Cole CR, Blackstone EH, Pashkow FJ et al. (1999). Heart-rate recovery immediately after exercise as a predictor of mortality. N Engl J Med 341: 1351–1357.

Cole CR, Foody JM, Blackstone EH et al. (2000). Heart rate recovery after submaximal exercise testing as a predictor of mortality in a cardiovascularly healthy cohort. Ann Intern Med 132: 552–555.

Craft LL, Perna FA, Freund KM et al. (2008). Psychosocial correlates of exercise in women with self-reported depressive symptoms. J Phys Act Health 5: 469–480.

Crisafulli A, Milia R, Vitelli S et al. (2009). Hemodynamic responses to metaboreflex activation: insights from spinal cord-injured humans. Eur J Appl Physiol 106: 525–533.

Dalgas U, Stenager E, Jakobsen J et al. (2009). Resistance training improves muscle strength and functional capacity in multiple sclerosis. Neurology 73: 1478–1484.

de Goede CJ, Keus SH, Kwakkel G et al. (2001). The effects of physical therapy in Parkinson's disease: a research synthesis. Arch Phys Med Rehabil 82: 509–515.

DeSouza CA, Shapiro LF, Clevenger CM et al. (2000). Regular aerobic exercise prevents and restores age-related declines in endothelium-dependent vasodilation in healthy men. Circulation 102: 1351–1357.

Dettmers C, Sulzmann M, Ruchay-Plossl A et al. (2009). Endurance exercise improves walking distance in MS patients with fatigue. Acta Neurol Scand 120: 251–257.

Dewland TA, Androne AS, Lee FA et al. (2007). Effect of acetylcholinesterase inhibition with pyridostigmine on cardiac parasympathetic function in sedentary adults and trained athletes. Am J Physiol Heart Circ Physiol 293: H86–H92.

DiCarlo SE, Bishop VS (1992). Onset of exercise shifts operating point of arterial baroreflex to higher pressures. Am J Physiol Heart Circ Physiol 262: H303–H307.

Dinenno FA, Joyner MJ (2003). Blunted sympathetic vasoconstriction in contracting skeletal muscle of healthy humans: is nitric oxide obligatory? J Physiol 553: 281–292.

Dorfman TA, Levine BD, Tillery T et al. (2007). Cardiac atrophy in women following bed rest. J Appl Physiol 103: 8–16.

Duncan JJ, Farr JE, Upton SJ et al. (1985). The effects of aerobic exercise on plasma catecholamines and blood pressure in patients with mild essential hypertension. JAMA 254: 2609–2613.

El-Sayed MS, Ali N, El-Sayed Ali Z (2005). Haemorheology in exercise and training. Sports Med 35: 649–670.

Fraga R, Franco FG, Roveda F et al. (2007). Exercise training reduces sympathetic nerve activity in heart failure patients treated with carvedilol. Eur J Heart Fail 9: 630–636.

Fu Q, Levine BD (2005). Cardiovascular response to exercise in women. Med Sci Sports Exerc 37: 1433–1435.

Fu Q, Vongpatanasin W, Levine BD (2008). Neural and nonneural mechanisms for sex differences in elderly hypertension: can exercise training help? Hypertension 52: 787–794.

Fu Q, Vangundy TB, Galbreath MM et al. (2010). Cardiac origins of the postural orthostatic tachycardia syndrome. J Am Coll Cardiol 55: 2858–2868.

Fu Q, VanGundy TB, Shibata S et al. (2011). Exercise training versus propranolol in the treatment of the postural orthostatic tachycardia syndrome. Hypertension 58: 167–175.

Fujimoto N, Prasad A, Hastings JL et al. (2010). Cardiovascular effects of 1 year of progressive and vigorous exercise training in previously sedentary individuals older than 65 years of age. Circulation 122: 1797–1805.

Galbreath MM, Shibata S, VanGundy TB et al. (2011). Effects of exercise training on arterial-cardiac baroreflex function in POTS. Clin Auton Res 21: 73–80.

Gardenghi G, Rondon MU, Braga AM et al. (2007). The effects of exercise training on arterial baroreflex sensitivity in neurally mediated syncope patients. Eur Heart J 28: 2749–2755.

Garssen MP, Bussmann JB, Schmitz PI et al. (2004). Physical training and fatigue, fitness, and quality of life in Guillain–Barre syndrome and CIDP. Neurology 63: 2393–2395.

Gates PE, Tanaka H, Graves J et al. (2003). Left ventricular structure and diastolic function with human ageing. Relation to habitual exercise and arterial stiffness. Eur Heart J 24: 2213–2220.

Goldberger JJ, Le FK, Lahiri M et al. (2006). Assessment of parasympathetic reactivation after exercise. Am J Physiol Heart Circ Physiol 290: H2446–H2452.

Goodman JM, Liu PP, Green HJ (2005). Left ventricular adaptations following short-term endurance training. J Appl Physiol 98: 454–460.

Goodson AR, Leibold JM, Gutterman DD (1994). Inhibition of nitric oxide synthesis augments centrally induced sympathetic coronary vasoconstriction in cats. Am J Physiol Heart Circ Physiol 267: H1272–H1278.

Grassi G, Seravalle G, Bertinieri G et al. (2000). Sympathetic and reflex alterations in systo-diastolic and systolic hypertension of the elderly. J Hypertens 18: 587–593.

Grassi G, Seravalle G, Calhoun D et al. (1992). Physical exercise in essential hypertension. Chest 101: 312S–314S.

Green DJ, Maiorana A, O'Driscoll G et al. (2004). Effect of exercise training on endothelium-derived nitric oxide function in humans. J Physiol 561: 1–25.

Hagberg JM, Hickson RC, McLane JA et al. (1979). Disappearance of norepinephrine from the circulation following strenuous exercise. J Appl Physiol 47: 1311–1314.

Hagberg JM, Hickson RC, Ehsani AA et al. (1980). Faster adjustment to and recovery from submaximal exercise in the trained state. J Appl Physiol 48: 218–224.

Hall JE, Hildebrandt DA, Kuo J (2001). Obesity hypertension: role of leptin and sympathetic nervous system. Am J Hypertens 14: 103S–115S.

Haller RG, Lewis SF (1984). Pathophysiology of exercise performance in muscle disease. Med Sci Sports Exerc 16: 456–459.

Haller RG, Henriksson KG, Jorfeldt L et al. (1991). Deficiency of skeletal muscle succinate dehydrogenase and aconitase. Pathophysiology of exercise in a novel human muscle oxidative defect. J Clin Invest 88: 1197–1206.

Hardman AE (2001). Physical activity and cancer risk. Proc Nutr Soc 60: 107–113.

Harrington R, Taylor G, Hollinghurst S et al. (2010). A community-based exercise and education scheme for stroke survivors: a randomized controlled trial and economic evaluation. Clin Rehabil 24: 3–15.

Harris SK, Petrella RJ, Overend TJ et al. (2003). Short-term training effects on left ventricular diastolic function and oxygen uptake in older and younger men. Clin J Sport Med 13: 245–251.

Henriksen EJ (2002). Invited review: effects of acute exercise and exercise training on insulin resistance. J Appl Physiol 93: 788–796.

Hespel P, Lijnen P, Van Hoof R et al. (1988). Effects of physical endurance training on the plasma renin-angiotensin-aldosterone system in normal man. J Endocrinol 116: 443–449.

Hicks AL, Martin KA, Ditor DS et al. (2003). Long-term exercise training in persons with spinal cord injury: effects on strength, arm ergometry performance and psychological well-being. Spinal Cord 41: 34–43.

Humm AM, Mason LM, Mathias CJ (2008). Effects of water drinking on cardiovascular responses to supine exercise and on orthostatic hypotension after exercise in pure autonomic failure. J Neurol Neurosurg Psychiatry 79: 1160–1164.

Ichinose M, Saito M, Kondo N et al. (2008). Baroreflex and muscle metaboreflex: control of muscle sympathetic nerve activity. Med Sci Sports Exerc 40: 2037–2045.

Imai K, Sato H, Hori M et al. (1994). Vagally mediated heart rate recovery after exercise is accelerated in athletes but blunted in patients with chronic heart failure. J Am Coll Cardiol 24: 1529–1535.

Iwasaki K, Zhang R, Zuckerman JH et al. (2003). Dose-response relationship of the cardiovascular adaptation to endurance training in healthy adults: how much training for what benefit? J Appl Physiol 95: 1575–1583.

Joosen M, Sluiter J, Joling C et al. (2008). Evaluation of the effects of a training programme for patients with prolonged fatigue on physiological parameters and fatigue complaints. Int J Occup Med Environ Health 21: 237–246.

Joyner MJ (2006). Baroreceptor function during exercise: resetting the record. Exp Physiol 91: 27–36.

Joyner MJ, Green DJ (2009). Exercise protects the cardiovascular system: effects beyond traditional risk factors. J Physiol 587: 5551–5558.

Joyner MJ, Masuki S (2008). POTS versus deconditioning: the same or different? Clin Auton Res 18: 300–307.

Joyner MJ, Thomas GD (2003). Having it both ways? Vasoconstriction in contracting muscles. J Physiol 550: 333.

Julius S, Gudbrandsson T, Jamerson K et al. (1991). The hemodynamic link between insulin resistance and hypertension. J Hypertens 9: 983–986.

Keller DM, Ogoh S, Greene S et al. (2004). Inhibition of KATP channel activity augments baroreflex-mediated vasoconstriction in exercising human skeletal muscle. J Physiol 561: 273–282.

Khan N, Afaq F, Mukhtar H (2010). Lifestyle as risk factor for cancer: evidence from human studies. Cancer Lett 293: 133–143.

Kingwell BA (2000). Nitric oxide as a metabolic regulator during exercise: effects of training in health and disease. Clin Exp Pharmacol Physiol 27: 239–250.

Knowler WC, Barrett-Connor E, Fowler SE et al. (2002). Reduction in the incidence of type 2 diabetes with lifestyle intervention or metformin. N Engl J Med 346: 393–403.

Kohno K, Matsuoka H, Takenaka K et al. (2000). Depressor effect by exercise training is associated with amelioration of hyperinsulinemia and sympathetic overactivity. Intern Med 39: 1013–1019.

Kraus WE, Levine BD (2007). Exercise training for diabetes: the "strength" of the evidence. Ann Intern Med 147: 423–424.

Krogh A, Lindhard J (1913). The regulation of respiration and circulation during the initial stages of muscular work. J Physiol 47: 112–136.

La Rovere MT, Pinna GD, Hohnloser SH et al. (2001). Baroreflex sensitivity and heart rate variability in the identification of patients at risk for life-threatening arrhythmias: implications for clinical trials. Circulation 103: 2072–2077.

Laterza MC, de Matos LD, Trombetta IC et al. (2007). Exercise training restores baroreflex sensitivity in never-treated hypertensive patients. Hypertension 49: 1298–1306.

Levine BD (1993). Regulation of central blood volume and cardiac filling in endurance athletes: the Frank–Starling mechanism as a determinant of orthostatic tolerance. Med Sci Sports Exerc 25: 727–732.

Levine BD (2001). Exercise physiology for the clinician. In: Thompson PD (Ed.), Exercise and Sports Cardiology. McGraw-Hill, Medical Publishing Division, New York, pp. 3–29.

Levine BD (2008). VO_2max: what do we know, and what do we still need to know? J Physiol 586: 25–34.

Levine BD, Buckey JC, Fritsch JM et al. (1991a). Physical fitness and cardiovascular regulation: mechanisms of orthostatic intolerance. J Appl Physiol 70: 112–122.

Levine BD, Lane LD, Buckey JC et al. (1991b). Left ventricular pressure-volume and Frank–Starling relations in endurance athletes. Implications for orthostatic tolerance and exercise performance. Circulation 84: 1016–1023.

Levy WC, Cerqueira MD, Harp GD et al. (1998). Effect of endurance exercise training on heart rate variability at rest in healthy young and older men. Am J Cardiol 82: 1236–1241.

Lindstrom J, Ilanne-Parikka P, Peltonen M et al. (2006). Sustained reduction in the incidence of type 2 diabetes by lifestyle intervention: follow-up of the Finnish Diabetes Prevention Study. Lancet 368: 1673–1679.

Lloyd PG, Yang HT, Terjung RL (2001). Arteriogenesis and angiogenesis in rat ischemic hindlimb: role of nitric oxide. Am J Physiol Heart Circ Physiol 281: H2528–H2538.

Marcus BH, Williams DM, Dubbert PM et al. (2006). Physical activity intervention studies: what we know and what we need to know: a scientific statement from the American Heart Association Council on Nutrition, Physical Activity, and Metabolism (Subcommittee on Physical Activity); Council on Cardiovascular Disease in the Young; and the Interdisciplinary Working Group on Quality of Care and Outcomes Research. Circulation 114: 2739–2752.

Martin WH 3rd, Montgomery J, Snell PG et al. (1987). Cardiovascular adaptations to intense swim training in sedentary middle-aged men and women. Circulation 75: 323–330.

Martin WH 3rd, Berman WI, Buckey JC et al. (1989). Effects of active muscle mass size on cardiopulmonary responses to exercise in congestive heart failure. J Am Coll Cardiol 14: 683–694.

Masuki S, Eisenach JH, Schrage WG et al. (2007a). Arterial baroreflex control of heart rate during exercise in postural tachycardia syndrome. J Appl Physiol 103: 1136–1142.

Masuki S, Eisenach JH, Schrage WG et al. (2007b). Reduced stroke volume during exercise in postural tachycardia syndrome. J Appl Physiol 103: 1128–1135.

Meredith IT, Friberg P, Jennings GL et al. (1991). Exercise training lowers resting renal but not cardiac sympathetic activity in humans. Hypertension 18: 575–582.

Michelini LC, Stern JE (2009). Exercise-induced neuronal plasticity in central autonomic networks: role in cardiovascular control. Exp Physiol 94: 947–960.

Millar PJ, Rakobowchuk M, Adams MM et al. (2009). Effects of short-term training on heart rate dynamics in individuals with spinal cord injury. Auton Neurosci 150: 116–121.

Mitchell JH (1990). J.B. Wolffe memorial lecture. Neural control of the circulation during exercise. Med Sci Sports Exerc 22: 141–154.

Monahan KD, Dinenno FA, Tanaka H et al. (2000). Regular aerobic exercise modulates age-associated declines in cardiovagal baroreflex sensitivity in healthy men. J Physiol 529: 263–271.

Mora S, Cook N, Buring JE et al. (2007). Physical activity and reduced risk of cardiovascular events: potential mediating mechanisms. Circulation 116: 2110–2118.

Mousa TM, Liu D, Cornish KG et al. (2008). Exercise training enhances baroreflex sensitivity by an angiotensin II-dependent mechanism in chronic heart failure. J Appl Physiol 104: 616–624.

Mtinangi BL, Hainsworth R (1998). Increased orthostatic tolerance following moderate exercise training in patients with unexplained syncope. Heart 80: 596–600.

Nakamura T, Hirayama M, Yamashita F et al. (2010). Lowered cardiac sympathetic nerve performance in response to exercise in Parkinson's disease. Mov Disord 25: 1183–1189.

Nelson AJ, Juraska JM, Musch TI et al. (2005). Neuroplastic adaptations to exercise: neuronal remodeling in cardiorespiratory and locomotor areas. J Appl Physiol 99: 2312–2322.

Niebauer J, Cooke JP (1996). Cardiovascular effects of exercise: role of endothelial shear stress. J Am Coll Cardiol 28: 1652–1660.

Nishime EO, Cole CR, Blackstone EH et al. (2000). Heart rate recovery and treadmill exercise score as predictors of mortality in patients referred for exercise ECG. JAMA 284: 1392–1398.

Okazaki K, Iwasaki K, Prasad A et al. (2005). Dose-response relationship of endurance training for autonomic circulatory control in healthy seniors. J Appl Physiol 99: 1041–1049.

Olsen DB, Orngreen MC, Vissing J (2005). Aerobic training improves exercise performance in facioscapulohumeral muscular dystrophy. Neurology 64: 1064–1066.

Ozkor MA, Murrow JR, Rahman AM et al. (2011). Endothelium-derived hyperpolarizing factor determines resting and stimulated forearm vasodilator tone in health and in disease. Circulation 123: 2244–2253.

Patel KP, Zhang K, Zucker IH et al. (1996). Decreased gene expression of neuronal nitric oxide synthase in hypothalamus and brainstem of rats in heart failure. Brain Res 734: 109–115.

Pierpont GL, Stolpman DR, Gornick CC (2000). Heart rate recovery post-exercise as an index of parasympathetic activity. J Auton Nerv Syst 80: 169–174.

Piko BF, Keresztes N (2006). Physical activity, psychosocial health, and life goals among youth. J Community Health 31: 136–145.

Pitetti KH, Barrett PJ, Abbas D (1993). Endurance exercise training in Guillain–Barre syndrome. Arch Phys Med Rehabil 74: 761–765.

Prior BM, Lloyd PG, Yang HT et al. (2003). Exercise-induced vascular remodeling. Exerc Sport Sci Rev 31: 26–33.

Raglin JS (1990). Exercise and mental health. Beneficial and detrimental effects. Sports Med 9: 323–329.

Raven PB (2008). Recent advances in baroreflex control of blood pressure during exercise in humans: an overview. Med Sci Sports Exerc 40: 2033–2036.

Raven PB, Fadel PJ, Ogoh S (2006). Arterial baroreflex resetting during exercise: a current perspective. Exp Physiol 91 (1): 37–49.

Ray CA, Rea RF, Clary MP et al. (1992). Muscle sympathetic nerve responses to static leg exercise. J Appl Physiol 73: 1523–1529.

Remensnyder JP, Mitchell JH, Sarnoff SJ (1962). Functional sympatholysis during muscular activity. Observations on influence of carotid sinus on oxygen uptake. Circ Res 11: 370–380.

Rennie KL, Hemingway H, Kumari M et al. (2003). Effects of moderate and vigorous physical activity on heart rate variability in a British study of civil servants. Am J Epidemiol 158: 135–143.

Robertson D, Hollister AS, Biaggioni I et al. (1993). The diagnosis and treatment of baroreflex failure. N Engl J Med 329: 1449–1455.

Rosenwinkel ET, Bloomfield DM, Arwady MA et al. (2001). Exercise and autonomic function in health and cardiovascular disease. Cardiol Clin 19: 369–387.

Routledge FS, Campbell TS, McFetridge-Durdle JA et al. (2010). Improvements in heart rate variability with exercise therapy. Can J Cardiol 26: 303–312.

Rowell LB (1993a). Arterial baroreflexes, central command, and muscle chemoreflexes: a synthesis. In: LB Rowell (Ed.), Human Cardiovascular Control. Oxford University Press, New York, pp. 441–483.

Rowell LB (1993b). Central circulatory adjustments to dynamic exercise. In: LB Rowell (Ed.), Human Cardiovascular Control. Oxford University Press, New York, pp. 162–203.

Rowell LB, O'Leary DS (1990). Reflex control of the circulation during exercise: chemoreflexes and mechanoreflexes. J Appl Physiol 69: 407–418.

Rush JW, Aultman CD (2008). Vascular biology of angiotensin and the impact of physical activity. Appl Physiol Nutr Metab 33: 162–172.

Saltin B, Blomqvist G, Mitchell JH et al. (1968). Response to exercise after bed rest and after training. Circulation 38: VII1–VII78.

Schmid A, Huonker M, Barturen JM et al. (1998). Catecholamines, heart rate, and oxygen uptake during exercise in persons with spinal cord injury. J Appl Physiol 85: 635–641.

Schrage WG, Eisenach JH, Dinenno FA et al. (2004). Effects of midodrine on exercise-induced hypotension and blood pressure recovery in autonomic failure. J Appl Physiol 97: 1978–1984.

Shetler K, Marcus R, Froelicher VF et al. (2001). Heart rate recovery: validation and methodologic issues. J Am Coll Cardiol 38: 1980–1987.

Shibata S, Hastings JL, Prasad A et al. (2008). "Dynamic" Starling mechanism: effects of ageing and physical fitness on ventricular-arterial coupling. J Physiol 586: 1951–1962.

Shibata S, Perhonen M, Levine BD (2010). Supine cycling plus volume loading prevent cardiovascular deconditioning during bed rest. J Appl Physiol 108: 1177–1186.

Shibata S, Fu Q, Bivens TB et al. (2012). Short-term exercise training improves the cardiovascular response to exercise in the postural orchostatic tachycardia syndrome. J Physiol 590: 3495–3505.

Shook RP, Fu Q, Vangundy TB et al. (2007). Patients with postural orthostatic tachycardia syndrome (POTS) have a lower level of fitness compared to healthy but sedentary females. Med Sci Sports Exerc 39: S169.

Smit AA, Timmers HJ, Wieling W et al. (2002). Long-term effects of carotid sinus denervation on arterial blood pressure in humans. Circulation 105: 1329–1335.

Smith GD, Mathias CJ (1996). Differences in cardiovascular responses to supine exercise and to standing after exercise in two clinical subgroups of Shy–Drager syndrome (multiple system atrophy). J Neurol Neurosurg Psychiatry 61: 297–303.

Smith GD, Watson LP, Pavitt DV et al. (1995). Abnormal cardiovascular and catecholamine responses to supine exercise in human subjects with sympathetic dysfunction. J Physiol 484: 255–265.

Smith GD, Watson LP, Mathias CJ (1996a). Cardiovascular and catecholamine changes induced by supine exercise and upright posture in vasovagal syncope. Comparisons with normal subjects and subjects with sympathetic denervation. Eur Heart J 17: 1882–1890.

Smith GD, Watson LP, Mathias CJ (1996b). Neurohumoral, peptidergic and biochemical responses to supine exercise in two groups with primary autonomic failure: Shy–Drager syndrome/multiple system atrophy and pure autonomic failure. Clin Auton Res 6: 255–262.

Snell PG, Martin WH, Buckey JC et al. (1987). Maximal vascular leg conductance in trained and untrained men. J Appl Physiol 62: 606–610.

Snook EM, Motl RW (2009). Effect of exercise training on walking mobility in multiple sclerosis: a meta-analysis. Neurorehabil Neural Repair 23: 108–116.

Sveen ML, Jeppesen TD, Hauerslev S et al. (2008). Endurance training improves fitness and strength in patients with Becker muscular dystrophy. Brain 131: 2824–2831.

Taivassalo T, Shoubridge EA, Chen J et al. (2001). Aerobic conditioning in patients with mitochondrial myopathies: physiological, biochemical, and genetic effects. Ann Neurol 50: 133–141.

Taivassalo T, Jensen TD, Kennaway N et al. (2003). The spectrum of exercise tolerance in mitochondrial myopathies: a study of 40 patients. Brain 126: 413–423.

Tanaka H, Dinenno FA, Monahan KD et al. (2000). Aging, habitual exercise, and dynamic arterial compliance. Circulation 102: 1270–1275.

Timmers HJ, Karemaker JM, Wieling W et al. (2003a). Baroreflex and chemoreflex function after bilateral carotid body tumor resection. J Hypertens 21: 591–599.

Timmers HJ, Karemaker JM, Wieling W et al. (2003b). Baroreflex control of muscle sympathetic nerve activity after carotid body tumor resection. Hypertension 42: 143–149.

Tuomilehto J, Lindstrom J, Eriksson JG et al. (2001). Prevention of type 2 diabetes mellitus by changes in lifestyle among subjects with impaired glucose tolerance. N Engl J Med 344: 1343–1350.

Valent L, Dallmeijer A, Houdijk H et al. (2007). The effects of upper body exercise on the physical capacity of people with a spinal cord injury: a systematic review. Clin Rehabil 21: 315–330.

Valent LJ, Dallmeijer AJ, Houdijk H et al. (2009). Effects of hand cycle training on physical capacity in individuals with tetraplegia: a clinical trial. Phys Ther 89: 1051–1060.

Voet NB, van der Kooi EL, Riphagen II et al. (2010). Strength training and aerobic exercise training for muscle disease. Cochrane Database Syst Rev CD003907.

Watanabe J, Thamilarasan M, Blackstone EH et al. (2001). Heart rate recovery immediately after treadmill exercise and left ventricular systolic dysfunction as predictors of mortality: the case of stress echocardiography. Circulation 104: 1911–1916.

Wenz T, Diaz F, Hernandez D et al. (2009). Endurance exercise is protective for mice with mitochondrial myopathy. J Appl Physiol 106: 1712–1719.

Werner WG, DiFrancisco-Donoghue J, Lamberg EM (2006). Cardiovascular response to treadmill testing in Parkinson disease. J Neurol Phys Ther 30: 68–73.

Winker R, Barth A, Bidmon D et al. (2005). Endurance exercise training in orthostatic intolerance: a randomized, controlled trial. Hypertension 45: 391–398.

Wolf AM, Woodworth KA (2009). Obesity prevention: recommended strategies and challenges. Am J Med 122: S19–S23.

Yousefi B, Tadibi V, Khoei AF et al. (2009). Exercise therapy, quality of life, and activities of daily living in patients with Parkinson disease: a small scale quasi-randomised trial. Trials 10: 67.

Ziegler MG, Ruiz-Ramon P, Shapiro MH (1993). Abnormal stress responses in patients with diseases affecting the sympathetic nervous system. Psychosom Med 55: 339–346.

Zucker IH, Patel KP, Schultz HD et al. (2004). Exercise training and sympathetic regulation in experimental heart failure. Exerc Sport Sci Rev 32: 107–111.

Handbook of Clinical Neurology, Vol. 117 (3rd series)
Autonomic Nervous System
R.M. Buijs and D.F. Swaab, Editors

Chapter 14

Autonomic control of bone formation: its clinical relevance

JI-YE HE, XIN-FENG ZHENG, AND LEI-SHENG JIANG*
Department of Orthopaedic Surgery, Xinhua Hospital, Shanghai Jiaotong University School of Medicine, Shanghai, China

INTRODUCTION

Bone is a dynamic living tissue that undergoes constant reshaping through the interaction of the anabolic (osteoblasts) and catabolic (osteoclasts) pathways in human bone cells. Under normal physiological conditions, the two processes are precisely balanced to maintain healthy bones. Once this balance is disturbed, bone loss or excessive bone formation eventually occurs. Traditionally, bone remodeling is thought to be regulated by nutritional status (Nakamura, 2006), humoral factors (Clowes et al., 2005), and biomechanical stress (Burr et al., 2002; Klein-Nulend et al., 2005). However, the involvement of the autonomic nervous system (ANS), mainly the sympathetic nervous system (SNS), in the modulation of bone remodeling is receiving increasing attention. In this chapter, we review current knowledge about the part played by the autonomic nervous system in abnormal bone formation and its clinical associations. A synopsis of the role of the SNS in some abnormal bone formation diseases is illustrated by a flow diagram (see Fig. 14.1).

EVIDENCE OF AUTONOMIC NERVOUS SYSTEM INVOLVEMENT IN BONE REMODELING

Both experimental and clinical evidence have demonstrated the neural innervation of bone (Serre et al., 1999; Mach et al., 2002). Sympathetic and sensory nerve fibers are identified in the periosteum and bone by immunohistochemical studies. These nerve fibers form neural networks that accompany blood vessels adjacent to the bone trabeculae, with the greatest density found around the growth plates and in the metaphysis of long bones. These neural networks are in close contact with bone cells, strongly indicating a physiological role in the innervation of bone tissue (Hohmann et al., 1986; Hill and Elde, 1991; Asmus et al., 2000; Togari, 2002; Burt-Pichat et al., 2005; Martin et al., 2007;). β-adrenergic and neuropeptide receptors have been detected on osteoblastic and osteoclastic cells (Togari, 2002). A direct neurite–osteoclastic cell communication via the adrenergic receptors was also reported by *in vitro* co-culture of murine superior cervical ganglia and osteoclast-like cells (Suga et al., 2010). Chemical sympathectomy with guanethidine or excision of the superior cervical ganglion resulted in bone loss and a deterioration in bone structure in rats (Haug et al., 2003; Pagani et al., 2008; Kim et al., 2009), which may be attributable to an increase in the surface and number of osteoclasts and abnormal bone deposition (Cherruau et al., 2003; Ladizesky et al., 2003; Pagani et al., 2008). These results suggest that there is neuroendocrine regulation in bone remodeling.

Autonomic dysfunction induced by genetic and pharmacological intervention offered deeper insight into the SNS mediation in bone remodeling. Dopamine β-hydroxylase (DβH)-deficient mice, deprived of norepinephrine production, exhibited a high bone mass phenotype in vertebrae and long bones despite hypogonadism and hypercortisolism (Li et al., 2001; Takeda et al., 2002). β_2-adrenergic receptor-deficient mice had high bone mass phenotype (Elefteriou et al., 2005), whereas β_1, $_2$-adrenergic receptor-deficient mice displayed low bone mass; the discrepancy between the two models indicated higher bone formation in Adr2R$^{-/-}$ requiring β_1-adrenergic signaling (Pierroz et al., 2005). A recent study with β_1, $_2$, $_3$-adrenergic receptor-deficient mice found that a high bone mass occurred in male mice (Bouxsein et al., 2009). Knockout of adenylyl cyclase 5, a downstream mediator of β_2-AR signaling, preserved bone quality in elderly mice (Yan et al., 2007).

*Correspondence to: Lei-Sheng Jiang, Department of Orthopaedic Surgery, Xinhua Hospital, Shanghai Jiaotong University School of Medicine, Shanghai 200092, China. E-mail: jiangleisheng@126.com

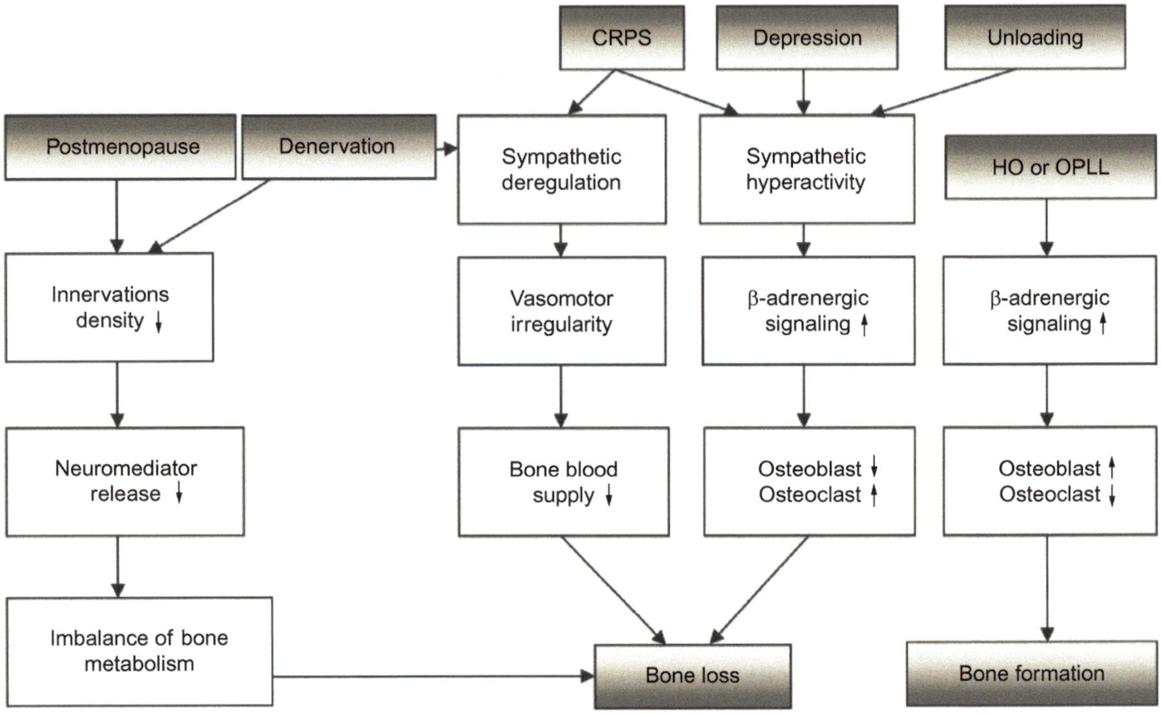

Fig. 14.1. Synopsis of the roles played by the sympathetic nervous system in a number of disorders involving abnormal bone remodeling. CRPS, complex regional pain syndrome; HO, heterotopic ossification; OPLL, ossification of the posterior longitudinal ligament.

Similar results were found with pharmacological interventions. Mice and rats treated with the nonselective β-adrenergic blocker propranolol exhibited an increased bone mass (Minkowitz et al., 1991; Takeda et al., 2002; Levasseur et al., 2003; Bonnet et al., 2006; Pierroz et al., 2006; Yirmiya et al., 2006; Zhang et al., 2007), while mice treated with the nonselective β agonist isoproterenol or the β_2-AR selective agonists clenbuterol and salbutamol showed a low bone mass (Takeda et al., 2002; Bonnet et al., 2005; Kondo et al., 2005; Pierroz et al., 2005). Intermittent parathyroid hormone (PTH) treatment has no osteoanabolic activity in mice lacking the β_2-adrenergic receptors due to the suppressed iPTH-induced bone formation and resorption (Hanyu et al., 2012). These results suggest that the integrity of the SNS and its transmitters are important in the maintenance and repair processes of bone tissue.

β-blockers, also known as β-adrenergic blocking agents, are widely prescribed for elderly patients with cardiovascular disease. Fortunately, we can evaluate the role of β-adrenergic signaling on bone remodeling from a clinical perspective. The findings from case-control analyses or cohort studies of patients suggested that use of β-blockers led to a higher bone mineral density (BMD) or a reduced risk of fracture (Pasco et al., 2004; Schlienger et al., 2004; Schoofs et al., 2005; Gage et al., 2006; Rejnmark et al., 2006; Turket er al., 2006; de Vries et al., 2007; Meisinger et al., 2007; Sosa et al., 2011). But definite conclusions need more prospective, randomized controlled trials.

THE SYMPATHETIC NERVOUS SYSTEM PLAYS A PIVOTAL ROLE IN LEPTIN-DEPENDENT REGULATION OF BONE

Leptin is an adipocyte-derived hormone that regulates reproduction and energy metabolism through its cytokine receptor ObRb located in the ventromedial hypothalamic neurons. Recently, its potent function in regulating bone remodeling has attracted some interest. Leptin is a powerful inhibitor of bone formation *in vivo*. Leptin receptor knockout mice displayed a high bone mass. Intracerebroventricular (ICV) infusion of leptin with no leakage into the general circulation caused bone loss in both wild-type and leptin-deficient mice by inhibiting the osteoblastic function (Ahima and Flier, 2000).

The link between a leptin-dependent central control and bone remodeling deserves exploration. Leptin receptor knockout mice have a low sympathetic tone, and ablation of adrenergic signaling yields a leptin-resistant high bone mass, suggesting that the central regulation of leptin on bone tissue is dependent on an intact SNS (Takeda et al., 2002). In agreement with this, leptin did not increase bone resorption in dopamine β-hydroxylase-deficient mice, adrb2-deficient mice or β-blocker pretreated mice, even with a high dose

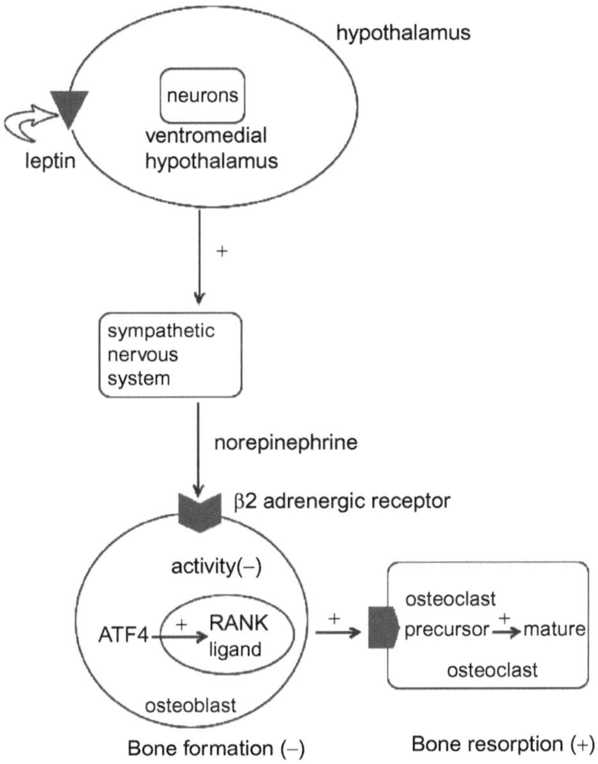

Fig. 14.2. Action of sympathetic nervous system on leptin-dependent regulation of bone formation.

(Elefteriou et al., 2005). Dense and intimate nerve endings have been detected in close contact with bone cells (Serre et al., 1999). Of note, the β_2-adrenergic receptor is so far the only adrenergic receptor known to be expressed in osteoblasts (Moore et al., 1993; Togari, 2002). These findings indicate that the antiosteogenic function of leptin is mediated by the SNS through the osteoblastic β_2-adrenergic receptor. Taken all together, the completed puzzle could potentially appear as follows (Fig. 14.2): leptin acts on a population of neurons located in the ventromedial hypothalamus which subsequently stimulates the noradrenergic sympathetic nerve fibers to release norepinephrine. The norepinephrine then specifically binds to β_2-adrenergic receptors expressed on the osteoblasts to inhibit osteoblast activity, and thus inhibits bone formation. At the same time, the adrenergic nerves may activate the receptor activator of the NF-κB ligand (RANKL) in osteoblasts and trigger RANKL-mediated osteoclastogenesis and bone resorption (Elefteriou et al., 2005).

ASSOCIATIONS WITH CLINICAL DISORDERS

Increased bone formation

Heterotopic ossification

Heterotopic ossification (HO), defined as excessive formation of bone in abnormal anatomical locations, commonly occurs after various types of trauma, such as fracture, dislocation, or burns, especially in patients with injuries to the central nervous system, including the spinal cord and brain.

Studies have confirmed that the neuroinflammatory factor substance P (SP) is involved in HO (Kan et al., 2011; Salisbury et al., 2011). Neurohumoral factors also play an important role in the pathogenesis of HO. Both increased active transforming growth factors (TGF-β1, TGF-β2, TGF-β3) and bone morphogenetic protein-2 (BMP-2) are found at the site of HO (Toom et al., 2007). The overexpression of BMPs is involved in the pathogenesis of HO in rectal cancers (Kypson et al., 2003) and fibrodysplasia ossificans progressiva (Kaplan et al., 2010). TGF-β1 directs the migration of Sca-1+CD29+CD45-CD11b- bone marrow-derived stroma cells (BMSCs), and is identified as a key factor in coupling bone resorption and formation (Tang et al., 2009). Antagonistic interaction of TGF-β and BMP signaling was demonstrated on chondrocyte proliferation and differentiation *in vivo*, in which BMP signaling was enhanced by TGF-β while TGF signaling was suppressed by BMP-2. Regardless of the specific mechanism, the superfamily of TGFβ and BMP signaling play a vital role in the pathogenesis of HO (Keller et al., 2011). On the other hand, BMP is a key signal in the cell-autonomous requirement of SNS development (Morikawa et al., 2009). Therefore, BMP expression is bound to be associated with all kinds of cells involved in bone formation under the control of sympathetic signaling.

Leptin-dependent regulation of bone formation has also been proved to be one of the possible causes of HO. An increased mineralization of the extracellular matrix in cells isolated from heterotopic bone biopsies was induced by exposure to exogenous leptin (Handschin et al., 2007). Strong expressions of both leptin mRNA and protein had been proved in HO-isolated tissue from rats with Achilles' tenotomy (Xu et al., 2009). Moreover, a study on leptin receptors and β_2-adrenergic receptor mRNA expression in brain injury-related HO revealed the mRNA expression of the leptin receptor isoforms in osteocytes isolated from all biopsies, and a significant decrease in the level of β_2-adrenergic receptor mRNA in cells isolated from heterotopic bone biopsies. These results indicated an association between hypothalamic leptin signaling and brain injury-related HO via the β_2-adrenergic system (Chauveau et al., 2008).

Ossification of the posterior longitudinal ligament

Ossification of the posterior longitudinal ligament (OPLL) of the spine is considered as a special type of HO. Contrary to decreased serum leptin levels in typical

HO patients (Chauveau et al., 2008), increased serum leptin levels were found in OPLL females. Serum leptin levels were significantly higher in patients in whom OPLL extended to the thoracic and/or lumbar spine than in patients in whom OPLL was limited to the cervical spine, and correlated positively with the number of involved vertebrae. These results suggest that hyperleptinemia may contribute to the development of HO of the spinal ligament in female patients with OPLL (Ikeda et al., 2011).

Loss of bone mass

Postmenopausal osteoporosis

Postmenopausal osteoporosis is characterized by progressive loss of bone mass and increased risk of fracture. The drastic changes of estrogen and cytokine levels within the body were considered the main contributors to the bone loss in postmenopausal osteoporosis (Kimble et al., 1996; Rossouw et al., 2002). However, rodent experiments with ovariectomy (OVX) highlighted the intervention of the SNS in the development of postmenopausal osteoporosis. β_2-adrenergic receptor-deficient mice displayed a high bone mass phenotype, whereby bone resorption was ameliorated even after OVX (Elefteriou et al., 2005). Dramatic decrease of innervation density in bone was found in OVX rats, proving the interconnection between the nervous system and bone loss after OVX (Burt-Pichat et al., 2005). The release of neuromediators was influenced by nerve fiber density, which is essential for bone cell function and for bone remodeling (Bliziotes et al., 2001; Itzstein et al., 2001).

Adrenergic antagonists have been used in OVX murine models to investigate the role of the SNS on postmenopausal osteoporosis. β-adrenergic blockade partially rescued bone loss following OVX in adult mice by preventing a general BMD decrease (Pierroz et al., 2006). Propranolol failed to improve the microarchitectural parameters in the lumbar vertebrae, but increased the femur cross-sectional and medullary areas. Propranolol treatment improved the decreased BMD and trabecular area, increased trabecular number (TbN), and lowered trabecular separation to some extent in OVX rats (Zhang et al., 2007). A dose-dependent effect of propranolol on bone tissue was also demonstrated (Bonnet et al., 2006).

Clinical observations on postmenopausal women yielded similar findings. A study of 1344 postmenopausal women who were β-blocker users revealed that β-blockers were positively correlated with a higher BMD at the total hip and ultradistal forearm (Pasco et al., 2004). Another study of 158 postmenopausal women who were taking β-blockers demonstrated that β-blocker use was associated with better trabecular microarchitecture at the femoral neck and lumbar spine (Bonnet et al., 2007).

However, there were controversial results from both animal experiments and clinical observations. Mice lacking β-adrenergic receptors had an increased bone mass, though they were not protected from the deleterious effects of estrogen deficiency on the trabecular bone microarchitecture (Bouxsein et al., 2009). A β-adrenergic blockade could only partially salvage the bone loss induced by OVX in adult mice (15 weeks old) (Pierroz et al., 2006), which is inconsistent with the reported full preventative effect against hypogonadism in young OVX mice (4 weeks old) (Takeda et al., 2002). The discrepancy may indicate that the effects of a β-adrenergic blockade on bone remodeling were modulated by other humoral or endocrine factors influenced by different developmental stages of the mice. At the same time, some prospective or observational studies failed to identify an association between β-blocker treatment and fracture risk in perimenopausal or older women (Rejnmark et al., 2004; Levasseur et al., 2005; Reid et al., 2005; Sosa et al., 2011).

Adolescent idiopathic scoliosis

Adolescent idiopathic scoliosis (AIS) is defined as a lateral curvature of the spine of more than 10° with vertebral rotation; it can occur in children aged from 10 years to maturity. Theories on its etiology include genetic predisposition (Ogilvie, 2011), biomechanical and neuromuscular factors (Fidler and Jowett, 1976; Veldhuizen et al., 2000), hormonal disturbances (Leboeuf et al., 2009), and neurological mechanisms (Repko et al., 2008). The autonomic nervous system, through its hypothalamic neuroendocrine control of puberty, skeletal growth, and menarche, contributes importantly to the pathogenesis of AIS (Fig. 14.3). Melatonin dysfunction detected in AIS subjects also involves the autonomic nervous system (Patel and Elefteriou, 2007).

Following the first report of the association of osteopenia with AIS in 1982 (Burner et al., 1982), an increasing number of investigations demonstrated a higher prevalence of osteoporosis in AIS patients (Cook et al., 1987; Cheng et al., 2006; Li et al., 2008). However, no evidence was found to support osteoporosis as an etiological factor for AIS (Lowe et al., 2000; Li et al., 2008). Recently, abnormal leptin bioavailability was found in girls with AIS (Qiu et al., 2007; Liu et al., 2012), suggesting that the leptin–hypothalamus–SNS system may possibly play a role in etiological cross-talk between osteoporosis and scoliosis.

A collective model of pathogenesis, a neuro-osseous timing of maturation (NOTOM) system, operating in a child's internal world during growth and maturation,

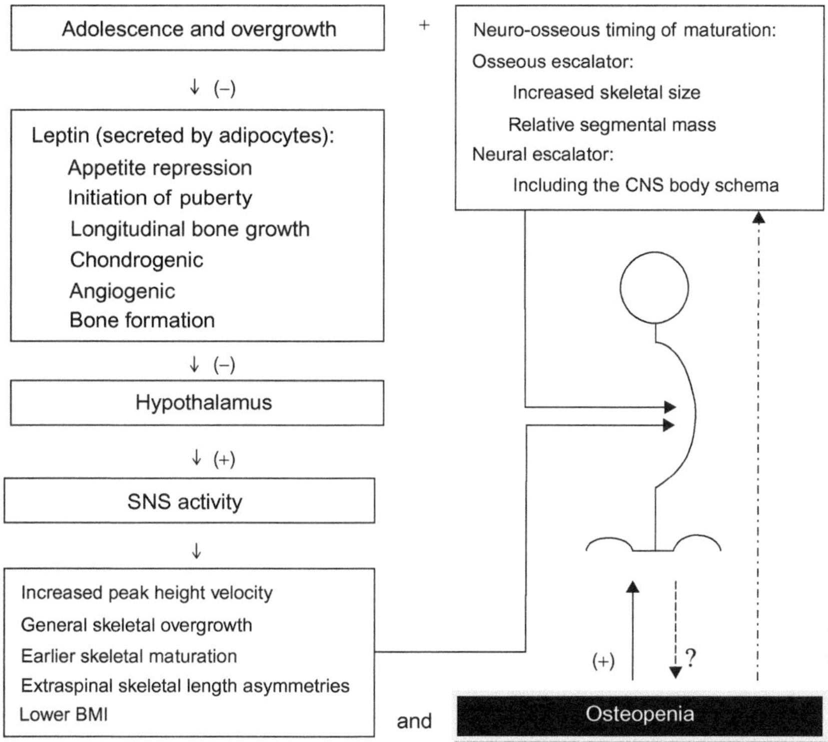

Fig. 14.3. Involvement of the autonomic nervous system in the pathogenesis of adolescent idiopathic scoliosis.

has been proposed for AIS (Burwell et al., 2008). This model provides a dynamic physiological balance of postural equilibrium through two escalators: (1) an osseous escalator – increasing skeletal size and relative segmental mass; (2) a neural escalator – including the CNS body schema. When the balance between the two escalators is disturbed in timing and/or space, asymmetric growth in the spine may lead to spinal three-dimensional deformity. The leptin–hypothalamus–SNS system, serving as a complement to the NOTOM escalators, regulates growth and reproduction through appetite repression, initiation of puberty, longitudinal bone growth, chondrogenic, angiogenic and bone formation. In AIS susceptible girls, given adequate nutrition and energy stores, circulating leptin talks to the hypothalamus, where dysfunction leads to an altered sensitivity to leptin resulting in increased SNS activity contributing by neuroendocrine mechanisms to: (1) earlier age at, and increased peak height velocity; (2) general skeletal overgrowth; (3) earlier skeletal maturation; (4) extraspinal skeletal length asymmetries, including periapical ribs and ilia; (5) generalized osteopenia; (6) lower body mass index.

Complex regional pain syndrome

Complex regional pain syndrome (CRPS), also known as reflex sympathetic dystrophy syndrome (RSDS) or Sudeck atrophy, was first reported when bone dystrophy secondary to trauma was observed clinically. Symptoms of diffuse limb pain and swelling are accompanied by autonomic dysfunction, increased bone resorption, and patchy osteoporosis (Adami et al., 1997). Among the various etiological theories, including inflammation (Kingery, 2010), hypoxia, autoimmune pathogenesis (Kohr et al., 2011), and the dysfunction of sympathetic or autonomic pathways (Cohen et al., 2012), the last still plays a important, if not a crucial role.

Regional dysfunction of sympathetic or autonomic nervous system is considered as a pathomechanism for patchy osteopenia in CRPS for the following reasons: (1) Increased SNS activity is demonstrated in CRPS patients by the observation that central sympathetic stimulation provokes pain, abnormal vasoconstriction, and sympathetic skin reflexes (Drummond and Finch, 2004). (2) An abnormal pain network of central somatomotor and autonomic integration pathways is supported by the observation of asymmetric vasomotor responses and enhanced pain within seconds of viewing the ambiguous visual stimuli (Cohen et al., 2012). (3) Autoimmune mechanisms for sympathetic and myenteric autonomic nervous system neurons and/or autonomic-differentiated cells is another possible explanation. In a subset of CRPS patients, the presence of autoantibodies to β_2-adrenergic receptor and/or the muscarinic-2 receptor supports an autoimmune pathogenesis of CRPS (Kohr et al., 2011). (4) The hypothalamic–pituitary–adrenal (HPA) axis, as a negative feedback for inflammatory processes, may play an important role in CRPS. Investigations have demonstrated inflammatory factors, such as the neuropeptide substance P and interleukin-1 family to be involved in CRPS (Kingery et al., 2003a; Li et al., 2009; Kingery, 2010).

Charcot neuro-osteoarthropathy

Charcot neuro-osteoarthropathy is a disorder characterized by destruction of bones and joints, and denervation of the limb (Jeffcoate, 2008). Its main clinical features are painless swelling, deformity, and instability in the affected joint. It may result from myelomeningocele, traumatic denervation, multiple local microinjuries, poliomyelitis, syringomyelia, HIV infection, leprosy, alcohol abuse, or diabetes mellitus (Koeck et al., 2009). Osteoporosis, a frequent phenomenon in most Charcot patients, has also been considered as a cause in the development of Charcot neuro-osteoarthropathy (Jirkovska et al., 2001).

The association between Charcot neuro-osteoarthropathy and peripheral autonomic neuropathy has yet to be determined (Shapiro et al., 1998; Koeck et al., 2009; Christensen et al., 2011). Compared with osteoarthritis, a significantly lower density of sympathetic nerve fibers was found in diabetic patients with Charcot foot (Koeck et al., 2009). Severe sensory nerve impairment with intact peripheral vasomotor and blood flow was demonstrated in diabetic patients with Charcot osteoarthropathy (Shapiro et al., 1998). Definitely opposite results showing weakened venoarteriolar sympathetic axon reflex with an increased blood flow were also reported in the patients with acute Charcot foot and first toe amputation (Christensen et al., 2011). Both neuropathy and diabetes are closely associated with lower bone mass; minor trauma induced by osteopenia combined with abnormal autonomic signals from neuropathy may together result in Charcot neuro-osteoarthropathy (Jones et al., 2004).

Unloading-induced osteoporosis

The regulation of bone remodeling is positively influenced by mechanical loading. Although the mechanisms underlying immobility-induced pathological osteopenia have not yet been fully understood, the role that the SNS plays in regulating bone mass has been investigated in bone mechanoadaptive responses. Recent investigations have demonstrated that a β-adrenergic blocker mitigated unloading-induced bone loss (Kondo et al., 2005; Baek and Bloomfield, 2009), and guanethidine, administered to selectively deplete noradrenaline/norepinephrine in the postganglionic adrenergic nerves, suppressed disuse osteopenia (Kondo et al., 2005). While isoproterenol, as a β-adrenergic agonist, decreased bone mass in loading mice, it did not do so in unloading ones (Kondo et al., 2005). Meanwhile, a direct effect of β-adrenergic blockade mitigating the unloading-induced bone loss may be enhanced by an indirect effect from reduced circulating leptin (Baek and Bloomfield, 2009). These findings indicate that a leptin–SNS concept might have a crucial role in the unloading-induced bone loss signaling pathway, and the signaling could be suppressed by sympathetic blockers. In addition, a downregulation of *N*-methyl-D-aspartate (NMDA) receptor with decreased osteogenesis in a disuse osteopenia rat model was observed (Ho et al., 2005), which suggested cross-talk between central mediation and mechanical loading response in bone health.

Central or peripheral nervous system damage

Denervation osteoporosis is one of the inevitable complications of spinal cord injury (SCI) and peripheral nerve injury. Rapid bone resorption and severe panskeletal bone loss in the pelvis and the lower extremities occurred in patients with SCI (Morse et al., 2008). More remarkable deterioration of bone trabecular microarchitecture and decrease of mechanical properties were seen in neurectomized rats than in OVX rats (Brouwers et al., 2009). Obviously, unloading has an important role in the development of denervation osteoporosis, so denervation osteoporosis was thought to be a special kind of disuse or unloading osteoporosis with a similar pathogenesis (Uebelhart et al., 1995). However, lower BMD in the proximal tibiae and decreased cortical thickness and cortical area of the tibial midshafts were shown in SCI rats than in hindlimb cast immobilization rats (Liu et al., 2008b), and weight-bearing exercises were ineffective in preventing osteoporosis or restoring bone mineral in SCI patients (Pacy et al., 1988; Leeds et al., 1990; BeDell et al., 1996). Decreased BMD in the upper extremities in paraplegics also implicated different mechanisms involved in denervation osteoporosis (Frey-Rindova et al., 2000).

Partial or complete disrupted neural signals from the central nervous system to the peripheral SNS after SCI induced an abnormal sympathetic innervation pathway (Munakata et al., 1997; Schmid et al., 1998; Krassioukov et al., 1999). An increased innervation density of substance P-immunoreactivity nerve fiber combined with a decreased neurofilament 200-immunoreactivity was found in an early stage SCI-induced osteopenia in young rats (Liu et al., 2008a). Blockade of the SNS using propranolol prevented neurectomy-induced bone loss (Huang et al., 2011). Increased plasma concentration of leptin with augmented body fat accumulation was seen in SCI patients, suggesting that these individuals may have become insensitive to leptin (Rayner and Trayhurn, 2001).

As mentioned above, the SNS–leptin system may play an important role in the pathomechanism of denervation osteoporosis. Meanwhile, denervation after SCI could also impair vasoregulation, affecting the otherwise intact blood supply to the sublesional bones.

Depression

Depression is a common psychological disorder that affects more than 5% of the population (Blazer et al., 1994), and severe depression could cloud a person's thinking and also induce osteoporosis. Depression-induced osteoporosis is site-specific, with a remarkable decrease in the bone mass of the hip and the spine (Jacka et al., 2005). The correlation between depression and osteoporosis has been widely investigated. Studies have shown that BMD decreases by 6–15% in major depressive disorder (MDD) compared to healthy controls (Jacka et al., 2005; Mussolino, 2005; Kahl et al., 2006; Altindag et al., 2007; Petronijevic et al., 2008). But controversy exists. Some studies reported no such correlation or no significant effect on BMD and bone turnover markers (Kavuncu et al., 2002; Whooley et al., 2004; Yazici et al., 2005).

Body weight is known to be a protective factor against bone loss. Women suffering from depression tend to weigh more than those free from depression; however, the greater body mass of these women did not help them to resist bone loss (Henry et al., 2000; Onyike et al., 2003). Depression was associated with a pronounced and enduring central and peripheral hypernoradrenergic state (Wong et al., 2000); the remarkable increases in norepinephrine (NE) levels in those with depression may contribute to an accrual of lower peak bone mass. Hence, depression-induced bone loss possibly occurs through activation of the SNS (Bab and Yirmiya, 2010). This hypothesis was supported by an experimental study using a chronic mild stress (CMS) murine model, in which there was a dramatic drop to almost half of bone mass compared with controls in the fourth week (Yirmiya et al., 2006). In order to confirm that the inhibition of bone formation is mediated by the activation of the SNS, a number of additional observational studies using stress models found that bone formation was inhibited, while bone resorption remained unaffected and propranolol attenuated the deterioration in bone mass and structure. Importantly, osteoblast numbers were reduced in the CMS murine model, which is consistent with previous studies which found that sympathetic antagonists ameliorate bone loss by enhancing osteoblast numbers (Takeda et al., 2002; Elefteriou et al., 2005).

Depression is a complex condition, interweaving the central nervous system disorder and poor lifestyle choices. Smoking, drinking, and an unhealthy diet, as well as reduced physical activity resulting from depression and the consumption of specific psychotropic drugs also contribute to the development of osteoporosis. However, to what extent osteoporosis is related to lifestyle changes in depressed patients remains unclear. The mechanisms involved need to be unraveled by further in-depth research.

CONCLUSIONS

There are increasing numbers of studies on autonomic nervous system regulation in bone metabolism. The sympathetic neural pathway is so far the only identified link between the potent leptin-dependent central control and bone cells. Clinical observations have also suggested that pharmacological intervention on the β-adrenergic system could be a novel way to treat osteoporosis. Despite conflicting reports, more inspiring and promising results have been demonstrated. Therefore, understanding the mechanism of sympathetic neural signaling on bone remodeling may shed light on a potential treatment avenue for the prevention or reversal of bone loss.

Numerous questions about sympathetic neural signaling in bone remodeling still await answers. Specific gene deletion studies for the SNS are helpful in unraveling the functions of specific neuroendocrine factors. However, generalized knockout rodent studies still suffer from many puzzling aspects. Consequently, conditional site-specific β_2-AR knockout in different animal models, such as OVX, depressed, unloading, and others, will be critical in addressing these questions.

So far, most experimental studies of the SNS in bone remodeling are confined to murine models not identical to human beings, and most clinical observations about the effect of β-blockers against osteoporosis are retrospective. Therefore, more scrupulously designed, prospective, randomized, and controlled clinical trials are crucial to obtain solid results.

References

Adami S, Fossaluzza V, Gatti D et al. (1997). Bisphosphonate therapy of reflex sympathetic dystrophy syndrome. Ann Rheum Dis 56: 201–204.

Ahima RS, Flier JS (2000). Leptin. Annu Rev Physiol 62: 413–437.

Altindag O, Altindag A, Asoglu M et al. (2007). Relation of cortisol levels and bone mineral density among premenopausal women with major depression. Int J Clin Pract 61: 416–420.

Asmus SE, Parsons S, Landis SC (2000). Developmental changes in the transmitter properties of sympathetic neurons that innervate the periosteum. J Neurosci 20: 1495–1504.

Bab IA, Yirmiya R (2010). Depression and bone mass. Ann N Y Acad Sci 1192: 170–175.

Baek K, Bloomfield SA (2009). Beta-adrenergic blockade and leptin replacement effectively mitigate disuse bone loss. J Bone Miner Res 24: 792–799.

BeDell KK, Scremin AM, Perell KL et al. (1996). Effects of functional electrical stimulation-induced lower extremity cycling on bone density of spinal cord-injured patients. Am J Phys Med Rehabil 75: 29–34.

Blazer DG, Kessler RC, McGonagle KA et al. (1994). The prevalence and distribution of major depression in a national community sample: the National Comorbidity Survey. Am J Psychiatry 151: 979–986.

Bliziotes MM, Eshleman AJ, Zhang XW et al. (2001). Neurotransmitter action in osteoblasts: expression of a functional system for serotonin receptor activation and reuptake. Bone 29: 477–486.

Bonnet N, Benhamou CL, Brunet-Imbault B et al. (2005). Severe bone alterations under beta2 agonist treatments: bone mass microarchitecture and strength analyses in female rats. Bone 37: 622–633.

Bonnet N, Laroche N, Vico L et al. (2006). Dose effects of propranolol on cancellous and cortical bone in ovariectomized adult rats. J Pharmacol Exp Ther 318: 1118–1127.

Bonnet N, Gadois C, McCloskey E et al. (2007). Protective effect of beta blockers in postmenopausal women: influence on fractures bone density micro and macroarchitecture. Bone 40: 1209–1216.

Bouxsein ML, Devlin MJ, Glatt V et al. (2009). Mice lacking beta-adrenergic receptors have increased bone mass but are not protected from deleterious skeletal effects of ovariectomy. Endocrinology 150: 144–152.

Brouwers JE, Lambers FM, van Rietbergen B et al. (2009). Comparison of bone loss induced by ovariectomy and neurectomy in rats analyzed by in vivo micro-CT. J Orthop Res 27: 1521–1527.

Burner WL 3rd, Badger VM, Sherman FC (1982). Osteoporosis and acquired back deformities. J Pediatr Orthop 2: 383–385.

Burr DB, Robling AG, Turner CH (2002). Effects of biomechanical stress on bones in animals. Bone 30: 781–786.

Burt-Pichat B, Lafage-Proust MH, Duboeuf F et al. (2005). Dramatic decrease of innervation density in bone after ovariectomy. Endocrinology 146: 503–510.

Burwell RG, Dangerfield PH, Freeman BJ (2008). Etiologic theories of idiopathic scoliosis. Somatic nervous system and the NOTOM escalator concept as one component in the pathogenesis of adolescent idiopathic scoliosis. Stud Health Technol Inform 140: 208–217.

Chauveau C, Devedjian JC, Delecourt C et al. (2008). Leptin receptors and beta2-adrenergic receptor mRNA expression in brain injury-related heterotopic ossification. J Recept Signal Transduct Res 28: 347–359.

Cheng JC, Hung VW, Lee WT et al. (2006). Persistent osteopenia in adolescent idiopathic scoliosis – longitudinal monitoring of bone mineral density until skeletal maturity. Stud Health Technol Inform 123: 47–51.

Cherruau M, Morvan FO, Schirar A et al. (2003). Chemical sympathectomy-induced changes in TH- VIP- and CGRP-immunoreactive fibers in the rat mandible periosteum: influence on bone resorption. J Cell Physiol 194: 341–348.

Christensen TM, Simonsen L, Holstein PE et al. (2011). Sympathetic neuropathy in diabetes mellitus patients does not elicit Charcot osteoarthropathy. J Diabetes Complications 25: 320–324.

Clowes JA, Khosla S, Eastell R (2005). Potential role of pancreatic and enteric hormones in regulating bone turnover. J Bone Miner Res 20: 1497–1506.

Cohen HE, Hall J, Harris N et al. (2012). Enhanced pain and autonomic responses to ambiguous visual stimuli in chronic complex regional pain syndrome (CRPS) type I. Eur J Pain 16: 182–195.

Cook SD, Harding AF, Morgan EL et al. (1987). Trabecular bone mineral density in idiopathic scoliosis. J Pediatr Orthop 7: 168–174.

de Vries F, Souverein PC, Cooper C et al. (2007). Use of β-blockers and the risk of hip/femur fracture in the United Kingdom and the Netherlands. Calcif Tissue Int 80: 69–75.

Drummond PD, Finch PM (2004). Persistence of pain induced by startle and forehead cooling after sympathetic blockade in patients with complex regional pain syndrome. J Neurol Neurosurg Psychiatry 75: 98–102.

Elefteriou F, Ahn JD, Takeda S et al. (2005). Leptin regulation of bone resorption by the sympathetic nervous system and CART. Nature 434: 514–520.

Fidler MW, Jowett RL (1976). Muscle imbalance in the aetiology of scoliosis. J Bone Joint Surg Br 58: 200–201.

Frey-Rindova P, de Bruin ED, Stussi E et al. (2000). Bone mineral density in upper and lower extremities during 12 months after spinal cord injury measured by peripheral quantitative computed tomography. Spinal Cord 38: 26–32.

Gage BF, Birman-Deych E, Radford MJ et al. (2006). Risk of osteoporotic fracture in elderly patients taking warfarin: results from the National Registry of Atrial Fibrillation 2. Arch Intern Med 166: 241–246.

Handschin AE, Trentz OA, Hemmi S et al. (2007). Leptin increases extracellular matrix mineralization of human osteoblasts from heterotopic ossification and normal bone. Ann Plast Surg 59: 329–333.

Hanyu R, Wehbi VL, Hayata T et al. (2012). Anabolic action of parathyroid hormone regulated by the β2-adrenergic receptor. Proc Natl Acad Sci U S A 109: 7433–7438.

Haug SR, Brudvik P, Fristad I et al. (2003). Sympathectomy causes increased root resorption after orthodontic tooth movement in rats: immunohistochemical study. Cell Tissue Res 313: 167–175.

Henry MJ, Pasco JA, Nicholson GC et al. (2000). Prevalence of osteoporosis in Australian women: Geelong Osteoporosis Study. J Clin Densitom 3: 261–268.

Hill EL, Elde R (1991). Distribution of CGRP- VIP-D beta H- SP- and NPY-immunoreactive nerves in the periosteum of the rat. Cell Tissue Res 264: 469–480.

Ho ML, Tsai TN, Chang JK et al. (2005). Down-regulation of N-methyl D-aspartate receptor in rat-modeled disuse osteopenia. Osteoporos Int 16: 1780–1788.

Hohmann EL, Elde RP, Rysavy JA et al. (1986). Innervation of periosteum and bone by sympathetic vasoactive intestinal peptide-containing nerve fibers. Science 232: 868–871.

Huang TH, Lin HS, Chen HI et al. (2011). The effects of systemic chemical sympathectomy on local bone loss induced by sciatic neurectomy. J Orthop Sci 16: 629–637.

Ikeda Y, Nakajima A, Aiba A et al. (2011). Association between serum leptin and bone metabolic markers and the development of heterotopic ossification of the spinal ligament in female patients with ossification of the posterior longitudinal ligament. Eur Spine J 20: 1450–1458.

Itzstein C, Cheynel H, Burt-Pichat B et al. (2001). Molecular identification of NMDA glutamate receptors expressed in bone cells. J Cell Biochem 82: 134–144.

Jacka FN, Pasco JA, Henry MJ et al. (2005). Depression and bone mineral density in a community sample of perimenopausal women: Geelong Osteoporosis Study. Menopause 12: 88–91.

Jeffcoate WJ (2008). Charcot neuro-osteoarthropathy. Diabetes Metab Res Rev 24: S62–S65.

Jirkovska A, Kasalicky P, Boucek P et al. (2001). Calcaneal ultrasonometry in patients with Charcot osteoarthropathy and its relationship with densitometry in the lumbar spine and femoral neck and with markers of bone turnover. Diabet Med 18: 495–500.

Jones KB, Mollano AV, Morcuende JA et al. (2004). Bone and brain: a review of neural hormonal and musculoskeletal connections. Iowa Orthop J 24: 123–132.

Kahl KG, Greggersen W, Rudolf S et al. (2006). Bone mineral density bone turnover and osteoprotegerin in depressed women with and without borderline personality disorder. Psychosom Med 68: 669–674.

Kan L, Lounev VY, Pignolo RJ et al. (2011). Substance P signaling mediates BMP-dependent heterotopic ossification. J Cell Biochem 112: 2759–2772.

Kaplan FS, Seemann P, Haupt J et al. (2010). Investigations of activated ACVR1/ALK2, a bone morphogenetic protein type I receptor that causes fibrodysplasia ossificans progressiva. Methods Enzymol 484: 357–373.

Kavuncu V, Kuloglu M, Kaya A et al. (2002). Bone metabolism and bone mineral density in premenopausal women with mild depression. Yonsei Med J 43: 101–108.

Keller B, Yang T, Chen Y et al. (2011). Interaction of TGFbeta and BMP signaling pathways during chondrogenesis. PLoS One 6: e16421.

Kim Y, Hamada N, Takahashi Y et al. (2009). Cervical sympathectomy causes alveolar bone loss in an experimental rat model. J Periodontal Res 44: 695–703.

Kimble RB, Srivastava S, Ross FP et al. (1996). Estrogen deficiency increases the ability of stromal cells to support murine osteoclastogenesis via an interleukin-1 and tumor necrosis factor-mediated stimulation of macrophage colony-stimulating factor production. J Biol Chem 271: 28890–28897.

Kingery WS (2010). Role of neuropeptide cytokine and growth factor signaling in complex regional pain syndrome. Pain Med 11: 1239–1250.

Kingery WS, Davies MF, Clark JD (2003a). A substance P receptor (NK1) antagonist can reverse vascular and nociceptive abnormalities in a rat model of complex regional pain syndrome type II. Pain 104: 75–84.

Klein-Nulend J, Bacabac RG, Mullender MG (2005). Mechanobiology of bone tissue. Pathol Biol (Paris) 53: 576–580.

Koeck FX, Bobrik V, Fassold A et al. (2009). Marked loss of sympathetic nerve fibers in chronic Charcot foot of diabetic origin compared to ankle joint osteoarthritis. J Orthop Res 27: 736–741.

Kohr D, Singh P, Tschernatsch M et al. (2011). Autoimmunity against the beta(2) adrenergic receptor and muscarinic-2 receptor in complex regional pain syndrome. Pain 152: 2690–2700.

Kondo H, Nifuji A, Takeda S et al. (2005). Unloading induces osteoblastic cell suppression and osteoclastic cell activation to lead to bone loss via sympathetic nervous system. J Biol Chem 280: 30192–30200.

Krassioukov AV, Bunge RP, Pucket WR et al. (1999). The changes in human spinal sympathetic preganglionic neurons after spinal cord injury. Spinal Cord 37: 6–13.

Kypson AP, Morphew E, Jones R et al. (2003). Heterotopic ossification in rectal cancer: rare finding with a novel proposed mechanism. J Surg Oncol 82: 132–136 disccussion 137.

Ladizesky M, Lama MA, Roldan EJ et al. (2003). Effect of unilateral superior cervical ganglionectomy on mandibular bone in rats. Neuro Endocrinol Lett 24: 314–320.

Leboeuf D, Letellier K, Alos N et al. (2009). Do estrogens impact adolescent idiopathic scoliosis? Trends Endocrinol Metab 20: 147–152.

Leeds EM, Klose KJ, Ganz W et al. (1990). Bone mineral density after bicycle ergometry training. Arch Phys Med Rehabil 71: 207–209.

Levasseur R, Sabatier JP, Potrel-Burgot C et al. (2003). Sympathetic nervous system as transmitter of mechanical loading in bone. Joint Bone Spine 70: 515–519.

Levasseur R, Dargent-Molina P, Sabatier JP et al. (2005). Beta-blocker use bone mineral density and fracture risk in older women: results from the Epidemiologie de l'Osteoporose prospective study. J Am Geriatr Soc 53: 550–552.

Li J, Ahmad T, Spetea M et al. (2001). Bone reinnervation after fracture: a study in the rat. J Bone Miner Res 16: 1505–1510.

Li X-F, Li H, Liu Z-D et al. (2008). Low bone mineral status in adolescent idiopathic scoliosis. Eur Spine J 17: 1431–1440.

Li WW, Guo TZ, Liang D et al. (2009). The NALP1 inflammasome controls cytokine production and nociception in a rat fracture model of complex regional pain syndrome. Pain 147: 277–286.

Liu D, Li H, Zhao CQ et al. (2008a). Changes of substance P-immunoreactive nerve fiber innervation density in the sublesional bones in young growing rats at an early stage after spinal cord injury. Osteoporos Int 19: 559–569.

Liu D, Zhao C-Q, Li H et al. (2008b). Effects of spinal cord injury and hindlimb immobilization on sublesional and supralesional bones in young growing rats. Bone 43: 119–125.

Liu Z, Tam EM, Sun GQ et al. (2012). Abnormal leptin bioavailability in adolescent idiopathic scoliosis: an important new finding. Spine (Phila Pa 1976) 37: 599–604.

Lowe TG, Edgar M, Margulies JY et al. (2000). Etiology of idiopathic scoliosis: current trends in research. J Bone Joint Surg Am 82-A: 1157–1168.

Mach DB, Rogers SD, Sabino MC et al. (2002). Origins of skeletal pain: sensory and sympathetic innervation of the mouse femur. Neuroscience 113: 155–166.

Martin CD, Jimenez-Andrade JM, Ghilardi JR et al. (2007). Organization of a unique net-like meshwork of CGRP+sensory fibers in the mouse periosteum: implications for the generation and maintenance of bone fracture pain. Neurosci Lett 427: 148–152.

Meisinger C, Heier M, Lang O et al. (2007). Beta-blocker use and risk of fractures in men and women from the general population: the MONICA/KORA Augsburg cohort study. Osteoporos Int 18: 1189–1195.

Minkowitz B, Boskey AL, Lane JM et al. (1991). Effects of propranolol on bone metabolism in the rat. J Orthop Res 9: 869–875.

Moore RE, Smith CK 2nd, Bailey CS et al. (1993). Characterization of beta-adrenergic receptors on rat and human osteoblast-like cells and demonstration that beta-receptor agonists can stimulate bone resorption in organ culture. Bone Miner 23: 301–315.

Morikawa Y, Zehir A, Maska E et al. (2009). BMP signaling regulates sympathetic nervous system development through Smad4-dependent and -independent pathways. Development 136: 3575–3584.

Morse L, Teng YD, Pham L et al. (2008). Spinal cord injury causes rapid osteoclastic resorption and growth plate abnormalities in growing rats (SCI-induced bone loss in growing rats). Osteoporos Int 19: 645–652.

Munakata M, Kameyama J, Kanazawa M et al. (1997). Circadian blood pressure rhythm in patients with higher and lower spinal cord injury: simultaneous evaluation of autonomic nervous activity and physical activity. J Hypertens 15: 1745–1749.

Mussolino ME (2005). Depression and hip fracture risk: the NHANES I epidemiologic follow-up study. Public Health Rep 120: 71–75.

Nakamura K (2006). Vitamin D insufficiency in Japanese populations: from the viewpoint of the prevention of osteoporosis. J Bone Miner Metab 24: 1–6.

Ogilvie JW (2011). Update on prognostic genetic testing in adolescent idiopathic scoliosis (AIS). J Pediatr Orthop 31: S46–S48.

Onyike CU, Crum RM, Lee HB et al. (2003). Is obesity associated with major depression? Results from the Third National Health and Nutrition Examination Survey. Am J Epidemiol 158: 1139–1147.

Pacy PJ, Hesp R, Halliday DA et al. (1988). Muscle and bone in paraplegic patients and the effect of functional electrical stimulation. Clin Sci (Lond) 75: 481–487.

Pagani F, Sibilia V, Cavani F et al. (2008). Sympathectomy alters bone architecture in adult growing rats. J Cell Biochem 104: 2155–2164.

Pasco JA, Henry MJ, Sanders KM et al. (2004). Beta-adrenergic blockers reduce the risk of fracture partly by increasing bone mineral density: Geelong Osteoporosis Study. J Bone Miner Res 19: 19–24.

Patel MS, Elefteriou F (2007). The new field of neuroskeletal biology. Calcif Tissue Int 80: 337–347.

Petronijevic M, Petronijevic N, Ivkovic M et al. (2008). Low bone mineral density and high bone metabolism turnover in premenopausal women with unipolar depression. Bone 42: 582–590.

Pierroz D, Muzzin P, Bouxsein M et al. (2005). Mice null for β1β2-adrenergic receptors have low bone mass and architecture and are resistant to isoproterenol-induced inhibition of bone growth. Bone 36: S130.

Pierroz D, Bouxsein M, Rizzoli R et al. (2006). Combined treatment with a beta-blocker and intermittent PTH improves bone mass and microarchitecture in ovariectomized mice. Bone 39: 260–267.

Qiu Y, Sun X, Qiu X et al. (2007). Decreased circulating leptin level and its association with body and bone mass in girls with adolescent idiopathic scoliosis. Spine (Phila Pa 1976) 32: 2703–2710.

Rayner DV, Trayhurn P (2001). Regulation of leptin production: sympathetic nervous system interactions. J Mol Med 79: 8–20.

Reid IR, Gamble GD, Grey AB et al. (2005). Beta-blocker use BMD and fractures in the study of osteoporotic fractures. J Bone Miner Res 20: 613–618.

Rejnmark L, Vestergaard P, Kassem M et al. (2004). Fracture risk in perimenopausal women treated with beta-blockers. Calcif Tissue Int 75: 365–372.

Rejnmark L, Vestergaard P, Mosekilde L (2006). Treatment with beta-blockers ACE inhibitors and calcium-channel blockers is associated with a reduced fracture risk: a nationwide case-control study. J Hypertens 24: 581–589.

Repko M, Horky D, Krbec M et al. (2008). The role of the autonomic nervous system in the etiology of idiopathic scoliosis: prospective electron microscopic and morphometric study. Childs Nerv Syst 24: 731–734.

Rossouw JE, Anderson GL, Prentice RL et al. (2002). Risks and benefits of estrogen plus progestin in healthy postmenopausal women: principal results. From the Women's Health Initiative randomized controlled trial. JAMA 288: 321–333.

Salisbury E, Rodenberg E, Sonnet C et al. (2011). Sensory nerve induced inflammation contributes to heterotopic ossification. J Cell Biochem 112: 2748–2758.

Schlienger RG, Kraenzlin ME, Jick SS et al. (2004). Use of beta-blockers and risk of fractures. JAMA 292: 1326–1332.

Schmid A, Huonker M, Stahl F et al. (1998). Free plasma catecholamines in spinal cord injured persons with different injury levels at rest and during exercise. J Auton Nerv Syst 68: 96–100.

Schoofs M, Sturkenboom MCM, Leeuwen JV et al. (2005). Use of beta-blockers is associated with BMD and fracture risk. Bone 36: S129–S130.

Serre CM, Farlay D, Delmas PD et al. (1999). Evidence for a dense and intimate innervation of the bone tissue including glutamate-containing fibers. Bone 25: 623–629.

Shapiro SA, Stansberry KB, Hill MA et al. (1998). Normal blood flow response and vasomotion in the diabetic Charcot foot. J Diabetes Complications 12: 147–153.

Sosa M, Saavedra P, Gómez de Tejada MJ et al. (2011). Beta-blocker use is associated with fragility fractures in post-menopausal women with coronary heart disease. Aging Clin Exp Res 23: 112–117.

Suga S, Goto S, Togari A (2010). Demonstration of direct neurite-osteoclastic cell communication in vitro via the adrenergic receptor. J Pharmacol Sci 112: 184–191.

Takeda S, Elefteriou F, Levasseur R et al. (2002). Leptin regulates bone formation via the sympathetic nervous system. Cell 111: 305–317.

Tang Y, Wu X, Lei W et al. (2009). TGF-β1-induced migration of bone mesenchymal stem cells couples bone resorption with formation. Nat Med 15: 757–765.

Togari A (2002). Adrenergic regulation of bone metabolism: possible involvement of sympathetic innervation of osteoblastic and osteoclastic cells. Microsc Res Tech 58: 77–84.

Toom A, Arend A, Gunnarsson D et al. (2007). Bone formation zones in heterotopic ossifications: histologic findings and increased expression of bone morphogenetic protein 2 and transforming growth factors beta2 and beta3. Calcif Tissue Int 80: 259–267.

Turker S, Karatosun V, Gunal I (2006). Beta-blockers increase bone mineral density. Clin Orthop Relat Res 443: 73–74.

Uebelhart D, Demiaux-Domenech B, Roth M et al. (1995). Bone metabolism in spinal cord injured individuals and in others who have prolonged immobilisation. A review. Paraplegia 33: 669–673.

Veldhuizen AG, Wever DJ, Webb PJ (2000). The aetiology of idiopathic scoliosis: biomechanical and neuromuscular factors. Eur Spine J 9: 178–184.

Whooley MA, Cauley JA, Zmuda JM et al. (2004). Depressive symptoms and bone mineral density in older men. J Geriatr Psychiatry Neurol 17: 88–92.

Wong ML, Kling MA, Munson PJ et al. (2000). Pronounced and sustained central hypernoradrenergic function in major depression with melancholic features: relation to hypercortisolism and corticotropin-releasing hormone. Proc Natl Acad Sci U S A 97: 325–330.

Xu JC, Wu T, Wu GH et al. (2009). Leptin expression by heterotopic ossification-isolated tissue in rats with Achilles' tenotomy. Saudi Med J 30: 605–610.

Yan L, Vatner DE, O'Connor JP et al. (2007). Type 5 adenylyl cyclase disruption increases longevity and protects against stress. Cell 130: 247–258.

Yazici AE, Bagis S, Tot S et al. (2005). Bone mineral density in premenopausal women with major depression. Joint Bone Spine 72: 540–543.

Yirmiya R, Goshen I, Bajayo A et al. (2006). Depression induces bone loss through stimulation of the sympathetic nervous system. Proc Natl Acad Sci U S A 103: 16876–16881.

Zhang W, Kanehara M, Zhang Y et al. (2007). Beta-blocker and other analogous treatments that affect bone mass and sympathetic nerve activity in ovariectomized rats. Am J Chin Med 35: 89–101.

Handbook of Clinical Neurology, Vol. 117 (3rd series)
Autonomic Nervous System
R.M. Buijs and D.F. Swaab, Editors

Chapter 15

The circadian system and the balance of the autonomic nervous system

RUUD M. BUIJS[1]*, CAROLINA ESCOBAR[2], AND DICK F. SWAAB[3]

[1]*Department of Cell Biology and Physiology, Institute for Biomedical Research, Universidad Nacional Autónoma de México, Mexico City, Mexico*

[2]*Department of Anatomy, Faculty of Medicine, Universidad Nacional Autónoma de México, Mexico City, Mexico*

[3]*Netherlands Institute for Neuroscience, an Institute of the Royal Netherlands Academy of Arts and Sciences, Amsterdam, The Netherlands*

INTRODUCTION

In the last 20 years, a great deal of effort has gone into the unraveling of the molecular mechanisms of the circadian clock, while the question of the functionality of this rhythmicity has been relatively neglected. The present review aims to highlight the functional aspect of circadian rhythms with regard to the physiology of our body, their role in disorders, and the potential of circadian therapies.

The biological clock, the suprachiasmatic nucleus (SCN), is situated at the base of the hypothalamus, immediately above the optic chiasm. It is essential for providing biological rhythms to our body (Buijs and Kalsbeek, 2001; Hastings et al., 2003). The SCN has an autonomous rhythmicity in the electrical activity of its neurons and receives light from the retina, which synchronizes that neuronal activity with the light–dark cycle (Gillette and Reppert, 1987). The SCN has neuronal connections to mainly hypothalamic target areas that receive the signal of the SCN and use this daily signal to adapt their output in agreement with the time of the day. The hypothalamus is an excellent target area for the SCN, since it is not only essential for the control of hormonal output but also for the autonomic control of body functions. In addition, many studies show the importance of the hypothalamus for the execution of fundamental behaviors such as sleep, temperature regulation, reproduction, and food intake. Herein lies – in our opinion – the essence of the function of the SCN: the organization of synchrony between hormonal, autonomic, and behavioral functions. This synchrony should be organized depending on the demands of the body, e.g., hunger or satiety, and should be organized in a timely was: e.g., so that food intake behavior is not initiated during the rest phase. After epidemiological studies had shown that, for example, obesity, diabetes, and hypertension are associated with less sleep in the night (Ayas et al., 2003; Gangwisch et al., 2006, 2007) and animal studies confirmed that food intake during the rest phase may induce obesity (Arble et al., 2009; Salgado-Delgado et al., 2010b), the central role of the biological clock in the organization of physiological functions is an aspect that is gaining more and more attention in the literature. This is also an important topic for the clinic, because it has become increasingly clear that a circadian misalignment between metabolic functions and the activity dictated by the SCN will result in disease, and circadian therapies may be effective. The strongest evidence yet for chronic diseases that might have their basis in a desynchronization of physiological functions exists for hypertension and type 2 diabetes (Knutsson, 2003; Karlsson et al., 2005; Gangwisch et al., 2006, 2007; Scheer et al., 2009), and is supported by findings that show that the biological clock is changed in the postmortem brain of hypertensive individuals (Goncharuk et al., 2001). The present chapter aims to review the mechanisms of rhythmicity and their importance for physiology and thus for health and disease. In order to appreciate the significance of the rhythmic organization of physiology we will first discuss the general principles of the hypothalamic organization of homeostasis.

*Correspondence to: Ruud M. Buijs, Ph.D., Department of Cell Biology and Physiology, Institute for Biomedical Research, Universidad Nacional Autónoma de México, Mexico City, DF, 04510, Mexico. E-mail: ruudbuijs@gmail.com

HOMEOSTASIS AND THE HYPOTHALAMUS

Claude Bernard envisioned that all physiological processes in the body need to be in perfect harmony with each other. Later, Walter Cannon coined the term "homeostasis" for this process. The hypothalamus is the core brain structure responsible for the organization and execution of homeostasis. Hypothalamic nuclei execute or influence basic behavioral functions such as food and water intake, sleep and reproduction, and integrate and synchronize these behaviors by controlling adequate levels of body temperature and hormone secretion by the pituitary gland (Swanson, 1987; Thompson and Swanson, 2003). Moreover, the expression of behavior and adjusted physiological functions depends on the time of day, on gender, on developmental stages, and, in many species, also on the season. All these functions interact and influence each other in such a way that when the required behavior is executed it is accompanied by adequate hormonal output and body functions at the right moment of the day or season. This necessitates all these functions in the hypothalamus to be integrated and translated into a homeostatic output. Thus an extensive exchange of information between the hypothalamus and the body and among the different hypothalamic nuclei is essential. In fact, all potentially life-threatening situations which may disrupt the homeostasis of the body, such as infections, hypertension, disturbance of the mineral balance, low circulating glucose levels, or food (i.e., fat) shortage, affect hypothalamic functioning in such a way that an adequate hormonal, autonomic, and behavioral response is the result. This means that the hypothalamus needs to be kept informed about the homeostasis of the body at all times and homeostasis can thus only be maintained if the organism can adequately register its internal as well as its external environment. In the course of evolution, various sensory organs have developed that transmit information about the body and about the environment to the central nervous system.

Hypothalamic hormonal and autonomic output maintains homeostasis

The tasks of the hypothalamus can be distinguished into an organizational one, inducing an during development and an immediate hormonal, behavioral and autonomic response (Swanson, 1987; Thompson and Swanson, 2003). For this it is important to take into consideration that recent studies have shown that the hypothalamus not only has the capacity to secrete hormones but that via its effect on the autonomic nervous system it may affect the sensitivity of the target endocrine organs for pituitary hormones (Engeland and Arnhold, 2005). The hypothalamus may thus, via the autonomic nervous system (ANS), target a selective set of organs and change its blood supply and the sensitivity for particular hormones. In this way the hypothalamus influences, for example, via the ANS, the glucose output of the liver (Kalsbeek et al., 2004), corticosterone secretion of the adrenal cortex (Buijs et al., 1999), and water filtration of the kidney (Deering and Coote, 2000). This capacity is not unique to the hypothalamus: brain areas such as the prefrontal cortex and amygdala also have the capacity to change the autonomic outflow to specific organs without interference from the hypothalamus (Rodrigues et al., 2009).

In order to maintain homeostasis, the hypothalamus not only needs to integrate information from the environment, body, and brain, but also to translate this information into a response that serves to maintain homeostasis. To this end an intricate network of connections between the different hypothalamic nuclei functions in such a way that information can be exchanged and shared.

Hypothalamic input

Information from the environment

The vasoactive intestinal polypeptide (VIP) neurons in the ventral SCN receive direct information about the light–dark situation from the ganglion cells of the retina. In addition, there are retinal projections to other hypothalamic areas close to the optic nerve, although the function of those projections is not yet clearly understood (Sollars et al., 2003; Brown et al., 2011). The light input to the SCN serves to synchronize to the light–dark cycle of the environment, as can be shown by the inhibition of the nightly melatonin secretion by light. Without the light input into the SCN, e.g., if there is damage to the optic nerves, animals and humans cannot synchronize their activity to the light–dark cycle, resulting in a free-running rhythm. This is different from lesions of the SCN itself, which result in a complete disturbance of the sleep–wake cycle. Suprasellar tumors involving compression of the optic chiasm, leading to permanent visual field defects (e.g., in case of craniopharyngioma, macroadenoma of the pituitary, or metastasis), are associated with an impaired sleep–wake rhythm (Borgers et al., 2011) and a reduction of the major SCN neuropeptide vasopressin (Swaab, 2003; Borgers et al., 2012). This shows that in humans too, damage to the SCN may lead to a disturbed organization of the sleep–wake cycle. This in contrast with a condition in which only the visual input is disturbed, such as in retinopathy, when humans remain rhythmic though not synchronized to the light–dark cycle (Czeisler et al., 1995).

Another sensory input from outside our body that is important for hypothalamic functioning is temperature. Although it is not yet known exactly how information about skin temperature reaches the hypothalamus, the first relay nucleus for temperature information to be transmitted to the hypothalamus is suggested to be the dorsomedial hypothalamic nucleus (DMH) (Morrison and Nakamura, 2011; Nakamura, 2011). The DMH transmits this information to the medial preoptic area (MPA). This hypothalamic region is the main area for the regulation of the temperature set point (Morrison and Nakamura, 2011; Nakamura and Morrison, 2011).

The importance of this function can be illustrated by the following observation. A young man with a craniopharyngioma which threatened to destroy his sight underwent surgery which caused serious damage to his hypothalamus. However, by far the greatest danger threatening him was his total lack of thermoregulation. If he engaged in physical activity and the outside temperature dropped somewhat, in a very short space of time he could develop life-threatening hypothermia; in contrast, if the sun came out or he warmed up by walking, e.g., within the hospital for his appointment with me (D.F. S.), his temperature very quickly rose to the level of a fever. Once, prior to an operation, the 15 minute wait he had wearing a hospital gown that tied down the back was enough to cause him to cool down so dramatically that he almost went into a coma, and he had to be given a hot shower to get his temperature back to normal before the operation could begin. His mother took over this function of the hypothalamus by following her son around with a thermometer, so that whenever the surrounding temperature changed she could take action.

Time of the day, hormones, inflammation, and feeding status influence the temperature set point (Liu et al., 2002; Scheer et al., 2005; Steiner et al., 2005). Furthermore, taste and smell information from, respectively, the nucleus of the solitary tract (NTS) and olfactory bulb also have access to hypothalamic nuclei; both change the temperature regulation and induce postprandial temperature increase and thus affect the MPA (Tentolouris et al., 2003).

Proprioceptive information from the body

Our perception of the surrounding world is limited to what our sensory organs tell us exists, but much more is going on in and around us than we realize consciously. In fact, most of the sensory information that reaches our brains does so without reaching our consciousness, a situation that is hardly surprising given the fact that most of this information deals with the homeostatic balance of our organs. A large part of the initial integration of homeostatic sensory information takes place in brain regions that are evolutionary older, such as the spinal cord, brainstem, and midbrain structures that are connected with the hypothalamus as the center where sensory information from body and environment converges. In principle, this sensory information (about pain or about the internal state changes of the body) may lead to homeostatic stress if a severe disbalance is noted. Initially this imbalance, at the level of the spinal cord and brainstem, leads, after it has been detected, to a reflex response using pathways from the sensory neurons in, for example, the NTS to the autonomic motor neurons in the dorsal motor nucleus of the vagus (DMV). This autonomic reflex adjusts, for example, blood pressure or the circulation in specific organs, by vasoconstriction or vasodilatation (Tjen-A-Looi et al., 1997; Craig, 2003). In addition, this sensory information is also transmitted higher up, to the hypothalamus or cerebral cortex, where feedback responses are also initiated and where the intensity of the response is evaluated, resulting in the appropriate behavioral and hormonal measures (Contreras et al., 2007; Kampe et al., 2009; Matyas et al., 2010).

Visceral sensory information entering the NTS or laminas 1–4 in the dorsal horn of the spinal cord reaches the hypothalamus directly via the NTS or indirectly via the parabrachial nucleus (PBN) (Craig, 2003). This type of sensory information is prone to modulation (gating) at the level of laminas 1–4 and at the level of the NTS. Since descending projections of the paraventricular nucleus (PVN), lateral hypothalamus, and arcuate nucleus (ARC; in human also called the infundibular nucleus) have terminations in the NTS, an interesting circuit is in place that allows the hypothalamus to modify its incoming information (Buijs et al., 2001). The same applies to laminae 1–4 of the spinal cord where only the PVN and lateral hypothalamus have terminations that may gate and modulate the incoming signals (Condes-Lara et al., 2007). The major areas where visceral sensory information via the NTS and parabrachial nucleus enters the hypothalamus are the ventromedial hypothalamus (VMH) the PVN, the MPA and the DMH (Swanson, 1987).

Circulating information from the bloodstream reaches the hypothalamus via the circumventricular organs (CVOs) such as the area postrema, the ARC, the subfornical nucleus, and the organum vasculosum of the lamina terminalis (OVLT). Hormones such as the pineal hormone, melatonin, can pass the blood–brain barrier and signal to those sites in the brain that are endowed with the appropriate receptors. If the substances in the circulation cannot pass the blood–brain barrier freely, such as insulin, angiotensin, glucose, or ghrelin, they signal to receptors on neurons that are situated outside the blood–brain barrier in the CVOs (Price et al., 2008; Smith and Ferguson, 2010). The CVOs have extensive connections with each other and with hypothalamic nuclei with a focus on the PVN; notably also the SCN receives input from the CVOs (Miselis, 1981; Lind et al., 1982; Yi et al., 2006; Trudel and Bourque,

2010). In addition, CVOs have extensive connections to autonomic motor neurons in the brainstem and spinal cord (Price et al., 2008a; Smith and Ferguson, 2010), ensuring rapid adjustment of autonomic functions. These circuits are comparable to the reflex response of the visceral neuronal sensory pathways that feed back to the autonomic motor neurons. In addition, the fact that these CVOs also receive information from the SCN and other hypothalamic nuclei indicates that gating of information may take place already at the level of the sensory structure.

Information from the cortex

It is clear that, in addition to information from the body and environment, the hypothalamus also needs information from the rest of the brain relating to the emotional state of the individual. Structures such as amygdala, prefrontal cortex, insular cortex and other cortical areas provide extensive input to the hypothalamus. Most of that input arrives at the lateral hypothalamus, which structure, not surprisingly, is also the main hypothalamic area that provides extensive input to the different cortical areas. For the purposes of this chapter it is sufficient for us to refer to some of the excellent reviews that have described these cortical-hypothalamic connections in depth (Risold et al., 1997; Swanson, 2000; Lambe et al., 2005; Swanson et al., 2005).

Hypothalamic integration

As has become clear from the summary of hypothalamic inputs given above, all this information is not only gated or modulated at the moment of entry to the nervous system, it also needs to be integrated with other hypothalamic information in order to lead to an adequate behavioral, autonomic, and hormonal response. To this end, hypothalamic nuclei establish extensive intrahypothalamic connections whereby the PVN is one of the main recipients of information (Canteras et al., 1994; Thompson et al., 1996; Thompson and Swanson, 2003). This anatomical organization allows an optimal exchange of information whereby the perception of a change in a specific physiological condition has repercussions for the regulation of other physiological functions. Consequently the hypothalamus does not respond to a stimulus in isolation or in a standard form, but modulates its response depending on the state of the individual. One of the structures that has a major impact on the state of the hypothalamus and its organization of the physiological response is the SCN. It has extensive projections mainly limited to within the hypothalamus. By transmitting its day–night signal it influences hypothalamic integration and output (Watts et al., 1987; Buijs et al., 1994; Buijs and Kalsbeek, 2001). The SCN provides the setting of the activity level of many hypothalamic circuits and nuclei and determines day and night temperature, day-night hormone levels, as well as blood pressure and activity levels (Hastings et al., 2003). In addition, the SCN takes care that the behavior is synchronized with the correct physiological conditions. For example, reproductive behavior needs to be coordinated in such a way that it is synchronized with ovulation and environmental conditions. Thus the time of the day, the nutritional status, and the environmental (stress) information are important variables that may determine whether reproduction will occur (Hill et al., 2008). The extensive intrahypothalamic connections that arise from the ARC will provide the information about the feeding status, while the projections from the SCN indicate the time of day information. Notably the ARC, the lateral hypothalamus, and the DMH, with their extensive intrahypothalamic connections, are crucial structures for receiving extrahypothalamic information. This anatomical construction provides the framework for the correct synchronization of hypothalamic output into behavior and physiology. The way the SCN modulates hypothalamic integration and output depending on the time of the day provides an interesting insight into how the hypothalamus integrates different types of information leading to an integrated response. In addition, the projections of the SCN to the DMH and ARC provide the SCN with the means to gate or modify the incoming or outgoing information. That this indeed takes place is demonstrated by the experiments showing that ghrelin, a hormone released by the stomach, activates the ARC more strongly at the beginning of the dark period than at the beginning of the light period (Yi et al., 2006).

Preautonomic neurons in the hypothalamus receive information about, e.g., metabolic conditions from the ARC, DMH, or VMH; about the time of day from the SCN; about temperature conditions from the MPO or DMH; and from the circumventricular organs, subfornical organ (SFO), ARC, and OVLT information about circulating sodium, glucose, insulin, and angiotensin II concentrations. Since areas that contain these preautonomic neurons, such as the PVN, MPO, and ARC, are also involved in the control of hormone secretion, this organization allows the brain to immediately synchronize autonomic output with hormonal secretion for optimal execution of functions. In addition to autonomic function or hormone secretion, several of these areas are known for their capacity to be able to induce certain types of behavior, such as food intake (PVN), water intake (OVLT and SFO), or temperature control (MPO). This indicates that these structures coordinate different tasks that support the same physiological function: to execute behavior and coordinate it with autonomic functions and hormone secretion. For example, food intake is stimulated by injections of neuropeptide

Y (NPY) into the PVN (Leibowitz and Wortley, 2004), although the PVN does not have projections to the cortex or nucleus accumbens; hence its projections to the NTS and the ascending projections of the NTS to the cortex. These may be essential for the final execution of food intake behavior. However, this has not been demonstrated and needs further investigation. Another example is that the PVN contains hormone-releasing neurons and has direct connections to autonomic motor nuclei that target the organs that will be reached by these hormones. This enables the PVN to influence the sensitivity of those organs for the hormones via the ANS (Buijs et al., 1999; Engeland and Arnhold, 2005; Bornstein et al., 2008). Since the PVN is also important for the execution of behavioral functions this provides a fascinating picture of the capacity of the hypothalamus to integrate and coordinate all these functions – especially within the PVN. In addition to all these intrahypothalamic connections that finally focus on the PVN, there are extensive intra-PVN connections (Van Den Pol, 1982) that form another possibility to exchange information between the different types of neurons to allow synchronization of functions. Thus the picture emerges of hypothalamic systems in which information at different levels is constantly compared in a circuit that integrates information from the body, the environment, and the brain. These circuits suggest a system that may operate so that behavior is executed, along with the associated hormonal and autonomic changes, only when all nodes in the loop signal GO. When the signal that arrives in one of the nodes does not agree with the behavior that is about to be executed then, at least momentarily, the behavior will not be executed.

The infundibular or arcuate nucleus and diabetes

As has become clear from the above summary of hypothalamic integration, there is a wealth of experimental data from animals obtained on the role of the ARC in monitoring the composition of the blood for metabolic substances such as, e.g., insulin, glucose, ghrelin, and leptin (Obici et al., 2002a, b, 2003; Horvath, 2005). Changing concentrations of insulin and glucose modify the activity of several neuronal populations within the ARC (Spanswick et al., 1997, 2000; van den Top et al., 2004). These nutrient/hormone-sensitive neurons project to preautonomic neurons located in the hypothalamus and brainstem, and modulate the autonomic output to liver and pancreas (Atasoy et al., 2012; Yi et al., 2012). At present, the ARC is mainly characterized by two antagonistic neuronal groups: the NPY and pro-opiomelanocortin (POMC)-expressing cells. In animals, NPY neurons in the ARC are activated by negative metabolic challenges, such as fasting and hypoglycemia (Watanabe et al., 2008), and NPY signaling is required to stimulate hepatic glucose production in response to a decrease in plasma glucose levels (Atasoy et al., 2012). POMC neurons, in contrast, are activated by positive energy conditions and satiety factors, and promote catabolism and energy expenditure (Cowley et al., 2001; Parton et al., 2007).

Notwithstanding the abundant data from animal experiments, little is known about the responses and changes of these peptide-containing neurons in the human brain. Recently we examined the changes of NPY neurons in the ARC in obese humans (Goldstone et al., 2002) and demonstrated that this condition did not affect the expression of NPY or POMC in the infundibular nucleus. However, this condition does not necessarily affect circulating levels of glucose or insulin, the two major players in the control of the activity in the ARC, in an important way. Therefore we investigated possible changes in NPY and α-MSH neurons in the infundibular nucleus in the diabetic condition (Saderi et al., 2012). The expression of NPY and α-MSH in the postmortem brain of type-2 diabetic individuals was compared with their expression in brains obtained from age- and sex-matched nondiabetic patients (Fig. 15.1).

In addition, since the NTS is an important area where sensory visceral information is integrated that reaches

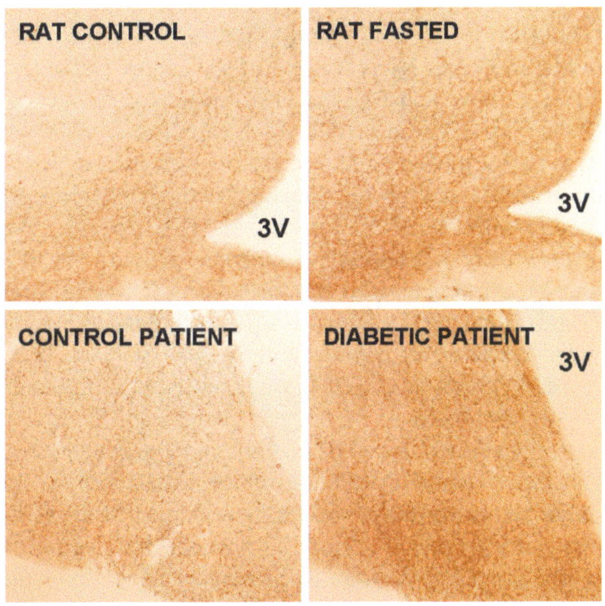

Fig. 15.1. Illustrates the NPY immunoreactivity in the rat arcuate nucleus and the human infundibular nucleus. The NPY immunoreactivity in the rat arcuate nucleus under 48 hour fasting condition is strongly increased as compared to the non-fasting condition. A similar increase in NPY immunoreactivity can be observed in the human diabetic infundibular nucleus suggesting a similar metabolic state.

this nucleus via the nervus vagus, the expression of NPY in the NTS was also monitored. A significant increase in the number of neurons expressing NPY was observed in the infundibular nucleus of the postmortem human diabetic brain. These changes, when extrapolated to animal studies, suggest that the human ARC in the diabetic state is reflecting a catabolic state and is possibly unable to sense high circulating glucose levels. Interestingly, the same brains showed a lower expression of NPY in the NTS of diabetic patients, demonstrating that the same neuronal phenotype in the NTS area responds quite differently to the diseased diabetic state. This shows that the decrease in staining in the infundibular nucleus is not caused by a general (nonspecific) change due to technical problems but may reflect that the NTS does indeed serve to detect different metabolic changes from the infundibular nucleus. Finally, the similarity between NPY expression in the ARC of a fasted animal and NPY expression in the postmortem brain of a diabetic patient (Fig. 15.1) strongly suggests that the diabetic brain perceives a hypoglycemic state in spite of the high circulating glucose levels. Consequently, the increased NPY signaling to second order neurons in the PVN that control the activity of the sympathetic system may contribute to the anomalous, unbalanced hepatic glucose production.

THE SUPRACHIASMATIC NUCLEUS AND BIOLOGICAL RHYTHMS IN HUMANS

The discovery in 1972 of the hypothalamic suprachiasmatic nuclei (SCN) as the seat of the central biological clock (Moore and Eichler, 1972; Stephan and Zucker, 1972) was followed in 1975 by the application of what was at that moment a new technique (i.e., immunocytochemistry) by our group, showing that the SCN produced vasopressin (Swaab et al., 1975). The first report on the presence of vasopressin-containing neurons in the human SCN appeared 2 years later (Dierickx and Vandesande, 1977). The first detailed description of peptide-containing neurons in the human SCN was published by Stopa et al. (1984). They described a predominantly ventral group of neurons containing VIP, and a dorsally adjacent group of neurons containing vasopressin (i.e., a spatial orientation similar to that previously described in the rat). As a striking example of evolutionary conservation, we found in postmortem material a pronounced daily rhythm in the number of vasopressin-expressing neurons over the 24 hour day/night cycle (Hofman and Swaab, 1993), which agrees with the dynamic of vasopressin in the rat brain (Gillette and Reppert, 1987; Reppert et al., 1987). Brain tissue from the Netherlands Brain Bank showed that if young donors (6–47 years of age) had died during the daytime (10:00–18:00 h), the SCN contained more than twice as many immunocytochemically stained vasopressin neurons than when death had occurred at night (22:00–06:00 h). Peak values in vasopressin cell number occurred in the early morning (06:00–10:00 h). This marked circadian fluctuation disappears in donors over 50 years of age, providing a biological basis for sleep disorders in the elderly (Hofman and Swaab, 1993) (Fig. 15.2). In addition to this diurnal variation, a marked seasonal fluctuation was observed in the number of vasopressin-immunoreactive neurons in young donors. The annual cycle of vasopressin abundance in the SCN showed a maximum in early autumn and a deep trough in late spring and early summer, with three times the number of vasopressin-immunoreactive neurons present during autumn than during summer. In donors over the age of 50 this seasonal rhythm was disrupted too, and had a reduced amplitude (Hofman and Swaab, 1995) (Fig. 15.3).

Interestingly, since the biological clock stops at the time of death, it was recently shown that it is possible to read the time of death in forensic tissues by measuring

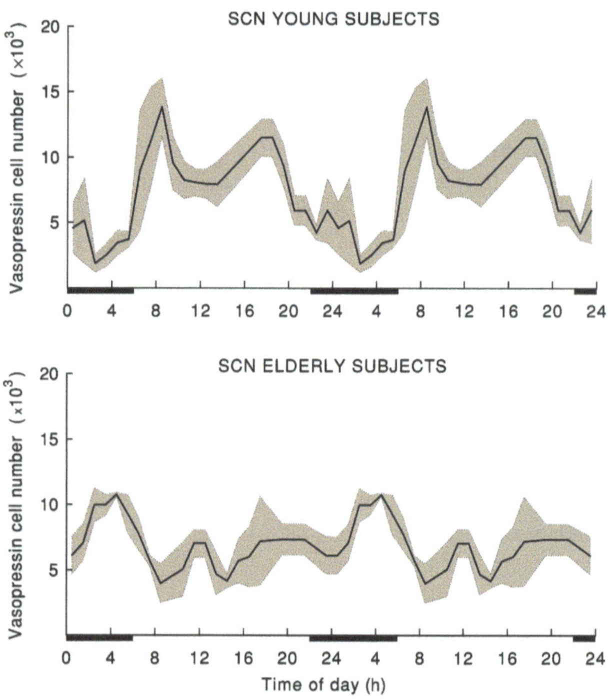

Fig. 15.2. Circadian rhythm in the number of vasopressin-containing neurons in the human suprachiasmatic nucleus (SCN) of (**A**) young subjects (<50 years of age) and (**B**) elderly subjects (≥50 years of age). The black bars indicate the night period (22:00–06:00 h). The general trend in the data is enhanced by using a smoothed double-plotted curve and is represented by mean ± S.E.M. values. Note the circadian rhythm in the SCN of young people with low values during the night period and peak values during the early morning. (Reproduced from Hofman and Swaab, 1994, Fig. 1, with permission.)

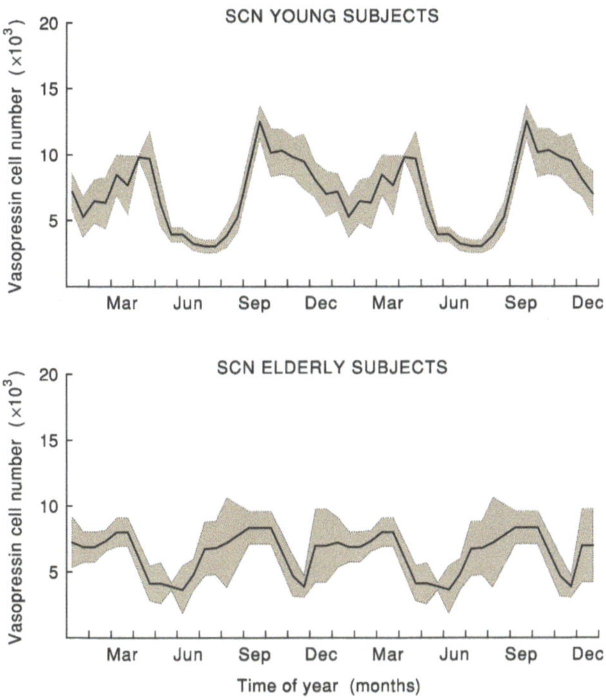

Fig. 15.3. Annual rhythm in the number of vasopressin-containing neurons in the human suprachiasmatic nucleus (SCN) of (**A**) young subjects (<50 years of age) and (B) elderly subjects (≥50 years of age). The general trend in the data is enhanced by using a smoothed, double-plotted curve and is represented by mean ± S.E.M. Note the circannual rhythm in the SCN of young people with low values during the summer and peak values in the autumn period. (Reproduced from Hofman and Swaab, 1995, Fig. 1, with permission.)

gene expression levels of various clock genes and their ratios (Kimura et al., 2011).

Hypothalamic output

In principle, there is just one area in the hypothalamus that has as its main task to provide information to the body, i.e., the PVN. Although the lateral hypothalamus (LH) also contains neurons that project to the autonomic motor nuclei, the PVN contains neurons that either provide hormonal output or autonomic output to sympathetic or parasympathetic motor neurons in the spinal cord or brainstem. Other nuclei, such as the LH or DMH, have a more mixed neuronal population with more intrahypothalamic connections and with limited autonomic connections (DMH) or with extensive projections to many brain areas, including hypothalamus and cortex (LH). Furthermore, as mentioned above, the same PVN neurons that provide autonomic output information have collaterals to NTS and spinal cord sensory laminae 1–4 that allow modulation of incoming sensory information at the same time. In addition, the retrochiasmatic area and the lateral hypothalamus (LH) have neurons that project directly to both sympathetic and parasympathetic motor nuclei (Buijs et al., 2003), but have for the most part projections to other brain regions and hypothalamic regions that are not principally involved in transmitting hypothalamic information to the rest of the body.

Tracing studies revealed that PVN and LH have different neurons that project to either sympathetic or parasympathetic motor neurons, and have shown that even brain nuclei like our biological clock or the ventromedial hypothalamus (VMH) contain neurons that project to preautonomic neurons in the PVN that are either parasympathetic or sympathetic (Buijs et al., 2003; Kreier et al., 2005). With respect to the SCN, this illustrates that even this relatively "simple" task (timing of the brain and body) requires a highly differentiated output whereby different organs receive information from different neurons in the same hypothalamic nucleus (Kreier et al., 2005). This differentiation in parasympathetic and sympathetic tasks for different SCN neurons provides the basis for an understanding of how, in humans, every 24 hours, a high sympathetic drive to the pineal at night, which is driven by the SCN, coincides with a higher parasympathetic drive to the heart, also driven by the SCN (Hu et al., 2011; Shea et al., 2011). In addition, it demonstrates that not all organs get similar (indirect) input from the SCN at all moments of the light–dark cycle: different organs receive different input from the brain and a differently timed input from the SCN.

Human suprachiasmatic nucleus output via neuronal connections

A detailed description of the efferent projections of the vasopressin and VIP neurons was created by our group (Dai et al., 1997) using a combination of immunohistochemistry and an *in vitro* postmortem anterograde tracing method (Dai et al., 1998a, b). It appeared that the main SCN projection areas were very similar to those described in rodents (i.e., the anteroventral hypothalamic area, the area below the paraventricular nucleus (i.e., sub-PVN), the ventral part of the PVN, and the DMH), indicating a striking evolutionary conservation of the SCN projections from rodent to human.

For blood pressure regulation it is important that vasopressin fibers, probably coming from the PVN, are most abundantly present in the human dorsal motor nucleus of the vagus nerve as well as in the adjacent medial and intermediate subnuclei of the solitary nucleus. Vasopressin-positive fibers were also found to greatly contact tyrosine hydroxylase-positive neuronal profiles in A2/C2, A2, and nucleus ambiguus (Fig. 15.4) (Goncharuk et al., 2011).

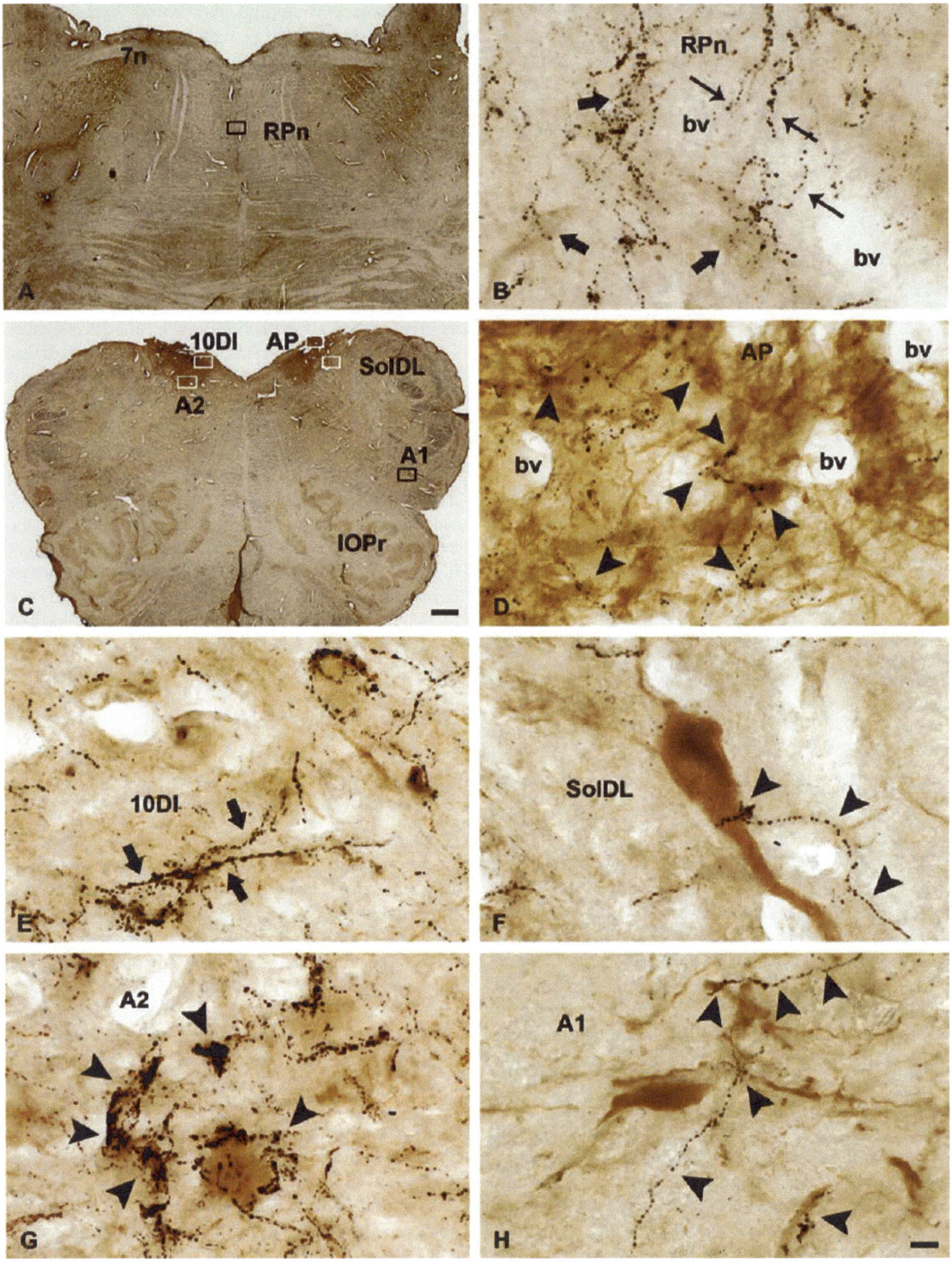
7n
RPn
A
RPn
bv
bv
B
10DI
AP
SolDL
A2
A1
IOPr
C
AP
bv
bv
bv
D
10DI
E
SolDL
F
A2
G
A1
H

Suprachiasmatic nucleus output via the human pineal gland

The pineal gland, or epiphysis cerebri, is a key structure of the circadian system. It contains pinealocytes, which produce melatonin (for references see Swaab, 2003). The pineal gland is innervated by a multisynaptic pathway from the SCN via the PVN, the intermediolateral column of the upper thoracic spinal cord and the superior cervical ganglion, which sends noradrenergic fibers to the pineal gland. The human pineal gland has a dense noradrenergic plexus. During darkness, due to stimulation arising from the SCN, noradrenaline is released from the sympathetic nerve endings in the pineal gland to activate N-acetyltransferase, the enzyme which catalyses the rate-limiting step of the synthesis of melatonin from serotonin. Serum and cerebrospinal fluid (CSF) melatonin levels in human are high at night and low during the day. Apart from the stimulatory input from the SCN, the main environmental stimulus that modifies the rhythmic melatonin fluctuations is light intensity. It is generally thought that melatonin reaches its targets in the brain via the peripheral circulation, although melatonin may also diffuse into the brain from the CSF. The SCN thus imposes circadian fluctuations indirectly on many more brain structures and functions by means of melatonin from the pineal gland – provided that melatonin receptors are present in those areas. In its turn, depending on the time of exposure, two distinct, separable, effects on the SCN, melatonin may elicit acute neuronal inhibition and phase shifting. We propose that this direct action of melatonin on the SCN is not only the neuroendocrine mediator of the circadian rhythm of sleep, but also influences arterial blood flow, decreases blood pressure, and blunts noradrenergic activation (Scheer et al., 2004). Melatonin is therefore clinically used in an increasing number of disorders (see below).

The suprachiasmatic nucleus synchronizes behavior, hormone secretion and autonomic output

The projections of the SCN reach different hypothalamic target areas where separate neurons of the SCN target separate parasympathetic or sympathetic preautonomic neurons in the PVN or LH. These autonomic connections serve to transmit the SCN message directly to the peripheral organs, allowing the SCN to modulate, e.g., blood pressure and heart rate via the sympathetic and parasympathetic system with a completely different timing and a different autonomic setting (Scheer et al., 2001, 2009; Kreier et al., 2005). Furthermore, these autonomic connections serve to time liver glucose production and insulin sensitivity, depending on the time of day (La Fleur et al., 2001; Kalsbeek et al., 2004; Yi et al., 2012). On the other hand, the SCN also sends projections to areas that serve to integrate information, such as the DMH and the paraventricular nucleus of the thalamus (PVT). The PVT, for example, sends information supposed to be essential for initiation of locomotor behavior to the amygdala, accumbens, lateral septum prefrontal cortex, and hypothalamus (Vertes and Hoover, 2008; Li et al., 2009). The DMH has extensive projections within the hypothalamus that serve, for example, to transmit the SCN signal to hormone producing neurons in the PVN, while its projections to the MPO are important for temperature regulation. A similar interaction of the SCN exists with the MPA, where gonadotropin-releasing hormone-producing neurons (GnRH) are located and where the temperature set point is organized. Also, here the SCN output serves to coordinate hormone secretion with autonomic functions and behavior (Van Der Beek et al., 1993; de la Iglesia et al., 1995, 2003; Vida et al., 2010). The more integrative structures in the hypothalamus that receive SCN projections distribute this time-related information throughout the whole hypothalamus and modulate a whole variety of functions according to the time of day. The picture emerges of a hypothalamic organization that provides an optimal basis for the synchronization and adaptation of functions such as behavior, hormone secretion, and control of organs that are fundamental for the homeostasis of the body.

By its hypothalamic projections the SCN organizes the onset of our locomotor activity, which is initiated with the beginning of the light period. However, before we wake up, the SCN has already prepared us for this activity,

Fig. 15.4. Vasopressin (VP) (blue) and tyrosine hydroxylase (TH) (brown) immunoreactive neuronal elements in frontal sections of the human brainstem (case #40) at the level of the facial nerve (7n) (**A, B**) and at the level of the middle part of the inferior olive (**C–H**). (**A**) Boxed area in the raphe pontis nucleus (RPn) is presented by high magnification in **B**. (**C**) Boxed areas in the area postrema (AP), dorsal motor nucleus of vagus, dorsointermediate part (10DI), solitary nucleus, dorsolateral subnucleus (SolDL), A2 noradrenaline cells (A2), and A1 noradrenaline cells (A1) are presented at high magnification in **D–H**, respectively. Note the chain-like series of vasopressin (VP) (blue) immunopositive punctuate varicosities that appear attached to either TH (brown) immunopositive neuronal profiles and fibers in the AP, SolDL, A2, and A1 (**D, F, G, H**, respectively, arrowheads) or to nonstained neuronal profiles in the RPn and 10DI (**B, E**, respectively, thick arrows). Note the VP fibers entering the blood vessel wall (**B**, thin arrows) in oblique sections of the blood vessel (bv). Note also the very high density of the VP fibers innervating the TH neuronal elements in A2 (**G**, arrowheads). IOPr, inferior olive, principal nucleus. Scale bars ¼ 1 mm in **A, C** and 20 μm in **B, D–H**. (Reproduced from Goncharuk et al., 2011, Fig. 2, with permission.)

by raising our heart rate, blood pressure, and sympathetic/parasympathetic balance (Scheer et al., 2003, 2010; Hu et al., 2011; Shea et al., 2011). At the same time, the SCN increases our body temperature, and cortisol and glucose levels (Buijs and Kalsbeek, 2001), in order to prepare us for the coming activity and feeding period.

The SCN optimizes its efficacy by affecting our peripheral organs via the autonomic nervous system and via hormonal secretion. The synchronization of these two different output systems of the hypothalamus by the SCN is well illustrated by the way the SCN influences cortisol secretion. Ample data illustrate the rhythm in blood cortisol levels, while the evidence for a rhythm in adrenocorticotropic hormone (ACTH) is limited (Kaneko et al., 1981), indicating that the sensitivity of the adrenal cortex for ACTH fluctuates. Yet textbook knowledge indicates generally that variations in blood cortisol levels depend on variations in ACTH levels. An investigation of how the SCN may affect the release of this hormone of the adrenal cortex revealed that there are only very limited SCN projections directly to corticotrophin-releasing hormone (CRH) neurons in the PVN, and also the rhythm of ACTH in the circulation is minimal (Buijs and Van Eden, 2000). Nevertheless, cortisol rhythms are pronounced, which is explained by the influence of the SCN on the sensitivity of the adrenal cortex via preautonomic sympathetic neurons in the PVN and via sympathetic input to the adrenal (Buijs et al., 1999; Engeland and Arnhold, 2005). This is a perfect example of how the SCN sensitizes our organs for hormones via the ANS, while at the same time targeting the same organs via the secretion of circulating hormones. Other evidence for such a mechanism can be seen in studies that investigated the importance of the autonomic innervation of the testes for the secretion of testosterone (Selvage and Rivier, 2003). The autonomic nervous system is also the main output pathway that the SCN uses to influence the liver and other organs, such as the kidney (Kalsbeek et al., 2004; Agarwal, 2010; Mahoney et al., 2010). The way the SCN influences our peripheral organs may indicate how the amygdala and the prefrontal cortex, too, may affect our physiology. These brain structures also contain preautonomic neurons, have extensive hypothalamic connections, and may thus directly and indirectly influence the autonomic tone of the organs of the body. This explains both why electrical stimulation of the amygdala induces changes in corticosterone secretion and emotional conditions change the sensitivity of the adrenal for ACTH. Indeed, changes in the ACTH/cortisol ratio are widely reported under many different conditions in humans (O'Brien et al., 1996; Bornstein et al., 2008; Wirtz et al., 2010). The basis for these changes is not very well understood, but the multisynaptic access to ANS motor neurons by the amygdala and prefrontal cortex may provide an excellent explanation for these phenomena. Consequently this may explain why in certain clinical conditions, such as, e.g., depression, there is such a discrepancy between the levels of ACTH and cortisol (Gotthardt et al., 1995; Petrowski et al., 2012).

Clock genes in the periphery: part of the circadian system

Recent research is aimed at understanding why the clock elements of the SCN, the clock genes, are present in all cells in all parts of the body. The current view is that these clock genes take part in organizing fundamental processes that are largely organized around metabolism or cell division (Matsuo et al., 2003; Green et al., 2008; Ramsey et al., 2009; Asher and Schibler, 2011). Apart from neurons in the olfactory bulb and in the retina, which have an autonomous rhythm, all cells in the body derive their 24 hour rhythmicity directly or indirectly from the SCN. Consequently the rhythmic expression of clock genes and metabolic genes in the periphery is tightly coupled to the rhythm of the SCN. Many studies have focused on how this SCN-induced rhythm is transferred to peripheral organs and how the metabolism of peripheral cells and their metabolic and other functional genes are dependent on the rhythm of the clock genes in peripheral cells, for example in the liver (Asher et al., 2010; Feng et al., 2011). For example, in theory, the SCN could drive the peripheral oscillators in the adrenal gland via the rhythm in ACTH secretion or via the ANS. The continuation of adrenal clock genes in hypophysectomized animals indicates that the rhythm of the adrenal clock genes is independent of SCN signaling via the pituitary gland (Fahrenkrug, 2008). In addition, experiments have shown that light influences not only the release of corticosterone but also the expression of certain clock genes in the adrenal (Ishida et al., 2005; Cailotto et al., 2009). Moreover, in a model of hamsters that show split rhythms of locomotor activity in constant light, a split rhythm in the SCN coincided with a split rhythm of clock genes in the adrenal, which also evidences the contribution of the ANS to determining the rhythm of the clock genes in the adrenal (Mahoney et al., 2010). These observations raise the question of whether there is a functional relationship between the rhythmic expression of the clock genes in the adrenal and the rhythm in glucocorticoids. Transplantation studies in which clock mutant adrenals are transplanted into wild-type mice and vice versa show that a corticosterone rhythmicity can be induced without autonomic connection to the brain, even in the absence of a functional SCN pacemaker in the adrenal (Oster et al., 2006), suggesting additional possible control mechanisms such as temperature and metabolites.

AUTONOMIC BALANCE AND THE SUPRACHIASMATIC NUCLEUS

All these examples emphasize that multiple mechanisms are in place that can drive rhythms in the peripheral organs, and most of these are related to the signal of the master clock, the SCN: food intake, temperature, activity, and corticosterone secretion are all immediately dependent on the output of the SCN. As we have seen above, the rhythm of clock gene expression in the periphery is normally tightly coupled to that of the SCN. However, changes in food availability, temperature, and corticosterone will alter the rhythm of these clock genes in the periphery but not in the SCN (Balsalobre et al., 2000; Damiola et al., 2000). In such a condition, a disbalance is induced between the rhythms of clock gene expression in the liver and in the SCN; at present, the question is whether such a disbalance may in time result in disease. This hypothesis is inspired by data on human shift workers who are prone to disease, especially diseases related to the digestive and circulatory system (Karlsson et al., 2005; Suwazono et al., 2009). In addition, an animal model for shift work has recently provided experimental evidence that dissociation between the rhythm of the SCN and the rhythm in activity and food intake induces a discrepancy between the rhythm within the hypothalamus as compared with the rhythm in physiological parameters such as temperature and glucose that are normally driven by the SCN. In addition, this animal model showed that under desynchronized food intake conditions the animal develops several aspects of the metabolic syndrome (Salgado-Delgado et al., 2010a, b). This observation suggests that in such a disbalance between the signals of the SCN and behavior, especially with respect to food intake, a pathological situation develops. Since clock genes are so closely linked to metabolic genes (Asher et al., 2010; Ramsey and Bass, 2011), it seems probable that the dissociation of their normal rhythm from the rhythm of the SCN is the crux of the process. With respect to the functionality of the clock genes in the liver, our observations indicate that a balanced autonomic drive to the liver is essential for a rhythm in glucose secretion and a rhythm in metabolic liver gene expression. However, complete sympathetic and parasympathetic denervation of the liver did not induce dramatic changes in glucose rhythm, liver clock gene expression, or metabolic liver gene expression. In contrast, only sympathetic or only parasympathetic denervation of the liver completely disturbed glucose production without altering, for example, corticosterone rhythm (Cailotto et al., 2007). Strikingly, despite the loss in glucose rhythm, such partial denervation of the liver did not disturb the rhythm of clock gene expression in the liver. These results illustrate the importance of a balance of the autonomic input to the liver and emphasize the clinical hazards of a disbalance.

AUTONOMIC DISBALANCE MAY LEAD TO DISEASE

Recently we investigated whether a disbalance in the autonomic output to different organs could explain the fact that the growth of intra-abdominal fat is associated with type 2 diabetes . The results showed that the subcutaneous and intra-abdominal fat compartments receive input from different sympathetic and parasympathetic autonomic neurons (Kreier et al., 2002, 2005). This somatotopy not only persists to the level of the PVN but also extends to the SCN, offering a breathtaking insight into the complexity of the regulation of the autonomic output. Vagal input to fat results in a higher uptake of glucose, suggesting that an accumulation of abdominal fat reflects a higher vagal drive. Since the pancreas and liver receive autonomic input from the same neurons that provide input to intra-abdominal fat, this observation suggests that these organs all receive a higher parasympathetic tone, which is reflected, for example, by more abdominal fat and higher insulin levels (Kreier et al., 2003). Since subcutaneous fat receives input from other neurons than abdominal fat, these fat tissues can grow at a different rate, which evidences the differentiated autonomic output (Fig. 15.5).

CIRCADIAN DESYNCHRONIZATION MAY INDUCE DISEASE

Since it is possible to override the signals of the SCN at any time and thus to be active during the sleep period or sleep during the active period, the timing of the physiological conditions of our body might not seem essential at first glance. However, taking into consideration that *night shift workers*, who by definition violate the signals of the biological clock, have a much bigger chance of developing chronic diseases (Karlsson et al., 2005; Suwazono et al., 2009), we hypothesize that diseases featuring a disbalance of the ANS, such as hypertension or type 2 diabetes, might be caused by changes in brain areas that drive the autonomic output. In addition, such a disbalance may be found in aging and dementia, induced by the marked decrease in the number of vasopressin-containing SCN neurons that occurs in subjects between 80 and 100 years of age and in patients with Alzheimer's disease (Swaab et al., 1985). Also, on the mRNA level, vasopressin was lower in the SCN of Alzheimer patients (Wu et al., 2006a), while the CRH neurons are activated in this condition (Raadsheer et al., 1995). Subsequently, it was shown that a flattening

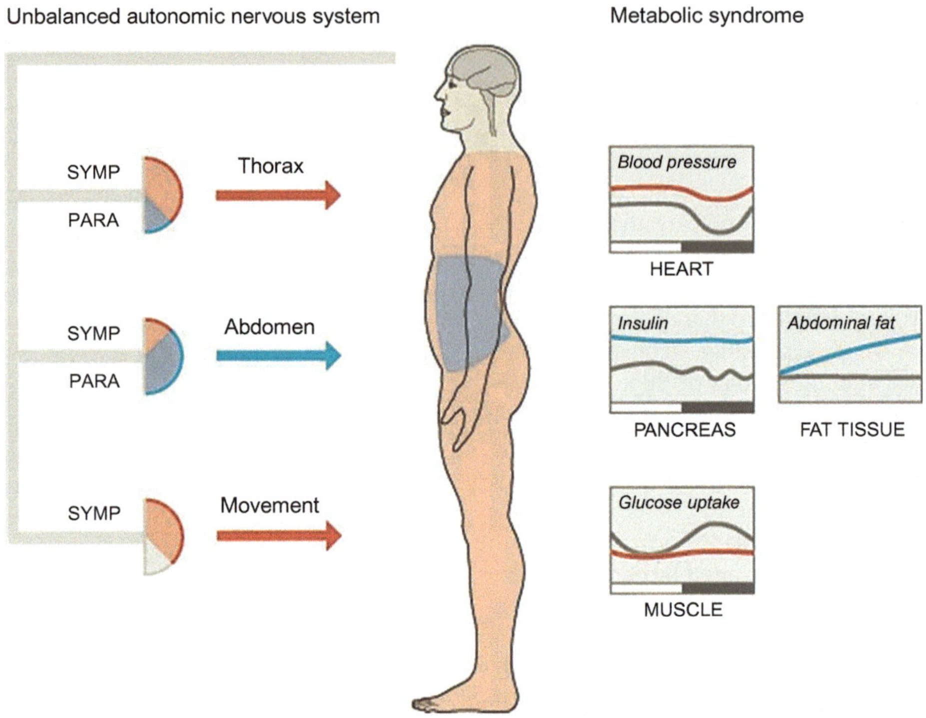

Fig. 15.5. Model of the metabolic syndrome caused by a central nervous deregulation. A disturbed output of the biological clock affects the selective balance of the autonomic nervous system (ANS) in different parts of the body. In the intra-abdominal compartment, the ANS is shifted in favor of the parasympathetic branch (blue), resulting in increased insulin secretion and growth of intra-abdominal fat tissue compared with normal values (gray). In contrast, in the thorax and movement compartment the sympathetic branch (red) prevails, leading to high blood pressure and impaired glucose uptake by the muscle compared with normal values. In this model, the symptoms of the metabolic syndrome are the result and not the cause of the disease. (Reproduced from Kreier et al., 2003, with permission.)

of the daily rhythm in vasopressin abundance is already observed in subjects >50 years of age (Fig. 15.2) (Hofman and Swaab, 1994). Because an age-related deterioration has been observed in many daily endocrine, physiological, and behavioral rhythms, it is tempting to hypothesize this to be related to the declining (vasopressinergic) output of the SCN. Indeed, recently, Harper et al. (2008) very elegantly demonstrated that the loss of SCN vasopressin-expressing neurons in the human postmortem brain significantly correlated with an increased deterioration of activity rhythms before death. Interestingly, the loss of vasopressin immunoreactivity during aging is possibly not the result of a loss of neurons, but of a decreased activity of these neurons. With the right stimulus, these neurons might be reactivated and thus rejuvenate the amplitude of circadian hormone and behavioral rhythms. Indeed, increased light intensities during the light period were able to prevent the aging-induced decrease in SCN vasopressin cell numbers in very old rats (Lucassen et al., 1995), as well as the deterioration of their behavioral rhythms (Witting et al., 1993). Moreover, in a recent double-blind, placebo-controlled randomized study with 189 elderly residents living in 12 different care facilities, long-term treatment with whole-day bright light significantly improved cognitive and non-cognitive symptoms of dementia (Riemersma-van der Lek et al., 2008).

A second example was found in the postmortem studies of human brains obtained after a long history of *hypertension* (Goncharuk et al., 2001). As was expected, the SCN was severely affected and showed decreased neuropeptide staining in the hypertensive brains. In agreement with the involvement of corticotropin releasing hormone (CRH) in the stress response, the CRH neurons in the PVN in the same hypertensive patients showed an increased activity in contrast to the SCN neurons (Goncharuk et al., 2002) (Fig. 15.6). The increased activity of CRH neurons in the PVN of hypertensive patients is probably not only responsible for an increased activity of the HPA axis, but also for an increased activity of the sympathetic branch of the ANS, which could very well be involved in the pathogenesis of hypertension. A major question that remains is whether these changes are a cause or a consequence of hypertension. A way to approach this question is to investigate whether a strengthening of the SCN signal can alleviate the

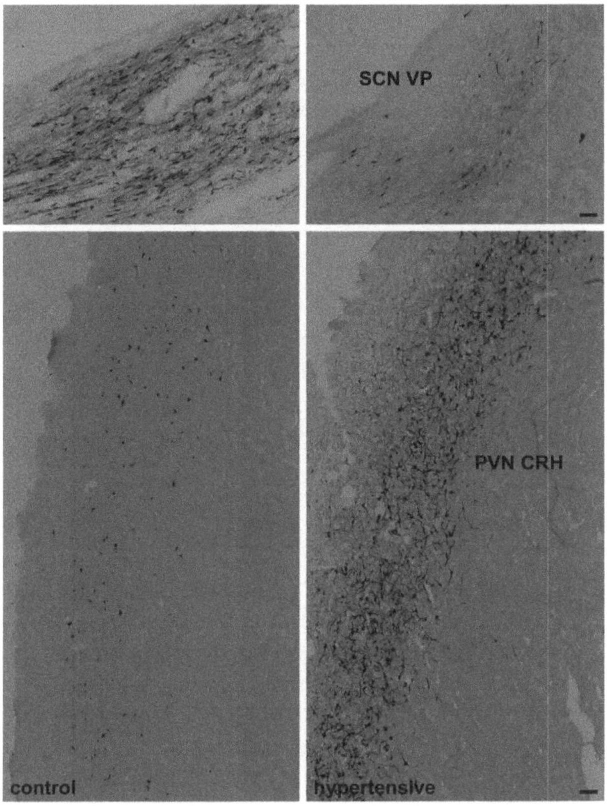

Fig. 15.6. Coronal sections of the human hypothalamus stained for vasopressin (VP) (upper images) (area of the suprachiasmatic nuclei (SCN)), or for corticotropin-releasing hormone (CRH) (lower images) (area of the paraventricular nuclei (PVN)). Sections on the left side of the figure are from the same control person, and those on the right from the same hypertensive person. It is evident that, although the vasopressin staining in the SCN of the hypertensive person is diminished compared to the control, CRH staining is enhanced. (With permission from Goncharuk et al., 2001, 2002.)

increased blood pressure. Accordingly, we conducted a randomized, double-blind, placebo-controlled, crossover study in which 16 men with untreated essential hypertension were treated with oral melatonin (2.5 mg daily, 1 hour before sleep) for 3 weeks. Repeated melatonin administration reduced ambulatory systolic and diastolic blood pressure by 6 and 4 mmHg, respectively (Scheer et al., 2004). Thus, as with the daily light treatment of the elderly, daily melatonin treatment, too, may help to improve circadian rhythms in behavior by an enhancement of biological clock functioning. This effect of melatonin does not seem to hold exclusively for this pineal hormone; corticosterone may also have such a feedback function on the SCN, probably via the serotonin system, while activity too has been shown to have a feedback role on the SCN (Buijs and Escobar, 2007; Malek et al., 2007; Van Oosterhout et al., 2012).

Changes in SCN vasopressin cell numbers and mRNA expression have also been observed in depressed patients. These changes are characterized by an increased number of vasopressin-immunoreactive neurons, a decreased amount of vasopressin mRNA, and a diminished amplitude of the daily fluctuation of vasopressin mRNA (Zhou et al., 2001). Also in depressed patients, a decreased release of SCN-derived vasopressin appears to go hand in hand with an increased activity of the HPA axis (Raadsheer et al., 1995). It has been proposed that the decreased activity of SCN vasopressin neurons in depressed patients is one of the deleterious consequences of the stress-induced increases in HPA activity. Indeed, increased circulating levels of glucocorticoids were shown to inhibit vasopressinergic activity in the SCN (Liu et al., 2006). On the other hand, polymorphisms of clock genes are risk factors for bipolar depression (Bao et al., 2008). Disturbed SCN function in depression may thus be related to both the cause and the effects of the depressive disorder. Light therapy in depressed patients may activate the SCN, directly inducing an increased synthesis and release of vasopressin that will inhibit the overactive CRH neurons (Bao et al., 2008).

We propose that the basis for pathologies such as type 2 diabetes might be a disbalance in communication between the periphery and the hypothalamus and SCN.

That the SCN might indeed be able to receive information about the physiology of our body has been shown in several recent studies. The SCN has reciprocal connections with the ARC and is sensitive to metabolic information (Yi et al., 2006, 2008). In addition, it has been long established that the SCN is also sensitive to information about locomotor activity via the lateral geniculate body (Biello et al., 1991). Projections from other circumventricular organs, OVLT, and the subfornical organ to the SCN are demonstrated and may signal similar or alternative types of circulating information to the SCN (Lind et al., 1982; Trudel and Bourque, 2010). When the feedback of metabolic information is out of phase with the signal provided from the SCN, for example, when an individual eats at the moment that the SCN signals the sleep phase, an aberrant physiological situation occurs that may result, in the long term, in the metabolic syndrome. It is possible that epigenetic alterations are involved or network changes in hypothalamic communication (Hu et al., 2012; Masri and Sassone-Corsi, 2012). This can only be understood if we consider the SCN not only as an autonomous clock but also as a structure that sets the level of physiological functions in different organs in a differential way. When the hypothalamus is challenged by nonhomeostatic conditions it reacts to that: we propose that the SCN, being the structure that sets the time for body functions, also receives via its extensive reciprocal connections with different hypothalamic structures this homeostatic information and will set the hypothalamic output according to the time of the day. Eating and/or activity at a time point that the SCN signals fasting or sleep disturbs, via its feedback, its rhythmic function and may lead to network failure within the SCN and/or within the hypothalamus (Hu et al., 2012). Support for this hypothesis is the recent observation that in patients with type 2 diabetes the infundibular nucleus shows an enhanced expression of NPY (Saderi et al., 2012), while in hypertensive individuals a strongly diminished activity of the SCN is evident (Goncharuk et al., 2001) and a higher activity in one of its target structures (Goncharuk et al., 2002). Such a mechanism may explain why individuals who sleep less than 5 hours per night have a greater chance of developing hypertension or diabetes than individuals who sleep longer, and why only with sleep times longer than 9 hours per day does the risk of developing these diseases increase (Gangwisch et al., 2006, 2007, 2008).

CIRCADIAN THERAPIES

The circadian system can be targeted at the level of the SCN by the input of bright light, by movement, or by administration of melatonin. Chronobiological treatments have proven to be effective in very diverse disorders, such as circadian sleep disorders, delirium, headache, obesity, and hypertension.

Bright light treatment has proven to be effective in dementia by improving circadian wake/sleep rhythms, performance, and mood (Riemersma-van der Lek et al., 2008). Bright light treatment also improves mood in seasonal affective disorder and major depression, and in bulimia nervosa it improves both the eating disorder and mood (Swaab, 2004).

Melatonin is rhythmically produced by the pineal gland under the control of the SCN. This hormonal output of the circadian system acts mainly via two G-coupled melatonin receptors, MT1 and MT2, on many brain areas, including the SCN (Wu et al., 2006b), and in this way is implicated in many physiological functions. In case of disorders of the circadian system, administration of exogenous melatonin can be effective with relatively minor side-effects if taken during the rest phase. Positive effects of melatonin administration have been reported in many disorders and in all phases of life, e.g., in circadian sleep disorders, ADHD, Rett syndrome, Angelman syndrome, septo-optic dysplasia, children with mental retardation and other developmental disabilities, craniopharyngioma, autism, Smith–Magenis syndrome, fragile-X syndrome, blindness, asthma, traumatic brain injury, whiplash syndrome, postoperative delirium, tardive dyskinesia, cluster headache, migraine, major depression, dementia, and glaucoma (Swaab, 2003; Riemersma-van der Lek et al., 2008; Braam et al., 2009; Hoebert et al., 2009; de Jonghe et al., 2011). A new development is the use of ramelteon and agomelatine, synthetic analogs of melatonin, with a longer half-life and a stronger affinity for MT1 and MT2 (Srinivasan et al., 2011; Serfaty and Raven, 2012).

Melatonin also influences arterial blood flow and decreases blood pressure. Repeated melatonin intake reduced systolic and diastolic blood pressure during sleep by 6 and 4 mmHg, respectively (Scheer et al., 2004).

The short-term positive effect of physical exercise on mood in depression (Krogh et al., 2011) may also act via its input on the SCN (Swaab, 2003).

CONCLUDING REMARKS

The hypothalamus plays a crucial role in the organization of homeostasis. It collects information from the environment and from the body in order to determine an optimal autonomic, hormonal, and behavioral response. The hypothalamus coordinates these three basic responses and fine-tunes them so that they support and amplify each other. The SCN, by putting a time stamp on each hypothalamic function, plays a crucial role in this organization. We propose that this crucial role of the hypothalamus in the homeostatic response is probably the reason why, e.g., in aging or depression, changes in the functioning of the biological clock, the SCN, might be the basis for the development of pathology. In addition, if by our behavior in our modern society we overrule the output of the biological clock by being active at moments that the SCN signals rest, in the long term this may lead to severe pathology such as diabetes or hypertension. This hypothesis is supported by observations of increased chances for the development of diabetes or hypertension in shift workers. This idea is also supported by experiments in animals in which an autonomic disbalance induced in the liver results in a loss of glucose rhythmicity and in obesity. Together, these data suggest that diseases of our modern society such as severe obesity, diabetes, and hypertension are induced by a process that is called internal desynchronization, implying a state whereby several rhythmic processes of our body are not in synchrony with each other anymore.

ACKNOWLEDGMENTS

The authors are grateful to Wilma Verweij for English text editing and acknowledge the support of Conacyt grant 79797 and PAPIIT IN 209711.

References

Agarwal R (2010). Regulation of circadian blood pressure: from mice to astronauts. Curr Opin Nephrol Hypertens 19: 51–58.

Arble DM, Bass J, Laposky AD et al. (2009). Circadian timing of food intake contributes to weight gain. Obesity (Silver Spring) 17: 2100–2102.

Asher G, Schibler U (2011). Crosstalk between components of circadian and metabolic cycles in mammals. Cell Metab 13: 125–137.

Asher G, Reinke H, Altmeyer M et al. (2010). Poly(ADP-ribose) polymerase 1 participates in the phase entrainment of circadian clocks to feeding. Cell 142: 943–953.

Atasoy D, Betley JN, Su HH et al. (2012). Deconstruction of a neural circuit for hunger. Nature 488: 172–177.

Ayas NT, White DP, Al Delaimy WK et al. (2003). A prospective study of self-reported sleep duration and incident diabetes in women. Diabetes Care 26: 380–384.

Balsalobre A, Brown SA, Marcacci L et al. (2000). Resetting of circadian time in peripheral tissues by glucocorticoid signaling. Science 289: 2344–2347.

Bao AM, Meynen G, Swaab DF (2008). The stress system in depression and neurodegeneration: focus on the human hypothalamus. Brain Res Rev 57: 531–553.

Biello SM, Harrington ME, Mason R (1991). Geniculo-hypothalamic tract lesions block chlordiazepoxide-induced phase advances in Syrian hamster. Brain Res 552: 47–52.

Borgers AJ, Romeijn N, van Someren E et al. (2011). Compression of the optic chiasm is associated with permanent shorter sleep duration in patients with pituitary insufficiency. Clin Endocrinol (Oxf) 75: 347–353.

Borgers AJ, Fliers E, Siljee JE et al. (2012). Arginine vasopressin immunoreactivity is decreased in the hypothalamic suprachiasmatic nucleus of subjects with suprasellar tumors. Brain Pathol http://dx.doi.org/10.1111/bpa.12016.

Bornstein SR, Engeland WC, Ehrhart-Bornstein M et al. (2008). Dissociation of ACTH and glucocorticoids. Trends Endocrinol Metab 19: 175–180.

Braam W, Smits MG, Didden R et al. (2009). Exogenous melatonin for sleep problems in individuals with intellectual disability: a meta-analysis. Dev Med Child Neurol 51: 340–349.

Brown TM, Wynne J, Piggins HD et al. (2011). Multiple hypothalamic cell populations encoding distinct visual information. J Physiol 589: 1173–1194.

Buijs RM, Escobar C (2007). Corticosterone and activity: the long arms of the clock talk back. Endocrinology 148: 5162–5164.

Buijs RM, Kalsbeek A (2001). Hypothalamic integration of central and peripheral clocks. Nat Rev Neurosci 2: 521–526.

Buijs RM, Van Eden CG (2000). The integration of stress by the hypothalamus, amygdala and prefrontal cortex: balance between the autonomic nervous system and the neuroendocrine system. Prog Brain Res 126: 117–132.

Buijs RM, Hou YX, Shinn S et al. (1994). Ultrastructural evidence for intra- and extranuclear projections of GABAergic neurons of the suprachiasmatic nucleus. J Comp Neurol 335: 42–54.

Buijs RM, Wortel J, Van Heerikhuize JJ et al. (1999). Anatomical and functional demonstration of a multisynaptic suprachiasmatic nucleus adrenal (cortex) pathway. Eur J Neurosci 11: 1535–1544.

Buijs RM, Chun SJ, Niijima A et al. (2001). Parasympathetic and sympathetic control of the pancreas: a role for the suprachiasmatic nucleus and other hypothalamic centers that are involved in the regulation of food intake. J Comp Neurol 431: 405–423.

Buijs RM, La Fleur SE, Wortel J et al. (2003). The suprachiasmatic nucleus balances sympathetic and parasympathetic output to peripheral organs through separate preautonomic neurons. J Comp Neurol 464: 36–48.

Cailotto C, Van Heijningen C, van der Vliet J et al. (2007). Daily rhythms in metabolic liver enzymes and plasma glucose require a balance in the autonomic output to the liver. Endocrinology 149: 1914–1925.

Cailotto C, Lei J, van der Vliet J et al. (2009). Effects of nocturnal light on (clock) gene expression in peripheral organs: a role for the autonomic innervation of the liver. PLoS One 4: e5650.

Canteras NS, Simerly RB, Swanson LW (1994). Organization of projections from the ventromedial nucleus of the hypothalamus: a *Phaseolus vulgaris*-leucoagglutinin study in the rat. J Comp Neurol 348: 41–79.

Condes-Lara M, Martinez-Lorenzana G, Rojas-Piloni G et al. (2007). Branched oxytocinergic innervations from the paraventricular hypothalamic nuclei to superficial layers in the spinal cord. Brain Res 1160: 20–29.

Contreras M, Ceric F, Torrealba F (2007). Inactivation of the interoceptive insula disrupts drug craving and malaise induced by lithium. Science 318: 655–658.

Cowley MA, Smart JL, Rubinstein M et al. (2001). Leptin activates anorexigenic POMC neurons through a neural network in the arcuate nucleus. Nature 411: 480–484.

Craig AD (2003). Interoception: the sense of the physiological condition of the body. Curr Opin Neurobiol 13: 500–505.

Czeisler CA, Shanahan TL, Klerman EIB et al. (1995). Suppression of melatonin secretion in some blind patients by exposure to bright light. N Engl J Med 332: 6–11.

Dai JP, Swaab DF, Buijs RM (1997). Distribution of vasopressin and vasoactive intestinal polypeptide (VIP) fibers in the human hypothalamus with special emphasis on suprachiasmatic nucleus efferent projections. J Comp Neurol 383: 397–414.

Dai J, Swaab DF, Buijs RM (1998a). Recovery of axonal transport in "dead neurons". Lancet 351: 499–500.

Dai JP, van der Vliet J, Swaab DF et al. (1998b). Postmortem tracing reveals the organization of hypothalamic projections of the suprachiasmatic nucleus in the human brain. J Comp Neurol 400: 87–102.

Damiola F, Le Minh N, Preitner N et al. (2000). Restricted feeding uncouples circadian oscillators in peripheral tissues from the central pacemaker in the suprachiasmatic nucleus. Genes Dev 14: 2950–2961.

de Jonghe A, van Munster BC, van Oosten HE et al. (2011). The effects of melatonin versus placebo on delirium in hip fracture patients: study protocol of a randomised, placebo-controlled, double blind trial. BMC Geriatr 11: 34.

de la Iglesia HO, Blaustein JD, Bittman EL (1995). The suprachiasmatic area in the female hamster projects to neurons containing estrogen receptors and GnRH. Neuroreport 6: 1715–1722.

de la Iglesia HO, Meyer J, Schwartz WJ (2003). Lateralization of circadian pacemaker output: activation of left- and right-sided luteinizing hormone-releasing hormone neurons involves a neural rather than a humoral pathway. J Neurosci 23: 7412–7414.

Deering J, Coote JH (2000). Paraventricular neurones elicit a volume expansion-like change of activity in sympathetic nerves to the heart and kidney in the rabbit. Exp Physiol 85: 177–186.

Dierickx K, Vandesande F (1977). Immunocytochemical localization of the vasopressinergic and the oxytocinergic neurons in the human hypothalamus. Cell Tissue Res 184: 15–27.

Engeland WC, Arnhold MM (2005). Neural circuitry in the regulation of adrenal corticosterone rhythmicity. Endocrine 28: 325–332.

Fahrenkrug J, Hannibal J, Georg B (2008). Diurnal rhythmicity of the canonical clock genes Per1, Per2 and Bmal1 in the rat adrenal gland is unaltered after hypophysectomy. J Neuroendocrinol 20: 323–329.

Feng D, Liu T, Sun Z et al. (2011). A circadian rhythm orchestrated by histone deacetylase 3 controls hepatic lipid metabolism. Science 331: 1315–1319.

Gangwisch JE, Boden-Albala B, Buijs RM et al. (2006). Short sleep duration as risk factor for hypertension: analysis of the NHANES I. Hypertension 47: 1–7.

Gangwisch JE, Heymsfield SB, Boden-Albala B et al. (2007). Sleep duration as a risk factor for diabetes incidence in a large US sample. Sleep 30: 1667–1673.

Gangwisch JE, Heymsfield SB, Boden-Albala B et al. (2008). Sleep duration associated with mortality in elderly, but not middle-aged, adults in a large US sample. Sleep 31: 1087–1096.

Gillette MU, Reppert SM (1987). The hypothalamic suprachiasmatic nuclei: circadian patterns of vasopressin secretion and neuronal activity in vitro. Brain Res Bull 19: 135–139.

Goldstone AP, Unmehopa UA, Bloom SR et al. (2002). Hypothalamic NPY and agouti-related protein are increased in human illness but not in Prader–Willi syndrome and other obese subjects. J Clin Endocrinol Metab 87: 927–937.

Goncharuk VD, Van Heerikhuize J, Dai JP et al. (2001). Neuropeptide changes in the suprachiasmatic nucleus in primary hypertension indicate functional impairment of the biological clock. J Comp Neurol 431: 320–330.

Goncharuk VD, Van Heerikhuize J, Swaab DF et al. (2002). Paraventricular nucleus of the human hypothalamus in primary hypertension: activation of corticotropin-releasing hormone neurons. J Comp Neurol 443: 321–331.

Goncharuk VD, Buijs RM, Jhamandas JH et al. (2011). Vasopressin (VP) and neuropeptide FF (NPFF) systems in the normal and hypertensive human brainstem. J Comp Neurol 519: 93–124.

Gotthardt U, Schweiger U, Fahrenberg J et al. (1995). Cortisol, ACTH, and cardiovascular response to a cognitive challenge paradigm in aging and depression. Am J Physiol Regul Integr Comp Physiol 37: R865–R873.

Green CB, Takahashi JS, Bass J (2008). The meter of metabolism. Cell 134: 728–742.

Harper DG, Stopa EG, Kuo-Leblanc V et al. (2008). Dorsomedial SCN neuronal subpopulations subserve different functions in human dementia. Brain 131: 1609–1617.

Hastings MH, Reddy AB, Maywood ES (2003). A clockwork web: circadian timing in brain and periphery, in health and disease. Nat Rev Neurosci 4: 649–661.

Hill J, Elmquist JK, Elias CF (2008). Hypothalamic pathways linking energy balance and reproduction. Am J Physiol Endocrinol Metab 294: E827–E832.

Hoebert M, van der Heijden KB, van Geijlswijk IM et al. (2009). Long-term follow-up of melatonin treatment in children with ADHD and chronic sleep onset insomnia. J Pineal Res 47: 1–7.

Hofman MA, Swaab DF (1993). Diurnal and seasonal rhythms of neuronal activity in the suprachiasmatic nucleus of humans. J Biol Rhythm 8: 283–295.

Hofman MA, Swaab DF (1994). Alterations in circadian rhythmicity of the vasopressin-producing neurons of the human suprachiasmatic nucleus (SCN) with aging. Brain Res 651: 134–142.

Hofman MA, Swaab DF (1995). Influence of aging on the seasonal rhythm of the vasopressin-expressing neurons in the human suprachiasmatic nucleus. Neurobiol Aging 16: 965–971.

Horvath TL (2005). The hardship of obesity: a soft-wired hypothalamus. Nat Neurosci 8: 561–565.

Hu K, Scheer FA, Laker M et al. (2011). Endogenous circadian rhythm in vasovagal response to head-up tilt. Circulation 123: 961–970.

Hu K, Meijer JH, Shea SA et al. (2012). Fractal patterns of neural activity exist within the suprachiasmatic nucleus and require extrinsic network interactions. PLoS One 7: e48927.

Ishida A, Mutoh T, Ueyama T et al. (2005). Light activates the adrenal gland: timing of gene expression and glucocorticoid release. Cell Metab 2: 297–307.

Kalsbeek A, la Fleur S, Van Heijningen C et al. (2004). Suprachiasmatic GABAergic inputs to the paraventricular nucleus control plasma glucose concentrations in the rat via sympathetic innervation of the liver. J Neurosci 24: 7604–7613.

Kampe J, Tschop MH, Hollis JH et al. (2009). An anatomic basis for the communication of hypothalamic, cortical and mesolimbic circuitry in the regulation of energy balance. Eur J Neurosci 30: 415–430.

Kaneko M, Kaneko K, Shinsako J et al. (1981). Adrenal sensitivity to adrenocorticotropin varies diurnally. Endocrinology 109: 70–75.

Karlsson B, Alfredsson L, Knutsson A et al. (2005). Total mortality and cause-specific mortality of Swedish shift- and dayworkers in the pulp and paper industry in 1952–2001. Scand J Work Environ Health 31: 30–35.

Kimura A, Ishida Y, Hayashi T et al. (2011). Estimating time of death based on the biological clock. Int J Legal Med 125: 385–391.

Knutsson A (2003). Health disorders of shift workers. Occup Med (Lond) 53: 103–108.

Kreier F, Fliers E, Voshol PJ et al. (2002). Selective parasympathetic innervation of subcutaneous and intra-abdominal fat – functional implications. J Clin Invest 110: 1243–1250.

Kreier F, Yilmaz A, Kalsbeek A et al. (2003). Hypothesis: shifting the equilibrium from activity to food leads to autonomic unbalance and the metabolic syndrome. Diabetes 52: 2652–2656.

Kreier F, Kap YS, Mettenleiter TC et al. (2005). Tracing from fat tissue, liver and pancreas: a neuroanatomical framework for the role of the brain in type 2 diabetes. Endocrinology 147: 1140–1147.

Krogh J, Nordentoft M, Sterne JA et al. (2011). The effect of exercise in clinically depressed adults: systematic review and meta-analysis of randomized controlled trials. J Clin Psychiatry 72: 529–538.

La Fleur SE, Kalsbeek A, Wortel J et al. (2001). A daily rhythm in glucose tolerance: a role for the suprachiasmatic nucleus. Diabetes 50: 1237–1243.

Lambe EK, Olausson P, Horst NK et al. (2005). Hypocretin and nicotine excite the same thalamocortical synapses in prefrontal cortex: correlation with improved attention in rat. J Neurosci 25: 5225–5229.

Leibowitz SF, Wortley KE (2004). Hypothalamic control of energy balance: different peptides, different functions. Peptides 25: 473–504.

Li Y, Li S, Sui N et al. (2009). Orexin-A acts on the paraventricular nucleus of the midline thalamus to inhibit locomotor activity in rats. Pharmacol Biochem Behav 93: 506–514.

Lind RW, Van Hoesen GW, Johnson AK (1982). An HRP study of the connections of the subfornical organ of the rat. J Comp Neurol 210: 265–277.

Liu S, Chen XM, Yoda T et al. (2002). Involvement of the suprachiasmatic nucleus in body temperature modulation by food deprivation in rats. Brain Res 929: 26–36.

Liu RY, Unmehopa UA, Zhou JN et al. (2006). Glucocorticoids suppress vasopressin gene expression in human suprachiasmatic nucleus. J Steroid Biochem Mol Biol 98: 248–253.

Lucassen PJ, Hofman MA, Swaab DF (1995). Increased light intensity prevents the age related loss of vasopressin-expressing neurons in the rat suprachiasmatic nucleus. Brain Res 693: 261–266.

Mahoney CE, Brewer D, Costello MK et al. (2010). Lateralization of the central circadian pacemaker output: a test of neural control of peripheral oscillator phase. Am J Physiol Regul Integr Comp Physiol 299: R751–R761.

Malek ZS, Sage D, Pevet P et al. (2007). Daily rhythm of tryptophan hydroxylase-2 messenger ribonucleic acid within raphe neurons is induced by corticoid daily surge and modulated by enhanced locomotor activity. Endocrinology 148: 5165–5172.

Masri S, Sassone-Corsi P (2012). The circadian clock: a framework linking metabolism, epigenetics and neuronal function. Nat Rev Neurosci 14: 69–75.

Matsuo T, Yamaguchi S, Mitsui S et al. (2003). Control mechanism of the circadian clock for timing of cell division in vivo. Science 302: 255–259.

Matyas F, Sreenivasan V, Marbach F et al. (2010). Motor control by sensory cortex. Science 330: 1240–1243.

Miselis RR (1981). The efferent projections of the subfornical organ of the rat: a circumventricular organ within a neural network subserving water balance. Brain Res 230: 1–23.

Moore RY, Eichler VB (1972). Loss of a circadian adrenal corticosterone rhythm following suprachiasmatic lesions in the rat. Brain Res 42: 201–206.

Morrison SF, Nakamura K (2011). Central neural pathways for thermoregulation. Front Biosci 16: 74–104.

Nakamura K (2011). Central circuitries for body temperature regulation and fever. Am J Physiol Regul Integr Comp Physiol 301: R1207–R1228.

Nakamura K, Morrison SF (2011). Central efferent pathways for cold-defensive and febrile shivering. J Physiol 589: 3641–3658.

O'Brien JT, Ames D, Schweitzer I et al. (1996). Enhanced adrenal sensitivity to adrenocorticotrophic hormone (ACTH) is evidence of HPA axis hyperactivity in Alzheimer's disease. Psychol Med 26: 7–14.

Obici S, Feng Z, Karkanias G et al. (2002a). Decreasing hypothalamic insulin receptors causes hyperphagia and insulin resistance in rats. Nat Neurosci 5: 566–572.

Obici S, Zhang BB, Karkanias G et al. (2002b). Hypothalamic insulin signaling is required for inhibition of glucose production. Nat Med 8: 1376–1382.

Obici S, Feng Z, Arduini A et al. (2003). Inhibition of hypothalamic carnitine palmitoyltransferase-1 decreases food intake and glucose production. Nat Med 9: 756–761.

Oster H, Damerow S, Kiessling S et al. (2006). The circadian rhythm of glucocorticoids is regulated by a gating mechanism residing in the adrenal cortical clock. Cell Metab 4: 163–173.

Parton LE, Ye CP, Coppari R et al. (2007). Glucose sensing by POMC neurons regulates glucose homeostasis and is impaired in obesity. Nature 449: 228–232.

Petrowski K, Wintermann GB, Kirschbaum C et al. (2012). Dissociation between ACTH and cortisol response in DEX-CRH test in patients with panic disorder. Psychoneuroendocrinology 37: 1199–1208.

Price CJ, Hoyda TD, Ferguson AV (2008a). The area postrema: a brain monitor and integrator of systemic autonomic state. Neuroscientist 14: 182–194.

Raadsheer FC, Van Heerikhuize JJ, Lucassen PJ et al. (1995). Corticotropin-releasing hormone mRNA levels in the paraventricular nucleus of patients with Alzheimer's disease and depression. Am J Psychiatry 152: 1372–1376.

Ramsey KM, Bass J (2011). Circadian clocks in fuel harvesting and energy homeostasis. Cold Spring Harb Symp Quant Biol 76: 63–72.

Ramsey KM, Yoshino J, Brace CS et al. (2009). Circadian clock feedback cycle through NAMPT-mediated NAD + biosynthesis. Science 324: 651–654.

Reppert SM, Schwartz WJ, Uhl GR (1987). Arginine vasopressin: a novel peptide rhythm in cerebrospinal fluid. Trends Neurosci 10: 76–80.

Riemersma-van der Lek RF, Swaab DF, Twisk J et al. (2008). Effect of bright light and melatonin on cognitive and noncognitive function in elderly residents of group care facilities: a randomized controlled trial. JAMA 299: 2642–2655.

Risold PY, Thompson RH, Swanson LW (1997). The structural organization of connections between hypothalamus and cerebral cortex. Brain Res Brain Res Rev 24: 197–254.

Rodrigues SM, Le Doux JE, Sapolsky RM (2009). The influence of stress hormones on fear circuitry. Annu Rev Neurosci 32: 289–313.

Saderi N, Salgado-Delgado R, Avendano-Pradel R et al. (2012). NPY and VGF immunoreactivity increased in the arcuate nucleus, but decreased in the nucleus of the tractus solitarius, of type-II diabetic patients. PLoS One 7: e40070.

Salgado-Delgado R, Angeles-Castellanos M, Saderi N et al. (2010a). Food intake during the normal activity phase prevents obesity and circadian desynchrony in a rat model of night work. Endocrinology 151: 1019–1029.

Salgado-Delgado R, Nadia S, Angeles-Castellanos M et al. (2010b). In a rat model of night work, activity during the normal resting phase produces desynchrony in the hypothalamus. J Biol Rhythms 25: 421–431.

Scheer FA, Ter Horst GJ, van der Vliet J et al. (2001). Physiological and anatomic evidence for regulation of the heart by suprachiasmatic nucleus in rats. Am J Physiol 280: H1391–H1399.

Scheer FA, Kalsbeek A, Buijs RM (2003). Cardiovascular control by the suprachiasmatic nucleus: neural and neuroendocrine mechanisms in human and rat. Biol Chem 384: 697–709.

Scheer FA, van Montfrans GA, van Someren EJ et al. (2004). Daily nighttime melatonin reduces blood pressure in male patients with essential hypertension. Hypertension 43: 192–197.

Scheer FA, Pirovano C, van Someren EJ et al. (2005). Environmental light and suprachiasmatic nucleus interact in the regulation of body temperature. Neuroscience 132: 465–477.

Scheer FA, Hilton MF, Mantzoros CS et al. (2009). Adverse metabolic and cardiovascular consequences of circadian misalignment. Proc Natl Acad Sci U S A 106: 4453–4458.

Scheer FA, Hu K, Evoniuk H et al. (2010). Impact of the human circadian system, exercise, and their interaction on cardiovascular function. Proc Natl Acad Sci U S A 107: 20541–20546.

Selvage DJ, Rivier C (2003). Importance of the paraventricular nucleus of the hypothalamus as a component of a neural pathway between the brain and the testes that modulates testosterone secretion independently of the pituitary. Endocrinology 144: 594–598.

Serfaty M, Raven PW (2012). Novel melatonin-based treatments for major depression. Lancet 379: 217–219.

Shea SA, Hilton MF, Hu K et al. (2011). Existence of an endogenous circadian blood pressure rhythm in humans that peaks in the evening. Circ Res 108: 980–984.

Smith PM, Ferguson AV (2010). Circulating signals as critical regulators of autonomic state – central roles for the subfornical organ. Am J Physiol Regul Integr Comp Physiol 299: R405–R415.

Sollars PJ, Smeraski CA, Kaufman JD et al. (2003). Melanopsin and non-melanopsin expressing retinal ganglion cells innervate the hypothalamic suprachiasmatic nucleus. Vis Neurosci 20: 601–610.

Spanswick D, Smith MA, Groppi VE et al. (1997). Leptin inhibits hypothalamic neurons by activation of ATP- sensitive potassium channels. Nature 390: 521–525.

Spanswick D, Smith MA, Mirshamsi S et al. (2000). Insulin activates ATP-sensitive K + channels in hypothalamic neurons of lean, but not obese rats. Nat Neurosci 3: 757–758.

Srinivasan V, Brzezinski A, Pandi-Perumal SR et al. (2011). Melatonin agonists in primary insomnia and depression-associated insomnia: are they superior to sedative-hypnotics?. Prog Neuropsychopharmacol Biol Psychiatry 35: 913–923.

Steiner AA, Chakravarty S, Robbins JR et al. (2005). Thermoregulatory responses of rats to conventional preparations of lipopolysaccharide are caused by lipopolysaccharide per se – not by lipoprotein contaminants. Am J Physiol Regul Integr Comp Physiol 289: R348–R352.

Stephan FK, Zucker I (1972). Circadian rhythms in drinking behavior and locomotor activity of rats are eliminated by hypothalamic lesions. Proc Natl Acad Sci U S A 69: 1583–1586.

Stopa EG, King JC, Lydic R et al. (1984). Human brain contains vasopressin and vasoactive intestinal polypeptide neuronal subpopulations in the suprachiasmatic region. Brain Res 297: 159–163.

Suwazono Y, Dochi M, Oishi M et al. (2009). Shiftwork and impaired glucose metabolism: a 14-year cohort study on 7104 male workers. Chronobiol Int 26: 926–941.

Swaab DF (2003). The Human Hypothalamus. Basic and Clinical Aspects. Part I. Nuclei of the Hypothalamus Handbook of Clinical Neurology, vol. 79: Elsevier, Amsterdam, p.vii.

Swaab DF (2004). Neuropeptides in hypothalamic neuronal disorders. Int Rev Cytol 240: 305–375.

Swaab DF, Pool CW, Nijveldt F (1975). Immunofluorescence of vasopressin and oxytocin in the rat hypothalamo-neurohypophyseal system. J Neural Transm 36: 195–215.

Swaab DF, Fliers E, Partiman TS (1985). The suprachiasmatic nucleus of the human brain in relation to sex, age and senile dementia. Brain Res 342: 37–44.

Swanson LW (1987). The hypothalamus. In: A Björklund, T Hökfelt, LW Swanson (Eds.), Integrated Systems of the CNS. Part I. Hypothalamus, Hippocampus, Amygdala, Retina, 5th edn. Elsevier, Amsterdam, pp. 1–124.

Swanson LW (2000). Cerebral hemisphere regulation of motivated behavior. Brain Res 886: 113–164.

Swanson LW, Sanchez-Watts G, Watts AG (2005). Comparison of melanin-concentrating hormone and hypocretin/orexin mRNA expression patterns in a new parceling scheme of the lateral hypothalamic zone. Neurosci Lett 387: 80–84.

Tentolouris N, Tsigos C, Perea D et al. (2003). Differential effects of high-fat and high-carbohydrate isoenergetic meals on cardiac autonomic nervous system activity in lean and obese women. Metabolism 52: 1426–1432.

Thompson RH, Swanson LW (2003). Structural characterization of a hypothalamic visceromotor pattern generator network. Brain Res Brain Res Rev 41: 153–202.

Thompson RH, Canteras NS, Swanson LW (1996). Organization of projections from the dorsomedial nucleus of the hypothalamus: a PHA-L study in the rat. J Comp Neurol 376: 143–173.

Tjen-A-Looi S, Bonham A, Longhurst J (1997). Interactions between sympathetic and vagal cardiac afferents in nucleus tractus solitarii. Am J Physiol Heart Circ Physiol 41: H2843–H2851.

Trudel E, Bourque CW (2010). Central clock excites vasopressin neurons by waking osmosensory afferents during late sleep. Nat Neurosci 13: 467–474.

Van Den Pol AN (1982). The magnocellular and parvocellular paraventricular nucleus of rat: intrinsic organization. J Comp Neurol 206: 317–345.

van den Top M, Lee K, Whyment AD et al. (2004). Orexigen-sensitive NPY/AgRP pacemaker neurons in the hypothalamic arcuate nucleus. Nat Neurosci 7: 493–494.

Van Der Beek EM, Wiegant VM, Van Der Donk HA et al. (1993). Lesions of the suprachiasmatic nucleus indicate the presence of a direct vasoactive intestinal polypeptide-containing projection to gonadotrophin-releasing hormone neurons in the female rat. J Neuroendocrinol 5: 137–144.

van Oosterhout F, Lucassen EA, Houben T et al. (2012). Amplitude of the SCN clock enhanced by the behavioral activity rhythm. PLoS One 7: e39693.

Vertes RP, Hoover WB (2008). Projections of the paraventricular and paratenial nuclei of the dorsal midline thalamus in the rat. J Comp Neurol 508: 212–237.

Vida B, Deli L, Hrabovszky E et al. (2010). Evidence for suprachiasmatic vasopressin neurons innervating kisspeptin neurons in the rostral periventricular area of the mouse brain: regulation by oestrogen. J Neuroendocrinol 22: 1032–1039.

Watanabe M, Arima H, Fukushima K et al. (2008). Direct and indirect modulation of neuropeptide Y gene expression in response to hypoglycemia in rat arcuate nucleus. FEBS Lett 582: 3632–3638.

Watts AG, Swanson LW, Sanchez-Watts G (1987). Efferent projections of the suprachiasmatic nucleus: I. Studies using anterograde transport of *Phaseolus vulgaris* leucoagglutinin in the rat. J Comp Neurol 258: 204–229.

Wirtz PH, Siegrist J, Schuhmacher A et al. (2010). Higher overcommitment to work is associated with higher plasma cortisol but not ACTH responses in the combined dexamethasone/CRH test in apparently healthy men and women. Psychoneuroendocrinology 35: 536–543.

Witting W, Mirmiran M, Bos NP et al. (1993). Effect of light intensity on diurnal sleep-wake distribution in young and old rats. Brain Res Bull 30: 157–162.

Wu YH, Fischer DF, Kalsbeek A et al. (2006a). Pineal clock gene oscillation is disturbed in Alzheimer's disease, due to functional disconnection from the "master clock". FASEB J 20: 1874–1876.

Wu YH, Zhou JN, Balesar R et al. (2006b). Distribution of MT1 melatonin receptor immunoreactivity in the human hypothalamus and pituitary gland: colocalization of MT1 with vasopressin, oxytocin, and corticotropin-releasing hormone. J Comp Neurol 499: 897–910.

Yi CX, van der Vliet J, Dai J et al. (2006). Ventromedial arcuate nucleus communicates peripheral metabolic information to the suprachiasmatic nucleus. Endocrinology 147: 283–294.

Yi CX, Challet E, Pevet P et al. (2008). A circulating ghrelin mimetic attenuates light-induced phase delay of mice and light-induced Fos expression in the suprachiasmatic nucleus of rats. Eur J Neurosci 27: 1965–1972.

Yi CX, Foppen E, Abplanalp W et al. (2012). Glucocorticoid signaling in the arcuate nucleus modulates hepatic insulin sensitivity. Diabetes 61: 339–345.

Zhou JN, Riemersma RF, Unmehopa UA et al. (2001). Alterations in arginine vasopressin neurons in the suprachiasmatic nucleus in depression. Arch Gen Psychiatry 58: 655–662.

Handbook of Clinical Neurology, Vol. 117 (3rd series)
Autonomic Nervous System
R.M. Buijs and D.F. Swaab, Editors

Chapter 16

Autonomic nervous system control of the cerebral circulation

PETER J. GOADSBY*
Department of Neurology, University of California, San Francisco, CA, USA

INTRODUCTION

The classic view of the cerebral circulation has been that blood flow and cerebral metabolism are tightly coupled under the influence of substances, such as H^{+}, adenosine, nitric oxide and K^{+}, which ensure a rapid and matched supply of blood when required without neural influence (Purves, 1972; Kuschinsky, 1989). Most recently, and particularly driven by the use of cerebral blood flow measurements in functional brain imaging (Frackowiak and Friston, 1994), it has become clear that astrocytes play a role in modulating functionally associated changes in cerebral blood flow (Attwell et al., 2010). The concept of neural control or neurogenically mediated changes in cerebral blood flow is relatively newer (Edvinsson and Krause, 2002), notwithstanding the fact that the observation that nerves exist on the vessels dates to Thomas Willis (1664).

The innervation of the cerebral circulation can be divided conveniently into intrinsic and extrinsic systems. The intrinsic systems originate within the central nervous system and, without exiting the brain, innervate cerebral parenchymal vessels (Hartman et al., 1980). They are not strictly autonomic in the classic sense of the word, although as their function is better understood such a distinction may eventually not be useful. They have been reviewed extensively elsewhere (Goadsby and Edvinsson, 2002). The extrinsic systems are those that commence within the central nervous system but exit and have an extra-axial synapse before innervating the cerebral vessels. The autonomic innervation of the cerebral circulation is a subset of these extrinsic nerves (Table 16.1), the sympathetic and parasympathetic nerves. Since it is relevant clinically the chapter will highlight areas where classic cerebrovascular physiology is impacted by these nerves, such as in cerebral autoregulation, and in clinical areas where the cranial autonomic system is clearly active. The sensory innervation (Table 16.1) from the trigeminovascular system is crucial in understanding the effects of the cranial autonomic system, as it acts in a feedback role, although this is outside the scope of this chapter (May and Goadsby, 1999).

PARASYMPATHETIC INFLUENCES ON THE CEREBRAL CIRCULATION

The parasympathetic innervation of the cerebral circulation represents the most powerful of the neural vasodilator influences upon that bed. The parasympathetic system is capable of altering brain blood flow independent of the prevailing metabolic demand and perfusion pressure. It is a potential reserve system that is well characterized in experimental animals but lacks detailed analysis in humans.

Anatomy

The parasympathetic system is that system arising from the superior salivatory nucleus and passing out of the brain in the facial (VIIth cranial) nerve, distributing fibers through the pterygopalatine (sphenopalatine) and otic ganglia and the carotid miniganglia to dilate vessels (Gibbins et al., 1983, 1984), almost certainly by way of a peptidergic transmitter. The terms "pterygopalatine" and "sphenopalatine" imply the same structure, with the former being the correct term in humans because of its relationship to the pterygopalatine fossa (Warwick and Williams, 1973). Pharmacologically, the system is characterized by the presence of one or more substances, such as acetylcholine, vasoactive intestinal polypeptide (VIP), and peptide histidine methionine (isoleucine in the rat, thus PHI).

Origin

Chorobski and Penfield (1932) were the first to describe in detail the anatomy of the facial nerve dilator pathway

*Correspondence to: Professor P.J. Goadsby, UCSF Headache Center, 1701 Divisadero Street, San Francisco CA 94115, USA. E-mail: pgoadsby@headache.ucsf.edu

Table 16.1

Extrinsic innervation of the cerebral circulation

System	Ganglion	Transmitter
	Autonomic innervation	
Parasympathetic	Pterygopalatine* Otic Carotid mini	Vasoactive intestinal polypeptide (VIP) Peptide histidine methionine (PHM)† Pituitary adenylate cyclase-activating peptide (PACAP) Helodermin Helospectin I and II Nitric oxide
Sympathetic	Superior cervical	Noradrenaline Neuropeptide Y (NPY)
	Sensory innervation	
Trigeminal	Trigeminal	Calcitonin gene-related peptide (CGRP) Substance P Neurokinin A Amylin Cholecystokinin-8

*Sphenopalatine in experimental animals.
†Peptide histidine isoleucine (PHI) in experimental animals.

that runs from the medulla via the greater superficial petrosal (GSP) (Cobb and Finesinger, 1932) to the pial vessels of the cat and monkey. McNaughton (1966) confirmed in humans that the GSP innervated the carotids although the density is less than that seen in the monkey (Fang, 1961). Vasquez and Purves (1979) showed that at least some GSP axons directly innervate the carotid plexus in rat and rabbit although the number with direct connections to the carotid artery are small (Ruskell, 1970). Studies using electron microscopy have also shown that fibers from the GSP innervate choroidal vessels via the sphenopalatine ganglion (Ruskell, 1971). Walters and colleagues (1986) traced the fibers from the GSP to the carotid plexus directly and to the carotid via the sphenopalatine and otic ganglia, thus clearly establishing the anatomical course of the GSP cerebral dilator pathway. The course that this innervation takes thereafter is via the ethmoidal nerve in the rat (Suzuki et al., 1988) and this distribution is similar in species that have been studied (Hardebo et al., 1991; Suzuki and Hardebo, 1993). Furthermore, using the transganglionic tracer pseudorabies virus it has been established that the central nucleus for these fibers is in the superior salivatory nucleus of the pons (Spencer et al., 1990).

Initially, studies examining the parasympathetic innervation of the cerebral vessels defined them as cholinergic. This led to considerable confusion as, although these nerves have cholinergic markers useful for anatomical studies, there was no apparent vasomotor function for the acetylcholine putatively stored in these nerves. It was thus concluded that the nerves had no physiological effect when, in fact, the data, as they were, only demonstrated that acetylcholine plays no immediate role in the effect of activation of parasympathetic nerves. Biochemical (Florence and Bevan, 1979) and histochemical (Bevan et al., 1982) evidence of choline acetyltransferase (ChAT) activity and the presence of a high affinity choline uptake system on the vessels (Duckles, 1981) have also provided markers for the parasympathetic system. Ultrastructural studies have noted a small number of agranular vesicles that store nonadrenergic transmitters (Nielsen et al., 1973). The innervation with cholinergic nerves is regionally variable anterior to posterior and the concentration of nerves along any vessel varies, with the densest innervation being at branching points, particularly in the pial vessels. In the cat and the rat, where detailed studies are available using acetylcholinesterase as a marker, the densest innervation is that of the anterior vessels (Lavrentieva et al., 1968). Notable variations in ChAT activity are seen between species, with the middle cerebral artery density being less than the basilar in the cat, and other variables such as age being recognized.

Transmitters and modulators

The parasympathetic system contains a number of transmitter or neuromodulator substances which are often colocalized in the same neuron. All the substances known to exist in the parasympathetic nerves are vasodilator (Table 16.2) but their precise role and, in particular, relationship to one another remains unclear. Whether these anatomical subgroups with one, two, three, or even four colocalized transmitters have functional correlates remains to be established.

Acetycholine

The cerebral vessels in isolation bind tritiated cholinergic ligands and thus various muscarinic receptors have been shown to exist on them. Acetylcholine dilates most of the cerebral vessels either *in vitro* or *in vivo* by an atropine-sensitive receptor, an effect that is mediated via the release of an endothelium-derived relaxing factor, nitric oxide. Local application of a cholinomimetic (carbachol) into the cerebrospinal fluid causes pial vessel dilatation, which is again antagonized by atropine (Kuschinsky and Wahl, 1973). Parenteral administration of acetylcholine increases cerebral blood flow, an effect again inhibited by atropine (Heistad et al., 1980). Curiously, then, it will be seen below that nicotinic antagonists are completely ineffective when tested during parasympathetic stimulation. The role of acetylcholine in these nerves remains unresolved.

Vasoactive intestinal polypeptide

The characterization of neuropeptides in the 1980s and 1990s led to a substantial re-evaluation of virtually all neuronal systems with respect to transmitter content. Potent vasodilator peptides were characterized, and identified in the parasympathetic nerves of the cranium, particularly vasoactive intestinal polypeptide (VIP) and, coexistent with it, peptide histidine methionine (PHM). VIP is a 28-amino-acid basic polypeptide that was first isolated from the porcine duodenum (Said and Mutt, 1970, 1972; Mutt and Said, 1974). It belongs to a structural superfamily of peptides along with glucagon, secretin, and gastrin inhibitory peptide. The family is characterized by helodermin/helospectin-like peptides that are distributed in the central nervous system and in endocrine cells, such as the C cells of the thyroid. It has been shown that each of helodermin and the helospectins I and II are vasodilator in the cerebral circulation. There are as yet no data that address the interrelationship of these transmitters, although they can have added effects.

It has been established clearly that the large cerebral vessels and cortical pial vessels have a rich VIPergic innervation that can be immunohistochemically identified and measured by radioimmunoassay (Edvinsson and Ekman, 1984). Indeed, using ultrastructural techniques it can be seen that vasoactive intestinal polypeptide is found in large, dense-core neuronal vesicles in perivascular nerve terminals on the vessels. Two important features characterize this innervation: first, it is predominantly in the anterior segments of the circle of Willis (Matsuyama et al., 1983), and, secondly, the fibers may be seen to follow the vessels and penetrate into the parenchyma. However, it has been shown that VIP-immunoreactive nerves that innervate the pial vessels may arise, at least in part, from intracortical neurons. The pattern of innervation of the vessels has further been characterized as having a spiral distribution with respect to the lumen and, importantly, this innervation is seen in human vessels. The origin of the nerves is essentially as it is for the cholinergic system (Hara et al., 1985; Suzuki et al., 1988).

Table 16.2

Transmitters of the parasympathetic system

Level	Structure	Transmitter	Antagonist
Brainstem (pons)	Superior salivatory nucleus	Glutamate	?
course	Facial nerve†		
Peripheral ganglion	Pterygopalatine, otic, carotid minigangIia	Acetylcholine	Hexamethonium-sensitive nicotinic receptor
End-organ	Cerebral blood vessels	VIP	None available
		PHI	
		Nitric oxide	Nitric oxide synthase inhibitors
		Helospectin-related*	

*Helodermin, Helospectin I and II.
†The fibers traverse the facial nerve, passing through but not synapsing in the geniculate ganglion and emerging as the greater superficial petrosal nerve.
VIP, vasoactive intestinal polypeptide; PHI, peptide histidine isoleucine (methionine); ?, unknown (see also Table 16.1).

Several peptides with sequence homology to VIP have also been found in parts of the parasympathetic cranial dilator system. Pituitary adenylate cyclase activating polypeptide (PACAP) is a vasoactive peptide with a 68% homology to VIP (Arimura, 1992). PACAP-like immunoreactivity is to be found on the cerebral vessels and PACAP dilates cerebral vessels *in vitro* and *in vivo* (Uddman et al., 1993; Seki et al., 1995). It has similarly been shown that each of helodermin and helospectin I and II are to be found on cranial vessels and are vasodilators in the cerebral circulation (Uddman et al., 1999).

Peptide histidine methionine (isoleucine)

The third major marker for the parasympathetic system is peptide histidine methionine (PHM). This is cleaved from the same prepropeptide as VIP (Itoh et al., 1983) and there is at least a 50% sequence homology (Tatemoto and Mutt, 1981). PHM and PHI (peptide histidine isoleucine) are almost identical, with a difference in two amino acid residues (92, lysine for arginine; 107, methionine for isoleucine), with the latter being the rodent peptide. Immunohistochemical studies have confirmed the existence of PHI(M)-like immunoreactivity on cerebral vessels, and the distribution is essentially parallel to that of VIP. PHI(M) elicits a less potent dose-dependent vasodilatation than VIP *in vitro* and *in vivo* (Suzuki et al., 1984; Edvinsson and McCulloch, 1985). Microapplication of PHI(M) dilates both arteries and veins *in situ* (Edvinsson and McCulloch, 1985).

Nitric oxide

The most recent addition to the transmitters or modulators that are involved in parasympathetic cerebrovascular actions is nitric oxide (NO). A short-lived, highly reactive molecule (Moncada et al., 1991), NO can dilate cerebral vessels (Tanaka et al., 1991; Morikawa et al., 1992). Blockade of NO synthesis reduces the effect of parasympathetic stimulation in the cerebral circulation (Goadsby et al., 1996). This observation is important because NO can be both deleterious and protective during cerebral ischemia, thus providing an avenue by which the parasympathetic nervous innervation might mediate protective neurovascular dilatation in the cerebral circulation (Kano et al., 1991).

Physiological effects

Effect of parasympathetic blockade

Resting cerebral blood flow and autoregulation

Few experiments have addressed the question of the facial nerve and its role in autoregulation. Sectioning the facial nerve does not alter autoregulation in the cat, while resting cerebral blood flow or glucose utilization are also unaffected (Goadsby, 1991).

Hypercapnia and hypoxia

Studies in the baboon (Hoff et al., 1977), cat (Bates and Sundt, 1976; Busija and Heistad, 1981, 1982), dog (D'Alecy and Rose, 1977), rat, and rabbit (Pinard et al., 1979) all demonstrate no effect of sections of the facial nerve on either hypercapnic or hypoxic vasodilatation. Similarly, Seylaz and colleagues (1988) have demonstrated that the main parasympathetic outflow ganglia, the sphenopalatine ganglion, may be ablated without any alteration of hypercapnic vasodilatation.

Stimulation of the parasympathetic nerves

Direct stimulation of the facial nerve in humans leads to an increase in total cranial blood flow (Bridges et al., 1958), as does facial nerve nucleus stimulation in the monkey (Meyer et al., 1971) or stimulation of an area just dorsal to it in the cat (Kuo et al., 1987) or chemical stimulation of the superior salivatory nucleus (Nakai et al., 1993). Salanga and Waltz (1973) have shown that stimulation of the intact facial nerve reduces cerebral blood flow, while stimulation of the sectioned distal end increases it. James and colleagues (1969) observed increases in cerebral blood flow in the baboon using the technique of direct intracarotid injection of ^{133}Xenon. This was confirmed in the dog by D'Alecy and Rose (1977) and in the rat and rabbit (Pinard et al., 1979) using the cerebral venous outflow technique. Clearly stimulation of the facial nerve increases cerebral blood flow without altering metabolic activity, as reflected by stable glucose utilization (Goadsby, 1989) and sagittal sinus oxygen content (Goadsby, 1991). These responses are mediated through a classic parasympathetic ganglion as they can be blocked by hexamethonium (Goadsby et al., 1983) and probably use vasoactive intestinal polypeptide as the major transmitter of the system.

VIP is released when cerebral arterial nerves are stimulated. A specific VIP antiserum has been used to inhibit noncholinergic/nonadrenergic dilator responses in the cranial circulation resulting from both direct and distant nerve stimulation (Goadsby and Macdonald, 1985). It has also been shown that facial nerve stimulation results in local cortical release of VIP (Goadsby and Shelley, 1990). This release is blocked by hexamethonium (Goadsby and Shelley, 1990), demonstrating release of VIP in the context of activation of a classic nicotinic autonomic ganglion in the cerebral circulation.

Sphenopalatine (pterygopalatine) ganglion

Studies of the peripheral ganglion mediating facial nerve vasodilatation have included peripheral reflex (trigeminal ganglion) (Lambert et al., 1984) and central structure (locus coeruleus) (Goadsby et al., 1983) stimulation. It is clear that facial nerve mediated effects can be blocked by sphenopalatine ganglion removal (Goadsby et al., 1984) and that this same response is VIP mediated (Goadsby and Macdonald, 1985). Indeed, stimulation of the sphenopalatine ganglion in cat (Goadsby, 1990) and rat (Seylaz et al., 1988) have been shown to produce strong frequency-dependent cerebral vasodilator responses. Importantly, cerebral glucose utilization (Goadsby, 1990) and tissue PO_2 (Seylaz et al., 1988) are not affected representing a truly neurogenic response.

In summary, the cranial parasympathetic pathway to the cerebral vessels arises in the superior salivatory nucleus in the pons; it traverses the facial nerve, joining the greater superficial petrosal nerve to be distributed to the vessels after synapsing chiefly in the sphenopalatine or otic ganglia. A variable small number of fibers in different species (including humans) have this peripheral synapse located in microganglia on the wall of the internal carotid artery, particularly near the carotid siphon. The transmitters contained in this system are acetylcholine, VIP, PHM(I), nitric oxide, and helospectin-related peptides. The ganglionic transmission in the periphery is mediated by a classic parasympathetic nicotinic ganglion, while current data would suggest that VIP is the major neuroeffector substance at the nerve–smooth muscle junction. The pathway does not play a role in either hypercapnic or hypoxic vasodilator responses or autoregulatory responses to changes in arterial perfusion pressure. The system can be activated either by direct stimulation or via connections with other important central neural vasoactive nuclei to increase cerebral blood flow independent of cerebral metabolic needs.

Role and clinical implications

Although there have been considerable advances in understanding the basic capabilities and connectivity of the parasympathetic innervation of the cerebral circulation, there is no clear physiological role for the system. The nerves are not involved directly in the most basic cerebrovascular responses, such as hypoxic or hypercapnic vasodilatation, nor do they appear to play a role in autoregulation. Their effects are, however, independent of direct metabolic intervention (Goadsby, 1989). This latter fact suggests a role in times of threat, such as during ischemia or in vasospastic conditions, such as subarachnoid hemorrhage (Table 16.3). The parasympathetic system is ideally placed to be engaged to increase cerebral blood flow when ordinary metabolic driving factors are impaired. This protection may, however, be regionally variable since the posterior circulation innervation with VIP is much less than that seen anteriorly. This finding may have implications in situations where predominantly posterior changes are reported, such as migraine, which I will call out below in the context autonomic regulation more broadly. It may be too narrow a perspective to consider only a vasomotor function for these nerves as there is evidence that cholinergic mechanisms can alter capillary permeability, including the movement of amino acids.

Table 16.3

Clinical conditions in which parasympathetic activation may be implicated

Stroke
Subarachnoid hemorrhage
Migraine*
Cluster headache†
Paroxysmal hemicrania ‡
Short-lasting unilateral neuralgiform headache attacks with conjunctival injection and tearing (SUNCT)‡

*(Goadsby et al., 1990); †(Goadsby and Edvinsson, 1994); ‡(Goadsby and Lipton, 1997).

SYMPATHETIC INFLUENCES UPON THE CEREBRAL CIRCULATION

The sympathetic innervation of the cerebral circulation was the first to be characterized, with the development of the Falck–Hillarp histofluorescence technique in the early 1960s (Falck et al., 1962). The sympathetic innervation of the brain circulation has thus been studied longer and is better understood than the parasympathetic system.

Anatomy

The sympathetic nervous innervation of the cerebral circulation arises in the hypothalamus as first-order neurons and projects to the intermediolateral cell column of the spinal cord (Purves, 1972). Second-order neurons arise from the sympathetic chain and proceed to synapse with third-order neurons in the superior cervical ganglion (Goadsby, 2002b). The innervation is lateralized and largely respects the midline, with subsequent projection to the cerebral vessels being provided by sympathetic nerves that run rostral with the carotid artery. There is a dense innervation, particularly of more proximal large arteries, which follows vessels out to the pia and along the brain surface but generally only follow penetrating vessels for a short distance into the cerebral parenchyma. The innervation is established for the cat

(Nielsen and Owman, 1967), monkey (Hernandez-Perez et al., 1975), rat (Edvinsson, 1982), rabbit (Peerless and Yasargil, 1971), dog (D'Alecy and Feigl, 1972) and humans (Edvinsson et al., 1976b). For the large cerebral vessels and pial vessels the noradrenaline/norepinephrine-containing nerves arise from the superior cervical ganglion since extirpation of the ganglion eliminates almost all adrenergic fluorescence (Edvinsson et al., 1972). The noradrenaline-containing nerves that appear to innervate the cerebral parenchymal vessels do not arise from the superior cervical ganglion (Edvinsson et al., 1973) but arise centrally from the major central source of noradrenaline in the pons, the nucleus locus coeruleus (Raichle et al., 1976). The sympathetic nerve terminals are often located close to smooth muscle cells in the outer media. The terminals are characterized electron microscopically as small, electron-dense vesicles. The sympathetic innervation of the cerebral vessels accounts for about half the nerves observed in the vessels walls. The innervation is most dense anteriorly, with a relatively sparser supply in the vertebrobasilar territory that is considered to arise largely from the stellate ganglion.

Transmitters and neuromodulators

Noradrenaline

The main transmitter of the sympathetic innervation is the classic autonomic amine transmitter, noradrenaline. Extirpation of the superior cervical ganglion results in a mainly ipsilateral reduction in adrenergic fibers from the vessels. In addition, sympathetic nerve stimulation can cause perivascular release of noradrenaline, while labeled noradrenaline is taken up into the nerves. The effect of noradrenaline release is to constrict the vessels through activation of α-adrenoceptors in the vessels. In humans, the predominant receptor subtype is the α_1-adrenoceptor, while in some experimental animals α_2-adrenoceptors can be seen; both mediate vasoconstriction. In humans, both β_1- and β_2-adrenoceptors are located on cerebral vessels (Edvinsson and Krause, 2002).

Peptides

In addition to the classic constrictor aminergic transmitter, sympathetic nerves release the vasoconstrictor peptide neuropeptide Y (NPY). NPY is widely distributed in the brain and peripheral nervous system. Fibers that contain NPY form a dense network around cerebral arteries and veins and double stain for noradrenaline (Edvinsson et al., 1983). Retrograde tracing studies have shown NPY-positive fibers that have cell bodies in the superior cervical ganglion, and these fibers are substantially reduced by sympathectomy (Edvinsson et al., 1983, 1984). Furthermore, NPY colocalizes with dopamine β-hydroxylase in cell bodies in the superior cervical ganglion, demonstrating that the peptide exists in cells that synthesize noradrenaline. NPY is potent constrictor of cerebral vessels at the NPY Y_1 receptor. This receptor has been identified on cerebral vessels using PCR for the human NPY Y_1 receptor and constriction can be mimicked by the agonist NPY13-36 and blocked by the Y_1 antagonist BIBP.

Physiological effects

Stimulation

Direct stimulation of the cervical sympathetic nerves in experimental animals results in large vessel constriction but no change to small caliber vessels. The effect is more prominent on the cerebral veins and is blocked by α_1-adrenoceptor blockade. Consistent with these data, microapplication of noradrenaline reduces local vessel diameter. The details of this system have been further studied using local transmural nerve stimulation, although, because of species variation in nerve supply, the results have not always been consistent.

Physiological roles

Trophic effects

There are excellent data that have established a role for the sympathetic nerves in development, particularly of the muscular layer of the wall of cerebral vessels (Dimitriadou et al., 1988). The trophic effects of the sympathetic nerves deserve further investigation.

AUTOREGULATION

Sympathetic nerve stimulation extends the upper limit of autoregulation (Bill and Linder, 1976; Edvinsson et al., 1976a; MacKenzie et al., 1979), whereas acute sympathectomy, α-blockade (Fitch et al., 1975; Edvinsson et al., 1985) or inhibition of the effects of neuropeptide Y (Goadsby and Edvinsson, 1993) shifts the lower limit of autoregulation towards a lower blood pressure. α-Adrenoceptor blockade with phenoxybenzamine (Fitch et al., 1975) or with the α_2-adrenoceptor antagonist yohimbine (Edvinsson et al., 1985) resulted in positive effects. The pharmacological action of NPY is to constrict cerebral vessels as potently as noradrenaline but with a more prolonged time course (Edvinsson et al., 1984). This similar but more sustained action may have functional implications particularly since autoregulation has a time course of seconds.

IN CONCLUSION: CLINICAL IMPLICATIONS FOR PRIMARY HEADACHE DISORDERS

The extrinsic cranial autonomic neural systems are both well placed and certainly implicated in the pathophysiology of the primary headache disorders, such as migraine and cluster headache. A detailed description of the pathophysiology of these disorders can be found elsewhere (Goadsby, 2002a; Akerman et al., 2011), although some general principles may be worth considering. Taking migraine as the index condition since it is so common, it is reasonably clear from functional imaging studies that brainstem areas are involved (Sprenger and Goadsby, 2010). The clinical presentation has implicated parasympathetic cranial autonomic mechanisms (Goadsby, 2009), so that the superior salivatory nucleus must be active, based on what we know of the physiological anatomy. Indeed, given work again on imaging (Sprenger et al., 2012), and experimental studies (Malick et al., 2000), there is a clear role for hypothalamic modulation of the pain and indeed other aspects of migraine. Taken all together, the mysteries of these clinical syndromes will only be well characterized when the anatomy and physiology of brainstem cranial autonomic mechanisms are much better understood.

References

Akerman S, Holland P, Goadsby PJ (2011). Diencephalic and brainstem mechanisms in migraine. Nat Rev Neurosci 12: 570–584.

Arimura A (1992). Pituitary adenylate cyclase activating polypeptide (PACAP): discovery and current status of research. Regul Pept 37: 287–303.

Attwell D, Buchan AM, Charpak S et al. (2010). Glial and neuronal control of brain blood flow. Nature 468: 232–243.

Bates D, Sundt TM (1976). The relevance of peripheral baroreceptors and chemoreceptors to the regulations of cerebral blood flow in the cat. Circ Res 38: 488–493.

Bevan JA, Buga GM, Florence VM et al. (1982). Distribution of choline acetyltransferase in cerebral and extracerebral cranial arteries of the cat. Circ Res 50: 470–476.

Bill A, Linder J (1976). Sympathetic control of cerebral blood flow in acute arterial hypertension. Acta Physiol Scand 96: 114–121.

Bridges TJ, Clark K, Yahr MD (1958). Plethsymographic studies of the cerebral circulation evidence for cranial nerve vasomotor activity. J Clin Invest 37: 763–772.

Busija DW, Heistad DD (1981). Effects of cholinergic nerves on cerebral blood flow in cats. Circ Res 48: 62–69.

Busija DW, Heistad DD (1982). Atropine does not attenuate cerebral vasodilatation during hypercapnia. Am J Physiol 242: H683–H687.

Chorobski J, Penfield W (1932). Cerebral vasodilator nerves and their pathway from the medulla oblongata. Arch Neurol Psychiatry 28: 1257–1289.

Cobb S, Finesinger JE (1932). Cerebral circulation XIX. The vagal pathway of the vasodilator impulses. Arch Neurol Psychiatry 28: 1243–1256.

D'Alecy LG, Feigl EO (1972). Sympathetic control of cerebral blood flow in dogs. Circ Res 31: 267–283.

D'Alecy LG, Rose CJ (1977). Parasympathetic cholinergic control of cerebral blood flow in dogs. Circ Res 41: 324–331.

Dimitriadou V, Aubineau P, Taxi J et al. (1988). Ultrastructural changes in the cerebral artery wall induced by long-term sympathetic denervation. Blood Vessels 25: 122–143.

Duckles SP (1981). Evidence for a functional cholinergic innervation of cerebral arteries. J Pharmacol Exp Ther 217: 544–548.

Edvinsson L (1982). Sympathetic control of the cerebral circulation. Trends Neurosci 5: 425–429.

Edvinsson L, Ekman R (1984). Distribution and dilatory effect of vasoactive intestinal polypeptide (VIP) in human cerebral vessels. Peptides 5: 329–331.

Edvinsson L, Krause DN (2002). Cerebral Blood Flow and Metabolism. Lippincott Williams and Wilkins, Philadelphia.

Edvinsson L, McCulloch J (1985). Distribution and vasomotor effects of peptide HI (PHI) in feline cerebral blood vessels in vitro and in situ. Regul Pept 10: 345–356.

Edvinsson L, Owman C, Rosengren E et al. (1972). Concentration of noradrenaline in pial vessels, choroid plexus, and iris during two weeks after sympathetic ganglionectomy or decentralization. Acta Physiol Scand 85: 201–206, 1972.

Edvinsson L, Lindvall M, Nielsen KC et al. (1973). Are brain vessels innervated also by central (non-sympathetic) adrenergic neurones? Brain Res 63: 496–499.

Edvinsson L, Owman C, Siesjo B (1976a). Physiological role of cerebrovascular sympathetic nerves in the autoregulation of cerebral blood flow. Brain Res 117: 519–523.

Edvinsson L, Owman C, Sjoberg NO (1976b). Autonomic nerves mast cells and amine receptors in human brain vessels. A histochemical and pharmacological study. Brain Res 115: 377–393.

Edvinsson L, Emson P, McCulloch J (1983). Neuropeptide Y: cerebrovascular innervation and vasomotor effects in the cat. Neurosci Lett 43: 79–84.

Edvinsson L, Emson P, McCulloch J et al. (1984). Neuropeptide Y: immunocytochemical localization to and effect upon feline pial arteries and veins in vitro and in situ. Acta Physiol Scand 122: 155–163.

Edvinsson L, MacKenzie ET, Robert J-P et al. (1985). Cerebrovascular responses to haemorrhagic hypotension in anaesthetized cats. Effects of a-adrenoceptor antagonists. Acta Physiol Scand 123: 317–323.

Falck B, Hillarp NA, Thieme G et al. (1962). Fluorescence of catecholamines and related compounds condensed with formaldehyde. J Histochem Cytochem 10: 348–354.

Fang HCH (1961). Cerebral arterial innervations in man. Arch Neurol 4: 651–656.

Fitch W, MacKenzie ET, Harper AM (1975). Effects of decreasing arterial blood pressure on cerebral blood flow in the baboon. Influence of the sympathetic nervous system. Circ Res 37: 550–557.

Florence VM, Bevan JA (1979). Biochemical determinations of cholinergic innervation in cerebral cortex. Circ Res 45: 212–218.

Frackowiak RSJ, Friston KJ (1994). Functional neuroanatomy of the human brain: positron emission tomography – a new neuroanatomical technique. J Anat 184: 211–225.

Gibbins IL, Brayden JE, Bevan JA (1983). The origin and distribution of vasoactive intestinal polypeptide immunoreactive nerves supplying cat cerebral arteries. Fed Proc 1983: 484.

Gibbins IL, Brayden JE, Bevan JA (1984). Perivascular nerves with immunoreactivity to vasoactive intestinal polypeptide in cephalic arteries of the cat: distribution, possible origins and functional implications. Neuroscience 13: 1327–1346.

Goadsby PJ (1989). Effect of stimulation of the facial nerve on regional cerebral blood flow and glucose utilization in cats. Am J Physiol 257: R517–R521.

Goadsby PJ (1990). Sphenopalatine ganglion stimulation increases regional cerebral blood flow independent of glucose utilization in the cat. Brain Res 506: 145–148.

Goadsby PJ (1991). Characteristics of facial nerve elicited cerebral vasodilatation determined with laser Doppler flowmetry. Am J Physiol 260: R255–R262.

Goadsby PJ (2002a). Pathophysiology of cluster headache: a trigeminal autonomic cephalgia. Lancet Neurol 1: 37–43.

Goadsby PJ (2002b). Raeder's syndrome: "paratrigeminal" paralysis of oculo-pupillary sympathetic. J Neurol Neurosurg Psychiatry 72: 297–299.

Goadsby PJ (2009). Lacrimation, conjunctival injection, nasal symptoms . . . cluster headache, migraine and cranial autonomic symptoms in primary headache disorders – what's new? J Neurol Neurosurg Psychiatry 80: 1057–1058.

Goadsby PJ, Edvinsson L (1993). Examination of the involvement of neuropeptide Y (NPY) in cerebral autoregulation using the novel NPY antagonist PP56. Neuropeptides 24: 27–33.

Goadsby PJ, Edvinsson L (1994). Human *in vivo* evidence for trigeminovascular activation in cluster headache. Brain 117: 427–434.

Goadsby PJ, Edvinsson L (2002). Neurovascular control of the cerebral circulation. In: L Edvinsson, DN Krause (Eds.), Cerebral Blood Flow and Metabolism, Lipincott Williams and Wilkins, Philadelphia, pp. 172–188.

Goadsby PJ, Lipton RB (1997). A review of paroxysmal hemicranias, SUNCT syndrome and other short-lasting headaches with autonomic features, including new cases. Brain 120: 193–209.

Goadsby PJ, Macdonald GJ (1985). Extracranial vasodilatation mediated by VIP (vasoactive intestinal polypeptide). Brain Res 329: 285–288.

Goadsby PJ, Shelley S (1990). High frequency stimulation of the facial nerve results in local cortical release of vasoactive intestinal polypeptide in the anesthetised cat. Neurosci Lett 112: 282–289.

Goadsby PJ, Lambert GA, Lance JW (1983). Effects of locus coeruleus stimulation on carotid vascular resistance in the cat. Brain Res 278: 175–183.

Goadsby PJ, Lambert GA, Lance JW (1984). The peripheral pathway for extracranial vasodilatation in the cat. J Auton Nerv Syst 10: 145–155.

Goadsby PJ, Edvinsson L, Ekman R (1990). Vasoactive peptide release in the extracerebral circulation of humans during migraine headache. Ann Neurol 28: 183–187.

Goadsby PJ, Uddman R, Edvinsson L (1996). Cerebral vasodilatation in the cat involves nitric oxide from parasympathetic nerves. Brain Res 707: 110–118.

Hara H, Hamill GS, Jacobowitz DM (1985). Origin of cholinergic nerves to the rat major cerebral arteries: coexistence with vasoactive intestinal polypeptide. Brain Res Bull 14: 179–188.

Hardebo JE, Arbab M, Suzuki N et al. (1991). Pathways of parasympathetic and sensory cerebrovascular nerves in monkeys. Stroke 22: 331–342.

Hartman BK, Swanson LW, Raichle ME et al. (1980). Central adrenergic regulation of cerebral microvascular permeability and blood flow; anatomic and physiologic evidence. Cerebral Microvasculature. Adv Exp Med Biol 131: 113–126.

Heistad DD, Marcus ML, Said SI et al. (1980). Effect of acetylcholine and vasoactive intestinal polypeptide on cerebral blood flow. Am J Physiol 239: H73–H80.

Hernandez-Perez MJ, Raichle ME, Stone HL (1975). The role of the peripheral sympathetic nervous system in cerebral blood flow autoregulation. Stroke 6: 284–292.

Hoff JT, MacKenzie ET, Harper AM (1977). Responses of the cerebral circulation to hypercapnia and hypoxia after 7th cranial nerve transection in baboons. Circ Res 40: 258–262.

Itoh N, Obata K-I, Yanaihara N et al. (1983). Human preprovasoactive intestinal polypeptide contains a novel PHI-27-like peptide, PHM-27. Nature 304: 547–549.

James IM, Millar RA, Purves MJ (1969). Observations on the extrinsic neural control of cerebral blood flow in the baboon. Circ Res 25: 77–93.

Kano M, Moskowitz MA, Yokota M (1991). Parasympathetic dennervation of rat pial vessels significantly increases infarction volume following middle cerebral artery occlusion. J Cereb Blood Flow Metab 11: 628–637.

Kuo JS, Wang MR, Liu RH et al. (1987). Reduction of common carotid resistance upon stimulation of an area dorsal to the facial nucleus of cats. Brain Res 417: 181–184.

Kuschinsky W (1989). Coupling of blood flow and metabolism in the brain – the classical view. In: J Seylaz, R Sercombe (Eds.), Neurotransmission and Cerebrovascular Function, Vol. 2. Elsevier Science Publishers, Amsterdam, pp. 331–342.

Kuschinsky W, Wahl M (1973). Cholinergic receptors at pial arteries. A microapplication study. Stroke 4: 371.

Lambert GA, Bogduk N, Goadsby PJ et al. (1984). Decreased carotid arterial resistance in cats in response to trigeminal stimulation. J Neurosurg 61: 307–315.

Lavrentieva NB, Mchedishvili GI, Plechkova EK (1968). Distribution and activity of cholinesterase in nerve structures of the pial arteries: histochemical investigation. Bull Exp Biol Med 66: 1282–1285.

MacKenzie ET, McGeorge AP, Graham DI et al. (1979). Effects of increasing arterial pressure on cerebral blood flow in the baboon: influence of the sympathetic nervous system. Pflugers Arch 378: 189–195.

Malick A, Strassman AM, Burstein R (2000). Trigeminohypothalamic and reticulohypothalamic tract neurons in the upper cervical spinal cord and caudal medulla of the rat. J Neurophysiol 84: 2078–2112.

Matsuyama T, Shiosaka S, Matsumoto M et al. (1983). Overall distribution of vasoactive intestinal polypeptide-containing nerves on the wall of the cerebral arteries: an immunohistochemical study using whole-mounts. Neuroscience 10: 89–96.

May A, Goadsby PJ (1999). The trigeminovascular system in humans: pathophysiological implications for primary headache syndromes of the neural influences on the cerebral circulation. J Cereb Blood Flow Metab 19: 115–127.

McNaughton FL (1966). The innervation of the intracranial blood vessels and the dural sinuses. In: S Cobb, AM Frantz, W Penfield et al. (Eds.), The Circulation of the Brain and Spinal Cord. Hafner Publishing, New York, pp. 178–200.

Meyer JS, Teraura T, Sakamoto K et al. (1971). Central neurogenic control of cerebral blood flow. Neurology 21: 247–262.

Moncada S, Palmer RMJ, Higgs EA (1991). Nitric oxide: physiology, pathophysiology, and pharmacology. Pharmacol Rev 43: 109–142.

Morikawa E, Rosenblatt S, Moskowitz MA (1992). L-arginine dilates rat pial arterioles by nitric oxide-dependent mechanisms and increases cerebral blood flow duirng focal cerebral ischemia. Br J Pharmacol 107: 905–907.

Mutt V, Said SI (1974). Structure of the porcine vasoactive intestinal octacosapeptide. Eur J Biochem 42: 581–589.

Nakai M, Tamaki K, Ogata J et al. (1993). Parasympathetic cerebrovascular center of the facial nerve. Circ Res 72: 470–475.

Nielsen KC, Owman C (1967). Adrenergic innervation of pial arteries related to the circle of Willis in the cat. Brain Res 6: 773–776.

Nielsen KC, Edvinsson L, Owman C (1973). Cholinergic innervation of pial artery system. Stroke 4: 371.

Peerless SJ, Yasargil MG (1971). Adrenergic innervation of the cerebral blood vessels in the rabbit. J Neurosurg 35: 148–154.

Pinard E, Purves MJ, Seylaz J et al. (1979). The cholinergic pathway to cerebral blood vessels II. Physiological studies. Pflugers Arch 379: 165–172.

Purves MJ (1972). The Physiology of the Cerebral Circulation. Cambridge University Press, Cambridge.

Raichle ME, Eichling JO, Grubb RL et al. (1976). Central noradrenergic regulation of brain micro-circulation. In: HM Pappius, W Feindel (Eds.), Dynamics of Brain Edema. Springer-Verlag, New York, pp. 11–17.

Ruskell GL (1970). An ocular parasympathetic nerve pathway of facial nerve origin and its influence on intraocular pressure. Exp Eye Res 10: 319–330.

Ruskell GL (1971). Facial parasympathetic innervation of the choroidal blood vessels in monkeys. Exp Eye Res 12: 166–172.

Said SI, Mutt V (1970). Polypeptide with broad biological activity: isolation from small intestine. Science 169: 1217–1218.

Said SI, Mutt V (1972). Isolation from porcine-intestinal wall of a vasoactive octacosapeptide related to secretin and to glucagon. Eur J Biochem 28: 199–204.

Salanga VD, Waltz AG (1973). Regional cerebral blood flow during stimulation of the seventh cranial nerve. Stroke 4: 213–217.

Seki Y, Suzuki Y, Baskaya MK (1995). The effects of pituitary adenylate cyclase activating polypeptide cerebral arteries and vertebral artery blood flow in anaesthetized dogs. Eur J Pharmacol 275: 259–266.

Seylaz J, Hara H, Pinard E et al. (1988). Effect of stimulation of the sphenopalatine ganglion on cortical blood flow in the rat. J Cereb Blood Flow Metab 8: 875–878.

Spencer SE, Sawyer WB, Wada H et al. (1990). CNS projections to the pterygopalatine parasympathetic preganglionic neurons in the rat: a retrograde transneuronal viral cell body labeling study. Brain Res 534: 149–169.

Sprenger T, Goadsby PJ (2010). What has functional neuroimaging done for primary headache ... and for the clinical neurologist? J Clin Neurosci 17: 547–553.

Sprenger T, Maniyar FH, Monteith TS et al. (2012). Midbrain activation in the premonitory phase of migraine: a $H_2{}^{15}O$-positron emission tomography study. Headache 52: 863–864.

Suzuki N, Hardebo JE (1993). The cerebrovascular parasympathetic innervation. Cerebrovasc Brain Metab Rev 5: 33–46.

Suzuki Y, McMaster D, Lederis K et al. (1984). Characterization of the relaxant effects of vasoactive intestinal polypeptide (VIP) and PHI on isolated brain arteries. Brain Res 322: 9–16.

Suzuki N, Hardebo JE, Owman C (1988). Origins and pathways of cerebrovascular vasoactive intestinal polypeptide-positive nerves in rat. J Cereb Blood Flow Metab 8: 697–712.

Tanaka K, Gotoh F, Gomi S et al. (1991). Inhibition of nitric oxide synthesis induces a significant reduction in local cerebral blood flow in the rat. Neurosci Lett 127: 129–132.

Tatemoto K, Mutt V (1981). Isolation and characterization of the intestinal peptide porcine PHI (PHI-27), a new member of the glucagon-secretin family. Proc Natl Acad Sci U S A 78: 6603–6607.

Uddman R, Goadsby PJ, Jansen I et al. (1993). PACAP, a VIP-like peptide, immunohistochemical localization and effect upon cat pial arteries and cerebral blood flow. J Cereb Blood Flow Metab 13: 291–297.

Uddman R, Goadsby PJ, Jansen-Olesen I et al. (1999). Helospectin-like peptides: immunochemical localization, and effects on isolated cerebral arteries and on local cerebral blood flow in the cat. J Cereb Blood Flow Metab 19: 61–67.

Vasquez J, Purves MJ (1979). The cholinergic pathway to cerebral blood vessels. I. Morphological studies. Pflugers Arch 379: 153–163.

Walters DW, Gillespie SA, Moskowitz M (1986). Cerebrovascular projections from the sphenopalatine and otic ganglia to the middle cerebral artery of the cat. Stroke 17: 488–494.

Warwick R, Williams PL (1973). Gray's Anatomy, Longman, Edinburgh.

Willis T (1664). The Anatomy of the Brain and Nerves. McGill University Press, Montreal.

Handbook of Clinical Neurology, Vol. 117 (3rd series)
Autonomic Nervous System
R.M. Buijs and D.F. Swaab, Editors

Chapter 17

Autonomic regulation of kidney function

EDWARD J. JOHNS*
Department of Physiology, University College Cork, Cork, Republic of Ireland

INTRODUCTION

The kidney sits at a pivotal position in balancing the excretion of sodium and water with fluid reabsorption along the nephrons to ensure that cardiovascular homeostasis is maintained. The maintenance of extracellular fluid volume, and hence blood volume, is achieved by a complex neurohumoral regulation of kidney function. Long-term fluid regulation is exerted by hormonal influences, such as aldosterone on the fine control of sodium reabsorption, and antidiuretic hormone on water reabsorption. By contrast, short-term dynamic control in response to changes in sodium and water intake and loss through normal everyday activities are primarily determined by the direct actions of the autonomic nervous system, via the renal sympathetic nerves, and indirectly via the renin–angiotensin system. It is becoming increasingly evident that the kidney is also an important source of sensory information and that afferent renal nerve input into the central nervous system can be responsible for determining efferent renal sympathetic nerve activity which in physiological circumstances is the foundation of renorenal reflexes, but in pathophysiological states may be responsible for the sympathoexcitation widely observed in hypertension, heart failure, and chronic renal disease.

Some 25% of cardiac output is delivered to both kidneys and of this one fifth is filtered at the glomerulus and processed by the nephrons. This filtered load, of 180 L/day in man, is subject to processing, that is, secretion of metabolites and reabsorption of fluid, as it passes along the tubules, with the end result that some 98% of the fluid volume is reabsorbed. Following filtration, about two-thirds is reabsorbed iso-osmotically at the high permeability proximal tubule epithelial cells, and thereafter it passes to the loop of Henle, comprising descending and ascending limbs, dipping deeply into the medulla and papillary regions, and is responsible for a further 25% reabsorption of the filtered load. It is at the thick ascending limb of the loop of Henle that activity of the sodium-potassium-2 chloride transporter is responsible for fluid reabsorption but also initiates the generation of the medullary concentration gradients essential for the production of a concentrated urine. The later section of the nephron, the distal tubule and collecting duct, are low permeability epithelia which are responsible for the reabsorption of up to 7% of the filtered load. At these sites, sodium transport is regulated by aldosterone acting on the principal cells and antidiuretic hormone acting on water transport processes (Johns et al., 2011).

The renal sympathetic nerves exert their modulatory influence on this processing and reabsorption of tubular fluid in two major ways: first, via a direct action at the proximal tubule and thick limb of the loop of Henle; second, via stimulation of renin release and subsequent generation of angiotensin II, which itself increases proximal fluid reabsorption and also, via aldosterone, on distal sodium reabsorption. The way in which the fluid compartmental volumes link and may be regulated are illustrated in Figure 17.1.

RENAL SYMPATHETIC NERVES

ANATOMY

Sympathetic preganglionic nerve fibers originate from a number of autonomic control nuclei within the hypothalamus and brainstem, which include the rostral ventrolateral medulla (RVLM), paraventricular nucleus (PVN), and, to a lesser extent, fibers from the raphe nucleus and A5 regions, and project to the intermediolateral areas of the spinal cord. The postganglionic fibers exit the spinal cord at spinal levels T11–L3 and pass through the paravertebral ganglia, with some passing

*Correspondence to: Edward J. Johns, Department of Physiology, Western Gateway Building, University College Cork, Cork, Republic of Ireland. Tel: +21-420-5977, Fax: +21-420-5370, E-mail: e.j.johns@ucc.ie

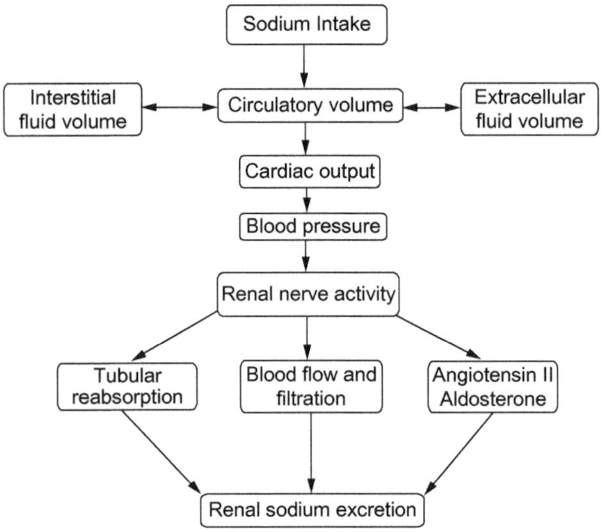

Fig. 17.1. This figure illustrates the relationship between sodium intake, its distribution across body fluid compartments, its impact on blood pressure and the role of the renal sympathetic nerves in regulation renal function to ensure an efficient mobilization of the sodium load.

through the aorticorenal, splenic, celiac, and mesenteric ganglia (Fig. 17.2). The fibers then align to the renal artery and enter the kidney at the hilus (Johns et al., 2011). Within the kidney, the fibers generally track with the arterial system, the interlobar, arcuate, interlobular, and then the arteriolar resistance vessels. Thereafter, the fibers pass alongside not only the afferent and efferent arterioles but also the different sections of the nephron, where they make close contact. The density of innervation is greatest in the cortex and outer medulla, and is much less in the inner medulla and papilla. The renal sympathetic nerve fibers are typical of autonomic nerves in that they have varicosities along their length which are in close proximity to the neuroeffector junctions at multiple sites along the resistance vessels and the nephrons. In terms of the nephrons, the greatest density of neuroeffector junctions occurs at the thick limb of the ascending loop of Henle and is relatively less along the proximal and distal tubules (Barajas et al., 1984, 1992). More recently, it has become evident that a small proportion of the sympathetic fibers penetrate as far as the pelvic wall where they make close connections with the sensory nerve endings present in this region (Kopp et al., 2007b).

Neural regulation of kidney function

Action potentials passing along the postganglionic fibers on reaching the varicosities lead to the exocytosis of norepinephrine into the neuroeffector junctions to activate adrenoceptors on the postsynaptic membrane. Neuroeffector junctions are present at the level of the arterioles, the granular renin containing cells, and along the nephron, and therefore there can be adrenergic regulation of renal hemodynamics, renin secretion, and the tubular fluid reabsorption.

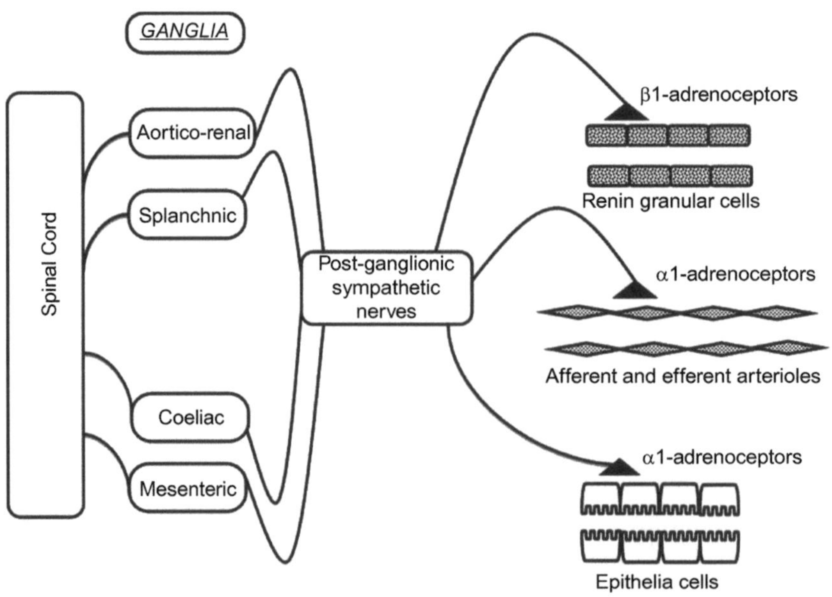

Fig. 17.2. This cartoon gives a general picture of the track of the sympathetic preganglionic fibers arising from the spinal cord which then traverse either directly or indirectly through different ganglia to give rise to the postganglionic sympathetic fibers which innervate the different functional components of the kidney. The different adrenoceptor subtypes regulating each cell type is indicated.

Neural regulation of renal blood flow and glomerular filtration

The rate of action potentials passing along the postganglionic fibers will have a frequency-dependent action on renal hemodynamics. Early studies reported that electrical depolarization of the renal nerves at high rates caused a marked reduction in both renal blood flow and glomerular filtration rate (Coote et al., 1972; Johns et al., 1976). These responses reflected a vasoconstriction of the renal resistance vessels which comprise the interlobular arteries and more importantly, the afferent and efferent arterioles. Activation of the renal nerves at somewhat lower rates, reducing renal blood flow by 15–20%, causes minimal change in glomerular filtration rate (Johns et al., 1976; Handa and Johns, 1985) which is likely to be due to a differential action at the resistance arterioles whereby the degree of constriction induced is relatively greater at the efferent than afferent arterioles with the result that glomerular filtration pressure remains unchanged.

There has been increasing attention devoted to investigating whether there is a differential neural control of blood flow through the cortex as against the medulla. Electrical stimulation of the renal sympathetic nerves reduces local cortical blood flow to approximately the same degree as total renal blood flow but the magnitudes of the reductions in the medulla are much less (Eppel et al., 2004; Walkowska et al., 2005). The reasons underlying this lack of sensitivity to neural regulation in the medulla may result from a relatively lower innervation density or the local presence or production of autocrine/paracrine factors, for example, nitric oxide, reactive oxygen species, arachidonic acid metabolites, or a combination of all these factors which may offset the vasoconstrictor action of norepinephrine (Evans et al., 2004).

Neural regulation of tubular reabsorption

A key finding (La Grange et al., 1973) was that electrical stimulation of the renal sympathetic nerves in the dog, at low levels that did not alter either renal blood flow or glomerular filtration rate, caused a marked and reversible reduction in urine flow and sodium excretion. These observations provided the first evidence of a direct neural influence on tubular fluid reabsorption and were supported by later findings in the rabbit (Hesse and Johns, 1984) and in the rat (Johns and Manitius, 1987). This concept was further supported by the micropuncture studies of Gottschalk and coworkers (Bello-Reuss et al., 1975), who demonstrated that proximal tubular fluid reabsorption was decreased following renal denervation and increased during low level renal sympathetic nerve stimulation at a time when neither renal blood flow nor glomerular filtration rates were changed. There has been relatively little investigation of how the renal nerves influence fluid reabsorption in other nephron segments, most likely due to technical challenges, but it has been reported (DiBona and Sawin, 1982) that low rates of electrical stimulation of the renal sympathetic nerves increased sodium, but not water, reabsorption in the thick ascending limb of the loop of Henle.

Release of norepinephrine from neuroeffector junctions at the proximal tubules activates adrenoceptors which stimulate two important transporters, the sodium/hydrogen exchanger, isoform 3 (NHE3) at the apical membranes of the microvillae to allow sodium to enter the epithelial cells (Liu et al., 1997; Gesek, 1999) and the sodium potassium ATPase (Na/KATPase) at the basolateral membrane (Aperia et al., 1992), which actively pumps sodium out of the cells into the interstitium. The NHE3 is the key regulator in that it is present at two sites, at the apical membrane where it is active, and within the subendosomal areas where it is inactive (McDonough et al., 2003; McDonough, 2010). There is evidence that reflex activation of the renal sympathetic nerves leads to a rise in the amount of NHE protein present in the apical membrane and a decrease in the subendosomal regions (Yang et al., 2002).

Neural regulations of renin release

The modified vascular smooth muscle cells containing renin granules are found at the afferent arteriole just at the entrance to the glomerulus. The granular content, as well as the number of cells recruited to contain renin, varies under different physiological conditions. Thus, following periods of elevated or reduced dietary sodium intake, there is a decrease or increase in renin content and/or number of cells, respectively, while in pathophysiological states where renal blood flow may be restricted, as with renal artery stenosis, granular content and number of granular cells will be elevated (Johns et al., 2011).

As with the other vascular smooth muscle cells, the renin-containing cells are innervated (Barajas and Wang, 1978; Barajas et al., 1992), and early functional studies (Coote et al., 1972) demonstrated that direct electrical stimulation of the renal sympathetic nerves not only markedly reduced renal blood flow but also caused large increases in renin secretion. Interpretation of these early reports regarding a direct neurally mediated renin secretion were confounded by the concomitant reductions in renal blood flow but this was clarified by later studies in the dog (Kopp et al., 1980; Osborn et al., 1981) in which much lower levels of electrical stimulation of the renal sympathetic nerves were used that did not change renal blood flow, and demonstrated that there were still large increases in renin secretion. Together, these studies provided strong support for the concept that the renal sympathetic nerves had a direct action to cause renin secretion from the granular cells which was independent of other renin secretion stimuli arising from alteration in renal blood flow or glomerular filtration rate.

Integration of the neural regulation of the kidney

It is now clear that neural control is exerted on all aspects of kidney function, renin secretion, tubular fluid reabsorption, and renal hemodynamics. What has become evident is that there is a progressive recruitment of these functional endpoints in a characteristic manner and this is illustrated in Figure 17.3.

Initially, at very low rates of renal nerve activity there is a direct nerve-induced renin secretion which takes place in the absence of changes in either fluid reabsorption or renal blood flow and glomerular filtration rate.

At slightly higher rates of electrical stimulation of the renal sympathetic nerves, not only is there a higher rate of renin secretion but this is accompanied by an increase in tubular fluid reabsorption, reflected as a reduction in urine flow and sodium excretion, even though there is no change in renal hemodynamics.

At high rates of renal nerve stimulation there is a constriction of the resistance vessels, the afferent and efferent arterioles, which will cause frequency dependent reductions in renal blood flow and glomerular filtration rate. At the same time there are marked increases in renin secretion and reductions in urine output.

It is important to appreciate that as the level of sympathetic nerve activity is increased there is a complex recruitment of functionalities that impact on cardiovascular homeostasis in different ways. Thus, under normal everyday conditions, there will be small but dynamic adjustments of renal sympathetic nerve activity in response to sodium and water intake as a result of feeding, fluid loss though sweating and respiration, to ensure that extracellular fluid volume, and hence blood volume, is maintained within appropriate limits. This will be achieved primarily though the release of renin and generation of angiotensin II and a direct action of the nerves on tubular fluid reabsorption without causing significant changes in renal hemodynamics. On the other hand, very high rates of renal sympathetic nerve activity, as occurs with severe exercise, exertion, or stress, when there is a necessity to redirect cardiac output to essential organs such as the skeletal muscles during the "fight-or-flight" response, will result in large reductions in renal blood flow and glomerular filtration rate. Nonetheless, such marked activation of the sympathetic nervous system only takes place over a relatively short time frame and will not have a major impact on cardiovascular homeostasis.

Effect of direct electrical stimulation of the renal nerves.

Stimulation rate	Low	Medium	High
Renin release	+	++	+++
Sodium excretion	-	+	++
RBF/GFR	-	-	+

Fig. 17.3. This table aims to relate the increasing frequency (low, medium and high) of electrical stimulation of the renal sympathetic nerves with the functional responses observed. It shows that at low stimulation frequencies, only renin release is increased and thereafter at medium frequencies there is a reduction in sodium excretion with an even larger release of renin. At the highest rates of stimulation, renal blood flow (RBF) and glomerular filtration rate (GFR) are reduced and with even greater responses in renin release and sodium excretion.

RENAL NEUROPHARMACOLOGY

Norepinephrine released from the varicosities of the sympathetic nerve terminals act on a range of adrenoceptors that have been shown to be present in the kidney. All adrenoceptors couple to G-proteins which activate selected signaling cascades dependent on the cell type involved, renin-containing, tubular epithelial, and vascular smooth muscle cells. Different adrenoceptors are present within the kidney (Johns, 1991) and have been assigned to the regulation of various functionalities.

Early investigations using direct electrical stimulation of the renal sympathetic nerves found that neurally mediated reductions in renal blood flow could be prevented by both the nonselective α-adrenoceptor antagonist phentolamine and the selective α_1-antagonist prazosin, but not by the α_2-adrenoceptor antagonist yohimbine (Johns, 1991). Later reports which attempted to examine in greater depth the subtypes of α_1-adrenoceptors involved in meditating the renal nerve-induced reductions in renal hemodynamics found that a large component required activation of the α_{1A}-adrenoceptor subtype (Witkowski et al., 2011). At the level of the tubular epithelia cells, a number of studies provided evidence that the antinatriuresis resulting from low-level stimulation of the renal sympathetic nerves could be blocked by prazosin, but not by the α_2-adrenoceptor antagonist idazoxan, consistent with the involvement of α_1-adrenoceptors (Johns and Manitius, 1986) which was later shown (Sattar and Johns, 1995) to primarily involve α_{1A}-adrenoceptors. It became evident at a very early stage that renal nerve-dependent renin release involved the activation of β-adrenoceptors. One of the first reports in the dog (Kopp et al., 1980) demonstrated that propranolol effectively blocked renin release in response to electrical stimulation of the renal sympathetic nerves at several different rates. Further investigations evaluating the subtype of β-adrenoceptor that activated renin secretion provided clear evidence that it was the β_1-adrenoceptor subtype that was involved, and this appears to be the case for most experimental species (Johns, 1981; Osborn et al., 1981) as well as in man. This diversity of adrenoceptors regulating function within the kidney is presented in Figure 17.2.

Together, this evidence makes it clear that the sympathetic innervation of the kidney exerts its effects by means of activating different adrenoceptor subtypes at different types of cells (Fig. 17.2). These findings would imply that in various pathophysiological conditions where there is overactivity of the sympathetic nervous system, administration of autonomic blocking drugs may differentially impact on kidney function. Consideration should be given to the fact that administration of β-blockers may suppress neurally mediated renin release and thereby angiotensin II generation which will be antihypertensive, but that α-adrenergically mediated sodium retention will persist. However, as fluid retention and renin secretion are the two primary endpoints of cardiovascular homeostasis, blockade of either arm may be sufficient to result in an effective reduction in blood pressure.

REFLEX REGULATION OF RENAL SYMPATHETIC NERVE ACTIVITY

Measurement of renal sympathetic nerve activity is relatively straightforward in experimental situations but is far more difficult in man. Often the alternative approach taken in humans is to record sympathetic nerve activity from the peroneal nerve which primarily innervates skeletal muscle blood vessels. The multifiber recordings comprise action potential arising from a large number of sympathetic nerve fibers giving a complex signal. The nerves do not fire continuously but are entrained to the changes in blood pressure over each heart cycle and are also modulated by respiration, changing with both inspiration and expiration.

Most sensory systems provide information to the hypothalamic and medullary areas of the brain which is then integrated in order to generate the appropriate sympathetic outflow to maintain homeostasis. Thus, there will be great variation in sensory input, dependent on whether an individual is asleep, physically active, or in a postprandial state. The major sensory areas providing essential afferent information to the brain are shown in Figure 17.4.

High pressure baroreceptors

There are mechanoreceptors embedded in the walls of the carotid sinus and aortic arch which are depolarized when these vessels are stretched. Thus, when blood pressure is increased, the rate of action potentials delivered

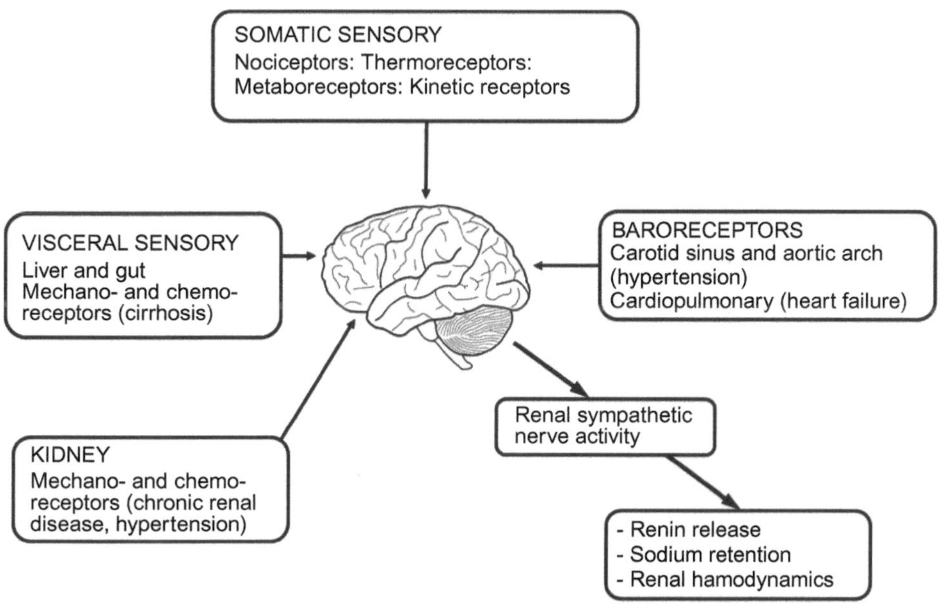

Fig. 17.4. An illustration of the integration of afferent inputs from various sensory systems of the body which determine the overall level of renal sympathetic nerve activity. The arrows indicate the direction of information flow into the general area of the brain. Associated disease states where neural control of the kidney is deranged are also indicated.

by the sensory nerves into the brain is elevated which results in a reflex decrease in sympathetic nerve activity and, conversely, as blood pressure falls at the carotid sinus and aortic arch, sympathetic nerve activity reflexly increases. The mechanoreceptors sense pressure in two modalities: first, by responding to increases and decrease in mean blood pressure and, second, over the course of a cardiac cycle, whereby as pressure increases from diastolic to systolic, there is a reduction in sympathetic outflow, while as pressure falls from systolic to diastolic, there is an increase in sympathetic nerve activity. Overall, this results in a bursting pattern which is entrained to the cardiac cycle, although the pattern is modulated by different conditions depending on whether the data are collected in the conscious or anesthetized state (Zhang et al., 1997; Huang et al., 2006; Li et al., 2008). The relationship between blood pressure and sympathetic nerve activity has been exploited experimentally to gain insight into physiological and pathophysiological mechanisms that may alter baroreflex sensitivity. This has been approached by generating baroreflex gain curves which allow the sensitivity of the relationship to be assessed and it has become apparent that the high pressure baroreceptors represent one of the most important regulators of renal sympathetic nerve activity (Johns, 2002, 2005).

Studies in the conscious chronically instrumented rat (Miki and Yoshimoto, 2005) demonstrated that the baroreflex gain curve and its sensitivity could change depending on the behavioral state of the animal. These authors found that the baroreflex gain curve for renal sympathetic nerve activity was shifted to operate over a higher pressure range with a greater sensitivity as physical activity levels were increased, from quiet awake to grooming, but that during rapid eye movement sleep, the curve was set at a lower blood pressure with less sensitivity. Interestingly, when the animals were faced with a threatening situation causing them to exhibit a “freezing” type behavior, blood pressure was unchanged while renal, but not lumbar, sympathetic nerve activity was increased, although baroreflex sensitivity was not altered (Miki and Yoshimoto, 2010; Yoshimoto et al., 2010).

The baroreceptor regulation of renal sympathetic nerve activity has a number of functional consequences. Reduction of pressure at the carotid sinuses in the dog, either via mechanical constriction of the carotid arteries or the use of head-up tilt, increased renal sympathetic nerve activity and plasma renin activity, and reduced sodium and water excretion in the absence of changes in renal hemodynamics, which were dependent on an intact renal innervation (Johns et al., 2011). These observations demonstrate that during normal physiological activities, the baroreflex regulation of blood pressure will modulate kidney function, mainly at the level of renin secretion and fluid balance rather than in the regulation of renal vascular resistance.

Low pressure baroreceptors

The heart and great veins contain sensory receptors which are sensitive to both mechanical stretch as well as the local chemical environment. In this part of the cardiovascular system, blood pressure is normally low but hydrostatic pressure will be increased if venous return is elevated, as may occur normally or as a result of a failing heart being unable to pump efficiently causing venous congestion. Under these conditions there will be a reflex suppression of sympathetic outflow to the peripheral organs.

There are convincing experimental data showing that stretching of the atria, using mechanical stretching via balloon inflation, causes a renal sympathetic nerve dependent natriuresis and diuresis (Johns et al., 2011). Indeed, studies in the conscious chronically instrumented dog (Miki et al., 1989, 2002) found that head-out total body water immersion (causing a shift of fluid from the periphery to the central compartments) resulted in a decrease in renal sympathetic nerve activity and a natriuresis in the intact dogs but not those subjected to a cardiac denervation. Moreover, studies in conscious and anesthetized rat preparations reported that an acute saline infusion to increase circulatory volume, and hence a stretching of the atrial chambers, caused a renal sympathoinhibition (Johns et al., 2011). In reports using the conscious dog (Miki et al., 2009), it was reported that following activation of the cardiopulmonary receptors using total body water immersion, there was a resetting of the baroreflex gain curves for renal sympathetic nerve activity which was shifted leftward and with a blunted operating range and increased gain coefficient. The important point arising from these observations is that there is a significant interaction between the high and low pressure baroreceptors in the regulation of sympathetic nerve activity passing to the kidney.

Somatosensory system

The musculoskeletal system is richly endowed with a sensory innervation sensitive to a range of stimuli, the metaboreceptors (chemoreceptors) within muscle tissue, mechanoreceptors in the tendons and joints, as well as the cutaneous receptors responsive to heat and cold (thermoreceptors) and pain (nociceptors). An early study in the anesthetized dog (Thames and Abboud, 1979) demonstrated that electrical stimulation of the sciatic nerve caused a renal nerve-dependent reduction in renal blood flow. Later studies in the anesthetized rat found that electrical stimulation of the brachial plexus or the local administration of capsaicin to activate cutaneous nociceptors had minimal effects on renal hemodynamics, but caused an antinatriuresis and antidiuresis which was mediated by the renal sympathetic nerves (Davis and Johns, 1991; Zhang and Johns, 1997; Huang and Johns, 1998). Interestingly, a thermal challenge to the tail of the anesthetized rat (DiBona and Sawin, 1999), an important organ of heat regulation in this species, elicited an increase in activity in a selected population of renal sympathetic nerve fibers. It is evident from all these studies that the wealth of sensory information arising from the soma provides an important input to the autonomic control centers. In this way, information regarding the status of the soma is continually transmitted to the brain to ensure that appropriate cardiovascular responses are obtained.

Visceral system

Also present within the visceral organs of the body are chemo- and mechanoreceptors which serve to transmit important sensory information to the autonomic control centers. At the level of the gastrointestinal tract (Weaver et al., 1987), it was reported that exposing the gut to exogenously applied bradykinin elicited a rise in blood pressure and a renal sympathoexcitation which occurred together with a renal nerve-dependent antinatriuresis and antidiuresis. The liver also provides significant sensory input to the central nervous system, thereby contributing to the maintenance of cardiovascular homeostasis. It has been found that lowering plasma glucose in the blood passing through the hepatic portal vein, but not the hepatic artery, caused an increase in plasma adrenaline concentrations which was prevented by deafferentation of the hepatic portal vein (Hevener et al., 2000). Thus, sensory receptors in this region can cause a sympathoactivation which may extend to the renal sympathetic nerves. Tonicity of the fluid in the hepatic portal vein also induces a sensory signal (Ishiki et al., 1991; Morita et al., 1991) as hepatic portal vein perfusions using a hypertonic solution resulted in a renal sympathoinhibition associated with a renal nerve-dependent increase in sodium and water excretion. Thus, it is clear that a reflex exists whereby reabsorption of sodium by the stomach would rapidly transfer to the hepatic portal vein eliciting a reduction in renal sympathetic nerve activity with a subsequent mobilization and excretion of the sodium load to reestablish sodium homeostasis.

Evidence is accruing that there are mechanoreceptors present within the splenic circulation which also provides input to the autonomic control centers of the brain. In anesthetized rat preparations (Hamza and Kaufman, 2004; Moncrief and Kaufman, 2006), it was observed that increasing splenic venous pressure raised splenic afferent nerve activity and reduced blood pressure and renal blood flow. The decrease in renal blood flow was a renal nerve-mediated event as it was prevented following renal denervation. Importantly, it was recognized that this "spleno-renal" reflex was activated by pressure within the splenic vein rather than blood flow though the spleen. What these findings highlight is that an increase in venous pressure in the splenic area, which may result from an increase in central venous pressure associated with cardiac failure, portal hypertension or liver disease, may reflexly activate the renal sympathetic nerves, cause fluid retention, and exacerbate the progression of the underlying disease.

Renal sensory innervation

The kidney itself has a sensory innervation which forms the afferent limb of renorenal reflexes. The region primarily innervated appears to be the pelvic wall, although functional studies suggest they are also present in the cortical and medullary regions. At the pelvic wall there is evidence for both mechanoreceptors sensitive to stretch resulting from increased hydrostatic pressure within the renal hilus, or chemoreceptors which respond to the urinary composition within the pelvis. The cell bodies of the afferent nerves are found in the ipsilateral dorsal root ganglia, from T6 to L4, and thereafter project axons to the ipsilateral dorsal horn of the spinal cord. At this point, they synapse to allow sensory information to be passed onto both spinal and supraspinal pathways.

Elevation of pelvic hydrostatic pressure over the range 3–20 mmHg gives rise to an approximate proportionate increase in afferent renal nerve activity which reflects a response of the mechanoreceptors. Two main categories of chemoreceptors have been identified, those which respond to ischemia and hypoxia/anoxia in the whole kidney, and those which respond to increased tonicity of fluid perfusing the renal pelvic areas. Importantly, activation of these mechano- and chemoreceptors is achieved by hydrostatic and urinary osmotic pressures that are within the normal physiological range (Kopp et al., 2007a). There is evidence for two types of renorenal reflexes, inhibitory and excitatory. The inhibitory reflex is one where activation of receptors results in a suppression of renal sympathetic nerve activity and excitatory renorenal reflexes where the sensory nerves increase activity of the renal sympathetic nerves.

The inhibitory renorenal reflex is well documented. A key observation in the anesthetized rat was that an acute surgical denervation of one kidney caused a diuresis and natriuresis of that kidney but a renal nerve-dependent antinatriuresis and antidiuresis in the contralateral kidney (Colindres et al., 1980; DiBona and Rios, 1980). Conversely, increased pelvic hydrostatic pressure in the ipsilateral kidney caused a renal nerve-mediated diuresis and natriuresis in the contralateral kidney. At a conceptual level, this neural interaction between the kidneys means that should there be either an increased fluid loss or an accumulation of urine in the pelvis, there will be a renal sympathetic nerve-mediated compensation in fluid excretion from the contralateral kidney.

Interestingly, there is a close structural apposition of the sympathetic and afferent nerve endings in the pelvis wall which underlies an important functional interaction whereby an increase in renal sympathetic nerve activity causes a concomitant rise in afferent nerve activity, which then further limits the renal sympathoexcitation (Kopp et al., 2007b). The importance of the efferent and afferent nerve interaction at the pelvis was reinforced by the report that rats on a high dietary sodium intake had elevated afferent renal nerve activity. Dorsal rhizotomy, to selectively remove the influence of the renal afferent nerves, resulted in an elevated blood pressure in the high sodium state together with a blunted sensitivity of the baroreflex regulation of renal sympathetic nerve activity. At the same time there was an enhanced renal sympathoexcitation in response to an air-jet stress challenge (Kopp et al., 2008). These particular studies highlight the importance of the afferent and efferent nerves in the long-term regulation of fluid balance and hence the level at which blood pressure is set.

The mechanisms underlying the sensory transduction processes to activate the renal sensory nerves have been explored in depth. There is good evidence that increased pelvic pressure causes the local release of bradykinin which acts on its receptors to initiate a PKA-mediated activation of COX-2, leading to the generation of PGE2. This signaling cascade leads to the production of cAMP and, via PKA, results in substance P generation which then depolarizes the terminal dendrites of the sensory nerves. The responsiveness of this pathway is modulated by both angiotensin II and locally produced endothelin (Kopp et al., 2003, 2007a).

An understanding of the physiology of the excitatory renorenal reflexes has been elusive but appears to be more associated with pathophysiological states. A series of reports by Campese and coworkers (Campese and Kogosov, 1995; Ye et al., 2002) found that renal injury or damage, resulting from either a 5/6th nephrectomy or the acute injection of a small bolus of phenol into the renal cortex, elicited a renal sympathoexcitation which was dependent on intact afferent renal nerves. These authors put forward the hypothesis that an inappropriate stimulation of the renal sensory nerves may occur in renal disease, giving rise to a sympathetically mediated hypertension that is resistant to most conventional antihypertensive drugs (Campese et al., 2006). There are experimental data demonstrating that intrarenal arterial infusion of adenosine in the conscious dog initiated a renal sympathoexcitation (Katholi et al., 1983) and increases in blood pressure and renal vascular resistance. All these responses were prevented by denervation of the infused (ipsilateral) kidney. Furthermore, it was found that intrarenal arterial infusion of bradykinin in the conscious rat elicited a marked rise in blood pressure and renal vascular resistance (Smits and Brody, 1984). This group observed that these responses required an intact renal innervation and could be prevented by ganglionic blockade which was consistent with the view that the bradykinin acted on sensory nerves to cause a reflex sympathoexcitation. The role of this excitatory renorenal reflex both normally and in pathophysiological states remains to be explored.

PATHOPHYSIOLOGICAL STATES

Diseases of the cardiovascular and renal systems are frequently characterized by deficits in autonomic control. In many situations, injury or inflammation of the blood vessels, the heart or the kidney initiates inappropriate activation of sensory nerves passing information to the brain, which causes an elevated sympathetic drive. This sympathoexcitation may in itself exacerbate the rate of disease progression (Fig. 17.4).

HYPERTENSION

There has been ongoing investigation into the contribution of the sympathetic nerves and the neural control of the kidney in the development and maintenance of hypertension. Animal models have been studied and these comprise the spontaneously hypertensive rat (SHR), a model of genetic or essential hypertension, the Goldblatt hypertensive rat (2K1C, 1K1C), which reflects renovascular hypertension, and the DOCA-salt model, which is a state of volume expansion-induced hypertension. An early study using the SHR found that renal denervation in young prehypertensive animals prevented or blunted the increase in blood pressure to hypertensive levels in the adult. Moreover, a slowing of the rate of hypertension development was also reported in the Goldblatt and DOCA-salt models of hypertension (DiBona and Kopp, 1997; Johns et al., 2011). Lundin and coworkers (1984) produced definitive evidence in the SHR showing that the rate of action potential firing in single renal sympathetic nerve fibers was over twice that recorded in normotensive control rats.

In terms of essential hypertensive patients, the issue of concern is how to establish if there is sympathoexcitation and whether this includes the kidney. There is significant difficulty in obtaining a measure of sympathetic activity in the renal nerves and two alternative approaches have been used. The first directly records peripheral sympathetic nerve activity, for example by placing electrodes onto the peroneal nerve to record primarily muscle sympathetic nerve activity which is then taken to reflect sympathetic outflow to other regions, including the kidney. This technique has been used to demonstrated that in young essential hypertensive patients the bursting pattern in muscle sympathetic nerve activity was much higher than that recorded in normotensive control subjects (Esler and Kaye, 2000). A second approach developed by Esler and colleagues (2003) has been the norepinephrine "spillover" methodology whereby norepinephrine concentrations in venous blood taken from specific organs is compared with that in arterial blood, which allows calculation of norepinephrine production by that organ. Using this approach, there is persuasive evidence showing that there is a marked increase in renal norepinephrine production in essential hypertensive patients (Esler et al., 2003).

The role of the renal sympathetic nerves in the genesis of hypertension has recently received renewed attention arising from a report in a patient group exhibiting resistant hypertension, that is uncontrolled hypertension in spite of receiving four to five antihypertensive drugs. Krum and coworkers (2009) utilized a catheter device at the end of which was a probe designed to emit bursts of high energy radio frequency waves. This probe was inserted into each renal artery in turn and the probe was set to deliver sufficient energy to disrupt the renal nerves running alongside the renal artery but without causing damage to the vessel wall. In this proof of concept study (Krum et al., 2009), there was a marked fall in blood pressure within 1 month which persisted for at least 12 months. Subsequently, a randomized double blind study was undertaken comparing two groups, a control group which was not subjected to the procedure, and the experimental group that was subjected to the bilateral renal denervation (Krum et al., 2011). Systolic blood pressure was reduced by some 20–30 mmHg in the renally denervated group which was maintained over a 24 month period, but was unchanged in the nonoperated group. These studies are now being extended, first, to determine if the reduction in blood pressure persists for even longer time periods, and second, to evaluate whether the drug regimen of these patients can be reduced. The question now arising is whether there are other patient cohorts who may benefit from bilateral renal denervation in this way.

There is now a growing view that hypertensive diseases arising from a genetic source are associated with an elevation in sympathetic nerve activity, including that to the kidney. Any excessive sympathetic drive to the kidney would impinge on function by elevating plasma renin activity and angiotensin II levels in the plasma as well as causing fluid retention. Together, these two factors could exacerbate the deterioration in cardiovascular homeostasis and cause blood pressure to rise further. The knowledge that the renal sympathetic nerves can importantly regulate renal function is now being translated into therapeutic approaches as exemplified by the successful renal denervation of the resistant hypertensive patients.

Heart failure

The failing heart is characterized as a state whereby it is unable to provide sufficient cardiac output to supply the metabolic needs of the body. One major cause is a narrowing of the coronary arteries due to atherosclerosis, with the result that there is insufficient perfusion, and hence oxygen supply, to the ventricular myocardium, which limits cardiac output.

The development and long-term presence of hypertension is associated with congestive heart failure where there is a chronic elevation in afterload, which eventually causes a deterioration of the cardiac pump. There is now clear evidence from experimental models (Zucker et al., 2004) and human studies (Esler et al., 2003) that heart failure is associated with a sympathoexcitation which is a response to further drive the failing heart to increase output. This elevation in sympathetic drive has the drawback that at the kidney it causes fluid retention, directly by the action of the nerves on the tubular reabsorptive processes, via angiotensin II, and indirectly as a result of the increased secretion of aldosterone.

It is also important to recognize in any pathophysiological state that there may be increased basal sympathetic drive as well as a derangement in the reflex regulation of sympathetic outflow. This issue was investigated in experimental models of heart failure where it was found that both the sensitivity of the renal sympathetic nerve baroreceptor gain curves and the renal sympathoinhibition resulting from an acute saline volume load were depressed (Zucker et al., 2004). Importantly, it would appear that in models of cardiac hypertrophy, which is a stage that often precedes overt heart failure, there is also a deranged baroreceptor control of renal sympathetic nerve activity. Two approaches were used to induce cardiac hypertrophy, one using caffeine/isoprenaline administration and the other using thyroxine treatment. Under these conditions it was found that there was a decreased sensitivity of the baroreceptor gain curve regulation of renal sympathetic nerve activity (Flanagan et al., 2008), while the renal sympathoinhibition in response to an acute saline volume load did not occur (Buckley and Johns, 2011; Johns et al., 2011). It becomes apparent from all these reports that in both cardiac hypertrophy and failure, the increase in sympathetic drive to the kidney alters function in two ways, by raising basal tone which will cause an underlying fluid retention, and the loss of reflex regulation in response to normal everyday activities. Together, these deficits will ensure an inappropriate fluid balance which will further impact on the deterioration in cardiovascular homeostasis associated with cardiac failure.

Cirrhosis

Cirrhosis of the liver occurs with edema formation as a consequence of fluid retention by the kidney. It was noted earlier that there are sensory receptors in the liver and spleen which, when activated, result in an increase in renal sympathetic nerve activity and neural regulation of kidney function. Experimental investigations in this area have utilized the procedure of common bile duct ligation in rodents where cirrhosis develops over several weeks. In this model, the renal sympathoinhibition resulting from an acute saline volume load is blunted (DiBona and Kopp, 1997; Johns et al., 2011). At a functional level, the excretion of an acute saline volume load is attenuated but following renal denervation, the proportion of the load excreted returns towards normal values (DiBona and Kopp, 1997; Johns et al., 2011). Findings from patient studies are compatible with there being a higher level of neural regulation of the kidney as administration of α-adrenoceptor antagonists or application of anesthetics to induce a bilateral spinal block caused a diuresis and natriuresis which would reflect loss of renal sympathetic tone. Together, these reports indicate that in hepatic cirrhosis there is a sympathoexcitation which contributes to the edema formation and is most likely due to an excessive activation of sensory receptors in the hepatic and splenic areas of the body.

Renal failure

Renal diseases are strongly associated with the development of secondary hypertension (Campese et al., 2006). The exact physiological mechanisms that are brought into play during the development of this renal dependent hypertension is uncertain, but findings from experimental and clinical investigations have indicated an important role for the sympathetic nervous system. It has been pointed out above that the kidney has a significant sensory innervation which, when activated physiologically, can enhance or suppress renal sympathetic as a component of the renorenal reflexes. Using a rat model of renal injury induced by the injection of a small volume of the irritant phenol into the pole of the kidney (Ye et al., 2002), there was a rapid and sustained increase in blood pressure, an elevation in plasma norepinephrine levels, together with a contralateral renal sympathoexcitation. Together, these findings pointed to an inappropriate stimulation of the renal sensory receptors causing a rise in renal sympathetic tone which would contribute to the chronic elevation in blood pressure.

The situation existing in human chronic renal failure has been investigated by measuring muscle sympathetic nerve activity. Hausberg and coworkers (2002) reported that the bursting rate in the muscle sympathetic nerve signal in renal transplantation patients in whom the diseased kidney was retained was approximately threefold greater that that obtained in normal subjects. By contrast, if the diseased native kidneys were removed the muscle sympathetic nerve bursting rate was indistinguishable from the normal group. These key reports focus attention on the possibility that in the diseased kidney tissue, injury or inflammatory changes cause an overactivation of the sensory nerves, which initiates a sympathoexcitation which will not only increase peripheral vascular resistance and ultimately blood pressure, but will also further impact on the kidney to impair its function (Ditting et al., 2009).

CONCLUSION

The renal sympathetic innervation serves to regulate the secretory, excretory, and hemodynamic functions of the organ, all of which impinge on cardiovascular homeostasis. Sensory information arising from the cardiovascular, somatic, and visceral systems of the body is delivered to the autonomic centers of the brain for integration and the generation of an appropriate sympathetic outflow which may be global or selected to specific organs. The outcome of a patterned sympathetic response is a normalization of function. It is often the case that in pathophysiological states, inappropriate activation of sensory receptors within different organs of the body leads to an altered sympathetic outflow and a defective reflex regulation of kidney function such that it contributes to further deterioration in cardiovascular homeostasis. Moreover, the recent demonstration that bilateral renal denervation in man, albeit in a selected cohort of hypertensive patients, serves to emphasize the importance of the kidney in cardiovascular diseases. This may point the way to further therapeutic opportunities targeting both the afferent and efferent neural control of the kidney.

References

Aperia A, Ibarra F, Svensson LB et al. (1992). Calcineurin mediates alpha-adrenergic stimulation of Na+, K(+)-ATPase activity in renal tubule cells. Proc Natl Acad Sci U S A 89: 7394–7397.

Barajas L, Wang P (1978). Myelinated nerves of the rat kidney. A light and electron microscopic autoradiographic study. J Ultrastruct Res 65: 148–162.

Barajas L, Powers K, Wang P (1984). Innervation of the renal cortical tubules: a quantitative study. Am J Physiol 247: F50–F60.

Barajas L, Liu L, Powers K (1992). Anatomy of the renal innervation: intrarenal aspects and ganglia of origin. Can J Physiol Pharmacol 70: 735–749.

Bello-Reuss E, Colindres RE, Pastoriza-Munoz E et al. (1975). Effects of acute unilateral renal denervation in the rat. J Clin Invest 56: 208–217.

Buckley MM, Johns EJ (2011). Impact of L-NAME on the cardiopulmonary reflex in cardiac hypertrophy. Am J Physiol 301: R1549–R1556.

Campese VM, Kogosov E (1995). Renal afferent denervation prevents hypertension in rats with chronic renal failure. Hypertension 25: 878–882.

Campese VM, Mitra N, Sandee D (2006). Hypertension in renal parenchymal disease: why is it so resistant to treatment? Kidney Int 69: 967–973.

Colindres RE, Spielman WS, Moss NG et al. (1980). Functional evidence for renorenal reflexes in the rat. Am J Physiol 239: F265–F270.

Coote JH, Johns EJ, Macleod VH et al. (1972). Effect of renal nerve stimulation, renal blood flow and adrenergic blockade on plasma renin activity in the cat. J Physiol 226: 15–36.

Davis G, Johns EJ (1991). Effect of somatic nerve stimulation on the kidney in intact, vagotomized and carotid sinus-denervated rats. J Physiol 432: 573–584.

DiBona GF, Kopp UC (1997). Neural control of renal function. Physiol Rev 77: 75–197.

DiBona GF, Rios LL (1980). Renal nerves in compensatory renal response to contralateral renal denervation. Am J Physiol 238: F26–F30.

DiBona GF, Sawin LL (1982). Effect of renal nerve stimulation on NaCl and H_2O transport in Henle's loop of the rat. Am J Physiol 243: F576–F580.

DiBona GF, Sawin LL (1999). Renal hemodynamic effects of activation of specific renal sympathetic nerve fiber groups. Am J Physiol 276: R539–R549.

Ditting T, Tiegs G, Veelken R (2009). Autonomous innervation in renal inflammatory disease-innocent bystander or active modulator? J Mol Med 87: 865–870.

Eppel GA, Malpas SC, Denton KM et al. (2004). Neural control of renal medullary perfusion. Clin Exp Pharmacol Physiol 31: 387–396.

Esler M, Kaye D (2000). Sympathetic nervous system activation in essential hypertension, cardiac failure and psychosomatic heart disease. J Cardiovasc Pharmacol 35: S1–S7.

Esler M, Lambert G, Brunner-La Rocca HP et al. (2003). Sympathetic nerve activity and neurotransmitter release in humans: translation from pathophysiology into clinical practice. Acta Physiol Scand 177: 275–284.

Evans RG, Eppel GA, Anderson WP et al. (2004). Mechanisms underlying the differential control of blood flow in the renal medulla and cortex. J Hypertens 22: 1439–1451.

Flanagan ET, Buckley MM, Aherne CM et al. (2008). Impact of cardiac hypertrophy on arterial and cardiopulmonary baroreflex control of renal sympathetic nerve activity in anaesthetized rats. Exp Physiol 93: 1058–1064.

Gesek FA (1999). Alpha1- and alpha2-adrenoceptor control of sodium transport reverses in developing hypertension. Hypertension 33: 524–529.

Hamza SM, Kaufman S (2004). Splenorenal reflex modulates renal blood flow in the rat. J Physiol 558: 277–282.

Handa RK, Johns EJ (1985). Interaction of the renin-angiotensin system and the renal nerves in the regulation of rat kidney function. J Physiol 369: 311–321.

Hausberg M, Kosch M, Harmelink P et al. (2002). Sympathetic nerve activity in end-stage renal disease. Circulation 106: 1974–1979.

Hesse IF, Johns EJ (1984). The effect of graded renal nerve stimulation on renal function in the anaesthetized rabbit. Comp Biochem Physiol A 79: 409–414.

Hevener AL, Bergman RN, Donovan CM (2000). Portal vein afferents are critical for the sympathoadrenal response to hypoglycemia. Diabetes 49: 8–12.

Huang C, Johns EJ (1998). Role of ANG II in mediating somatosensory-induced renal nerve-dependent antinatriuresis in the rat. Am J Physiol 275: R194–R202.

Huang C, Yoshimoto M, Miki K et al. (2006). The contribution of brain angiotensin II to the baroreflex regulation of renal sympathetic nerve activity in conscious normotensive and hypertensive rats. J Physiol 574: 597–604.

Ishiki K, Morita H, Hosomi H (1991). Reflex control of renal nerve activity originating from the osmoreceptors in the hepato-portal region. J Auton Nerv Syst 36: 139–148.

Johns EJ (1981). An investigation into the type of beta-adrenoceptor mediating sympathetically activated renin release in the cat. Br J Pharmacol 73: 749–754.

Johns EJ (1991). The physiology and pharmacology of the renal nerves. Pol Arch Med Wewn 85: 141–149.

Johns EJ (2002). The autonomic nervous system and pressure-natriuresis in cardiovascular-renal interactions in response to salt. Clin Auton Res 12: 256–263.

Johns EJ (2005). Angiotensin II in the brain and the autonomic control of the kidney. Exp Physiol 90: 163–168.

Johns EJ, Manitius J (1986). An investigation into the alpha-adrenoceptor mediating renal nerve-induced calcium reabsorption by the rat kidney. Br J Pharmacol 89: 91–97.

Johns EJ, Manitius J (1987). An investigation into the neural regulation of calcium excretion by the rat kidney. J Physiol 383: 745–755.

Johns EJ, Lewis BA, Singer B (1976). The sodium-retaining effect of renal nerve activity in the cat: role of angiotensin formation. Clin Sci Mol Med 51: 93–102.

Johns E, Kopp UC, DiBona GF (2011). Neural control of renal function. Compr Physiol 1: 731–767.

Katholi RE, Hageman GR, Whitlow PL et al. (1983). Hemodynamic and afferent renal nerve responses to intra-renal adenosine in the dog. Hypertension 5: I149–I154.

Kopp U, Aurell M, Nilsson IM et al. (1980). The role of beta-1-adrenoceptors in the renin release response to graded renal sympathetic nerve stimulation. Pflugers Arch 387: 107–113.

Kopp UC, Cicha MZ, Smith LA (2003). Angiotensin blocks substance P release from renal sensory nerves by inhibiting PGE2-mediated activation of cAMP. Am J Physiol Renal Physiol 285: F472–F483.

Kopp UC, Cicha MZ, Smith LA (2007a). Activation of endothelin-a receptors contributes to angiotensin-induced suppression of renal sensory nerve activation. Hypertension 49: 141–147.

Kopp UC, Cicha MZ, Smith LA et al. (2007b). Renal sympathetic nerve activity modulates afferent renal nerve activity by PGE2-dependent activation of alpha1- and alpha2-adrenoceptors on renal sensory nerve fibers. Am J Physiol Regul Integr Comp Physiol 293: R1561–R1572.

Kopp UC, Jones SY, DiBona GF (2008). Afferent renal denervation impairs baroreflex control of efferent renal sympathetic nerve activity. Am J Physiol Regul Integr Comp Physiol 295: R1882–R1890.

Krum H, Schlaich M, Whitbourn R et al. (2009). Catheter-based renal sympathetic denervation for resistant hypertension: a multicentre safety and proof-of-principle cohort study. Lancet 373: 1275–1281.

Krum H, Simplicity HTN-1 Investigators (2011). Catheter-based renal sympathetic denervation for resistant hypertension: durability of blood pressure reduction out to 24 months. Hypertension 57: 911–917.

La Grange RG, Sloop CH, Schmid HE (1973). Selective stimulation of renal nerves in the anesthetized dog. Effect on renin release during controlled changes in renal hemodynamics. Circ Res 33: 704–712.

Li Y, Qiu J, Yang Z et al. (2008). Long-range correlation of renal sympathetic nerve activity in both conscious and anesthetized rats. J Neurosci Methods 172: 131–136.

Liu F, Nesbitt T, Drezner MK et al. (1997). Proximal nephron Na+/H+exchange is regulated by alpha 1A- and alpha 1B-adrenergic receptor subtypes. Mol Pharmacol 52: 1010–1018.

Lundin S, Ricksten SE, Thoren P (1984). Renal sympathetic activity in spontaneously hypertensive rats and normotensive controls as studied by three different methods. Acta Physiol Scand 120: 265–272.

McDonough AA (2010). Mechanisms of proximal tubule sodium transport regulation that link extracellular fluid volume and blood pressure. Am J Physiol Regul Integr Comp Physiol 298: R851–R861.

McDonough AA, Leong PK, Yang LE (2003). Mechanisms of pressure natriuresis: how blood pressure regulates renal sodium transport. Ann N Y Acad Sci 986: 669–677.

Miki K, Yoshimoto M (2005). Differential effects of behaviour on sympathetic outflow during sleep and exercise. Exp Physiol 90: 155–158.

Miki K, Yoshimoto M (2010). Role of differential changes in sympathetic nerve activity in the preparatory adjustments of cardiovascular functions during freezing behaviour in rats. Exp Physiol 95: 56–60.

Miki K, Hayashida Y, Sagawa S et al. (1989). Renal sympathetic nerve activity and natriuresis during water immersion in conscious dogs. Am J Physiol 256: R299–R305.

Miki K, Hayashida Y, Shiraki K (2002). Role of cardiac-renal neural reflex in regulating sodium excretion during water immersion in conscious dogs. J Physiol 545: 305–312.

Miki K, Yoshimoto M, Hayashida Y et al. (2009). Role of cardiopulmonary and carotid sinus baroreceptors in regulating renal sympathetic nerve activity during water immersion in conscious dogs. Am J Physiol Regul Integr Comp Physiol 296: R1807–R1812.

Moncrief K, Kaufman S (2006). Splenic baroreceptors control splenic afferent nerve activity. Am J Physiol Regul Integr Comp Physiol 290: R352–R356.

Morita H, Nishida Y, Hosomi H (1991). Neural control of urinary sodium excretion during hypertonic NaCl load in conscious rabbits: role of renal and hepatic nerves and baroreceptors. J Auton Nerv Syst 34: 157–169.

Osborn JL, DiBona GF, Thames MD (1981). Beta-1 receptor mediation of renin secretion elicited by low-frequency renal nerve stimulation. J Pharmacol Exp Ther 216: 265–269.

Sattar MA, Johns EJ (1995). Alpha1-Adrenoceptor subtypes mediating antinatriuresis in Wistar and stroke-prone spontaneously hypertensive rats. Eur J Pharmacol 294: 727–736.

Smits JF, Brody MJ (1984). Activation of afferent renal nerves by intrarenal bradykinin in conscious rats. Am J Physiol 247: R1003–R1008.

Thames MD, Abboud FM (1979). Interaction of somatic and cardiopulmonary receptors in control of renal circulation. Am J Physiol 237: H560–H565.

Walkowska A, Badzynska B, Kompanowska-Jezierska E et al. (2005). Effects of renal nerve stimulation on intrarenal blood flow in rats with intact or inactivated NO synthases. Acta Physiol Scand 183: 99–105.

Weaver LC, Genovesi S, Stella A et al. (1987). Neural, hemodynamic, and renal responses to stimulation of intestinal receptors. Am J Physiol 253: H1167–H1176.

Witkowski A, Prejbisz A, Florczak E et al. (2011). Effects of renal sympathetic denervation on blood pressure, sleep apnea course, and glycemic control in patients with resistant hypertension and sleep apnea. Hypertension 58: 559–565.

Yang L, Leong PK, Chen JO et al. (2002). Acute hypertension provokes internalization of proximal tubule NHE3 without inhibition of transport activity. Am J Physiol Renal Physiol 282: F730–F740.

Ye S, Zhong H, Yanamadala V et al. (2002). Renal injury caused by intrarenal injection of phenol increases afferent and efferent renal sympathetic nerve activity. Am J Hypertens 15: 717–724.

Yoshimoto M, Nagata K, Miki K (2010). Differential control of renal and lumbar sympathetic nerve activity during freezing behavior in conscious rats. Am J Physiol Regul Integr Comp Physiol 299: R1114–R1120.

Zhang T, Johns EJ (1997). Somatosensory influences on renal sympathetic nerve activity in anesthetized Wistar and hypertensive rats. Am J Physiol 272: R982–R990.

Zhang T, Huang C, Johns EJ (1997). Neural regulation of kidney function by the somatosensory system in normotensive and hypertensive rats. Am J Physiol 273: R1749–R1757.

Zucker IH, Schultz HD, Li YF et al. (2004). The origin of sympathetic outflow in heart failure: the roles of angiotensin II and nitric oxide. Prog Biophys Mol Biol 84: 217–232.

Handbook of Clinical Neurology, Vol. 117 (3rd series)
Autonomic Nervous System
R.M. Buijs and D.F. Swaab, Editors

Chapter 18

Autonomic neural control of the airways

STUART B. MAZZONE[1] AND BRENDAN J. CANNING[2]*
[1]*School of Biomedical Sciences, University of Queensland, St Lucia, Queensland, Australia*
[2]*Johns Hopkins Asthma and Allergy Center, Baltimore, MD, USA*

INTRODUCTION

The airways are a conduit for air to and from the gas-exchanging regions of the lung with the cells and end organs that comprise the airway wall being dynamically regulated to optimize ventilation. Airway smooth muscle regulates the dimensions of the airways, while mucus glands and the vasculature of the bronchial mucosa regulate airway surface liquid composition and volume. Protective reflexes, including cough and mucus secretion, clear the airways of accumulated secretions, irritants, and pathogens, and thus subserve the essential role of maintaining airway patency and preserving lung capacity for gas exchange. This dynamic control of the airways occurs on a breath-by-breath basis, and is regulated in large part by the opposing actions of the divisions of the autonomic nervous system. Autonomic nerves are the primary regulators of airway smooth muscle tone (and hence airway caliber), mucus gland secretion, and bronchial vascular tone (Widdicombe, 1991; Coleridge and Coleridge, 1994; Barnes and Liu, 1995; Haxhiu et al., 2005; Canning, 2006; Wine, 2007). Autonomic inputs also match airway and pulmonary vascular smooth muscle tone with ventilation to meet the ever-changing physiological demands for gas exchange.

Dysfunction and/or dysregulation of airway autonomic nerves account, in part, for the symptoms of respiratory disorders, including asthma, chronic obstructive pulmonary disease, pulmonary hypertension, pulmonary embolism, neurogenic pulmonary edema, allergic rhinitis and possibly sleep-disordered breathing (Joos et al., 2000; Canning, 2002; Undem and Kollarik, 2005; Kim and Baraniuk, 2007; Sedý et al., 2008; Undem and Nassenstein, 2009). It follows logically from the above that understanding the intricacies of these neural circuits will provide insights into the symptoms of respiratory diseases and potential therapeutic strategies. This chapter focuses on our current understanding of the control of the airway and airway protective reflexes by the autonomic nervous system. An overview of the efferent and afferent innervation of the airways precedes discussions of their control by the central nervous system (CNS).

AIRWAY PARASYMPATHETIC AND SYMPATHETIC POSTGANGLIONIC NEURONS

Intrinsic, postganglionic parasympathetic neurons innervate the airways and lungs and contribute significantly to the regulation of mucus secretion, contractions and relaxations of airway smooth muscle as well as dilatation of bronchial and pulmonary blood vessels (Coleridge and Coleridge, 1994; Barnes and Liu, 1995; Haxhiu et al., 2005; Canning, 2006; Wine, 2007). The effects of airway parasympathetic nerves on airway and lung end organs have been attributed primarily to acetylcholine acting at muscarinic type 3 receptors but also to vasoactive intestinal peptide, pituitary adenylate cyclase-activating polypeptide, peptide histidine isoleucine or peptide histidine methionine, and/or the gaseous transmitter nitric oxide (Palmer et al., 1986; Canning and Undem, 1993; Lei et al., 1993; Ward et al., 1995; Canning et al., 1996; Fischer et al., 1996; Kinhult et al., 2000; Canning, 2006). Postganglionic sympathetic nerves also innervate the airways and lungs and arise from paravertebral and superior cervical ganglia. Stimulation of sympathetic nerves can evoke noradrenergic contractions of the pulmonary and bronchial vasculature and relaxations of airway smooth muscle via α- and β-adrenoceptors, respectively (Cabezas et al., 1971; Barnes and Liu., 1995;

*Correspondence to: Brendan J. Canning, Johns Hopkins Asthma and Allergy Center, Baltimore, MD, 21224, USA. Tel: +1-410-550-2156, Fax: +1 (410) 550 2130, E-mail: bjc@jhmi.edu

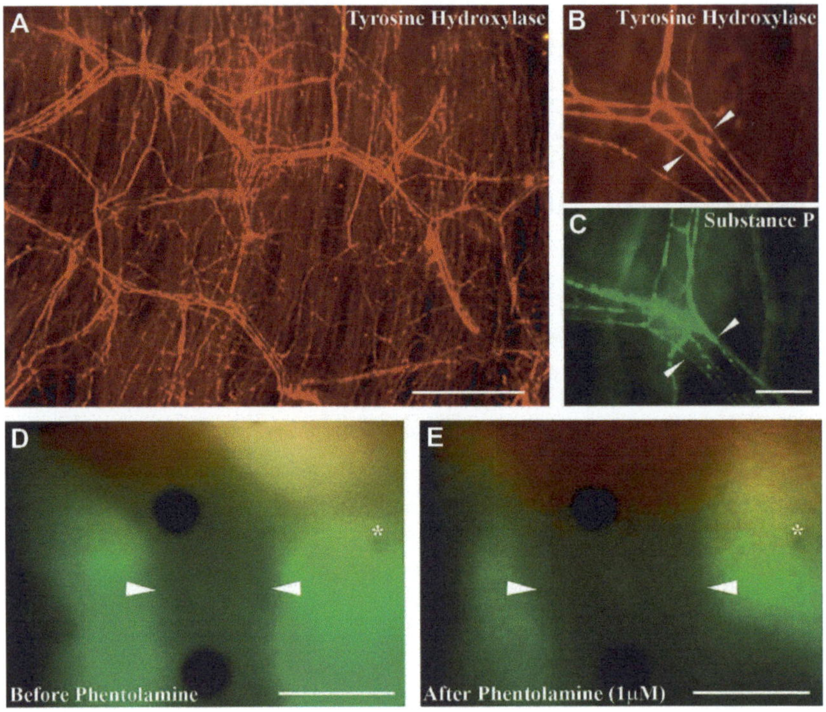

Fig. 18.1. Sympathetic control of airway vasculature in guinea pigs. (**A**, **B**) Immunohistochemical staining for tyrosine hydroxylase reveals sympathetic postganglionic nerve fibers innervating submucosal blood vessels in tracheal wholemounts. (**C**) Tracheal blood vessels are also densely innervated by substance P-expressing vagal sensory neurons. (**D**, **E**) Blockade of α-adrenoceptors with phentolamine in anesthetized guinea pigs evokes dilatation of submucosal arterioles, consistent with the presence of basal sympathetic vasomotor tone. (Reproduced from Mazzone et al., 2010.)

Canning, 2006; Oh et al., 2006; Mazzone et al., 2010) (Fig. 18.1). The smooth muscle of the bronchial and pulmonary vasculature may also contract in response to stimulation of sympathetic nerves due to the actions of neuropeptide Y, which is coreleased with noradrenaline/norepinephrine from postganglionic sympathetic nerves (Wahlestedt et al., 1990). Sympathetic innervation of mucus glands in the airways has been described anatomically, but its physiological function is unclear (Rogers, 2001).

Parasympathetic cholinergic innervation of the airways is well conserved amongst species, as is the sympathetic innervation of the pulmonary and bronchial vasculature. In contrast, the distribution and function of the sympathetic innervation and the noncholinergic parasympathetic of airway smooth muscle varies considerably between species. Sympathetic adrenergic nerves play a dominant inhibitory role in the airways of cats and dogs. In contrast, sympathetic adrenergic nerves do not innervate human and nonhuman primate airway smooth muscle (van der Velden and Hulsmann, 1999; Canning, 2006). In these species, circulating catecholamines provide the primary drive for β-adrenoceptor-mediated airway smooth muscle relaxations, whereas nonadrenergic, noncholinergic neurotransmitters, presumably released from postganglionic parasympathetic nerves, mediate nerve-dependent relaxations of human and nonhuman primate airway smooth muscle. Airway smooth muscle in rats and mice appears to be devoid of a direct, functional relaxant innervation (reviewed in Canning, 2006). Also in some species (dogs, rabbits, rats, mice) but not in others (humans, guinea pigs), stimulation of sympathetic nerves following blockade of β-adrenoceptors evokes contractions of airway smooth muscle through the effects of noradrenaline acting on α-adrenoceptors (Canning, 2006).

Anatomically and functionally distinct parasympathetic pathways regulate cholinergic and nonadrenergic, noncholinergic responses in the airways of cats and guinea pigs (Canning and Undem, 1993; Fischer et al., 1998; Canning, 2006). Reflexes differentially regulate these cholinergic and noncholinergic parasympathetic pathways (Mazzone and Canning, 2002a) and recent anatomical and functional studies confirm that distinct subtypes of preganglionic neurons drive parasympathetic contractile and relaxant responses in the airways (McGovern and Mazzone, 2010, discussed below). The parasympathetic innervation of human airways is probably most similar to that of cats and guinea pigs amongst the commonly studied animals (Canning, 2006).

Sensory innervation to the airways

Multiple vagal afferent nerve subtypes innervate the airways (reviewed in Mazzone, 2005; Canning et al., 2006).

Rapidly (RAR) and slowly (SAR) adapting stretch receptors, differentiated based on their responses to sustained lung inflation, are the two primary classes of mechanoreceptors innervating the airways and lungs. RARs, and to a lesser extent SARs, are also activated by other mechanical stimuli including airway smooth muscle contractions, lung deflation (or collapse), pulmonary edema, pulmonary vascular congestion, and punctate mechanical stimulation of the airway mucosa. Typically SARs are more active than RARs during eupnea (Fig. 18.2). Both vagal afferent nerve subtypes regulate respiration and autonomic tone at eupnea and in response to stimulation (Canning et al., 2001, 2006; Mazzone and Canning, 2002a, d) (Fig. 18.2).

More recently, mechanoreceptors not readily fitting into the classification scheme of RARs and SARs were shown to innervate the airways and lungs. These vagal afferents were found to play an essential role in the initiation of cough. The "cough receptors" are highly sensitive to punctate mechanical stimulation, acids and low chloride solutions but, unlike RARs and SARs, are not activated by distending pressures, airway collapse, or stretch (Canning et al., 2004; Mazzone, 2005; Mazzone and McGovern, 2006; Canning, 2010). Also unlike RARs and SARs, cough receptors are likely quiescent during eupnea, only becoming active in the presence of stimuli that evoke cough. Such vagal afferents likely innervate the airways of other species that cough such as dogs, cats, and humans (Widdicombe, 1954; Mortola et al., 1975; Yanaura et al., 1978), but have not yet been formally described in the published literature.

The structures of the peripheral terminals of bronchopulmonary mechanoreceptors are poorly described.

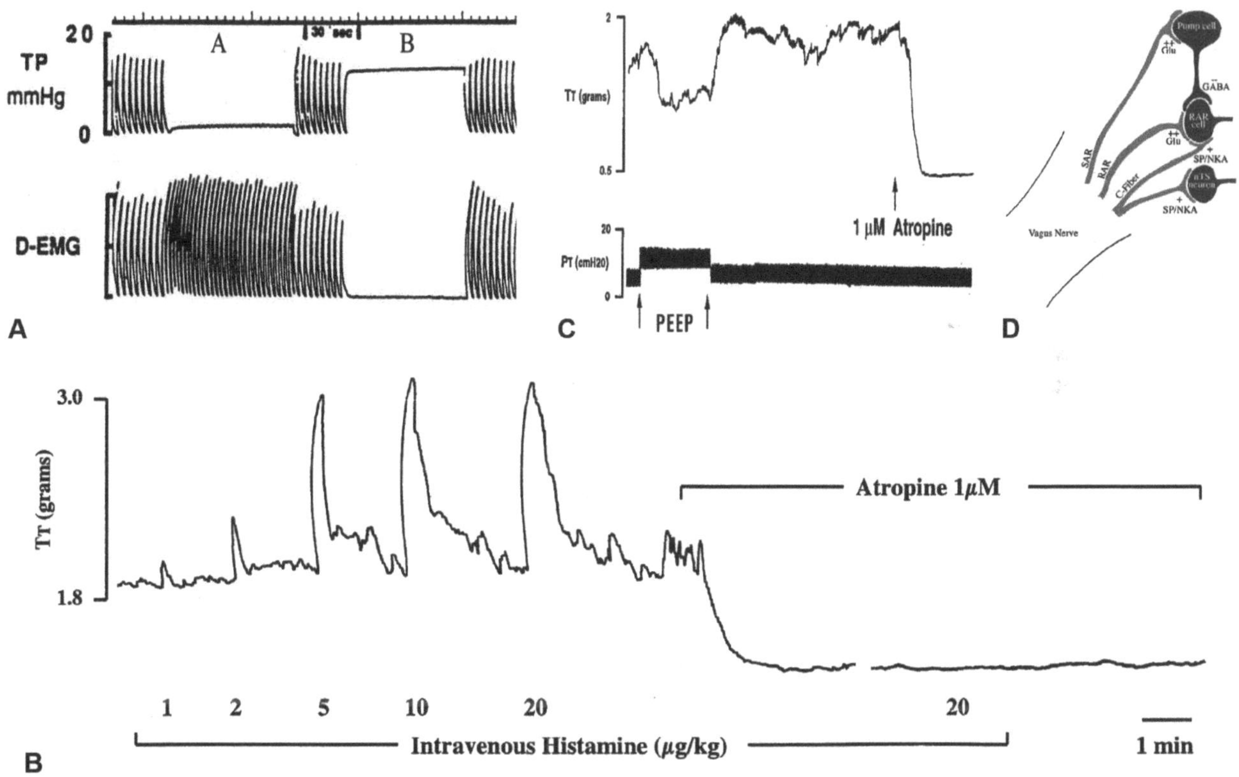

Fig. 18.2. Reflex regulation of breathing and airway smooth muscle tone by lung mechanoreceptors. In panel (**A**) lung collapse and sustained lung inflation are used to activate classic rapidly (RAR) and slowly (SAR) adapting receptors, respectively, in mechanically ventilated dogs. Rapidly adapting receptor activation increases respiratory muscle electromyographic activity (D-EMG) whereas slowly adapting receptor activation abolishes D-EMG. (**B**) RAR activation following bronchoconstriction of the intrapulmonary airways evokes atropine-sensitive, vagally mediated reflex contractions of the trachea in anesthetized guinea pigs. (**C**) By contrast, SAR activation during sustained lung stretch (induced by increasing positive end expiratory pressure (PEEP)) evokes a withdrawal of basal cholinergic tone in the guinea pig trachea. (**D**) A simplified diagram showing the brainstem circuitry which receives inputs from pulmonary mechanoreceptors. Note the excitatory glutamatergic (Glu) inputs of RARs onto second order neurons (RAR cells) which can be inhibited by SAR-mediated excitation of GABAergic pump cells. This circuitry explains the excitatory and inhibitory effects of RARs and SARs, respectively, on respiratory and airway smooth muscle control. (Data in panel (**A**) reproduced from Coon, 1994; panel (**B**) from Mazzone and Canning, 2002c; panel (**C**) from Kesler and Canning, 1999; panel (**D**) from Mazzone and Canning, 2002d.)

This is due in part to the fact that there is marked overlap in the termination sites of the various vagal afferent nerve subtypes innervating the airways but also because specific histochemical markers for these various subtypes are lacking. In most species, RARs and SARs are found primarily in the intrapulmonary airways and display baseline activity during tidal breathing. RAR terminations are more concentrated in the conducting airways, whereas SARs terminate in the conducting airways but also in and around the gas-exchanging units of the lung (Schelegle and Green, 2001; Widdicombe, 2003; Yu et al., 2003). However, species differences may exist. In dogs, for example, SARs are abundant in the trachea while few innervate the extrapulmonary airways of guinea pigs, rats, rabbits, or cats (Keller et al., 1989; Ho et al., 2001; Schelegle and Green, 2001; Canning et al., 2006). Based on their characteristic responses to stretching of the lung and smooth muscle contraction, however, it is reasonable to speculate that intrapulmonary RARs and SARs may directly innervate airway smooth muscle and/or align with the longitudinal or circumferential axis of the airways. Studies in rabbits have described such structures that correspond to the terminals of SARs (Yu et al., 2003). Comparable studies of RAR nerve terminals have not been reported. Recent studies in guinea pigs have described a population of highly organized neuronal structures in the airway mucosa that likely correspond to the terminals of cough receptors (Mazzone et al., 2009). These terminals were found between the epithelium and the smooth muscle layer, an observation that is consistent with the lack of effect of epithelial removal or airway smooth muscle contraction on the responsiveness or excitability of these afferent nerves (Canning et al., 2004; Mazzone et al., 2009).

Bronchopulmonary nociceptors are relatively less sensitive to stretch and touch, but respond vigorously to a variety of chemical irritants and proinflammatory molecules (Mazzone, 2005; Canning et al., 2006). They can be found innervating the airways from the larynx to the terminal bronchioles and are typically defined by their responsiveness to noxious endogenous (e.g., bradykinin) and exogenous (e.g., capsaicin) chemical stimuli. In general, nociceptors are likely quiescent in the normal, uninflamed airways, are relatively unresponsive to lung inflation, collapse, or stretch, and contribute little to the ongoing regulation of tidal breathing. Not surprisingly, nociceptor activation is associated with airway defensive reflexes, including cough, reflex bronchospasm, mucus secretion, and inhibition of sympathetic vasomotor outflow.

Bronchopulmonary nociceptors are a heterogenous population of sensory nerves. Although the axons of airway nociceptors are mostly unmyelinated, conducting action potentials in the C-fiber range ($\leq$1 m/sec), faster conducting Aδ nociceptors have been described (Riccio et al., 1996). Subtypes of airway afferent C fibers have also been identified based on their differential distribution in the airways and lungs and/or their differing embryological origins (Coleridge and Coleridge, 1984; Undem et al., 2004; Nassenstein et al., 2010; Lieu et al., 2011). Subtypes may also differ in their neurochemistry, responsiveness to chemical stimuli and the reflexes they initiate upon activation (Chou et al., 2008). The organization and anatomical attributes of these distinct nociceptor subpopulations in the airways is unknown.

In rats and in guinea pigs, a major subpopulation of afferent C fibers in the airways and lungs express the neuropeptides substance P, neurokinin A, and calcitonin gene-related peptide. Immunohistochemical staining for these neuropeptides reveals that nociceptors innervate structures throughout the airway walls and the lungs (e.g., Fig. 18.1), but are particularly prominent in the airway epithelium, where they form a dense, poorly organized plexus beneath and between the epithelial cell layers lining the airways (Baluk and Gabella, 1991; Mazzone et al., 2009). Structures comparable to the neuropeptide-containing fibers in the epithelium of guinea pigs and rats have been localized to the mucosa of other species, but in general, express far less neuropeptide immunoreactivity (Lamb and Sparrow, 2002). Whether these structures correspond to the peripheral terminals of C fibers in these species and where nociceptors that do not express the neuropeptides terminate in the airways of guinea pigs and rats is unknown. The poorly organized but extensive epithelial terminations of at least a subpopulation of nociceptors may place these vagal afferent nerves in an ideal position for detecting and responding to the external environment and to epithelial damage and inflammation, but renders them less sensitive than RARs and SARs to the internal environment and the mechanical forces associated with breathing. This is consistent with the primary role of airway nociceptors in defensive reflexes, their modest activity during eupnea, their relative insensitivity to lung volume changes, and thus their limited role in homeostatic reflexes.

Sensory neurons with cell bodies in the thoracic dorsal root ganglia also innervate the airways and lungs (Kummer et al., 1992; Groth et al., 2006; Oh et al., 2006). The functions of these primary afferent neurons, particularly in relation to neural control of the airways, remain poorly defined. Both mechanoreceptors and chemically sensitive nociceptors have been identified amongst the afferent nerves innervating the airways and associated structures (e.g., the pleura). Spinal terminations of afferent nerves innervating the airways and lungs have not been studied in any great detail.

Origin of airway parasympathetic preganglionic neurons

The existence of distinct postganglionic neurons controlling contractile and relaxant airway responses raises the question of whether these neurons are in turn regulated by common or distinct preganglionic inputs. Given that reflexes differentially regulate the parasympathetic postganglionic cholinergic and noncholinergic nerves that control airway smooth muscle tone (Mazzone and Canning, 2002a), it follows that distinct populations of preganglionic neurons likely control these opposing effects (McGovern and Mazzone, 2010). Functional and tract tracing studies have identified parasympathetic preganglionic neurons with cell bodies in the nucleus ambiguus that regulate cholinergic contractions of airway smooth muscle (Fig. 18.3) (Haxhiu and Loewy, 1996; Kc et al., 2004; Atoji et al., 2005; Chen et al., 2007; McGovern and Mazzone, 2010). The preganglionic nerves that innervate airway noncholinergic parasympathetic postganglionic neurons may be unmyelinated (Canning and Undem, 1993). However, anatomically these neurons are difficult to distinguish from those regulating cholinergic ganglia as they are intermingled within a common region of the nucleus ambiguus and express comparable neurochemical phenotypes (McGovern and Mazzone, 2010). Although subtypes of airway parasympathetic preganglionic neurons in the nucleus ambiguus have not been identified, preganglionic neurons can be differentiated from laryngeal and esophageal motor neurons within this nucleus (McGovern and Mazzone, 2010). Neurons in the DVN have been retrogradely labeled from the airways and from the neuronal plexuses containing the cholinergic and noncholinergic parasympathetic postganglionic neurons. However, airway dorsal vagal nucleus (DVN) neurons are far outnumbered by the neurons in the nucleus ambiguus that project to the airways (Haxhiu and Loewy, 1996; McGovern and Mazzone, 2010). To date, no function has been ascribed to the preganglionic neurons with cell bodies in the DVN. A relaxant innervation of the airways from the DVN would parallel a relaxant innervation of the stomach and esophagus from the DVN (Krowicki et al., 1999; Hornby et al., 2002). In the guinea pig at least, the noncholinergic parasympathetic postganglionic nerves regulating nonadrenergic relaxations of airway smooth muscle emanate from the adjacent esophagus (Canning and Undem, 1993); however, few DVN neurons project to the extrathoracic esophagus in this species (McGovern and Mazzone, 2010).

Parasympathetic reflexes may differentially regulate different airway segments and the right and left vagus nerves may also differentially regulate the airways (Phipps and Richardson, 1976; Canning and Undem, 1993; Kondo et al., 1995; Coon et al., 2000; McGovern

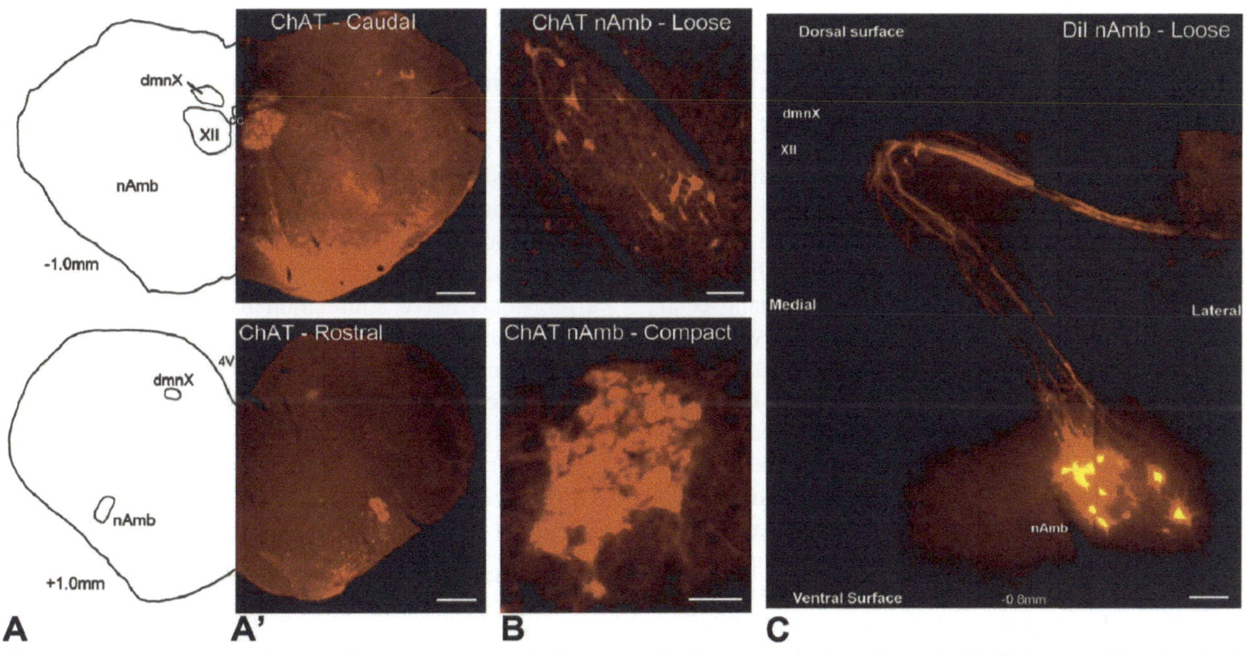

Fig. 18.3. Brainstem organization of vagal parasympathetic preganglionic neurons in the guinea pig. (**A, B**) Immunohistochemical staining for choline acetyltransferase in caudal and rostral medulla shows the location of vagal preganglionic neurons in the dorsal motor nucleus of the vagus nerve (dmnX) and the nucleus ambiguus (nAmb). (**C**) Retrograde tracing from the upper trachea using the fluorescent dye DiI shows the location of vagal motor neurons in the nAmb. Traced neurons in the dmnX are rarely seen. XII hypoglossal nucleus. (Reproduced from McGovern and Mazzone, 2010.)

and Mazzone, 2010). It is unclear what possible physiological role these differences would play in adult vertebrates, which have roughly symmetrical lungs that move air by creating pressure gradients in the airways secondary to changes in thoracic volume and not, as in amphibians, by peristaltic contractions of smooth muscle.

Gap junctions between neurons in the nucleus ambiguus that project to the airways or neuronal interconnections between postganglionic neurons in adjacent airway ganglia could coordinate muscle tone throughout the airways. There is conflicting evidence for electrical coupling of preganglionic neurons (Johnson and Getting, 1991; Lewis, 1994; Rekling and Feldman, 1997).

A novel class of airway preganglionic neurons, not originating from brainstem nuclei but rather derived from the adjacent esophageal myenteric plexus, has been identified in guinea pigs (Mazzone and McGovern, 2010) (Fig. 18.4). These neurons are characterized by their expression of the calcium-binding protein calretinin and the cholinergic marker choline acetyltransferase, and provide dense projections to airway cholinergic postganglionic neurons. Activation of these esophageal-derived preganglionic neurons evokes hexamethonium-sensitive contractions of the airway smooth muscle. This previously unrecognized complexity in the neural control of the airways may have important implications in the association between airway diseases and concomitant gastroesophageal reflux.

CENTRAL NERVOUS SYSTEM REGULATION OF AIRWAY SYMPATHETIC NEURONS

Several investigators have described recordings from pre- and postganglionic sympathetic nerves that may regulate airway smooth muscle and bronchial or pulmonary vascular tone (Bachoo and Polosa, 1987; Huangfu and Guyenet, 1991; Bartsch et al., 1999; Huang et al., 2000). The available evidence suggests that sympathetic outflow from the CNS to the airways is distinct from that of the sympathetic outflow to other organs. Pulmonary embolism, for example, alters pulmonary and systemic vascular tone secondary to the release of serotonin from platelets and the initiation of vasomotor reflexes (Leaños et al., 1995). Intravascularly administered serotonin or experimental pulmonary embolism both evoke a sympathetic reflex-dependent increase in pulmonary vascular tone but a decrease in renal sympathetic nerve activity and thus systemic blood pressure (Veelken et al., 1997; Stratmann and Gregory, 2003).

Some elements of CNS control of airway sympathetic nerves have been reported. Stimulating the posterior hypothalamus produces a sympathetic nerve-mediated decrease in tracheal smooth muscle tone with a concomitant increase in blood pressure and heart rate (Iwase et al., 1998). These responses were abolished by a histamine H_1-receptor antagonist administered into the fourth ventricle, suggesting a role for central histaminergic mechanisms in the evoked responses. Hyperthermia produces effects comparable to hypothalamic stimulation, with microdialysis confirming the release of histamine rostral ventrolateral medulla (RVLM), the nucleus of the solitary tract (NTS), and raphe magnus (Kanamaru et al., 2001). Microinjecting a histamine H_1-receptor antagonist into the RVLM inhibited the effects of hyperthermia on the airways and vasculature.

Reflexes may differentially regulate the sympathetic nerves innervating the airways and the extrapulmonary vasculature. Aerosol administration of capsaicin into the lower airways evoked a pronounced reversal of a histamine evoked contraction of the trachealis in vagotomized guinea pigs (Oh et al., 2006). Propranolol, hexamethonium, and tetrodotoxin applied topically to the tracheal mucosa abolished these reflexes, as did extrinsic (efferent) denervation of the trachea, or thoracic dorsal root transection. The tracheal smooth muscle relaxation proceeded without changes in heart rate or blood pressure.

Neurokinin receptor antagonists also abolished the airway sympathetic reflex evoked by aerosol administration of capsaicin (Oh et al., 2006), a result that was consistent with the observation that the majority of spinal sensory nerves supplying the airways and lungs of guinea pigs were primarily capsaicin-sensitive and expressed substance P. We speculate that tachykinins augment synaptic transmission in the spinal cord and thus augment basal sympathetic outflow to the airways. This could account for the selective effects of inhaled capsaicin on airway sympathetic outflow.

We also observed that electrical stimulation of the central cut end of the bilaterally transected vagi evokes a pronounced propranolol-sensitive relaxation of the precontracted trachealis *in situ*, which occurred only after vagus nerve stimulation had ceased (Oh et al., 2006). These relaxations were accompanied by dynamic changes in blood pressure that coincided with relaxation of the tracheal smooth muscle. These results are consistent with previous reports but also fail to identify the vagal afferent nerve subtypes involved (Huangfu and Guyenet, 1991; Veelken et al., 1997; Oh et al., 2006).

Brainstem integration of airway sensory nerves

The brainstem terminations of airway-specific vagal sensory nerve fibers have been characterized in some detail. Airway mechanoreceptors (RARs and SARs) and bronchopulmonary C fibers, originating in the nodose and jugular ganglia of the vagus nerves, terminate primarily in the caudal, medial, and ventromedial regions

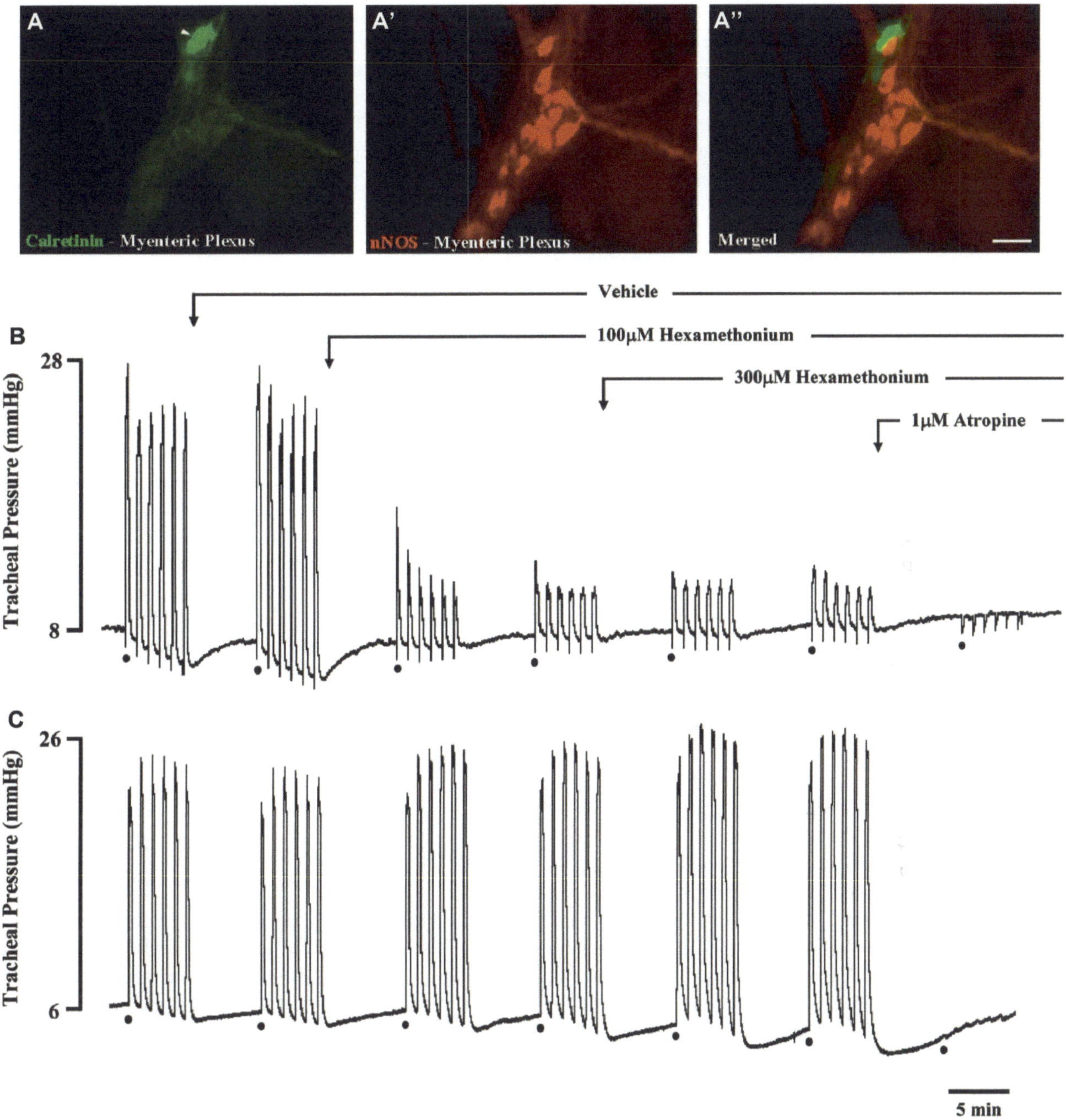

Fig. 18.4. Recent studies have revealed a previously unrecognized neural pathway for controlling airway smooth muscle tone. (**A**) Neurons in the guinea pig esophageal myenteric plexus, which are immunoreactive for the calcium-binding protein calretinin and choline acetyltransferase but not neuronal nitric oxide synthase, project axons to the adjacent tracheal cholinergic ganglia. (**B**) Electrical stimulation of the esophagus evokes contractions in the tracheal smooth muscle which are sensitive to the ganglionic blocker hexamethonium and the muscarinic receptor antagonist atropine. (**C**) By contrast, contractions of the trachea evoked by electric field stimulation are not blocked by hexamethonium, confirming that contractions evoked by esophageal stimulation are entirely dependent on ganglionic neurotransmission and are not due to aberrant spread of the electrical current to the tracheal ganglia. (Data modified from Mazzone and McGovern, 2010.)

of the nucleus of the solitary tract (NTS) (Kalia and Mesulam, 1980a, b; Miyazaki et al., 1999; Ezure and Tanaka, 2000; Otake et al., 2001). Although there is some topographical organization of these terminations, different afferent nerve subtypes from the airways also project directly or indirectly to adjacent and overlapping regions of the NTS. Electrophysiological studies have defined and characterized in detail the second order NTS neurons that receive input from vagal SARs and RARs (referred to as pump cells and RAR cells respectively) (Bonham and McCrimmon, 1990; Miyazaki et al., 1999; Ezure and Tanaka, 2000) (see Fig. 18.2D). Less is known about the NTS neurons receiving input from bronchopulmonary C fibers, or whether the different subtypes of bronchopulmonary C fibers that have been identified project to common or different NTS neurons. It is, however, interesting that the respiratory reflex responses that follow the activation of different airway C-fiber populations are quite distinct, suggesting that airway afferent subtypes likely synapse with distinct NTS neurons which in turn provide differential regulatory inputs to the brainstem autonomic and respiratory networks (Chou et al., 2008). The recent description of the novel airway mechanoreceptor (cough receptor) has precipitated investigation into where these afferent nerves terminate in the NTS. Functional and tracing studies indicate that cough receptor relay neurons are likely confined to a discrete region of the medial NTS (Ohi et al., 2005; Jakus et al., 2008; Canning and Mori, 2010). Microinjection of excitatory amino acid antagonists into this area abolishes cough receptor-evoked coughing but has negligible effects on basal respiratory rate or other airway afferent nerve evoked reflexes. Whether these findings reflect a unique termination locus for cough receptor afferents relative to other airway sensory nerves or a unique projection of cough receptor relay neurons to brainstem circuitry purely involved in cough is at present unknown.

Neurotransmission at the synapses between identified vagal sensory neurons innervating the airways and the second-order NTS neurons that receive their input has been studied using physiological and electrophysiological methods. Electrophysiological analyses identify glutamate acting at both *N*-methyl-D-aspartate (NMDA) and non-NMDA receptors as the primary transmitter of vagal afferents innervating the airways (Aylwin et al., 1997; Haxhiu et al., 2000; Canning and Mori, 2011). The actions of glutamate are mediated primarily through non-NMDA receptors at vagal afferent synapses in the NTS. Nitric oxide released in the NTS secondary to activation of glutamate receptors may enhance, while presynaptic metabotropic glutamate receptors may attenuate or modulate, the actions of glutamate during airway reflexes (Chianca et al., 2004). Tachykinins substance P and neurokinin A acting at neurokinin-1, neurokinin-2, and neurokinin-3 receptors may play a role in modulating or mediating synaptic transmission at synapses made by bronchopulmonary C fibers in the NTS (Mazzone and Geraghty, 1999; Canning et al., 2001). Other transmitters, including adenosine, adenosine triphosphate (ATP), angiotensin, calcitonin gene-related peptide, carbon monoxide, GABA, norepinephrine, serotonin, and ligands for the transient receptor potential vanilloid type 1 receptor, may also regulate synaptic transmission in the NTS and consequently airway autonomic reflexes (reviewed in Lawrence and Jarrott, 1996).

Second-order neurons in the NTS often receive input from multiple afferent nerve subtypes, including afferent nerve fibers that do not innervate the airways. These multiple and varying sensory inputs to the NTS help shape the reflex response to a given stimulus and may explain why visceral reflexes rarely proceed independently under physiological conditions (Mazzone and Canning, 2002b, d) (Fig. 18.5). Anatomical and functional studies have also shown considerable convergence of vagal afferents at sites of brainstem integration (Mifflin, 1996; Paton, 1998; Mazzone and Canning, 2002b). Through volume transmission or convergence, bronchopulmonary C fibers may augment reflexes evoked by airway mechanoreceptors (Mazzone and Canning, 2002b; Mazzone et al., 2005) (Fig. 18.2). Such sensitizing effects may be mediated by neurokinins (Sekizawa et al., 2008, 2011).

Central interactions between vagal afferents may not necessarily always be synergistic. Slowly adapting receptor activation attenuates rapidly adapting receptor dependent parasympathetic-cholinergic contractions and cough receptor-dependent coughing (Kesler and Canning, 1999; Canning et al., 2001; Mazzone and Canning, 2002a; Ezure and Tanaka, 2000, 2004; Canning and Mori, 2011) and reduces synaptic transmission in NTS neurons receiving input from rapidly adapting receptors (Ezure and Tanaka, 2000) (Fig. 18.2).

Brainstem pathways that regulate airway autonomic tone

There is limited information available on brainstem regulation of airway autonomic tone (Haxhiu et al., 1996, 2005). Tracing studies using pseudorabies virus have revealed a network of airway vagal afferent endings in the NTS and airway parasympathetic preganglionic neurons projecting primarily from the nucleus ambiguus (Haxhiu et al., 1993). Second-order labeling was found in multiple locations including the hypothalamus, ventral medullary raphe nuclei, the A5 noradrenergic cell group, the locus coeruleus, and the Kölliker-Fuse nucleus. Second-order NTS neurons receiving slowly and rapidly adapting receptor input from the airways project directly to the nucleus ambiguus and to adjacent ipsilateral and contralateral NTS subnuclei (Otake et al., 2001; Ezure et al., 2002). The projections of second-order NTS neurons receiving input from bronchopulmonary C fibers have not been described.

Airway smooth muscle tone is the consequence of what seems to be a complex, interconnected network of neural pathways. GABAergic periaqueductal gray neurons project to the nucleus ambiguus and their activation results in a withdrawal of airway cholinergic tone

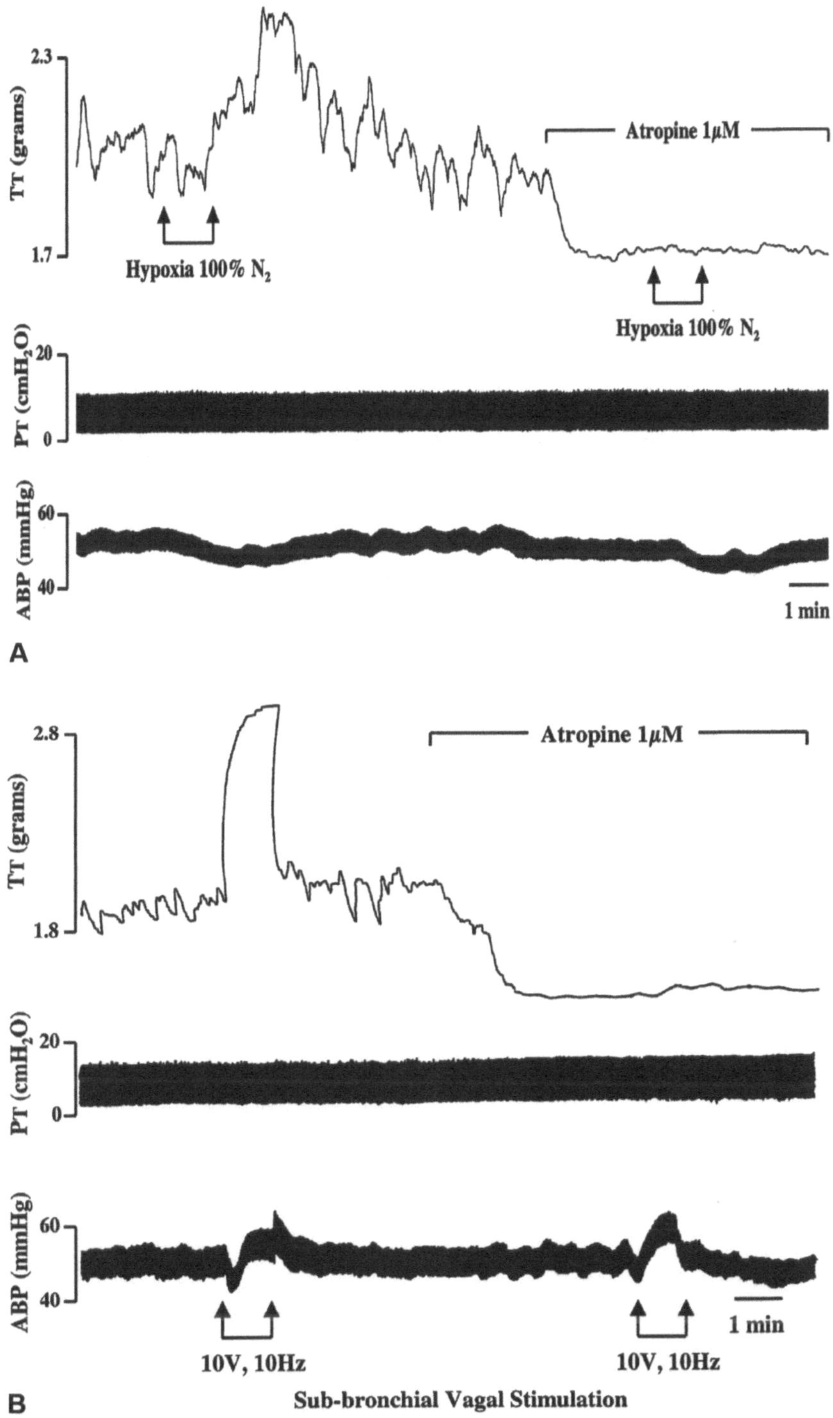

Fig. 18.5. Extrapulmonary afferent pathways regulate vagal cholinergic tone in the airways. Panels (**A**) and (**B**) show data obtained from anesthetized, paralyzed and mechanically ventilated guinea pigs. In each experiment, tracheal tone (TT), ventilator pressure (PT), and arterial blood pressure (ABP) are monitored. In (**A**), a brief period of hypoxic stimulation of arterial chemoreceptors, produced by temporarily replacing the ventilated air with 100% nitrogen, produces a substantial elevation in tracheal cholinergic tone. In (**B**), electrical stimulation of the vagi below the lungs (i.e., abdominal vagal afferents) produces a rapid rise in tracheal tone. Both responses are abolished by atropine pretreament confirming their reflex cholinergic nature.

(Haxhiu et al., 2002). Hypothalamic paraventricular nucleus neurons also project to the nucleus ambiguus and elsewhere in the brainstem (Kc et al., 2006). Hypothalamic stimulation either increases or decreases airway parasympathetic drive, depending on which subpopulation of neurons is recruited (Tatsuta and Arita, 1996; Beyaert et al., 1998). Serotonergic medullary raphe nuclei neurons and noradrenergic neurons arising from the locus coeruleus also regulate airway parasympathetic tone.

Ascending and descending suprapontine pathways

Airway afferent inputs to the NTS ascend the neuroaxis, presumably via relays in the parabrachial, midbrain, and thalamic nuclei (Fig. 18.6). These pathways are poorly described and their functional relevance is only beginning to be clarified. Stimulation of vagal afferents in rats modifies activity in several higher order nuclei including the ventral posteriomedial and ventral posteriolateral parvocellular nuclei of the thalamus, insula cortex, primary sensory cortex, and amygdala (Cechetto and Saper, 1987; Hanamori et al., 1998; Ito, 2002; Lyubashina and Panteleev, 2009). Functional brain imaging studies in humans have shown that sensory irritation of the airways recruits neuronal activity in the primary sensory, insula, cingulate, premotor, motor, and orbitofrontal cortices (Mazzone et al., 2007, 2011). These regions likely contribute to the perceptual awareness of airway irritations, and the resulting cognitive, emotional, and behavioral consequences that arise (unpleasantness, anxiety, spatial awareness, and the motivational drive to respond). This circuitry may be particularly relevant to respiratory responses that offer some level of voluntary control (e.g., coughing, speech, breath holding, etc.). Indeed, distinct motor control networks arising from the cortex in humans and animals provide descending drive for coughing and breathing, including the primary motor, insula, premotor, and anterior cingulate cortices, supplementary motor area, ventral thalamus, caudate nucleus, putamen, amygdala, and cerebellum (Kase et al., 1984; McKay et al., 2003; Simonyan et al., 2007; Mazzone et al., 2011). Furthermore, a higher order brain network that includes the inferior frontal gyrus, anterior insula, and prefrontal cortex provides descending inhibitory control of evoked respiratory responses, including the suppression of breathing and reflexively evoked coughing (McKay et al., 2008; Mazzone et al., 2011). How these higher order brain networks interact with the brainstem and spinal respiratory circuitry is at present unclear. Although not yet directly studied, cortical structures almost certainly also regulate airway autonomic tone.

AUTONOMIC REGULATION OF THE AIRWAYS AT REST

Basal cholinergic tone is demonstrable in all species, while basal noradrenergic and nonnoradrenergic relaxant tone is highly variable even within species (Kesler and Canning, 1999; Haxhiu et al., 2002; Kesler et al., 2002; Mazzone and Canning, 2002a, b, c). The mechanisms regulating basal autonomic tone in the airways are not fully understood but are likely attributable to ongoing vagal afferent drive, presumably from RARs and C fibers (Kesler and Canning, 1999; Kesler et al., 2002; Canning, 2006). CNS pathways also regulate autonomic outflow to the airways both at rest and upon reflex activation, and central regulation of airway autonomic tone may be closely coupled to central respiratory rhythm generation (Mitchell et al., 1985). The presence and level of basal tone in the airways of all species studied to date suggests a "set point" for smooth muscle contraction, perhaps optimized such that a withdrawal or augmentation of tone can be rapidly achieved in response to physiological and pathophysiological stimuli. The presence of this tone adds

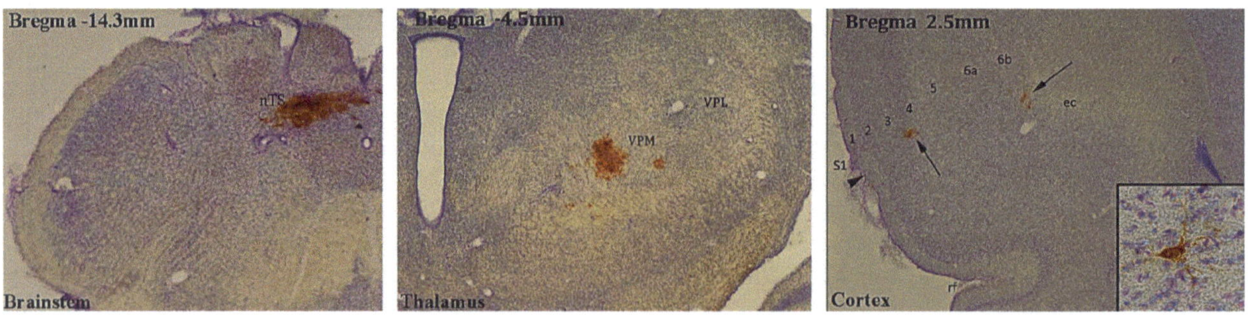

Fig. 18.6. Vagal afferent pathways project via the brainstem to higher order forebrain nuclei. The three panels show immunoreactivity for the herpes simplex virus (HSV1, strain H129) which has been transneuronally transported to the brainstem, thalamus, and primary sensory cortex, following inoculation of the virus into the trachea of guinea pigs. nTS, nucleus of the solitary tract; VPM, ventral posteriomedial nucleus; VPL, ventral posteriolateral nucleus; S1 primary sensory cortex; rh; rhinal fissure; numbers 1–6b = cortical layers. (Reproduced from McGovern et al., 2011.)

considerable complexity to the mechanisms underlying reflex regulation of airway caliber (Canning, 2006).

CONCLUSIONS

The airways are regulated by multiple autonomic efferent pathways with often opposing actions on the effector cells comprising the airway wall and with pathway-specific reflex mechanisms recruited by multiple vagal afferent nerve subtypes. These nerves play a central role in the pathophysiology of diseases of the airways and lung including asthma, chronic obstructive pulmonary disease, pulmonary hypertension, neurogenic pulmonary edema, and pulmonary thromboembolism. Understanding how central autonomic control of airway function affects disease outcomes and changes with chronic illness may lead to new therapeutic approaches to bronchopulmonary diseases.

ACKNOWLEDGMENTS

Dr. Mazzone's research is supported by grants from the National Health and Medical Research Council of Australia (566734, 454776). Dr. Canning's research is supported by a grant from the National Institutes of Health (HL083192).

References

Atoji Y, Kusindarta DL, Hamazaki N et al. (2005). Innervation of the rat trachea by bilateral cholinergic projections from the nucleus ambiguus and direct motor fibers from the cervical spinal cord: a retrograde and anterograde tracer study. Brain Res 1031: 90–100.

Aylwin ML, Horowitz JM, Bonham AC (1997). NMDA receptors contribute to primary visceral afferent transmission in the nucleus of the solitary tract. J Neurophysiol 77: 2539–2548.

Bachoo M, Polosa C (1987). Properties of the inspiration-related activity of sympathetic preganglionic neurones of the cervical trunk in the cat. J Physiol 385: 545–564.

Baluk P, Gabella G (1991). Afferent nerve endings in the tracheal muscle of guinea-pigs and rats. Anat Embryol (Berl) 183: 81–87.

Barnes PJ, Liu SF (1995). Regulation of pulmonary vascular tone. Pharmacol Rev 47: 87–131.

Bartsch T, Häbler HJ, Jänig W (1999). Hypoventilation recruits preganglionic sympathetic fibers with inspiration-related activity in the superior cervical trunk of the rat. J Auton Nerv Syst 77: 31–38.

Beyaert CA, Hill JM, Lewis BK et al. (1998). Effect on airway caliber of stimulation of the hypothalamic locomotor region. J Appl Physiol 84: 1388–1394.

Bonham AC, McCrimmon DR (1990). Neurones in a discrete region of the nucleus tractus solitarius are required for the Breuer–Hering reflex in rat. J Physiol 427: 261–280.

Cabezas GA, Graf PD, Nadel JA (1971). Sympathetic versus parasympathetic nervous regulation of airways in dogs. J Appl Physiol 31: 651–655.

Canning BJ (2002). Neurology of allergic inflammation and rhinitis. Curr Allergy Asthma Rep 2: 210–215.

Canning BJ (2006). Reflex regulation of airway smooth muscle tone. J Appl Physiol 101: 971–985.

Canning BJ (2010). Afferent nerves regulating the cough reflex: mechanisms and mediators of cough in disease. Otolaryngol Clin North Am 43: 15–25.

Canning BJ, Mori N (2010). An essential component to brainstem cough gating identified in anesthetized guinea pigs. FASEB J 24: 3916–3926.

Canning BJ, Mori N (2011). Encoding of the cough reflex in anesthetized guinea pigs. Am J Physiol Regul Integr Comp Physiol 300: R369–R377.

Canning BJ, Undem BJ (1993). Evidence that distinct neural pathways mediate parasympathetic contractions and relaxations of guinea-pig trachealis. J Physiol 471: 25–40.

Canning BJ, Mazzone SB, Meeker SN et al. (2004). Identification of the tracheal and laryngeal afferent neurones mediating cough in anaesthetized guinea-pigs. J Physiol 557: 543–558.

Canning BJ, Mori N, Mazzone SB (2006). Vagal afferent nerves regulating the cough reflex. Respir Physiol Neurobiol 152: 223–242.

Canning BJ, Undem BJ, Karakousis PC et al. (1996). Effects of organotypic culture on parasympathetic innervation of guinea pig trachealis. Am J Physiol 271: L698–L706.

Canning BJ, Reynolds SM, Mazzone SB (2001). Multiple mechanisms of reflex bronchospasm in guinea pigs. J Appl Physiol 91: 2642–2653.

Cechetto DF, Saper CB (1987). Evidence for a viscerotopic sensory representation in the cortex and thalamus in the rat. J Comp Neurol 262: 27–45.

Chen Y, Li M, Liu H et al. (2007). The airway-related parasympathetic motoneurones in the ventrolateral medulla of newborn rats were dissociated anatomically and in functional control. Exp Physiol 92: 99–108.

Chianca DA Jr, Lin LH, Dragon DN et al. (2004). NMDA receptors in nucleus tractus solitarii are linked to soluble guanylate cyclase. Am J Physiol Heart Circ Physiol 286: H1521–H1527.

Chou YL, Scarupa MD, Mori N et al. (2008). Differential effects of airway afferent nerve subtypes on cough and respiration in anesthetized guinea pigs. Am J Physiol Regul Integr Comp Physiol 295: R1572–R1584.

Coleridge JC, Coleridge HM (1984). Afferent vagal C fibre innervation of the lungs and airways and its functional significance. Rev Physiol Biochem Pharmacol 99: 1–110.

Coleridge HM, Coleridge JC (1994). Neural regulation of bronchial blood flow. Respir Physiol 98: 1–13.

Coon RL (1994). Reflex effects of lung inflation on tracheomotor tone observed during apnea produced by the Hering–Breuer reflex. J Appl Physiol 76: 2546.

Coon RL, Mueller PJ, Clifford PS (2000). Functional anatomy of the vagal innervation of the cervical trachea of the dog. J Appl Physiol 89: 139–142.

Ezure K, Tanaka I (2000). Lung inflation inhibits rapidly adapting receptor relay neurons in the rat. Neuroreport 11: 1709–1712.

Ezure K, Tanaka I (2004). GABA, in some cases together with glycine, is used as the inhibitory transmitter by pump cells in the Hering–Breuer reflex pathway of the rat. Neuroscience 127: 409–417.

Ezure K, Tanaka I, Saito Y et al. (2002). Axonal projections of pulmonary slowly adapting receptor relay neurons in the rat. J Comp Neurol 446: 81–94.

Fischer A, Canning BJ, Kummer W (1996). Correlation of vasoactive intestinal peptide and nitric oxide synthase with choline acetyltransferase in the airway innervation. Ann N Y Acad Sci 805: 717–722.

Fischer A, Canning BJ, Undem BJ et al. (1998). Evidence for an esophageal origin of VIP-IR and NO synthase-IR nerves innervating the guinea pig trachealis: a retrograde neuronal tracing and immunohistochemical analysis. J Comp Neurol 394: 326–334.

Groth M, Helbig T, Grau V et al. (2006). Spinal afferent neurons projecting to the rat lung and pleura express acid sensitive channels. Respir Res 7: 96.

Hanamori T, Kunitake T, Kato K et al. (1998). Responses of neurons in the insular cortex to gustatory, visceral, and nociceptive stimuli in rats. J Neurophysiol 79: 2535–2545.

Haxhiu MA, Loewy AD (1996). Central connections of the motor and sensory vagal systems innervating the trachea. J Auton Nerv Syst 57: 49–56.

Haxhiu MA, Jansen AS, Cherniack NS et al. (1993). CNS innervation of airway-related parasympathetic preganglionic neurons: a transneuronal labeling study using pseudorabies virus. Brain Res 618: 115–1134.

Haxhiu MA, Erokwu BO, Cherniack NS (1996). The brainstem network involved in coordination of inspiratory activity and cholinergic outflow to the airways. J Auton Nerv Syst 61: 155–161.

Haxhiu MA, Yamamoto B, Dreshaj IA et al. (2000). Involvement of glutamate in transmission of afferent constrictive inputs from the airways to the nucleus tractus solitarius in ferrets. J Auton Nerv Syst 80: 22–30.

Haxhiu MA, Yamamoto BK, Dreshaj IA et al. (2002). Activation of the midbrain periaqueductal gray induces airway smooth muscle relaxation. J Appl Physiol 93: 440–449.

Haxhiu MA, Kc P, Moore CT et al. (2005). Brain stem excitatory and inhibitory signaling pathways regulating bronchoconstrictive responses. J Appl Physiol 98: 1961–1982.

Ho CY, Gu Q, Lin YS et al. (2001). Sensitivity of vagal afferent endings to chemical irritants in the rat lung. Respir Physiol 127: 113–124.

Hornby PJ, Abrahams TP, Partosoedarso ER (2002). Central mechanisms of lower esophageal sphincter control. Gastroenterol Clin North Am 31 (4 Suppl): S11–S20.

Huang WX, Yu Q, Cohen MI (2000). Fast (3 Hz and 10 Hz) and slow (respiratory) rhythms in cervical sympathetic nerve and unit discharges of the cat. J Physiol 523: 459–477.

Huangfu DH, Guyenet PG (1991). Sympatholytic response to stimulation of superior laryngeal nerve in rats. Am J Physiol 260: R290–R297.

Ito S (2002). Visceral region in the rat primary somatosensory cortex identified by vagal evoked potential. J Comp Neurol 444: 10–24.

Iwase M, Kanamaru M, Kanamaru A et al. (1998). Central histaminergic neurons regulate rabbit tracheal tension through the cervical sympathetic nerve. J Auton Nerv Syst 74: 23–32.

Jakus J, Poliacek I, Halasova E et al. (2008). Brainstem circuitry of tracheal-bronchial cough: c-fos study in anesthetized cats. Respir Physiol Neurobiol 160: 289–300.

Johnson SM, Getting PA (1991). Electrophysiological properties of neurons within the nucleus ambiguus of adult guinea pigs. J Neurophysiol 66: 744–761.

Joos GF, Germonpré PR, Pauwels RA (2000). Neural mechanisms in asthma. Clin Exp Allergy 30 (Suppl 1): 60–65.

Kalia M, Mesulam MM (1980a). Brain stem projections of sensory and motor components of the vagus complex in the cat: I. The cervical vagus and nodose ganglion. J Comp Neurol 193: 435–465.

Kalia M, Mesulam MM (1980b). Brain stem projections of sensory and motor components of the vagus complex in the cat: II. Laryngeal, tracheobronchial, pulmonary, cardiac, and gastrointestinal branches. J Comp Neurol 193: 467–508.

Kanamaru M, Iwase M, Homma I (2001). Neuronal histamine release elicited by hyperthermia mediates tracheal dilation and pressor response. Am J Physiol Regul Integr Comp Physiol 280: R1748–R1754.

Kase Y, Kito G, Miyata T et al. (1984). Influence of cerebral cortex stimulation upon cough-like spasmodic expiratory response (SER) and cough in the cat. Brain Res 306: 293–298.

Kc P, Mayer CA, Haxhiu MA (2004). Chemical profile of vagal preganglionic motor cells innervating the airways in ferrets: the absence of noncholinergic neurons. J Appl Physiol 97: 1508–1517.

Kc P, Karibi-Ikiriko A, Rust CF et al. (2006). Phenotypic traits of the hypothalamic PVN cells innervating airway-related vagal preganglionic neurons. Respir Physiol Neurobiol 154: 319–330.

Keller E, Kohl J, Koller EA (1989). Location of pulmonary stretch receptors in the guinea-pig. Respir Physiol 76: 149–157.

Kesler BS, Canning BJ (1999). Regulation of baseline cholinergic tone in guinea-pig airway smooth muscle. J Physiol 518: 843–855.

Kesler BS, Mazzone SB, Canning BJ (2002). Nitric oxide-dependent modulation of smooth-muscle tone by airway parasympathetic nerves. Am J Respir Crit Care Med 165: 481–488.

Kim D, Baraniuk JN (2007). Neural aspects of allergic rhinitis. Curr Opin Otolaryngol Head Neck Surg 15: 268–273.

Kinhult J, Andersson JA, Uddman R et al. (2000). Pituitary adenylate cyclase-activating peptide 38 a potent endogenously produced dilator of human airways. Eur Respir J 15: 243–247.

Kondo T, Kobayashi I, Hirokawa Y et al. (1995). Differences in motor control in the bronchus and extrathoracic trachea. J Auton Nerv Syst 55: 1–8.

Krowicki KZ, Sivarao VD, Abrahams PT et al. (1999). Excitation of dorsal motor vagal neurons evokes non-nicotinic receptor-mediated gastric relaxation. J Auton Nerv Syst 77: 83–89.

Kummer W, Fischer A, Kurkowski R, Heym C (1992). The sensory and sympathetic innervation of guinea-pig lung and trachea as studied by retrograde neuronal tracing and double-labelling immunohistochemistry. Neuroscience 49: 715–737.

Lamb JP, Sparrow MP (2002). Three-dimensional mapping of sensory innervation with substance P in porcine bronchial mucosa: comparison with human airways. Am J Respir Crit Care Med 166: 1269–1281.

Lawrence AJ, Jarrott B (1996). Neurochemical modulation of cardiovascular control in the nucleus tractus solitarius. Prog Neurobiol 48: 21–53.

Leaños OL, Hong E, Amezcua JL (1995). Reflex circulatory collapse following intrapulmonary entrapment of activated platelets: mediation via 5-HT3 receptor stimulation. Br J Pharmacol 116: 2048–2052.

Lei YH, Barnes PJ, Rogers DF (1993). Regulation of NANC neural bronchoconstriction in vivo in the guinea-pig: involvement of nitric oxide, vasoactive intestinal peptide and soluble guanylyl cyclase. Br J Pharmacol 108: 228–235.

Lewis DI (1994). Dye-coupling between vagal motoneurones within the compact region of the adult rat nucleus ambiguus, in-vitro. J Auton Nerv Syst 47: 53–58.

Lieu T, Kollarik M, Myers AC et al. (2011). Neurotrophin and GDNF family ligand receptor expression in vagal sensory nerve subtypes innervating the adult guinea pig respiratory tract. Am J Physiol Lung Cell Mol Physiol 300: L790–L798.

Lyubashina O, Panteleev S (2009). Effects of cervical vagus nerve stimulation on amygdala-evoked responses of the medial prefrontal cortex neurons in rat. Neurosci Res 65: 122–125.

Mazzone SB (2005). An overview of the sensory receptors regulating cough. Cough 1: 2.

Mazzone SB, Canning BJ (2002a). Evidence for differential reflex regulation of cholinergic and noncholinergic parasympathetic nerves innervating the airways. Am J Respir Crit Care Med 165: 1076–1083.

Mazzone SB, Canning BJ (2002b). Synergistic interactions between airway afferent nerve subtypes mediating reflex bronchospasm in guinea pigs. Am J Physiol Regul Integr Comp Physiol 283: R86–R98.

Mazzone SB, Canning BJ (2002c). An in vivo guinea pig preparation for studying the autonomic regulation of airway smooth muscle tone. Auton Neurosci 99: 91–101.

Mazzone SB, Canning BJ (2002d). Central nervous system control of the airways: pharmacological implications. Curr Opin Pharmacol 2: 220.

Mazzone SB, Geraghty DP (1999). Respiratory action of capsaicin microinjected into the nucleus of the solitary tract: involvement of vanilloid and tachykinin receptors. Br J Pharmacol 127: 473–481.

Mazzone SB, McGovern AE (2006). Na+-K+-2Cl- cotransporters and Cl-channels regulate citric acid cough in guinea pigs. J Appl Physiol 101: 635–643.

Mazzone SB, McGovern AE (2010). Innervation of tracheal parasympathetic ganglia by esophageal cholinergic neurons: evidence from anatomic and functional studies in guinea pigs. Am J Physiol Lung Cell Mol Physiol 298: L404–L416.

Mazzone SB, Mori N, Canning BJ (2005). Synergistic interactions between airway afferent nerve subtypes regulating the cough reflex in guinea-pigs. J Physiol 569: 559–573.

Mazzone SB, McLennan L, McGovern AE et al. (2007). Representation of capsaicin-evoked urge-to-cough in the human brain using functional magnetic resonance imaging. Am J Respir Crit Care Med 176: 327–332.

Mazzone SB, Reynolds SM, Mori N et al. (2009). Selective expression of a sodium pump isozyme by cough receptors and evidence for its essential role in regulating cough. J Neurosci 29: 13662–13671.

Mazzone SB, Lim LH, Wagner EM et al. (2010). Sympathetic nerve-dependent regulation of mucosal vascular tone modifies airway smooth muscle reactivity. J Appl Physiol 109: 1292–1300.

Mazzone SB, Cole LJ, Ando A et al. (2011). Investigation of the neural control of cough and cough suppression in humans using functional brain imaging. J Neurosci 31: 2948–2958.

McGovern AE, Mazzone SB (2010). Characterization of the vagal motor neurons projecting to the guinea pig airways and esophagus. Front Neurol 1: 153.

McGovern AE, Davis-Poynter N, Mazzone SB (2011). Transneuronal tracing of airways related sensory circuitry using herpes simplex virus 1, strain H129. FASEB J 25: 1077.13.

McKay LC, Evans KC, Frackowiak RS et al. (2003). Neural correlates of voluntary breathing in humans. J Appl Physiol 95: 1170–1178.

McKay LC, Adams L, Frackowiak RS et al. (2008). A bilateral cortico-bulbar network associated with breath holding in humans, determined by functional magnetic resonance imaging. Neuroimage 40: 1824–1832.

Mifflin SW (1996). Convergent carotid sinus nerve and superior laryngeal nerve afferent inputs to neurons in the NTS. Am J Physiol 271: R870–R880.

Mitchell RA, Herbert DA, Baker DG (1985). Inspiratory rhythm in airway smooth muscle tone. J Appl Physiol 58: 911–920.

Miyazaki M, Tanaka I, Ezure K (1999). Excitatory and inhibitory synaptic inputs shape the discharge pattern of pump neurons of the nucleus tractus solitarii in the rat. Exp Brain Res 129: 191–200.

Mortola J, Sant'Ambrogio G, Clement MG (1975). Localization of irritant receptors in the airways of the dog. Respir Physiol 24: 107–114.

Nassenstein C, Taylor-Clark TE, Myers AC et al. (2010). Phenotypic distinctions between neural crest and placodal derived vagal C-fibres in mouse lungs. J Physiol 588: 4769–4783.

Oh EJ, Mazzone SB, Canning BJ et al. (2006). Reflex regulation of airway sympathetic nerves in guinea-pigs. J Physiol 573: 549–564.

Ohi Y, Yamazaki H, Takeda R et al. (2005). Functional and morphological organization of the nucleus tractus solitarius in the fictive cough reflex of guinea pigs. Neurosci Res 53: 201–209.

Otake K, Nakamura Y, Tanaka I et al. (2001). Morphology of pulmonary rapidly adapting receptor relay neurons in the rat. J Comp Neurol 430: 458–470.

Palmer JB, Cuss FM, Barnes PJ (1986). VIP and PHM and their role in nonadrenergic inhibitory responses in isolated human airways. J Appl Physiol 61: 1322–1328.

Paton JF (1998). Pattern of cardiorespiratory afferent convergence to solitary tract neurons driven by pulmonary vagal C-fiber stimulation in the mouse. J Neurophysiol 79: 2365–2373.

Phipps RJ, Richardson PS (1976). The effects of irritation at various levels of the airway upon tracheal mucus secretion in the cat. J Physiol 261: 563–581.

Rekling JC, Feldman JL (1997). Bidirectional electrical coupling between inspiratory motoneurons in the newborn mouse nucleus ambiguus. J Neurophysiol 78: 3508–3510.

Riccio MM, Kummer W, Biglari B et al. (1996). Interganglionic segregation of distinct vagal afferent fibre phenotypes in guinea-pig airways. J Physiol 496: 521–530.

Rogers DF (2001). Motor control of airway goblet cells and glands. Respir Physiol 125: 129–144.

Schelegle ES, Green JF (2001). An overview of the anatomy and physiology of slowly adapting pulmonary stretch receptors. Respir Physiol 125: 17–31.

Sedý J, Zicha J, Kunes J et al. (2008). Mechanisms of neurogenic pulmonary edema development. Physiol Res 57: 499–506.

Sekizawa S, Chen CY, Bechtold AG et al. (2008). Extended secondhand tobacco smoke exposure induces plasticity in nucleus tractus solitarius second-order lung afferent neurons in young guinea pigs. Eur J Neurosci 28: 771–781.

Sekizawa S, Joad JP, Pinkerton KE et al. (2011). Distinct tachykinin NK(1) receptor function in primate nucleus tractus solitarius neurons is dysregulated after second-hand tobacco smoke exposure. Br J Pharmacol 163: 782–791.

Simonyan K, Saad ZS, Loucks TM et al. (2007). Functional neuroanatomy of human voluntary cough and sniff production. Neuroimage 37: 401–409.

Stratmann G, Gregory GA (2003). Neurogenic and humoral vasoconstriction in acute pulmonary thromboembolism. Anesth Analg 97: 341–354.

Tatsuta H, Arita H (1996). Nonadrenergic relaxation of the cat cervical trachea evoked by stimulation in the lateral hypothalamic area. Neurosci Res 26: 315–322.

Undem BJ, Kollarik M (2005). The role of vagal afferent nerves in chronic obstructive pulmonary disease. Proc Am Thorac Soc 2: 355–360.

Undem BJ, Nassenstein C (2009). Airway nerves and dyspnea associated with inflammatory airway disease. Respir Physiol Neurobiol 167: 36–44.

Undem BJ, Chuaychoo B, Lee MG et al. (2004). Subtypes of vagal afferent C-fibres in guinea-pig lungs. J Physiol 556: 905–917.

van der Velden VH, Hulsmann AR (1999). Autonomic innervation of human airways: structure, function, and pathophysiology in asthma. Neuroimmunomodulation 6: 145–159.

Veelken R, Leonard M, Stetter A et al. (1997). Pulmonary serotonin 5-HT3-sensitive afferent fibers modulate renal sympathetic nerve activity in rats. Am J Physiol 272: H979–H986.

Wahlestedt C, Håkanson R, Vaz CA et al. (1990). Norepinephrine and neuropeptide Y: vasoconstrictor cooperation in vivo and in vitro. Am J Physiol 258: R736–R742.

Ward JK, Barnes PJ, Springall DR et al. (1995). Distribution of human i-NANC bronchodilator and nitric oxide-immunoreactive nerves. Am J Respir Cell Mol Biol 13: 175–184.

Widdicombe JG (1954). Receptors in the trachea and bronchi of the cat. J Physiol 123: 71–104.

Widdicombe JG (1991). Neural control of airway vasculature and edema. Am Rev Respir Dis 143: S18–S21.

Widdicombe J (2003). Functional morphology and physiology of pulmonary rapidly adapting receptors (RARs). Anat Rec A Discov Mol Cell Evol Biol 270: 2–10.

Wine JJ (2007). Parasympathetic control of airway submucosal glands: central reflexes and the airway intrinsic nervous system. Auton Neurosci 133: 35–54.

Yanaura S, Hosokawa T, Kitagawa H et al. (1978). Influence of tracheal muscular tone on the initiation of cough reflex. Jpn J Pharmacol 28: 447–455.

Yu J, Wang YF, Zhang JW (2003). Structure of slowly adapting pulmonary stretch receptors in the lung periphery. J Appl Physiol 95: 385–393.

Handbook of Clinical Neurology, Vol. 117 (3rd series)
Autonomic Nervous System
R.M. Buijs and D.F. Swaab, Editors

Chapter 19

Multiple system atrophy

GREGOR K. WENNING* AND FLORIAN KRISMER
Division of Clinical Neurobiology, Department of Neurology, Medical University, Innsbruck, Austria

INTRODUCTION

Multiple system atrophy (MSA) is a sporadic and fatal α-synucleinopathy characterized clinically by either levodopa-unresponsive parkinsonism or cerebellar ataxia accompanied by progressive autonomic failure (Wenning et al., 2004b; Gilman et al., 2008; Kollensperger et al., 2010). The term MSA was introduced by Graham and Oppenheimer in 1969 (Graham and Oppenheimer, 1969), recognizing the overlapping pathologies of olivopontocerebellar atrophy (OPCA) (Dejerine and Thomas, 1900), striatonigral degeneration (SND) (Adams et al., 1961), and Shy–Drager syndrome (SDS; parkinsonism with autonomic failure) (Shy and Drager, 1960). The notion of MSA was further supported by the discovery of argyrophilic glial cytoplasmic inclusions (GCI) in neuropathological specimens of patients with MSA irrespective of clinical presentation (Papp et al., 1989). Misfolded, hyperphosphorylated, fibrillar α-synuclein represents the principal component of GCIs, classifying MSA among the α-synucleinopathies together with Parkinson's disease (PD), dementia with Lewy bodies (DLB), and pure autonomic failure (PAF) (Spillantini et al., 1998; Trojanowski and Revesz, 2007).

Quinn (1989) was the first to propose a set of diagnostic criteria and distinguished two motor subtypes, namely MSA-SND and MSA-OPCA, with predominantly parkinsonism and cerebellar ataxia, respectively (Quinn, 1989). For ease of use, MSA-SND was reclassified as MSA-P and MSA-OPCA as MSA-C (Schulz et al., 1994). Due to unintended overlap and practical difficulties with the Quinn criteria, a consensus conference was convened to develop clinical guidelines for the diagnosis of MSA (Gilman et al., 1998). Ten years later, warning signs ("red flags") and neuroimaging features were included in the revised consensus diagnostic criteria (Gilman et al., 2008). Meanwhile, a unified MSA rating scale evaluating disease severity semiquantitatively was developed and validated by the European MSA study group (EMSA-SG) (Wenning et al., 2004b). Furthermore, animal models have become available as preclinical testbeds for candidate neuroprotection strategies (Shults et al., 2005; Stefanova et al., 2005; Yazawa et al., 2005). As a result, disease modification trials have been completed or launched based on our growing insights into the pathogenic events underlying MSA pathology (Holmberg et al., 2007; Bensimon et al., 2009; Dodel et al., 2010).

Over the last decade, multiple international academic networks have been established to promote genetic research into MSA. These intensified efforts have led to the identification of variants in the α-synuclein gene and their association with increased disease risk in a large population of MSA patients (Al-Chalabi et al., 2009; Scholz et al., 2009). In addition, several pedigrees suggesting familial MSA with undetermined locus have been reported (Wullner et al., 2004; Hara et al., 2007). More recently, functionally impaired variants of coenzyme Q2 were shown to be associated with an increased risk of MSA in multiplex families and patients with sporadic disease (The Multiple-System Atrophy Research Collaboration 2013; N Engl J Med 2013, Jun 12 [Epub ahed of print]. PMID: 23758206).

EPIDEMIOLOGY

The current knowledge relies on few descriptive epidemiological studies. MSA affects both sexes equally and the mean age at onset is around 55 years. This fatal disorder progresses relentlessly until death with a median survival of 9 years. However, there is considerable variation in disease progression, with individual cases surviving up to 15 years. Survival rates at 5 and 10 years are 83.5% and 39.9%, respectively (Wenning et al., 1994; Testa et al., 1996; Wenning et al., 1997; Watanabe

*Correspondence to: Professor Gregor K. Wenning, M.D., Ph.D., MSc, Division of Clinical Neurobiology, Department of Neurology, Medical University, Anichstrasse 35, 6020 Innsbruck, Austria. E-mail: gregor.wenning@i-med.ac.at

et al., 2002; O'Sullivan et al., 2008). In the Olmsted County study, the annual incidence rate was estimated to be 0.6 cases per 100 000, reflecting nine novel diagnosed MSA cases. Interestingly, none of them had an onset before 50 years of age. Indeed, in the elderly population > 50 years the incidence rate is greater (3 per 100 000)(Bower et al., 1997). Estimates on the age-adjusted prevalence range from 1.9 to 4.9 cases per 100 000 (Schrag et al., 1999; Tison et al., 2000).

MSA is commonly regarded a sporadic disorder. However, in recent years rare familial manifestations have been observed, suggesting the presence of familial monogenic MSA with autosomal dominant or recessive inheritance (Wullner et al., 2004; Hara et al., 2007). In one of these multiplex families, functionally impaired variants in the coenzyme Q2 gene, that is essential for the biosynthesis of coenzyme Q10, were detected (The Multiple-System Atrophy Research Collaboration 2013; N Engl J Med 2013, Jun 12 [Epub ahead of print]. PMID: 23758206). Moreover, were further supported by reports from two independent groups suggesting an association of single nucleotide polymorphisms within the SNCA locus and increased MSA disease risk (Al-Chalabi et al., 2009; Scholz et al., 2009).

A case-control study reported an increased MSA risk resulting from occupational exposure to organic solvents, plastic monomers and additives, pesticides, and metals (Nee et al., 1991). Furthermore, an Italian multicenter case-control study observed a significantly higher risk of developing MSA in subjects having worked in agriculture (Vanacore et al., 2001). Comparably to Parkinson's disease, smoking seems to be less frequent in MSA cases compared to healthy controls. The fact that the inverse association with smoking found previously in PD is shared by MSA but not by progressive supranuclear palsy (PSP) lends epidemiologic support to the notion that different smoking habits are associated with different groups of neurodegenerative disease (Vanacore et al., 2000). These findings were further underpinned by a case-control study suggesting that smoking and farming may influence MSA risk independently (Vanacore et al., 2005). By contrast, a French case-control study revealed that MSA was not associated with exposure to pesticides, solvents, and other occupational toxics in the assessed cohort. Further, this study showed that aspirin intake, alcohol consumption, and fish and seafood consumption was more common within the control group (Vidal et al., 2008). Furthermore, cholesterol in brain membranes may modulate the conformational state and accumulation of α-synuclein. Data from a case-control study suggest that lower levels of total cholesterol and high-density lipoprotein may be associated with an increased risk of MSA (Lee et al., 2009). Finally, in a Korean cohort, multivariate regression analysis revealed that use of antihypertensive medication, vitamins, and consumption of meat and poultry were associated with decreased MSA disease risk, whereas heavy smoking was more prevalent among MSA patients (Seo et al., 2010).

CLINICAL FEATURES AND DIAGNOSTIC CRITERIA

MSA is characterized clinically by any combination of parkinsonism, cerebellar dysfunction, autonomic failure, and pyramidal signs. According to clinical diagnostic criteria two major subtypes may be distinguished by their predominant motor feature. The variant characterized by parkinsonian features is designated MSA-P (Parkinson subtype) while cerebellar ataxia is the main motor presentation in MSA-C (Cerebellar subtype) (Gilman et al., 2008). In Europe, MSA-P accounts for up to 70% of cases (Kollensperger et al., 2010), whereas in Japan, MSA-C (almost 70% of cases) is more common (Watanabe et al., 2002). Nevertheless, autonomic failure is observed almost universally in both motor presentations.

MSA-associated parkinsonism is dominated by a progressive akinetic-rigid syndrome. In contrast to Parkinson's disease, pill-rolling rest tremor is uncommon. Frequent cerebellar features are gait ataxia, limb kinetic ataxia, scanning dysarthria, and cerebellar oculomotor disturbances. Urogenital dysfunction and orthostatic dysregulation are the two most common features of autonomic failure in MSA. Urogenital dysregulation includes urge incontinence, incomplete bladder emptying, and erectile dysfunction which is present in the majority of male patients at disease onset. Micturition disorders in MSA are caused by changes in the peripheral and central innervations of the bladder involving sacral autonomic nuclei such as the Onuf's nucleus (Yamamoto et al., 2005) and pontine micturition centers. Additionally, approximately 75% of patients suffer from orthostatic hypotension; however, recurrent syncope emerges only in 15% (Wenning et al., 1994, 1997; Gilman et al., 2008; Kollensperger et al., 2010). Conversely, a recent study (Kollensperger et al., 2010) revealed that less than 70% of patients with history of syncope had a documented blood pressure drop fulfilling orthostatic hypotension criteria.

Diagnostic criteria

Clinical diagnostic criteria have been developed within the last two decades. The work of Quinn (Quinn, 1989) served as a basis for the introduction of clinical consensus criteria (Gilman et al., 1998). Advances in clinical knowledge of MSA required a revision of the consensus criteria in 2008 (Gilman et al., 2008). The revised consensus criteria specify three levels of increasing certainty: probable, possible, and definite MSA. While definite MSA requires autopsy confirmation, probable and possible MSA rely on history and physical examination. Functional neuroimaging findings can be applied for a diagnosis of possible MSA. Probable MSA is defined as a sporadic, progressive, and adult-onset (30 years or older) disorder characterized by autonomic failure involving urinary incontinence (with erectile dysfunction in males) or severe orthostatic hypotension (decrease of blood pressure by at least 30 mmHg systolic or 15 mmHg diastolic within 3 minutes of standing) and either poorly levodopa-responsive parkinsonism or a cerebellar syndrome (Gilman et al., 2008). Possible MSA is characterized clinically by either parkinsonism or a cerebellar syndrome together with at least one feature suggesting autonomic dysfunction which cannot be explained otherwise. Moreover, one additional feature suggestive of MSA is required (Gilman et al., 2008).

Investigations

According to the consensus criteria, additional investigations such as autonomic function tests, external sphincter electromyography, and neuroimaging may be used to support the diagnosis and to exclude other disease entities (Gilman et al., 2008). However, the diagnostic accuracy in early MSA is unknown.

Urinary symptoms, being the most frequent feature of autonomic failure in MSA, may not only be examined by history (frequency of voiding, difficulties in initiating or suppressing voiding), but also by sonographic measurement of residual volume. Alternatively, residual volume after voiding can be determined by catheterization. In order to exclude the most frequent secondary cause of urinary symptoms, lower tract infections, a standard urine analysis may be performed (Wenning et al., 2004a). Further, in men, prostate hypertrophy with outflow obstruction needs to be considered.

As mentioned previously, symptoms of OH such as postural dizziness and postural visual disturbances are frequently present in MSA. To this end, cardiovascular autonomic function tests such as a standing test or tilt table testing may determine the presence and quantify objectively the magnitude of OH. According to international guidelines, OH is defined as systolic blood pressure drop of at least 20 mmHg and/or as diastolic blood pressure drop of at least 10 mmHg within 3 minutes of standing or tilt (Consensus Committee, 1996). This decline is usually accompanied by an inadequate rise of heart rate. However, the revised consensus criteria on the diagnosis of probable MSA require a more pronounced degree of OH by at least 30 mmHg systolic or 15 mmHg diastolic within 3 minutes of standing (Gilman et al., 2008). Interestingly, delayed OH (beyond 3 minutes of postural challenge) may occur in more than half of the patients, suggesting this to be a mild or early form of sympathetic adrenergic failure (Gibbons and Freeman, 2006). OH in MSA largely reflects degeneration of central autonomic pathways (Benarroch et al., 2006b), whereas cardiovascular autonomic failure in PD is characterized by cardiac sympathetic postganglionic denervation as shown by [^{123}I]metaiodobenzylguanidine (MIBG) scintigraphy (see below) (Rascol and Schelosky, 2009).

External sphincter electromyography (EMG) is frequently abnormal in MSA patients consistent with degeneration of Onuf's nucleus within the sacral cord. Interestingly, EMG abnormalities may even precede clinical symptoms and EMG differentiates MSA from Parkinson's disease within 5 years from disease onset (Vodusek, 2001; Paviour et al., 2005; Yamamoto et al., 2005).

In terms of neuroimaging, structural as well as functional methods can support the diagnosis of MSA. Ultrasound imaging of the basal ganglia and midbrain structures has recently been introduced to aid the early recognition of PD and to discriminate PD from MSA and related atypical parkinsonian disorders (Berg et al., 2008). Ultrasound imaging, however, does not reflect dopaminergic integrity and may be impossible or only partially assessable due to a missing temporal acoustic bone window (Walter et al., 2007). MRI demonstration of putaminal, pontine, and middle cerebellar peduncle atrophy as well as dilatation of IV ventricle is helpful in MSA. Additionally, intensity changes in T2-signal may be observed. In particular, the "putaminal slit" sign (hyperintensity in the dorsolateral margin of the putamen; Fig. 19.1) and the "hot cross bun" sign (cruciform hypointensity at pons level; Fig. 19.2) have high positive predictive value and reasonable sensitivity (Schrag et al., 2000; Horimoto et al., 2002; Seppi and Poewe, 2010). Furthermore, the assessment of the random movement of water molecules, which is the basis of diffusion-weighted imaging, reveals enhanced signal in degenerating tissue. Elevated regional apparent diffusion coefficients (rADC) in putaminal structures differentiate MSA-P from PD, even at early disease stages; however, they fail to discriminate MSA-P from PSP (Schocke et al., 2002; Seppi et al., 2003). In contrast, increased rADCs in the middle cerebellar peduncle have been shown to discriminate MSA-P from both PD and PSP (Nicoletti et al., 2006).

Patterns of resting regional cerebral glucose metabolism were investigated using fludeoxyglucose (^{18}F) positron emission tomography (FDG PET) in MSA patients, revealing decreased striatal glucose metabolism (Gilman et al., 1988). Concordance between clinical and imaging-based measurements can be used in diagnosis as observed in a study showing that image analysis using statistical parametric maps agreed with clinical diagnosis in 96% (Eckert et al., 2005). The assessment of brain

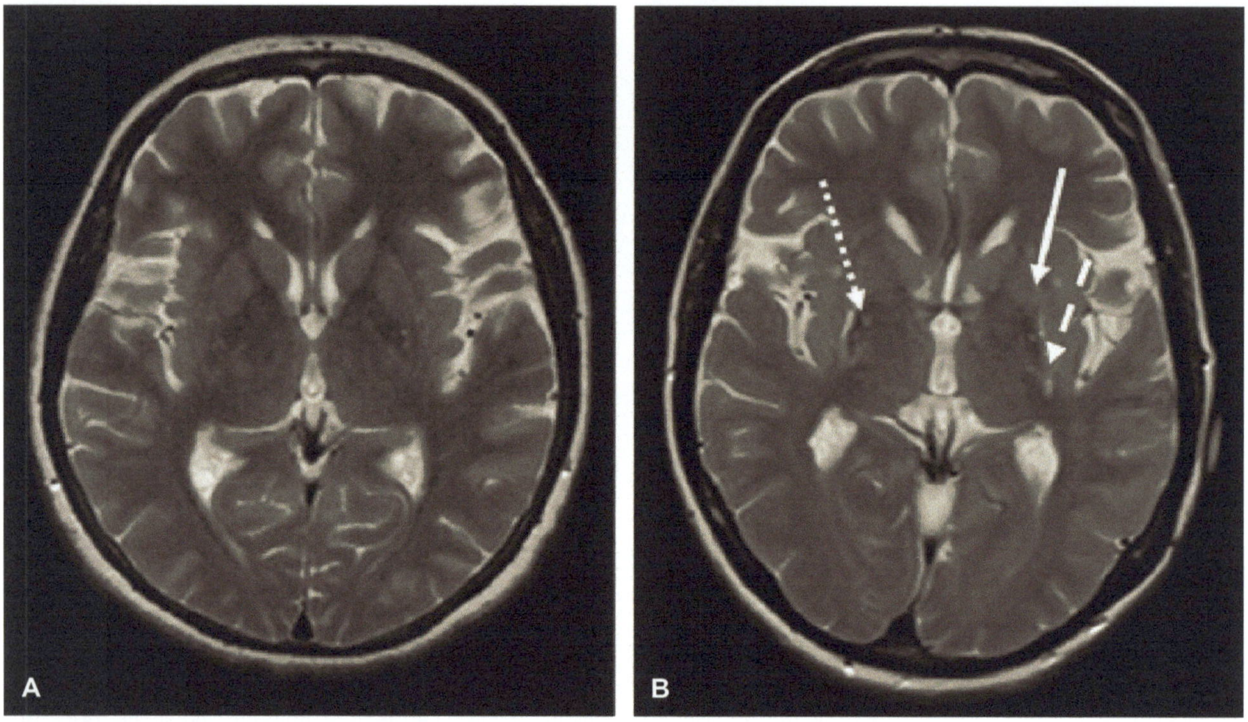

Fig. 19.1. T2-weighted MRI image (**A**) unaltered in PD, and (**B**) putaminal atrophy (arrow), putaminal hypointensity (dotted line) and hyperintense margin (dashed line) in a patient with MSA-P. (Reproduced from Mahlknecht et al., 2010, with kind permission from Springer Science + Business Media.)

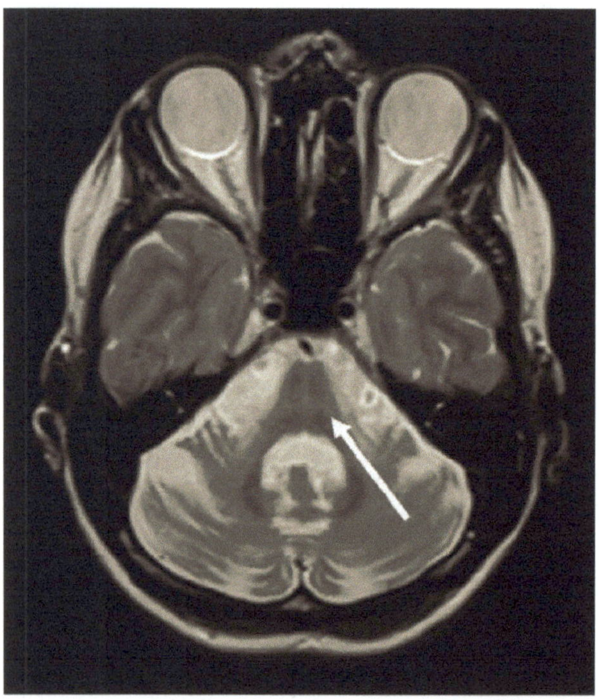

Fig. 19.2. "Hot cross bun" sign in a T2-weighted MRI image of a patient with MSA-P. (Reproduced from Mahlknecht et al., 2010, with kind permission from Springer Science + Business Media.)

perfusion using single-photon emission computed tomography (SPECT) reveals areas of significant hypoperfusion in the striatum, brainstem, and cerebellum in MSA compared to PD and age-matched controls (Cilia et al., 2005). In terms of dopaminergic imaging, presynaptic dopaminergic function ligands such as F-dopa-PET and β-CIT-SPECT may not satisfactorily separate PD from MSA. By contrast, ligands targeting postsynaptic dopamine D_2 receptor ligands are able to differentiate PD from MSA based on normal or increased (PD) or reduced binding signal, respectively (Brooks and Seppi, 2009). Finally, imaging of sympathetic innervation of the heart using metaiodobenzylguanidine (MIBG)-scintigraphy may help to discriminate PD from MSA (Fig. 19.3) (Rascol and Schelosky, 2009). However, recent evidence suggests that cardiac sympathetic denervation does occur in MSA more frequently than considered previously (Orimo et al., 2007).

Disease progression

MSA is a rapidly progressive disease with disability requiring walking aids at early disease stages. In a recent clinicopathological study patients were confined to a wheelchair at a median of 6.7 years after disease onset (O'Sullivan et al., 2008). Features predicting aggressive

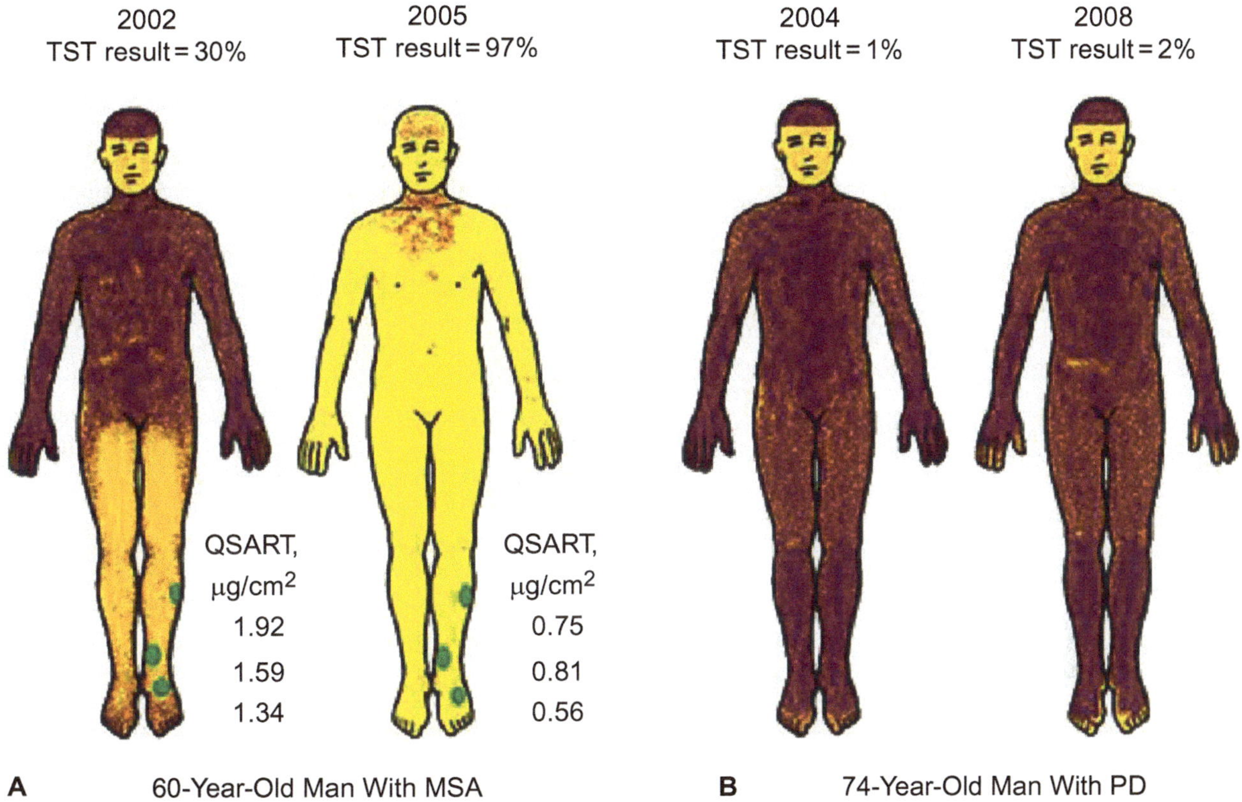

Fig. 19.3. Progressive hypohidrosis revealed by the thermoregulatory sweat test (TST) discriminates multiple system atrophy (MSA) (**A**) from Parkinson's disease (PD) (**B**). (Reproduced from Lipp et al., 2009, with kind permission from the American Medical Association. Copyright (©) (2010) American Medical Association. All rights reserved.)

disease with shorter disease duration are female gender, an older age at disease onset, a short interval from disease onset to the development of the first clinical milestone, and not being admitted to residential care (O'Sullivan et al., 2008). Furthermore, patients suffering from MSA-P experience a more rapid progression than MSA-C patients. MSA patients commonly die from cardiopulmonary arrest and urinary tract infections. Aspiration pneumonia and the wasting syndrome are other frequent causes of death (Papapetropoulos et al., 2007). A natural history study conducted by the European MSA study group (EMSA-SG) characterized the rapid progression using clinical rating scales including the unified MSA rating scale (UMSARS). In detail, the average increase in the UMSARS subscore on activities of daily living (UMSARS I) was 16.8% in the first 6 months from disease onset. Further, UMSARS II (motor examination) and UMSARS IV (global disability) scores increased by 26.1% and 12.5%, respectively (comparing baseline and 6 month assessment) (Geser et al., 2006). Not only motor symptoms progress over time, but also symptoms of autonomic failure deteriorate. A study assessing self-perceived autonomic failure showed a pronounced worsening on a self-completed- questionnaire comparing baseline and 6 month follow-up assessment (Kollensperger et al., 2007).

PATHOLOGY

Gross examination of postmortem brain from MSA patients frequently reveals atrophy within nigrostriatal and/or olivopontocerebellar structures (Trojanowski and Revesz, 2007). Microscopically, abundant and widespread argyrophilic (oligodendro-)glial cytoplasmic inclusions (GCI) composed predominantly of α-synuclein are widely regarded as cellular hallmark lesions (Spillantini et al., 1998; Trojanowski and Revesz, 2007). This finding classified MSA together with PD and dementia with Lewy bodies as α-synucleinopathies. Although the localization of GCIs is widespread, they appear to be more prominent in regions of pronounced neuronal cell loss (Ozawa et al., 2004). Additionally, GCIs correlate with disease duration and disease severity. Focal abnormal accumulation of fibrillar α-synuclein has been reported in neuronal cytoplasm as well as nuclei (NCI and NNI, respectively). However, those inclusions have not been accepted as a defining neuropathological criterion for MSA, but they are likely relevant in the disease process (Yoshida, 2007).

The clinical subtypes correlate with distinct neuropathological findings. In MSA-P, the striatonigral system is the major site of pathology, whereas MSA-C is characterized by involvement of the olivopontocerebellar system and relatively preserved striatonigral structures. Macroscopically, MSA-P patients frequently show an atrophic and green-gray discolored putamen as well as depigmentation of the substantia nigra (Wenning et al., 1997; Daniel, 1999). Semiquantitative assessment of lesion severity based on neuronal loss, astrogliosis, and the presence of GCI led to the proposal of a grading scale for striatonigral degeneration ranging from SND I ("degeneration of the substantia nigra pars compacta (SNC) with relative preservation of the striatum except for minimal gliosis and GCIs in the posterior putamen"; minimal change MSA) to SND III, showing severe striatal pathology and extension to anterior ventromedial subregions (Wenning et al., 2002).

By contrast, patients with MSA-C show varying degrees of atrophy within the paleocerebellum and neocerebellum, middle cerebellar peduncles, basis pontis, and the inferior olives. Microscopically, oligodendroglial pathology is predominantly present in the projections from the precerebellar nuclei to the cerebellum and the ascending or descending fiber tracts of the motor system. In line with MSA-P, a grading score ranging from OPCA 0/I (no or little degeneration, respectively) to OPCA III has been proposed (Jellinger et al., 2005).

Autonomic failure in MSA corresponds to dysfunction of central and preganglionic efferent autonomic neurons, neuronal networks in the brainstem that control cardiovascular and respiratory activity and degeneration of neuroendocrine pathways in the hypothalamopituitary axis (Ozawa, 2007). In detail, cell loss is present in the dorsal motor nucleus of the vagal nerve (Sung et al., 1979), in the locus coeruleus (Wenning et al., 1997), and in the catecholaminergic neurons of the ventrolateral medulla (Benarroch et al., 1998). Furthermore, Papp and Lantos suggested that the substantial involvement of pontomedullary reticular formation is the histological counterpart for visceral dysfunction (Papp and Lantos, 1994). Hypocretin neurons in the posterior lateral hypothalamus are markedly decreased in number and abundant GCI are observed in this area in patients with MSA, presumably influencing the regulation phase switch between wakefulness and sleep (Benarroch et al., 2007). Finally, the loss of autonomic neurons in both the paraventricular nucleus and the locus coeruleus may be the underlying cause of inadequate neurohormonal responses to various stressors (Benarroch et al., 2006a). Finally, MSA patients have diffuse and widespread anhidrosis largely caused by preganglionic sudomotor involvement (Fig. 19.3) (Lipp et al., 2009).

PATHOGENESIS

The underlying pathogenesis of MSA is not yet understood. However, in recent years, substantial progress has been achieved towards the clarification of the underlying pathogenesis. α-Synuclein-positive GCIs, being the obligatory subcellular feature of a definite diagnosis of MSA according to consensus criteria (Gilman et al., 2008), are closely linked to the development of MSA. Indeed, the density of GCIs correlates significantly with neuronal deterioration and disease duration (Ozawa et al., 2004). The crucial role of GCI has been elucidated further with the discovery of early phosphoprotein-25-α (p25α; also called tubulin polymerization promoting protein) accumulation in disease-related, but not healthy oligodendrocytes (Jellinger, 2006; Kovacs et al., 2007; Song et al., 2007). Myelination is regarded as the main function of p25α and the protein is present in myelin basic protein (MBP)-immunopositive sheaths. *In vitro* studies suggest that a number of events, such as translocation within oligodendroglial cell bodies, impede the normal cellular function of p25α and lead to subsequent deposition and fibrillation of α-synuclein (Hlavanda et al., 2002; Lindersson et al., 2005; Song et al., 2007). Accordingly, recent data in brain homogenates from adult wild-type mice indicate a specific association between α-synuclein and MBP. However, details remain to be resolved (Lindersson et al., 2005; Kovacs et al., 2007).

Furthermore, a recent study revealed a lack of detectable *SNCA* gene expression signals in oligodendroglia of either healthy adults or MSA patients (Miller et al., 2005), suggesting that accumulated α-synuclein derived from sources outside the oligodendroglial cell population. Comparably to findings in PD, striatal transplants in a rodent model of MSA were affected by α-synuclein pathology (Stefanova et al., 2009). In contrast to the findings of Lewy bodies in aged grafts of PD patients (Kordower et al., 2008a, b; Li et al., 2008), there was no evidence of graft-derived α-synuclein pathology in MSA mice; instead the aggregates within striatal grafts were shown to be of host origin, with deleterious effects on the anatomical graft integration. Furthermore, *in vitro* studies indicated prion-like seeding effects of oligomeric α-synuclein (Danzer et al., 2007, 2009; Luk et al., 2009). However, the basis of α-synuclein cell-to-cell transfer is yet to be discovered.

The majority of recent studies support the theory that GCI induce glial cell degeneration and subsequently promote neuronal cell death. The underlying mechanisms are still elusive. However, p25α and MBP misprocessing are the earliest detectable changes in MSA and therefore emphasize the importance of the oligodendroglia–myelin–axon interface. To sum up, MSA can be seen as primary oligodendrogliopathy featuring early myelin disruption, followed by synucleinopathy and subsequent axonal damage, and, finally, leading to secondary neurodegeneration (Wenning et al., 2008).

THERAPY

Due to the fatal prognosis and the early development of disability, neuroprotective and neuroregenerative therapies are urgently needed. Unfortunately, so far no single interventional trial has shown beneficial effects in this disease entity. Therefore, medical treatment rests largely on the alleviation of clinical symptoms, in particular parkinsonism and autonomic failure (Stefanova et al., 2009).

Neuroprotective and neurorestorative strategies

As mentioned previously, clinical trials targeting neuroprotective strategies failed to modify the course of MSA (Stefanova et al., 2009). Nevertheless, a placebo-controlled double-blind pilot trial treating patients with recombinant human growth hormone (r-HGH) showed a trend toward reduction in progression of motor symptoms that failed to reach significance (Holmberg et al., 2007). Furthermore, after confirming neuroprotective efficacy of minocycline in a transgenic mouse model of MSA (Stefanova et al., 2007) and in models of related neurodegenerative disorders (Zhu et al., 2002; Casarejos et al., 2006), a phase II clinical neuroprotection trial was conducted by EMSA-SG and the German Parkinson Network (KNP) proving minocycline to be ineffective in patients with MSA. In addition, the NNIPPS group (Neuroprotection and Natural History in Parkinson Plus Syndromes) assessed the disease-modifying efficacy of riluzole, which is an antiglutamatergic agent, in a double-blind placebo-controlled fashion. In accordance with results from a small randomized controlled trial (Seppi et al., 2006), this study also failed to demonstrate beneficial effects (Bensimon et al., 2009).

In contrast, a pilot study assessing the efficacy and feasibility of autologous mesenchymal stem cell therapy in human MSA patients revealed a delay in progression of neurological deficits in the treatment group and a functional improvement in the follow-up period (Lee et al., 2008). A subsequent randomized controlled trial conducted by the same group confirmed the results of the pilot study (Lee et al., 2012), however, these results have not been verified by other groups. Recent preclinical evidence supported a neuroprotective efficacy of these stem cells and hypothesized transdifferentiation or induction of nonspecific immunological responses as underlying mechanisms (Park et al., 2011; Stemberger et al., 2011).

Finally, rifampicin and rasagiline are currently under investigation in randomized controlled multisite studies. The rationale for these two trials is based on preclinical findings in transgenic MSA mouse models. In particular, the irreversible inhibitor of type B monoamine oxidase rasagiline showed neuroprotective properties in a transgenic mouse model overexpressing α-synuclein under the oligodendroglial-specific proteolipid protein promoter (Stefanova et al., 2008). Comparably, rifampicin has been shown to remove aggregated α-synuclein and to mediate neuroprotective effects in a mouse model overexpressing α-synuclein under a MBP-promoter (Ubhi et al., 2008).

Symptomatic treatment

Autonomic symptoms

Therapy of autonomic symptoms is the only branch in MSA management where substantial improvement can be achieved. Such intervention is important since untreated MSA patients frequently suffer from ascending urinary tract infections (Ito et al., 2006) and may sustain serious trauma from falls due to fainting caused by postural hypotension. Needless to say, this autonomic dysfunction has a huge impact on the quality of life of both patient and caretaker(s) (Schrag et al., 2006; Kollensperger et al., 2007).

Urinary symptoms

Postvoid residual volume in MSA is usually unaltered by procholinergic agents whereas anticholinergic drugs may improve symptoms of sphincter-detrusor dyssynergy and detrusor hyperreflexia early in the course of the disease, although central nervous adverse effects may limit doses (Beck et al., 1994). More recently, a peripherally acting anticholinergic, trospium chloride, was shown to be equally effective in patients with detrusor hyperreflexia, but with a better tolerability (Halaska et al., 2003); however, this agent has not been investigated in MSA. Furthermore β-adrenergic receptor antagonists mediate improvements in voiding with reduction of residual volumes in MSA patients (Sakakibara et al., 2000). Alternatively, injection of botulinum toxin type A may be helpful in the treatment of detrusor overactivity (Giannantoni et al., 2009) and urethral hypertonia (Apostolidis et al., 2009).

Finally, postmicturition volumes > 150 mL require clean intermittent catheterization three to four times a day to prevent secondary consequences. If mechanical obstructions within the urethra or motor symptoms prevent uncomplicated catheterization, a permanent transcutaneuous suprapubic catheter may become necessary (Fowler and O'Malley, 2003).

Orthostatic hypotension

Orthostatic hypotension and subsequent fainting is frequently the cause of recurrent falls which can lead to serious trauma. However, treatment of orthostatic hypotension requires increasing intravascular volume and reducing orthostatic pooling of venous blood. Mechanical and pharmacological therapies are currently being used to reduce orthostatic hypotension symptoms in MSA.

In detail, nonpharmacological treatments should be considered first. These options include compression stockings or custom-made elastic body garments, sufficient fluid intake, high-salt diet, and spreading of total carbohydrate intake in order to reduce postprandial hypotension (Young and Mathias, 2004; Deguchi et al., 2007; Freeman, 2008). Furthermore, during sleep, head-up tilt reduces hypertensive cerebral perfusion pressure and ameliorates nocturia by increasing intravasal volume which is particularly helpful in improving early morning orthostatic tolerance (Fan et al., 2009).

Midodrine is a prodrug forming an active metabolite which activates α-adrenergic receptors of the arteriolar and venous vasculature resulting in an increased vascular tone. In randomized placebo-controlled trials, midodrine mediated beneficial effects in patients with neurogenic orthostatic hypotension including MSA (Jankovic et al., 1993; Low et al., 1997). Adverse effects such as exacerbation of urinary retention, pruritus, and supine hypertension are usually mild and rarely require withdrawal of treatment. Another promising drug seems to be the norepinephrine precursor L-threodihydroxyphenylserine (L-threo-DOPS or droxidopa) which has been frequently used in Japan to ameliorate symptoms of orthostatic hypotension. Efficacy was proven by two double-blind placebo-controlled trials (Kaufmann, 2008; Mathias, 2008). Finally, the somatostatin analog octreotide mediated beneficial effects regarding postprandial hypotension in patients with pure autonomic failure, presumably due to inhibition of the release of vasodilatory gastrointestinal peptides (Raimbach et al., 1989; Alam et al., 1995).

Constipation

One-third of MSA patients complain of chronic constipation (Kollensperger et al., 2010). Nonpharmacological treatment is crucial, including daily exercise, high fluid and high fiber intake, and the consumption of fruit juices, particularly prune juice. If laxative therapy becomes necessary, polycarbophil and mosapride citrate may be prescribed as suggested by two small open label trials (Sakakibara et al., 2007).

Erectile dysfunction

Erectile dysfunction is almost universal in male MSA patients. The need for medical treatment should be discussed with each patient individually. The phosphodiesterase type 5 inhibitor sildenafil improved erectile function in a placebo-controlled double-blind study (Hussain et al., 2001), but should only be prescribed after exclusion of severe postural hypotension as sildenafil may unmask or exacerbate orthostatic hypotension. Erectile dysfunction may also be treated by oral yohimbine, intracavernosal injection of papaverine (Papatsoris et al., 2008), subcutaneous apomorphine (O'Sullivan, 2002) or a penis implant (Beck et al., 1994).

Inspiratory stridor

Inspiratory stridor may be observed in up to 30% at late disease stages. Therapeutic approaches include nighttime continuous positive airway pressure (CPAP) in order to attenuate nocturnal stridor (Iranzo et al., 2004) and, in patients with relatively preserved mobility and severe stridor, tracheostomy.

Parkinsonism

There are many different classes of drugs used in the therapy of parkinsonism. Although no randomized controlled trial evaluating the efficacy of L-dopa in MSA patients has been conducted, L-dopa is widely regarded the first choice.

Dopaminergics

Despite the fact that MSA patients are commonly regarded as L-dopa unresponsive, a recent clinical series documented beneficial L-dopa response in up to 40% of MSA patients. However, these effects are usually transient with benefits lasting only few years (O'Sullivan et al., 2008; Kollensperger et al., 2010). Further, according to the revised consensus criteria L-dopa responsiveness should be tested over 3 months with escalating doses up to 1000 mg per day if needed and tolerated (Gilman et al., 2008). Unfortunately, up to half of the L-dopa-responsive MSA patients will develop motor fluctuations and L-dopa-induced dyskinesias, particularly affecting orofacial and neck muscles.

Other dopaminergic agents including dopamine agonists have not been evaluated in controlled clinical trials. Case reports suggest that dopamine agonists do not mediate benefits beyond L-dopa treatment (Lees and Bannister, 1981; Goetz et al., 1984). In contrast, adverse effects such as the development or exacerbation of postural hypotension, leg edema, and impulse control disorders including hypersexuality may require agonist dose reduction or even withdrawal (Klos et al., 2005).

N-methyl-D-aspartate receptor antagonists

Amantadine was proven ineffective in a short-term open label trial in L-dopa-unresponsive MSA patients (Colosimo et al., 1996). This finding was confirmed by a randomized placebo-controlled trial (Wenning, 2005).

Selective serotonin reuptake inhibitors

Paroxetine improved motor abilities of the upper limbs and speech in a double-blind placebo-controlled trial suggesting modifications in serotonergic neurotransmissions in basal ganglia circuitry (Friess et al., 2006). Interestingly, the degree of depressive symptoms was not significantly influenced by this selective serotonin reuptake inhibitor.

Surgical therapy

Ablative neurosurgical interventions such as medial pallidotomy did not improve parkinsonism in MSA patients (Lang et al., 1997) and studies on bilateral subthalamic stimulation revealed controversial results (Visser-Vandewalle et al., 2003; Lezcano et al., 2004). Therefore, surgical therapy is currently irrelevant in the routine management of MSA.

Cerebellar ataxia

Cerebellar ataxia, being the defining criterion of MSA-C (Gilman et al., 2008), is present in two-thirds of MSA patients (Kollensperger et al., 2010). Thus far, despite anecdotal reports of the beneficial effects of multiple drugs, no single medication has shown beneficial effects in larger patient series (Wenning et al., 2003). Consequently, physical therapy and mechanical support devices remain the only therapeutic options available.

CONCLUSION

In recent years, substantial progress has been achieved towards the understanding of underlying pathogenic mechanisms of MSA (Stefanova et al., 2009), particularly due to the development of animal models replicating the human phenotype (Stefanova et al., 2009). Furthermore, clinical criteria for the diagnosis of MSA have been defined and specific rating scales assessing clinical features and disability have been developed (Wenning et al., 2004b; Schrag et al., 2007; Gilman et al., 2008). Finally, candidate single nucleotide polymorphism (SNP) studies have identified variants of the α-synuclein gene with significantly increased disease risk (Al-Chalabi et al., 2009; Scholz et al., 2009). This progress has allowed researchers to develop novel therapeutic approaches and to perform multicenter randomized clinical trials to test the efficacy of candidate drugs. Recent advances have only been possible thanks to the combined efforts of international academic networks in Europe (EMSA-SG – www.emsa-sg.org), North America (NAMSA), and Japan (JAMSAC).

Unfortunately, thus far we have been unable to modify the rapid course of this fatal disease. Hence, future research is urgently needed to answer questions regarding the pathogenesis of the disease and to provide for patients a better prognosis offering them powerful management strategies.

References

Adams MR, Van Bogaert L, van der Eecken H (1961). Nigrostriate and cerebello-nigro-striate degeneration (Clinical uniqueness and pathological variability of presenile degeneration of the extrapyramidal rigidity type). Psychiatr Neurol (Basel) 142: 219–259.

Alam M, Smith G, Bleasdale-Barr K et al. (1995). Effects of the peptide release inhibitor octreotide on daytime hypotension and on nocturnal hypertension in primary autonomic failure. J Hypertens 13: 1664–1669.

Al-Chalabi A, Durr A, Wood NW et al. (2009). Genetic variants of the alpha-synuclein gene SNCA are associated with multiple system atrophy. PLoS One 4: e7114.

Apostolidis A, Dasgupta P, Denys P et al. (2009). Recommendations on the use of botulinum toxin in the treatment of lower urinary tract disorders and pelvic floor dysfunctions: a European consensus report. Eur Urol 55: 100–119.

Beck RO, Betts CD, Fowler CJ (1994). Genitourinary dysfunction in multiple system atrophy: clinical features and treatment in 62 cases. J Urol 151: 1336–1341.

Benarroch EE, Smithson IL, Low PA et al. (1998). Depletion of catecholaminergic neurons of the rostral ventrolateral medulla in multiple systems atrophy with autonomic failure. Ann Neurol 43: 156–163.

Benarroch EE, Schmeichel AM, Sandroni P et al. (2006a). Differential involvement of hypothalamic vasopressin neurons in multiple system atrophy. Brain 129: 2688–2696.

Benarroch EE, Schmeichel AM, Sandroni P et al. (2006b). Involvement of vagal autonomic nuclei in multiple system atrophy and Lewy body disease. Neurology 66: 378–383.

Benarroch EE, Schmeichel AM, Sandroni P et al. (2007). Involvement of hypocretin neurons in multiple system atrophy. Acta Neuropathol 113: 75–80.

Bensimon G, Ludolph A, Agid Y et al. (2009). Riluzole treatment survival and diagnostic criteria in Parkinson plus disorders: the NNIPPS study. Brain 132: 156–171.

Berg D, Godau J, Walter U (2008). Transcranial sonography in movement disorders. Lancet Neurol 7: 1044–1055.

Bower JH, Maraganore DM, McDonnell SK et al. (1997). Incidence of progressive supranuclear palsy and multiple system atrophy in Olmsted County Minnesota 1976 to 1990. Neurology 49: 1284–1288.

Brooks DJ, Seppi K (2009). Proposed neuroimaging criteria for the diagnosis of multiple system atrophy. Mov Disord 24: 949–964.

Casarejos MJ, Menendez J, Solano RM et al. (2006). Susceptibility to rotenone is increased in neurons from parkin null mice and is reduced by minocycline. J Neurochem 97: 934–946.

Cilia R, Marotta G, Benti R et al. (2005). Brain SPECT imaging in multiple system atrophy. J Neural Transm 112: 1635–1645.

Colosimo C, Merello M, Pontieri FE (1996). Amantadine in parkinsonian patients unresponsive to levodopa: a pilot study. J Neurol 243: 422–425.

Consensus Committee of the American Autonomic Society and the American Academy of Neurology [No authors listed] (1996). Consensus statement on the definition of orthostatic hypotension pure autonomic failure and multiple system atrophy. Neurology 46: 1470.

Daniel S (1999). The neuropathology and neurochemistry of multiple system atrophy. In: CJ Mathias, R Bannister (Eds.), Autonomic Failure: A Textbook of Clinical Disorders of the Autonomic Nervous System, Oxford University Press, Oxford.

Danzer KM, Haasen D, Karow AR et al. (2007). Different species of alpha-synuclein oligomers induce calcium influx and seeding. J Neurosci 27: 9220–9232.

Danzer KM, Krebs SK, Wolff M et al. (2009). Seeding induced by alpha-synuclein oligomers provides evidence for spreading of alpha-synuclein pathology. J Neurochem 111: 192–203.

Deguchi K, Ikeda K, Shimamura M et al. (2007). Assessment of autonomic dysfunction of multiple system atrophy with laryngeal abductor paralysis as an early manifestation. Clin Neurol Neurosurg 109: 892–895.

Dejerine J, Thomas A (1900). L'atrophie olivo-ponto-cerebelleuse. Nouvelle Iconographie de la Salpêtrière: Clinique des Maladies du Système Nerveux 13: 330–370.

Dodel R, Spottke A, Gerhard A et al. (2010). Minocycline 1-year therapy in multiple-system-atrophy: effect on clinical symptoms and [(11).C] (R).-PK11195 PET (MEMSA-trial). Mov Disord 25: 97–107.

Eckert T, Barnes A, Dhawan V et al. (2005). FDG PET in the differential diagnosis of parkinsonian disorders. Neuroimage 26: 912–921.

Fan CW, Gasparro D, Crowley V et al. (2009). Acute haemodynamic response to sleeping head-up at 6 inches in older inpatients. Clin Auton Res 19: 51–57.

Fowler CJ, O'Malley KJ (2003). Investigation and management of neurogenic bladder dysfunction. J Neurol Neurosurg Psychiatry 74 (Suppl 4): iv27–iv31.

Freeman R (2008). Current pharmacologic treatment for orthostatic hypotension. Clin Auton Res 18 (Suppl 1): 14–18.

Friess E, Kuempfel T, Modell S et al. (2006). Paroxetine treatment improves motor symptoms in patients with multiple system atrophy. Parkinsonism Relat Disord 12: 432–437.

Geser F, Wenning GK, Seppi K et al. (2006). Progression of multiple system atrophy (MSA): a prospective natural history study by the European MSA Study Group (EMSA SG). Mov Disord 21: 179–186.

Giannantoni A, Rossi A, Mearini E et al. (2009). Botulinum toxin A for overactive bladder and detrusor muscle overactivity in patients with Parkinson's disease and multiple system atrophy. J Urol 182: 1453–1457.

Gibbons CH, Freeman R (2006). Delayed orthostatic hypotension: a frequent cause of orthostatic intolerance. Neurology 67: 28–32.

Gilman S, Markel DS, Koeppe RA et al. (1988). Cerebellar and brainstem hypometabolism in olivopontocerebellar atrophy detected with positron emission tomography. Ann Neurol 23: 223–230.

Gilman S, Low P, Quinn N et al. (1998). Consensus statement on the diagnosis of multiple system atrophy. American Autonomic Society and American Academy of Neurology. Clin Auton Res 8: 359–362.

Gilman S, Wenning GK, Low PA et al. (2008). Second consensus statement on the diagnosis of multiple system atrophy. Neurology 71: 670–676.

Goetz CG, Tanner CM, Klawans HL (1984). The pharmacology of olivopontocerebellar atrophy. Adv Neurol 41: 143–148.

Graham JG, Oppenheimer DR (1969). Orthostatic hypotension and nicotine sensitivity in a case of multiple system atrophy. J Neurol Neurosurg Psychiatry 32: 28–34.

Halaska M, Ralph G, Wiedemann A et al. (2003). Controlled double-blind multicentre clinical trial to investigate long-term tolerability and efficacy of trospium chloride in patients with detrusor instability. World J Urol 20: 392–399.

Hara K, Momose Y, Tokiguchi S et al. (2007). Multiplex families with multiple system atrophy. Arch Neurol 64: 545–551.

Hlavanda E, Kovacs J, Olah J et al. (2002). Brain-specific p25 protein binds to tubulin and microtubules and induces aberrant microtubule assemblies at substoichiometric concentrations. Biochemistry 41: 8657–8664.

Holmberg B, Johansson JO, Poewe W et al. (2007). Safety and tolerability of growth hormone therapy in multiple system atrophy: a double-blind placebo-controlled study. Mov Disord 22: 1138–1144.

Horimoto Y, Aiba I, Yasuda T et al. (2002). Longitudinal MRI study of multiple system atrophy – when do the findings appear and what is the course? J Neurol 249: 847–854.

Hussain IF, Brady CM, Swinn MJ et al. (2001). Treatment of erectile dysfunction with sildenafil citrate (Viagra) in parkinsonism due to Parkinson's disease or multiple system atrophy with observations on orthostatic hypotension. J Neurol Neurosurg Psychiatry 71: 371–374.

Iranzo A, Santamaria J, Tolosa E et al. (2004). Long-term effect of CPAP in the treatment of nocturnal stridor in multiple system atrophy. Neurology 63: 930–932.

Ito T, Sakakibara R, Yasuda K et al. (2006). Incomplete emptying and urinary retention in multiple-system atrophy:

when does it occur and how do we manage it? Mov Disord 21: 816–823.

Jankovic J, Gilden JL, Hiner BC et al. (1993). Neurogenic orthostatic hypotension: a double-blind placebo-controlled study with midodrine. Am J Med 95: 38–48.

Jellinger KA (2006). P25alpha immunoreactivity in multiple system atrophy and Parkinson disease. Acta Neuropathol 112: 112.

Jellinger KA, Seppi K, Wenning GK (2005). Grading of neuropathology in multiple system atrophy: proposal for a novel scale. Mov Disord 20 (Suppl 12): S29–S36.

Kaufmann H (2008). L-dihydroxyphenylserine (Droxidopa): a new therapy for neurogenic orthostatic hypotension: the US experience. Clin Auton Res 18 (Suppl 1): 19–24.

Klos KJ, Bower JH, Josephs KA et al. (2005). Pathological hypersexuality predominantly linked to adjuvant dopamine agonist therapy in Parkinson's disease and multiple system atrophy. Parkinsonism Relat Disord 11: 381–386.

Kollensperger M, Stampfer-Kountchev M, Seppi K et al. (2007). Progression of dysautonomia in multiple system atrophy: a prospective study of self-perceived impairment. Eur J Neurol 14: 66–72.

Kollensperger M, Geser F, Ndayisaba JP et al. (2010). Presentation, diagnosis and management of multiple system atrophy in Europe: final analysis of the European multiple system atrophy registry. Mov Disord 25: 2604–2612.

Kordower JH, Chu Y, Hauser RA et al. (2008a). Lewy body-like pathology in long-term embryonic nigral transplants in Parkinson's disease. Nat Med 14: 504–506.

Kordower JH, Chu Y, Hauser RA et al. (2008b). Transplanted dopaminergic neurons develop PD pathologic changes: a second case report. Mov Disord 23: 2303–2306.

Kovacs GG, Gelpi E, Lehotzky A et al. (2007). The brain-specific protein TPPP/p25 in pathological protein deposits of neurodegenerative diseases. Acta Neuropathol 113: 153–161.

Lang AE, Lozano A, Duff J et al. (1997). Medial pallidotomy in late-stage Parkinson's disease and striatonigral degeneration. Adv Neurol 74: 199–211.

Lee PH, Kim JW, Bang OY et al. (2008). Autologous mesenchymal stem cell therapy delays the progression of neurological deficits in patients with multiple system atrophy. Clin Pharmacol Ther 83: 723–730.

Lee PH, Lim TS, Shin HW et al. (2009). Serum cholesterol levels and the risk of multiple system atrophy: a case-control study. Mov Disord 24: 752–758.

Lee PH, Lee JE, Kim HS et al. (2012). A randomized trial of mesenchymal stem cells in multiple system atrophy. Ann Neurol. 72: 32–40.

Lees AJ, Bannister R (1981). The use of lisuride in the treatment of multiple system atrophy with autonomic failure (Shy–Drager syndrome). J Neurol Neurosurg Psychiatry 44: 347–351.

Lezcano E, Gomez-Esteban JC, Zarranz JJ et al. (2004). Parkinson's disease-like presentation of multiple system atrophy with poor response to STN stimulation: a clinicopathological case report. Mov Disord 19: 973–977.

Li J, Englund YE, Holton JL et al. (2008). Lewy bodies in grafted neurons in subjects with Parkinson's disease suggest host-to-graft disease propagation. Nat Med 14: 501–503.

Lindersson E, Lundvig D, Petersen C et al. (2005). p25alpha Stimulates alpha-synuclein aggregation and is co-localized with aggregated alpha-synuclein in alpha-synucleinopathies. J Biol Chem 280: 5703–5715.

Lipp A, Sandroni P, Ahlskog JE et al. (2009). Prospective differentiation of multiple system atrophy from Parkinson disease with and without autonomic failure. Arch Neurol 66: 742–750.

Low PA, Gilden JL, Freeman R et al. (1997). Efficacy of midodrine vs placebo in neurogenic orthostatic hypotension. A randomized double-blind multicenter study. Midodrine Study Group. JAMA 277: 1046–1051.

Luk KC, Song C, O'Brien P et al. (2009). Exogenous alpha-synuclein fibrils seed the formation of Lewy body-like intracellular inclusions in cultured cells. Proc Natl Acad Sci U S A 106: 20051–20056.

Mahlknecht P, Schocke M, Seppi K (2010). Differenzialdiagnose der Parkinson-Syndrome mittels MRT. Nervenarzt 81: 1168–1179.

Mathias CJ (2008). L-dihydroxyphenylserine (Droxidopa) in the treatment of orthostatic hypotension: the European experience. Clin Auton Res 18 (Suppl 1): 25–29.

Miller DW, Johnson JM, Solano SM et al. (2005). Absence of alpha-synuclein mRNA expression in normal and multiple system atrophy oligodendroglia. J Neural Transm 112: 1613–1624.

Nee LE, Gomez MR, Dambrosia J et al. (1991). Environmental-occupational risk factors and familial associations in multiple system atrophy: a preliminary investigation. Clin Auton Res 1: 9–13.

Nicoletti G, Lodi R, Condino F et al. (2006). Apparent diffusion coefficient measurements of the middle cerebellar peduncle differentiate the Parkinson variant of MSA from Parkinson's disease and progressive supranuclear palsy. Brain 129: 2679–2687.

O'Sullivan JD (2002). Apomorphine as an alternative to sildenafil in Parkinson's disease. J Neurol Neurosurg Psychiatry 72: 681.

O'Sullivan SS, Massey LA, Williams DR et al. (2008). Clinical outcomes of progressive supranuclear palsy and multiple system atrophy. Brain 131: 1362–1372.

Orimo S, Kanazawa T, Nakamura A et al. (2007). Degeneration of cardiac sympathetic nerve can occur in multiple system atrophy. Acta Neuropathol 113: 81–86.

Ozawa T (2007). Morphological substrate of autonomic failure and neurohormonal dysfunction in multiple system atrophy: impact on determining phenotype spectrum. Acta Neuropathol 114: 201–211.

Ozawa T, Paviour D, Quinn NP et al. (2004). The spectrum of pathological involvement of the striatonigral and olivopontocerebellar systems in multiple system atrophy: clinicopathological correlations. Brain 127: 2657–2671.

Papapetropoulos S, Tuchman A, Laufer D et al. (2007). Causes of death in multiple system atrophy. J Neurol Neurosurg Psychiatry 78: 327–329.

Papatsoris AG, Papapetropoulos S, Singer C et al. (2008). Urinary and erectile dysfunction in multiple system atrophy (MSA). Neurourol Urodyn 27: 22–27.

Papp MI, Lantos PL (1994). The distribution of oligodendroglial inclusions in multiple system atrophy and its relevance to clinical symptomatology. Brain 117: 235–243.

Papp MI, Kahn JE, Lantos PL (1989). Glial cytoplasmic inclusions in the CNS of patients with multiple system atrophy (striatonigral degeneration olivopontocerebellar atrophy and Shy–Drager syndrome). J Neurol Sci 94: 79–100.

Park HJ, Bang G, Lee BR et al. (2011). Neuroprotective effect of human mesenchymal stem cells in an animal model of double toxin-induced multiple system atrophy-parkinsonism. Cell Transplant 20: 827–835.

Paviour DC, Williams D, Fowler CJ et al. (2005). Is sphincter electromyography a helpful investigation in the diagnosis of multiple system atrophy? A retrospective study with pathological diagnosis. Mov Disord 20: 1425–1430.

Quinn N (1989). Multiple system atrophy – the nature of the beast. J Neurol Neurosurg Psychiatry (Suppl): 78–89.

Raimbach SJ, Cortelli P, Kooner JS et al. (1989). Prevention of glucose-induced hypotension by the somatostatin analogue octreotide (SMS 201-995) in chronic autonomic failure: haemodynamic and hormonal changes. Clin Sci (Lond) 77: 623–628.

Rascol O, Schelosky L (2009). ^{123}I-metaiodobenzylguanidine scintigraphy in Parkinson's disease and related disorders. Mov Disord 24 (Suppl 2): S732–S741.

Sakakibara R, Hattori T, Uchiyama T et al. (2000). Urinary dysfunction and orthostatic hypotension in multiple system atrophy: which is the more common and earlier manifestation? J Neurol Neurosurg Psychiatry 68: 65–69.

Sakakibara R, Yamaguchi T, Uchiyama T et al. (2007). Calcium polycarbophil improves constipation in primary autonomic failure and multiple system atrophy subjects. Mov Disord 22: 1672–1673.

Schocke MF, Seppi K, Esterhammer R et al. (2002). Diffusion-weighted MRI differentiates the Parkinson variant of multiple system atrophy from PD. Neurology 58: 575–580.

Scholz SW, Houlden H, Schulte C et al. (2009). SNCA variants are associated with increased risk for multiple system atrophy. Ann Neurol 65: 610–614.

Schrag A, Ben Shlomo Y, Quinn NP (1999). Prevalence of progressive supranuclear palsy and multiple system atrophy: a cross-sectional study. Lancet 354: 1771–1775.

Schrag A, Good CD, Miszkiel K et al. (2000). Differentiation of atypical parkinsonian syndromes with routine MRI. Neurology 54: 697–702.

Schrag A, Geser F, Stampfer-Kountchev M et al. (2006). Health-related quality of life in multiple system atrophy. Mov Disord 21: 809–815.

Schrag A, Selai C, Mathias C et al. (2007). Measuring health-related quality of life in MSA: the MSA-QoL. Mov Disord 22: 2332–2338.

Schulz JB, Klockgether T, Petersen D et al. (1994). Multiple system atrophy: natural history MRI morphology and dopamine receptor imaging with 123IBZM-SPECT. J Neurol Neurosurg Psychiatry 57: 1047–1056.

Seo JH, Yong SW, Song SK et al. (2010). A case-control study of multiple system atrophy in Korean patients. Mov Disord 25: 1953–1959.

Seppi K, Poewe W (2010). Brain magnetic resonance imaging techniques in the diagnosis of parkinsonian syndromes. Neuroimaging Clin N Am 20: 29–55.

Seppi K, Schocke MF, Esterhammer R et al. (2003). Diffusion-weighted imaging discriminates progressive supranuclear palsy from PD but not from the parkinson variant of multiple system atrophy. Neurology 60: 922–927.

Seppi K, Peralta C, Diem-Zangerl A et al. (2006). Placebo-controlled trial of riluzole in multiple system atrophy. Eur J Neurol 13: 1146–1148.

Shults CW, Rockenstein E, Crews L et al. (2005). Neurological and neurodegenerative alterations in a transgenic mouse model expressing human alpha-synuclein under oligodendrocyte promoter: implications for multiple system atrophy. J Neurosci 25: 10689–10699.

Shy GM, Drager GA (1960). A neurological syndrome associated with orthostatic hypotension: a clinical-pathologic study. Arch Neurol 2: 511–527.

Song YJ, Lundvig DM, Huang Y et al. (2007). p25alpha relocalizes in oligodendroglia from myelin to cytoplasmic inclusions in multiple system atrophy. Am J Pathol 171: 1291–1303.

Spillantini MG, Crowther RA, Jakes R et al. (1998). Filamentous alpha-synuclein inclusions link multiple system atrophy with Parkinson's disease and dementia with Lewy bodies. Neurosci Lett 251: 205–208.

Stefanova N, Tison F, Reindl M et al. (2005). Animal models of multiple system atrophy. Trends Neurosci 28: 501–506.

Stefanova N, Reindl M, Neumann M et al. (2007). Microglial activation mediates neurodegeneration related to oligodendroglial alpha-synucleinopathy: implications for multiple system atrophy. Mov Disord 22: 2196–2203.

Stefanova N, Poewe W, Wenning GK (2008). Rasagiline is neuroprotective in a transgenic model of multiple system atrophy. Exp Neurol 210: 421–427.

Stefanova N, Bucke P, Duerr S et al. (2009). Multiple system atrophy: an update. Lancet Neurol 8: 1172–1178.

Stemberger S, Jamnig A, Stefanova N et al., 2011, PLoS One. 2011;6:e19808; http://dx.doi.org/10.1371/journal.pone.0019808.

Sung JH, Mastri AR, Segal E (1979). Pathology of Shy–Drager syndrome. J Neuropathol Exp Neurol 38: 353–368.

Testa D, Filippini G, Farinotti M et al. (1996). Survival in multiple system atrophy: a study of prognostic factors in 59 cases. J Neurol 243: 401–404.

Tison F, Yekhlef F, Chrysostome V et al. (2000). Prevalence of multiple system atrophy. Lancet 355: 495–496.

Trojanowski JQ, Revesz T (2007). Proposed neuropathological criteria for the post mortem diagnosis of multiple system atrophy. Neuropathol Appl Neurobiol 33: 615–620.

Ubhi K, Rockenstein E, Mante M et al. (2008). Rifampicin reduces alpha-synuclein in a transgenic mouse model of multiple system atrophy. Neuroreport 19: 1271–1276.

Vanacore N, Bonifati V, Fabbrini G et al. (2000). Smoking habits in multiple system atrophy and progressive supranuclear palsy. European Study Group on Atypical Parkinsonisms. Neurology 54: 114–119.

Vanacore N, Bonifati V, Fabbrini G et al. (2001). Epidemiology of multiple system atrophy. ESGAP Consortium. European Study Group on Atypical Parkinsonisms. Neurol Sci 22: 97–99.

Vanacore N, Bonifati V, Fabbrini G et al. (2005). Case-control study of multiple system atrophy. Mov Disord 20: 158–163.

Vidal JS, Vidailhet M, Elbaz A et al. (2008). Risk factors of multiple system atrophy: a case-control study in French patients. Mov Disord 23: 797–803.

Visser-Vandewalle V, Temel Y, Colle H et al. (2003). Bilateral high-frequency stimulation of the subthalamic nucleus in patients with multiple system atrophy – parkinsonism. Report of four cases. J Neurosurg 98: 882–887.

Vodusek DB (2001). Sphincter EMG and differential diagnosis of multiple system atrophy. Mov Disord 16: 600–607.

Walter U, Behnke S, Eyding J et al. (2007). Transcranial brain parenchyma sonography in movement disorders: state of the art. Ultrasound Med Biol 33: 15–25.

Watanabe H, Saito Y, Terao S et al. (2002). Progression and prognosis in multiple system atrophy: an analysis of 230 Japanese patients. Brain 125: 1070–1083.

Wenning GK (2005). Placebo-controlled trial of amantadine in multiple-system atrophy. Clin Neuropharmacol 28: 225–227.

Wenning GK, Ben Shlomo Y, Magalhaes M et al. (1994). Clinical features and natural history of multiple system atrophy. An analysis of 100 cases. Brain 117: 835–845.

Wenning GK, Tison F, Ben Shlomo Y et al. (1997). Multiple system atrophy: a review of 203 pathologically proven cases. Mov Disord 12: 133–147.

Wenning GK, Seppi K, Tison F et al. (2002). A novel grading scale for striatonigral degeneration (multiple system atrophy). J Neural Transm 109: 307–320.

Wenning GK, Geser F, Stampfer-Kountchev M et al. (2003). Multiple system atrophy: an update. Mov Disord 18 (Suppl 6): S34–S42.

Wenning GK, Colosimo C, Geser F et al. (2004a). Multiple system atrophy. Lancet Neurol 3: 93–103.

Wenning GK, Tison F, Seppi K et al. (2004b). Development and validation of the Unified Multiple System Atrophy Rating Scale (UMSARS). Mov Disord 19: 1391–1402.

Wenning GK, Stefanova N, Jellinger KA et al. (2008). Multiple system atrophy: a primary oligodendrogliopathy. Ann Neurol 64: 239–246.

Wullner U, Abele M, Schmitz-Huebsch T et al. (2004). Probable multiple system atrophy in a German family. J Neurol Neurosurg Psychiatry 75: 924–925.

Yamamoto T, Sakakibara R, Uchiyama T et al. (2005). When is Onuf's nucleus involved in multiple system atrophy? A sphincter electromyography study. J Neurol Neurosurg Psychiatry 76: 1645–1648.

Yazawa I, Giasson BI, Sasaki R et al. (2005). Mouse model of multiple system atrophy alpha-synuclein expression in oligodendrocytes causes glial and neuronal degeneration. Neuron 45: 847–859.

Yoshida M (2007). Multiple system atrophy: alpha-synuclein and neuronal degeneration. Neuropathology 27: 484–493.

Young TM, Mathias CJ (2004). The effects of water ingestion on orthostatic hypotension in two groups of chronic autonomic failure: multiple system atrophy and pure autonomic failure. J Neurol Neurosurg Psychiatry 75: 1737–1741.

Zhu S, Stavrovskaya IG, Drozda M et al. (2002). Minocycline inhibits cytochrome c release and delays progression of amyotrophic lateral sclerosis in mice. Nature 417: 74–78.

Handbook of Clinical Neurology, Vol. 117 (3rd series)
Autonomic Nervous System
R.M. Buijs and D.F. Swaab, Editors

Chapter 20

Pure autonomic failure

EMILY M. GARLAND[1], WILLIAM B. HOOPER[2], AND DAVID ROBERTSON[1]*
[1]*Autonomic Dysfunction Center, Vanderbilt University, Nashville, TN, USA*
[2]*Department of Internal Medicine, Greenwich Hospital, Greenwich, CT, USA*

INTRODUCTION

Pure autonomic failure (PAF), or Bradbury–Eggleston syndrome, is a rare, primary disease of autonomic nervous system degeneration. It generally presents in middle to late life, progresses slowly, and has a relatively good prognosis. In contrast to the central neurodegenerative disease multiple system atrophy (MSA), PAF is characterized by lesions in the peripheral autonomic nervous system and deficits in norepinephrine synthesis and release from sympathetic nerve terminals. Disabling orthostatic hypotension and symptoms of cerebral hypoperfusion most frequently bring these patients to physicians specializing in autonomic dysfunction. The orthostatic hypotension is often accompanied by supine hypertension, making blood pressure management more difficult. With denervation hypersensitivity of adrenergic receptors and impaired baroreflex function, patients with PAF experience exaggerated responses to drugs and interventions that affect blood pressure, a consequence that can be useful in both the diagnosis and treatment of the disorder. Studies continue on the underlying pathology of PAF, which is likely related to α-synuclein-containing neuronal cytoplasmic inclusions called Lewy bodies that have been found in ganglia, peripheral autonomic nerves, and the central nervous system (CNS).

BACKGROUND

In 1925, Bradbury and Eggleston first described three patients with extreme posture-related falls in blood pressure (Bradbury and Eggleston, 1925). Their evaluation and conclusions were so sophisticated for the era in which they practiced, it is worthwhile to consider their work in some detail. In addition to standing- and exercise-induced syncope, the patients experienced "a slow unchanging pulse rate, incapacity to perspire, a lowered basal metabolism and signs of slight and indefinite changes in the nervous system." Nocturia and erectile dysfunction were prominent. All three cases had a slight anemia and high normal blood urea nitrogen.

Precipitous 50% drops in blood pressure with low, relatively steady heart rates occurred with standing or head-up tilt. Head-down tilt increased blood pressure, again with little effect on heart rate. Acute epinephrine dosing raised systolic blood pressure and heart rate in supine and standing positions without a consistent effect on diastolic blood pressure, indicating that epinephrine produced little or no vasoconstriction, and the investigators surmised that the patients were hypersensitive to epinephrine's vasodilatory effects. Atropine did not change blood pressure or heart rate. Low autonomic tone was inferred from the absence of a heart rate rise during orthostatic decreases in blood pressure or after atropine. Pilocarpine stimulated sweating but did not affect orthostatic blood pressure changes. Bandaging the legs or wearing an abdominal binder provided benefit in case 2, allowing him to walk about more freely.

Bradbury and Eggleston concluded that these patients had lost "the normal mechanism by which blood pressure is maintained in the different body positions," including an impaired ability to increase heart rate via sympathetic activation and "decreased responsiveness of the vagus to pronounced elevation of blood pressure." They proposed "paralysis of the sympathetic vasoconstrictor endings" as the underlying pathology. Subsequent to this publication, similar cases were termed Bradbury–Eggleston syndrome or idiopathic orthostatic hypotension.

By the mid 1950s, a number of tests had been developed to assess sympathetic and parasympathetic nervous system function, and it was established that severe

*Correspondence to: David Robertson, M.D., Autonomic Dysfunction Center, AA3228 Medical Center North, Vanderbilt University, Nashville, TN 37232-2195, USA. Tel: +1-615-343-6499, E-mail: davidrobertson@vanderbilt.edu

orthostatic hypotension associated with impaired sympathetic reflexes could be related to an underlying disease, such as diabetes or cancer, or it could be a primary disorder without an identified structural or biochemical cause. Interest in the primary cases focused on the site of the lesion, with evidence presented in 1946 (Briggs, 1946) and 1955 (Barnett et al., 1955; Barnett and Wagner, 1958) that primary orthostatic hypotension could be attributed to either a central or a peripheral lesion in the autonomic nervous system.

Patients with orthostatic hypotension and central nervous system signs were described in 1960 by Shy and Drager, leading to the designation "Shy–Drager syndrome" (Shy, 1960). In addition to the autonomic symptoms, similar to those described by Bradbury and Eggleston, i.e., postural falls in blood pressure, anhidrosis, and genitourinary and gastrointestinal defects, the patients of Shy and Drager also demonstrated symptoms associated with a Parkinson-like syndrome, such as tremor, rigidity, bradykinesia, and slurred speech. An autopsy of one patient revealed widespread neuronal cell loss and gliosis in the intermediolateral cell column, medulla, cerebellum, midbrain, hypothalamus, and basal ganglia. Shy and Drager suggested that a primary degenerative nervous system disorder can underlie some cases of orthostatic hypotension. In recognition of the numerous sites of cell loss in the central nervous systems of these types of patients, the term "multiple system atrophy" was proposed in 1969 by Graham and Oppenheimer (Graham and Oppenheimer, 1969).

Further evidence accumulated over the next two decades for multiple distinct clinical entities under the umbrella of orthostatic hypotension and for different patterns of neurodegeneration. A link between idiopathic orthostatic hypotension and decreased norepinephrine synthesis in sympathetic nerves was demonstrated by Goodall et al. (1968). They proposed that the decreased urinary output of radioactive norepinephrine in three patients infused with labeled dopamine was related to the amyloid deposits and degenerative changes in sympathetic nerves and ganglia observed in the one autopsy. Hughes described eight cases of idiopathic orthostatic hypotension, six of whom underwent extensive autonomic function testing and one of whom was autopsied. Problems were restricted to the autonomic nervous system in three patients. They experienced orthostatic hypotension, anhidrosis, and nocturia but none of the rigidity, tremor and gait abnormalities observed in the other five patients. Both groups had abnormal Valsalva maneuvers, negative cold pressor tests, and no heart rate change with atropine. The authors noted that prognosis worsens when systems other than the autonomic nervous system are involved (Hughes et al., 1970). Kontos et al. reported in 1975 on five patients in whom idiopathic postural hypotension was associated with loss of noradrenergic nerve terminals. Not only did these patients fail to show a decrease in forearm blood flow when given an infusion of tyramine and enhanced vasoconstriction in response to a norepinephrine infusion, but they also lacked catecholamine-specific fluorescence around blood vessels from deltoid muscle biopsies (Kontos et al., 1975). A distinction between disrupted central autonomic pathways and isolated peripheral autonomic dysfunction was confirmed in a study of 10 cases of idiopathic orthostatic hypotension studied by Ziegler et al. (1977). Six had symptoms of a central nervous system disorder, including parkinsonism and cerebellar dysfunction. These had normal plasma norepinephrine levels while supine, in contrast to patients without signs of a central neurologic lesion, who had low plasma norepinephrine levels.

In 1979, Bannister compared results of cardioreflex testing and responses to pressor agents in 10 patients with chronic autonomic failure and postural hypotension. Patients were classified as autonomic failure in isolation or pure autonomic failure (PAF), autonomic failure associated with multiple system atrophy (MSA), or autonomic failure associated with parkinsonism. Results for the Valsalva maneuver, mental arithmetic test, and sweat test confirmed that cardioreflexes were impaired in all three patient groups (Bannister, 1979). The differences in the levels of the lesions have more recently been associated with the diverse pathophysiological findings in PAF, MSA, and Parkinson's disease (PD) with autonomic failure. Patients with MSA show an impairment of central autonomic cardiovascular pathways with intact peripheral postganglionic noradrenergic fibers. Although the tonic discharge of sympathetic neurons is intact, baroreflex modulation of sympathetic tone is affected in MSA. In patients with PAF and PD with autonomic failure, central pathways are intact but lesions are found in the sympathetic nerve tracts in the spinal cord and postganglionic noradrenergic neurons (Shibao et al., 2012). In addition, the movement disorder of PD relates to a loss of dopaminergic innervation in the striatum (Goldstein et al., 2008a). Patients with a central lesion synthesize norepinephrine in postganglionic neurons and retain normal pressor sensitivity to directly acting agents, while norepinephrine synthesis is decreased in patients with a peripheral lesion and the pressor response is exaggerated.

In a consensus statement published in 1996 (Consensus Committee of the American Autonomic Society and the American Academy of Neurology, 1996), orthostatic hypotension was defined as a decrease in systolic blood pressure ≥ 20 mmHg or diastolic pressure fall ≥ 10 mmHg within 3 minutes of standing or head-up tilt. The document went on to define PAF as an idiopathic sporadic disorder with orthostatic hypotension and frequently other signs of autonomic failure but no additional neurological features. It was recognized that supine plasma norepinephrine is low in these patients.

ORTHOSTASIS: THE CHALLENGE OF UPRIGHT POSTURE

In a person lying flat, blood is distributed fairly evenly throughout the body. Standing is accompanied by a shift of approximately 500–1000 mL of blood from the upper body to the legs, buttocks, and abdominal cavity within 3–5 minutes. Furthermore, plasma volume decreases by an average of 14% over approximately 30 minutes of standing motionless as plasma moves out of the circulation and into the tissues, and this is reflected in a transient increase in hematocrit of 3–4% over a 15–30 minute timeframe (Jacob et al., 2005; Gupta and Nair, 2008; Robertson, 2008). As a result, venous return to the heart decreases, reducing stroke volume and cardiac output. In individuals with a healthy autonomic nervous system, baroreceptors detect these changes and signal the central nervous system, which activates the sympathetic nervous system and suppresses vagal activity. As a result of vasoconstriction and positive inotropic and chronotropic effects on the heart, peripheral resistance increases, venous return improves, and cardiac output normalizes to maintain blood pressure and prevent a fall in cerebral perfusion. In addition to norepinephrine release, activation of the renin–angiotensin–aldosterone system and stimulation of arginine vasopressin contribute to the vasoconstriction and help the body retain fluid (Gupta and Nair, 2008). The homeostatic mechanisms that are needed to limit blood pressure fluctuations are curtailed in patients with PAF. Following changes in blood volume and blood pressure, stimulation of the sympathetic nervous system and norepinephrine release are limited. Compensatory vasoconstriction is inadequate or absent, as is the reflex increase in heart rate. In addition, the loss of sympathetic innervation to juxtaglomerular cells in PAF has been proposed to cause a deficit of renin-producing cells, so that the postural increase in plasma renin activity is blunted, although baseline aldosterone and the aldosterone response to standing are maintained (Biaggioni et al., 1993).

In patients with autonomic failure, the decrease in blood pressure and inadequate compensatory heart rate increase while upright are accompanied by a proportionate drop in middle cerebral artery flow. As blood pressure falls to a certain level, the accompanying decrease in cerebral blood flow may lead to orthostatic symptoms. Without intervention, symptoms may be associated with syncope, a sudden, transient loss of consciousness with spontaneous recovery. A shift in the cerebral autoregulatory curve may enable some patients to tolerate a substantial fall in blood pressure without orthostatic symptoms (Gupta and Nair, 2008).

CLINICAL PRESENTATION

PAF occurs sporadically, without evidence of a familial or environmental etiology, but it is possible that some genetic contribution to PAF may have eluded investigators. It is slightly more prevalent in men than women. Symptoms of autonomic dysfunction begin gradually in middle age and progress slowly. Orthostatic hypotension is prominent without cognitive, sensory or motor deficits (Freeman, 2004). Early symptoms may include unsteadiness, lightheadedness or faintness, as well as neck pain and headache (Robertson et al., 1994) while standing. The neck pain has a coathanger distribution and might result from hypoperfusion of the neck muscles during periods of low blood pressure (Bleasdale-Barr and Mathias, 1998; Mathias et al., 1999). Symptoms are generally worse in the morning, after meals, and during hot weather, and can often be relieved by lying down. Given their insidious nature and gradual onset, symptoms may be experienced for many years before patients seek medical care. Patients frequently present initially with symptoms of urinary and sexual dysfunction. Nocturia may cause the patient to get up as many as five times per night. Men may describe urinary hesitancy, urgency, dribbling, and incontinence, as well as erectile and ejaculatory dysfunction. Likewise, women may experience urinary retention or incontinence. Other symptoms related to autonomic dysfunction occur more variably. These include a decreased ability to sweat, dry mouth, diarrhea, constipation and other gastrointestinal problems (Kaufmann and Schatz, 2012).

At least a 5 year history of such complaints without development of neurological signs provides some assurance that a diagnosis of MSA, Parkinson's disease, or dementia with Lewy bodies can be excluded. In most but not all cases of Parkinson's disease, extrapyramidal symptoms develop years before autonomic symptoms. However, autonomic dysfunction is often the initial presenting symptom of MSA, and patients must be carefully evaluated for MSA before being given a diagnosis of PAF. Hoarseness is frequently an early sign of MSA, and both MSA and PD patients may suffer from sleep apnea and REM disturbances. The clinical condition can be relatively stable over many years in PAF, and prognosis is relatively good with occasional survival into the 90s (Mabuchi et al., 2005; Robertson, 2008). Yet, it has been proposed that, in some cases, PAF is merely the initial presentation of a central neurodegenerative disorder (Kaufmann et al., 2004).

AUTONOMIC ASSESSMENT OF CARDIOVASCULAR REFLEXES

In patients presenting with orthostatic hypotension, causes other than a primary autonomic failure must be ruled out. Dehydration, poor nutrition, and deconditioning can contribute to orthostatic hypotension. Impaired baroreflex function and less responsive vasculature raise the prevalence of orthostatic hypotension to 15% in individuals aged 64–69 years and higher in older people (Low, 2008). Predisposing medications include tricyclic antidepressants with norepinephrine transporter (NET)-blocking properties, antihypertensives, diuretics, vasodilators, and antiparkinsonian agents, which can lower sympathetic tone and impair compensatory changes in vascular resistance (Metzler et al., 2012). A complete history and physical examination may identify a primary cardiac or neurological cause. Rarely, an immune-mediated process or a congenital disorder can be recognized. When these possibilities have been eliminated, autonomic assessment of the cardiovascular system is required to determine the severity of the impairment, the involvement of the sympathetic and parasympathetic nervous systems, and the participation of afferent and efferent neuronal pathways.

During a head-up tilt test, the patient is tilted gradually with continuous blood pressure and EKG monitoring. Efforts are made to reproduce the patient's orthostatic symptoms to determine the blood pressure and heart rate changes that accompany presyncope or syncope. Patients with PAF and MSA experience a decrease in blood pressure during head-up tilt. Chandler (Chandler and Mathias, 2002) reported a 41 ± 4 and 19 ± 4 mmHg decrease in mean arterial pressure in PAF and MSA, respectively, in contrast to a 12 ± 4 mmHg increase in controls. Others have found no difference in the blood pressure decrease between PAF and MSA (Asahina et al., 2005; Donadio et al., 2010). Between 5 and 11 minutes after return to the supine position (tilt reversal), patients with PAF are more likely than those with MSA to increase systolic blood pressure above the pretilt level (Chandler and Mathias, 2002; Asahina et al., 2005). This blood pressure overshoot does not generally occur in healthy individuals, and its cause is unclear. It is associated with increased total peripheral resistance, suggesting that vasoconstrictor factors are responsible. Because of the peripheral rather than central lesion in PAF, vasopressin release is stimulated by hypotension in PAF, but not in MSA, and could contribute to the vasoconstriction (Asahina et al., 2005).

Whereas patients with vasovagal syncope generally have normal blood pressure while recumbent, sitting, or standing, they occasionally experience fainting associated with brief hypotension and bradycardia. A head-up tilt table test can be valuable for differentiating neurally mediated syncope from PAF. Other tests described below also confirm normal autonomic function in neurally mediated syncope.

Lesions in the adaptive mechanisms in the cardiovascular system can be determined by simple physiological tests of autonomic function performed with blood pressure and heart rate monitoring. These tests commonly include a 10 minute stand test, hyperventilation, the Valsalva maneuver, sinus arrhythmia, a cold pressor test, and a handgrip test.

Orthostatic responses to standing are determined following at least 30 minutes rest in the supine position. Participants stand for 10 minutes or until symptoms require them to lie down. Although orthostatic hypotension occurs in essentially all patients with PAF, it can also occur in 16–58% of patients with Parkinson's disease with autonomic failure and in approximately 75% of patients with MSA (Metzler et al., 2012). An orthostatic decrease in systolic blood pressure of 50 mmHg or greater is common in untreated patients with PAF. However, a diagnosis cannot be excluded based on a single normal upright blood pressure measurement; several measurements of orthostatic blood pressure should be made. With parasympathetic and sympathetic dysfunction, heart rate will show little increase with hypotension.

Rapid, shallow breathing (hyperventilation) for 30 seconds minimally affects blood pressure and raises heart rate in healthy controls. Hypocapnia-induced vasodilatation is counteracted by sympathetically mediated reflex vasoconstriction and increases in cardiac output (Onrot et al., 1991; Jordan et al., 1997). In the absence of the reflex vasoconstriction, hyperventilation powerfully decreases blood pressure in patients with autonomic failure (Shibao et al., 2010).

Sinus arrhythmia (SA) is assessed during controlled breathing (pattern of 5 seconds inhalation and 5 seconds exhalation repeated over 90 seconds). During inspiration, heart rate increases, whereas it slows during expiration. The SA ratio is calculated from the highest to lowest heart rate during this 90 second period. Subjects under the age of 40 years generally have a 10–30 beats per minute (bpm) difference between the maximum and minimum heart rates. The variation in heart rate is mainly under vagal control, and blunted sinus arrhythmia in autonomic failure indicates a defect in the parasympathetic nervous system (Shibao and Biaggioni, 2010).

The Valsalva maneuver tests several of the components of the baroreflex arc. Individuals exhale forcibly against pressure, increasing intrathoracic and intra-abdominal pressures. In normal subjects, mechanical compression of the aorta in phase I leads to a transient increase in blood pressure. With continued straining during phase II, venous return and cardiac output fall, causing a drop in blood pressure. Reflex sympathetic stimulation and parasympathetic withdrawal produce a tachycardia (at least 20–25 bpm above baseline). With release of strain in phase III, blood pressure abruptly falls as intrathoracic pressure decreases. During phase IV, blood pressure then rises above basal levels (by at least 10 mmHg) as sympathetic tone increases and venous return and cardiac output normalize. The increase in blood pressure triggers a reflex bradycardia. The Valsalva ratio is calculated from the maximum heart rate during or shortly after straining and the minimum heart rate in phase IV. Blood pressure responses during the Valsalva maneuver reflect sympathetic nervous system function. In patients with autonomic failure, the decrease in systolic blood pressure during phase II is magnified whereas the phase IV blood pressure overshoot is absent. Accordingly, heart rate does not increase during phase IV and the Valsalva ratio is lowered. Parasympathetic lesions also decrease the Valsalva ratio.

Placing the hand in ice water during a cold pressor test normally elicits a rapid reduction in forearm and skin blood flow with a concomitant increase in blood pressure. Increases in heart rate and blood pressure are also observed after fatiguing muscle exercise, such as the handgrip test. A normal response to both the cold pressor test and the handgrip test averages a 20 mmHg increment in systolic blood pressure and a 10 bpm increase in heart rate. A negative cold pressor or handgrip test is consistent with a lesion in sympathetic efferent activity. Since it increases blood pressure by cerebral stimulation, a mental arithmetic test can help differentiate a peripheral from a central lesion.

Unfortunately, blood pressure and heart rate responses during these autonomic function tests are not always helpful in differentiating the different classes of chronic autonomic failure (Jordan et al., 1998). Because of the difficulty distinguishing between PAF and MSA, Asahina et al. (2009) looked at the sweat response and skin blood flow during deep inspiration and mental arithmetic. A decrease in skin blood flow is mediated by vasoconstrictor sympathetic efferent fibers (Young et al., 2006). Control subjects increased sweat output and decreased skin blood flow during testing, MSA patients had no sweat response but decreased skin blood flow, and PAF patients lacked both the sweat and skin blood flow responses.

Because PAF is a less progressive and generally less disabling disorder, it is important for the clinicians, patients and caregivers to be able to distinguish it from MSA and PD with autonomic failure (Table 20.1). Results of pharmacological testing, measurement of

Table 20.1

Comparison of multiple system atrophy, pure autonomic failure and Parkinson's disease (modified from Garland and Robertson, 2009)

Characteristic	Multiple system atrophy (MSA)	Pure autonomic failure (PAF)	Parkinson's disease (PD)
Site of autonomic impairment	Mainly preganglionic, central	Mainly postganglionic; peripheral	Mainly postganglionic; peripheral
Prognosis	Poor; median survival 5–9 years	Good; slow disease progression with survival > 10–20 years	Intermediate prognosis between MSA and PAF; variable progression
Parkinsonian or cerebellar symptoms	Common	Rare	Always present
Neuropathological marker	Glial cytoplasmic inclusions in oligodendroglia and neurons	Lewy bodies in neurons	Lewy bodies in neurons
Orthostatic hypotension	Common	Common	Less common than in MSA or PAF
Sleep apnea and respiratory disturbances	Common	Uncommon	Common
Urinary tract disturbances	Early in disease	Late in disease	Later than extrapyramidal symptoms
Supine plasma norepinephrine	Normal	Very low	Normal
Orthostatic rise in norepinephrine	Subnormal	Subnormal	Normal
Postsynaptic adrenoreceptor sensitivity	Mildly increased	Increased	Mildly increased

norepinephrine and vasoactive hormones, imaging studies, sweat assessments, and odor identification tests contribute to the complete clinical picture and facilitate the differential diagnosis.

SENSITIVITY TO NEUROTRANSMITTER AGONISTS AND ANTAGONISTS

PAF can be distinguished from MSA with acute drug studies that compare blood pressure responses to agents that act either directly or indirectly on adrenergic receptors. Pressor agents that act directly on the α-adrenergic receptors, such as norepinephrine, midodrine, and phenylephrine, produce an exaggerated blood pressure response in PAF, consistent with peripheral denervation and receptor hypersensitivity (Polinsky et al., 1981; Freeman, 1995). Biaggioni et al. (1987) reported that extremely low doses of phenylpropanolamine produce dramatic increases in blood pressure in patients with autonomic failure. They underscored the fact that while this excessive pressor effect could be exploited in the treatment of autonomic failure, the safety of similarly acting drugs could be a concern in individuals with compromised autonomic function. Accordingly, application of a phenylephrine ophthalmic solution raises blood pressure in patients with idiopathic orthostatic hypotension but has no such pressor effect in healthy subjects (Robertson, 1979). Similarly, direct β-adrenergic receptor agonists such as isoprenaline (isoproterenol) stimulate greater tachycardia in PAF (Robertson et al., 1984; Sharabi et al., 2006) and have an even more profound depressor effect via vasodilatation unopposed by compensatory tachycardia (Robertson et al., 1984).

Tyramine is an indirectly acting sympathetic agent that requires uptake by the norepinephrine transporter. Like the α_2-adrenergic receptor antagonist yohimbine, it causes norepinephrine release from sympathetic nerves. These drugs cause a large pressor response in MSA but not in PAF, consistent with the postganglionic lesion in PAF. Clonidine is an α_2-adrenergic agonist that affects blood pressure through central and peripheral effects. It usually reduces blood pressure but in severe PAF, clonidine can exert a pressor effect at doses above 0.3 mg and improves standing time in some patients with PAF, possibly by peripheral actions on postjunctional α_2-adrenergic receptors (Robertson et al., 1983). This response is greater in patients with lower supine plasma norepinephrine (Robertson et al., 1986), but clonidine is an unreliable means to raise blood pressure in all but the most severely affected PAF patients; the 0.4 mg pressor dose of clonidine in PAF may elicit CNS side-effects such as altered perception.

Acting in the lower brainstem, clonidine suppresses sympathetically mediated elevations in plasma norepinephrine in healthy individuals and lowers blood pressure by decreasing sympathetic outflow. Although clonidine should not produce a normal catecholamine response in PAF patients, this pathway also leads to stimulation of growth hormone release by actions on the hypothalamus and anterior pituitary. The clonidine-induced increase in serum growth hormone levels is maintained in patients with PD and PAF, but not in patients with MSA, due to their central lesion (Gupta and Nair, 2008; Mathias, 2009).

Ganglionic blockade with trimetaphan is a useful test for differentiating MSA from Parkinson's disease with autonomic failure and PAF. Whereas trimetaphan causes a profound decrease in blood pressure in MSA, the response is smaller in PAF and more closely resembles the normal response when assessed in patients with Parkinson's disease and autonomic failure (Goldstein et al., 2002b; Sharabi et al., 2006).

A significant increase in blood pressure occurs after indometacin but not after another nonsteroidal anti-inflammatory drug, ibuprofen (Jordan et al., 1998).

Additional information on autonomic cardiovascular function is provided by spectral analysis of spontaneous blood pressure and heart rate variability and baroreflex testing. Spectral analysis confirms the impaired sympathetic efferent function in PAF and also demonstrates deficient baroreflex mechanisms. Low frequency distribution for RR interval and blood pressure are absent in PAF, and the high frequency distribution for RR interval does not show the decrease with tilt that is part of the normal compensatory vagal withdrawal (Furlan et al., 1995). Impaired baroreflexes are also evident in PAF in the absence of compensatory heart rate changes to agents that raise or lower blood (Mathias, 2009). Baroreflex-mediated release of vasopressin acts as a useful test of afferent autonomic pathway integrity. Patients with PAF will demonstrate normal baroreceptor-mediated vasopressin release, whereas patients with MSA will show a diminished response since vasopressin is produced centrally (Kaufmann et al., 1992).

CATECHOLAMINES

Human plasma contains not only norepinephrine, epinephrine, and dopamine, but also the catecholamine precursor, 3,4-L-dihydroxyphenylalanine or dopa, and neuronal metabolites of norepinephrine, dihydroxyphenylglycol (DHPG), and of dopamine, 3,4-dihydroxyphenylacetic acid or DOPAC. Measurement of these additional catechols and their labeled counterparts provides information about norepinephrine synthesis and turnover. Resting plasma concentrations of norepinephrine, epinephrine, and their metabolites are reduced in PAF consistent with decreased production or release due to the postganglionic lesion (Yamamoto et al., 1996; Goldstein et al., 2003a). Levels in MSA patients are normal. Neither patient group has a normal baroreflex-mediated increase in plasma norepinephrine with standing (Ziegler et al., 1977; Yamamoto et al., 1996). Plasma dopamine and DOPAC are also low in PAF due to impaired synthesis and metabolism in sympathetic neurons (Yamamoto et al., 1996). In contrast, cerebrospinal fluid (CSF) levels of norepinephrine and DHPG are low in both PAF and MSA (Goldstein et al., 2003a). The value of measures of plasma, CSF, and urinary catechols in the differential diagnosis for PAF and MSA is limited by the overlap between patient groups (Goldstein et al., 1989). Furthermore, since venous norepinephrine concentration is modulated by neuronal reuptake, α_1-adrenoreceptor activity, blood flow, and clearance from the circulation (Goldstein et al., 2002b), plasma venous norepinephrine concentration is sometimes problematic as a reflection of sympathetic activity.

Kinetic studies based on the dilution of tracer ^{3}H-norepinephrine by unlabeled endogenous norepinephrine are employed to determine norepinephrine spillover, that is, the movement of norepinephrine from various organs to the bloodstream (Esler et al., 1982, 1988). Patients with Parkinson's disease and autonomic failure have decreased or absent norepinephrine spillover in the heart (Goldstein et al., 2002a). Patients with PAF have normal cerebrovascular norepinephrine spillover but decreased total body spillover and essentially absent cardiac spillover (Esler et al., 1988; Yamamoto et al., 1996).

MICRONEUROGRAPHY

Intraneuronal microneurography enables the direct recording from postganglionic sympathetic nerves. A tungsten electrode can be placed into the peroneal nerve with the aid of an electrical stimulator to confirm placement in a muscle rather than skin nerve. Action potentials from muscle sympathetic fibers are coupled with cardiac rhythm and diastolic blood pressure (Delius et al., 1972) and can be evaluated at rest and after maneuvers expected to stimulate efferent sympathetic activity. Because of the degeneration of postganglionic neurons in PAF, this technique may not be very informative (Freeman, 1995).

SWEAT TESTING AND SYMPATHETIC CHOLINERGIC INNERVATION

Patients with chronic autonomic failure often have impaired sweating, and tests of sudomotor function can be useful in differentiating preganglionic from postganglionic sympathetic lesions (Mathias, 2009). The thermoregulatory sweat test (TST) assesses the sweating response to an increase in the core body temperature and can be impaired by central or peripheral lesions. A more localized response, focusing on the sweat glands, can be examined with the quantitative sudomotor axonal reflex test (Q-SART) (Low et al., 1983) or quantitative direct and indirect reflex testing of sudomotor function (QDIRT) (Gibbons et al., 2008). This evaluates the sweat gland response to the iontophoresis of acetylcholine at peripheral test sites and thus assesses postganglionic sympathetic sudomotor function. The sympathetic skin response detects changes in electrical potential in the skin in response to physiological and electrical stimuli and is used to study sympathetic cholinergic activity to sweat glands. An abnormal TST with a normal Q-SART would be consistent with a preganglionic lesion whereas an abnormal Q-SART would indicate postganglionic failure and is found in most patients with PAF (Cohen et al., 1987).

Donadio and colleeagues (2010) used immunofluorescent techniques to assess skin innervation by noradrenergic and sudomotor cholinergic fibers. Antibodies to dopamine β-hydroxylase and vasoactive intestinal peptide demonstrated a loss of noradrenergic and cholinergic autonomic skin fibers in PAF, relative to controls and patients with MSA. These results are consistent with a postganglionic denervation in PAF.

IMAGING

MSA and PAF produce distinct patterns of neurodegeneration which may be detected in imaging studies. Cardiac sympathetic nerves use the uptake-1 process to take up ^{123}I-metaiodobenzylguanidine (^{123}I-MIBG) and 6-[^{18}F]fluorodopamine, and vesicles in cardiac sympathetic neurons can be visualized by single-photon emission computed tomography (SPECT) or positron emission tomography (PET), respectively. Cardiac sympathetic innervation is generally intact in MSA but absent in Parkinson's disease with autonomic failure (Goldstein et al., 2002a). Myocardial radioactivity is also decreased in PAF. Cardiac sympathetic denervation precedes the loss of neurons in the sympathetic ganglia (Kaufmann and Benarroch, 2008).

6-[^{18}F]fluorodopa-derived radioactivity visualized through PET scans is decreased in the putamen, caudate, and substantia nigra of the brains of patients with Parkinson's disease (Fig. 20.1). The ratio of radioactivity in the putamen relative to occipital cortex is low in Parkinson's but normal in PAF. Both patient groups have similarly decreased ratios of radioactivity for substantia nigra relative to occipital cortex (Goldstein et al., 2008b).

Magnetic resonance imaging (MRI) and positron emission tomography may help differentiate MSA from other chronic primary autonomic disorders. MRI of the

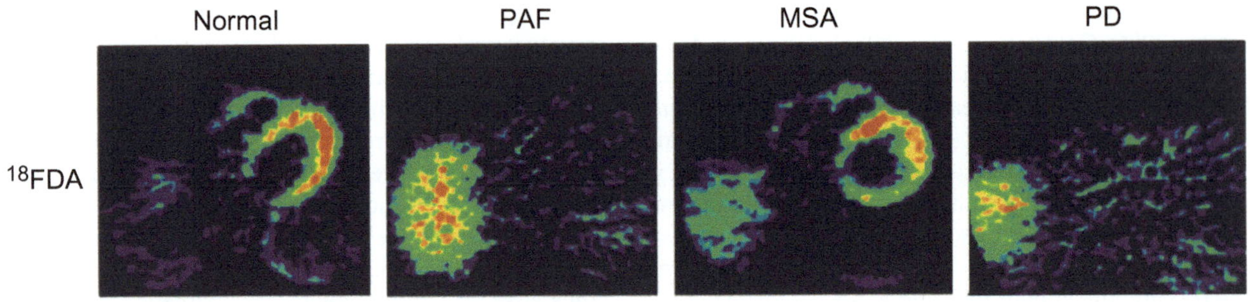

Fig. 20.1. Cardiac PET scan after intravenous 6-[^{18}F]Fluorodopamine (FDA) in a normal control individual and in patients with pure autonomic failure (PAF), multiple system atrophy (MSA) and Parkinson's disease (PD). Radioactivity was absent in the patients with PAF and PD, indicating loss of sympathetic nerve terminals. 6-[^{18}F]Fluorodopamine-derived radioactivity in the patient with MSA indicates intact sympathetic terminals. (Modified from an image provided by David S. Goldstein, Clinical Neurocardiology Section, National Institute of Neurological Disorders and Stroke, National Institutes of Health, Bethesda, MD 20892, USA.)

brains of patients with MSA demonstrates characteristic hypointensity of the putamen relative to the pallidum on T2 images and atrophy of the brainstem and cerebellum, whereas MRI of patients with PAF is normal. Likewise, PET imaging of MSA patients shows generalized reduction in glucose metabolism, particularly in the cerebellum, brainstem, striatum, and frontal and motor cortex (Fulham et al., 1991).

UPSIT ODOR IDENTIFICATION TEST

An odor identification test (UPSIT; University of Pennsylvania Smell Identification Test) has been proposed to differentiate MSA from PD (Goldstein et al., 2008a; Goldstein and Sewell, 2009). UPSIT may also aid in the differential diagnosis of PAF vs. MSA. Whereas odor identification is comparatively normal in patients with MSA, it is impaired in patients with PAF, as well as in patients with Parkinson's disease (Goldstein and Sewell, 2009; Garland et al., 2011). Goldstein and Sewell reported a correlation between olfactory dysfunction and neuroimaging evidence of cardiac noradrenergic denervation, consistent with a common disease process in PAF and Parkinson's disease. The olfactory bulb and several cortical areas involved in olfactory processing are innervated by noradrenergic afferents from the locus coeruleus (Doucette et al., 2007; Veyrac et al., 2007) and noradrenergic signaling in the olfactory bulb appears to play a role in odor discrimination (Doucette et al., 2007). Goldstein proposed that noradrenergic terminals may be lost in the olfactory bulb, as well as in the heart, in PAF and PD, disrupting noradrenergic-mediated olfactory processing in these disorders (Goldstein and Sewell, 2009). UPSIT scores in Parkinson's disease were independent of the central dopamine deficiency. However, a similar study found that the UPSIT score does not differ between PAF and MSA, with both patient groups scoring higher than patients with Parkinson's disease (Silveira-Moriyama et al., 2009).

PUPILLARY RESPONSES

Drugs can be instilled onto the conjunctiva to test autonomic pupillary responses. Cocaine normally blocks neuronal uptake of norepinephrine and causes pupillary dilation. This response is attenuated in patients with sympathetic failure. On the other hand, hydroxyamphetamine causes pupil dilation by stimulating norepinephrine release, and similar to the blood pressure response to drugs acting by this mechanism, pupillary dilation does not occur in patients with a postganglionic lesion. Dilator supersensitivity to directly acting agents is found with sympathetic denervation whereas constrictor supersensitivity is found with parasympathetic denervation. Pupil abnormalities are more prevalent in PAF than in MSA, including bilateral Horner's syndrome in approximately 50% (Bremner and Smith, 2006).

SUPINE HYPERTENSION

Approximately 50% of patients with PAF also experience marked supine hypertension with systolic blood pressures as high as 200 mmHg (Shannon et al., 1997; Garland et al., 2009). In PAF, the high blood pressures are not related to sympathetic nervous system activation (Shannon et al., 2000; Goldstein et al., 2003b), despite increased systemic vascular resistance (Kronenberg et al., 1990). Mechanisms may relate to the inadequate baroreflex buffering of blood pressure, the increased sensitivity of adrenergic receptors, or to medication effects. The long-term consequences of supine hypertension are unknown, although end-organ damage, including left ventricular hypertrophy (Vagaonescu et al., 2000) and renal function impairment (Garland et al., 2009), have been reported. The supine hypertension can also contribute to pressure natriuresis (Jordan et al., 1999b) and exacerbate orthostatic hypotension in the morning.

At this point, treating the disabling orthostatic hypotension of PAF and preventing associated falls takes precedence over treatment of the supine hypertension because the enduring benefit of antihypertensive therapy is not known; a need for such treatment must first be established by overnight blood pressure monitoring, and less direct approaches such as decreasing nocturnal polyuria should be tested first. Patients can minimize the amount of time in a recumbent position and tilt the head of the bed up for sleeping. Although blood pressure remains high throughout the night in some patients, others experience a "dipping" phenomenon with the blood pressure dropping to normal levels (Okamoto et al., 2009). It is important to avoid nighttime antihypertensives in patients who are "dippers" and to remember that with impaired baroreflex function, patients with PAF will be more sensitive to antihypertensive agents. Uncontrolled supine hypertension has been treated with transdermal nitroglycerin applied at bedtime and removed before arising, although this intervention does not prevent nighttime pressure natriuresis (Jordan et al., 1999b). Short-acting nifedipine decreases blood pressure during the night but worsens nocturia and orthostatic hypotension in the morning. Hydralazine and minoxidil less effectively lower nighttime blood pressure. Unlike other antihypertensive agents, clonidine lowers blood pressure in both MSA and PAF and also reduces nocturia. Yet, morning orthostatic tolerance does not improve.

EXERCISE-INDUCED HYPOTENSION

In contrast to the increases in blood pressure, heart rate, and plasma norepinephrine in exercising healthy individuals, patients with PAF have exercise-induced hypotension, an attenuated heart rate increase, and no change in norepinephrine (Akinola et al., 2001). Orthostatic hypotension may be exaggerated postexercise. Hyperventilation reduces blood carbon dioxide, and this may contribute to vasodilatation in exercising muscles and the hypotension experienced during exercise (Onrot et al., 1991; Mathias, 2009).

SIGNAL AMPLIFICATION: EATING AND DRINKING

With the loss of baroreflex function in PAF comes signal amplification in that stimuli that have only mild or no effects on blood pressure in healthy individuals can have profound blood pressure effects in patients with PAF (Robertson, 2008). For example, patients may experience a 40 mmHg drop in blood pressure after a meal and a 40 mmHg increase after 16 ounces of water. Pressures generally remain stable after eating and water ingestion in healthy people.

Release of vasodilatory gastrointestinal peptides following a meal results in dilation of vessels and blood pooling in the splanchnic region. As a result of compensatory increases in cardiac output, stroke volume, and heart rate, and vasoconstriction of vessels in the skeletal muscles, blood pressure does not drop. Since these responses are mediated by the sympathetic nervous system, they are absent in patients with chronic autonomic failure, leading to postprandial hypotension (Mathias, 2009; Shibao et al., 2012). Systolic blood pressure decreases about 50 mmHg in patients with severe autonomic failure, starting about 15 minutes after the meal and reaching a nadir between 30 and 60 minutes (Robertson et al., 1981). The severity of postprandial hypotension rises with a carbohydrate-rich meal. Since it increases the risk of falls, postprandial hypotension should be prevented when possible by consuming small, more frequent meals and limiting carbohydrate intake. Patients can drink coffee with their meal, taking advantage of the pressor effect of caffeine to attenuate postprandial hypotension (Onrot et al., 1985). Acarbose was shown in a randomized, double-blind crossover study to diminish postprandial hypotension in patients with PAF (Shibao et al., 2007), possibly by decreasing glucose absorption and inhibiting release of gastrointestinal peptides.

The pressor effect of water in autonomic failure patients begins after about 5 minutes and peaks around 30 minutes. The pressor response is sympathetically mediated and occurs independently of a change in plasma volume (Jordan et al., 1999a). Evidence indicates that it is related to decreased osmolality sensed in the gut (Jordan et al., 1999a; McHugh et al., 2010). Water can provide rapid relief from orthostatic symptoms, and patients are advised to drink two glasses before engaging in upright activities.

IMPAIRED RENAL FUNCTION

Diminished renal function has been reported in some forms of autonomic dysfunction characterized by profound orthostatic hypotension. Patients with familial dysautonomia are much more likely to develop chronic kidney disease than the US population as a whole (Elkayam et al., 2006), and serum urea nitrogen and serum creatinine are elevated in patients with dopamine β-hydroxylase deficiency (Garland et al., 2009). In a group of 64 patients with PAF having a mean supine plasma norepinephrine concentration of 0.62 ± 0.32 nmol/L and an average systolic blood pressure drop of 67 mmHg with standing, serum urea nitrogen and serum creatinine were elevated relative to a matched group of healthy controls. Some 48% of the patients had supine hypertension, and these patients experienced a greater orthostatic decrease in blood pressure than patients without supine hypertension. Serum creatinine was higher and estimated glomerular filtration rate was lower in those positive for supine hypertension. There is therefore evidence that renal function is diminished in patients with PAF and that the compromised renal function is associated with supine hypertension (Garland et al., 2009).

PATHOPHYSIOLOGY

PAF is one of a group of neurodegenerative diseases classified as α-synucleinopathies that share the characteristic histopathological finding of insoluble deposits of the protein α-synuclein. In MSA, the protein precipitates are found in the cytoplasm and nuclei of glial cells, while they take the form of neuronal cytoplasmic inclusions called Lewy bodies in PAF, Parkinson's disease, and dementia with Lewy bodies. The exact relationship between α-synuclein deposits and the pathology remains undetermined. It is not entirely clear whether cell death and the resulting pathological condition stem directly from the accumulation of α-synuclein or if these accumulations are merely a secondary effect of the pathogenic process.

Histofluorescence and electron microscopy of vasomotor nerves have generally shown decreased catecholamine-specific histofluorescence and loss of noradrenergic nerves in patients with PAF (Kontos et al., 1975; Bannister et al., 1981). In PAF, the largest concentrations of Lewy bodies are found in the sympathetic and parasympathetic ganglia and pre- and postganglionic autonomic neurons (Freeman, 2004), corresponding to the site of greatest cell loss. These include sympathetic neurons in epicardial fat, periadrenal adipose tissue and nerves innervating the bladder wall (Kaufmann and Benarroch, 2008). Lewy bodies are also found to a lesser extent in the substantia nigra, the locus coeruleus, and the thoracolumbar and sacral spinal cord, as well as in unmyelinated neuronal fibers in the skin.

It has been suggested that skin biopsy with α-synuclein immunostaining may soon become an important method of aiding in the diagnosis of Lewy body diseases (Fig. 20.2). A 2008 study demonstrated α-synuclein immunoreactivity in the skin only in patients who were positive for Lewy body pathology in the central nervous system. None of three patients with MSA had positive skin immunoreactivity for α-synuclein (Ikemura et al., 2008). Igari et al. (2011) reported intense staining for α-synuclein in axons of cutaneous nerves from a patient with severe orthostatic hypotension, low plasma norepinephrine, and reduced cardiac MIBG imaging.

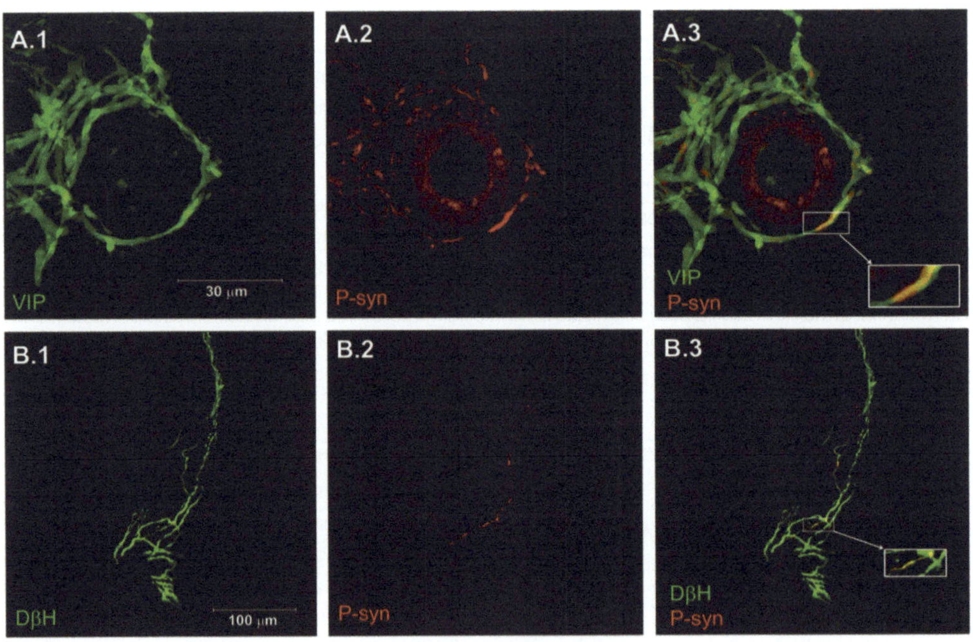

Fig. 20.2. Thigh skin biopsy from a patient with pure autonomic failure. Sections were immunostained for vasoactive intestinal peptide (VIP) as a marker of sudomotor cholinergic fibers (A1) and dopamine-β-hydroxylase (DβH) as a marker of noradrenergic fibers (B1). Middle panels show staining for phosphorylated α-synuclein which was demonstrated by colocalization studies to occur in both cholinergic (A3) and postganglionic sympathetic adrenergic (B3) cholinergic fibers. (Reproduced from Figure 2 in Donadio et al., 2013.)

TREATMENT/MANAGEMENT

Patients with PAF have a generally good outlook; many live for 20 years or more after the onset of their disease. Treatment of patients with PAF should be directed at relieving symptoms, improving daily functioning, and preventing falls, rather than at meeting specific blood pressure goals (Shibao et al., 2012).

Nonpharmacological interventions are initially recommended to patients with PAF to improve functional capacity and improve their quality of life. Abrupt changes in posture, large meals, and hot environments should be avoided. Eating several small meals a day with a dietary sodium intake of 6–10 g per day and fluid intake ≥ 2 L per day might improve orthostatic symptoms but can exacerbate supine hypertension. Patients can take advantage of the amplification of food and water effects on blood pressure by consuming carbohydrates to lower pressure and drinking 500 mL of water to raise pressure depending on their needs throughout the day. Efforts should be made to participate in some form of physical exercise to avoid deconditioning. Compression stockings and an abdominal binder restrict the redistribution of blood that occurs with standing and can attenuate postural changes in blood pressure. Other physical maneuvers like leg-crossing, squatting, and contraction of leg muscles can stimulate skeletal muscle pump activity and defer the need to lie down. Sleeping with the head of the bed tilted up can lessen nocturia and supine hypertension.

Medications producing volume expansion and vasoconstriction are cornerstones of treatment for PAF. Fludrocortisone is a synthetic mineralocorticoid that can increase plasma volume via stimulating sodium retention in the kidney. Its mechanism may also include an increase in peripheral vascular resistance, since the pressor effect of fludrocortisone outlasts its actions on sodium reabsorption (Metzler et al., 2012). The onset of its treatment effect is also delayed for 1–2 weeks after initiation of treatment (Chobanian et al., 1979). Although there have been no randomized, controlled trials to test its efficacy, fludrocortisone has been used as treatment for orthostatic hypotension for several decades (Shibao et al., 2012). Nevertheless, care must be taken when using this drug in patients with supine hypertension which can be exacerbated.

Desmopressin (DDAVP), a vasopressin analog, reduces nocturia and can expand plasma volume. Unlike vasopressin, DDAVP does not have a pressor effect. Patients treated with desmopressin must be carefully monitored for hyponatremia and water intoxication (Mathias and Young, 2003).

Although sympathomimetic agents would be expected to help patients with PAF by causing vasoconstriction, the only agent approved by the US Food and Drug Administration (FDA) for PAF treatment is the peripheral direct α_1-adrenergic agonist midodrine. Its beneficial effects on orthostatic hypotension and symptoms have been demonstrated in double-blind, placebo-controlled trials (Low et al., 1997; Wright et al., 1998). Its duration of action is short enough so that it can be taken early in the day to allow patients to attend to their activities without worsening supine hypertension at night. Droxidopa, used very effectively to restore norepinephrine and improve orthostatic hypotension in dopamine β-hydroxylase deficiency (Robertson and Garland, 2013), has not yet received approval in the USA. Improvement of orthostatic hypotension and symptoms without changes in supine systolic blood pressure was demonstrated in patients with PAF and MSA in a European trial. It has been used to treat chronic autonomic failure in Japan since 1989 (Kaufmann et al., 2003; Kaufmann, 2008; Mathias, 2008). Other drugs have been prescribed off-label in the treatment of PAF. Pyridostigmine, a cholinesterase inhibitor, enhances ganglionic transmission and improves sympathetic tone during standing. It thus enables the normal physiological response to orthostatic stress, somewhat increasing upright diastolic blood pressure and improving orthostatic symptoms without worsening supine hypertension (Singer et al., 2006). Pyridostigmine may be of limited benefit in patients with severe autonomic failure (Shibao and Biaggioni, 2010). The α_2- adrenoreceptor antagonist yohimbine stimulates neuronal release of norepinephrine and increases muscle sympathetic nerve activity, norepinephrine spillover, and blood pressure in healthy volunteers (Goldberg et al., 1983; Grossman et al., 1991). Acute doses diminish the hypotensive response to upright posture in patients with autonomic failure (Biaggioni et al., 1994b). Likely due to their receptor hypersensitivity, patients with PAF are able to respond to yohimbine despite denervation of the peripheral sympathetic nervous system. The orthostatic hypotension in PAF can be aggravated by anemia, due in part to impaired erythropoiesis (Shibao et al., 2012). Lower hemoglobin is associated with lower upright plasma norepinephrine. Erythropoietin has been used to treat the anemia with some increase in upright blood pressure (Biaggioni et al., 1994a). It has been proposed to increase intravascular volume and thereby improve venous return and blood pressure (Shibao et al., 2012). Octreotide, which stimulates somatostatin receptors, may reduce release of vasodilatory gastrointestinal peptides after parenteral administration and may be beneficial in preventing postprandial hypotension. It constricts the splanchnic circulation and has been demonstrated in an acute study to prolong standing time (Hoeldtke et al., 1998). Treatment with these agents can contribute to patients' ability to tolerate standing and therefore can result in substantial improvement in quality of life.

Because the liver is a highly vascular organ, the liver blood flow and the clearance of drugs by the liver can be influenced by orthostatic hypotension. Patients with orthostatic hypotension experience a greater than normal decrease in hepatic blood flow when they assume an upright posture. As a result, drugs that are extracted by the liver, such as lidocaine, have a lower volume of distribution and elevated plasma concentration. These findings have implications for drug toxicity in patients with PAF (Feely et al., 1982; Hoeldtke et al., 1998).

Since PAF can produce problems with urinary retention, intermittent catheterization may be needed. Cholinergic agents may also be helpful. Changes in diet, laxatives, and antidiarrheals may be needed to treat gastrointestinal symptoms of PAF. Patient education about the nature of the disease is also an extremely important component of treatment.

FUTURE DIRECTIONS

It is important to identify the factors underlying the progression in some patients from a presumed diagnosis of PAF to a disease that involves the central nervous system. Environmental, genetic and immunological elements must be considered. Since Lewy body diseases share a common histopathological finding which may suggest a similar pathogenesis, it remains difficult to understand why only some patients with PAF go on to develop parkinsonism or dementia, while almost all patients with Parkinson's disease will develop some degree of autonomic dysfunction, and patients with dementia with Lewy bodies may develop both parkinsonism and autonomic dysfunction. Patients with PAF and those with Parkinson's disease plus orthostatic hypotension not only share neuroimaging evidence of cardiac noradrenergic denervation, but also have similar baroreflex abnormalities and abnormal cardiovascular responses to agents acting through the sympathetic nervous system (Goldstein et al., 2008a). Therefore, it remains unclear as to whether PAF, Parkinson's disease, and dementia with Lewy bodies represent discrete diseases, or if they are part of one continuum of disease with the phenotype determined by diverse neurodegenerative processes.

An alternative possibility is that patients with PAF have some neuroprotective factor that prevents them from developing degeneration in striatal terminals as seen in Parkinson's disease but rather directs denervation to postganglionic sympathetic neurons. An interesting finding possibly supportive of this theory is that while patients with PAF and Parkinson's demonstrate similarly low dopamine concentrations in their substantia nigra, striatal dopamine concentrations are 10-fold higher in PAF than in Parkinson's (Goldstein et al., 2008b). Whether this represents the effect of some as yet undetermined neuroprotective factor will remain an active area of research in the coming years.

References

Akinola AB, Smith GD, Mathias CJ et al. (2001). The metabolic, catecholamine and cardiovascular effects of exercise in human sympathetic denervation. Clin Auton Res 11: 251–257.

Asahina M, Young TM, Bleasdale-Barr K et al. (2005). Differences in overshoot of blood pressure after head-up tilt in two groups with chronic autonomic failure: pure autonomic failure and multiple system atrophy. J Neurol 252: 72–77.

Asahina M, Akaogi Y, Yamanaka Y et al. (2009). Differences in skin sympathetic involvements between two chronic autonomic disorders: multiple system atrophy and pure autonomic failure. Parkinsonism Relat Disord 15: 347–350.

Bannister R (1979). Chronic autonomic failure with postural hypotension. Lancet 2: 404–406.

Bannister R, Crowe R, Eames R et al. (1981). Adrenergic innervation in autonomic failure. Neurology 31: 1501–1506.

Barnett AJ, Wagner GR (1958). Severe orthostatic hypotension: case report and description of response to sympatheticomimetic drugs. Am Heart J 56: 412–424.

Barnett AJ, Hamilton MD, Kay HB (1955). Severe orthostatic hypotension. Australas Ann Med 4: 183–194.

Biaggioni I, Onrot J, Stewart CK et al. (1987). The potent pressor effect of phenylpropanolamine in patients with autonomic impairment. JAMA 258: 236–239.

Biaggioni I, Garcia F, Inagami T et al. (1993). Hyporeninemic normoaldosteronism in severe autonomic failure. J Clin Endocrinol Metab 76: 580–586.

Biaggioni I, Robertson D, Krantz S et al. (1994a). The anemia of primary autonomic failure and its reversal with recombinant erythropoietin. Ann Intern Med 121: 181–186.

Biaggioni I, Robertson RM, Robertson D (1994b). Manipulation of norepinephrine metabolism with yohimbine in the treatment of autonomic failure. J Clin Pharmacol 34: 418–423.

Bleasdale-Barr KM, Mathias CJ (1998). Neck and other muscle pains in autonomic failure: their association with orthostatic hypotension. J R Soc Med 91: 355–359.

Bradbury S, Eggleston C (1925). Postural hypotension: a report of three cases. Am Heart J 1: 73–86.

Bremner FD, Smith SE (2006). Pupil abnormalities in selected autonomic neuropathies. J Neuroophthalmol 26: 209–219.

Briggs NM (1946). Neurological Manifestations Associated with Postural Hypotension. University of Minnesota, Minneapolis, MN.

Chandler MP, Mathias CJ (2002). Haemodynamic responses during head-up tilt and tilt reversal in two groups with chronic autonomic failure: pure autonomic failure and multiple system atrophy. J Neurol 249: 542–548.

Chobanian AV, Volicer L, Tifft CP et al. (1979). Mineralocorticoid-induced hypertension in patients with orthostatic hypotension. N Engl J Med 301: 68–73.

Cohen J, Low P, Fealey R et al. (1987). Somatic and autonomic function in progressive autonomic failure and multiple system atrophy. Ann Neurol 22: 692–699.

Consensus Committee of the American Autonomic Society and the American Academy of Neurology [no authors listed] (1996). Consensus statement on the definition of orthostatic hypotension, pure autonomic failure, and multiple system atrophy. The Consensus Committee of the American Autonomic Society and the American Academy of Neurology. Neurology 46: 1470.

Delius W, Hagbarth KE, Hongell A et al. (1972). General characteristics of sympathetic activity in human muscle nerves. Acta Physiol Scand 84: 65–81.

Donadio V, Cortelli P, Elam M et al. (2010). Autonomic innervation in multiple system atrophy and pure autonomic failure. J Neurol Neurosurg Psychiatry 81: 1327–1335.

Donadio V, Incensi A, Cortelli P et al. (2013). Skin sympathetic fiber alpha-synuclein deposits: a potential biomarker for pure autonomic failure. Neurology 80: 725–732.

Doucette W, Milder J, Restrepo D (2007). Adrenergic modulation of olfactory bulb circuitry affects odor discrimination. Learn Mem 14: 539–547.

Elkayam L, Matalon A, Tseng CH et al. (2006). Prevalence and severity of renal disease in familial dysautonomia. Am J Kidney Dis 48: 780–786.

Esler M, Leonard P, O'Dea K et al. (1982). Biochemical quantification of sympathetic nervous activity in humans using radiotracer methodology: fallibility of plasma noradrenaline measurements. J Cardiovasc Pharmacol 4 (Suppl 1): S152–S157.

Esler M, Jennings G, Korner P et al. (1988). Assessment of human sympathetic nervous system activity from measurements of norepinephrine turnover. Hypertension 11: 3–20.

Feely J, Wade D, McAllister CB et al. (1982). Effect of hypotension on liver blood flow and lidocaine disposition. N Engl J Med 307: 866–869.

Freeman R (1995). Pure autonomic failure. In: D Robertson, I Biaggioni (Eds.), Disorders of the Autonomic Nervous System. Harwood Academic Publishers, Luxembourg, pp. 83–105.

Freeman R (2004). Pure autonomic failure: an immaculate misconception? Neurology 63: 953–954.

Fulham MJ, Dubinsky RM, Polinsky RJ et al. (1991). Computed tomography, magnetic resonance imaging and positron emission tomography with [^{18}F]fluorodeoxyglucose in multiple system atrophy and pure autonomic failure. Clin Auton Res 1: 27–36.

Furlan R, Piazza S, Bevilacqua M et al. (1995). Pure autonomic failure: complex abnormalities in the neural mechanisms regulating the cardiovascular system. J Auton Nerv Syst 51: 223–235.

Garland EM, Robertson D (2009). Autonomic failure. In: LR Squire (Ed.), Encyclopedia of Neuroscience. Academic Press, London, pp. 825–832.

Garland EM, Gamboa A, Okamoto L et al. (2009). Renal impairment of pure autonomic failure. Hypertension 54: 1057–1061.

Garland EM, Raj SR, Peltier AC et al. (2011). A cross-sectional study contrasting olfactory function in autonomic disorders. Neurology 76: 456–460.

Gibbons CH, Illigens BM, Centi J et al. (2008). QDIRT: quantitative direct and indirect test of sudomotor function. Neurology 70: 2299–2304.

Goldberg MR, Hollister AS, Robertson D (1983). Influence of yohimbine on blood pressure, autonomic reflexes, and plasma catecholamines in humans. Hypertension 5: 772–778.

Goldstein DS, Sewell L (2009). Olfactory dysfunction in pure autonomic failure: implications for the pathogenesis of Lewy body diseases. Parkinsonism Relat Disord 15: 516–520.

Goldstein DS, Polinsky RJ, Garty M et al. (1989). Patterns of plasma levels of catechols in neurogenic orthostatic hypotension. Ann Neurol 26: 558–563.

Goldstein DS, Holmes CS, Dendi R et al. (2002a). Orthostatic hypotension from sympathetic denervation in Parkinson's disease. Neurology 58: 1247–1255.

Goldstein DS, Robertson D, Esler M et al. (2002b). Dysautonomias: clinical disorders of the autonomic nervous system. Ann Intern Med 137: 753–763.

Goldstein DS, Holmes C, Sharabi Y et al. (2003a). Plasma levels of catechols and metanephrines in neurogenic orthostatic hypotension. Neurology 60: 1327–1332.

Goldstein DS, Pechnik S, Holmes C et al. (2003b). Association between supine hypertension and orthostatic hypotension in autonomic failure. Hypertension 42: 136–142.

Goldstein DS, Holmes C, Bentho O et al. (2008a). Biomarkers to detect central dopamine deficiency and distinguish Parkinson disease from multiple system atrophy. Parkinsonism Relat Disord 14: 600–607.

Goldstein DS, Holmes C, Sato T et al. (2008b). Central dopamine deficiency in pure autonomic failure. Clin Auton Res 18: 58–65.

Goodall MC, Harlan WR Jr, Alton H (1968). Decreased noradrenaline (norepinephrine) synthesis in neurogenic orthostatic hypotension. Circulation 38: 592–603.

Graham JG, Oppenheimer DR (1969). Orthostatic hypotension and nicotine sensitivity in a case of multiple system atrophy. J Neurol Neurosurg Psychiatry 32: 28–34.

Grossman E, Rea RF, Hoffman A et al. (1991). Yohimbine increases sympathetic nerve activity and norepinephrine spillover in normal volunteers. Am J Physiol 260: R142–R147.

Gupta D, Nair MD (2008). Neurogenic orthostatic hypotension: chasing "the fall". Postgrad Med J 84: 6–14.

Hoeldtke RD, Horvath GG, Bryner KD et al. (1998). Treatment of orthostatic hypotension with midodrine and octreotide. J Clin Endocrinol Metab 83: 339–343.

Hughes RC, Cartlidge NE, Millac P (1970). Primary neurogenic orthostatic hypotension. J Neurol Neurosurg Psychiatry 33: 363–371.

Igari R, Wada M, Wakabayashi K et al. (2011). alpha-Synuclein in the skin nerve of pure autonomic failure. Intern Med 50: 3049–3050.

Ikemura M, Saito Y, Sengoku R et al. (2008). Lewy body pathology involves cutaneous nerves. J Neuropathol Exp Neurol 67: 945–953.

Jacob G, Raj SR, Ketch T et al. (2005). Postural pseudoanemia: posture-dependent change in hematocrit. Mayo Clin Proc 80: 611–614.

Jordan J, Shannon J, Robertson D (1997). The physiological conundrum of hyperadrenergic orthostatic intolerance. Chin J Physiol 40: 1–8.

Jordan J, Shannon JR, Biaggioni I et al. (1998). Contrasting actions of pressor agents in severe autonomic failure. Am J Med 105: 116–124.

Jordan J, Shannon JR, Grogan E et al. (1999a). A potent pressor response elicited by drinking water. Lancet 353: 723.

Jordan J, Shannon JR, Pohar B et al. (1999b). Contrasting effects of vasodilators on blood pressure and sodium balance in the hypertension of autonomic failure. J Am Soc Nephrol 10: 35–42.

Kaufmann H (2008). L-dihydroxyphenylserine (Droxidopa): a new therapy for neurogenic orthostatic hypotension: the US experience. Clin Auton Res 18 (Suppl 1): 19–24.

Kaufmann HC, Benarroch EE (2008). Degenerative autonomic disorders (autonomic synucleinopathies). In: PA Low, EE Benarroch (Eds.), Clinical Autonomic Disorders. Lippincott Williams and Wilkins, Baltimore, pp. 287–306.

Kaufmann H, Schatz I (2012). Pure autonomic failure. In: D Robertson, I Biaggioni, G Burnstock et al. (Eds.), Primer on the Autonomic Nervous System. 3rd edn. Academic Press, New York, pp. 467–469.

Kaufmann H, Oribe E, Miller M et al. (1992). Hypotension-induced vasopresson release distinguishes between pure autonomic failure and multiple system atrophy with autonomic failure. Neurology 42: 590–593.

Kaufmann H, Saadia D, Voustianiouk A et al. (2003). Norepinephrine precursor therapy in neurogenic orthostatic hypotension. Circulation 108: 724–728.

Kaufmann H, Nahm K, Purohit D et al. (2004). Autonomic failure as the initial presentation of Parkinson disease and dementia with Lewy bodies. Neurology 63: 1093–1095.

Kontos HA, Richardson DW, Norvell JE (1975). Norepinephrine depletion in idiopathic orthostatic hypotension. Ann Intern Med 82: 336–341.

Kronenberg MW, Forman MB, Onrot J et al. (1990). Enhanced left ventricular contractility in autonomic failure: assessment using pressure-volume relations. J Am Coll Cardiol 15: 1334–1342.

Low PA (2008). Prevalence of orthostatic hypotension. Clin Auton Res 18 (Suppl 1): 8–13.

Low PA, Caskey PE, Tuck RR et al. (1983). Quantitative sudomotor axon reflex test in normal and neuropathic subjects. Ann Neurol 14: 573–580.

Low PA, Gilden JL, Freeman R et al. (1997). Efficacy of midodrine for neurogenic orthostatic hypotension. Reply. JAMA 278: 388.

Mabuchi N, Hirayama M, Koike Y et al. (2005). Progression and prognosis in pure autonomic failure (PAF): comparison with multiple system atrophy. J Neurol Neurosurg Psychiatry 76: 947–952.

Mathias CJ (2008). L-dihydroxyphenylserine (Droxidopa) in the treatment of orthostatic hypotension: the European experience. Clin Auton Res 18 (Suppl 1): 25–29.

Mathias CJ (2009). Autonomic nervous system: clinical testing. In: LR Squire (Ed.), Encyclopedia of Neuroscience. Academic Press, London, pp. 911–928.

Mathias CJ, Young TM (2003). Plugging the leak – benefits of the vasopressin-2 agonist, desmopressin in autonomic failure. Clin Auton Res 13: 85–87.

Mathias CJ, Mallipeddi R, Bleasdale-Barr K (1999). Symptoms associated with orthostatic hypotension in pure autonomic failure and multiple system atrophy. J Neurol 246: 893–898.

McHugh J, Keller NR, Appalsamy M et al. (2010). Portal osmopressor mechanism linked to transient receptor potential vanilloid 4 and blood pressure control. Hypertension 55: 1438–1443.

Metzler M, Duerr S, Granata R et al. (2012). Neurogenic orthostatic hypotension: pathophysiology, evaluation, and management. J Neurol, ePub ahead of print.

Okamoto LE, Gamboa A, Shibao C et al. (2009). Nocturnal blood pressure dipping in the hypertension of autonomic failure. Hypertension 53: 363–369.

Onrot J, Goldberg MR, Biaggioni I et al. (1985). Hemodynamic and humoral effects of caffeine in autonomic failure. Therapeutic implications for postprandial hypotension. N Engl J Med 313: 549–554.

Onrot J, Bernard GR, Biaggioni I et al. (1991). Direct vasodilator effect of hyperventilation-induced hypocarbia in autonomic failure patients. Am J Med Sci 301: 305–309.

Polinsky RJ, Kopin IJ, Ebert MH et al. (1981). Pharmacologic distinction of different orthostatic hypotension syndromes. Neurology 31: 1–7.

Robertson D (1979). Contraindication to the use of ocular phenylephrine in idiopathic orthostatic hypotension. Am J Ophthalmol 87: 819–822.

Robertson D (2008). The pathophysiology and diagnosis of orthostatic hypotension. Clin Auton Res 18 (Suppl 1): 2–7.

Robertson D, Garland EM (2013). Dopamine Beta-Hydroxylase Deficiency. GeneReviews at GeneTests: Medical Genetics Information Resource, University of Washington, Seattle. http://www.ncbi.nlm.nih.gov/books/NBK1474/.

Robertson D, Wade D, Robertson RM (1981). Postprandial alterations in cardiovascular hemodynamics in autonomic dysfunction states. Am J Cardiol 48: 1048–1052.

Robertson D, Goldberg MR, Hollister AS et al. (1983). Clonidine raises blood pressure in severe idiopathic orthostatic hypotension. Am J Med 74: 193–200.

Robertson D, Hollister AS, Carey EL et al. (1984). Increased vascular beta2-adrenoceptor responsiveness in autonomic dysfunction. J Am Coll Cardiol 3: 850–856.

Robertson D, Goldberg MR, Tung CS et al. (1986). Use of alpha 2 adrenoreceptor agonists and antagonists in the functional assessment of the sympathetic nervous system. J Clin Invest 78: 576–581.

Robertson D, Kincaid DW, Haile V et al. (1994). The head and neck discomfort of autonomic failure: an unrecognized aetiology of headache. Clin Auton Res 4: 99–103.

Shannon J, Jordan J, Costa F et al. (1997). The hypertension of autonomic failure and its treatment. Hypertension 30: 1062–1067.

Shannon JR, Jordan J, Diedrich A et al. (2000). Sympathetically mediated hypertension in autonomic failure. Circulation 101: 2710–2715.

Sharabi Y, Eldadah B, Li ST et al. (2006). Neuropharmacologic distinction of neurogenic orthostatic hypotension syndromes. Clin Neuropharmacol 29: 97–105.

Shibao C, Biaggioni I (2010). Orthostatic hypotension and cardiovascular risk. Hypertension 56: 1042–1044.

Shibao C, Gamboa A, Diedrich A et al. (2007). Acarbose, an alpha-glucosidase inhibitor, attenuates postprandial hypotension in autonomic failure. Hypertension 50: 54–61.

Shibao C, Okamoto L, Biaggioni I (2012). Pharmacotherapy of autonomic failure. Pharmacol Ther 134: 279–286.

Shibao C, Okamoto LE, Gamboa A et al. (2010). Comparative efficacy of yohimbine against pyridostigmine for the treatment of orthostatic hypotension in autonomic failure. Hypertension 56: 847–851.

Shy G (1960). A neurological syndrome associated with orthostatic hypotension: a clinical-pathologic study. AMA Arch Neurol 2: 511–527.

Silveira-Moriyama L, Mathias C, Mason L et al. (2009). Hyposmia in pure autonomic failure. Neurology 72: 1677–1681.

Singer W, Opfer-Gehrking TL, Nickander KK et al. (2006). Acetylcholinesterase inhibition in patients with orthostatic intolerance. J Clin Neurophysiol 23: 476–481.

Vagaonescu TD, Saadia D, Tuhrim S et al. (2000). Hypertensive cardiovascular damage in patients with primary autonomic failure. Lancet 355: 725–726.

Veyrac A, Nguyen V, Marien M et al. (2007). Noradrenergic control of odor recognition in a nonassociative olfactory learning task in the mouse. Learn Mem 14: 847–854.

Wright RA, Kaufmann HC, Perera R et al. (1998). A double-blind, dose-response study of midodrine in neurogenic orthostatic hypotension. Neurology 51: 120–124.

Yamamoto T, Polinsky RJ, Goldstein DS et al. (1996). Plasma sulfoconjugated dopamine levels are normal in patients with autonomic failure. J Lab Clin Med 128: 488–491.

Young TM, Asahina M, Nicotra A et al. (2006). Skin vasomotor reflex responses in two contrasting groups of autonomic failure: multiple system atrophy and pure autonomic failure. J Neurol 253: 846–850.

Ziegler MG, Lake CR, Kopin IJ (1977). The sympathetic-nervous-system defect in primary orthostatic hypotension. N Engl J Med 296: 293–297.

Handbook of Clinical Neurology, Vol. 117 (3rd series)
Autonomic Nervous System
R.M. Buijs and D.F. Swaab, Editors

Chapter 21

Autonomic dysfunction in Parkinson disease

HORACIO KAUFMANN[1]* AND DAVID S. GOLDSTEIN[2]

[1]*Dysautonomia Center, Department of Neurology, New York University School of Medicine, New York, NY, USA*

[2]*Clinical Neurocardiology Section, Clinical Neurosciences Program, Division of Intramural Research, National Institute of Neurological Disorders and Stroke, National Institutes of Health, Bethesda, MD, USA*

INTRODUCTION

Autonomic dysfunction occurs commonly in patients with Parkinson disease (PD). Indeed, the original report by James Parkinson in 1817 noted prominent symptoms of constipation and urinary incontinence. The well-known movement disorder usually dominates the clinical picture and has occupied the attention of clinicians and researchers. Nevertheless, a substantial minority of parkinsonian patients have severe and disabling symptoms of autonomic impairment, several of which are treatable.

Autonomic disturbances in PD can manifest as dysphagia, constipation, urinary urgency, incontinence, erectile dysfunction, orthostatic and postprandial hypotension, dyshidrosis, and impaired thermoregulation. It has been difficult to quantify the prevalence of autonomic dysfunction in PD. First, antiparkinsonian medication with levodopa can decrease blood pressure while standing and delay gastric emptying. Anticholinergics further decrease gastrointestinal motility. Until relatively recently, these abnormalities were believed, incorrectly, to reflect side-effects of the drugs, whereas we now know that the drugs interact importantly with the dysautonomia that is part of the disease process itself. Second, the parkinsonian form of multiple system atrophy (MSA-P), which always features signs and symptoms of autonomic dysfunction, can resemble PD clinically, so that studies can overestimate or underestimate the frequency of autonomic dysfunction by misdiagnosis.

In a retrospective study, almost one-third of patients with pathologically proven PD had autonomic dysfunction documented in the medical record (Rogers et al., 1980). This retrospective approach most likely underestimates the frequency of autonomic failure. Compared to age-matched control subjects, PD patients have higher frequencies of constipation, erectile dysfunction, urinary urgency, incomplete bladder emptying, dysphagia, and orthostatic lightheadedness. Indeed, about 9 in 10 patients with PD have one or more of these autonomic symptoms (Singer et al., 1991). Autonomic problems increase significantly with increasing disease severity (Visser et al., 2004).

Here we review components of the autonomic nervous system and clinical manifestations, diagnosis, and treatment of autonomic abnormalities in PD. We also note similarities and differences between autonomic abnormalities in MSA and PD. MSA is covered elsewhere in this volume (Ch. 19).

COMPONENTS OF THE AUTONOMIC NERVOUS SYSTEM

The autonomic nervous system has five components: enteric, parasympathetic cholinergic, sympathetic cholinergic, sympathetic noradrenergic, and adrenomedullary hormonal.

Langley, who introduced the term "autonomic nervous system" almost a century ago (Langley, 1903), referred to neurons in ganglia outside the brain and spinal cord that seemed to function independently, or autonomously, of the central nervous system. Now we know that the components of the autonomic nervous system do not function independently of the central nervous system, but the phraseology stuck. Langley identified enteric, sympathetic, and parasympathetic components. In the early 20th century, Cannon (1939) discovered and emphasized the adrenal hormonal component. The autonomic nervous system, therefore, is not only neuronal but also neurohormonal. One might instead

*Correspondence to: Horacio Kaufmann, M.D., Dysautonomia Center, Department of Neurology, New York, NY 10016, USA. Tel: +1-212-263-7225, E-mail: Horacio.Kaufmann@nyumc.org

refer to the components as "automatic" neuroendocrine systems, along with, e.g., the vasopressin system, renin–angiotensin–aldosterone system, and hypothalamo–pituitary–adrenocortical system.

Failure of a particular component of the autonomic nervous system produces characteristic clinical manifestations. Parasympathetic cholinergic failure presents as constipation, dry mouth, a constant pulse rate, urinary retention, and erectile failure in men. Sweating, whether thermoregulatory, gustatory, or emotional, depends on delivery of acetylcholine, not catecholamines, from sympathetic nerves. Sympathetic cholinergic failure therefore manifests as decreased sweating. Due to the absolute requirement of intact sympathetic neurocirculatory function for humans to tolerate standing up, sympathetic noradrenergic failure presents as orthostatic intolerance and orthostatic hypotension.

CARDIOVASCULAR AUTONOMIC DYSFUNCTION

Orthostatic and postprandial hypotension

The sympathetic nervous system is essential for maintaining blood pressure during orthostasis. Orthostatic hypotension (OH) is the cardinal manifestation of sympathetic neurocirculatory failure and occurs in 30–40% of patients with PD. OH is often defined as a decrease in systolic blood pressure of at least 20 mmHg or diastolic blood pressure of at least 10 mmHg within 3 minutes of standing (Kaufmann, 1996). Symptoms, which occur only when standing or sitting and are relieved when lying down, include light-headedness, blurred vision, generalized weakness, fatigue, cognitive impairment, and pain in the shoulders and back of the neck.

A related problem is postprandial hypotension, a fall in blood pressure after meals, which occurs even in the supine position and can be very pronounced. Postprandial hypotension is common in PD, even in patients who do not experience OH (Micieli et al., 1987; Hasegawa and Okamoto, 1992; Nozaki et al., 1993). Postprandial hypotension frequently underlies the apparent worsening of parkinsonian symptoms after meals. Patients appear dazed or "frozen," suggesting an "off" state related to fluctuating levodopa responses.

The lack of a reliable method for the differential diagnosis between PD and the parkinsonian variant of multiple system atrophy (MSA-P) during life, and the widespread assumption that early severe OH is diagnostic of MSA, complicate estimating the frequency of OH in PD. Indeed, in small postmortem studies (Saito et al., 1992; Benarroch et al., 2000), four out of five patients with prominent OH and pathologically proven PD had carried a diagnosis of MSA during life. Meanwhile, one-third of patients with pathologically proven MSA have been reported to die misdiagnosed with PD (Magalhaes et al., 1995).

Among 24 studies (Rajput and Rozdilsky, 1976; Kuroiwa et al., 1983; Micieli et al., 1987; Briebach et al., 1990; Awerbuch and Sandyk, 1992; Sandyk and Awerbuch, 1992; Hubble et al., 1995; Loew et al., 1995; Magalhaes et al., 1995; Bellon et al., 1996; Hillen et al., 1996; Senard et al., 1997; Turkka et al., 1997; Wenning et al., 1999; Kujawa et al., 2000; Tranchant et al., 2000; Holmberg et al., 2001; Papapetropoulos et al., 2001; Krygowska-Wajs et al., 2002; Bhattacharya et al., 2003; Bonuccelli et al., 2003; Allcock et al., 2004; Korchounov et al., 2004; Thaisetthawatkul et al., 2004), the reported frequency of OH in PD varied widely from 5% to 100%, but the variability decreased noticeably as the size of the study increased. All studies involving more than 80 patients reported an OH frequency between 20% and 60%. Across all studies, the mean percentage of PD patients with OH was 41%. In a recent meta-analysis (Velseboer et al., 2011), the pooled estimate of the point prevalence of OH in PD was 30.1% (95% CI: 22.9% to 38.4%). The authors found a large statistical heterogeneity between studies, which could not be reduced by subgroup analyses, and concluded that this pooled estimate should be interpreted with caution. More data from unselected population-based cohorts are needed (Velseboer et al., 2011).

In most clinical series, OH is believed to be a late complication of PD. Few studies have actually analyzed formally the timing of onset of OH in relation to the movement disorder. In an analysis of historical data from patients with PD and OH who were evaluated at the NIH, about 65% of patients had evidence that OH had developed early in their disease (Goldstein, 2006). In four patients, OH dominated the clinical picture, before the onset of a symptomatic movement disorder, and the patients were thought to have pure autonomic failure (PAF). The diagnosis of PD in these patients was based on typical clinical features, including resting tremor and good motor response to levodopa, as well as neuroimaging evidence of cardiac sympathetic denervation. In another study, OH was present in 14% of patients with early, newly diagnosed, untreated PD who were followed for 7 years to ascertain their diagnosis (Bonuccelli et al., 2003). In that study, 15% of the original cohort of patients turned out to have other types of parkinsonism during the follow up period.

The notion that OH can precede the movement disorder was suggested half a century ago (Nylin and Levander, 1948) based on the case of a patient who developed OH at the age of 75 and a year later developed a unilateral resting tremor, a masked face, and "cogwheel" rigidity, findings recognized by the authors as typical of PD. This notion was confirmed in 2004 by a postmortem pathology report of a patient with typical PD who had detailed historical data conclusively showing that symptomatic OH had occurred before the onset of the motor abnormalities (Kaufmann et al., 2004).

Studies have noted higher frequencies of OH in older patients, in patients with more severe disease, or in patients with a longer duration of disease at the time of evaluation.

The role of levodopa

Contrary to a long-held notion, treatment with levodopa does not cause OH in PD (Hoehn, 1975). If levodopa was the cause, then a higher proportion of patients with than without OH would be on levodopa therapy, and this is not the case (Goldstein et al., 2002; Bhattacharya et al., 2003). Patients with PD + OH do not differ from those without OH in levodopa treatment or actual plasma levodopa concentrations. Even more convincingly, OH can occur in patients with PD who have never taken levodopa or discontinued levodopa treatment in the remote past (Martignoni et al., 1995). As discussed below, such patients have physiological evidence of decreased cardiovascular innervation by sympathetic nerves, which at least partly explains the OH.

It is important to consider, however, that even with concomitant carbidopa treatment, which attenuates conversion of levodopa to dopamine outside the central nervous system, levodopa increases plasma levels of both dopamine and its deaminated metabolite, dihydroxyphenylacetic acid (Kaakkola et al., 1985; Rose et al., 1988; Myllyla et al., 1993; Tohgi et al., 1995). Exogenously administered dopamine at relatively low doses produces vasodilatation, by stimulating dopamine receptors on vascular smooth muscle cells and possibly by inhibiting norepinephrine release from sympathetic nerves (Yeh et al., 1969; Lokhandwala and Hegde, 1990; Durrieu et al., 1991). Dopamine also augments natriuresis and diuresis, which promotes depletion of extracellular fluid and blood volumes.

In patients with PD and decreased cardiovascular sympathetic innervation and baroreflex abnormalities (see below), vasodilatation and hypovolemia elicited by dopamine produced from levodopa could decrease blood pressure both during supine rest and during standing. Thus, symptomatic OH may occur in patients with PD while taking levodopa/carbidopa or dopamine receptor agonists, not from effects of these drugs alone but from interactions with baroreflex and sympathoneural pathophysiological abnormalities occurring as part of the disease process.

Cardiac sympathetic denervation

A great many studies over the past several years have shown that virtually all patients with PD have loss of sympathetic innervation of the heart. This is indicated by low myocardial concentrations of radioactivity after injection of the sympathoneural imaging agents ^{123}I-metaiodobenzylguanidine (Satoh et al., 1997, 1999; Braune et al., 1998, 1999; Yoshita et al., 1998; Orimo et al., 1999; Druschky et al., 2000; Ohmura, 2000; Reinhardt et al., 2000; Takatsu et al., 2000a, b) and 6-[^{18}F] fluorodopamine (Goldstein et al., 1997, 2000, 2002), by neurochemical assessments during right heart catheterization (Goldstein et al., 2000), and by postmortem pathological studies (Herrmann et al., 1999; Orimo et al., 2001; Amino et al., 2005).

Sympathetic cardiac denervation in PD can occur without OH. About half of the patients with PD without OH had a loss of 6-[^{18}F]fluorodopamine-derived radioactivity diffusely in the left ventricular myocardium, and a bit less than half had loss localized to the lateral or inferior walls, with relative preservation in the septum or anterior wall. Only a very small minority had entirely normal cardiac 6-[^{18}F] fluorodopamine-derived radioactivity. Even in PD patients with neuroimaging evidence for normal cardiac sympathetic innervation, eventually denervation occurs (Fig. 21.1).

Neuropathological data support this *in vivo* observation: Lewy bodies have been reported in cardiac plexus of patients with PD (Iwanaga et al., 1999) and tyrosine hydroxylase (TH) immunoreactive axons had nearly disappeared in the epicardium of the left ventricular anterior wall from specimens with PD. Moreover, the numbers of neurofilament (NF) and S-100 protein immunoreactive axons were also drastically decreased. Triple immunofluorolabeling for NF, TH, and myelin basic protein showed profound involvement of cardiac sympathetic axons in PD (Amino et al., 2005).

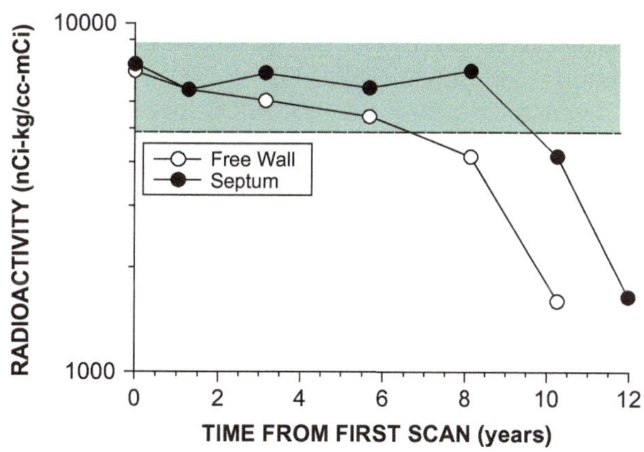

Fig. 21.1. Serial ^{18}F-dopamine-derived radioactivity in the cardiac septum and free wall of a patient with established PD. It took several years before there was any evidence of loss of sympathetic nerves. Free wall radioactivity declined first, followed soon after by septal radioactivity. All PD patients eventually have cardiac sympathetic denervation, but as illustrated here. the denervation does not necessarily precede the movement disorder.

Neuroimaging and neuropathological evidence of cardiac sympathetic denervation has also been shown in most patients with pure autonomic failure (PAF), a Lewy body (LB) disorder likely related to PD. PAF is a neurodegenerative disorder of peripheral autonomic neurons without motor abnormalities (Kaufmann, 1996). Postmortem studies of patients with PAF have shown Lewy bodies in autonomic ganglia, distal sympathetic axons and epicardial nerves (Hague et al., 1997). However, despite the absence of clinical parkinsonism in PAF, LB were also found in substantia nigra of these patients, and PAF patients have evidence of central dopamine deficiency (Goldstein et al., 2008). Most neuropathological studies have shown that patients with PD have Lewy bodies, not only in the substantia nigra but in sympathetic ganglia as well. The overlapping pathological findings suggest that PAF and PD may lie along a spectrum of Lewy body disorders that affect peripheral autonomic neurons (Kaufmann and Biaggioni, 2003). Mechanisms for different extents of involvement of central and peripheral catecholamine lesions in Lewy body disorders remain poorly understood (Fig. 21.2).

Absence of postganglionic lesion in multiple system atrophy

Just as the literature about cardiac sympathetic denervation in PD is consistent, the literature is also consistent about intact cardiac sympathetic innervation in most patients with multiple system atrophy (MSA), ascertained by neuroimaging (Yoshita et al., 1997; Braune et al., 1999; Druschky et al., 2000; Reinhardt et al., 2000), normal or even increased rates of entry of norepinephrine and other catechols into coronary sinus plasma (Goldstein et al., 2000), and postmortem pathology (Orimo et al., 2001; Yelnik, 2002).

In contrast to PAF and PD, MSA does not involve Lewy bodies, either in the substantia nigra or sympathetic neurons. Glial and neuronal cells of MSA patients contain cytoplasmic inclusions that, similar to Lewy bodies, contain α-synuclein (Wakabayashi et al., 1998; Dickson et al., 1999; Kaufmann et al., 2001). Thus, these disorders may be considered α-synucleinopathies (Jellinger, 2003). A minority of MSA patients have cardiac sympathetic denervation (Orimo et al., 2007); however, whether such patients have a “hybrid” condition, involving both Lewy bodies in catecholamine neurons and glial cytoplasmic inclusions, remains unknown.

In sum, compelling neuroimaging, neurochemical, and postmortem pathological evidence indicates that PD (especially PD + OH) features cardiac sympathetic denervation, while MSA features generally intact cardiac sympathetic innervation. This difference is consistent with a postganglionic sympathetic lesion in PD and PAF but not in MSA. Neuroimaging evidence of cardiac sympathetic denervation is a useful tool for distinguishing PD from other disorders with clinical parkinsonism, such as MSA or progressive supranuclear palsy.

Vascular sympathetic denervation

The extent to which sympathetic denervation in the peripheral vasculature contributes to OH in PD is unclear. The degree of loss of sympathetic innervation in PD seems to vary among organs. Normal tissue concentrations of 6-[^{18}F] fluorodopamine-derived radioactivity have been noted in the liver, spleen, salivary glands, and nasopharyngeal mucosa, but decreased concentrations in the thyroid gland and renal cortex (Goldstein et al., 2002). Findings based on ^{123}I-metaiodobenzylguanidine scanning have led to the view that in PD cardiac sympathetic denervation occurs independently of OH or other manifestations of autonomic failure and that the denervation is more pronounced in the heart (Yoshita et al., 1998; Braune et al., 1999; Satoh et al., 1999; Takatsu et al., 2000a).

Consistent with more generalized sympathetic denervation in PD + OH than in PD without OH is the finding that patients with PD + OH have lower mean plasma levels of norepinephrine, the sympathetic neurotransmitter, and of dihydroxyphenylglycol, the main neuronal metabolite of norepinephrine during supine rest, than do patients without OH (Senard et al., 1990, 1993; Niimi et al., 1999; Goldstein et al., 2002).

Plasma norepinephrine

Concentrations of norepinephrine in antecubital venous plasma provide a means – albeit indirect – to detect sympathetic denervation in the body as a whole. Thus, patients with OH due to PAF have low plasma norepinephrine levels during supine rest (Ziegler et al., 1977; Goldstein et al., 1989). Patients with PD + OH have lower plasma norepinephrine concentrations than patients with PD without OH (Senard et al., 1990, 1993). In patients with PD + OH, plasma norepinephrine levels, while significantly lower than in PD patients without OH, are not particularly low for healthy people of similar age and are clearly higher than in patients with PAF. It is possible that partial loss of sympathetic fibers leads to augmented traffic in the remaining fibers, resulting in increased proportionate release of norepinephrine from the reduced vesicular stores. Moreover, because denervation would produce concurrent decreases in both release and reuptake of norepinephrine, plasma norepinephrine levels might fail to detect a real decrease in norepinephrine release. Levodopa/carbidopa treatment may also increase individual variability in plasma norepinephrine levels. Why sympathetic nerves in the heart seem particularly susceptible in PD remains unknown.

Baroreflex abnormalites

A particular pattern of beat-to-beat blood pressure responses to the Valsalva maneuver can detect sympathetic neurocirculatory failure, including that in PD + OH (Goldstein and Tack, 2000). During phase II of the maneuver the blood pressure decreases progressively, because reflex, sympathetically mediated cardiovascular stimulation is deficient in response to reduced cardiac filling. During phase IV, the pressure fails to exceed the baseline value (Fig. 21.1). All patients with unequivocal PD + OH who are able to perform a technically adequate Valsalva maneuver show this abnormal pattern, regardless of levodopa/carbidopa treatment. Blunted baroreflex mediated sympathetic activation, in response to acutely decreased venous return to the heart, characterizes OH in PD. Similar abnormalities occur in MSA and PAF.

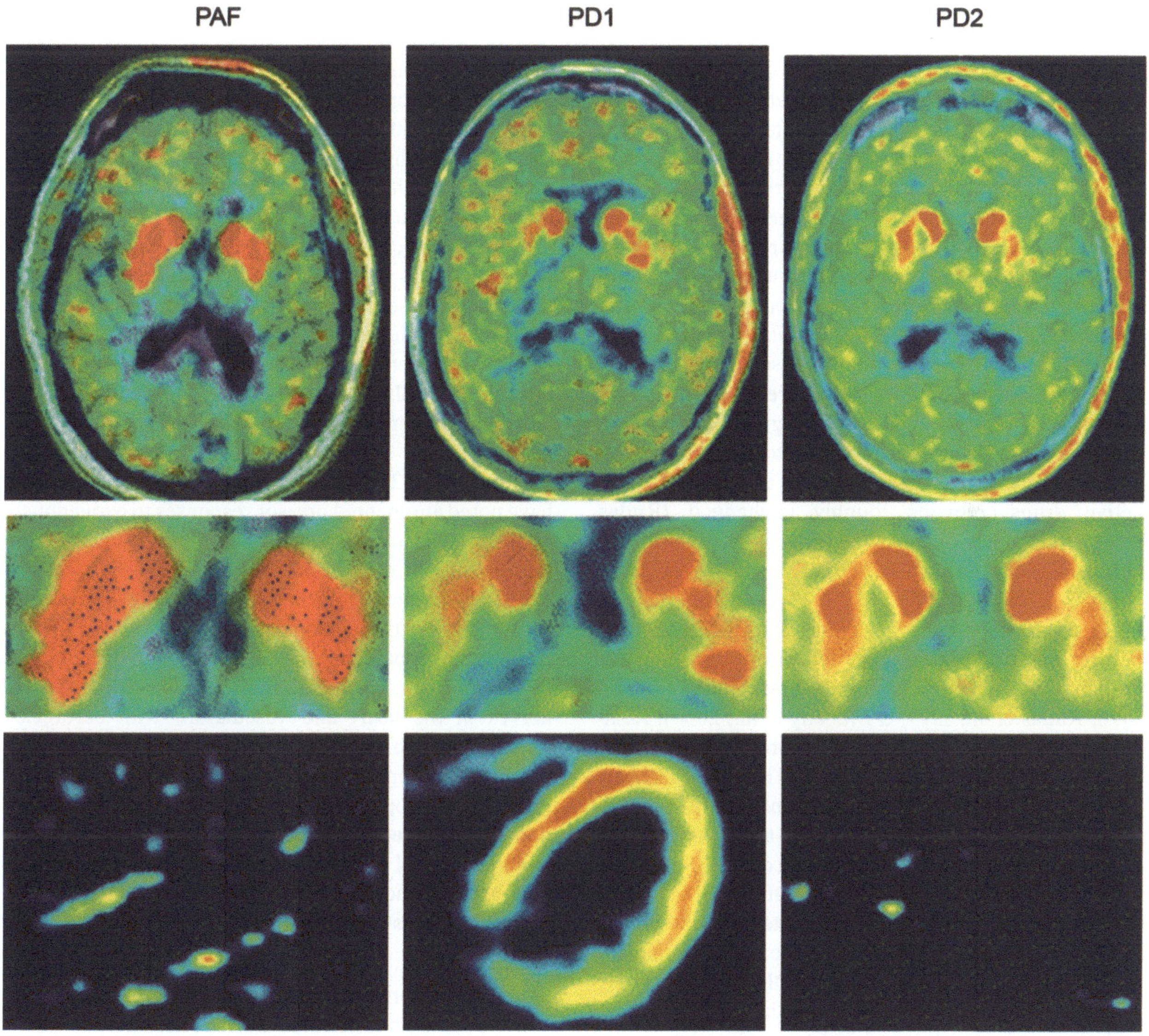

Fig. 21.2. Cardiac sympathetic denervation occurs independently of the loss of striatal dopaminergic innervation. The figure shows low- and high-resolution ^{18}F-DOPA PET/MRI brain scans and ^{18}F-dopamine PET cardiac scans in (**A**), a patient with pure autonomic failure (PAF), (**B**) a patient with Parkinson disease (PD1), and (**C**) another PD patient (PD2). The PAF patient has normal striatal ^{18}F-DOPA-derived radioactivity and decreased cardiac ^{18}F-dopamine-derived radioactivity, as expected. The two PD patients have similar losses of striatal ^{18}F-DOPA-derived radioactivity. PD1 has completely normal ^{18}F-dopamine-derived radioactivity, whereas PD2 has undetectable ^{18}F-dopamine-derived radioactivity.

Normally, plasma norepinephrine levels approximately double within 5 minutes of standing from the supine position (Lake et al., 1976). Most PD + OH patients have an attenuated increase in plasma norepinephrine levels during orthostasis. The finding that OH in PD is associated with failure to increase norepinephrine levels appropriately during orthostasis is consistent with decreased baroreflex-sympathoneural function.

Studies have disagreed about whether baroreflex-sympathoneural gain changes as a function of aging (Shimada et al., 1985, 1986; Ebert et al., 1992; Matsukawa et al., 1996; Davy et al., 1997, 1998; Rudas et al., 1999; Tanaka et al., 1999; Niimi et al., 2000; O'Mahony et al., 2000; Seals et al., 2001; Ferrari, 2002). Some of this inconsistency may have resulted from the different types of measures used – direct indices, such as peroneal muscle sympathetic activity, or indirect indices, such as limb vascular resistance. When both direct and indirect measurements have been applied in the same subjects, cardiopulmonary baroreflex control of sympathetic outflow, assessed by exposure of subjects to lower body negative pressure, has been found to be augmented rather than impaired with age in healthy humans; meanwhile reflexive limb vasoconstriction is attenuated (Davy et al., 1998). The ability to inhibit sympathetic outflow in response to increased cardiac filling, from head-down tilt, does not decrease with normal human aging (Tanaka et al., 1999). Regulation of sympathetic outflow by arterial baroreceptors, measured by sympathetic microneurography after injection of vasoactive drugs, remains roughly unchanged (Rudas et al., 1999), even with lower body negative pressure applied concurrently to keep central venous pressure constant (Davy et al., 1997).

In contrast, studies have consistently found that baroreflex-cardiovagal gain decreases with normal human aging (Bristow et al., 1969; Matsukawa et al., 1996; Rudas et al., 1999; O'Mahony et al., 2000). Relatively few studies have assessed baroreflex-cardiovagal gain in PD (Szili-Torok et al., 2001). The extent of heart rate change with the Valsalva maneuver is blunted (Camerlingo et al., 1987), but this might reflect the advanced age of PD patients (van Dijk et al., 1993). When estimating baroreflex-cardiovagal gain from the slope of the relationship between interbeat interval and systolic blood pressure during phase II of the Valsalva maneuver, almost all patients with PD + OH have markedly decreased baroreflex-cardiovagal gain. In PD lacking OH, baroreflex-cardiovagal gain may be statistically decreased from normal, but in PD + OH, baroreflex-cardiovagal gain is virtually always low (Goldstein et al., 2005).

A combination of baroreflex impairment, cardiac sympathetic denervation, and at least some loss of sympathetic innervation of the vasculature are likely causing the OH in PD.

Denervation supersensitivity

Clinical and preclinical studies of chronic autonomic failure have consistently noted increased blood pressure or vasoconstrictor responses to exogenously administered adrenoceptor agonists in patients with PD + OH. This finding would be consistent with "denervation supersensitivity," as described classically by Cannon (1939). At least part of this supersensitivity may result from increased expression of α- or β-adrenoceptors or increased intracellular signaling after receptor occupation (Davies et al., 1982; Vatner et al., 1985; Baser et al., 1991; Warner et al., 1993; Kurvers et al., 1998). Moreover, theoretically, cardiac sympathetic denervation supersensitivity might predispose to the development of arrhythmias (Inoue and Zipes, 1987). Molecular mechanisms of denervation supersensitivity remain poorly understood.

Augmented cardiovascular responsiveness to adrenoceptor agonists can have other explanations, however, such as decreased baroreflex buffering of sympathetic outflows, which, as noted above, seems to characterize PD + OH. Structural adaptations of vascular walls, with increases in wall:lumen ratios, occur commonly in hypertension, and supine hypertension often attends OH in patients with autonomic failure (Biaggioni and Robertson, 2002). Thus, although studies of PD patients have noted augmented pressor responses to exogenously administered norepinephrine, with the augmentation seen mainly or only in patients with PD + OH (Senard et al., 1990; Niimi et al., 1999), the results do not necessarily lead to the conclusion that PD + OH features denervation supersensitivity.

In summary, a combination of loss of sympathetic nerves in the heart and vasculature with baroreflex impairment can explain OH in PD and worsening of orthostatic symptoms during treatment with levodopa/carbidopa or dopamine receptor agonists. Cardiac sympathetic denervation characterizes most patients with PD and virtually all patients with PD + OH. These findings contrast with normal cardiac sympathetic innervation in most patients with MSA.

Functional consequences of cardiac sympathetic denervation

Limited evidence based on studies in transplanted hearts (Bengel et al., 2001) suggests that cardiac denervation impairs cardiac performance during exercise. Indeed, during exercise stress testing, patients with PD reached lower heart rates and blood pressures than controls

(DiFrancisco-Donoghue et al., 2009). Also suggesting cardiac β_1-adrenergic impairment in patients with PD was the hyperdynamic cardiac contractility shown during dobutamine stress testing (Nakamura et al., 2007). Moreover, a recent small study showed that in patients with PD who complained of fatigue, (123)I-metaiodobenzylguanidine heart-to-mediastinal uptake ratio was lower than in a matched group of PD patients who were nonfatigued, a finding suggesting that cardiac sympathetic denervation is associated with fatigue in patients with PD (Nakamura et al., 2011).

Treatment of orthostatic hypotension

The first step to avoid or minimize OH is to identify and discontinue drugs that can cause OH, such as antihypertensive agents and diuretics. Levodopa and dopamine receptor agonists may exacerbate OH, especially during the first weeks of treatment. Gradual dosage increases when initiating therapy or dose reductions in established patients can minimize this adverse effect. Sodium and water intake should be increased in these patients with liberal use of table salt or administration of sodium tablets. Patients should also be instructed not to lie flat, even during the daytime. Lying flat results in accelerated sodium loss from effects of increased cardiac filling and from pressure-natriuresis, leading to loss of intravascular volume. Overnight volume depletion can explain the typical finding of worse OH in the morning in patients with autonomic failure. Elevating the head of the bed on blocks is recommended (MacLean et al., 1944).

Patients and their families should be educated about the hypotensive effects of meal ingestion, exposure to environmental heat, and prolonged physical exertion. Isotonic exercise produces less hypotension than does isometric exercise. Exercise in a pool prevents blood pressure reductions during the exercise; however, OH can be worse after exiting the pool. In patients with autonomic failure, eating a meal can significantly lower blood pressure, because vasoconstriction in other vascular beds fails to compensate adequately for splanchnic vasodilatation induced by meal ingestion. In some patients, hypotension only occurs postprandially. Thus, patients should eat frequent, small meals with a low carbohydrate content, and alcohol intake should be minimized. Caffeine taken with breakfast may be helpful. Hot baths should be avoided, and patients should be especially careful during warm weather. This is because heat-induced vasodilatation and perspiration still occur, but sympathetic vasoconstriction is impaired. Straining at stool with a closed glottis (i.e., producing a Valsalva maneuver), playing wind instruments, and singing can be particularly dangerous for patients with PD + OH. A high-fiber diet is encouraged to prevent constipation, and singing or playing wind instruments should be undertaken only when sitting.

The use of knee-high compressive stockings is not effective, but waist-high stockings or abdominal binders may be an effective, albeit poorly tolerated, countermeasure for OH.

OH should only be treated pharmacologically in patients who are symptomatic, because the treatments usually worsen supine hypertension. Perhaps because of adaptive cerebral autoregulatory changes, some patients with autonomic failure tolerate very low arterial pressures when standing without experiencing symptoms of cerebral hypoperfusion. Blood pressure levels change throughout the day and from one day to another (Vagaonescu et al., 2000). Thus, the patient's normal cycle of blood pressure and orthostatic symptoms should be identified before treatment is initiated. If the patient has worse symptoms in the morning upon arising from bed, then treatment with short-acting drugs may be more effective in the morning and not needed later in the day.

The physiological mechanisms of OH, if identified, can guide its management. Measures include drugs to increase intravascular volume, increase peripheral vascular resistance, and correct anemia if present. Fludrocortisone, a synthetic mineralocorticoid, is widely used to increase intravascular volume in PD patients with symptomatic OH (Hickler, 1959). Therapy with fludrocortisone is initiated at a dose of 0.1 mg per day. The daily dose can be increased but more than 0.3 mg per day is rarely effective. Maximal clinical response occurs after approximately a week; dosage adjustments should take into consideration this delayed onset. Pedal edema and weight gain of 5–7 pounds are expected consequences of fludrocortisone therapy. For fludrocortisone to work effectively requires that the patient be on a high salt diet. Because of the potential for potassium wasting, serum potassium should be monitored in patients during initiation of fludrocortisone treatment for OH and at frequent intervals thereafter.

DDAVP, a synthetic vasopressin analog that acts at V2 receptors on distal renal tubules, altering its permeability to enhance water reabsorption, thus would be expected to work adjunctively with fludrocortisone to expand intravascular volume. DDAVP is administered intranasally in doses of 5–40 micrograms at bedtime (Mathias et al., 1986). Since DDAVP can induce hyponatremia, careful monitoring of serum sodium, preferably during a brief inpatient stay, is necessary during the first 4–5 days of treatment and at monthly intervals thereafter. Indometacin, a prostaglandin inhibitor, has been used to treat OH, especially in combination with fludrocortisone, but the lack of rigorous clinical data supporting the efficacy of this combination precludes a recommendation for its use (Crook et al., 1981).

Sympathomimetic agents increase peripheral vascular resistance and are useful in the treatment of symptomatic OH in PD. Midodrine is an orally active, selective α_1-adrenoceptor agonist that does not cross the blood–brain barrier and does not cause central excitatory effects (Kaufmann et al., 1988; Wright et al., 1998). The pressor response to midodrine begins within about an hour, making this agent potentially useful in treating patients who benefit from on-demand increases in blood pressure (e.g., for postprandial and morning hypotension). Midodrine therapy is started with a dose of 2.5 mg and increased to no more than 10 mg three times a day. A typical daily regimen includes a dose before breakfast, a dose before lunch, and a third dose midafternoon. Theoretically, routine administration of midodrine at high, fixed doses might downregulate α-adrenoceptors and mitigate the pressor effect. Midodrine should not be administered before bedtime, as the blood pressure typically is high when the patient is lying down. Theoretically, midodrine would be especially effective in patients with documented sympathetic denervation, because of denervation supersensitivity.

Erythropoietin increases red blood cell mass and blood viscosity. In addition, it also increases plasma endothelin, inhibits nitric oxide, and increases renal sodium reabsorption (Perera et al., 1994). Hypotensive parkinsonian patients with anemia may benefit from a 6 week course of subcutaneously administered recombinant erythropoietin (4000 units twice weekly) (Perera et al., 1994). Other treatments for OH should be continued during erythropoietin therapy.

A number of investigational agents are currently being studied. L-threo-DOPS (the biologically active stereoisomer of the amino acid 3,4-dihydroxyphenylserine) is a precursor of norepinephrine that has shown promise in the treatment of OH in large phase 3 clinical trials (Kaufmann et al., 2003).

Acarbose, an α-glucosidase inhibitor that decreases glucose absorption in the small intestine, significantly attenuates postprandial hypotension in patients with autonomic failure, and we have used it successfully on patients with PD and severe symptomatic postprandial hypotension (Shibao et al., 2007).

GASTROINTESTINAL DYSFUNCTION

James Parkinson's 1817 description of gastrointestinal problems in the patients in his case series was lastingly accurate: "food is with difficulty retained in the mouth until masticated; and then as difficultly swallowed ... the saliva fails of being directed to the back part of the fauces, and hence is continually draining from the mouth ... the bowels which all along had been torpid, now in most cases, demand stimulating medicines of very considerable power: the expulsion of the faeces from the rectum sometimes requiring mechanical aid" (Parkinson, 1817). While dysphagia, drooling, and constipation are the most common gastrointestinal abnormalities, early satiety, epigastrial distention, and nausea caused by delayed gastric emptying are also very frequent in PD patients.

Neuropathology

PD affects both the extrinsic and intrinsic innervation of the gut, which explains the prominent motility disturbances. Lewy bodies have been found in enteric neurons, in the Auerbach and Meissner plexuses along the entire GI tract, including the esophagus, stomach, small intestine, and colon, particularly in neurons of the Auerbach plexus in the lower esophagus (Wakabayashi et al., 1988). A study using histochemistry showed that LB are found mostly in vasoactive intestinal peptide (VIP)-containing neurons in the enteric plexus (Wakabayashi et al., 1990). Interestingly, transgenic mice expressing mutant α-synuclein from insertions of an entire human *SNCA* gene showed marked abnormalities in enteric nervous system function and synuclein-immunoreactive aggregates in enteric ganglia by 3 months of age (Kuo et al., 2010).

The extrinsic parasympathetic innervation of the gut originates in neurons of the dorsal motor nucleus of the vagus in the medulla. These neurons innervate the entire gastrointestinal tract with the exception of the proximal esophagus (innervated by the glossopharyngeal nerve) and the distal colon and rectum (innervated by the sacral parasympathetic nerves), and are severely affected early in PD. Vagal activity increases propulsive motility and relaxation of sphincters, and stimulates secretions of the exocrine and endocrine glands of the stomach, intestine, pancreas, and liver.

Early involvement of vagal or enteric neurons promoting peristalsis may explain the finding that constipation is a predictor of later development of PD (Abbott et al., 2001). According to Braak (Braak and Braak, 2000), PD develops in a sequence of six neuropathologic stages, with the earliest change being in the dorsal nucleus of the vagus nerve. The vulnerability of these neurons may be related to their having long unmyelinated axons that project to postganglionic neurons of the enteric nervous system. Braak has also proposed that a neurotoxic pathogen, gaining entry to the body via the gastrointestinal tract, might ascend via retrograde and transneuronal transport in vagal post- and then preganglionic fibers, to harm vulnerable neurons of the dorsal motor nucleus of the vagus nerve (Braak et al., 2003).

In contrast to the involvement of dorsal vagal neurons, neurons in the nucleus ambiguus are not directly affected in PD. Nucleus ambiguus neurons innervate the muscles of the palate, pharynx, and larynx through myelinated axons in the vagus nerve. The reason for abnormal swallowing in PD is likely to be abnormal supranuclear control of oropharyngeal muscles. This is suggested by the observation that many patients suffer severe dysphagia only when "off," and improve as soon as a dose of levodopa becomes effective.

Sympathetic outflow to the gastrointestinal tract, which arises from preganglionic neurons at the T1–L1 segments of the spinal cord and relays, via the splanchnic nerves, in the celiac and mesenteric ganglia, is involved in reflexes that decrease gut motility.

A substantial proportion of dopamine production in the body takes place in non-neuronal cells of the gut (Eisenhofer et al., 1997) that express tyrosine hydroxylase (Mezey et al., 1996, 1998, 1999). Whether PD involves altered non-neuronal dopamine production in the gut remains unknown. Singaram et al. (1995) reported decreased dopaminergic myenteric neurons in patients with PD and chronic constipation. The number of neurons containing immunoreactive tyrosine hydroxylase, however, was normal. PD patients also had decreased dopamine in the external muscular layer but not in the mucosa. No evidence has accrued for improvement in constipation by levodopa treatment.

Clinical gastrointestinal problems

Drooling

Excessive drooling is a distressing and frequent problem in patients with PD, but it is not due to excessive saliva production. On the contrary, salivation is reduced in patients with PD (Bagheri et al., 1999; Proulx et al., 2005). Drooling is due to reduced swallowing frequency allowing excess saliva to accumulates in the mouth.

Treatment with oral anticholinergics is ineffective. Botulinum toxin injected in the parotid and submandibulary glands has been used successfully (Jost, 1999; Pal et al., 2000; Friedman and Potulska, 2001; Dogu et al., 2004), but dysphagia is a potential adverse effect of the diffusion of botulinum toxin into nearby muscles.

Dysphagia

Dysphagia in patients with PD is related to the severity of the disease and may occur in up to 50% of patients (Bushmann et al., 1989; Edwards et al., 1992; Johnston et al., 1995). In general, abnormalities of swallowing are mild (Wintzen et al., 1994). PD patients who experience significant swallowing dysfunction should be evaluated by a speech and swallowing expert. Swallowing studies may help to define the nature of the dysphagia and the presence or absence of silent aspiration.

The three phases of swallowing, buccal, pharyngeal and esophageal, may be disrupted in PD. Abnormal lingual control (lingual festination) can impair the ability to pass a bolus of food backward into the pharynx. The pharyngeal swallow reflex may be disturbed as well (Born et al., 1996). Normally, the soft palate moves to prevent the bolus from entering the nasal cavity, the hyoid bone rises; the larynx prevents the bolus from entering the trachea, the true and false vocal cords close and the epiglottis lowers; the cricopharyngeal sphincter opens, and food or liquid moves to the esophagus. In PD patients, abnormalities in the pharyngeal phase can lead to silent aspiration. There have been reports of repetitive reflux of food from the vallecula and pyriform sinuses into the oral cavity (Wintzen et al., 1994). In the esophageal phase, the smooth muscles of the esophagus move the bolus in rhythmic, wavelike contractions into the stomach. Esophageal dysmotility occurs in as many as 70% of PD patients, with nonperistaltic swallows, belching, segmental spasms, esophageal dilatation, and gastroesophageal reflux (Ertekin et al., 2002). Repetitive, spontaneous contractions of the proximal esophagus have been described in patients with PD, a finding similar to that in achalasia (Johnston et al., 2001). Vocal cord palsy is frequent in patients with MSA and may lead to aspiration (Simpson et al., 1992; Wu et al., 1996).

Treatment

Soft diets help most types of dysphagia by making it easier to move food in the mouth and esophagus. Soft food also decreases aspiration by reducing the need for separate fluid intake, which is a potential source of aspiration. Patients with motor fluctations should be instructed to eat only during "on" times when dysphagia is less pronounced. Some patients suffer from achalasia, which can be treated with botulinum toxin injection into the cardia (Gui et al., 2003). Feeding gastrostomies or jejunostomies are a last resort and are rarely necessary for patients with PD. However, these procedures can provide the benefit of allowing more normal food and medication intake.

Delayed gastric emptying

Gastric retention due to delayed gastric emptying is a common problem in PD and results in nausea, early satiety, and abdominal distension. Levodopa, as a large neutral amino acid, is absorbed relatively little in the stomach and mainly in the small bowel, mostly the duodenum, by an active transport mechanism (Wade et al., 1973). Because of the high capacity of the transporter, competition between levodopa and other dietary neutral amino acids (e.g., valine, leucine, and isolucine) is not common but may occur.

Delayed gastric emptying retards delivery of levodopa to the absorptive sites in the duodenum. Reduced bioavailability of levodopa explains some of the response fluctuations that develop after long-term levodopa therapy (Kurlan et al., 1988b). Studies have shown higher gastric retention 1 hour after a meal in patients with PD who experience motor fluctuations, compared with patients without fluctuations (Djaldetti et al., 1996). Factors that slow gastric emptying also delay and blunt peak plasma levodopa levels and may cause a delay or a complete failure of the clinical response to the dose. With direct delivery of L-dopa into the duodenum, motor fluctuations can improve markedly (Kurlan et al., 1988a).

Treatment

Timing of gastric emptying is related to meal characteristics such as bulk, tonicity, and composition. Lipid and carbohydrates and excessive gastric acidity delay gastric emptying. Small frequent meals are recommended. Prokinetic drugs that accelerate gastric emptying include muscarinic agents, peripheral dopamine blockers, and serotonin-4 receptor agonists. The muscarinic receptor agonists bethanechol and carbachol have long been used for the treatment of markedly delayed gastric emptying or paralytic ileus. These agents exert a prokinetic effect by stimulating M_3 receptors in intestinal smooth muscle. Unfortunately, bethanechol increases the amplitude of gastric contractions in an uncoordinated manner, with little improvement in coordinated peristalsis. Additionally, bethanechol can elicit nausea and vomiting. Other common side-effects are diaphoresis, flushing, salivation, and abdominal cramping. The typical dosage is 25 mg orally four times daily or 2.5–5 mg subcutaneously four times daily.

Dopamine D_2-receptor blockers stimulate gastric motility. Metoclopramide, the most effective, cannot be used in parkinsonian patients because it blocks central dopamine receptors and worsens parkinsonism. Domperidone acts mostly on peripheral dopamine receptors and is an effective prokinetic agent (10–40 mg orally four times daily) in PD (Soykan et al., 1997). It is not yet available in the US.

Agonists at serotonin $(5HT)_4$ receptors stimulate release of acetylcholine from enteric neurons, activating prokinetic pathways. The first available $5HT_4$ agonist was cisapride (Jost and Schimrigk, 1993; Katayama et al., 1995). Unfortunately, cisapride prolongs the cardiac QT interval, predisposing to the ventricular arrhythmia *torsades de pointes*, which may cause hypotension, syncope, and sudden death. The agent was withdrawn from the market in the US. The proarrhythmic action of cisapride is due to its ability to block the myocyte cell membrane potassium channel, an effect that is independent of its prokinetic action. Other $5HT_4$ agonists with gastrointestinal prokinetic effects and low affinity for the cardiac potassium channel have been developed, such as mosapride and tegaserod. These drugs do not prolong the QT interval, and, in theory, are not arrhythmogenic (Potet et al., 2001). However, after initial approval by the US Food and Drug Administration in 2002, tegaserod was withdrawn from the market in 2007 due to concerns about a possible increased risk of heart attack or stroke (http://www.fda.gov/NewsEvents/Newsroom/PressAnnouncements/2007/ucm108879.htm). The issue remains controversial, as other large surveillance studies have not shown increased cardiovascular risk (Loughlin et al., 2010).

Nausea and vomiting as a side-effect of levodopa and dopamine agonists

The most prominent toxic effect of levodopa and dopamine agonists (particularly apomorphine) is nausea and vomiting. A proposed mechanism for this toxicity is increased occupation of dopamine receptors in the area postrema of the dorsal medulla, which lacks an efficient blood–brain barrier. Dopamine produced from levodopa outside the central nervous system could occupy area postrema receptors, evoking nausea and vomiting. Carbidopa inhibits the enzymatic conversion of levodopa to dopamine. As a catechol, carbidopa has little ability to penetrate the blood–brain barrier. The combination of levodopa with carbidopa therefore attenuates conversion of levodopa to dopamine outside the brain, augmenting entry of levodopa to the central nervous system, where enzymatic conversion to dopamine can proceed. The combination of levodopa with carbidopa decreases dopaminergic occupation of receptors in the area postrema, resulting in less nausea and vomiting. The brand name for levodopa/carbidopa, Sinemet™, comes from the Latin words for "without vomiting."

Trimethobenzamide is a dopamine receptor blocker with some prokinetic effects commonly used for the prevention of nausea associated with the use of apomorphine (at a dose of 300 mg orally three times daily) (Bowron, 2004).

Constipation

Normal defecation requires two separate processes: movement of stool along the colon by peristaltic waves of contracting smooth muscle and then expulsion of feces through the anal canal by the coordinated action of voluntary and involuntary muscles. In patients with PD, stool transit time is prolonged because colonic motility is reduced due to abnormal intrinsic and extrinsic vagal innervation. Degeneration of intrinsic enteric neurons and extrinsic parasympathetic efferent fibers that regulate contractility of colonic muscle underlie slow transit time, resulting in reduced frequency of defecation.

Frequently, defecation is also abnormal due to pelvic floor dyssynergia. Defecography and anal sphincter EMG in some PD patients showed paradoxical contraction of the puborectalis muscle (Mathers et al., 1988, 1989).

The puborectalis muscle, which is one of the muscles that comprise the pelvic floor and plays an important role in both fecal continence and defecation, is tonically contracted and maintains the anorectal angle at rest. Contraction of the internal and external anal sphincters contributes to continence. During defecation, the puborectalis muscle relaxes, opening the anorectal angle, the internal anal sphincter opens reflexively, and the external anal sphincter is voluntarily relaxed, thus allowing normal expulsion of rectal stools. In a patient with pelvic floor dyssynergia, the puborectalis muscle fails to relax, or contracts, increasing the anorectal angle. This accentuates its flap valve action. Moreover, anal sphincters paradoxically contract during attempted defecation. This results in outlet obstruction, dyschezia (straining to start or finish a bowel movement), and constipation. It has been suggested that this paradoxical contraction of the pelvic musculature is dystonic in nature (Stocchi et al., 2000). In support of this argument, apomorphine has been shown to alleviate this defecatory problem in some patients with PD. Similarly, injection of botulinum toxin in the puborectalis muscle or in the external anal sphincter has been reported as helpful. Other disorders associated with constipation in PD patients include megacolon (Kupsky et al., 1987) and sigmoid volvulus (Lewitan et al., 1951).

Treatment

The management of constipation in PD consists of dietary changes, exercise, and pharmacotherapy. Dietary modifications are aimed at increasing bulk and softening the stool. Patients should be encouraged to drink at least eight glasses of water each day and to increase the bulk and fiber content of their diet. Low-fiber foods, such as many baked goods, should be eaten infrequently and bananas should be avoided altogether. At least two meals per day should include high-fiber raw vegetables, to stimulate the gastrocolic reflex. Increasing physical activity can also be helpful. If stools remain hard, stool softeners (e.g., docusate) given with meals can be used. Lactulose in doses of 10–20 g per day may benefit some patients. Patients should be educated about the delayed onset of effect of stool softeners and encouraged to continue with fluids, increased bulk, high-fiber diet, and exercise. Discontinuing anticholinergic agents may increase bowel motility. Milk of magnesia, other mild laxatives, or enemas should be reserved for patients who do not respond to other interventions. Laxatives or enemas may be useful once weekly as part of an overall bowel regimen.

Prokinetic agents. Mosapride citrate, a 5-HT_4 agonist and partial 5-HT_3 antagonist (Yoshida et al., 1993), an effect that makes it also antiemetic blocking vagal serotonin receptors in the chemoreceptor trigger zone, ameliorates constipation in parkinsonian patients. In a recent study of parkinsonian patients with constipation and difficulty in defecation, mosapride treatment for 3 months improved bowel frequency and difficult defecation (Liu et al., 2005). Mosapride shortened colonic transit time, augmented the amplitude of rectal contraction during defecation and lessened the volume of postdefecation residuals.

BLADDER DYSFUNCTION

Three types of neuronal outflow regulate urinary bladder function: sacral parasympathetic, lumbar sympathetic, and somatic. The pelvic nerves carry the sacral parasympathetic (S2–S4) output to the bladder. Activation of muscarinic cholinergic receptors promotes bladder emptying (micturition) through contraction of the detrusor muscle and relaxation of the bladder neck. The lumbar sympathetic (T11–L2) output, carried via the hypogastric nerves, relaxes the detrusor muscle via β-adrenoceptors and contracts the bladder neck via α-adrenoceptors, thus promoting urinary retention. The sacral somatomotor output arises from motor neurons of the nucleus of Onuf (S2–S4) and is carried by the pudendal nerve. Stimulation of the motor neurons augments contraction of the external sphincter via nicotinic cholinergic receptors and promotes storage of urine.

Micturition involves a spinopontospinal reflex that is initiated by bladder tension receptors and integrated in pontine micturition centers. Extensive research on central neural pathways underlying reflexive micturition has pointed to glutamic acid as the major excitatory transmitter, with other transmitters, including norepinephrine, dopamine, and GABA, modulating the glutamatergic transmission (de Groat, 1998).

In the cat, four brainstem regions appear to regulate micturition: Barrington's nucleus (or the pontine micturition center) in the dorsomedial pons, the periaqueductal gray, the preoptic area of the hypothalamus, and an area in the ventrolateral pons called the L-region. It has been suggested that cells in Barrington's nucleus directly excite bladder motor neurons and indirectly inhibit internal urethral sphincter motor neurons, preoptic hypothalamic cells regulate initiation of micturition, L-region cells control motor neurons innervating the pelvic floor (including the external urethral sphincter), and periaqueductal gray cells receive afferent input about bladder filling (Blok, 2002). Studies using positron emission tomography (PET) and functional magnetic resonance imaging (MRI) scanning in humans have indicated activation of the same regions associated with urination or the attempt to urinate (Kershen et al., 2003). Afferent information for the micturition reflex comes from bladder distention. A balance between cortical stimulation and supraspinal inhibition determines a setpoint for reflexive responses as the bladder fills (Hebjorn et al., 1976).

In PD, cell loss in the substantia nigra, which normally has an inhibitory effect on the micturition reflex (Lewin et al., 1967; Yoshimura et al., 2003), leads to hyperreflexia of the detrusor muscle, with involuntary or uninhibited contractions and an urge to urinate (Fitzmaurice et al., 1985; Araki et al., 2000). The animal model produced by MPTP (1-methyl-4-phenyl-1, 2,3,6-tetrahydropyridine) suggests an inhibitory role of the substantia nigra (STN) on the micturition reflex (Albanese et al., 1988; Yoshimura et al., 1998). Further evidence that the basal ganglia affect the micturition reflex comes from a recent report in patients with PD receiving deep brain stimulation (DBS) of the STN (Seif et al., 2004). With the stimulator off, urodynamic parameters showed detrusor hyperreflexia. When turning the stimulator on, induction of the micturition reflex was delayed towards normalization, with the initial desire to void at higher bladder volumes and an increment of the maximal bladder capacity. During the voiding phase, DBS STN stimulation induced a small, nonsignificant increase in the pressure of the detrusor, maximum urinary flow, and reduction in residual urine.

Urinary problems are common and afflict both women and men with PD (Dmochowski, 1999). A comprehensive questionnaire and urodynamic evaluation in patients with PD and urinary symptoms revealed detrusor hyperreflexia in 79% of patients, while 16% had detrusor hyporeflexia (Araki et al., 2000). Impaired contractile function occurred in 9% of patients and detrusor-sphincter dyssynergia in 3%. Scores derived from questionnaires of irritative and obstructive symptom were fairly accurate in predicting urodynamic abnormalities. Bladder function deteriorated and postvoid residual urine volume increased with advancing disease severity. In another study, sphincter electromyography revealed pseudodyssynergia or bradykinesia in 50% of female PD patients (Dmochowski, 1999).

In MSA there is also degeneration of neurons in Onuf's nucleus. In later stages of the disease there is neuronal loss affecting parasympathetic innervation of the detrusor producing detrusor hypocontractility or arreflexia, with increased residual urine and overflow incontinence.

Prostate surgery

In men with PD, bladder outflow obstruction due to benign prostatic hyperplasia results in urinary hesitancy and low urine flow. Obstruction can cause detrusor overactivity and urinary urgency as well. Therefore, surgery of the prostate is considered in PD patients in the hope that detrusor hyperactivity results from bladder outlet obstruction rather than PD. Unfortunately, surgery frequently worsens symptoms and results in overt urinary incontinence.

A recent study found that bladder filling during urodynamic evaluation occurs earlier when detrusor hyperactivity is due to PD and that urge incontinence is rare in men with detrusor hyperactivity due to bladder outlet obstruction (Defreitas et al., 2003). Urological intervention is not contraindicated in men with PD, but patients should try anticholinergic medication first if urge incontinence is prominent. If conservative measures fail, a voiding cystometrogram to demonstrate obstructed voiding should be performed before transurethral resection of the prostate is considered (Chandiramani et al., 1997).

It has been suggested that in addition to detrusor hyperreflexia, patients with PD may have impaired relaxation or "bradykinesia" of the urethral sphincter, resulting in bladder outflow obstruction and difficulty in micturition, with similar symptoms as in prostatic hypertrophy. A study of subcutaneous apomorphine in patients with PD showed that apomorphine reduced bladder outflow resistance and improved voiding. It was proposed that this intervention be used to demonstrate the reversibility of outflow obstruction in men with PD before prostatic surgery be undertaken (Christmas et al., 1988; Aranda and Cramer, 1993).

Onset of urinary symptoms is earlier in MSA

In MSA, urinary symptoms, like OH, are typically present before the onset of the motor symptoms. Urinary symptoms in PD tend to occur later. Also characteristic of MSA are early urinary incontinence due to Onuf's nucleus involvement, postmicturition residual volume more than 100 mL, loss of the bulbocavernosus reflex, and denervation indicated by sphincter EMG (Chandiramani et al., 1997).

Worsening urinary control after transurethral resection of the prostate in men or anti-incontinence procedures in women is typical with MSA, immediately after or within a year after surgery (Beck et al., 1994). Since in PD the anterior horn cells of Onuf's nucleus are not affected, sphincter EMG was proposed as a means of distinguishing between PD and MSA. Both the anal and urethral sphincters are innervated by the anterior horn cells in Onuf's nucleus, leading to changes of chronic reinnervation, with prolongation of the mean duration of motor units in patients with MSA (Eardley et al., 1989; Fowler, 1996).

Treatment

Many patients can reduce nocturnal frequency by restricting fluid intake after the evening meal. In PD patients with autonomic dysfunction and supine hypertension, nocturia may also result from pressure natriuresis (see Orthostatic hypotension) and improves by sleeping with the head and torso elevated. Pharmacological treatments include peripherally acting anticholinergics, such as oxybutynin (5–10 mg at bedtime or three times daily), propantheline (7.5–15 mg at bedtime or three times daily), or tolterodine tartrate (1–2 mg twice daily based on individual response and tolerability).

Anticholinergic agents reduce detrusor contractions and are useful in the treatment of detrusor hyperactivity but may worsen voiding problems and even produce urinary retention in patients with detrusor hypoactivity or outlet obstruction. Therefore, before starting treatment, it is important to measure postvoid residual volume with an ultrasound study or urodynamic evaluation. If postvoid residual volume is less than 100 mL, then treatment with anticholinergics may provide benefit (Fowler, 1999). It is important to reevaluate the patient if there is no improvement after pharmacological therapy. Increased residual urine can stimulate detrusor contractions.

Anticholinergic drugs should also be administered with caution, as they may also aggravate gastrointestinal motility disorders and increase gastric retention. α_1-Adrenoceptor antagonists can decrease tone in the bladder neck and may be helpful for patients with a hypoactive detrusor; however, these agents worsen OH. If the patient has residual volume more than 100 mL, then self-catheterization is indicated. This can also be combined with anticholinergic therapy to enhance continence between catheterizations. If the patient or relative cannot perform catheterization, surgical management of the problem may be needed.

If daytime frequency or urgency precede nocturia, mechanical outlet obstruction should be ruled out. Any deterioration in voiding pattern (even in the absence of dysuria) should raise concern about a urinary tract infection, and this should be treated promptly.

SEXUAL DYSFUNCTION

Dopaminergic mechanisms seem to participate in libido and arousal-related vasodilatation of penile erectile tissue. The cause of erectile dysfunction (ED; difficulty achieving or sustaining an erection) in PD is unknown but may reflect dopamine deficiency. About 60% of male PD patients have ED (Singer et al., 1992). Impaired sexual arousal, behavior, orgasm, and drive are also frequent. Sexual fantasy, however, seems to remain normal in most patients (Yu et al., 2004). ED, almost invariably, is the earliest symptom in men with MSA and can precede other symptoms by several years.

Treatment

Many drugs can cause male sexual dysfunction, and a thorough medication history often uncovers causative agents. Propranolol and other β-adrenergic blockers, which are sometimes used to treat tremor or hypertension in PD, should be discontinued if possible. Other drugs that may cause sexual dysfunction include α_1-adrenoceptor blockers, guanethidine, thiazide diuretics, anxiolytics, digoxin, cimetidine, and some antidepressants.

Depression is a common cause of impotence and can respond to antidepressants, though some antidepressants themselves can cause impotence (e.g., selective serotonin reuptake inhibitors, tricyclic antidepressants, monoamine oxidase inhibitors). Some patients with anxiety- or stress-associated sexual dysfunction may benefit from low-dose anxiolytics.

If no medical or psychological reason appears to be causing impotence, several options are available. Intracavernous injections or transurethral suppositories of alprostadil, a synthetic prostaglandin E_1, induce penile erection, but their use is cumbersome. Sildenafil, an orally active inhibitor of the type V cGMP specific phosphodiesterase (the predominant isoenzyme in the human corpus cavernosum) has improved ED in small clinical trials of PD patients (Zesiewicz et al., 2000; Hussain et al., 2001). Patients with MSA, however, developed severe hypotension (Saadia et al., 2002). A report on men using subcutaneous injections of apomorphine to treat motor fluctuations in PD noted that the treatment benefited their sexual function and induced penile erection (O'Sullivan and Hughes, 1998). Drug trials to assess the effect of sublingual apomorphine to treat ED had promising results, although nausea occurs in a proportion of the patients (Dula et al., 2000; Perimenis et al., 2004). Some patients on high doses of antiparkinsonian therapy become hypersexual, even in the face of inability to perform.

THERMOREGULATION AND SWEATING ABNORMALITIES

Preoptic and hypothalamic neurons are important for thermoregulatory function and may be affected in PD. During normal human aging, the ability to tolerate swings of environmental temperature declines. Elderly healthy volunteers have a decreased setpoint for mounting sympathetic noradrenergic responses to core hypothermia (Frank et al., 1997).

Thermoregulatory sweating is a sympathetic nervous system function where the effector neurotransmitter is acetylcholine, in contrast to noradrenergic mediation of sympathetic cardiovascular functions. Many studies assessing cutaneous sympathetic cholinergic function in PD have relied on measurements of skin humidity or electrical conductance as indices of sweat production; results have been variable (Wang et al., 1993; Jost et al., 1995; Denislic and Meh, 1996; Braune et al., 1997; De Marinis et al., 2000; Haapaniemi et al., 2000). PD patients can have increased, decreased, or normal sweating.

In PD patients with sympathetic neurocirculatory failure and cardiac sympathetic noradrenergic denervation, sympathetic cholinergic innervation of sweat glands appears to remain intact, since such patients have normal sweating during the quantitative sudomotor axon reflex test (Sharabi et al., 2003). Turkka et al. (1997) reported increased sweating in PD patients, both before and after heat exposure; however, whether PD patients have appropriate thermoregulatory sweating responses in terms of maintaining core temperature is unknown. Abnormal sensations of heat or cold, and hypothermia can occur in the PD patient.

Excessive sweating of the head and neck in response to external heat has been associated with poor heat dissipation in the rest of the body. Some of these phenomena disappear with levodopa treatment. Severe drenching sweats can also occur as an end-of-dose "off" phenomenon in patients with motor fluctuations (Sage and Mark, 1995). In contrast, some patients will experience sweating during "on" responses following levodopa administration, frequently in association with dyskinesia (Swinn et al., 2003), although it is rarely as severe as that seen in the "off" state. Severe hyperpyrexia after levodopa withdrawal can represent a form of neuroleptic malignant syndrome (Cao and Katz, 1999). Ethanol and aspirin in high doses can cause intermittent sweating. Thyrotoxicosis, chronic infections, and the postmenopausal state should also be considered in the differential diagnosis.

Almost all research on thermoregulatory function in PD has focused on responses to increased temperature. It is well known that cold exposure markedly increases sympathetic noradrenergic outflows. Whether PD entails deficient cold-induced sympathetic stimulation remains unknown.

CONCLUSIONS

Symptoms and signs of autonomic failure occur commonly in patients with PD. These include cardiovascular abnormalities, constipation, urinary frequency and urgency, erectile failure in men, and altered sweating. Most patients with PD have neuroimaging evidence of cardiac sympathetic denervation which occurs independently of the loss of striatal dopamine terminals and appears to result from a dying-back pathogenetic process. Patients with PD and orthostatic hypotension have reduced baroreflex-mediated cardiovagal and sympathoneural responses accounting for their cardiovascular abnormalities.

References

Abbott RD, Petrovitch H, White LR et al. (2001). Frequency of bowel movements and the future risk of Parkinson's disease. Neurology 57: 456–462.

Albanese A, Jenner P, Marsden CD et al. (1988). Bladder hyperreflexia induced in marmosets by 1-methyl-4-phenyl-1,2,3,6-tetrahydropyridine. Neurosci Lett 87: 46–50.

Allcock LM, Ullyart K, Kenny RA et al. (2004). Frequency of orthostatic hypotension in a community based cohort of patients with Parkinson's disease. J Neurol Neurosurg Psychiatry 75: 1470–1471.

Amino T, Orimo S, Itoh Y et al. (2005). Profound cardiac sympathetic denervation occurs in Parkinson disease. Brain Pathol 15: 29–34.

Araki I, Kitahara M, Oida T et al. (2000). Voiding dysfunction and Parkinson's disease: urodynamic abnormalities and urinary symptoms. J Urol 164: 1640–1643.

Aranda B, Cramer P (1993). Effects of apomorphine and L-dopa on the parkinsonian bladder. Neurourol Urodyn 12: 203–209.

Awerbuch GI, Sandyk R (1992). Autonomic functions in the early stages of Parkinson's disease. Int J Neurosci 64: 7–14.

Bagheri H, Damase-Michel C, Lapeyre-Mestre M et al. (1999). A study of salivary secretion in Parkinson's disease. Clin Neuropharmacol 22: 213–215.

Baser SM, Brown RT, Curras MT et al. (1991). Beta-receptor sensitivity in autonomic failure. Neurology 41: 1107–1112.

Beck RO, Betts CD, Fowler CJ (1994). Genitourinary dysfunction in multiple system atrophy: clinical features and treatment in 62 cases. J Urol 151: 1336–1341.

Bellon AK, Jost WH, Schimrigk K et al. (1996). Blood pressure adaptation and hormone regulation in Parkinson patients following orthostasis. Dtsch Med Wochenschr 121: 1077–1083.

Benarroch EE, Schmeichel AM, Parisi JE (2000). Involvement of the ventrolateral medulla in parkinsonism with autonomic failure. Neurology 54: 963–968.

Bengel FM, Ueberfuhr P, Schiepel N et al. (2001). Effect of sympathetic reinnervation on cardiac performance after heart transplantation. N Engl J Med 345: 731–738.

Bhattacharya KF, Nouri S, Olanow CW et al. (2003). Selegiline in the treatment of Parkinson's disease: its impact on orthostatic hypotension. Parkinsonism Relat Disord 9: 221–224.

Biaggioni I, Robertson RM (2002). Hypertension in orthostatic hypotension and autonomic dysfunction. Cardiol Clin 20: 291–301, vii.

Blok BF (2002). Central pathways controlling micturition and urinary continence. Urology 59 (Suppl 1): 13–17.
Bonuccelli U, Lucetti C, Del Dotto P et al. (2003). Orthostatic hypotension in de novo Parkinson disease. Arch Neurol 60: 1400–1404.
Born LJ, Harned RH, Rikkers LF et al. (1996). Cricopharyngeal dysfunction in Parkinson's disease: role in dysphagia and response to myotomy. Mov Disord 11: 53–58.
Bowron A (2004). Practical considerations in the use of apomorphine injectable. Neurology 62 (Suppl 4): S32–S36.
Braak H, Braak E (2000). Pathoanatomy of Parkinson's disease. J Neurol 247 (Suppl 2): II3–II10.
Braak H, Rub U, Gai WP et al. (2003). Idiopathic Parkinson's disease: possible routes by which vulnerable neuronal types may be subject to neuroinvasion by an unknown pathogen. J Neural Transm 110: 517–536.
Braune HJ, Korchounov AM, Schipper HI (1997). Autonomic dysfunction in Parkinson's disease assessed by sympathetic skin response: a prospective clinical and neurophysiological trial on 50 patients. Acta Neurol Scand 95: 293–297.
Braune S, Reinhardt M, Bathmann J et al. (1998). Impaired cardiac uptake of meta-[123I]iodobenzylguanidine in Parkinson's disease with autonomic failure. Acta Neurol Scand 97: 307–314.
Braune S, Reinhardt M, Schnitzer R et al. (1999). Cardiac uptake of [123I]MIBG separates Parkinson's disease from multiple system atrophy. Neurology 53: 1020–1025.
Briebach T, Baas H, Fischer PA (1990). Orthostatic dysregulation in Parkinson syndrome. Results of a study of 250 patients. Nervenarzt 61: 491–494.
Bristow JD, Honour AJ, Pickering GW et al. (1969). Diminished baroreflex sensitivity in high blood pressure. Circulation 39: 48–54.
Bushmann M, Dobmeyer SM, Leeker L et al. (1989). Swallowing abnormalities and their response to treatment in Parkinson's disease. Neurology 39: 1309–1314.
Camerlingo M, Aillon C, Bottacchi E et al. (1987). Parasympathetic assessment in Parkinson's disease. Adv Neurol 45: 267–269.
Cannon WB (1939). A law of denervation. Am J Med Sci 198: 737–750.
Cao L, Katz RH (1999). Acute hypernatremia and neuroleptic malignant syndrome in Parkinson disease. Am J Med Sci 318: 67–68.
Chandiramani VA, Palace J, Fowler CJ (1997). How to recognize patients with parkinsonism who should not have urological surgery. Br J Urol 80: 100–104.
Christmas TJ, Kempster PA, Chapple CR et al. (1988). Role of subcutaneous apomorphine in parkinsonian voiding dysfunction. Lancet 2: 1451–1453.
Crook JE, Robertson D, Whorton AR (1981). Prostaglandin suppression: inability to correct severe idiopathic orthostatic hypotension. South Med 73: 318–320.
Davies B, Sudera D, Sagnella G et al. (1982). Increased numbers of alpha receptors in sympathetic denervation supersensitivity in man. J Clin Invest 69: 779–784.
Davy KP, Jones PP, Seals DR (1997). Influence of age on the sympathetic neural adjustments to alterations in systemic oxygen levels in humans. Am J Physiol 273: R690–R695.
Davy KP, Seals DR, Tanaka H (1998). Augmented cardiopulmonary and integrative sympathetic baroreflexes but attenuated peripheral vasoconstriction with age. Hypertension 32: 298–304.
de Groat WC (1998). Anatomy of the central neural pathways controlling the lower urinary tract. Eur Urol 34 (Suppl 1): 2–5.
De Marinis M, Stocchi F, Gregori B et al. (2000). Sympathetic skin response and cardiovascular autonomic function tests in Parkinson's disease and multiple system atrophy with autonomic failure. Mov Disord 15: 1215–1220.
Defreitas GA, Lemack GE, Zimmern PE et al. (2003). Distinguishing neurogenic from non-neurogenic detrusor overactivity: a urodynamic assessment of lower urinary tract symptoms in patients with and without Parkinson's disease. Urology 62: 651–655.
Denislic M, Meh D (1996). Sympathetic skin response in parkinsonian patients. Electromyogr Clin Neurophysiol 36: 231–235.
Dickson DW, Liu W, Hardy J et al. (1999). Widespread alterations of alpha-synuclein in multiple system atrophy. Am J Pathol 155: 1241–1251.
DiFrancisco-Donoghue J, Elokda A, Lamberg EM et al. (2009). Norepinephrine and cardiovascular responses to maximal exercise in Parkinson's disease on and off medication. Mov Disord 24: 1773–1778.
Djaldetti R, Baron J, Ziv I et al. (1996). Gastric emptying in Parkinson's disease: patients with and without response fluctuations. Neurology 46: 1051–1054.
Dmochowski RR (1999). Female voiding dysfunction and movement disorders. Int Urogynecol J Pelvic Floor Dysfunct 10: 144–151.
Dogu O, Apaydin D, Sevim S et al. (2004). Ultrasound-guided versus "blind" intraparotid injections of botulinum toxin-A for the treatment of sialorrhoea in patients with Parkinson's disease. Clin Neurol Neurosurg 106: 93–96.
Druschky A, Hilz MJ, Platsch G et al. (2000). Differentiation of Parkinson's disease and multiple system atrophy in early disease stages by means of I-123-MIBG-SPECT. J Neurol Sci 175: 3–12.
Dula E, Keating W, Siami PF et al. (2000). Efficacy and safety of fixed-dose and dose-optimization regimens of sublingual apomorphine versus placebo in men with erectile dysfunction. The Apomorphine Study Group. Urology 56: 130–135.
Durrieu G, Senard JM, Tran MA et al. (1991). Effects of levodopa and bromocriptine on blood pressure and plasma catecholamines in parkinsonians. Clin Neuropharmacol 14: 84–90.
Eardley I, Quinn NP, Fowler CJ et al. (1989). The value of urethral sphincter electromyography in the differential diagnosis of parkinsonism. Br J Urol 64: 360–362.
Ebert TJ, Morgan BJ, Barney JA et al. (1992). Effects of aging on baroreflex regulation of sympathetic activity in humans. Am J Physiol 263: H798–H803.
Edwards LL, Quigley EM, Pfeiffer RF (1992). Gastrointestinal dysfunction in Parkinson's disease: frequency and pathophysiology [see comments]. Neurology 42: 726–732.
Eisenhofer G, Aneman A, Friberg P et al. (1997). Substantial production of dopamine in the human gastrointestinal tract. J Clin Endocrinol Metab 82: 3864–3871.

Ertekin C, Tarlaci S, Aydogdu I et al. (2002). Electrophysiological evaluation of pharyngeal phase of swallowing in patients with Parkinson's disease. Mov Disord 17: 942–949.

Ferrari AU (2002). Modifications of the cardiovascular system with aging. Am J Geriatr Cardiol 11: 30–33.

Fitzmaurice H, Fowler CJ, Rickards D et al. (1985). Micturition disturbance in Parkinson's disease. Br J Urol 57: 652–656.

Fowler CJ (1996). Investigation of the neurogenic bladder. J Neurol Neurosurg Psychiatry 60: 6–13.

Fowler CJ (1999). Neurological disorders of micturition and their treatment. Brain 122: 1213–1231.

Frank SM, Higgins MS, Fleisher LA et al. (1997). Adrenergic, respiratory, and cardiovascular effects of core cooling in humans. Am J Physiol 272: R557–R562.

Friedman A, Potulska A (2001). Botulinum toxin for treatment of parkinsonian sialorrhea. Neurol Neurochir Pol 35 (Suppl 3): 23–27.

Goldstein DS (2006). Orthostatic hypotension as an early finding in Parkinson's disease. Clin Auton Res 16: 46–54.

Goldstein DS, Tack C (2000). Noninvasive detection of sympathetic neurocirculatory failure. Clin Auton Res 10: 285–291.

Goldstein DS, Polinsky RJ, Garty M et al. (1989). Patterns of plasma levels of catechols in neurogenic orthostatic hypotension. Ann Neurol 26: 558–563.

Goldstein DS, Holmes C, Cannon RO 3rd et al. (1997). Sympathetic cardioneuropathy in dysautonomias. N Engl J Med 336: 696–702.

Goldstein DS, Holmes C, Li ST et al. (2000). Cardiac sympathetic denervation in Parkinson disease. Ann Intern Med 133: 338–347.

Goldstein DS, Holmes CS, Dendi R et al. (2002). Orthostatic hypotension from sympathetic denervation in Parkinson's disease. Neurology 58: 1247–1255.

Goldstein DS, Eldadah BA, Holmes C et al. (2005). Neurocirculatory abnormalities in Parkinson disease with orthostatic hypotension: independence from levodopa treatment. Hypertension 46: 1–7.

Goldstein DS, Holmes C, Sato T et al. (2008). Central dopamine deficiency in pure autonomic failure. Clin Auton Res 18: 58–65.

Gui D, Rossi S, Runfola M et al. (2003). Botulinum toxin in the therapy of gastrointestinal motility disorders. Review article. Aliment Pharmacol Ther 18: 1–16.

Haapaniemi TH, Korpelainen JT, Tolonen U et al. (2000). Suppressed sympathetic skin response in Parkinson disease. Clin Auton Res 10: 337–342.

Hague K, Lento P, Morgello S et al. (1997). The distribution of Lewy bodies in pure autonomic failure: autopsy findings and review of the literature. Acta Neuropathol (Berl) 94: 192–196.

Hasegawa Y, Okamoto S (1992). Parkinson's disease with syncope as a chief complaint induced by prominent postprandial hypotension. Rinsho Shinkeigaku 32: 1017–1021.

Hebjorn S, Andersen JT, Walter S et al. (1976). Detrusor hyperreflexia. A survey on its etiology and treatment. Scand J Urol Nephrol 10: 103–109.

Herrmann DN, Griffin JW, Hauer P et al. (1999). Epidermal nerve fiber density and sural nerve morphometry in peripheral neuropathies [see comments]. Neurology 53: 1634–1640.

Hickler RB (1959). Successful treatment of orthostatic hypotension with 9-alpha fluorohydrocortisone. N Eng J Med 261: 788–791.

Hillen ME, Wagner ML, Sage JI (1996). "Subclinical" orthostatic hypotension is associated with dizziness in elderly patients with Parkinson disease. Arch Phys Med Rehabil 77: 710–712.

Hoehn MM (1975). Levodopa-induced postural hypotension. Treatment with fludrocortisone. Arch Neurol 32: 50–51.

Holmberg B, Kallio M, Johnels B et al. (2001). Cardiovascular reflex testing contributes to clinical evaluation and differential diagnosis of Parkinsonian syndromes. Mov Disord 16: 217–225.

Hubble JP, Koller WC, Cutler NR et al. (1995). Pramipexole in patients with early Parkinson's disease. Clin Neuropharmacol 18: 338–347.

Hussain IF, Brady CM, Swinn MJ et al. (2001). Treatment of erectile dysfunction with sildenafil citrate (Viagra) in parkinsonism due to Parkinson's disease or multiple system atrophy with observations on orthostatic hypotension. J Neurol Neurosurg Psychiatry 71: 371–374.

Inoue H, Zipes DP (1987). Results of sympathetic denervation in the canine heart: supersensitivity that may be arrhythmogenic. Circulation 75: 877–887.

Iwanaga K, Wakabayashi K, Yoshimoto M et al. (1999). Lewy body-type degeneration in cardiac plexus in Parkinson's and incidental Lewy body diseases. Neurology 52: 1269–1271.

Jellinger KA (2003). Neuropathological spectrum of synucleinopathies. Mov Disord 18 (Suppl 6): S2–S12.

Johnston BT, Li Q, Castell JA et al. (1995). Swallowing and esophageal function in Parkinson's disease. Am J Gastroenterol 90: 1741–1746.

Johnston BT, Colcher A, Li Q et al. (2001). Repetitive proximal esophageal contractions: a new manometric finding and a possible further link between Parkinson's disease and achalasia. Dysphagia 16: 186–189.

Jost WH (1999). Treatment of drooling in Parkinson's disease with botulinum toxin. Mov Disord 14: 1057.

Jost WH, Schimrigk K (1993). Cisapride treatment of constipation in Parkinson's disease. Mov Disord 8: 339–343.

Jost WH, Kirchhofer U, Houy S et al. (1995). Sympathetic skin response in Parkinson syndrome. Nervenarzt 66: 777–780.

Kaakkola S, Mannisto PT, Nissinen E et al. (1985). The effect of an increased ratio of carbidopa to levodopa on the pharmacokinetics of levodopa. Acta Neurol Scand 72: 385–391.

Katayama K, Morio Y, Haga K et al. (1995). Cisapride, a gastroprokinetic agent, binds to 5-HT4 receptors. Nippon Yakurigaku Zasshi 105: 461–468.

Kaufmann H (1996). Consensus statement on the definition of orthostatic hypotension, pure autonomic failure and multiple system atrophy. Clin Auton Res 6: 125–126.

Kaufmann H, Biaggioni I (2003). Autonomic failure in neurodegenerative disorders. Semin Neurol 23: 351–363.

Kaufmann H, Brannan T, Krakoff L et al. (1988). Treatment of orthostatic hypotension due to autonomic failure with a peripheral alpha-adrenergic agonist (midodrine). Neurology 38: 951–956.

Kaufmann H, Hague K, Perl D (2001). Accumulation of alpha-synuclein in autonomic nerves in pure autonomic failure. Neurology 56: 980–981.

Kaufmann H, Saadia D, Voustianiouk A et al. (2003). Norepinephrine precursor therapy in neurogenic orthostatic hypotension. Circulation 108: 724–728.

Kaufmann H, Nahm K, Purohit D et al. (2004). Autonomic failure as the initial presentation of Parkinson disease and dementia with Lewy bodies. Neurology 63: 1093–1095.

Kershen RT, Kalisvaart J, Appell RA (2003). Functional brain imaging and the bladder: new insights into cerebral control over micturition. Curr Urol Rep 4: 344–349.

Korchounov A, Kessler KR, Schipper HI (2004). Differential effects of various treatment combinations on cardiovascular dysfunction in patients with Parkinson's disease. Acta Neurol Scand 109: 45–51.

Krygowska-Wajs A, Furgala A, Laskiewicz J et al. (2002). Early diagnosis of orthostatic hypotension in idopathic Parkinson's disease. Folia Med Cracov 43: 59–67.

Kujawa K, Leurgans S, Raman R et al. (2000). Acute orthostatic hypotension when starting dopamine agonists in Parkinson's disease. Arch Neurol 57: 1461–1463.

Kuo YM, Li Z, Jiao Y et al. (2010). Extensive enteric nervous system abnormalities in mice transgenic for artificial chromosomes containing Parkinson disease-associated alpha-synuclein gene mutations precede central nervous system changes. Hum Mol Genet 19: 1633–1650.

Kupsky WJ, Grimes MM, Sweeting J et al. (1987). Parkinson's disease and megacolon: concentric hyaline inclusions (Lewy bodies) in enteric ganglion cells. Neurology 37: 1253–1255.

Kurlan R, Nutt JG, Woodward WR et al. (1988a). Duodenal and gastric delivery of levodopa in parkinsonism. Ann Neurol 23: 589–595.

Kurlan R, Rothfield KP, Woodward WR et al. (1988b). Erratic gastric emptying of levodopa may cause "random" fluctuations of parkinsonian mobility. Neurology 38: 419–421.

Kuroiwa Y, Shimada Y, Toyokura Y (1983). Postural hypotension and low R-R interval variability in parkinsonism, spino-cerebellar degeneration, and Shy–Drager syndrome. Neurology 33: 463–467.

Kurvers H, Daemen M, Slaaf D et al. (1998). Partial peripheral neuropathy and denervation induced adrenoceptor supersensitivity. Functional studies in an experimental model. Acta Orthop Belg 64: 64–70.

Lake CR, Ziegler MG, Kopin IJ (1976). Use of plasma norepinephrine for evaluation of sympathetic neuronal function in man. Life Sci 18: 1315–1325.

Langley JN (1903). The autonomic nervous system. Brain 26: 1–26.

Lewin RJ, Dillard GV, Porter RW (1967). Extrapyramidal inhibition of the urinary bladder. Brain Res 4: 301–307.

Lewitan A, Nathanson L, Slade WR Jr (1951). Megacolon and dilatation of the small bowel in parkinsonism. Gastroenterology 17: 367–374.

Liu Z, Sakakibara R, Odaka T et al. (2005). Mosapride citrate, a novel 5-HT4 agonist and partial 5-HT3 antagonist, ameliorates constipation in parkinsonian patients. Mov Disord 20: 680–686.

Loew F, Gauthey L, Koerffy A et al. (1995). Postprandial hypotension and orthostatic blood pressure responses in elderly Parkinson's disease patients. J Hypertens 13: 1291–1297.

Lokhandwala MF, Hegde SS (1990). Cardiovascular pharmacology of dopamine receptor agonists. In: F Amenta (Ed.), Peripheral Dopamine Pathophysiology. CRC Press, Boca Raton, FL, pp. 63–77.

Loughlin J, Quinn S, Rivero E et al. (2010). Tegaserod and the risk of cardiovascular ischemic events: an observational cohort study. J Cardiovasc Pharmacol Ther 15: 151–157.

MacLean AR, Allen EV, Magath TB (1944). Orthostatic tachycardia and orthostatic hypotension: defects in the return of venous blood to the heart. Am Heart J 27: 145–163.

Magalhaes M, Wenning GK, Daniel SE et al. (1995). Autonomic dysfunction in pathologically confirmed multiple system atrophy and idiopathic Parkinson's disease – a retrospective comparison. Acta Neurol Scand 91: 98–102.

Martignoni E, Pacchetti C, Godi L et al. (1995). Autonomic disorders in Parkinson's disease. J Neural Transm Suppl 45: 11–19.

Mathers SE, Kempster PA, Swash M et al. (1988). Constipation and paradoxical puborectalis contraction in anismus and Parkinson's disease: a dystonic phenomenon? J Neurol Neurosurg Psychiatry 51: 1503–1507.

Mathers SE, Kempster PA, Law PJ et al. (1989). Anal sphincter dysfunction in Parkinson's disease. Arch Neurol 46: 1061–1064.

Mathias CJ, Fosbraey P, da Costa DF et al. (1986). The effect of desmopressin on nocturnal polyuria overnight weight loss and morning postural hypotension in patients with autonomic failure. Br Med J (Clin Res Ed) 293: 353–354.

Matsukawa T, Sugiyama Y, Mano T (1996). Age-related changes in baroreflex control of heart rate and sympathetic nerve activity in healthy humans. J Auton Nerv Syst 60: 209–212.

Mezey E, Eisenhofer G, Harta G et al. (1996). A novel non-neuronal catecholaminergic system: exocrine pancreas synthesizes and releases dopamine. Proc Natl Acad Sci U S A 93: 10377–10382.

Mezey E, Eisenhofer G, Hansson S et al. (1998). Dopamine produced by the stomach may act as a paracrine/autocrine hormone in the rat. Neuroendocrinology 67: 336–348.

Mezey E, Eisenhofer G, Hansson S et al. (1999). Non-neuronal dopamine in the gastrointestinal system. Clin Exp Pharmacol Physiol Suppl 26: S14–S22.

Micieli G, Martignoni E, Cavallini A et al. (1987). Postprandial and orthostatic hypotension in Parkinson's disease. Neurology 37: 386–393.

Myllyla VV, Sotaniemi KA, Illi A et al. (1993). Effect of entacapone a COMT inhibitor on the pharmacokinetics of levodopa and on cardiovascular responses in patients with Parkinson's disease. Eur J Clin Pharmacol 45: 419–423.

Nakamura T, Hirayama M, Ito H et al. (2007). Dobutamine stress test unmasks cardiac sympathetic denervation in Parkinson's disease. J Neurol Sci 263: 133–138.

Nakamura T, Hirayama M, Hara T et al. (2011). Does cardiovascular autonomic dysfunction contribute to fatigue in Parkinson's disease? Mov Disord 26: 1869–1874.

Niimi Y, Ieda T, Hirayama M et al. (1999). Clinical and physiological characteristics of autonomic failure with Parkinson's disease. Clin Auton Res 9: 139–144.

Niimi Y, Iwase S, Fu Q et al. (2000). Effect of aging on muscle sympathetic nerve activity and peripheral venous pressure in humans. Environ Med 44: 56–59.

Nozaki S, Kang J, Miyai I et al. (1993). Postprandial hypotension in Parkinson's disease – the incidence and risk factor. Rinsho Shinkeigaku 33: 1135–1139.

Nylin G, Levander M (1948). Studies on the circulation with the aid of tagged erythrocytes in a case of orthostatic hypotension (asympathicotomic hypotension). Ann Intern Med 28: 723–746.

O'Mahony D, Bennett C, Green A et al. (2000). Reduced baroreflex sensitivity in elderly humans is not due to efferent autonomic dysfunction. Clin Sci (Lond) 98: 103–110.

O'Sullivan JD, Hughes AJ (1998). Apomorphine-induced penile erections in Parkinson's disease. Mov Disord 13: 536–539.

Ohmura M (2000). Loss of 123I-MIBG uptake by the heart in Parkinson's disease: assessment of cardiac sympathetic denervation and diagnostic value. J Nucl Med 41: 1594–1595.

Orimo S, Ozawa E, Nakade S et al. (1999). (123)I-metaiodobenzylguanidine myocardial scintigraphy in Parkinson's disease. J Neurol Neurosurg Psychiatry 67: 189–194.

Orimo S, Ozawa E, Oka T et al. (2001). Different histopathology accounting for a decrease in myocardial MIBG uptake in PD and MSA. Neurology 57: 1140–1141.

Orimo S, Kanazawa T, Nakamura A et al. (2007). Degeneration of cardiac sympathetic nerve can occur in multiple system atrophy. Acta Neuropathol 113: 81–86.

Pal PK, Calne DB, Calne S et al. (2000). Botulinum toxin A as treatment for drooling saliva in PD. Neurology 54: 244–247.

Papapetropoulos S, Paschalis C, Athanassiadou A et al. (2001). Clinical phenotype in patients with alpha-synuclein Parkinson's disease living in Greece in comparison with patients with sporadic Parkinson's disease. J Neurol Neurosurg Psychiatry 70: 662–665.

Parkinson J (1817). An Essay on the Shaking Palsy. Reprinted in James Parkinson (1755–1824), Macmillan, London (1955).

Perera R, Isola L, Kaufmann H (1994). Erythropoietin improves orthostatic hypotension in primary autonomic failure. Neurology 44 (Suppl 2): A363.

Perimenis P, Markou S, Gyftopoulos K et al. (2004). Efficacy of apomorphine and sildenafil in men with nonarteriogenic erectile dysfunction. A comparative crossover study. Andrologia 36: 106–110.

Potet F, Bouyssou T, Escande D et al. (2001). Gastrointestinal prokinetic drugs have different affinity for the human cardiac human ether-a-gogo K(+) channel. J Pharmacol Exp Ther 299: 1007–1012.

Proulx M, de Courval FP, Wiseman MA et al. (2005). Salivary production in Parkinson's disease. Mov Disord 20: 204–207.

Rajput AH, Rozdilsky B (1976). Dysautonomia in Parkinsonism: a clinicopathological study. J Neurol Neurosurg Psychiatry 39: 1092–1100.

Reinhardt MJ, Jungling FD, Krause TM et al. (2000). Scintigraphic differentiation between two forms of primary dysautonomia early after onset of autonomic dysfunction: value of cardiac and pulmonary iodine-123 MIBG uptake. Eur J Nucl Med 27: 595–600.

Rogers RC, Kita H, Butcher LL et al. (1980). Afferent projections to the dorsal motor nucleus of the vagus. Brain Res Bull 5: 365–373.

Rose S, Jenner P, Marsden CD (1988). The effect of carbidopa on plasma and muscle levels of L-dopa dopamine and their metabolites following L-dopa administration to rats. Mov Disord 3: 117–125.

Rudas L, Crossman AA, Morillo CA et al. (1999). Human sympathetic and vagal baroreflex responses to sequential nitroprusside and phenylephrine. Am J Physiol 276: H1691–H1698.

Saadia D, Voustianiouk A, Martinelli G et al. (2002). Hemodynamic effects of sildenafil in primary autonomic failure. Neurology 58 (Suppl 3): A88–A89, S14.006.

Sage JI, Mark MH (1995). Drenching sweats as an off phenomenon in Parkinson's disease: treatment and relation to plasma levodopa profile. Ann Neurol 37: 120–122.

Saito F, Tsuchiya K, Kotera M (1992). An autopsied case of Parkinson's disease manifesting Shy–Drager syndrome. Rinsho Shinkeigaku 32: 1238–1244.

Sandyk R, Awerbuch GI (1992). Dysautonomia in Parkinson's disease: relationship to motor disability. Int J Neurosci 64: 23–31.

Satoh A, Serita T, Tsujihata M (1997). Total defect of metaiodobenzylguanidine (MIBG) imaging on heart in Parkinson's disease: assessment of cardiac sympathetic denervation. Nippon Rinsho 55: 202–206.

Satoh A, Serita T, Seto M et al. (1999). Loss of 123I-MIBG uptake by the heart in Parkinson's disease: assessment of cardiac sympathetic denervation and diagnostic value. J Nucl Med 40: 371–375.

Seals DR, Monahan KD, Bell C et al. (2001). The aging cardiovascular system: changes in autonomic function at rest and in response to exercise. Int J Sport Nutr Exerc Metab 11 (Suppl): S189–S195.

Seif C, Herzog J, van der Horst C et al. (2004). Effect of subthalamic deep brain stimulation on the function of the urinary bladder. Ann Neurol 55: 118–120.

Senard JM, Valet P, Durrieu G et al. (1990). Adrenergic supersensitivity in parkinsonians with orthostatic hypotension. Eur J Clin Invest 20: 613–619.

Senard JM, Rascol O, Durrieu G et al. (1993). Effects of yohimbine on plasma catecholamine levels in orthostatic hypotension related to Parkinson disease or multiple system atrophy. Clin Neuropharmacol 16: 70–76.

Senard JM, Rai S, Lapeyre-Mestre M et al. (1997). Prevalence of orthostatic hypotension in Parkinson's disease. J Neurol Neurosurg Psychiatry 63: 584–589.

Sharabi Y, Li ST, Dendi R et al. (2003). Neurotransmitter specificity of sympathetic denervation in Parkinson's disease. Neurology 60: 1036–1039.

Shibao C, Gamboa A, Diedrich A et al. (2007). Acarbose an alpha-glucosidase inhibitor attenuates postprandial hypotension in autonomic failure. Hypertension 50: 54–61.

Shimada K, Kitazumi T, Sadakane N et al. (1985). Age-related changes of baroreflex function plasma norepinephrine and blood pressure. Hypertension 7: 113–117.

Shimada K, Kitazumi T, Ogura H et al. (1986). Effects of age and blood pressure on the cardiovascular responses to the Valsalva maneuver. J Am Geriatr Soc 34: 431–434.

Simpson DM, Kaufmann H, Sanders I et al. (1992). Laryngeal dystonia in multiple system atrophy. Muscle Nerve 15: 1213–1215.

Singaram C, Ashraf W, Gaumnitz EA et al. (1995). Dopaminergic defect of enteric nervous system in Parkinson's disease patients with chronic constipation. Lancet 346: 861–864.

Singer C, Weiner WJ, Sanchez-Ramos J et al. (1991). Sexual function in patients with Parkinson's disease. J Neurol Neurosurg Psychiatry 54: 942.

Singer C, Weiner WJ, Sanchez Ramos JR (1992). Autonomic dysfunction in men with Parkinson's disease. Eur Neurol 32: 134–140.

Soykan I, Sarosiek I, Shifflett J et al. (1997). Effect of chronic oral domperidone therapy on gastrointestinal symptoms and gastric emptying in patients with Parkinson's disease. Mov Disord 12: 952–957.

Stocchi F, Badiali D, Vacca L et al. (2000). Anorectal function in multiple system atrophy and Parkinson's disease. Mov Disord 15: 71–76.

Swinn L, Schrag A, Viswanathan R et al. (2003). Sweating dysfunction in Parkinson's disease. Mov Disord 18: 1459–1463.

Szili-Torok T, Kalman J, Paprika D et al. (2001). Depressed baroreflex sensitivity in patients with Alzheimer's and Parkinson's disease. Neurobiol Aging 22: 435–438.

Takatsu H, Nagashima K, Murase M et al. (2000a). Differentiating Parkinson disease from multiple-system atrophy by measuring cardiac iodine-123 metaiodobenzylguanidine accumulation. JAMA 284: 44–45.

Takatsu H, Nishida H, Matsuo H et al. (2000b). Cardiac sympathetic denervation from the early stage of Parkinson's disease: clinical and experimental studies with radiolabeled MIBG. J Nucl Med 41: 71–77.

Tanaka H, Davy KP, Seals DR (1999). Cardiopulmonary baroreflex inhibition of sympathetic nerve activity is preserved with age in healthy humans. J Physiol 515: 249–254.

Thaisetthawatkul P, Boeve BF, Benarroch EE et al. (2004). Autonomic dysfunction in dementia with Lewy bodies. Neurology 62: 1804–1809.

Tohgi H, Abe T, Yamazaki K et al. (1995). Effects of the catechol-*O*-methyltransferase inhibitor tolcapone in Parkinson's disease: correlations between concentrations of dopaminergic substances in the plasma and cerebrospinal fluid and clinical improvement. Neurosci Lett 192: 165–168.

Tranchant C, Guiraud-Chaumeil C, Echaniz-Laguna A et al. (2000). Is clonidine growth hormone stimulation a good test to differentiate multiple system atrophy from idiopathic Parkinson's disease? J Neurol 247: 853–856.

Turkka J, Suominen K, Tolonen U et al. (1997). Selegiline diminishes cardiovascular autonomic responses in Parkinson's disease. Neurology 48: 662–667.

Vagaonescu TD, Saadia D, Tuhrim S et al. (2000). Hypertensive cardiovascular damage in patients with primary autonomic failure. Lancet 355: 725–726.

van Dijk JG, Haan J, Zwinderman K et al. (1993). Autonomic nervous system dysfunction in Parkinson's disease: relationships with age medication duration and severity. J Neurol Neurosurg Psychiatry 56: 1090–1095.

Vatner DE, Lavallee M, Amano J et al. (1985). Mechanisms of supersensitivity to sympathomimetic amines in the chronically denervated heart of the conscious dog. Circ Res 57: 55–64.

Velseboer DC, de Haan RJ, Wieling W et al. (2011). Prevalence of orthostatic hypotension in Parkinson's disease: a systematic review and meta-analysis. Parkinsonism Relat Disord 17: 724–729.

Visser M, Marinus J, Stiggelbout AM et al. (2004). Assessment of autonomic dysfunction in Parkinson's disease: the SCOPA-AUT. Mov Disord 19: 1306–1312.

Wade DN, Mearrick PT, Morris JL (1973). Active transport of L-dopa in the intestine. Nature 242: 463–465.

Wakabayashi K, Takahashi H, Takeda S et al. (1988). Parkinson's disease: the presence of Lewy bodies in Auerbach's and Meissner's plexuses. Acta Neuropathol (Berl) 76: 217–221.

Wakabayashi K, Takahashi H, Ohama E et al. (1990). Parkinson's disease: an immunohistochemical study of Lewy body-containing neurons in the enteric nervous system. Acta Neuropathol (Berl) 79: 581–583.

Wakabayashi K, Yoshimoto M, Tsuji S et al. (1998). Alpha-synuclein immunoreactivity in glial cytoplasmic inclusions in multiple system atrophy. Neurosci Lett 249: 180–182.

Wang SJ, Fuh JL, Shan DE et al. (1993). Sympathetic skin response and R-R interval variation in Parkinson's disease. Mov Disord 8: 151–157.

Warner MR, Wisler PL, Hodges TD et al. (1993). Mechanisms of denervation supersensitivity in regionally denervated canine hearts. Am J Physiol 264: H815–H820.

Wenning GK, Scherfler C, Granata R et al. (1999). Time course of symptomatic orthostatic hypotension and urinary incontinence in patients with postmortem confirmed parkinsonian syndromes: a clinicopathological study. J Neurol Neurosurg Psychiatry 67: 620–623.

Wintzen AR, Badrising UA, Roos RA et al. (1994). Dysphagia in ambulant patients with Parkinson's disease: common not dangerous. Can J Neurol Sci 21: 53–56.

Wright RA, Kaufmann HC, Perera R et al. (1998). A double-blind dose-response study of midodrine in neurogenic orthostatic hypotension. Neurology 51: 120–124.

Wu YR, Chen CM, Ro LS et al. (1996). Vocal cord paralysis as an initial sign of multiple system atrophy in the central nervous system. J Formos Med Assoc 95: 804–806.

Yeh BK, McNay JL, Goldberg LI (1969). Attenuation of dopamine renal and mesenteric vasodilation by haloperidol: evidence for a specific dopamine receptor. J Pharmacol Exp Ther 168: 303–309.

Yelnik J (2002). Functional anatomy of the basal ganglia. Mov Disord 17 (Suppl 3): S15–S21.

Yoshida N, Omoya H, Kato S et al. (1993). Pharmacological effects of the new gastroprokinetic agent mosapride citrate and its metabolites in experimental animals. Arzneimittelforschung 43: 1078–1083.

Yoshimura N, Mizuta E, Yoshida O et al. (1998). Therapeutic effects of dopamine D1/D2 receptor agonists on detrusor hyperreflexia in 1-methyl-4-phenyl-1236-tetrahydropyridine-lesioned parkinsonian cynomolgus monkeys. J Pharmacol Exp Ther 286: 228–233.

Yoshimura N, Kuno S, Chancellor MB et al. (2003). Dopaminergic mechanisms underlying bladder hyperactivity in rats with a unilateral 6-hydroxydopamine (6-OHDA) lesion of the nigrostriatal pathway. Br J Pharmacol 139: 1425–1432.

Yoshita M, Hayashi M, Hirai S (1997). Iodine 123-labeled meta-iodobenzylguanidine myocardial scintigraphy in the cases of idiopathic Parkinson's disease multiple system atrophy and progressive supranuclear palsy. Rinsho Shinkeigaku 37: 476–482.

Yoshita M, Hayashi M, Hirai S (1998). Decreased myocardial accumulation of 123I-meta-iodobenzyl guanidine in Parkinson's disease. Nucl Med Commun 19: 137–142.

Yu M, Roane DM, Miner CR et al. (2004). Dimensions of sexual dysfunction in Parkinson disease. Am J Geriatr Psychiatry 12: 221–226.

Zesiewicz TA, Sullivan KL, Arnulf I et al. (2000). Sildenafil citrate (Viagra) for the treatment of erectile dysfunction in men with Parkinson's disease. Mov Disord 15: 305–308.

Ziegler MG, Lake CR, Kopin IJ (1977). The sympathetic-nervous-system defect in primary orthostatic hypotension. N Engl J Med 296: 293–297.

Handbook of Clinical Neurology, Vol. 117 (3rd series)
Autonomic Nervous System
R.M. Buijs and D.F. Swaab, Editors

Chapter 22

Diabetic autonomic neuropathy

AARON I. VINIK[1]* AND TOMRIS ERBAS[2]

[1]*Eastern Virginia Medical School, Strelitz Diabetes Center, Division of Endocrinology and Metabolism, Eastern Virginia Medical School, Norfolk, VA, USA*

[2]*Department of Endocrinology and Metabolism, Hacettepe University Medical School, Ankara, Turkey*

DIABETIC AUTONOMIC NEUROPATHY

Introduction

Diabetic autonomic neuropathy (DAN) is among the least recognized and least understood complications of diabetes, despite its significant negative impact on survival and quality of life in people with diabetes (Vinik and Erbas, 2001). It is also a major source of increased cost of caring for the diabetic patient. The metabolic disorders of diabetes lead to diffuse and widespread damage to peripheral and autonomic nerves and small vessels. When diabetic neuropathy affects the autonomic nervous system, it can damage the cardiovascular, gastrointestinal, genitourinary, and neurovascular systems and impair metabolic functions such as glucose counterregulation.

The autonomic nervous system (ANS) is primarily efferent, transmitting impulses from the central nervous system to peripheral organs. However, it also has an afferent component. Its two divisions, the parasympathetic and the sympathetic nervous systems, work in balanced opposition to control heart rate, force of cardiac contraction, dilatation and constriction of blood vessels, contraction and relaxation of smooth muscle in the digestive and urogenital systems, secretions of glands, and pupillary size. Diabetes can cause dysfunction of any or every part of the autonomic nervous system, leading to a wide range of disorders.

Pathogenesis

Hypotheses concerning the multiple etiologies of diabetic neuropathy include a metabolic insult to nerve fibers, neurovascular insufficiency, autoimmune damage, and neurohormonal growth factor deficiency.

Several different factors have been implicated in this pathogenic process. Hyperglycemic activation of the polyol pathway leading to accumulation of sorbitol and potential changes in the NAD:NADH ratio may cause direct neuronal damage and/or decreased nerve blood flow (Greene and Lattimer, 1983; Greene et al., 1988; Cameron and Cotter, 1997). Activation of protein kinase C pathways induces vasoconstriction and reduces neuronal blood flow. Increased oxidative stress, with increased free radical production, causes vascular endothelium damage, and reduces nitric oxide bioavailability. Alternately, excess nitric oxide production may result in formation of peroxynitrite and damage endothelium and neurons, a process referred to as nitrosative stress (Low et al., 1997; Veves and King, 2001; Hoeldtke et al., 2002). In a subpopulation of individuals with neuropathy, immune mechanisms may also be involved (Sundkvist et al., 1991). Reduction in neurotrophic growth factors, deficiency of essential fatty acids, and formation of advanced glycosylation end products (localized in endoneurial blood vessels) also result in reduced endoneurial blood flow and nerve hypoxia with altered nerve function (Brownlee, 1992; Horrobin, 1997; Pittenger et al., 1999; Obrosova, 2002). The result of this multifactorial process may be activation of polyADP ribosylation depletion of ATP, resulting in cell necrosis and activation of genes involved in neuronal damage (Pacher et al., 2002).

Prevalence

Reported prevalence of DAN varies, depending on whether studies have been carried out in the community, clinic, or tertiary referral center. This variance also reflects the type and number of tests performed and

*Correspondence to: Aaron I. Vinik, MD, PhD, FCP, MACP, Eastern Virginia Medical School, Strelitz Diabetes Center, 855 West Brambleton Avenue, Norfolk, VA 23510, USA. E-mail: vinikai@evms.edu

the presence or absence of signs and symptoms of autonomic neuropathy. Other factors that account for the marked variability in reported prevalence rates include the lack of a standard accepted definition of DAN, different diagnostic methods, variable study selection criteria, and referral bias.

For example, in a community-based population study of diabetic neuropathy in Oxford, England, the prevalence of autonomic neuropathy, as defined by one or more abnormal heart rate variability (HRV) test results, was 16.7% (Neil et al., 1989).

In a further study, Ziegler et al. evaluated the prevalence of cardiovascular autonomic neuropathy (CAN) in 1171 diabetic patients randomly recruited from 22 diabetes centers in Germany, Austria, and Switzerland. The study found that 25.3% of patients with type 1 diabetes and 34.3% of patients with type 2 diabetes had abnormal findings in more than two of six autonomic function tests. If more strict criteria were used (i.e., abnormalities present in at least three of six autonomic function tests), the prevalence of CAN was 16.8% for individuals with type 1 diabetes and 22.1% for individuals with type 2 diabetes (Ziegler et al., 1992). Another study group observed nearly an identical prevalence rate (16.6%) for individuals with type 1 diabetes (O'Brien et al., 1986).

Clinical manifestations

Clinical manifestations of autonomic dysfunction and other microvascular complications frequently occur concurrently but in inconsistent patterns. The ubiquitous distribution of the ANS renders virtually all organs susceptible to autonomic dysfunction. Therefore, a patient diagnosed with diabetes should be suspected of having at least subclinical disturbances of the ANS. Clinical symptoms of autonomic neuropathy generally do not occur until long after the onset of diabetes. However, subclinical autonomic dysfunction can occur within a year of diagnosis in type 2 diabetic patients and within 2 years in type 1 diabetic patients (Pfeifer et al., 1984).

The most important etiological factors that have been associated with autonomic neuropathy are poor glycemic control, diabetes duration, age, female sex, and higher body mass index. Of patients with symptomatic autonomic dysfunction, 25–50% die within 5–10 years of diagnosis (Ewing et al., 1991; Rathmann et al., 1993).

The 5 year mortality rate in patients with diabetic autonomic neuropathy is three times higher than in diabetic patients without autonomic involvement (O'Brien et al., 1991). Leading causes of death in diabetic patients with either symptomatic or asymptomatic autonomic neuropathy are heart disease and nephropathy. Increased urinary albumin excretion is related to autonomic neuropathy in diabetic patients. Impairments in the autonomic nervous system may also contribute to the pathogenesis of diabetic nephropathy and cardiovascular disease. Autonomic neuropathy is also an independent risk factor for stroke.

CARDIOVASCULAR AUTONOMIC NEUROPATHY

Perhaps one of the most overlooked of all serious complications of diabetes is cardiovascular autonomic neuropathy (CAN) (Maser et al., 2000). CAN results from damage to the autonomic nerve fibers that innervate the heart and blood vessels and causes abnormalities in heart rate control and vascular dynamics.

CAN has been linked to resting tachycardia, postural hypotension, exercise intolerance, enhanced intraoperative or perioperative cardiovascular lability, increased incidence of asymptomatic ischemia, myocardial infarction, and decreased rate of survival after myocardial infarction. Autonomic dysfunction can affect daily activities of individuals with diabetes and may invoke potentially life-threatening outcomes. Advances in technology, built on decades of research and clinical testing, now make it possible to objectively identify early stages of CAN with the use of careful measurement of autonomic function (Levitas and Kristal, 1972; Margolis et al., 1973; Storstein and Jervell, 1979; Burgos et al., 1989; Roy et al., 1989; Langer et al., 1991; Purewal and Watkins, 1995). Little information exists as to the frequency of CAN in representative diabetic populations. Fifteen studies using different end points reported prevalence rates of between 1% and 90% (Maser et al., 2003). The Diabetes Control and Complications Trial (DCCT) found that 1.65% had abnormal heart rate variability (HRV) at baseline for less than 5 years duration of diabetes. The rate rose to 6.2% among those with less than 9 but more than 5 years duration of diabetes, and by 9 years the prevalence increased to 12.2% (Diabetes Control and Complications Trial Research Group, 1998).

The heterogenous methodology makes it difficult to compare epidemiology across different studies. CAN may be present at diagnosis, and prevalence increases with age, duration of diabetes, and poor glycemic control. CAN also cosegregates with distal symmetric polyneuropathy, microangiopathy, and macroangiopathy. Age, diabetes, obesity, and smoking are risk factors for reduced HRV in type 2 diabetes. Thus, there may be selectivity and sex-related differences among the various cardiovascular risk factors as to their influence on autonomic dysfunction. A1c, hypertension, distal symmetrical polyneuropathy, retinopathy, and exposure to hyperglycemia were shown to be risk factors for developing CAN in type 1 diabetes (Ziegler et al., 2004, 2006; Witte et al., 2005).

Cardiovascular symptoms and signs

Resting tachycardia

Resting tachycardia and a fixed heart rate are characteristic late findings in diabetic patients with vagal impairment. Resting heart rates of 90–100 beats per minute and occasional heart rate increments up to 130 beats per minute occur. The highest resting heart rates have been found in patients with parasympathetic damage, occurring earlier in the course of CAN than sympathetic nerve function; in those with evidence for combined vagal and sympathetic involvement, the rate returns toward normal but remains elevated. A fixed heart rate that is unresponsive to moderate exercise, stress, or sleep indicates almost complete cardiac denervation (Ewing and Clarke, 1986).

Exercise intolerance

Autonomic dysfunction impairs exercise tolerance, reduces response in heart rate and blood pressure (BP), and blunts increases in cardiac output in response to exercise. Diabetic patients who are likely to have CAN should be tested for cardiac stress before undertaking an exercise program. Patients with CAN need to rely on their perceived exertion, not heart rate, to avoid hazardous levels of intensity of exercise (Vinik and Erbas, 2002; Vinik et al., 2003a; Colberg et al., 2003; Albers et al., 2006).

Intraoperative cardiovascular instability

Hemodynamic changes occur during surgery for individuals without diabetes. Perioperative cardiovascular morbidity and mortality are increased two- to threefold in patients with diabetes. Compared with nondiabetic subjects, diabetic patients undergoing general anesthesia may experience a greater degree of decline in heart rate and BP during induction of anesthesia and less of an increase after tracheal intubation and extubation. The normal autonomic response of vasoconstriction and tachycardia did not completely compensate for the vasodilating effects of anesthesia. There is an association between CAN and more severe intraoperative hypothermia that results in decreased drug metabolism and impaired wound healing. Preoperative cardiovascular autonomic screening may provide useful information for anesthesiologists planning the anesthetic management of diabetic patients and identify those at greater risk for intraoperative complications.

Orthostatic hypotension

Orthostatic hypotension is defined as a fall in blood pressure (i.e., >20 mmHg for systolic or >10 mmHg for diastolic blood pressure) in response to postural change, from supine to standing. In patients with diabetes, orthostatic hypotension is usually due to damage to the efferent sympathetic vasomotor fibers, particularly in the splanchnic vasculature. In addition, there is a decrease in cutaneous, splanchnic, and total vascular resistance that occurs in the pathogenesis of this disorder.

Normally, in response to postural change there is an increase in plasma norepinephrine. For individuals with orthostatic hypotension, there may be a reduction in this response relative to the fall in blood pressure. Diminished cardiac acceleration and cardiac output, particularly in association with exercise, may also be important in the presentation of this disorder. Less frequently, there is a rise in norepinephrine that may be due to low blood volume or reduced red cell mass. Frequently, there are fluctuations in the degree of orthostatic hypotension. This may reflect postprandial blood pooling, the hypotensive role of insulin, and changing patterns of fluid retention due to renal failure or congestive heart failure.

Patients with orthostatic hypotension typically present with lightheadedness and presyncopal symptoms. Symptoms such as dizziness, weakness, fatigue, visual blurring, and neck pain also may be due to orthostatic hypotension. Many patients, however, remain asymptomatic despite significant falls in blood pressure. If the cause of orthostatic hypotension is CAN, treatment goals should not only consist of therapies to increase the standing blood pressure, balanced against preventing hypertension in the supine position, but should also provide education to patients so that they avoid situations (e.g., vasodilatation from hot showers) that result in the creation of symptoms (i.e., syncopal episodes). Such symptoms can result in injuries from falling (Page and Watkins, 1976; Hilsted et al., 1981; Winocour et al., 1986; Mathias et al., 1987; Freeman, 1998, 1999).

Symptoms compatible with orthostasis may occur on changes from a supine to an erect position and may be caused by postural tachycardia syndrome (POTS), inappropriate sinus tachycardia, neurocardiogenic syncope, or abnormalities in baroreceptor function. The hallmark of these abnormalities is the absence of a fall in BP with standing, but a tachycardia or bradycardia with the change in posture. The pathogenesis of POTS is obscure. Some patients have defective peripheral vasoconstriction and an increase in calf blood flow, whereas others have increased peripheral arterial resistance and decreased blood flow. POTS is associated with a selective defect in intraepidermal nerve fiber in the skin. Norepinephrine concentrations have been significantly related to the estimate of the severity of autonomic neuropathy, and loss of peripheral sympathetic C fiber tone seems to translate to inadequate cardiac venous return with thoracic hypovolemia. POTS patients have paradoxically unchanged plasma renin activity and low aldosterone, given their marked reduction in plasma volume. POTS patients have exaggerated muscle sympathetic nerve activity with baroreceptor-reflex challenges (Stewart et al., 2002, 2003; Singer et al., 2004; Stewart and Montgomery, 2004; Muenter et al., 2005; Vinik and Ziegler, 2007).

Silent myocardial ischemia/cardiac denervation syndrome

The presence of both symptomatic and asymptomatic coronary artery disease (CAD) is increased in diabetic patients and subclinical neuropathy is an important cause of silent ischemia in patients with diabetes (Airaksinen and Koistinen, 1992; Marchant et al., 1993).

Features of myocardial infarction (MI) in patients with CAN are silence, cough, nausea and vomiting, dyspnea, tiredness, and ECG changes. Reduced appreciation for ischemic pain can impair timely recognition of myocardial ischemia or infarction, thereby delaying appropriate therapy. In the ECGs of diabetic patients with exertional chest pain, a prolonged anginal perceptual threshold (i.e., the time from onset of 0.1 mV ST depression to the onset of angina pectoris during exercise) was associated with the presence of CAN (Ambepityia et al., 1990).

Hence, patients with CAN and coronary artery disease are jeopardized because the longer threshold permits them to continue exercising despite increasing ischemia. Silent ischemia in diabetic patients may result either from CAN, or from autonomic dysfunction attributable to coronary artery disease itself, or from both. In the Framingham Study, the rates of unrecognized MI were 39% in diabetic patients and 22% in nondiabetic subjects, but the difference was not significant (Margolis et al., 1973).

In a survey from the National Registry of Myocardial Infarction 2 (NRMI-2), of 434 877 patients presenting with MI, 33% did not have chest pain. Among those presenting without chest pain, 32% had diabetes versus 25.4% in the group with chest pain (Canto et al., 2000).

The mechanisms of painless myocardial ischemia are, however, complex and not fully understood. Altered pain thresholds, subthreshold ischemia not sufficient to induce pain, and dysfunction of the afferent cardiac autonomic nerve fibers have all been suggested as possible mechanisms (Shakespeare et al., 1994).

Positron emission tomography to measure regional cerebral blood flow as an index of regional neuronal activation has shown that impaired afferent signaling resulting from autonomic dysfunction is associated with failed signal transmission from the thalamus to the frontal cortex (Rosen and Camici, 2000).

In the Detection of Ischemia in Asymptomatic Diabetics (DIAD) study of 1123 patients with type 2 diabetes, cardiac autonomic dysfunction was a strong predictor of ischemia (Wackers et al., 2004). Thus, patients with CAN warrant more careful attention, and cardiovascular autonomic function testing may be an important component in the risk assessment of diabetic patients with coronary artery disease.

Increased risk of mortality

The mechanisms by which CAN leads to increased mortality remain obscure. A number of studies have shown a 2.3-fold increased risk of CAN in diabetic patients showing a prolonged QT interval, leading to speculation that CAN might also predispose to malignant ventricular arrhythmias and to sudden death from cardiac arrest caused by torsades de pointes, as in long QT syndrome. It is difficult to determine the independent effects of CAN on mortality because of the coexistence of cardiovascular disease (CVD). Symptomatic patients with CAN have increased mortality from renal failure and a variety of other causes. However, several studies showing increased mortality from CAN have excluded CHD. If duration, renal disease, hypertension, and CVD were controlled for, the relative risk would decrease from 4.03 to 1.39, but it would remain significant (Orchard et al., 1996).

Furthermore, symptomatic CAN at 5 years of diabetes predicted mortality at 10 years, even after adjusting for conventional CVD risk factors. CAN remained a significant predictor of death after Cox regression adjustment for perfusion defects (Lee et al., 2003). Stepwise Cox regression with backward selection for smoking, sex, age, CVD, cholesterol, triglyceride, albumin excretion rate, systolic blood pressure, glycemic control, and HRV still gave a hazard ratio of 6.4 for CAN (Astrup et al., 2006). Finally, metaiodobenzylguanidine (MIBG) scan and CAN were independent predictors, according to multivariate analysis, for a cardiac event, suggesting that maldistribution of sympathetic innervention may contribute (Nagamachi et al., 2006).

HRV was found to be an independent predictor of all-cause mortality during a period of 9 years, in a population-based study using Cox proportional hazard models including dyslipidemia, diabetes, age, physical activity, alcohol intake, smoking, and CVD as independent variables (Wirta et al., 1997).

Moreover, the Hoorn study by Gerritsen et al. demonstrated that impaired autonomic function is associated with increased all-cause and cardiovascular mortality, and that CAN in patients already at risk (diabetes, hypertension, or history of CVD) may be especially hazardous (Gerritsen et al., 2001).

Results from the Action to Control Cardiovascular Risk in Diabetes (ACCORD) trial again confirmed the association of CAN and mortality. These investigators showed that the individuals in this trial with baseline CAN were between 1.55 and 2.14 times as likely to die as individuals without CAN. The investigators did not find that the presence of CAN appeared to contribute to the increased mortality observed in the intensive versus standard glycemic therapy group (Pop-Busui et al., 2010).

Increased mortality after myocardial infarction

Mortality rates after an MI are higher for diabetic compared with nondiabetic patients (Vinik and Ziegler, 2007). A simple bedside test that measured 1 minute HRV during deep breathing was a good predictor of all-cause mortality for 185 patients (17.8% with diabetes) after a first MI (Stewart et al., 2003). Autonomic function testing is a valuable tool in identifying a subgroup of post-MI patients who are at high risk for death (Miettinen et al., 1998; Katz et al., 1999).

Cardiovascular autonomic neuropathy and sudden death

Sudden, unexpected deaths occur among subjects with CAN. One potential cause may be severe but asymptomatic ischemia, which can induce lethal arrhythmias. QT prolongation may also predispose individuals to life-threatening cardiac arrhythmias and sudden death. Results from the European Diabetes Insulin-Dependent Diabetes Mellitus (IDDM) Complications Study showed that male patients with impaired HRV had a higher corrected QT prolongation than males without this complication (Veglio et al., 1999).

The question of more intensive therapy and reduction in cardiovascular complications was addressed for people with type 2 diabetes of long duration in three other studies (i.e., the Action to Control Cardiovascular Risk in Diabetes (ACCORD) trial (Buse et al., 2007), the Action in Diabetes and Vascular Disease: Preterax and Diamicron Modified Release Controlled Evaluation (ADVANCE) trial (Patel et al., 2008), and the Veterans Affairs Diabetes Trial (VADT) (Duckworth et al., 2009)). Although the three studies utilized different patient cohorts, with varying durations of diabetes (ACCORD, 10 years; VADT, 11.5 years; ADVANCE, 8 years) and had different treatment regimens, the results of these trials indicated that intensive glucose control did not reduce CVD events. In fact, the ACCORD trial was terminated early because an increased rate of mortality was found to be associated with intensive control of hyperglycemia (Gerstein et al., 2008). A higher rate of mortality for those on intensive treatment was not, however, found in ADVANCE and VADT. Recent articles from the ACCORD trial, by Pop-Busui et al. (2010) and Calles-Escandon et al. (2010), examined whether the effects of cardiovascular autonomic neuropathy (CAN) or self-reported history of neuropathy at baseline could have been a contributor to the higher mortality risk in the intensive glycemic arm.

Physiological activities of the cardiovascular system are under the control of the autonomic nervous system. Damage to the autonomic nerves that innervate the heart and blood vessels results in dysfunction in heart rate control and vascular dynamics (Vinik and Ziegler, 2007). Autonomic imbalance between the sympathetic and parasympathetic nervous systems' regulation of cardiovascular function contributes to metabolic abnormalities (Licht et al., 2010) and significant morbidity and mortality for individuals with diabetes (Maser et al., 2003; Vinik et al., 2003b; Ziegler et al., 2008). Clinical manifestations of cardiovascular autonomic dysfunction (e.g., exercise intolerance, intraoperative cardiovascular liability, orthostatic tachycardia and bradycardia syndromes, silent myocardial ischemia) can result in life-threatening outcomes (Maser et al., 2003; Vinik et al., 2003b). In fact, the ultimate outcome of increased risk of mortality is clearly associated with the presence of autonomic dysfunction (Maser et al., 2003). Results from the ACCORD trial again confirmed the association of CAN and mortality. These investigators showed that the individuals in this trial with baseline CAN were between 1.55 and 2.14 times as likely to die as individuals without CAN (Pop-Busui et al., 2010). Furthermore, CAN in the presence of peripheral neuropathy was the highest predictor of CVD mortality (i.e., hazard ratio (HR) 2.95, $p = 0.008$). Indeed, combining indexes of autonomic dysfunction have been shown to be associated with the risk of mortality (Maser et al., 2003; Lykke et al., 2008; Ziegler et al., 2008). Results from the Detection of Ischemia in Asymptomatic Diabetics (DIAD) study showed that the presence of cardiac autonomic dysfunction (defined via change in heart rate from lying to standing, a measure reflecting parasympathetic dysfunction) was among the highest HR (4.33) associated with the primary events (i.e., cardiac death or nonfatal MI) (Young et al., 2009). Just as the number and type of assessment modalities used to identify the presence of CAN affects prevalence rates and the association with mortality (Maser et al., 2003), the methods in the ACCORD trial used to assess CAN may have affected the ability to determine an association with intensive glucose therapy and mortality (Vinik et al., 2010).

The pathogenesis of autonomic imbalance between the sympathetic and parasympathetic nervous systems' regulation of cardiovascular function has not been established. Obesity and the overproduction of inflammatory cytokines have been implicated in the induction of numerous pathogenic mechanisms that can be responsible for autonomic imbalance impacting deleterious processes that increase and perpetuate cardiovascular risk. In addition, hyperleptinemia and a deficiency of adiponectin favor sympathetic overactivity and the inflammatory cascade is enhanced in the presence of sleep apnea, a frequent partner in diabetes and the metabolic syndrome. Restoration of autonomic balance is possible and has been shown with therapeutic lifestyle changes, increased physical activity, β-adrenergic blockers, aldose reductase inhibitors, ACE inhibitors, angiotensin-receptor blockers, potent antioxidants such as α-lipoic acid, and in animal models using inhibitors of peroxynitrite formation and its decomposition. There are exciting new prospects for pathogenesis-oriented intervention (Vinik et al., 2011).

Imaging of myocardial sympathetic innervation with various radiotracers (e.g., MIBG) has shown that predisposition to arrhythmias and an association with mortality may also be related to intracardiac sympathetic imbalance (Kahn et al., 1988; Stevens et al., 1998a).

The significance of CAN as an independent cause of sudden death has, however, been questioned recently. In the Rochester Diabetic Neuropathy Study, the investigators found that all cases of sudden death in individuals with and without diabetes had severe coronary artery disease or left ventricular (LV) dysfunction (Suarez et al., 2005).

They suggested that although CAN could be a contributing factor, it was not a significant independent cause of sudden death. Heart failure is, however, common in individuals with diabetes; it is identified in these patients by the presence of neuropathy, even in those without evidence of coronary artery disease or LV dysfunction.

Autonomic cardiopathy

CAN may be associated with abnormalities in LV systolic and particularly diastolic function in the absence of cardiac disease in diabetic patients. Echocardiographic studies have shown a significant correlation of the severity of CAN with reduced peak diastolic filling rate and with an augmented atrial contribution to diastolic filling as assessed by Doppler echocardiography. It is difficult to judge, however, whether CAN is an independent contributor to these abnormalities, because other factors such as interstitial myocardial fibrosis and microangiopathic or metabolic changes may also be responsible for LV dysfunction.

CAN is associated with LV diastolic dysfunction (LVDD) at rest in patients with long-term type 2 or type 1 diabetes. LVDD may progress to heart failure, mainly with preserved LV systolic function (diastolic heart failure), which is also related to high morbidity and mortality rates.

The pathophysiology of LVDD includes delayed relaxation, impaired LV filling, and/or increased stiffness. In patients with CAN, vagal impairment can lead to a relative predominance of sympathetic activity in the sympathovagal balance. Sympathetic overactivity stimulates the renin–angiotensin–aldosterone system and increases heart rate, stroke volume, and peripheral vascular resistance, thus contributing to LV dysfunction. The latter is promoted by the deleterious effects of the renin–angiotensin—aldosterone and adrenergic systems on systemic and coronary hemodynamics, myocyte hypertrophy and fibroblast growth, and myocyte necrosis and apoptosis.

Such a sympathetic hyperactivity, in combination with regional myocardial sympathetic denervation, has been shown recently to lead to diminished coronary blood flow reserve and diastolic dysfunction in diabetic patients with early microangiopathy. Analogous to the situation with neurohormones, the overexpression of cytokines is sufficient to contribute to LV dysfunction and, eventually, to heart failure.

Recently, Rennings et al. showed that glitazones increase vascular leakage and render the patient susceptible to the development of edema in insulin-treated type 2 diabetic patients with autonomic neuropathy. For this reason, autonomic neuropathy should be considered as a risk factor for glitazone-induced edema (Rennings et al., 2010).

Diagnosis of cardiovascular autonomic neuropathy

There are simple bedside tests to diagnose CAN using HRV, responses to breathing, the Valsalva maneuver, and standing (Table 22.1). Functional abnormalities and imbalance between the sympathetic and parasympathetic nervous systems are discerned with respiratory modulation of different-frequency oscillations in HRV.

The power spectrum reflects the amplitude of the heart rate fluctuations present at different oscillation frequencies. The power spectrum of HRV has been shown to consist of three major peaks:

1. very-low-frequency component (below 0.04 Hz), which is related to fluctuations in vasomotor tone associated with thermoregulation
2. low-frequency (LF) component (around 0.1 Hz), which represents the so-called 10 s rhythm (Mayer waves) associated with the baroreceptor reflex
3. high-frequency (HF) component (around 0.25 Hz), which is related to respiratory activity.

The very-low-frequency heart rate fluctuations are thought to be mediated primarily by the sympathetic system, and the LF fluctuations are predominantly under

Table 22.1

Diagnostic assessment of cardiovascular autonomic function (Vinik and Ziegler, 2007)

Parasympathetic	Sympathetic
Resting heart rate	Resting heart rate
Beat-to-beat variation with deep breathing (E:1 ratio)	Spectral analysis of heart rate variation, very-low-frequency power (VLFP; 0.003–0.04 Hz)
30:15 Heart rate ratio with standing	Orthostasis blood pressure
Valsalva ratio	Hand grip blood pressure
Spectral analysis of heart rate variation, high-frequency	Cold pressor response power (HFP; 0.15–0.40 Hz)
	Sympathetic skin galvanic response (cholinergic)
	Sudorometry (cholinergic)
	Cutaneous blood flow (peptidergic)

sympathetic control with vagal modulation, whereas the HF fluctuations are under parasympathetic control. Because spectral analysis is carried out under resting conditions, it has the advantage that active cooperation of the patient is not required.

Numerous factors may influence the test results: age, heart rate, respiratory rate, BP, eating, drinking coffee, smoking, body position, volume status, mental stress, drugs, exercise, and time of day.

Measurement of 24 hour HRV may be more sensitive and reliable in detecting CAN than single tests. Additionally, 24 hour recording of HRV may provide insights into abnormal patterns of circadian rhythms modulated by sympathovagal activity. Diabetic patients with autonomic neuropathy display an impairment in absolute values of both HF and LF oscillations. Blunting of nocturnal increases in the HF component, however, which expresses vagal modulation of the heart, seems to be the earliest and most prominent event. This leads to a relative predominance of sympathetic activity during the night.

Simultaneous beat-to-beat measurement of R-R intervals and BP is useful to detect spontaneous baroreceptor–cardiac reflex sensitivity and to evaluate the relationship between spontaneous changes in BP and R-R interval in the time domain (sequence method) and in the frequency domain (cross-spectral method). These estimates can be obtained under conditions suitable for routine outpatient evaluation.

Cardiac radionuclide imaging

Radionuclide techniques for cardiac mapping have recently been used to directly quantify cardiac sympathetic innervation in CAN. The nonmetabolized norepinephrine analog MIBG participates in norepinephrine uptake in postganglionic sympathetic neurons. Several studies have demonstrated decreased myocardial MIBG uptake in patients with CAN as assessed by autonomic reflex tests (Stevens et al., 1998a, b; Ziegler et al., 1998).

There is evidence to suggest that scintigraphic assessment using MIBG and single-photon emission computed tomography is more sensitive in detecting CAN than indirect autonomic reflex testing, because MIBG uptake is reduced in patients with normal autonomic tests. The norepinephrine analog [^{11}C] hydroxyephedrine (HED) has also been employed to examine cardiac innervation defects. In diabetic patients, attenuated HED retention is related to the severity of CAN and is most pronounced in the inferior, apical, and lateral segments. In severe CAN, the myocardial retention of HED was remarkably heterogenous, and as the extent of distal deficits increased, HED retention became paradoxically increased in the proximal myocardial segments, which showed the highest deficits in coronary blood flow reserve (Stevens et al., 1999).

Thus, cardiac radionuclide imaging provides a unique, sensitive tool for direct assessment of the pathophysiology and progression of early sympathetic innervation defects not accessible by indirect autonomic function testing. This asymmetry of innervation is thought to be responsible for the predisposition to arrhythmias, and it provides a tool for the determination of the success of intervention on reinnervation of the myocardium.

Treatment of orthostatic hypotension

Treatment of orthostatic hypotension comprises nonpharmacological and pharmacological measures. Nonpharmacological measures such as increasing consumption of water and the use of lower-extremity stockings to reduce symptoms (e.g., dizziness, dyspnea) should be encouraged when treating orthostatic hypotension attributable to autonomic dysfunction. Recently, some novel approaches using other pharmacological agents have been investigated in nondiabetic individuals with orthostatic symptoms. Enhancement of ganglionic transmission via the use of pyridostigmine (inhibitor of acetylcholinesterase) improved symptoms and orthostatic BP with only modest effects in supine BP for patients with POTS. Similarly, the use of β-adrenergic blockers may benefit the tachycardia, and anticholinergics, the orthostatic bradycardia. Pyridostigmine has also been shown to improve HRV in healthy young adults. In patients with pooling of blood in the splanchnic bed, somatostatin may be of value, and in patients with contracted plasma volume, treatment with erythropoietin is recommended.

Potential for reversal of CAN

Several agents have become available for the correction of functional defects in the autonomic nervous system. In patients with minimal abnormalities, endurance training under strict supervision and lifestyle intervention associated with weight loss improve HRV. The Diabetes Complications and Control Trial (DCCT) group has shown without question the importance of attention to intensive insulin therapy in preventing the onset and progression of diabetic complications (Table 22.2). The DCCT demonstrated that autonomic dysfunction was reduced by 53% in type 1 diabetic patients with very intensive glycemic control (Diabetes Control and Complications Trial Research Group, 1995).

The EDIC (Epidemiology of Diabetes Intervention and Complication study) is a longitudinal cohort follow-up study for the DCCT, in which patients with type 1 diabetes were randomized to conventional or intensive glycemic control. The EDIC study has shown persistent beneficial effects of past glucose control on microvascular complications, despite the loss of glycemic separation in type 1 diabetic patients (DCCT EDIC Research Group, 2003).

Table 22.2

Progression of cardiac autonomic dysfunction (Ziegler, 1994; Vinik and Ziegler, 2007)

- It can be detected at the time of diagnosis
- Neither age nor type of diabetes is a limiting factor in its emergence; it has been found both in young people with newly diagnosed type 1 diabetes and in older people newly diagnosed with type 2 diabetes
- Poor glycemic control plays a central role in development and progression
- Intensive therapy can slow the progression and delay the appearance of abnormal autonomic function tests
- Subclinical autonomic neuropathy can be detected early using autonomic function tests
- Autonomic features that are associated with sympathetic nervous system dysfunction (e.g., orthostatic hypotension) are relatively late complications of diabetes
- There is an association between CAN and diabetic nephropathy that contributes to high mortality rates
- Some individuals with symptoms associated with autonomic neuropathy die suddenly and unexpectedly
- Clinical signs and symptoms of autonomic dysfunction do not always progress. This underscores the need for performance of quantitative autonomic function tests to identify individuals at risk for premature death
- Type 1 and type 2 diabetes may have different progression paths
- The relationship between autonomic damage and the duration of diabetes is not clear, although numerous studies support an association
- Prevalence and mortality rates may be higher among individuals with type 2 diabetes, possibly because of the longer duration of glycemic abnormalities before diagnosis

Pop-Busui et al. (2009) showed that during EDIC follow-up, CAN progressed in both treatment groups, but the incidence and prevalence of CAN remained lower in the former intensive group than in the former conventional group, despite similar levels of glycemic control in the EDIC study. To minimize the development of autonomic neuropathy, intensive glucose control of type 1 diabetes should be started as early as possible (Pop-Busui et al., 2009). On the other hand, glycemic control with a reduction of HbA1c from 9.5 to 8.4 has also been shown to improve HRV with mild autonomic abnormalities; this was not so in type 1 diabetics of advanced autonomic abnormalities (Burger et al., 1999).

The effects of glycemic control in type 2 diabetics are less definite. The VA Cooperative Study showed no difference in the prevalence of autonomic neuropathy after 2 years of intensive glycemic control in type 2 diabetic patients (Azad et al., 1999). In the Steno-2 Study, where people with type 2 diabetes received intensive multifactorial treatment (e.g., targeting hyperglycemia, hypertension, dyslipidemia, and microalbuminuria, along with secondary prevention of CVD with aspirin), the approach reduced autonomic dysfunction by 63%. In terms of the multifactorial approach, the glucose-lowering agents appeared to have the least effect when compared with antihypertensive treatment, lipid-lowering agents, aspirin, and vitamin-mineral supplements (Gaede et al., 1999). In addition, a survey of evidence from clinical trials shows that early identification of autonomic neuropathy permits timely initiation of therapy with antioxidant α-lipoic acid, which slows or reverses progression of CAN (Ziegler et al., 1997).

Aldose reductase inhibitors reduce the flux of glucose through the polyol pathway, inhibiting tissue accumulation of sorbitol and fructose. The use of aldose reductase inhibitors such as sorbinil improved resting and maximum cardiac output, and epalrestat improved MIBG uptake and HRV in patients with mild abnormalities but not in those with advanced CAN.

β-Blockers such as bisoprolol improved HRV in heart failure. The addition of spironolactone to enalapril, furosemide, and digoxin in patients with heart failure improved sympathovagal balance. Angiotensin-converting enzyme (ACE) inhibition with quinapril increases total HRV and improves the parasympathetic/sympathetic balance in patients with mild but not advanced autonomic neuropathy.

To investigate the effect of ACE inhibition or angiotensin-receptor blockade and their combination on both DAN and LVDD in asymptomatic patients with diabetes, Didangelos et al. examined patients with long-term diabetes mellitus and DAN. Early ACE inhibition or angiotensin-receptor blockade improved both DAN and LVDD after 1 year of treatment in asymptomatic patients with diabetes mellitus. The combination may be slightly better than monotherapies on DAN and LVDD, auguring well for the patient with established CAN (Didangelos et al., 2006).

Matsuki et al. showed that fluvastatin improves cardiac sympathetic neuropathy in the diabetic rat heart in association with attenuation of increased cardiac oxidative stress (Matsuki et al., 2010). On the other hand, Kellogg et al. showed that in experimental diabetes, selective COX-2 inactivation confers protection against sympathetic denervation by reducing intramyocardial oxidative stress and inflammation (Kellogg et al., 2009). Therefore, statins and COX-2 inactivation might help to attenuate cardiac sympathetic neural dysfunction.

GASTROINTESTINAL AUTONOMIC NEUROPATHY

Gastrointestinal disturbances caused by autonomic neuropathy are common and are often a disabling complication of diabetes. However, gastrointestinal symptoms are frequently overlooked and untreated. Autonomic neuropathy can affect every part of the gastrointestinal tract, i.e., the esophagus, stomach, small intestine, and colon. Thus, the gastrointestinal manifestations are quite variable and include dysphagia, abdominal pain, nausea, vomiting, malabsorption, fecal incontinence, diarrhea, and constipation. Diabetic patients may present with a spectrum of manifestations from mild gastrointestinal symptoms to severe clinical disease. Gastrointestinal tract complications of diabetes appear to be more common in patients with longstanding disease and poorly controlled blood glucose levels.

Esophageal dysfunction

Despite the frequent motility abnormalities described, symptoms of esophageal disease are rather uncommon in diabetics and only approximately 30% experience symptoms. Motor abnormalities include impairment of peristaltic activity with double peak and tertiary contractions or impaired peristalsis and diminished lower esophageal sphincteric pressures. These factors may further predispose to gastroesophageal reflux disease, particularly in the setting of impaired gastric emptying. Esophageal dysfunction is detectable through esophageal motility test and esophageal scintigraphy in diabetic patients (Sundkvist et al., 1989).

Gastroparesis diabeticorum

Gastroparesis diabeticorum is usually clinical silent, although severe diabetic gastroparesis is one of the most debilitating of all the gastrointestinal complications of diabetes. It has been demonstrated that gastric emptying is delayed in 30–50% of longstanding type 1 and type 2 diabetic patients (Ma et al., 2009).

DAN can impair gastric acid secretion and gastrointestinal motility. Gastric dysrhythmias known as functional pathology are defined as bradygastria and tachygastria. Gastric dysrhythmias interfere with the normal gastric peristaltic contractions. Loss of vagal tone and increased sympathetic nervous system activity have been associated with gastric dysrhythmias. The typical symptoms of diabetic gastroparesis are early satiety, nausea, vomiting, postprandial fullness, abdominal bloating, epigastric pain, and anorexia. Patients with gastroparesis have emesis of undigested food consumed many hours or even days previously. Episodes of nausea and vomiting may last days to months or occur in cycles (Horowitz et al., 1991).

Even with mild symptoms, gastroparesis interferes with nutrient delivery to the small bowel and therefore disrupts the relationship between glucose absorption and exogenous insulin administration. This may result in wide swings of glucose levels and unexpected episodes of postprandial hypoglycemia and apparent "brittle diabetes." Gastroparesis should therefore always be suspected in patients with erratic glucose control. Upper gastrointestinal symptoms should not be attributed to gastroparesis until conditions such as gastric ulcer, duodenal ulcer, gastritis, and gastric cancer have been excluded.

Delayed gastric emptying influences the absorption of orally administered drugs in the intestine.

Evaluation of the patient with suspected diabetic gastroparesis includes manometery to detect antral hypomotility or pylorospasm, and double-isotope scintigraphy to measure solid- and liquid-phase gastric emptying. Another method for assessing gastric emptying is the stable isotope breath test. Since gastric emptying slows by hyperglycemia, it should be measured during euglycemia.

The evaluation of gastric mucosal nerve fiber density by endoscopic biopsies is another clinical method to histologically confirm the presence of gastric autonomic neuropathy (Jin et al., 2009; Selim et al., 2010).

Treatment

Control of blood glucose levels may improve gastric motor dysfunction. The management of gastroparesis should include multiple small feedings with a reduction in the fat content (<40 g/day), which tends to delay gastric emptying. A low-fiber diet is also advised in diabetic gastroparesis to avoid bezoar formation. Gastrokinetic drugs used to treat diabetic gastropathy are metoclopramide, domperidone, and erythromycin (Sturm et al., 1999). The gastrokinetic efficacy of metoclopramide, domperidone, and erythromycin diminishes by tachyphylaxis during prolonged administration.

Metoclopramide is a cholinergic and dopaminergic agent with central antiemetic activity (Weber and McCallum, 1996). The recommended dosage is 10 mg four times a day, 30–60 minutes before the meal and bedtime. Metoclopramide may be given intravenously. Side-effects observed with metoclopramide are hyperprolactinemia, galactorrhea, tremor, restlessness, tardive dyskinesia, and drowsiness.

Domperidone is a peripherally acting dopamine 2-receptor antagonist without cholinergic activity. Domperidone improves gastrointestinal motility by enhancing antral concentrations. The oral dose of domperidone for the treatment of gastroparesis is 20–40 mg, three times daily, taken 30 minutes before meals, if necessary, at bedtime. Central nervous sideeffects are observed less frequently with domperidone than with metoclopramide.

Erythromycin and its derivatives improved gastric emptying of solids and liquids and increased antral contraction (Janssens et al., 1990; Erbas et al., 1993). The oral dose of erythromycin is 250 mg three times daily, taken 30 minutes before meals. Intravenous erythromycin (3 mg/kg/every 8 hours by infusion) is a useful drug for clearance of gastric bezoars. Erythromycin may cause nausea, abdominal cramps and diarrhea, rash, and allergic manifestation.

Several novel drugs, including the ghrelin (orexigenic hormone) and ghrelin receptor agonists, motilin agonist (mitemcinal), 5-HT4-receptor agonists and the muscarinic antagonist are being investigated for their prokinetic effects. If medications fail and severe gastroparesis persists, jejunostomy placement into normally functioning bowel may be needed. Feedings can be given at night, freeing patients during the day, and insulin can accordingly be adjusted to accommodate the feeding schedule (Vinik and Milicevic, 1996). However, even resting the stomach with intravenous alimentation and nasogastric suction can return gastric function within a few days. Gastric electrical stimulation also shows promise in the treatment of refractory gastroparesis.

Diabetic diarrhea

Diarrhea may be evident in 20% of diabetic patients, particularly those with known autonomic neuropathy. A number of factors may contribute to the diarrhea, which include stasis of bowel contents with bacterial overgrowth, bile acid-induced intestinal hurry and malabsorption, and a diminution in pancreatic exocrine secretions, presumably due to the need for insulin to maintain exocrine pancreatic function and the vagal pancreatic neuropathy. Autonomic neuropathy-associated diarrhea can be sudden, explosive, and paroxysmal. Fecal incontinence may be associated with severe diabetic diarrhea or constitute an independent disorder of anorectal dysfunction. Incontinent diabetic have decreased basal sphincteric pressure, suggesting abnormal internal and external anal sphincter function.

Treatment

The severe and intermittent nature of diabetic diarrhea makes treatment and assessment difficult. Initial therapy of diabetic diarrhea should be directed toward correction of electrolyte/water disturbances and nutrition. Good control of glucose levels is helpful. A broad-spectrum antibiotic (doxycyline and metranidazole) is usually the treatment of choice for bacterial overgrowth. Retention of bile sometimes occurs, which may be highly irritant to the gut and chelation of bile salts with 4 g cholestyramine mixed with fluid and given orally three times daily may be of considerable help. Antidiarrheal agents (loperamide and diphenoxylate) can reduce the number of stools. Clonidine may reserve adrenergic nerve dysfunction and improve diarrhea. Initial treatment should begin with 0.1 mg twice a day. Clonidine may cause orthostatic hypotension. Octreotide inhibits the release of gastroenteropancreatic endocrine peptides, which may be a pathogenetic factor responsible for diarrhea and electrolyte imbalance in diabetic patients. Octreotide has been shown to be effective in improving diarrhea at a dose of 50–75 micrograms administered subcutaneously twice a day.

Constipation

Constipation is the most common gastrointestinal complication, affecting nearly 25% of diabetic patients. Severe constipation may be complicated by ulceration, perforation, and fecal impaction (Battle et al., 1980). Anorectal manometry is suggested for evaluating the rectal anal inhibitory reflex to distinguish colonic hypomotility from rectosigmoid dysfunction causing outlet obstructive symptoms.

Treatment

Treatment for constipation should begin with emphasis on good bowel habits, which include regular exercise and maintenance of adequate hydration and fiber consumption. Many constipated patients respond to a high-soluble-fiber diet supplemented with daily hydrophilic colloid. Sorbitol and lactulose may be helpful. The intermittent use of saline or osmotic laxatives may be required for patients with more severe symptoms.

GENITOURINARY AUTONOMIC NEUROPATHY

Erectile dysfunction

The most common form of organic sexual dysfunction in male diabetics is erectile dysfunction (ED). ED is defined as the consistent inability to attain and maintain an erection adequate for sexual intercourse, usually qualified by being present for several months and occurring at least half the time. The incidence of ED in diabetic men has been estimated to be between 35% and 75% (McCulloch et al., 1980). Importantly, ED is a marker for the development of generalized vascular disease and for premature death from MI, and ED may presage future cardiovascular events.

The symptoms of organic ED are gradual in onset and progress with time. The earliest complaints are decreased rigidity with incomplete tumescence before total failure. It occurs with all partners, and there is no loss of libido. Morning erections are lost.

Treatment

A number of treatment modalities are available and each treatment has positive and negative effects, therefore patients must be made aware of both aspects before a therapeutic decision is made. Before considering any form of treatment, every effort should be made to have the patient withdraw from alcohol and eliminate smoking. The patient should be removed, if possible, from drugs that are known to cause erectile dysfunction. Metabolic control should be optimized. Direct injection of prostacylin (epoprostenol) into the corpus cavernosum will induce satisfactory erections in a significant number of men. Also, surgical implantation of a penile prosthesis may be appropriate.

Relaxation of the corpus cavernosus smooth-muscle cells is caused by NO and cGMP, and the ability to have and maintain an erection depends on NO and cGMP. The peripherally acting oral phosphodiesterase type 5 inhibitors are the mainstay of oral medical treatment of ED in diabetic patients. The PDE5 inhibitors block the action of PDE5, and cGMP accumulates. This class of agents consists of sildenafil, vardenafil, and tadalafil. These agents enhance blood flow to the corpora cavernosae with sexual stimulation and have been evaluated in diabetics with similar levels of efficacy of about 70%. Generally, diabetic patients require the maximum dose of each agent, sildenafil 100 mg, tadalafil 20 mg, and vardenafil 20 mg. These are absolutely contraindicated in patients being treated with nitroglycerine or other nitrate-containing drugs. Severe hypotension and fatal cardiac events can occur (Rendell et al., 1999). Side-effects are headache, flushing, dyspepsia, and muscle pain.

Retrograde ejaculation

Retrograde ejaculation is caused by damage to efferent sympathetic innervation, which coordinates the simultaneous closure of the internal vesicle sphincter and relaxation of the external vesicle sphincter during ejaculation. Retrograde ejaculation is reported to occur in up to 32% of diabetic men. Lack of spermatozoa in the semen and presence of motile sperm in a postcoital specimen of urine confirm the diagnosis.

Female sexual dysfunction

Women with diabetes mellitus may experience decreased sexual desire and more pain on sexual intercourse, but they are at risk of decreased sexual arousal, with inadequate lubrication. Diagnosis of female sexual dysfunction using vaginal plethysmography to measure lubrication and vaginal flushing has not been well established.

Cystopathy

Neural impairment of the bladder includes sympathetic, parasympathetic, and pudendal nerves. In diabetic autonomic neuropathy, the motor function of the bladder is unimpaired, but afferent fiber damage results in diminished bladder sensation. The urinary bladder can be enlarged to more than three times its normal size. Patients are seen with bladders filled to their umbilicus, yet they feel no discomfort. Loss of bladder sensation occurs early in DAN, with diminished voiding frequency and the patient is no longer able to void completely. Consequently, dribbling and overflow incontinence are common complaints. Autonomic neuropathy associated neurogenic bladder dysfunction may manifest itself as partial or complete urinary retention, incontinence, or frequent urination including nocturia. Chronic dysfunction causes inadequate emptying with consequent urinary infection. In addition, chronic neurogenic bladder can cause vesicoureteric reflux with renal damage.

Evaluation of diabetic bladder dysfunction should be done for any diabetic patient with recurrent urinary tract infection, pyelonephritis, incontinence, or a palpable bladder. Following autonomic neuropathy testing (e.g., HRV assessment) that implicates a neuropathic basis for cystopathy, the clinician is in a position to further evaluate the condition and determine the appropriate treatment. Evaluation should include postvoid ultrasound to assess residual volume and upper urinary tract dilatation, and cystometry or voiding cystometrogram to measure bladder sensation and volume pressure changes associated with bladder filling with known volumes of water and voiding.

Treatment

The principal aim of the treatment should be to improve bladder emptying and to reduce the risk of urinary tract infection. Patients with cystopathy should be instructed to palpate their bladder and, if they are unable to initiate micturition when their bladders are full, use Credé's maneuver to start the flow of urine. Parasympathomimetics such as bethanechol are sometimes helpful, although frequently they do not help to fully empty the bladder. Extended sphincter relaxation can be achieved with an α_1-blocker, such as doxazosin 1–2 mg two to three times daily. Self-catheterization can be particularly useful in this setting, with the risk of infection generally being low.

NEUROVASCULAR DYSFUNCTION

Impaired skin blood flow

Skin blood flow is important in maintaining nutrition, regional and whole body temperature, and healing traumatized skin. There are two different types of skin. The first is apical (glabrous) skin, which is present on the palmar surface of the hand, the plantar surface of the foot, and the face. Apical skin contains a large number of arteriovenous anastomoses or AV shunts; and functions in thermoregulation. In contrast, nonapical (hairy) skin is present over most of the body surface. There are relatively few AV shunt vessels, and blood flow is primarily nutritive in function. Microvascular skin flow is under the control of the ANS and is regulated by both the central and peripheral components of the ANS. Arteriovenous shunts provide a potential low-resistance pathway by which blood flow can be diverted from the arteriolar to venular circulations, bypassing the capillary bed. The shunts are maintained in a constricted state by sympathetic tone. Loss of this tone due to neuropathy causes the shunt to open, thus deviating blood flow from the skin. Increased venular oxygen levels, apparent ischemic lesions despite the presence of palpable pulses, and raised skin temperatures in the distal extremities are consistent with arteriovenous shunting.

In diabetes, the rhythmic contraction of arterioles and small arteries is disordered. Defective blood flow in the small capillary circulation is found with decreased responsiveness to mental arithmetic, cold pressor, hand grip, and heating. The defect is associated with a reduction in the amplitude of vasomotion and resembles premature aging (Stansberry et al., 1996, 1997). There are differences in the glabrous and hairy skin circulations. In hairy skin, a functional defect is found prior to the development of neuropathy (Stansberry et al., 1999). The clinical counterpart is dry skin, loss of sweating, and the development of fissures and cracks that are portals of entry for microorganisms, leading to infectious ulcers and ultimately gangrene.

Evaluation

The development of neurovascular dysfunction in diabetic patients involves complex and multifactorial processes. There are three key elements: endothelium, smooth muscle, and nerve supply. Evaluation of each component is essential to understanding the neurovascular dysfunction of diabetes mellitus. The vascular endothelium has a key role in the pathogenesis of microangiopathy and neuropathy by producing important chemicals such as nitric oxide (NO), endothelin, and prostacylin. The advantage of noninvasive methods for assessing skin blood flow has enabled clinical measurements of the effects of diabetes on neurovascular dysfunction. Microvascular blood flow can be accurately measured noninvasively using laser Doppler flowmetry under basal and stimulated conditions. Smooth-muscle microvasculature in the periphery reacts sympathetically to a number of stressor tasks. These may be divided into those dependent upon the integrity of the central nervous system – orienting response and mental arithmetic – and those dependent upon the distal sympathetic axon – handgrip and cold pressor tests:

Orienting response

Vasoconstriction and resulting drop in peripheral (index finger, pulp surface) skin blood flow when a subject engages in speech after several minutes of relaxation with music.

Mental arithmetic

Mental arithmetic as a serial subtraction task typically results in a 30% reduction in peripheral (index finger, pulp surface) skin blood flow. There is no response in the presence of either a proximal or distal ANS lesion.

Hand grip

Peripheral contralateral (index finger, pulp surface) response to sustained 40% maximum grip on a dynamometer is biphasic over 60 seconds. The initial normal response is a 40–50% reduction of flow from basal during the initial 20–30 seconds, followed by a dilation resulting in a return to typically suprabasal levels; there is no response if the peripheral ANS is damaged.

Cold pressure

Immersion of the contralateral hand in cold (ice) water typically results in a 50–60% reduction in peripheral skin blood flow at the contralateral pulp index surface. In some individuals this response becomes biphasic after prolonged exposure (30 seconds) to such intense cold, as it is extremely uncomfortable. There is a predominantly peripheral component but pain generates a centrally mediated response.

Heating and gravity

Heating the limb to 44°C and dropping it below the level of the heart results in a marked increase in blood flow in normal subjects. The response is a measure of autonomic microvascular integrity and is markedly depressed in patients with autonomic neuropathy (Vinik and Erbas, 2001).

References

Airaksinen KEJ, Koistinen MJ (1992). Association between silent coronary artery disease, diabetes, and autonomic neuropathy: fact or fallacy? Diabetes Care 15: 288–292.

Albers AR, Krichavsky MZ, Balady GJ (2006). Stress testing in patients with diabetes mellitus: diagnostic and prognostic value. Circulation 113: 583–592.

Ambepityia G, Kopelman PG, Ingram D et al. (1990). Exertional myocardial ischemia in diabetes: a quantitative analysis of anginal perceptual threshold and the influence of autonomic function. J Am Coll Cardiol 15: 72–77.

Astrup AS, Tarnow L, Rossing P et al. (2006). Cardiac autonomic neuropathy predicts cardiovascular morbidity and mortality in type 1 diabetic patients with diabetic nephropathy. Diabetes Care 29: 334–339.

Azad N, Emanuele NV, Abraira C et al. (1999). The effects of intensive glycemic control on neuropathy in the VA cooperative study on type II diabetes mellitus (VA CSDM). J Diabetes Complications 13: 307–313.

Battle WM, Snape WJ Jr, Alavi A et al. (1980). Colonic dysfunction in diabetes mellitus. Gastroenterology 79: 1217–1221.

Brownlee M (1992). Glycation products and the pathogenesis of diabetic complications. Diabetes Care 15: 1835–1843.

Burger AJ, Weinrauch LA, D'Elia JA et al. (1999). Effects of glycemic control on heart rate variability in type I diabetic patients with cardiac autonomic neuropathy. Am J Cardiol 84: 687–691.

Burgos LG, Ebert TJ, Assiddao C et al. (1989). Increased intraoperative cardiovascular morbidity in diabetics with autonomic neuropathy. Anesthesiology 70: 591–597.

Buse JB, Bigger JT, Byington RP et al. (2007). Action to Control Cardiovascular Risk in Diabetes (ACCORD) trial: design and methods. Am J Cardiol 99: 21i–33i.

Calles-Escandon J, Lovato L, Simons-Morton D et al. (2010). Effect of intensive compared with standard glycemia treatment strategies on mortality by baseline subgroup characteristics. Diabetes Care 33: 721–727.

Cameron N, Cotter M (1997). Metabolic and vascular factors in the pathogenesis of diabetic neuropathy. Diabetes 46: S31–S37.

Canto JG, Shlipak MG, Rogers WJ et al. (2000). Prevalence, clinical characteristics, and mortality among patients with myocardial infarction presenting without chest pain. JAMA 283: 3223–3229.

Colberg S, Swain D, Vinik A (2003). Use of heart rate reserve and rating of perceived exertion to prescribe exercise intensity in diabetic autonomic neuropathy. Diabetes Care 26: 986–990.

DCCT EDIC Research Group (2003). Sustained effect of intensive treatment of type 1 diabetes mellitus on development and progression of diabetic nephropathy. JAMA 290: 2159–2167.

Diabetes Control and Complications Trial Research Group (1995). The effect of intensive diabetes therapy on the development and progression of neuropathy. Ann Intern Med 122: 561–568.

Diabetes Control and Complications Trial Research Group (1998). The effect of intensive diabetes therapy on measures of autonomic nervous system function in the Diabetes Control and Complications Trial (DCCT). Diabetologia 41: 416–423.

Didangelos TP, Arsos GA, Karamitsos DT et al. (2006). Effect of quinapril or losartan alone and in combination on left ventricular systolic and diastolic functions in asymptomatic patients with diabetic autonomic neuropathy. J Diabetes Complications 20: 1–7.

Duckworth W, Abraira C, Moritz T et al. (2009). Glucose control and vascular complications in veterans with type 2 diabetes. N Engl J Med 360: 129–139.

Erbas T, Varoglu E, Erbas B et al. (1993). Comparison of metoclopramide and erythromycin in the treatment of diabetic gastroparesis. Diabetes Care 16: 1511–1514.

Ewing DJ, Clarke BF (1986). Diabetic autonomic neuropathy: present insights and future prospects. Diabetes Care 9: 648–665.

Ewing DJ, Boland O, Neilson JM et al. (1991). Autonomic neuropathy, QT interval lengthening, and unexpected deaths in male diabetic patient. Diabetologia 34: 182–185.

Freeman R (1998). Diabetic autonomic neuropathy: an overview. In: A Veves (Ed.), Clinical Management of Diabetic Neuropathy. Humana Press, Totowa, NJ, pp. 181–208.

Freeman R (1999). Cardiovascular autonomic neuropathy. In: PJ Dyck, PK Thomas (Eds.), Diabetic Neuropathy. 2nd edn. WB Saunders, Philadelphia, pp. 541–554.

Gaede P, Vedel P, Parving HH et al. (1999). Intensified multifactorial intervention in patients with type 2 diabetes mellitus and microalbuminuria: the Steno type 2 randomized study. Lancet 353: 617–622.

Gerritsen J, Dekker JM, TenVoorde BJ et al. (2001). Impaired autonomic function is associated with increased mortality, especially in subjects with diabetes, hypertension, or a history of cardiovascular disease: the Hoorn Study. Diabetes Care 24: 1793–1798.

Gerstein HC, Miller ME, Byington RP et al. (2008). Effects of intensive glucose lowering in type 2 diabetes. N Engl J Med 358: 2545–2559.

Greene DA, Lattimer SA (1983). Impaired rat sciatic nerve sodium-potassium adenosine triphosphatase in acute streptozocin diabetes and its correction by dietary myo-inositol supplementation. J Clin Invest 72: 1058–1063.

Greene DA, Lattimer SA, Sima AA (1988). Are disturbances of sorbitol, phosphoinositide, and Na+-K+-ATPase regulation involved in pathogenesis of diabetic neuropathy? Diabetes 37: 688–693.

Hilsted J, Parving HH, Christensen NJ et al. (1981). Hemodynamics in diabetic orthostatic hypotension. J Clin Invest 68: 1427–1434.

Hoeldtke RD, Bryner KD, McNeil DR et al. (2002). Nitrosative stress, uric acid, and peripheral nerve function in early type 1 diabetes. Diabetes 51: 2817–2825.

Horowitz M, Edelbroek M, Fraser R (1991). Disordered gastric motor function in diabetes mellitus. Recent insights into prevalence, pathophysiology, clinical relevance and treatment. Scand J Gastroenterol 26: 673–684.

Horrobin DF (1997). Essential fatty acids in the management of impaired nerve function in diabetes. Diabetes 46 (Suppl 2): S90–S93.

Janssens J, Peeters TL, Vantrappen G (1990). Improvement of gastric emptying in diabetic gastroparesis by erthromycin. N Engl J Med 322: 1028–1031.

Jin HY, Kang YM, Kim CY et al. (2009). Morphological comparison of small nerve fibres in gastric mucosa in non-diabetic and type 2 diabetic subjects. Diabet Med 26: 943–946.

Kahn JK, Sisson JC, Vinik AI (1988). Prediction of sudden cardiac death in diabetic autonomic neuropathy. J Nucl Med 29: 1605–1606.

Katz A, Liberty IF, Porath A et al. (1999). A simple bedside test of 1-minute heart rate variability during deep breathing as a prognostic index after myocardial infarction. Am Heart J 138: 32–38.

Kellogg AP, Converso K, Wiggin T et al. (2009). Effects of cyclooxygenase-2 gene inactivation on cardiac autonomic and left ventricular function in experimental diabetes. Am J Physiol Heart Circ Physiol 296: H453–H461.

Langer A, Freeman MR, Josse RG et al. (1991). Detection of silent myocardial ischemia in diabetes mellitus. Am J Cardiol 67: 1073–1078.

Lee KH, Jang HJ, Kim YH et al. (2003). Prognostic value of cardiac autonomic neuropathy independent and incremental to perfusion defects in patients with diabetes and suspected coronary artery disease. Am J Cardiol 92: 1458–1461.

Levitas IM, Kristal JJ (1972). Stress exercise testing of the young diabetic for the detection of unknown coronary artery disease. Isr J Med Sci 8: 845–847.

Licht CM, Vreeburg SA, van Reedt Dortland AK et al. (2010). Increased sympathetic and decreased parasympathetic activity rather than changes in hypothalamic–pituitary–adrenal axis activity is associated with metabolic abnormalities. J Clin Endocrinol Metab 95: 2458–2466.

Low PA, Nickander KK, Tritschler HJ (1997). The roles of oxidative stress and antioxidant treatment in experimental diabetic neuropathy. Diabetes 46 (Suppl 2): S38–S42.

Lykke JA, Tarnow L, Parving HH et al. (2008). A combined abnormality in heart rate variation and QT corrected interval is a strong predictor of cardiovascular death in type 1 diabetes. Scand J Clin Lab Invest 68: 654–659.

Ma J, Rayner CK, Jones KL et al. (2009). Diabetic gastroparesis: diagnosis and management. Drugs 69: 971–986.

Marchant B, Umachandran V, Stevenson R et al. (1993). Silent myocardial ischemia: role of subclinical neuropathy in patients with and without diabetes. J Am Coll Cardiol 22: 1433–1437.

Margolis JR, Kannel WS, Feinleib M et al. (1973). Clinical features of unrecognized myocardial infarction – silent and symptomatic. Eighteen year follow-up: the Framingham Study. Am J Cardiol 32: 1–7.

Maser R, Lenhard M, DeCherney G (2000). Cardiovascular autonomic neuropathy: the clinical significance of its determination. Endocrinologist 10: 27–33.

Maser RE, Mitchell BD, Vinik AI et al. (2003). The association between cardiovascular autonomic neuropathy and mortality in individuals with diabetes: a meta-analysis. Diabetes Care 26: 1895–1901.

Mathias CJ, Da Coata DF, Fosbraey P (1987). Hypotensive and sedative effects of insulin in autonomic failure. Br Med J 295: 161–163.

Matsuki A, Nozawa T, Igarashi N et al. (2010). Fluvastatin attenuates diabetes-induced cardiac sympathetic neuropathy in association with a decrease in oxidative stress. Circ J 74: 468–475.

McCulloch DK, Campbell IW, Wu FC et al. (1980). The prevalence of diabetic impotence. Diabetologia 18: 279–283.

Miettinen H, Lehto S, Salomaa V et al. (1998). Impact of diabetes on mortality after the first myocardial infarction. The FINMONICA Myocardial Infarction Register Study Group. Diabetes Care 21: 69–75.

Muenter SN, Charkoudian N, Dotson RM et al. (2005). Baroreflex control of muscle sympathetic nerve activity in postural orthostatic tachycardia syndrome. Am J Physiol Heart Circ Physiol 289: H1226–H1233.

Nagamachi S, Fujita S, Nishii R et al. (2006). Prognostic value of cardiac I–123 metaiodobenzylguanidine imaging in patients with non-insulin-dependent diabetes mellitus. J Nucl Cardiol 13: 34–42.

Neil HA, Thompson AV, John S et al. (1989). Diabetic autonomic neuropathy: the prevalence of impaired heart rate variability in a geographically define population. Diabet Med 6: 20–24.

O'Brien IA, O'Hare JP, Lewin IG et al. (1986). The prevalence of autonomic neuropathy in insulin-dependent diabetes mellitus: a controlled study based on heart rate variability. Q J Med 61: 957–967.

O'Brien IA, McFadden JP, Corrall RJ (1991). The influence of autonomic neuropathy on mortality in insulin-dependent diabetes. Q J Med 79: 495–502.

Obrosova IG (2002). How does glucose generate oxidative stress in peripheral nerve? Int Rev Neurobiol 50: 3–35.

Orchard TJ, Lloyd CE, Maser RE et al. (1996). Why does diabetic autonomic neuropathy predict IDDM mortality? An analysis from the Pittsburgh Epidemiology of Diabetes Complications Study. Diabetes Res Clin Pract 34: S165–S171.

Pacher P, Liaudet L, Soriano F et al. (2002). The role of poly (ADP-ribose) polymerase activation in the development of myocardial and endothelial dysfunction in diabetes. Diabetes 51: 514–521.

Page MM, Watkins PJ (1976). Provocation of postural hypotension by insulin in diabetic autonomic neuropathy. Diabetes 25: 90–95.

Patel A, MacMahon S, Chalmers J et al. (2008). Intensive blood glucose control and vascular outcomes in patients with type 2 diabetes. N Engl J Med 358: 2560–2572.

Pfeifer MA, Weinberg CR, Cook DL et al. (1984). Autonomic neural dysfunction in recently diagnosed diabetic subjects. Diabetes Care 7: 447–453.

Pittenger GL, Malik RA, Burcus N et al. (1999). Specific fiber deficits in sensorimotor diabetic polyneuropathy correspond to cytotoxicity against neuroblastoma cells of sera from patients with diabetes. Diabetes Care 22: 1839–1844.

Pop-Busui R, Low PA, Waberski BH et al. (2009). Effects of prior intensive insulin therapy on cardiac autonomic nervous system function in type 1 diabetes mellitus: the Diabetes Control and Complications Trial/Epidemiology of Diabetes Interventions and Complications study (DCCT/EDIC). Circulation 119: 2886–2893.

Pop-Busui R, Evans G, Gerstein H et al.; ACCORD Study Group (2010). Effects of cardiac autonomic dysfunction on mortality risk in the action to control cardiovascular risk in diabetes (ACCORD) trial. Diabetes Care 33: 1578–1584.

Purewal T, Watkins PJ (1995). Postural hypotension in diabetic neuropathy: a review. Diabetic Med 12: 192–200.

Rathmann W, Ziegler D, Jahnke M et al. (1993). Mortality in diabetic patients with cardiovascular autonomic neuropathy. Diabet Med 10: 820–824.

Rendell MS, Rajfer J, Wicker PA et al. (1999). Sildenafil for treatment of erectile dysfunction in men with diabetes: a randomized controlled trial. JAMA 281: 421–426.

Rennings AJ, Smits P, Stewart MW et al. (2010). Autonomic neuropathy predisposes to rosiglitazone-induced vascular leakage in insulin-treated patients with type 2 diabetes: a randomised, controlled trial on thiazolidinedione-induced vascular leakage. Diabetologia 53: 1856–1866.

Rosen SD, Camici PG (2000). The brain–heart axis in the perception of cardiac pain: the elusive link between ischaemia and pain. Ann Med 32: 350–364.

Roy TM, Peterson HR, Snider HL (1989). Autonomic influence on cardiovascular performance in diabetic subjects. Am J Med 87: 382–388.

Selim MM, Wendelschafer-Crabb G, Redmon JB et al. (2010). Gastric mucosal nerve density: a biomarker for diabetic autonomic neuropathy? Neurology 75: 973–981.

Shakespeare CF, Katritsis D, Crowther A et al. (1994). Differences in autonomic nerve function in patients with silent and symptomatic myocardial ischaemia. Br Heart J 71: 22–29.

Singer W, Spies JM, McArthur J et al. (2004). Prospective evaluation of somatic and autonomic small fibers in selected autonomic neuropathies. Neurology 62: 612–618.

Stansberry KB, Shapiro SA, Hill MA et al. (1996). Impaired peripheral vasomotion in diabetes. Diabetes Care 19: 715–721.

Stansberry KB, Hill MA, Shapiro SA et al. (1997). Impairment of peripheral blood flow responses in diabetes resembles an enhanced aging effect. Diabetes Care 20: 1711–1716.

Stansberry KB, Peppard HR, Babyak LM et al. (1999). Primary nociceptive afferents mediate the blood flow dysfunction in non-glabrous (hairy) skin of type 2 diabetes: a new model for the pathogenesis of microvascular dysfunction. Diabetes Care 22: 1549–1554.

Stevens M, Dayanikli F, Raffel D et al. (1998a). Scintigraphic assessment of regionalized defects in myocardial sympathetic innervation and blood flow regulation in diabetic patients with autonomic neuropathy. J Am Coll Cardiol 31: 1575–1584.

Stevens M, Raffel D, Allman KC et al. (1998b). Cardiac sympathetic dysinnervation in diabetes implications for enhanced cardiovascular risk. Circulation 98: 961–968.

Stevens M, Raffel D, Allman K et al. (1999). Regression and progression of cardiac sympathetic dysinnervation complicating diabetes: an assessment by C-11 hydroxyephedrine and positron emission tomography. Metabolism 48: 92–101.

Stewart JM, Munoz J, Weldon A (2002). Clinical and physiological effects of an acute alpha-1 adrenergic agonist and a beta-1 adrenergic antagonist in chronic orthostatic intolerance. Circulation 106: 2946–2954.

Stewart JM, Medow MS, Montgomery LD (2003). Local vascular responses affecting blood flow in postural tachycardia syndrome. Am J Physiol Heart Circ Physiol 285: H2749–H2756.

Stewart JM, Montgomery LD (2004). Regional blood volume and peripheral blood flow in postural tachycardia syndrome. Am J Physiol Heart Circ Physiol 287: H1319–H1327.

Storstein L, Jervell J (1979). Response to bicycle exercise testing in long-standing juvenile diabetes. Acta Med Scand 205: 227–230.

Sturm A, Holtmann G, Goebell H et al. (1999). Prokinetics in patients with gastroparesis: a systematic analysis. Digestion 60: 422–427.

Suarez GA, Clark VM, Norell JE et al. (2005). Sudden cardiac death in diabetes mellitus: risk factors in the Rochester Diabetic Neuropathy Study. J Neurol Neurosurg Psychiatry 76: 240–245.

Sundkvist G, Hillarp B, Lilja B et al. (1989). Esophageal motor function evaluated by scintigraphy, video-radiography and manometry in diabetic patients. Acta Radiol 30: 17–19.

Sundkvist G, Lind P, Bergstrom B et al. (1991). Autonomic nerve antibodies and autonomic nerve function in type 1 and type 2 diabetic patients. J Intern Med 229: 505–510.

Veglio M, Borra M, Stevens LK et al. (1999). The relation between QTc interval prolongation and diabetic complications. The EURODIAB IDDM Complications Study Group. Diabetologia 42: 68–75.

Veves A, King G (2001). Can VEGF reverse diabetic neuropathy in human subjects? J Clin Invest 107: 1215–1218.

Vinik AI, Erbas T (2001). Recognizing and treating diabetic autonomic neuropathy. Cleve Clin J Med 68: 928–944.

Vinik A, Erbas T (2002). Neuropathy. In: N Ruderman, JT Devlin, S Schneider et al. (Eds.), Handbook of Exercise in Diabetes. American Diabetes Association, Alexandria, VA, pp. 463–495.

Vinik AI, Milicevic Z (1996). Recent advances in the diagnosis and treatment of diabetic neuropathy. Endocrinologist 6: 443–461.

Vinik AI, Ziegler D (2007). Diabetic cardiovascular autonomic neuropathy. Circulation 115: 387–397.

Vinik A, Erbas T, Pfeifer MA (2003a). Diabetic autonomic neuropathy. In: D Porte, RS Sherwin, A Baron (Eds.), Diabetes Mellitus, Ellenberg & Rifkin's. 6th edn. Mc Graw Hill, 2002, pp. 789–804.

Vinik AI, Maser RE, Mitchell BD et al. (2003b). Diabetic autonomic neuropathy. Diabetes Care 26: 1553–1579.

Vinik AI, Maser RE, Ziegler D (2010). Neuropathy: the crystal ball for cardiovascular disease? Diabetes Care 33: 1688–1690.

Vinik A, Maser R, Ziegler D (2011). Autonomic imbalance: prophet of doom or scope for hope? Diabetic Med 28: 643–651.

Wackers FJ, Young LH, Inzucchi SE et al. (2004). Detection of silent myocardial ischemia in asymptomatic diabetic subjects: the DIAD Study. Diabetes Care 27: 1954–1961.

Weber F, McCallum R (1996). Gastric motor disorders. In: WJ Snape Jr. (Ed.), Consultations in Gastroentelogy. WB Saunders, Philadelphia, pp. 247–259.

Winocour PH, Dhar H, Anderson DC (1986). The relationship between autonomic neuropathy and urinary sodium and albumin excretion in insulin-treated diabetics. Diabet Med 3: 436–440.

Wirta O, Pasternack A, Mustonen J et al. (1997). Renal and cardiovascular predictors of 9 year total and sudden cardiac mortality in non-insulin-dependent diabetic subjects. Nephrol Dial Transplant 12: 2612–2617.

Witte DR, Tesfaye S, Chaturvedi N et al. (2005). Risk factors for cardiac autonomic neuropathy in type 1 diabetes mellitus. Diabetologia 48: 164–171.

Young LH, Wackers FJ, Chyun DA et al. (2009). Cardiac outcomes after screening for asymptomatic coronary artery disease in patients with type 2 diabetes: the DIAD Study: a randomized controlled trial. JAMA 301: 1547–1555.

Ziegler D (1994). Diabetic cardiovascular autonomic neuropathy: prognosis, diagnosis and treatment. Diabetes Metab Rev 10: 339–383.

Ziegler D, Gries FA, Spuler M et al. (1992). The epidemiology of diabetic neuropathy. Diabetic Cardiovascular Autonomic Neuropathy Multicenter Study Group. J Diabetes Complications 6: 49–57.

Ziegler D, Schatz H, Conrad F et al. (1997). Effects of treatment with the antioxidant alpha-lipoic acid on cardiac autonomic neuropathy in NIDDM patients. A 4-month randomized controlled multicenter trial (DEKAN Study). Deutsche Kardiale Autonome Neuropathie. Diabetes Care 20: 369–373.

Ziegler D, Weise F, Langen KJ et al. (1998). Effect of glycaemic control on myocardial sympathetic innervation assessed by metaiodobenzylguanidine scintigraphy: a 4-year prospective study in IDDM patients. Diabetologia 41: 443–451.

Ziegler D, Zentai C, Perz S et al. (2004). Diminished heart rate variability (HRV) and prolonged QTc interval, but not increased QT dispersion (QTD) are predicators of mortality in the diabetic population. Diabetes 53: A57.

Ziegler D, Zentai C, Perz S et al. (2006). Selective contribution of diabetes and other cardiovascular risk factors to cardiac autonomic dysfunction in the general population. Exp Clin Endocrinol Diabetes 114: 153–159.

Ziegler D, Zentai CP, Perz S et al. (2008). Prediction of mortality using measures of cardiac autonomic dysfunction in the diabetic and nondiabetic population: the MONICA/KORA Augsburg Cohort Study. Diabetes Care 31: 556–561.

Handbook of Clinical Neurology, Vol. 117 (3rd series)
Autonomic Nervous System
R.M. Buijs and D.F. Swaab, Editors

Chapter 23

Hypoglycemia-associated autonomic failure in diabetes

PHILIP E. CRYER*
Department of Medicine, Washington University in St. Louis and Barnes-Jewish Hospital, St. Louis, Missouri, USA

INTRODUCTION

Diabetes mellitus is common. It is estimated that 285 million people worldwide had diabetes in 2010; this number is projected to increase by ~50%, to 438 million people, by the year 2030 (International Diabetes Federation, 2009). Glycemic control – lowering plasma glucose concentrations closer to the nondiabetic range – delays the microvascular complications of diabetes (retinopathy, nephropathy, and neuropathy) in type 1 diabetes mellitus (T1DM) (Diabetes Control and Complications Trial (DCCT), 1993) and in type 2 diabetes (T2DM) (UK Prospective Diabetes Study (UKPDS) Group, 1998a, b). It may also delay the macrovascular complications of diabetes (coronary, cerebral, and peripheral atherosclerosis) (Diabetes Control and Complications Trial/Epidemiology of Diabetes Interventions and Complications (DCCT/EDIC) Study Research Group, 2005; Holman et al., 2008). However, glucose-lowering with a sulfonylurea or a glinide or with insulin introduces the risk of iatrogenic hypoglycemia, the limiting factor in the glycemic management of diabetes (Cryer, 2009a). Iatrogenic hypoglycemia causes recurrent morbidity in most people with T1DM and many with advanced T2DM, and is sometimes fatal. In addition, it generally precludes maintenance of euglycemia over a lifetime of diabetes, and thus, full realization of the vascular benefits of glycemic control. Finally, it impairs defenses against subsequent hypoglycemia and thus causes a vicious cycle of recurrent hypoglycemia.

Hypoglycemia in diabetes is fundamentally iatrogenic, the result of therapeutic hyperinsulinemia caused by treatment with a sulfonylurea or a glinide or with insulin. But, because of the effectiveness of the normal defenses against hypoglycemia – the glucose counterregulatory mechanisms – hypoglycemia in diabetes is typically the result of the interplay of therapeutic hyperinsulinemia and compromised physiological and behavioral defenses against falling plasma glucose concentrations (Cryer, 2009a). Attenuated adrenomedullary epinephrine and sympathetic neural responses to falling glucose levels play a key role in that pathophysiology, albeit in the unique setting of absent insulin and glucagon responses to falling glucose levels in T1DM and advanced T2DM (Cryer, 2009a). (In the presence of normal insulin and glucagon responses, attenuated sympathoadrenal responses to falling plasma glucose concentrations would not cause clinical problems.) This functional and at least partially reversible sympathoadrenal failure, termed hypoglycemia-associated autonomic failure (HAAF) in diabetes (Cryer, 2009a), is distinct from classic diabetic autonomic neuropathy which is structural and largely irreversible (Tesfaye et al., 2010).

EPIDEMIOLOGY OF HYPOGLYCEMIA IN DIABETES

Type 1 diabetes

Hypoglycemia is a fact of life for most people with T1DM (Cryer, 2009a). They suffer untold numbers of episodes of asymptomatic hypoglycemia, an average of two episodes of symptomatic hypoglycemia each week – thousands of such episodes over a lifetime of diabetes – and an average of one episode of severe hypoglycemia, one requiring the assistance of another person, each year.

If it is shown to accurately reflect plasma glucose concentrations and is linked to continuous recording of symptoms, continuous subcutaneous glucose monitoring may soon lead to a more detailed knowledge of the prevalence and incidence of hypoglycemia in diabetes (Juvenile Diabetes Research Foundation (JDRF) Continuous Glucose Monitoring Study Group, 2008; Cooke et al., 2009). Pending that, we must rely on

*Correspondence to: Philip E. Cryer, M.D., Campus Box 8127, Washington University School of Medicine, 660 South Euclid Avenue, St. Louis, MO 63110, USA. Tel: +1-314-362-7635, Fax: +1-314-362-7989, E-mail: pcryer@wustl.edu

estimates. Those estimates are most reliably based on episodes of severe hypoglycemia, that is, one requiring the assistance of another person, since these are dramatic events that are likely to be recalled and reported (Pedersen-Bjergaard et al., 2003), even though they represent only a very small fraction of the total hypoglycemic experience (Cryer, 2009a). The estimates are also most reliably based on prospective studies focused on hypoglycemia (Cryer, 2009a).

The problem of hypoglycemia during aggressive glycemic therapy of T1DM was highlighted by the report of the Diabetes Control and Complications Trial (DCCT) in 1993. Perhaps because patients at high risk for hypoglycemia were excluded, the event rates of 62 episodes of severe hypoglycemia per 100 patient years during intensive therapy in the DCCT provided a minimum estimate; rates of 110 episodes per 100 patient-years were reported in a contemporary study (Reichard and Pihl, 1994), and in subsequent studies, rates of 115 episodes per 100 patient-years (Donnelly et al., 2005) and 110–320 episodes per 100 patient-years (UK Hypoglycaemia Study Group, 2007) have been reported. Clearly, the clinical problem has not gone away. Prospective, population-based data indicate that the prevalence and incidence of hypoglycemia increase with a longer duration of T1DM (UK Hypoglycaemia Study Group, 2007). For example, event rates for severe hypoglycemia increased from 110 to 320 episodes per 100 patient-years in patients with T1DM for <5 and >15 years, respectively.

Type 2 diabetes

Among the approved drug classes used to treat T2DM, thiazolidinediones, α-glucosidase inhibitors, dipeptidyl peptidase-IV inhibitors, and glucagon-like peptide-1 receptor agonists should not, and probably do not, cause hypoglycemia (Cryer, 2009a; Phung et al., 2010). However, these drugs increase the risk of hypoglycemia if used with an insulin secretagogue or with insulin. On the other hand, sulfonylureas (e.g., glyburide (INN glibenclamide), glipizide, gliclazide, and glimepiride) and glinides (e.g., repaglinide and nateglinide) and any insulin preparation can cause therapeutic hyperinsulinemia independent of the plasma glucose concentration and, therefore, hypoglycemia (Cryer, 2009a; Phung et al., 2010). Among the as yet unapproved drug classes, inhibitors of renal glucose reabsorption should not produce hypoglycemia (Ferrannini et al., 2010) but glucokinase activators might well do so (Bonadonna et al., 2010).

Hypoglycemia is relatively infrequent early in the course of T2DM treated with a sulfonylurea or even with insulin (UK Hypoglycaemia Study Group, 2007; Cryer, 2009a). Prospective, population-based data indicate that the overall incidence of any hypoglycemia and of severe hypoglycemia in insulin-treated T2DM is about one-third of that in T1DM. For example, Donnelly et al. (2005) reported event rates of severe hypoglycemia of 35 and 115 episodes per 100 patient-years in insulin-treated T2DM and T1DM, respectively. However, as insulin secretion decreases further (UK Prospective Diabetes Study (UKPDS) Group, 1998b), hypoglycemia becomes progressively more limiting to glycemic control over time in T2DM (UK Prospective Diabetes Study Group, 1998c). Indeed, the frequency of hypoglycemia is similar in T2DM and T1DM matched for duration of insulin therapy (Hepburn et al., 1993). Prospective data indicate that the frequency of severe hypoglycemia increases with the duration of insulin therapy in T2DM. For example, the UK Hypoglycaemia Study Group (2007) found the proportion of patients affected by severe hypoglycemia to increase from 7% to 25% and the event rates for severe hypoglycemia to increase from 10 to 70 episodes per 100 patient years in patients with T2DM treated with insulin for < 2 years and for > 5 years, respectively.

Because T2DM is ~20-fold more prevalent than T1DM, and most people with T2DM ultimately require treatment with insulin, the available data suggest that most episodes of hypoglycemia, including severe hypoglycemia, occur in people with T2DM (Cryer, 2009a).

PHYSIOLOGY OF DEFENSE AGAINST HYPOGLYCEMIA

As plasma glucose levels fall, the prevention or correction of clinical hypoglycemia normally involves both physiological and behavioral defenses (Cryer, 2001, 2006) (Fig. 23.1). The physiological component of glucose counterregulation includes: (1) decreased insulin secretion as plasma glucose levels decline within the physiological range and (2) increased glucagon secretion and, absent that, increased epinephrine secretion as glucose levels fall just below the physiological range. These favor increased hepatic (and renal) glucose production into the circulation and decreased glucose clearance from the circulation by insulin-sensitive tissues such as muscle and, therefore, increased plasma glucose concentrations. Increased epinephrine secretion is physiologically important only in the setting of an absent decrease in insulin secretion and increase in glucagon secretion. The behavioral component of glucose counterregulation is carbohydrate ingestion prompted by symptoms of hypoglycemia. It is symptoms – largely neurogenic symptoms mediated by sympathetic neural activation – that allow people to recognize that their plasma glucose concentrations are low (Towler et al., 1993; DeRosa and Cryer, 2004).

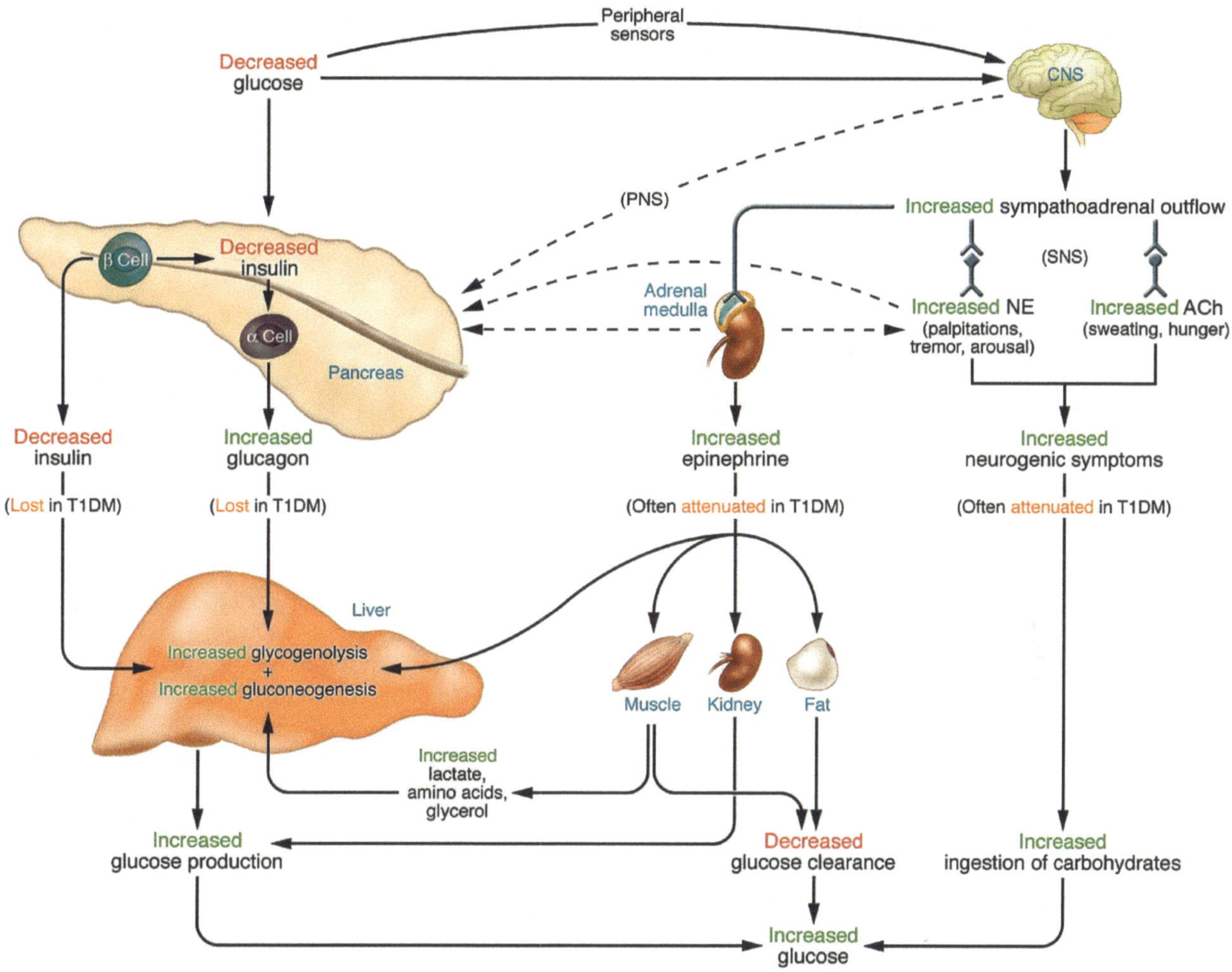

Fig. 23.1. Schematic diagram of glucose counterregulation – physiological defenses (decrements in β-cell insulin and increments in α-cell glucagon and adrenomedullary epinephrine) and behavioral defenses (sympathoadrenal-induced awareness of hypoglycemia prompting carbohydrate ingestion) against falling plasma glucose concentrations – in people without diabetes. All four of these, not just insulin secretion, are compromised in people with T1DM and those with advanced T2DM. (Reproduced from Cryer, 2006, with permission of the American Society for Clinical Investigation.)

PATHOPHYSIOLOGY OF DEFENSE AGAINST HYPOGLYCEMIA IN DIABETES

Defective glucose counterregulation and hypoglycemia unawareness

Although hypoglycemia can be caused by marked therapeutic insulin excess, iatrogenic hypoglycemia is typically the result of the interplay of mild to moderate absolute or even relative therapeutic insulin excess and compromised glucose counterregulation in T1DM and in advanced, absolutely endogenous insulin-deficient T2DM (Cryer, 2009a). As plasma glucose concentrations fall in nondiabetic individuals, insulin secretion decreases and glucagon and epinephrine secretion increases (Cryer, 2001). As plasma glucose concentrations fall in people with established T1DM, insulin secretion does not decrease, glucagon secretion does not increase, and the increase in epinephrine secretion is often attenuated (Dagogo-Jack et al., 1993). The latter patients have defective glucose counterregulation and a 25-fold (White et al., 1983), or perhaps even greater (Bolli et al., 1984), increased risk of severe hypoglycemia during aggressive glycemic therapy of their T1DM compared with those with comparable impairments of insulin and glucagon secretion but a normal epinephrine secretory response. Further, the attenuated epinephrine response is a marker of an attenuated sympathoadrenal response to hypoglycemia (Paramore et al., 1998; Goldstein and Kopin, 2008). The latter, largely the attenuated sympathetic neural response (DeRosa and Cryer, 2004), causes hypoglycemia unawareness, loss of the warning symptoms that previously allowed the individual to recognize hypoglycemia (Cryer, 2009a). Patients with T1DM and impaired awareness of hypoglycemia have a sixfold increased risk of severe hypoglycemia during aggressive glycemic therapy of their diabetes compared with those with normal awareness (Geddes et al., 2008). Those with insulin-treated T2DM and impaired awareness have been reported to have a 17-fold increased risk (Schopman et al., 2010).

Early in T2DM, insulin secretion decreases and glucagon and epinephrine secretion increase normally as glucose levels fall. Glucose counterregulation is intact. Later, however, as the patient approaches the absolute endogenous insulin-deficient end of the spectrum of T2DM, insulin secretion does not decrease and glucagon secretion does not increase as glucose levels fall (Segel et al., 2002); if the sympathoadrenal response becomes attenuated, defective glucose counterregulation and hypoglycemia unawareness develop in such patients with advanced T2DM, just as they do earlier in T1DM (Cryer, 2009a).

A decrease in intraislet insulin, perhaps among other β-cell secretory products, is normally a signal to increased α-cell glucagon secretion during hypoglycemia (Banarer et al., 2002; Gosmanov et al., 2005; Israelian et al., 2005; Raju and Cryer, 2005; Cooperberg and Cryer, 2010). Thus, β-cell failure plausibly explains both loss of the decrement in insulin secretion and loss of the increment in glucagon secretion as plasma glucose concentrations fall (Cryer, 2009a). That, like the high risk of hypoglycemia, develops early in T1DM but only later in T2DM. Loss of central nervous system (CNS)-mediated signaling of glucagon secretion (Taborsky et al., 2009; McCrimmon and Sherwin, 2010; Paranjape et al., 2010) cannot be the sole explanation since the denervated (transplanted) human pancreas (Diem et al., 1990) and the denervated dog pancreas (Sherck et al., 2001), as well as the perfused rat pancreas (Gerich et al., 1974) and perifused mouse, rat, and human islets (MacDonald et al., 2007), release glucagon in response to low glucose levels in the absence of neural connections to the CNS (Cryer, 2009a). Furthermore, the magnitude of the glucagon secretory response to low plasma glucose concentrations *in vivo* is the same in the presence and in the absence of islet innervation (Diem et al., 1990; Sherck et al., 2001).

An increase in intraislet insulin, perhaps among other β-cell secretory products, is normally a signal to decreased α-cell glucagon secretion following a mixed meal (Cooperberg and Cryer, 2009). Thus, β-cell failure may also contribute to the loss of postprandial suppression of glucagon secretion in T1DM and advanced T2DM.

In summary, available evidence suggests that the metabolic pathophysiology, and thus the pathogenesis of iatrogenic hypoglycemia, is the same in T1DM and advanced T2DM, although it develops rapidly – shortly after diagnosis – in T1DM but much more gradually – as absolute insulin deficiency develops – in T2DM (Cryer, 2009a). This pathophysiology explains why iatrogenic hypoglycemia becomes a barrier to glycemic control early in T1DM but only later in T2DM.

Hypoglycemia-associated autonomic failure in diabetes

Recent antecedent hypoglycemia shifts the glycemic thresholds for sympathoadrenal (among other) responses to hypoglycemia to lower plasma glucose concentrations. The pivotal finding was that a 2 hour episode of afternoon hypoglycemia reduced the neuroendocrine and symptomatic responses to hypoglycemia the following morning in healthy humans (Heller and Cryer, 1991). That finding led to the concept of hypoglycemia-associated autonomic failure (Cryer, 2009a) in T1DM (Dagogo-Jack et al., 1993) and T2DM (Segel et al., 2002).

The concept of hypoglycemia-associated autonomic failure (HAAF) in diabetes (Fig. 23.2) posits that recent antecedent hypoglycemia, as well as sleep or prior exercise, causes both defective glucose counterregulation (by attenuating the adrenomedullary epinephrine response, in the setting of absent insulin and glucagon responses, during subsequent hypoglycemia) and hypoglycemia unawareness (by attenuating the sympathoadrenal, largely the sympathetic neural, response during subsequent hypoglycemia) and thus a vicious cycle of recurrent hypoglycemia (Cryer, 2009a) in T1DM (Dagogo-Jack et al., 1993) and T2DM (Segel et al., 2002). Perhaps the most compelling support of the clinical impact of HAAF is the finding, originally reported from three independent laboratories (Fanelli et al., 1993, 1994; Cranston et al., 1994; Dagogo-Jack et al., 1994), that as little as 2–3 weeks of scrupulous avoidance of hypoglycemia reverses hypoglycemia unawareness and improves the reduced epinephrine component of defective glucose counterregulation in most affected patients.

Mechanism of hypoglycemia-associated autonomic failure

In contrast to those of absent insulin and glucagon secretory responses to hypoglycemia, discussed earlier, the mechanism of the key component of HAAF in diabetes – the attenuated CNS-mediated sympathoadrenal response to falling plasma glucose concentrations that causes hypoglycemia unawareness and, in the setting of absent insulin and glucagon responses, defective glucose counterregulation – is not known (Cryer, 2009a). The alteration could be in the afferent or efferent components of the sympathoadrenal system. Indeed, a reduced plasma metanephrine response to hypoglycemia in patients with T1DM and reduced plasma epinephrine responses to that stimulus has been interpreted to

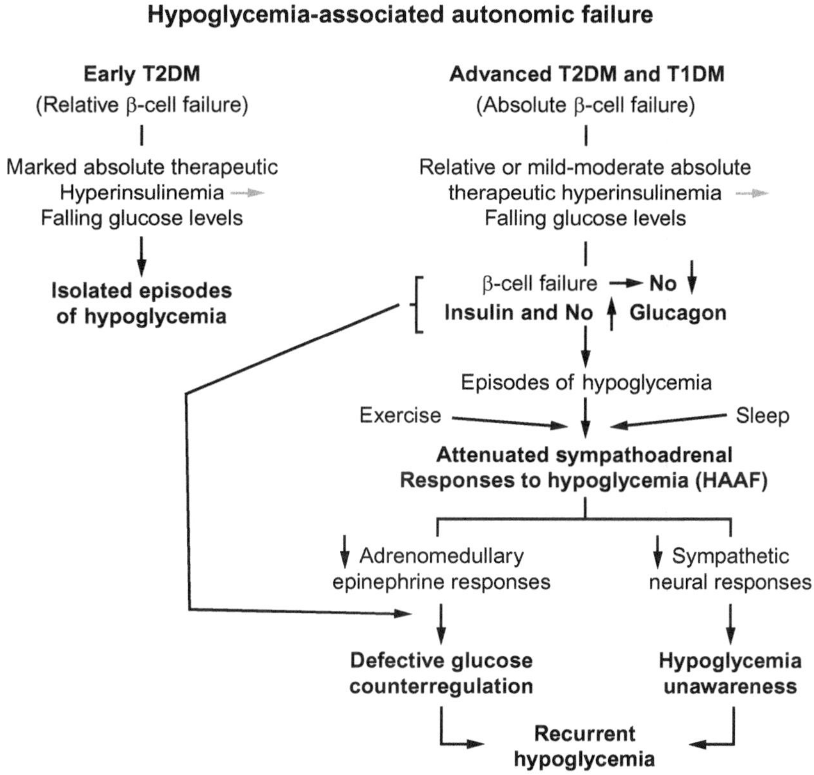

Fig. 23.2. The concept of hypoglycemia-associated autonomic failure (HAAF) in advanced (absolutely endogenous insulin deficient) T2DM and T1DM (right) in the pathogenesis of defective glucose counterregulation and hypoglycemia unawareness and thus recurrent hypoglycemia compared with that of isolated episodes of hypoglycemia early in T2DM (left). (Reproduced from Cryer, 2009a, with permission of the American Diabetes Association.)

indicate a decreased adrenomedullary capacity to secrete epinephrine (de Galan et al., 2004). Nonetheless, the primary alteration is widely assumed to reside in the brain. The various theories, none of which have been established, have been reviewed (Cryer, 2009a; McCrimmon and Sherwin, 2010). They include the systemic mediator (cortisol) hypothesis, the brain fuel (glucose, other?) transport hypothesis, the brain metabolism hypothesis, and, most recently, the cerebral network hypothesis.

The cerebral network hypothesis is based on the finding that hypoglycemia increases synaptic activity – assessed by measurement of regional cerebral blood flow (rCBF) with [^{15}O]water positron emission tomography (PET) – in a discrete system of interconnected brain regions including the medial prefrontal cortex, the lateral orbital prefrontal cortex, the thalamus, the globus pallidus, and the periaqueductal gray in humans (Teves et al., 2004). We studied a model of HAAF – hypoglycemia on two consecutive days with interval hypoglycemia between the two studies – which reduces sympathoadrenal and symptomatic responses to hypoglycemia on the second day in healthy individuals. Interval hypoglycemia did not alter brain synaptic activity patterns during euglycemia but resulted in enhanced synaptic activation during hypoglycemia in the dorsal midline thalamus and only in that region (Arbelaez et al., 2008). We therefore suggested that increased thalamic synaptic activation, induced by recent antecedent hypoglycemia, might inhibit the hypothalamic glucose sensor-initiated sympathoadrenal response to subsequent hypoglycemia, the key feature of HAAF (Arbelaez et al., 2008).

Hypoglycemia-associated autonomic failure: adaptive or maladaptive?

McCrimmon and Sherwin (2010) have expressed the belief that changes in brain metabolism and downregulation of the stress response following recurrent hypoglycemia "are adaptive and not maladaptive and, to that extent this would not be consistent with the current description of this phenomenon as hypoglycemia-associated autonomic failure." Rather, their view is that repeated hypoglycemia induces tolerance of hypoglycemia through preconditioning. That is a plausible view from an evolutionary perspective, and is consistent with: (1) the reasoning that it allows individuals to be less troubled by symptoms during hypoglycemia; (2) the evidence of habituation of the hypothalamic–pituitary–adrenocortical response to recurrent restraint stress in rats (Jaferi and Bhatnagar, 2006), which could also apply to habituation of the sympathoadrenal response to recurrent hypoglycemic stress in humans (Arbelaez et al., 2008); (3) the finding that rats subjected to recurrent moderate hypoglycemia had less brain cell death following marked hypoglycemia (Puente et al., 2010). However, the concept is not plausible from the perspective of people with diabetes.

Recurrent hypoglycemia causes HAAF in diabetes (Cryer, 2009a). HAAF is the sum of defective glucose counterregulation (historically inadequate glucose counterregulation) and hypoglycemia unawareness. Patients with T1DM with defective glucose counterregulation – defined as failure to defend the plasma glucose concentration during low-dose insulin infusion ($\sim$0.7 mU $kg^{-1} \cdot min^{-1}$) and associated with attenuated plasma epinephrine responses to hypoglycemia – are at 25-fold (White et al., 1983), or perhaps even greater (Bolli et al., 1984), increased risk of severe iatrogenic hypoglycemia compared with those with normal glycemic defense against such hyperinsulinemia and normal epinephrine responses. Furthermore, defective glucose counterregulation – again defined as failure to defend plasma glucose during low-dose insulin infusion – and attenuated plasma epinephrine responses to hypoglycemia characterize patients with a history of severe iatrogenic hypoglycemia (Sjöbom et al., 1989). In addition, patients with T1DM and impaired awareness of hypoglycemia are at about sixfold increased risk of severe iatrogenic hypoglycemia (Geddes et al., 2008). Those with insulin-treated T2DM and impaired awareness of hypoglycemia have been reported to be at 17-fold increased risk of severe iatrogenic hypoglycemia (Schopman et al., 2010). Finally, defective glucose counterregulation – defined as failure to defend plasma glucose during low-dose insulin infusion and associated with a history of serious clinical hypoglycemia – and hypoglycemic unawareness segregate together, but not with objective evidence of classic diabetic autonomic neuropathy (Ryder et al., 1990). Thus, HAAF is clearly maladaptive from the perspective of people with absolute endogenous insulin deficiency, those with T1DM and advanced T2DM. In such individuals HAAF substantially increases the risk of severe iatrogenic hypoglycemia, with its associated morbidity and potential mortality (White et al., 1983; Bolli et al., 1984; Sjöbom et al., 1989; Geddes et al., 2008; Schopman et al., 2010).

Nonetheless, the findings of Puente et al. (2010) indicate that recurrent hypoglycemia reduces seizures during, and necrosis of brain neurons and cognitive dysfunction following, marked hypoglycemia in rats. Thus, it may be that recurrent iatrogenic hypoglycemia is both adaptive, in that it renders patients less vulnerable to the devastating effects of subsequent episodes of hypoglycemia, and maladaptive, in that it renders patients more prone to subsequent hypoglycemia with its morbidity and potential mortality.

Summary

The metabolic pathophysiology that underlies HAAF in T1DM and advanced T2DM (Fig. 23.2) is summarized in Figures 23.3 (physiology) and 23.4 (pathophysiology in HAAF). In the setting of absent decrements in β-cell insulin and the resulting absent increments in α-cell glucagon, an attenuated adrenomedullary epinephrine response to hypoglycemia causes defective glucose counterregulation and an attenuated sympathetic neural response to hypoglycemia largely causes hypoglycemia unawareness. Recent antecedent hypoglycemia, as well as sleep and prior exercise, cause an attenuated sympathoadrenal response to subsequent hypoglycemia, and thus HAAF.

Insight into this pathophysiology has led to a comprehensive set of risk factors for hypoglycemia in diabetes (Table 23.1) and a systematic clinical approach to the problem (Table 23.2). These have been reviewed (Cryer, 2009a). Nonetheless, the clinical approach remains suboptimal (Cryer, 2011). It has also led to a definition of hypoglycemia in diabetes (all episodes of an abnormally low plasma glucose concentration that expose the individual to potential harm, with or without symptoms) and a classification of hypoglycemia in diabetes along with a recommended alert value of $\leq$ 70 mg/dL (3.9 mmol/L) (American Diabetes Association Work Group on Hypoglycemia, 2005; Cryer 2009b).

IMPACT OF HYPOGLYCEMIA

Morbidity and mortality

The morbidity of iatrogenic hypoglycemia ranges from nuisance to life-threatening. Hypoglycemia is common in T1DM and advanced T2DM but the vast majority of episodes, including those that cause functional brain failure – decreased cognition, aberrant behavior, even seizure or loss of consciousness – are corrected after the plasma glucose concentration is raised. Prolonged, profound hypoglycemia can cause brain death, but that is very rare, and most fatal episodes are the result of other mechanisms, presumably cardiac arrhythmias (Cryer, 2007).

Older estimates were that 2–4% of people with T1DM died from hypoglycemia (Cryer, 2009a). More recent reports include hypoglycemic mortality rates of 6% (DCCT/EDIC, 2007), 7% (Feltbower et al., 2008), and 10% (Skrivarhaug et al., 2006). Whatever the precise

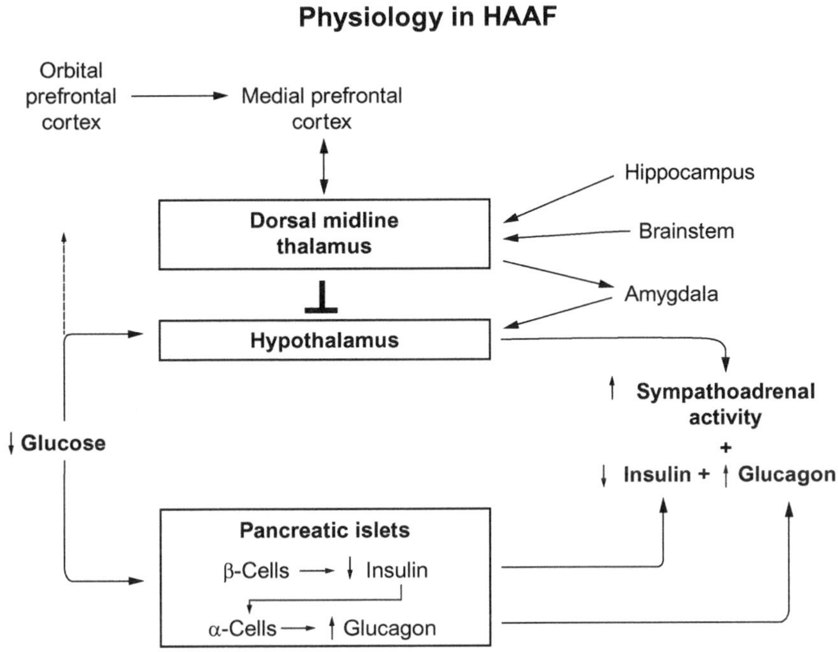

Fig. 23.3. Schematic diagram of the mechanisms of the physiology of glucose counterregulation, including decrements in insulin and increments in glucagon mediated primarily within the pancreatic islets and increments in sympathoadrenal activity – adrenomedullary epinephrine and sympathetic neural – mediated in the brain. (Reproduced from Cryer, 2009a, with permission of the American Diabetes Association.)

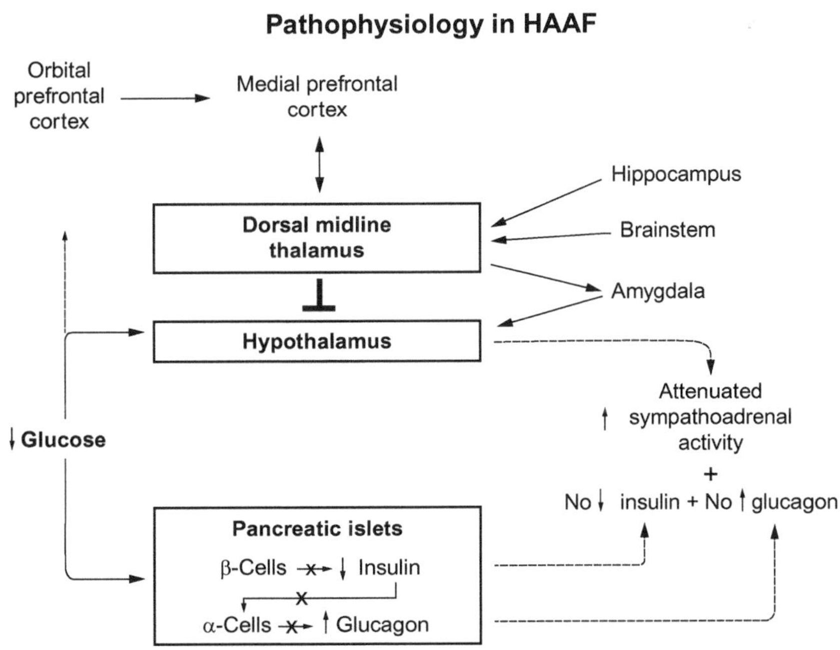

Fig. 23.4. Schematic diagram of the mechanisms of the pathophysiology of glucose counterregulation in HAAF in diabetes including absent decrements in insulin and the resulting absent increments in glucagon at the pancreatic islet level and attenuated increments in adrenomedullary epinephrine and sympathetic neural activity, perhaps because of inhibition of glucose-sensor initiated hypothalamic imitation of sympathoadrenal activity mediated by a cerebral network operating through the thalamus. (Adapted from Cryer, 2008; Cryer, 2009a, with permission of the American Diabetes Association.)

Table 23.1

Risk factors for hypoglycemia in diabetes

Absolute or relative insulin excess
- Insulin (or other drug) doses excessive, ill-timed or of the wrong type
- Decreased exogenous glucose delivery (missed meals, overnight fast, gastroparesis, celiac disease)
- Increased glucose utilization (exercise)
- Decreased endogenous glucose production (alcohol)
- Increased sensitivity to insulin (late after exercise, middle of the night, weight loss, glycemic control)
- Decreased insulin clearance (renal failure)

Compromised glucose counterregulation
- Absolute endogenous insulin deficiency
- History of severe hypoglycemia, hypoglycemia unawareness, or both (as well as recent antecedent hypoglycemia, prior exercise, or sleep)
- Aggressive glycemic therapy *per se* (lower HbA_{1C}, lower glycemic goals)

Table 23.2

Hypoglycemia risk reduction in diabetes

1. Address the issue of hypoglycemia in every contact with a patient at risk
2. Apply the principles of aggressive therapy
   - Patient education and empowerment
   - Frequent self monitoring of plasma glucose
   - Flexible insulin (or other drug) regimens
   - Rational individualized glycemic goals
   - Ongoing professional guidance and support
3. Consider the conventional risk factors, and adjust the regimen accordingly
4. Consider the possibility of compromised glucose counterregulation and seek a history of hypoglycemia unawareness
   - Consider 2–3 weeks of scrupulous avoidance of iatrogenic hypoglycemia

frequency, there is an iatrogenic hypoglycemia mortality rate in T1DM. An example, documented with continuous subcutaneous glucose monitoring, has been reported (Tanenberg et al., 2010).

Sulfonylureas and insulin can cause fatal hypoglycemia in T2DM but hypoglycemic mortality rates are as yet unknown (Cryer, 2009a). Deaths of up to 10% of patients with severe sulfonylurea-induced hypoglycemia have been reported (Ferner and Neil, 1988; Gerich, 1989; Holstein and Egberts, 2003).

Red flags

Increased mortality has been reported in randomized controlled trials during more aggressive, compared with less aggressive, glucose-lowering therapy in patients with T2DM (Action to Control Cardiovascular Risk in Diabetes (ACCORD) Study Group, 2008) and in intensive care unit patients with hyperglycemia (NICE-SUGAR Study Investigators et al., 2009). Whether that excess mortality is (as seems likely), or is not the direct result of iatrogenic hypoglycemia is debated. Absent measurements of plasma glucose concentrations at the time of death, this can neither be confirmed nor denied categorically. However, the finding of excess mortality in patients with T2DM with HbA_{1C} levels in the lower, as well as the higher, deciles (Currie et al., 2010) would seem to imply a direct effect of hypoglycemia. In any event, some aspect of aggressive glycemic therapy must be responsible. The clinical implication seems clear: overly aggressive glucose-lowering therapy, with current available methods, appears to cause excess mortality in diabetes.

Potential mechanisms of hypoglycemic mortality

It seems likely that most episodes of sudden death of people with diabetes are the result of ischemic heart disease and unrelated to hypoglycemia. However, hypoglycemia can be fatal (Nordin, 2010). One mechanism is impaired ventricular repolarization, reflected in a prolonged corrected QT interval (QT_C) in the electrocardiogram, which is associated with potentially lethal ventricular arrhythmias. Epinephrine infusion increases the QT_C interval (Lee et al., 2003). Insulin-induced hypoglycemia, which causes catecholamine, including epinephrine, release, increases the QT_C interval (Robinson et al., 2003); that effect is blunted by β_1-adrenergic blockade with atenolol in nondiabetic persons (Robinson et al., 2003) and patients with T1DM (Lee et al., 2005). QT_C prolongation occurs during iatrogenic hypoglycemia in people with T1DM (Murphy et al., 2004; Robinson et al., 2004; Gill et al., 2009). This might occur in patients without, or with classical autonomic neuropathy (structural autonomic failure) or with cardiac HAAF (functional autonomic failure). With respect to the latter, (Adler et al., 2009) found that, compared with hyperinsulinemic euglycemia, hyperinsulinemic hypoglycemia reduced baroreflex sensitivity, muscle sympathetic nerve activity responses to nitroprusside-induced hypotension, and plasma norepinephrine responses to lower body negative pressure the following day in nondiabetic humans. That cardiovascular HAAF is entirely analogous to the metabolic HAAF discussed earlier.

Given these findings, the following construct is plausible (Fig. 23.5). Recent antecedent hypoglycemia causes cardiovascular HAAF including reduced baroreflex sensitivity and increased vulnerability to a

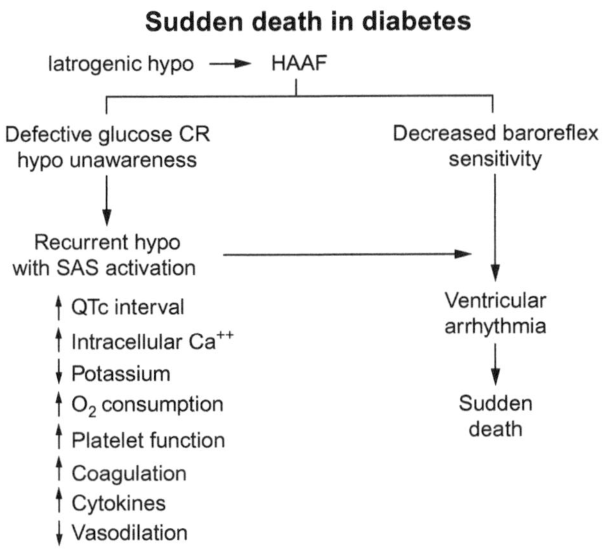

Fig. 23.5. Potential mechanism of iatrogenic hypoglycemia-induced hypoglycemia-associated autonomic failure (HAAF) mediated sudden death in diabetes – cardiovascular HAAF causing reduced baroreflex sensitivity and metabolic HAAF leading to an episode of hypoglycemia that increases sympathoadrenal system activity which triggers a fatal ventricular arrhythmia in the setting of reduced baroreflex sensitivity.

ventricular arrhythmia. It also causes metabolic HAAF with an increased risk for hypoglycemia with sympathoadrenal activation that could, through an array of mechanisms including abnormal cardiac repolarization, trigger a ventricular arrhythmia and sudden death.

GLYCEMIC CONTROL IN DIABETES

The conundrum

Albeit only one component of the management of diabetes, glycemic control delays the microvascular complications, and may delay the macrovascular complications, of T1DM and T2DM (DCCT, 1993; UKPDS, 1998a, b; DCCT/EDIC, 2005; Holman et al., 2008). On the other hand, aggressive glucose-lowering therapy with a sulfonylurea or a glinide or with insulin causes excess mortality, as well as morbidity, in T1DM and T2DM (Ferner and Neil, 1988; Gerich, 1989; Holstein and Egberts, 2003; Skrivarhaug et al., 2006; Cryer, 2007, 2009a; DCCT/EDIC, 2007; ACCORD, 2008; Feltbower et al., 2008; NICE-SUGAR Study Investigators et al., 2009; Currie et al., 2010; Tanenberg et al., 2010). Pending the elimination of hypoglycemia from the lives of people affected by diabetes at some unknown time in the future (Cryer, 2011), how do we balance the pros and cons of glycemic therapy in diabetes (Skyler et al., 2009; Pogach and Aron, 2010; Schernthaner, 2010; Yudkin et al., 2010)? It would seem that individualized glycemic goals would be in the patient's best interest.

Variances from the general glycemic goal of a HbA_{1C} of < 7% in nonpregnant adults have been suggested in a position statement of the American Diabetes Association and a scientific statement of the American College of Cardiology Foundation and the American Heart Association (Skyler et al., 2009). A lower glycemic goal might be selected for patients with "short duration of diabetes, long life expectancy and no significant cardiovascular disease." Conversely, a higher glycemic goal might be selected for patients with "a history of severe hypoglycemia, limited life expectancy, advanced microvascular or macrovascular complications, or extensive comorbid conditions or those with longstanding diabetes in whom the general goal is difficult to attain despite diabetes self-management education, appropriate glucose monitoring, and effective doses of multiple glucose-lowering agents including insulin." It would seem that this judgment would rest largely on the stage of diabetes, the required therapy, and the relative risk of hypoglycemia.

Generic glycemic goal

A reasonable generic glycemic goal is the lowest HbA_{1C} that does not cause severe hypoglycemia (that requiring the assistance of another person), preserves awareness of hypoglycemia, and causes, at worst, an acceptable number of documented episodes of symptomatic hypoglycemia at a given stage in the evolution of the individual's diabetes (Cryer, 2011).

Individualized glycemic goals

During effective therapy of early T2DM with lifestyle changes or with glucose-lowering therapy other than a sulfonylurea, a glinide or insulin, a reasonable glycemic goal might be a normal HbA_{1C}. That might be beneficial over a substantial portion of the course of T2DM. But, such therapies are seldom effective over a lifetime of T2DM, and are not effective in T1DM.

During therapy of T2DM with a sulfonylurea, a glinide, or with insulin, or of T1DM with insulin the glycemic goal might be a $HbA_{1C} < 7\%$ with the above provisos. Sometimes this can be accomplished safely relatively early in T2DM or shortly after diagnosis in T1DM. But, it is often not possible later.

If a $HbA_{1C} < 7\%$ is not achievable safely, there is demonstrable benefit from reducing HbA_{1C} from high to lower, albeit still above optimal, levels (Lachin et al., 2008). Indeed, a mean HbA_{1C} of 7.5% was associated with an all-cause mortality risk ratio of 1.0 in the UK General Practice Research Database (Currie et al., 2010).

Glucose levels low enough to prevent symptoms of hyperglycemia become a reasonable glycemic goal in individuals with a limited life expectancy or functional capacity in whom glycemic control is unlikely to be beneficial.

SUMMARY

Diabetes is common and glycemic control is desirable, but iatrogenic hypoglycemia is the limiting factor in the glycemic management of diabetes. Hypoglycemia becomes a major problem early in T1DM but later in T2DM. That parallels absolute endogenous insulin deficiency, early in T1DM but later in T2DM. Since a decrease in β-cell insulin secretion is normally a signal to increase α-cell glucagon secretion during hypoglycemia, absolute β-cell failure – early in T1DM but later in T2DM – plausibly explains loss of glucagon secretion in response to hypoglycemia in T1DM and advanced T2DM. In the setting of absent decrements in insulin and absent increments in glucagon as plasma glucose concentrations fall, which is unique to absolute endogenous insulin-deficient diabetes, an attenuated sympathoadrenal response to falling plasma glucose concentrations becomes key to the pathogenesis of iatrogenic hypoglycemia. An attenuated adrenomedullary epinephrine response causes defective glucose counterregulation and an attenuated sympathetic neural response largely causes hypoglycemia unawareness. Thus, these syndromes, both associated with a substantially increased risk of iatrogenic hypoglycemia, are the components of hypoglycemia-associated autonomic failure (HAAF) in diabetes. While this key feature of HAAF, an attenuated sympathoadrenal response, can be caused by recent antecedent iatrogenic hypoglycemia, or sleep or prior exercise, its mechanism is not known. Recent evidence indicates that this metabolic HAAF is paralleled by cardiovascular HAAF. Recent antecedent hypoglycemia also reduces baroreflex sensitivity, and thus increases the risk of ventricular arrhythmia and sudden death, perhaps triggered by the sympathoadrenal response to a subsequent episode of hypoglycemia in the setting of metabolic HAAF. Given the conundrum of the vascular benefits of glycemic control and the mortality risk of aggressive glycemic control, individualized glycemic goals are in the best interest of people with diabetes pending the future elimination of hypoglycemia from the lives of people affected by diabetes.

ACKNOWLEDGMENTS

The author's original research cited was supported by NIH grants R01/R37 DK27085 and MO1 RR00036 (now UL1 RR24992) and a fellowship award from the American Diabetes Association. The author acknowledges the substantial contributions of his collaborators, postdoctoral fellows, and technicians. Ms. Janet Dedeke prepared this manuscript.

DISCLOSURES

P.E.C. has served as a consultant to Merck & Co., MannKind Corp., Bristol-Myers Squibb/AstraZeneca and Novo Nordisk in the past year.

References

Action to Control Cardiovascular Risk in Diabetes Study Group ACCORD (2008). Effects of intensive glucose lowering in type 2 diabetes. N Engl J Med 358: 2545–2559.

Adler GK, Bonyhay I, Failing H et al. (2009). Antecedent hypoglycemia impairs autonomic cardiovascular function. Implications for rigorous glycemic control. Diabetes 58: 360–366.

American Diabetes Association Workgroup on Hypoglycemia (2005). Defining and reporting hypoglycemia in diabetes. Diabetes Care 28: 1245–1249.

Arbelaez AM, Powers WJ, Videen TO et al. (2008). Attenuation of counterregulatory responses to recurrent hypoglycemia by active thalamic inhibition: a mechanism for hypoglycemia-associated autonomic failure. Diabetes 57: 470–475.

Banarer S, McGregor VP, Cryer PE (2002). Intraislet hyperinsulinemia prevents the glucagon response to hypoglycemia despite an intact autonomic response. Diabetes 51: 958–965.

Bolli GB, De Feo P, De Cosmo S et al. (1984). A reliable and reproducible test for adequate glucose counterregulation in type I diabetes mellitus. Diabetes 33: 732–737.

Bonadonna RC, Heise T, Arbet-Engels C et al. (2010). Piragliatin (RO4389620), a novel glucokinase activator, lowers plasma glucose both in the postabsorptive state and after a glucose challenge in patients with type 2 diabetes mellitus: a mechanistic study. J Clin Endocrinol Metab 95: 5028–5036.

Cooke D, Hurel SJ, Casbardt A et al. (2009). Randomized controlled trial to assess the impact of continuous glucose monitoring on HbA(1c) in insulin-treated diabetes (MITRE Study). Diabet Med 26: 540–547.

Cooperberg BA, Cryer PE (2009). Beta-cell-mediated signaling predominates over direct alpha-cell signaling in the regulation of glucagon secretion in humans. Diabetes Care 32: 2275–2280.

Cooperberg BA, Cryer PE (2010). Insulin reciprocally regulates glucagon secretion in humans. Diabetes 59: 2936–2940.

Cranston I, Lomas J, Maran A et al. (1994). Restoration of hypoglycaemia awareness in patients with long-duration insulin-dependent diabetes. Lancet 344: 283–287.

Cryer PE (2001). The prevention and correction of hypoglycemia (Ch. 35). In: HM Goodman (Ed.), Handbook of Physiology, Section 7, The Endocrine System. In: LS Jefferson, AD Cherrington (Eds.), The Endocrine Pancreas and Regulation of Metabolism. Vol. II. Oxford University Press, New York, pp. 1057–1092.

Cryer PE (2006). Mechanisms of sympathoadrenal failure and hypoglycemia in diabetes. J Clin Invest 116: 1470–1473.

Cryer PE (2007). Hypoglycemia, functional brain failure, and brain death. J Clin Invest 117: 868–870.

Cryer PE (2008). The barrier of hypoglycemia in diabetes. Diabetes 57: 3169–3176.

Cryer PE (2009a). Hypoglycemia in Diabetes. Pathophysiology, Prevalence and Prevention American Diabetes Association, Alexandria, VA.

Cryer PE (2009b). Preventing hypoglycaemia: what is the appropriate glucose alert value? Diabetologia 52: 35–37.

Cryer PE (2011). Elimination of hypoglycemia from the lives of people affected by diabetes. Diabetes 60: 24–27.

Currie CJ, Peters JR, Tynan A et al. (2010). Survival as a function of HbA(1c) in people with type 2 diabetes: a retrospective cohort study. Lancet 375: 481–489.

Dagogo-Jack SE, Craft S, Cryer PE (1993). Hypoglycemia-associated autonomic failure in insulin-dependent diabetes mellitus. Recent antecedent hypoglycemia reduces autonomic responses to, symptoms of, and defense against subsequent hypoglycemia. J Clin Invest 91: 819–828.

Dagogo-Jack S, Rattarasarn C, Cryer PE (1994). Reversal of hypoglycemia unawareness, but not defective glucose counterregulation, in IDDM. Diabetes 43: 1426–1434.

de Galan BE, Tack CJ, Willemsen JJ et al. (2004). Plasma metanephrine levels are decreased in type 1 diabetic patients with a severely impaired epinephrine response to hypoglycemia, indicating reduced adrenomedullary stores of epinephrine. J Clin Endocrinol Metab 89: 2057–2061.

DeRosa MA, Cryer PE (2004). Hypoglycemia and the sympathoadrenal system: neurogenic symptoms are largely the result of sympathetic neural, rather than adrenomedullary, activation. Am J Physiol Endocrinol Metab 287: E32–E41.

Diabetes Control and Complications Trial Research Group (DCCT) (1993). The effect of intensive treatment of diabetes on the development and progression of long-term complications in insulin-dependent diabetes mellitus. N Engl J Med 329: 977–986.

Diabetes Control and Complications Trial/Epidemiology of Diabetes Interventions and Complications (DCCT/EDIC) Study Research Group (2005). Intensive diabetes treatment and cardiovascular disease in patients with type 1 diabetes. N Engl J Med 353: 2643–2653.

Diabetes Control and Complications Trial/Epidemiology of Diabetes Interventions and Complications (DCCT/EDIC) (2007). Long-term effect of diabetes and its treatment on cognitive function. N Engl J Med 356: 1842–1852.

Diem P, Redmon JB, Abid M et al. (1990). Glucagon, catecholamine and pancreatic polypeptide secretion in type I diabetic recipients of pancreas allografts. J Clin Invest 86: 2008–2013.

Donnelly LA, Morris AD, Frier BM et al. (2005). Frequency and predictors of hypoglycaemia in type 1 and insulin-treated type 2 diabetes: a population-based study. Diabet Med 22: 749–755.

Fanelli CG, Epifano L, Rambotti AM et al. (1993). Meticulous prevention of hypoglycemia normalizes the glycemic thresholds and magnitude of most of neuroendocrine responses to, symptoms of, and cognitive function during hypoglycemia in intensively treated patients with short-term IDDM. Diabetes 42: 1683–1689.

Fanelli C, Epifano L, Rambotti AM et al. (1994). Long-term recovery from unawareness, deficient counterregulation and lack of cognitive dysfunction during hypoglycaemia, following institution of rational, intensive insulin therapy in IDDM. Diabetologia 37: 1265–1276.

Feltbower RG, Bodansky HJ, Patterson CC et al. (2008). Acute complications and drug misuse are important causes of death for children and young adults with type 1 diabetes: results from the Yorkshire Register of Diabetes in Children and Young Adults. Diabetes Care 31: 922–926.

Ferner RE, Neil HA (1988). Sulphonylureas and hypoglycaemia. Br Med J (Clin Res Ed) 296: 949–950.

Ferrannini E, Ramos SJ, Salsali A et al. (2010). Dapagliflozin monotherapy in type 2 diabetic patients with inadequate glycemic control by diet and exercise: a randomized, double-blind, placebo-controlled, phase 3 trial. Diabetes Care 33: 2217–2224.

Geddes J, Schopman JE, Zammitt NN et al. (2008). Prevalence of impaired awareness of hypoglycaemia in adults with type 1 diabetes. Diabet Med 25: 501–504.

Gerich JE (1989). Oral hypoglycemic agents. N Engl J Med 321: 1231–1245.

Gerich JE, Charles MA, Grodsky GM (1974). Characterization of the effects of arginine and glucose on glucagon and insulin release from the perfused rat pancreas. J Clin Invest 54: 833–841.

Gill GV, Woodward A, Casson IF et al. (2009). Cardiac arrhythmia and nocturnal hypoglycaemia in type 1 diabetes – the "dead in bed" syndrome revisited. Diabetologia 52: 42–45.

Goldstein DS, Kopin IJ (2008). Adrenomedullary, adrenocortical, and sympathoneural responses to stressors: a meta-analysis. Endocr Regul 42: 111–119.

Gosmanov NR, Szoke E, Israelian Z et al. (2005). Role of the decrement in intraislet insulin for the glucagon response to hypoglycemia in humans. Diabetes Care 28: 1124–1131.

Heller SR, Cryer PE (1991). Reduced neuroendocrine and symptomatic responses to subsequent hypoglycemia after 1 episode of hypoglycemia in nondiabetic humans. Diabetes 40: 223–226.

Hepburn DA, MacLeod KM, Pell AC et al. (1993). Frequency and symptoms of hypoglycaemia experienced by patients with type 2 diabetes treated with insulin. Diabet Med 10: 231–237.

Holman RR, Paul SK, Bethel MA et al. (2008). 10-year follow-up of intensive glucose control in type 2 diabetes. N Engl J Med 359: 1577–1589.

Holstein A, Egberts EH (2003). Risk of hypoglycaemia with oral antidiabetic agents in patients with type 2 diabetes. Exp Clin Endocrinol Diabetes 111: 405–414.

International Diabetes Federation (2009). IDF Diabetes Atlas. 4th edn. International Diabetes Federation, Brussels, Belgium.

Israelian Z, Gosmanov NR, Szoke E et al. (2005). Increasing the decrement in insulin secretion improves glucagon responses to hypoglycemia in advanced type 2 diabetes. Diabetes Care 28: 2691–2696.

Jaferi A, Bhatnagar S (2006). Corticosterone can act at the posterior paraventricular thalamus to inhibit hypothalamic–pituitary–adrenal activity in animals that habituate to repeated stress. Endocrinology 147: 4917–4930.

Juvenile Diabetes Research Foundation Continuous Glucose Monitoring Study Group (2008). Continuous glucose monitoring and intensive treatment of type 1 diabetes. N Engl J Med 359: 1464–1476.

Lachin JM, Genuth S, Nathan DM et al. (2008). Effect of glycemic exposure on the risk of microvascular complications in the diabetes control and complications trial – revisited. Diabetes 57: 995–1001.

Lee S, Harris ND, Robinson RT et al. (2003). Effects of adrenaline and potassium on QT_C interval and QT dispersion in man. Eur J Clin Invest 33: 93–98.

Lee SP, Harris ND, Robinson RT et al. (2005). Effect of atenolol on QTc interval lengthening during hypoglycaemia in type 1 diabetes. Diabetologia 48: 1269–1272.

MacDonald PE, De Marinis YZ, Ramracheya R et al. (2007). A K_{ATP} channel-dependent pathway within alpha cells regulates glucagon release from both rodent and human islets of Langerhans. PLoS Biol 5: e143.

McCrimmon RJ, Sherwin RS (2010). Hypoglycemia in type 1 diabetes. Diabetes 59: 2333–2339.

Murphy NP, Ford-Adams ME, Ong KK et al. (2004). Prolonged cardiac repolarisation during spontaneous nocturnal hypoglycaemia in children and adolescents with type 1 diabetes. Diabetologia 47: 1940–1947.

NICE-SUGAR Study Investigators, Finfer S, Chittock DR et al. (2009). Intensive versus conventional glucose control in critically ill patients. N Engl J Med 360: 1283–1297.

Nordin C (2010). The case for hypoglycaemia as a proarrhythmic event: basic and clinical evidence. Diabetologia 53: 1552–1561.

Paramore DS, Fanelli CG, Shah SD et al. (1998). Forearm norepinephrine spillover during standing, hyperinsulinemia, and hypoglycemia. Am J Physiol 275: E872–E881.

Paranjape SA, Chan O, Zhu W et al. (2010). Influence of insulin in the ventromedial hypothalamus on pancreatic glucagon secretion in vivo. Diabetes 59: 1521–1527.

Pedersen-Bjergaard U, Pramming S, Thorsteinsson B (2003). Recall of severe hypoglycaemia and self-estimated state of awareness in type 1 diabetes. Diabetes Metab Res Rev 19: 232–240.

Phung OJ, Scholle JM, Talwar M et al. (2010). Effect of non-insulin antidiabetic drugs added to metformin therapy on glycemic control, weight gain, and hypoglycemia in type 2 diabetes. JAMA 303: 1410–1418.

Pogach L, Aron D (2010). Balancing hypoglycemia and glycemic control: a public health approach for insulin safety. JAMA 303: 2076–2077.

Puente EC, Silverstein J, Bree AJ et al. (2010). Recurrent moderate hypoglycemia ameliorates brain damage and cognitive dysfunction induced by severe hypoglycemia. Diabetes 59: 1055–1062.

Raju B, Cryer PE (2005). Loss of the decrement in intraislet insulin plausibly explains loss of the glucagon response to hypoglycemia in insulin-deficient diabetes: documentation of the intraislet insulin hypothesis in humans. Diabetes 54: 757–764.

Reichard P, Pihl M (1994). Mortality and treatment side-effects during long-term intensified conventional insulin treatment in the Stockholm Diabetes Intervention Study. Diabetes 43: 313–317.

Robinson RT, Harris ND, Ireland RH et al. (2003). Mechanisms of abnormal cardiac repolarization during insulin-induced hypoglycemia. Diabetes 52: 1469–1474.

Robinson RT, Harris ND, Ireland RH et al. (2004). Changes in cardiac repolarization during clinical episodes of nocturnal hypoglycaemia in adults with type 1 diabetes. Diabetologia 47: 312–315.

Ryder RE, Owens DR, Hayes TM et al. (1990). Unawareness of hypoglycaemia and inadequate hypoglycaemic counter-regulation: no causal relation with diabetic autonomic neuropathy. BMJ 301: 783–787.

Schernthaner G (2010). Diabetes and cardiovascular disease: is intensive glucose control beneficial or deadly? Lessons from ACCORD, ADVANCE, VADT, UKPDS, PROactive, and NICE-SUGAR. Wien Med Wochenschr 160: 8–19.

Schopman JE, Geddes J, Frier BM (2010). Prevalence of impaired awareness of hypoglycaemia and frequency of hypoglycaemia in insulin-treated type 2 diabetes. Diabetes Res Clin Pract 87: 64–68.

Segel SA, Paramore DS, Cryer PE (2002). Hypoglycemia-associated autonomic failure in advanced type 2 diabetes. Diabetes 51: 724–733.

Sherck SM, Shiota M, Saccomando J et al. (2001). Pancreatic response to mild non-insulin-induced hypoglycemia does not involve extrinsic neural input. Diabetes 50: 2487–2496.

Sjöbom NC, Adamson U, Lin PE (1989). The prevalence of impaired glucose counter-regulation during an insulin-infusion test in insulin-treated diabetic patients prone to severe hypoglycaemia. Diabetologia 32: 818–825.

Skrivarhaug T, Bangstad HJ, Stene LC et al. (2006). Long-term mortality in a nationwide cohort of childhood-onset type 1 diabetic patients in Norway. Diabetologia 49: 298–305.

Skyler JS, Bergenstal R, Bonow RO et al. (2009). Intensive glycemic control and the prevention of cardiovascular events: implications of the ACCORD, ADVANCE, and VA diabetes trials. Diabetes Care 32: 187–192.

Taborsky GJ Jr, Mei Q, Hackney DJ et al. (2009). Loss of islet sympathetic nerves and impairment of glucagon secretion in the NOD mouse: relationship to invasive insulitis. Diabetologia 52: 2602–2611.

Tanenberg RJ, Newton CA, Drake AJ (2010). Confirmation of hypoglycemia in the "dead-in-bed" syndrome, as captured by a retrospective continuous glucose monitoring system. Endocr Pract 16: 244–248.

Tesfaye S, Boulton AJ, Dyck P et al. (2010). Diabetic neuropathies: update on definitions, diagnostic criteria, estimation of severity, and treatments. Diabetes Care 33: 2285–2293.

Teves D, Videen TO, Cryer PE et al. (2004). Activation of human medial prefrontal cortex during autonomic responses to hypoglycemia. Proc Natl Acad Sci U S A 101: 6217–6221.

Towler DA, Havlin CE, Craft S et al. (1993). Mechanism of awareness of hypoglycemia. Perception of neurogenic (predominantly cholinergic) rather than neuroglycopenic symptoms. Diabetes 42: 1791–1798.

UK Hypoglycaemia Study Group (2007). Risk of hypoglycaemia in types 1 and 2 diabetes: effects of treatment modalities and their duration. Diabetologia 50: 1140–1147.

UK Prospective Diabetes Study (UKPDS) Group (1998a). Effect of intensive blood-glucose control with metformin on complications in overweight patients with type 2 diabetes (UKPDS 34). Lancet 352: 854–865.

UK Prospective Diabetes Study (UKPDS) Group (1998b). Intensive blood-glucose control with sulphonylureas or insulin compared with conventional treatment and risk of complications in patients with type 2 diabetes (UKPDS 33). Lancet 352: 837–853.

UK Prospective Diabetes Study (UKPDS) Group (1998c). United Kingdom Prospective Diabetes Study 24: A 6-year, randomized, controlled trial comparing sulfonylurea, insulin, and metformin therapy in patients with newly diagnosed type 2 diabetes that could not be controlled with diet therapy. Ann Intern Med 128: 165–175.

White NH, Skor DA, Cryer PE et al. (1983). Identification of type I diabetic patients at increased risk for hypoglycemia during intensive therapy. N Engl J Med 308: 485–491.

Yudkin JS, Richter B, Gale EA (2010). Intensified glucose lowering in type 2 diabetes: time for a reappraisal. Diabetologia 53: 2079–2085.

Handbook of Clinical Neurology, Vol. 117 (3rd series)
Autonomic Nervous System
R.M. Buijs and D.F. Swaab, Editors

Chapter 24

Sensory–autonomic interactions in health and disease

PETER D. DRUMMOND*
School of Psychology and Exercise Science, Murdoch University, Perth, Western Australia

Under normal conditions, cortical and subcortical regulatory processes integrate many sources of sensory input to drive autonomic effector nuclei in the brainstem and to moderate spinal autonomic reflexes. However, in disease states associated with dysfunction or injury to peripheral nerves, abnormal interactions sometimes develop between sensory and autonomic nerves. Putative mechanisms in disease states such as chronic neuropathic pain are reviewed in this chapter, with particular focus on changes in sweating and cutaneous blood flow.

NORMAL CONTROL OF SWEATING

Sweating greatly enhances evaporative heat loss from the skin during exercise and in hot environments, thereby helping to maintain body temperature within an optimal range. Sweating also precedes increases in body temperature during the early stages of exercise (van Beaumont and Bullard, 1963), and in glabrous skin is evoked by emotions that prepare the body for action (Bradley, 2009).

In humans, eccrine sweat glands are innervated chiefly by sympathetic cholinergic nerve fibers that induce sweating via release of acetylcholine and possibly cotransmitters such as nitric oxide, calcitonin gene-related peptide and vasoactive intestinal polypeptide (Shibasaki and Crandall, 2010). Sweating is driven primarily by internal body temperature (detected by thermosensors in the preoptic hypothalamus) and secondarily by mean skin temperature. Sweating can also be evoked by stimulation of axonal nicotinic receptors on cholinergic efferent fibers (Low et al., 1983). Thus, local mechanisms (e.g., endogenous acetylcholine release) might build upon effects of central thermoregulatory drive to modulate sweating.

Primary (essential) hyperhidrosis involves excessive sweating, mainly in the palms, soles, axillae and face, to emotional factors and thermal stimuli (Kreyden and Scheidegger, 2004). Individual differences in the rate of sweating are proportional to the size of sweat glands and to cholinergic sensitivity (Sato and Sato, 1983). However, the morphology of sweat glands appears to be normal in people with this disorder (Swartling et al., 2004). Spontaneous activity was detected in single-unit microelectrode recordings of individual sudomotor neurons supplying the glabrous skin of the hand of patients with hyperhidrosis, whereas activity is essentially absent in healthy individuals while relaxing in thermoneutral conditions (Macefield et al., 2008). Thus, an increase in central sympathetic drive to sweat glands may increase spontaneous sweat release in patients with hyperhidrosis.

REGULATION OF CUTANEOUS BLOOD FLOW

Blood flow through the skin is regulated both by sympathetic vasoconstrictor and vasodilator reflexes. These reflexes help to maintain body temperature within a narrow range, stabilize blood pressure during changes of posture, prepare the body for "fight or flight," and help to disperse body heat produced during exercise. In addition, parasympathetic vasodilator reflexes around the mouth and eyes supply the salivary and lacrimal glands with the raw materials needed to protect these sensitive regions against infection and injury.

Sympathetic vasoconstrictor reflexes

In the hands, feet, and face, sympathetic vasoconstrictor discharge closes high-capacity arteriovenous shunts during pain, cold, and stress. In other parts of the body, sympathetic vasoconstrictor nerves apply a weak tonic vasoconstrictor influence on arterioles and veins via corelease of noradrenaline/norepinephrine and neuropeptide Y (Stephens et al., 2004). Vasoconstrictor tone intensifies when ambient temperature falls below thermoneutral levels of around 33°C (Hodges and Johnson, 2009). Vasoconstriction is attenuated in elderly

*Correspondence to: Professor Peter Drummond, School of Psychology and Exercise Science, Murdoch University, Perth, 6150 Western Australia. Tel: +61-8-9360 2415, Fax: +61-8-9360-6492, E-mail: P.Drummond@murdoch.edu.au

people through a variety of mechanisms, including loss of the noradrenergic component of the response, a probable reduction in sympathetic outflow and neurotransmitter release, and reduced vascular responsiveness (Holowatz and Kenney, 2010).

Although sympathetic vasoconstrictor reflexes generally form part of a mass discharge to all regions of the body, there is some evidence of lateral organization of these reflexes. For example, application of mustard oil or histamine to the human forearm elicits sustained vasoconstriction in the ipsilateral hand (Magerl et al., 1996), possibly due to a spinal reflex comprising nociceptive fibers in the afferent limb and sympathetic vasoconstrictor fibers in the efferent limb. Curiously, however, vasoconstrictor responses to limb pain are greater in the contralateral than the ipsilateral temple (Drummond and Granston, 2003) due to ipsilateral *release* of sympathetic vasoconstrictor tone (Drummond, 2006a) (Fig. 24.1). The functional significance of this response is uncertain but could be worth exploring, as loss of sympathetic vasoconstrictor tone may aggravate vascular distension and pain in disorders such as cluster headache and migraine (Drummond, 1991, 2006b).

Sympathetic vasodilator reflexes

Increases in skin and internal body temperature trigger active sympathetic vasodilatation in nonglabrous (hairy) skin (Kamijo et al., 2005; Kellogg, 2006). This reflex appears to involve a complex combination of mediators, including vasoactive intestinal polypeptide, nitric oxide, acetylcholine, prostanoids, and histamine (Hodges and Johnson, 2009; Kellogg et al., 2010). Hormones may modulate this vasodilator reflex. For example, fluctuations in sex hormones during menopause may lower the threshold for active sympathetic vasodilatation, evoking menopausal hot flashes (Charkoudian, 2010).

Parasympathetic vasodilator reflexes

Irritation of orofacial tissues not only evokes lacrimation, salivation and rhinorrhea but also triggers parasympathetic vasodilatation, thereby increasing the blood supply to actively secreting glands and downstream tissues and skin around the mouth and eyes (Drummond, 1992, 1995, 1997). These trigeminal-nociceptive parasympathetic reflexes become active during attacks of cluster headache and other trigeminal autonomic cephalalgias, and could aggravate pain by distending tender intracranial and scalp arteries (Drummond, 2006b).

Axon reflexes

Activation of unmyelinated and thinly myelinated nociceptive fibers evokes the local release of neuropeptides such as substance P and calcitonin gene-related peptide from the activated nerve terminals. These neuropeptides

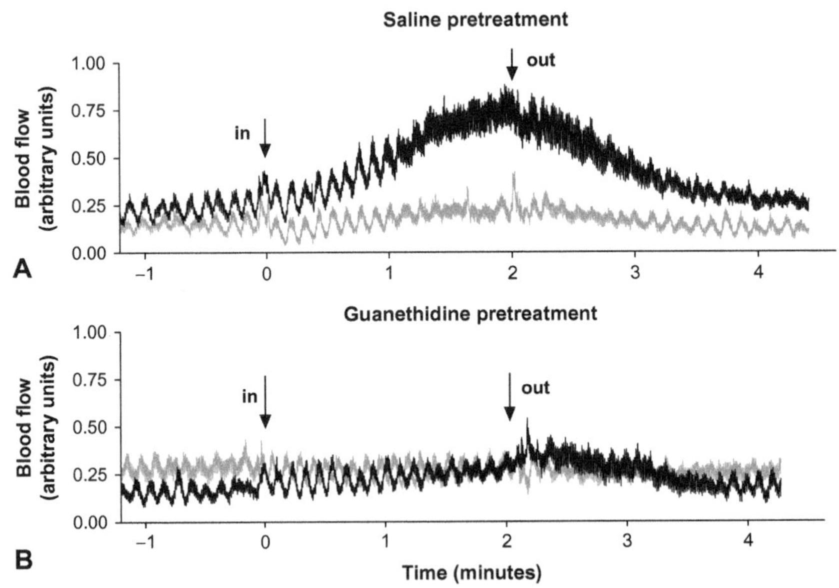

Fig. 24.1. Effect of immersing the right hand in 2°C water for 2 minutes on blood flow in the right temple (black waveform) and left temple (gray waveform) after bilateral saline pretreatment (**A**), and after guanethidine pretreatment to the right temple (to block adrenergic vasoconstriction) and saline pretreatment to the left temple (**B**). The right hand was immersed in the water at the arrow marked "in," and removed from the water at the arrow marked "out." The guanethidine pretreatment inhibited increases in blood flow in the right temple during the cold pressor test, indicating that the vasodilator response was mediated by release of adrenergic vasoconstrictor tone. Pulse amplitude (indicated by the thickness of the blood flow signal) also increased in the right temple after saline pretreatment, and this response was inhibited by guanethidine pretreatment. (Reprinted with permission from Drummond, 2006a.)

increase the diameter and permeability of cutaneous blood vessels and may also provoke mast cell degranulation, thus amplifying local changes in blood flow and vessel permeability (Drummond, 2004). The axon reflex is the first stage of a neuroinflammatory response that involves chemotaxis of white blood cells and extravasation of plasma proteins from blood vessels into the extracellular fluid. The axon reflex may help to protect the skin by dispersing heat or by diluting harmful substances that have entered the skin (Magerl and Treede, 1996), and may also modulate autoregulatory responses in the vascular endothelium that fine-tune local blood flow and inflammatory processes (Lindsey et al., 2000).

In addition to the axon reflex, resident immune and epidermal cells make up the major elements of a neuroimmune defense system in the skin. Epidermal keratinocytes and melanocytes synthesize adrenaline and noradrenaline, which they employ as paracrine signals in a transduction network involving α_1- and β_2-adrenoceptors expressed on a range of epidermal cells (Grando et al., 2006). Among other functions, this signaling network controls cell growth, differentiation, cell motility, and immune responses, with α_1-adrenoceptors generally playing a stimulatory role and β_2-adrenoceptors an inhibitory role (Maestroni, 2006). Catecholamines normally act via β_2-adrenoceptors to inhibit proinflammatory cytokine release. However, disruption of the adrenergic signaling network may contribute to chronic inflammation because a relative increase in α_1-adrenoceptors on immune cells facilitates proinflammatory cytokine release (Kavelaars, 2002).

ADRENERGIC INVOLVEMENT IN AXON REFLEXES

Introduction of α-adrenergic agonists into the skin of the human forearm by iontophoresis evokes vasoconstriction at the site of administration (Drummond, 2002). Surprisingly, however, blood flow increases in surrounding skin (Drummond and Lipnicki, 1999; Drummond, 2009a). As this response can be blocked by pretreatment with local anesthetic cream, it appears to be an axon reflex. The flare is evoked by stimulation of α_1-adrenoceptors but not α_2-adrenoceptors (Fig. 24.2), and depends partly on prostaglandin synthesis at the site of iontophoresis (Drummond, 2011).

Animal studies have confirmed that α_1-adrenoceptors contribute to the flare that develops around the site of inflammation and injury. For example, Lin et al. (2003) reported that sympathectomy inhibited the flare generated by intradermal injection of capsaicin in the rat hindlimb, whereas intra-arterial injection of the α_1-adrenoceptor agonist phenylephrine restored the capsaicin-induced flare. In sympathetically intact rats,

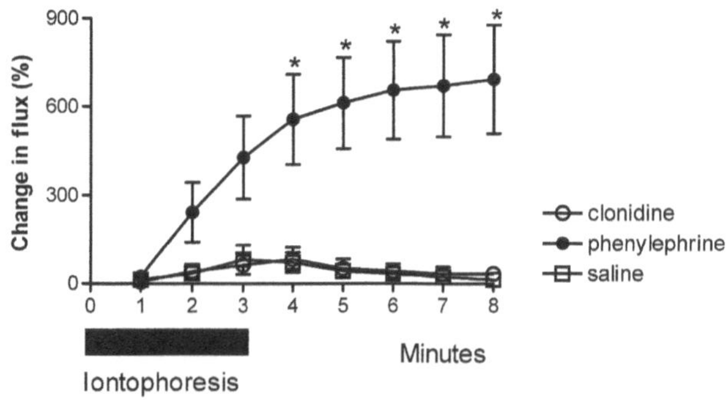

Fig. 24.2. Change in flux (± S.E.) 8 mm from the site of iontophoresis (350 μA for 3 minutes) of phenylephrine, clonidine, and saline-control. Responses are expressed as the percentage change from levels recorded immediately before the iontophoreses. Axon reflex vasodilatation to phenylephrine was significantly greater than the response to saline (*$p < 0.05$). (Reprinted with permission from Drummond, 2011.)

blockade of peripheral α_1-adrenoceptors with the α_1-adrenoceptor antagonist terazosin inhibited the flare to capsaicin. Similarly, sympathectomy and pretreatment with terazosin inhibited sensitization of nociceptive C fibers to intradermal injection of capsaicin in the rat hindpaw, whereas intra-arterial injection of noradrenaline restored capsaicin-induced sensitization after sympathectomy (Ren et al., 2005). These effects may involve an adrenergic release of inflammatory mediators such as prostaglandins, or could entail direct stimulation of α_1-adrenoceptors on primary nociceptive afferents.

α-Adrenergic receptors may also contribute to vasodilatation evoked by heating the skin. This vasodilator response initially involves an axon reflex, followed by non-neural vasodilatation. Presynaptic adrenergic blockade with bretylium tosylate abolishes the initial axon reflex to gradual or rapid local heating of the skin, and greatly attenuates the peak vasodilator response to prolonged skin heating (Houghton et al., 2006; Hodges et al., 2008, 2009). Both noradrenaline and its cotransmitter neuropeptide Y are implicated in the vasodilator response to local heating, possibly via release of nitric oxide from the vascular endothelium. That is, heat may evoke the release of noradrenaline and neuropeptide Y from vasoconstrictor nerves which, in turn, triggers vasodilatation mediated by the vascular endothelium.

Alternatively, noradrenaline may bind directly to α_1-adrenoceptors on cutaneous nociceptive afferents (Dawson et al., 2011) (Fig. 24.3), thereby evoking antidromic release of substance P and calcitonin gene-related peptide from peripheral nerve terminals. One difficulty with this premise is that microdialysis of noradrenaline on the dorsal aspect of the human hand does not induce axon reflex vasodilatation (Zahn et al., 2004). However,

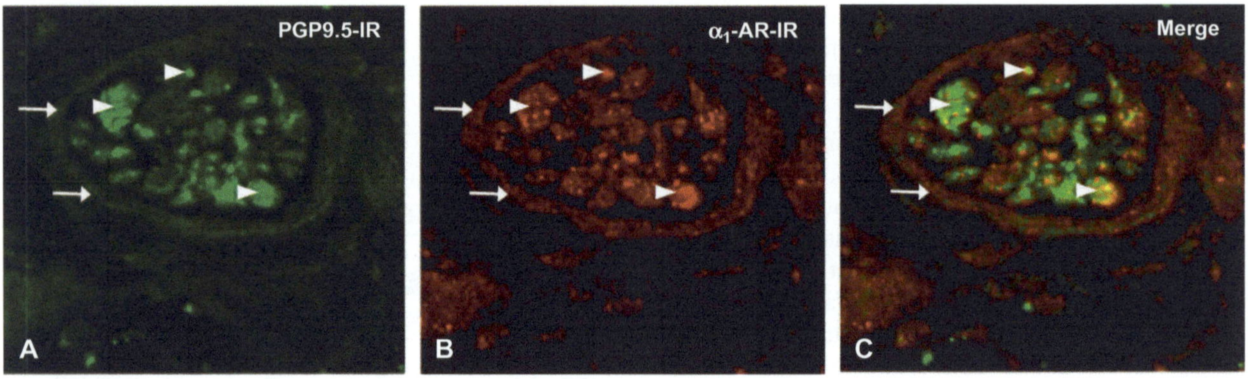

Fig. 24.3. Protein gene product 9.5 (PGP9.5: a general marker for all categories of nerve fiber) and α_1-adrenoceptor double-labeling illustrating α_1-adrenoceptor expression in a cutaneous nerve fascicle. In panels (**A**) and (**B**), paired confocal images illustrate PGP9.5 immunoreactivity (panel **A**) and α_1-adrenoceptor immunoreactivity (panel **B**). The merged image (panel **C**) illustrates colocalization of PGP9.5 immunoreactivity with α_1-adrenoceptor immunoreactivity. Arrows highlight α_1-adrenoceptors in the perineurium which was reactive for α_1-adrenoceptors but not PGP9.5, while arrow heads highlight examples of colocalization of PGP9.5 and α_1-adrenoceptors within nerve fibers in the fascicle. Scale bar in panel **B** = 50 μm. (Reprinted with permission from Dawson et al., 2011.)

under normal conditions, the perineurial sheath, which acts as a blood–nerve barrier, may prevent binding between α_1-adrenoceptors on cutaneous nociceptive afferents and noradrenaline in the extracellular fluid. If so, axon reflexes or inflammatory responses, evoked by heating the skin or electrical stimulation during iontophoresis, could disrupt the blood–nerve barrier and permit access to protected α_1-adrenoceptors.

An interaction between sympathetic and sensory nerves might play a normal physiological role in renal function. These fibers run close to each other and sometimes intertwine in the renal pelvic wall (Kopp et al., 2007). Reflex increases in efferent sympathetic nerve activity are associated with increases in afferent renal nerve activity. This relationship may be mediated by α_1-adrenoceptors on renal sensory nerves that release prostaglandin E_2 when activated; in turn, prostaglandin E_2 increases afferent renal nerve activity and evokes release of substance P (Kopp et al., 2007). The relationship between efferent sympathetic nerve activity and afferent renal nerve activity could be important in conditions that involve decreased responsiveness of the renal sensory nerves, such as sodium retention, hypertension, and heart failure.

ADRENERGIC INVOLVEMENT IN INFLAMMATORY PAIN

Sympathetic neural discharge does not normally activate nociceptors. Nevertheless, pharmacological release of adrenergic stores evokes thermal hyperalgesia in skin inflamed by topically applied capsaicin, possibly because of disruption to the perineurial blood–nerve barrier (Drummond, 1998). The hyperalgesic effect of noradrenaline is blocked by α-adrenergic antagonists, persists after arterial occlusion, and is independent of prostaglandin synthesis (Drummond, 1999; Schattschneider et al., 2007). The α_1-adrenoceptor agonist phenylephrine provokes thermal hyperalgesia in the mildly burnt skin of human volunteers, and this response is blocked by the α_1-antagonist terazosin (Drummond, 2009b). In contrast, α_2 agonists and antagonists have no effect on thermal hyperalgesia in this experimental paradigm. Guanethidine displaces noradrenaline from storage vesicles in sympathetic varicosities, thus evoking prolonged sympathetic blockade. To investigate effects of this treatment on sensitivity to heat, guanethidine was administered by iontophoresis to a 2 cm patch of skin four times over a 2 week period (Lipnicki and Drummond, 2001). The guanethidine treatment induced vascular signs of adrenergic supersensitivity and intensified sensitivity to noxious heat in inflamed skin in the presence of noradrenaline. Together, these findings indicate that adrenergic hyperalgesia is mediated by α_1-adrenoceptors in the inflamed skin of human subjects.

Pharmacological experiments in animals have also shown that α_1-adrenoceptors are involved in inflammatory pain, but this appears to depend on the type of pain model employed. For example, α_1-adrenoceptors mediated pain behaviors after subcutaneous injection of formalin or α-methyl-5-hydroxytryptamine plus noradrenaline into the plantar surface of rat paws (Hong and Abbott, 1996), whereas both α-adrenoceptor subtypes contributed to mechanical hyperalgesia after inflammation of the L5 spinal nerve of rats with complete Freund's adjuvant (Baik et al., 2003). However, α_2-adrenoceptors mediated the excitatory effect of lumbar sympathetic chain stimulation and close arterial injection of noradrenaline on nociceptive afferent discharge in the saphenous nerve of adjuvant-inflamed rats (Sato et al., 1993). Ren et al. (2005) investigated the role of sympathetic efferent fibers in sensitization of A-δ and C primary nociceptive afferents in the tibial nerve to mechanical stimulation following intradermal injection of capsaicin in the hindpaw of rats. Sensitization of C fibers decreased after postganglionic sympathectomy and after α_1-adrenoceptor blockade with terazosin, but was restored by intra-arterial injection of noradrenaline; however, blocking α_2-adrenergic receptors had little effect on sensitivity of C-fiber nociceptors. Similar findings have been reported by other research groups (Dogrul et al., 2006; Meisner et al., 2007). Thus, there seems to be some consensus that α_1-adrenoceptors augment hyperalgesia in inflamed skin.

An adrenergic component of hyperalgesia might be important in a range of chronic neuroinflammatory diseases. For example, activation of α_1-adrenoceptors on sensory afferents in the lower urinary tract may contribute to pain in patients with chronic pelvic symptoms (Trevisani et al., 2007). In support of this possibility, activation of α_1-adrenoceptors with phenylephrine evoked a calcium-dependent release of substance P from nociceptive afferents in the rat urinary bladder, in association with plasma extravasation (Trevisani et al., 2007). Conversely, blockade of α_1-adrenoceptors reduced signs of nociceptor activation to the chemotherapy agent cyclophosphamide, both in the urinary bladder and the lumbar spinal cord. Evidence that α_1-adrenoceptors play a role in chronic neuropathic pain syndromes is discussed in more detail below.

EFFECTS OF NERVE INJURY

After peripheral nerve injury, the high concentration of trophic agents produced by the injured and denervated tissue initiates regeneration of axotomized fibers and sprouting of collateral twigs from nearby surviving fibers (Diamond et al., 1992; Hendrix and Peters, 2007; Navarro et al., 2007). The regenerating fibers and collateral sprouts grow down vacated perineurial sheaths and may eventually make functional connections with denervated end-organs. Unfortunately, however, this regenerative process sometimes results in cross-innervation that is quite inappropriate. As sympathetic, parasympathetic, and sensory nerve fibers compete for trophic agents such nerve growth factor (Kessler, 1985), damage to one nerve fiber category is accompanied by disproportionate growth in other nerve fiber categories. For example, if sympathetic nerve fibers are selectively destroyed, they may be replaced by collateral sprouts from nearby sensory or parasympathetic nerve fibers. This could result in paradoxical sweating or flushing if messages destined for parasympathetic end-organs inadvertently find their way to sympathetic targets. Conversely, cross-innervation of sympathetic and parasympathetic end-organs by nociceptive afferents might sensitize these tissues to pain.

A second compensatory mechanism involves heightened sensitivity to local neurotransmitters, either through reductions in neurotransmitter uptake and/or enzymatic degradation, or through the development of postjunctional supersensitivity (Fleming and Westfall, 1988; Tripovic et al., 2010). Whether neurotransmitter depletion evokes specific or nonspecific postjunctional supersensitivity seems to depend on the target receptor and tissue. For example, after injury to sympathetic vasoconstrictor fibers, postjunctional supersensitivity develops not only to adrenergic agonists but also to vasoconstrictors such as vasopressin and angiotensin II, due to loss of the noradrenaline transporter and a selective increase in vascular reactivity to circulating vasoconstrictor agents (Tripovic et al., 2010). However, within the central nervous system, injury to adrenergic nerve fibers is associated with signs of specific adrenergic denervation supersensitivity, mediated by an increased density of postsynaptic adrenergic receptors (Hanft and Gross, 1990; Nowak et al., 1991; Roudet et al., 1993; Roudet et al., 1994; Giroux et al., 1999).

Nerve injury may also provoke expression of adrenergic receptors within the peripheral nervous system. Indeed, in rats, messenger RNA for α_{1B}-adrenoceptors increases in dorsal root ganglia following peripheral nerve section or ligation of spinal nerves supplying those ganglia (Xie et al., 2001; Maruo et al., 2006). Using whole cell patch clamp recordings, Petersen et al. (1996) showed that the proportion of DRG neurons that were activated by noradrenaline increased markedly after chronic nerve injury induced by loose or tight ligation of the sciatic nerve in the rat. Conversely, the α_1-adrenoceptor antagonist prazosin was found to be analgesic and to inhibit nociceptive fiber discharge in rat models of neuropathy when given systemically (Nam et al., 2000; Hord et al., 2001; Kim et al., 2005). In the spinal nerve ligation model, excitatory effects of noradrenaline on surviving nociceptive afferents were mediated by α_1-adrenoceptors in monkeys (Ali et al., 1999) and rats (Lee et al., 1999; Nam et al., 2000), but were mediated by α_2-adrenoceptors after partial lesion of the auricular nerve of rabbits (Sato and Perl, 1991) and after sciatic nerve transection in rats (Chen et al., 1996). Thus, the expression and excitability of α-adrenoceptor subtypes after nerve injury may vary across animal species or the type of nerve injury.

SENSORY–AUTONOMIC INTERACTIONS IN NEUROPATHIC PAIN

Lesions of the peripheral nervous system, induced by injury or disease, can trigger chronic neuropathic pain.

Paradoxically, such syndromes are often characterized by an analgesic response to sympathetic blockade, implying that the pain is maintained by activity in a surviving population of sympathetic fibers (Birklein, 2005). The pathophysiology of this sympathetically maintained pain is not known, but may involve an aberrant excitation of α_1-adrenergic receptors on blood vessels and pain-signaling (nociceptive) afferent fibers (Gibbs et al., 2008). There is also substantial evidence that immune cells and proinflammatory mediators play a critical role in the development and maintenance of neuropathic pain, and that this is also modulated by adrenergic activity (Maestroni, 2006; Thacker et al., 2007). Thus, the sympathetic nervous system may contribute to chronic neuropathic pain via several key interlinked mechanisms. In particular, injury to nociceptive afferent fibers, together with sympathetic nerve dysfunction, could trigger a compensatory increase in sensitivity to adrenergic agents that heightens the excitability of intact or regenerating nociceptive neurons, alters blood flow, and intensifies inflammation. These disturbances may then build upon each other, resulting in clinical deterioration (Fig. 24.4).

Complex regional pain syndrome (CRPS) type I (previously known as reflex sympathetic dystrophy) and CRPS type II (causalgia) are considered to be chronic neuropathic pain conditions – CRPS type I is associated with peripheral small-fiber neuropathy (Oaklander et al., 2006) whereas CRPS type II develops after major peripheral nerve injury (e.g., surgery or a crush injury). Pain is disproportionate to the initial injury, and often spreads. During the early stages of CRPS, affected skin typically is flushed, hot and dry, suggestive of vasodilatation and loss of sympathetic sudomotor outflow; but later on the skin becomes cold, clammy, and cyanotic, suggestive of excessive sympathetic neural discharge (Birklein, 2005). Despite this clinical picture, there is no direct evidence of an increase in sympathetic outflow in chronically affected limbs (Casale and Elam, 1992), no increase in venous noradrenaline levels (Drummond et al., 1991; Harden et al., 1994), and no increase in reflex vasoconstrictor responses (Rosen et al., 1988).

One explanation for this paradox is that an underlying sympathetic deficit initiates compensatory increases in the sensitivity or expression of adrenergic receptors in affected tissues, specifically that of blood vessels, sweat glands and sensory nerves. Indeed, venous noradrenaline and its metabolites are lower in the affected than the unaffected limbs of CRPS patients (Drummond et al., 1991; Harden et al., 1994), and the density of α_1-adrenoceptors is greater in the affected skin of CRPS patients than in controls (Drummond et al., 1996). Moreover, in patients with longstanding pain, cutaneous blood vessels (Arnold et al., 1993), sweat glands (Chemali et al., 2001), and nociceptive afferents (Jorum et al., 2007) are supersensitive to adrenergic agents. The CRPS-affected limb is also characterized by enhanced axon-reflex vasodilatation, elevated levels of inflammatory mediators, and cutaneous hypoxia (Birklein, 2005). Thus, heightened sensitivity to adrenergic agents appears to be associated with chronic inflammation, which may cause the condition to persist.

Nociceptive afferents in conditions such as postherpetic neuralgia and CRPS acquire an abnormal excitability to adrenergic agents. Injection of the α_1-adrenoceptor agonist phenylephrine into zoster-affected skin in humans evokes abnormal pain (Choi and Rowbotham, 1997), and cells in rat dorsal root ganglia infected with the varicella-zoster virus gain an unusual sensitivity to phenylephrine (Schmidt et al., 2003). Similarly, intradermal

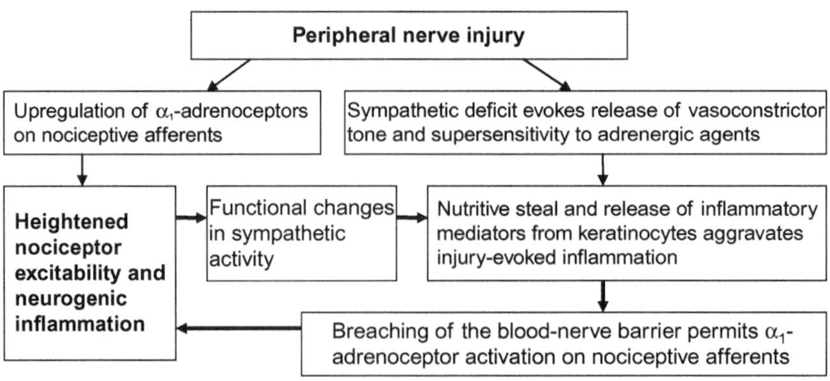

Fig. 24.4. Putative model of sensory–sympathetic interaction in neuropathic pain syndromes. A peripheral nerve injury that involves loss of sensory and sympathetic nerve fibers may initially evoke signs of sympathetic deficit (a warm dry limb) but later on the limb becomes cool and moist due to the development of adrenergic supersensitivity. Blood bypasses the superficial layers of skin by flowing through sympathetically denervated arteriovenous shunts, thereby aggravating inflammation, disrupting the blood–nerve barrier and exposing excitatory α_1-adrenoceptors on nociceptive afferents. This heightened nociceptor excitability not only contributes to pain but may also facilitate axon reflexes and further aggravate inflammation.

injection of adrenergic agonists into the CRPS-affected limb evokes pain in a subgroup of patients (Ali et al., 2000). Pain also increases during sympathetic arousal and body cooling (strong stimuli for increased sympathetic vasoconstrictor activity) (Drummond et al., 2001; Baron et al., 2002), and hyperalgesia spreads. In a recent case report, nociceptive fibers in the affected limb of a patient with neuropathic pain were excited by sympathetic activity and by intradermal injection of noradrenaline (Jorum et al., 2007). Although neurophysiological interaction between nociceptive afferents and sympathetic efferents may be the exception rather than the rule in CRPS (Campero et al., 2010), nonetheless, sympathetic agents appear to evoke pain in certain neuropathic pain syndromes, possibly due to direct excitation of adrenergic receptors on sensory nerves.

Vasoconstrictor reflexes are diminished in the early stages of CRPS (Wasner, 2010), consistent with peripheral nerve deficit. Nevertheless, vasoconstrictor responses to directly applied adrenergic agents are increased in the affected limb of CRPS patients (Arnold et al., 1993), in animal models of neuropathic pain (Kurvers et al., 1998), in animals with chronic post-ischemia pain (an animal model of CRPS type I) (Xanthos et al., 2008), and in patients with chronic limb ischemia associated with peripheral artery disease (Jarajapu et al., 2001). Thus, CRPS and related conditions may be associated with a supersensitivity to adrenergic agents. By limiting blood flow through already ischemic tissue, adrenergic supersensitivity in vascular smooth muscle may promote acidosis, hypoxia and pain.

Sympathetic involvement in pain mechanisms has been studied in greater detail in CRPS than in other chronic pain syndromes. However, some important similarities have begun to emerge across a spectrum of chronic pain disorders. In erythromelalgia, the painfully hot and swollen limbs result, in part, from chronic inflammation due to deficient nutritive perfusion of affected tissues (Cohen, 2000). Sympathetic vasoconstrictor reflexes are impaired, implying that sympathetic dysfunction contributes to microvascular insufficiency and pain (Mork et al., 2002). Similarly, in patients with painful diabetic neuropathy, sympathetic denervation and vascular endothelial dysfunction are associated with reduced dermal nutritive flow, hypoxia, and pain (Tack et al., 2002; Quattrini et al., 2007). Experimental findings have linked these clinical observations with an upregulation of adrenoceptors in peripheral nerves. For example, in an animal model of diabetic neuropathy, stimulation of α_1-adrenoceptors aggravated allodynia; moreover, messenger RNA and binding sites for α_1-adrenoceptors were increased in dorsal root ganglia (Lee et al., 2000). Thus, each of these conditions presents a similar clinical picture: an underlying sympathetic deficit is associated with altered blood flow, chronic inflammation, and pain. The inflammation and associated nutritive insufficiency may aggravate the primary disorder and contribute to clinical deterioration (Fig. 24.4).

A sympathetic deficit associated with adrenergic supersensitivity may also mediate abnormal sweating in the affected limb of patients with CRPS. In support of this possibility, iontophoresis of the α_1-adrenoceptor agonist phenylephrine evoked greater sweating in CRPS-affected skin than in unaffected skin or in healthy volunteers or patients with resolved CRPS (Chemali et al., 2001). Although the sympathetic innervation of eccrine sweat glands is primarily cholinergic in human skin, these glands also receive a rudimentary adrenergic supply (Donadio et al., 2006; Morgan et al., 2006). Thus, an increased expression of α_1-adrenoceptors on sudomotor fibers might contribute to hyperhidrosis in neuropathic pain syndromes such as CRPS.

SENSORY–AUTONOMIC INTERACTIONS IN CRANIAL DYSAUTONOMIAS

Most people sweat and flush when eating highly spiced food, particularly in warm environments (Lee, 1954). Gustatory sweating is mediated by increases in sympathetic sudomotor activity, whereas gustatory flushing is mediated by trigeminal-parasympathetic vasodilatation and is opposed by sympathetic vasoconstrictor tone (Drummond, 1995). In contrast, after peripheral nerve injury, misdirected regeneration and collateral sprouting of salivatory fibers in the cervical sympathetic chain or cranial nerves mediate aberrant sweating and flushing in areas of sympathetic denervation. This has been noted in a spectrum of autonomic disorders including Frey's syndrome, harlequin syndrome, and Ross's syndrome (Lance et al., 1988; Drummond and Edis, 1990; Drummond and Lance 1993; Shin et al., 2000; Kalapesi et al., 2005; Willaert et al., 2009), and in trigeminal autonomic cephalalgias such as cluster headache (Drummond and Lance, 1992; Drummond, 2006b).

Cluster headaches typically cycle in bouts that last weeks or months, followed by periods of complete freedom from headache. During bouts, intense unilateral headaches lasting minutes to hours recur once or more per day. The headaches may be associated with transient or persistent signs of cervical sympathetic deficit on the side of headache (miosis, ptosis, and loss of thermoregulatory sweating and flushing), and attacks are accompanied by signs of ipsilateral parasympathetic activity (lacrimation, rhinorrhea, and facial flushing) (Drummond, 2006b). Attacks may also be associated with paradoxical sweating in sympathetically denervated regions of the face. The most likely explanation for attack-related disturbances is pain-related activation of trigeminal-parasympathetic reflexes involving the lacrimal glands and the blood vessels that supply those glands. The internal carotid artery and its ophthalmic branch appear to swell during attacks of cluster headache (Ekbom and Greitz, 1970; Waldenlind et al., 1993; May et al., 2000), possibly due to an axon reflex. Thus, vascular compression of the sympathetic nerve plexus that surrounds the internal carotid artery could account for signs of postganglionic sympathetic deficit. In addition, cross-innervation of sympathetically denervated forehead sweat glands and blood vessels by parasympathetic lacrimal fibers may induce sweating and flushing in this region during attacks (Drummond and Lance, 1992). The attack itself is likely to originate in the brainstem (Goadsby, 2002); nevertheless, secondary vascular disturbances could amplify pain. In particular, trigeminal-parasympathetic vasodilator reflexes might activate sensitized vascular nociceptive afferents which, in turn, intensifies trigeminal-parasympathetic vasodilatation and possibly also initiates axon-reflex vasodilatation. Loss of sympathetic vasoconstrictor tone could further compound vasodilatation and pain (Drummond, 2006b).

CONCLUSIONS

Although the sensory and autonomic nervous systems generally are regarded as discrete structures, aberrant interactions between sensory and autonomic nerves mediate disturbances in sweating and cutaneous blood flow in certain disease states. An upregulation of α_1-adrenoceptors on intact or regenerating nociceptive neurons in injured or chronically inflamed tissue may be particularly important in patients with chronic neuropathic pain, as this might not only aggravate symptoms but also contribute to clinical deterioration by altering local circulatory dynamics. Despite substantial evidence of involvement of the sympathetic nervous system in neuropathic and inflammatory pain models, including evidence of α_1-adrenoceptor upregulation (Xie et al., 2001; Maruo et al., 2006), this has yet to be incorporated into guidelines for the assessment and treatment of patients with chronic neuropathic pain. Thus, priority should be given to research that evaluates the therapeutic benefits α_1-adrenoceptor blockade for patients with these intractable syndromes.

References

Ali Z, Ringkamp M, Hartke TV et al. (1999). Uninjured C-fiber nociceptors develop spontaneous activity and alpha-adrenergic sensitivity following L6 spinal nerve ligation in monkey. J Neurophysiol 81: 455–466.

Ali Z, Raja SN, Wesselmann U et al. (2000). Intradermal injection of norepinephrine evokes pain in patients with sympathetically maintained pain. Pain 88: 161–168.

Arnold JM, Teasell RW, MacLeod AP et al. (1993). Increased venous alpha-adrenoceptor responsiveness in patients with reflex sympathetic dystrophy. Ann Intern Med 118: 619–621.

Baik E, Chung JM, Chung K (2003). Peripheral norepinephrine exacerbates neuritis-induced hyperalgesia. J Pain 4: 212–221.

Baron R, Schattschneider J, Binder A et al. (2002). Relation between sympathetic vasoconstrictor activity and pain and hyperalgesia in complex regional pain syndromes: a case-control study. Lancet 359: 1655–1660.

Birklein F (2005). Complex regional pain syndrome. J Neurol 252: 131–138.

Bradley MM (2009). Natural selective attention: orienting and emotion. Psychophysiology 46: 1–11.

Campero M, Bostock H, Baumann TK et al. (2010). A search for activation of C nociceptors by sympathetic fibers in complex regional pain syndrome. Clin Neurophysiol 121: 1072–1079.

Casale R, Elam M (1992). Normal sympathetic nerve activity in a reflex sympathetic dystrophy with marked skin vasoconstriction. J Auton Nerv Syst 41: 215–219.

Charkoudian N (2010). Mechanisms and modifiers of reflex induced cutaneous vasodilation and vasoconstriction in humans. J Appl Physiol 109: 1221–1228.

Chemali KR, Gorodeski R, Chelimsky TC (2001). Alpha-adrenergic supersensitivity of the sudomotor nerve in complex regional pain syndrome. Ann Neurol 49: 453–459.

Chen Y, Michaelis M, Janig W et al. (1996). Adrenoreceptor subtype mediating sympathetic-sensory coupling in injured sensory neurons. J Neurophysiol 76: 3721–3730.

Choi B, Rowbotham MC (1997). Effect of adrenergic receptor activation on post-herpetic neuralgia pain and sensory disturbances. Pain 69: 55–63.

Cohen JS (2000). Erythromelalgia: new theories and new therapies. J Am Acad Dermatol 43: 841–847.

Dawson LF, Phillips JK, Finch PM et al. (2011). Expression of alpha(1)-adrenoceptors on peripheral nociceptive neurons. Neuroscience 175: 300–314.

Diamond J, Holmes M, Coughlin M (1992). Endogenous NGF and nerve impulses regulate the collateral sprouting of sensory axons in the skin of the adult rat. J Neurosci 12: 1454–1466.

Dogrul A, Coskun I, Uzbay T (2006). The contribution of alpha-1 and alpha-2 adrenoceptors in peripheral imidazoline and adrenoceptor agonist-induced nociception. Anesth Analg 103: 471–477, table of contents.

Donadio V, Nolano M, Provitera V et al. (2006). Skin sympathetic adrenergic innervation: an immunofluorescence confocal study. Ann Neurol 59: 376–381.

Drummond PD (1991). Cervical sympathetic deficit in unilateral migraine headache. Headache 31: 669–672.

Drummond PD (1992). The mechanism of facial sweating and cutaneous vascular responses to painful stimulation of the eye. Brain 115: 1417–1428.

Drummond PD (1995). Mechanisms of physiological gustatory sweating and flushing in the face. J Auton Nerv Syst 52: 117–124.

Drummond PD (1997). Photophobia and autonomic responses to facial pain in migraine. Brain 120: 1857–1864.

Drummond PD (1998). Enhancement of thermal hyperalgesia by alpha-adrenoceptors in capsaicin-treated skin. J Auton Nerv Syst 69: 96–102.

Drummond PD (1999). Nitroprusside inhibits thermal hyperalgesia induced by noradrenaline in capsaicin-treated skin. Pain 80: 405–412.

Drummond PD (2002). Prior iontophoresis of saline enhances vasoconstriction to phenylephrine and clonidine in the skin of the human forearm. Br J Clin Pharmacol 54: 45–50.

Drummond PD (2004). The effect of cutaneous mast cell degranulation on sensitivity to heat. Inflamm Res 53: 309–315.

Drummond PD (2006a). Immersion of the hand in ice water releases adrenergic vasoconstrictor tone in the ipsilateral temple. Auton Neurosci 128: 70–75.

Drummond PD (2006b). Mechanisms of autonomic disturbance in the face during and between attacks of cluster headache. Cephalalgia 26: 633–641.

Drummond PD (2009a). Alpha-1 adrenoceptor stimulation triggers axon-reflex vasodilatation in human skin. Auton Neurosci 151: 159–163.

Drummond PD (2009b). alpha(1)-Adrenoceptors augment thermal hyperalgesia in mildly burnt skin. Eur J Pain 13: 273–279.

Drummond PD (2011). Inflammation contributes to axon reflex vasodilatation evoked by iontophoresis of an alpha-1 adrenoceptor agonist. Auton Neurosci 159: 90–97.

Drummond PD, Edis RH (1990). Loss of facial sweating and flushing in Holmes–Adie syndrome. Neurology 40: 847–849.

Drummond PD, Granston A (2003). Facilitation of extracranial vasodilatation to limb pain in migraine sufferers. Neurology 61: 60–63.

Drummond PD, Lance JW (1992). Pathological sweating and flushing accompanying the trigeminal lacrimal reflex in patients with cluster headache and in patients with a confirmed site of cervical sympathetic deficit. Evidence for parasympathetic cross-innervation. Brain 115: 1429–1445.

Drummond PD, Lance JW (1993). Site of autonomic deficit in harlequin syndrome: local autonomic failure affecting the arm and the face. Ann Neurol 34: 814–819.

Drummond PD, Lipnicki DM (1999). Noradrenaline provokes axon reflex hyperaemia in the skin of the human forearm. J Auton Nerv Syst 77: 39–44.

Drummond PD, Finch PM, Smythe GA (1991). Reflex sympathetic dystrophy: the significance of differing plasma catecholamine concentrations in affected and unaffected limbs. Brain 114: 2025–2036.

Drummond PD, Skipworth S, Finch PM (1996). alpha 1-adrenoceptors in normal and hyperalgesic human skin. Clin Sci (Lond) 91: 73–77.

Drummond PD, Finch PM, Skipworth S et al. (2001). Pain increases during sympathetic arousal in patients with complex regional pain syndrome. Neurology 57: 1296–1303.

Ekbom K, Greitz T (1970). Carotid angiography in cluster headache. Acta Radiol Diagn (Stockh) 10: 177–186.

Fleming W, Westfall DP (1988). Adaptive supersensitivity. In: U Trendelenberg, N Weiner (Eds.), Catecholamines. I Handbook of Experimental Pharmacology, Vol. 90. Springer-Verlag, Berlin/Heidelberg, pp. 509–559.

Gibbs GF, Drummond PD, Finch PM et al. (2008). Unravelling the pathophysiology of complex regional pain syndrome: focus on sympathetically maintained pain. Clin Exp Pharmacol Physiol 35: 717–724.

Giroux N, Rossignol S, Reader TA (1999). Autoradiographic study of alpha1- and alpha2-noradrenergic and serotonin1A receptors in the spinal cord of normal and chronically transected cats. J Comp Neurol 406: 402–414.

Goadsby PJ (2002). Pathophysiology of cluster headache: a trigeminal autonomic cephalgia. Lancet Neurol 1: 251–257.

Grando SA, Pittelkow MR, Schallreuter KU (2006). Adrenergic and cholinergic control in the biology of epidermis: physiological and clinical significance. J Invest Dermatol 126: 1948–1965.

Hanft G, Gross G (1990). The effect of reserpine, desipramine and thyroid hormone on alpha 1a- and alpha 1b-adrenoceptor binding sites: evidence for a subtype-specific regulation. Br J Clin Pharmacol 30 (Suppl 1): 125S–127S.

Harden RN, Duc TA, Williams TR et al. (1994). Norepinephrine and epinephrine levels in affected versus unaffected limbs in sympathetically maintained pain. Clin J Pain 10: 324–330.

Hendrix S, Peters EM (2007). Neuronal plasticity and neuroregeneration in the skin – the role of inflammation. J Neuroimmunol 184: 113–126.

Hodges GJ, Johnson JM (2009). Adrenergic control of the human cutaneous circulation. Appl Physiol Nutr Metab 34: 829–839.

Hodges GJ, Kosiba WA, Zhao K et al. (2008). The involvement of norepinephrine, neuropeptide Y, and nitric oxide in the cutaneous vasodilator response to local heating in humans. J Appl Physiol 105: 233–240.

Hodges GJ, Kosiba WA, Zhao K et al. (2009). The involvement of heating rate and vasoconstrictor nerves in the cutaneous vasodilator response to skin warming. Am J Physiol Heart Circ Physiol 296: H51–H56.

Holowatz LA, Kenney WL (2010). Peripheral mechanisms of thermoregulatory control of skin blood flow in aged humans. J Appl Physiol 109: 1538–1544.

Hong Y, Abbott FV (1996). Contribution of peripheral alpha 1A-adrenoceptors to pain induced by formalin or by alpha-methyl-5-hydroxytryptamine plus noradrenaline. Eur J Pharmacol 301: 41–48.

Hord AH, Denson DD, Stowe B et al. (2001). alpha-1 and alpha-2 Adrenergic antagonists relieve thermal hyperalgesia in experimental mononeuropathy from chronic constriction injury. Anesth Analg 92: 1558–1562.

Houghton BL, Meendering JR, Wong BJ et al. (2006). Nitric oxide and noradrenaline contribute to the temperature threshold of the axon reflex response to gradual local heating in human skin. J Physiol 572: 811–820.

Jarajapu YP, Coats P, McGrath JC et al. (2001). Increased alpha(1)- and alpha(2)-adrenoceptor-mediated contractile responses of human skeletal muscle resistance arteries in chronic limb ischemia. Cardiovasc Res 49: 218–225.

Jorum E, Orstavik K, Schmidt R et al. (2007). Catecholamine-induced excitation of nociceptors in sympathetically maintained pain. Pain 127: 296–301.

Kalapesi FB, Krishnan AV, Kiernan MC (2005). Segmental facial anhidrosis and tonic pupils with preserved deep tendon reflexes: a novel autonomic neuropathy. J Neuroophthalmol 25: 5–8.

Kamijo Y, Lee K, Mack GW (2005). Active cutaneous vasodilation in resting humans during mild heat stress. J Appl Physiol 98: 829–837.

Kavelaars A (2002). Regulated expression of alpha-1 adrenergic receptors in the immune system. Brain Behav Immun 16: 799–807.

Kellogg DL Jr (2006). In vivo mechanisms of cutaneous vasodilation and vasoconstriction in humans during thermoregulatory challenges. J Appl Physiol 100: 1709–1718.

Kellogg DL Jr, Zhao JL, Wu Y et al. (2010). VIP/PACAP receptor mediation of cutaneous active vasodilation during heat stress in humans. J Appl Physiol 109: 95–100.

Kessler JA (1985). Parasympathetic, sympathetic, and sensory interactions in the iris: nerve growth factor regulates cholinergic ciliary ganglion innervation in vivo. J Neurosci 5: 2719–2725.

Kim SK, Min BI, Kim JH et al. (2005). Effects of alpha1- and alpha2-adrenoreceptor antagonists on cold allodynia in a rat tail model of neuropathic pain. Brain Res 1039: 207–210.

Kopp UC, Cicha MZ, Smith LA et al. (2007). Renal sympathetic nerve activity modulates afferent renal nerve activity by PGE2-dependent activation of alpha1- and alpha2-adrenoceptors on renal sensory nerve fibers. Am J Physiol Regul Integr Comp Physiol 293: R1561–R1572.

Kreyden OP, Scheidegger EP (2004). Anatomy of the sweat glands, pharmacology of botulinum toxin, and distinctive syndromes associated with hyperhidrosis. Clin Dermatol 22: 40–44.

Kurvers H, Daemen M, Slaaf D et al. (1998). Partial peripheral neuropathy and denervation induced adrenoceptor supersensitivity. Functional studies in an experimental model. Acta Orthop Belg 64: 64–70.

Lance JW, Drummond PD, Gandevia SC et al. (1988). Harlequin syndrome: the sudden onset of unilateral flushing and sweating. J Neurol Neurosurg Psychiatry 51: 635–642.

Lee TS (1954). Physiological gustatory sweating in a warm climate. J Physiol 124: 528–542.

Lee DH, Liu X, Kim HT et al. (1999). Receptor subtype mediating the adrenergic sensitivity of pain behavior and ectopic discharges in neuropathic Lewis rats. J Neurophysiol 81: 2226–2233.

Lee YH, Ryu TG, Park SJ et al. (2000). Alpha1-adrenoceptors involvement in painful diabetic neuropathy: a role in allodynia. Neuroreport 11: 1417–1420.

Lin Q, Zou X, Fang L et al. (2003). Sympathetic modulation of acute cutaneous flare induced by intradermal injection of capsaicin in anesthetized rats. J Neurophysiol 89: 853–861.

Lindsey KQ, Caughman SW, Olerud JE et al. (2000). Neural regulation of endothelial cell-mediated inflammation. J Investig Dermatol Symp Proc 5: 74–78.

Lipnicki DM, Drummond PD (2001). Vascular and nociceptive effects of localized prolonged sympathetic blockade in human skin. Auton Neurosci 88: 86–93.

Low PA, Caskey PE, Tuck RR et al. (1983). Quantitative sudomotor axon reflex test in normal and neuropathic subjects. Ann Neurol 14: 573–580.

Macefield VG, Sverrisdottir YB, Elam M et al. (2008). Firing properties of sudomotor neurones in hyperhidrosis and thermal sweating. Clin Auton Res 18: 325–330.

Maestroni GJ (2006). Sympathetic nervous system influence on the innate immune response. Ann N Y Acad Sci 1069: 195–207.

Magerl W, Koltzenburg M, Schmitz JM et al. (1996). Asymmetry and time-course of cutaneous sympathetic reflex responses following sustained excitation of chemosensitive nociceptors in humans. J Auton Nerv Syst 57: 63–72.

Magerl W, Treede RD (1996). Heat-evoked vasodilatation in human hairy skin: axon reflexes due to low-level activity of nociceptive afferents. J Physiol 497: 837–848.

Maruo K, Yamamoto H, Yamamoto S et al. (2006). Modulation of P2X receptors via adrenergic pathways in rat dorsal root ganglion neurons after sciatic nerve injury. Pain 120: 106–112.

May A, Bahra A, Buchel C et al. (2000). PET and MRA findings in cluster headache and MRA in experimental pain. Neurology 55: 1328–1335.

Meisner JG, Waldron JB, Sawynok J (2007). Alpha1-adrenergic receptors augment P2X3 receptor-mediated nociceptive responses in the uninjured state. J Pain 8: 556–562.

Morgan CJ, Friedmann PS, Church MK et al. (2006). Cutaneous microdialysis as a novel means of continuously stimulating eccrine sweat glands in vivo. J Invest Dermatol 126: 1220–1225.

Mork C, Kalgaard OM, Kvernebo K (2002). Impaired neurogenic control of skin perfusion in erythromelalgia. J Invest Dermatol 118: 699–703.

Nam TS, Yeon DS, Leem JW et al. (2000). Adrenergic sensitivity of uninjured C-fiber nociceptors in neuropathic rats. Yonsei Med J 41: 252–257.

Navarro X, Vivo M, Valero-Cabre A (2007). Neural plasticity after peripheral nerve injury and regeneration. Prog Neurobiol 82: 163–201.

Nowak G, Zak J, Superata J (1991). Role of dopaminergic neurons in denervation-induced alpha 1-adrenergic up-regulation in the rat cerebral cortex. J Neurochem 56: 914–916.

Oaklander AL, Rissmiller JG, Gelman LB et al. (2006). Evidence of focal small-fiber axonal degeneration in complex regional pain syndrome-I (reflex sympathetic dystrophy). Pain 120: 235–243.

Petersen M, Zhang J, Zhang JM et al. (1996). Abnormal spontaneous activity and responses to norepinephrine in dissociated dorsal root ganglion cells after chronic nerve constriction. Pain 67: 391–397.

Quattrini C, Harris ND, Malik RA et al. (2007). Impaired skin microvascular reactivity in painful diabetic neuropathy. Diabetes Care 30: 655–659.

Ren Y, Zou X, Fang L et al. (2005). Sympathetic modulation of activity in Adelta- and C-primary nociceptive afferents after intradermal injection of capsaicin in rats. J Neurophysiol 93: 365–377.

Rosen L, Ostergren J, Fagrell B et al. (1988). Skin microvascular circulation in the sympathetic dystrophies evaluated by videophotometric capillaroscopy and laser Doppler fluxmetry. Eur J Clin Invest 18: 305–308.

Roudet C, Savasta M, Feuerstein C (1993). Normal distribution of alpha-1-adrenoceptors in the rat spinal cord and its modification after noradrenergic denervation: a quantitative autoradiographic study. J Neurosci Res 34: 44–53.

Roudet C, Mouchet P, Feuerstein C et al. (1994). Normal distribution of alpha 2-adrenoceptors in the rat spinal cord and its modification after noradrenergic denervation: a quantitative autoradiographic study. J Neurosci Res 39: 319–329.

Sato J, Perl ER (1991). Adrenergic excitation of cutaneous pain receptors induced by peripheral nerve injury. Science 251: 1608–1610.

Sato K, Sato F (1983). Individual variations in structure and function of human eccrine sweat gland. Am J Physiol 245: R203–R208.

Sato J, Suzuki S, Iseki T et al. (1993). Adrenergic excitation of cutaneous nociceptors in chronically inflamed rats. Neurosci Lett 164: 225–228.

Schattschneider J, Zum Buttel I, Binder A et al. (2007). Mechanisms of adrenosensitivity in capsaicin induced hyperalgesia. Eur J Pain 11: 756–763.

Schmidt M, Kress M, Heinemann S et al. (2003). Varicella-zoster virus isolates, but not the vaccine strain OKA, induce sensitivity to alpha-1 and beta-1 adrenergic stimulation of sensory neurones in culture. J Med Virol 70 (Suppl 1): S82–S89.

Shibasaki M, Crandall CG (2010). Mechanisms and controllers of eccrine sweating in humans. Front Biosci (Schol Ed) 2: 685–696.

Shin RK, Galetta SL, Ting TY et al. (2000). Ross syndrome plus: beyond Horner, Holmes–Adie, and harlequin. Neurology 55: 1841–1846.

Stephens DP, Saad AR, Bennett LA et al. (2004). Neuropeptide Y antagonism reduces reflex cutaneous vasoconstriction in humans. Am J Physiol Heart Circ Physiol 287: H1404–H1409.

Swartling C, Naver H, Pihl-Lundin I et al. (2004). Sweat gland morphology and periglandular innervation in essential palmar hyperhidrosis before and after treatment with intradermal botulinum toxin. J Am Acad Dermatol 51: 739–745.

Tack CJ, van Gurp PJ, Holmes C et al. (2002). Local sympathetic denervation in painful diabetic neuropathy. Diabetes 51: 3545–3553.

Thacker MA, Clark AK, Marchand F et al. (2007). Pathophysiology of peripheral neuropathic pain: immune cells and molecules. Anesth Analg 105: 838–847.

Trevisani M, Campi B, Gatti R et al. (2007). The influence of alpha1-adrenoreceptors on neuropeptide release from primary sensory neurons of the lower urinary tract. Eur Urol 52: 901–908.

Tripovic D, Pianova S, McLachlan EM et al. (2010). Transient supersensitivity to alpha-adrenoceptor agonists, and distinct hyper-reactivity to vasopressin and angiotensin II after denervation of rat tail artery. Br J Pharmacol 159: 142–153.

Van Beaumont W, Bullard RW (1963). Sweating: its rapid response to muscular work. Science 141: 643–646.

Waldenlind E, Ekbom K, Torhall J (1993). MR-angiography during spontaneous attacks of cluster headache: a case report. Headache 33: 291–295.

Wasner G (2010). Vasomotor disturbances in complex regional pain syndrome – a review. Pain Med 11: 1267–1273.

Willaert WI, Scheltinga MR, Steenhuisen SF et al. (2009). Harlequin syndrome: two new cases and a management proposal. Acta Neurol Belg 109: 214–220.

Xanthos DN, Bennett GJ, Coderre TJ (2008). Norepinephrine-induced nociception and vasoconstrictor hypersensitivity in rats with chronic post-ischemia pain. Pain 137: 640–651.

Xie J, Ho Lee Y, Wang C et al. (2001). Differential expression of alpha1-adrenoceptor subtype mRNAs in the dorsal root ganglion after spinal nerve ligation. Brain Res Mol Brain Res 93: 164–172.

Zahn S, Leis S, Schick C et al. (2004). No alpha-adrenoreceptor-induced C-fiber activation in healthy human skin. J Appl Physiol 96: 1380–1384.

Handbook of Clinical Neurology, Vol. 117 (3rd series)
Autonomic Nervous System
R.M. Buijs and D.F. Swaab, Editors

Chapter 25

Autoimmune autonomic failure

SRIKANTH MUPPIDI AND STEVEN VERNINO*
Department of Neurology and Neurotherapeutics, University of Texas Southwestern Medical Center, Dallas, TX, USA

INTRODUCTION

There are many etiologies of autonomic failure, including degenerative, toxic, metabolic, and autoimmune causes. In many instances, acute or subacute autonomic failure can be attributed to an autoimmune etiology. Slowly progressive chronic autoimmune autonomic failure is also described, and this can be difficult to differentiate from chronic degenerative causes of autonomic failure. Among the autoimmune causes of autonomic failure, idiopathic autoimmune autonomic ganglionopathy (AAG) is unique because the pathophysiology of the disease is well understood, and it is one of a very few antibody-mediated neurological disorders. Autoimmune autonomic failure can also be seen in association with systemic rheumatological disorders (e.g., Sjögren's syndrome) or in association with cancer as a paraneoplastic syndrome. In this chapter, we will cover clinical manifestations, pathophysiology, and treatment considerations for idiopathic AAG and other autoimmune causes of autonomic failure.

CLINICAL MANIFESTATIONS OF AUTOIMMUNE AUTONOMIC GANGLIONOPATHY

Autoimmune autonomic ganglionopathy (AAG) was originally described as "pure pan-dysautonomia with recovery" (Young et al., 1975). In this original report, it was postulated to be a form of acute polyneuritis (Guillian–Barré syndrome), restricted to the autonomic nerves. Since then, many individual case reports and case series have shed light on this entity. In 1994, a case series of 27 patients described the clinical manifestations, electrodiagnostic features, and follow-up data in patients with this syndrome and used the term "idiopathic autonomic neuropathy" (Suarez et al., 1994). In 2000, the syndrome was associated with the identification of autoantibodies to ganglionic nicotinic acetylcholine receptors (gnAChR) (Vernino et al., 2000). As there is now clear evidence that these antibodies cause autonomic dysfunction by affecting synaptic transmission in the autonomic ganglia, the term autoimmune autonomic ganglionopathy (AAG) is preferred.

Classically, AAG has a subacute onset (onset to peak symptoms within 6 weeks), frequently preceded by a viral prodrome, immunization, or minor surgical procedure in a previously healthy individual. Patients may experience either an upper respiratory tract infection or gastroenteritis with fever although no pathogen has been consistently identified. There is a slight female predominance as is common with other autoimmune conditions, and the mean age of AAG patients is 61 years (Klein et al., 2003). Subacute or acute autonomic failure with prominent gastrointestinal dysmotility may occur in children, but antibodies to gnAChR have not been reported in children younger than 12 years.

Patients usually experience symptoms attributable to failure of all three limbs of the peripheral autonomic nervous system. Enteric autonomic failure (seen in up to 70% of the cases) presents with manifestations of impaired gastrointestinal motility including gastroparesis (early satiety, decreased appetite, and vomiting) and severe constipation. Severe hypomotility may cause intestinal pseudoobstruction with pain, abdominal distention, and evidence of dilated loops of bowel on abdominal imaging studies. Overall, the extent and severity of the enteric nervous system involvement helps to differentiate AAG from other degenerative causes of autonomic failure such as multiple system atrophy (MSA) or pure autonomic failure (PAF).

Sympathetic failure manifests as severe orthostatic intolerance, syncope, and reduced or absent sweating.

*Correspondence to: Steven Vernino, M.D., Ph.D., Professor and Vice chair of Neurology, Department of Neurology and Neurotherapeutics, University of Texas Southwestern Medical Center, 5323 Harry Hines Boulevard, Dallas, TX 75390, USA. Tel: +1-214-648-3802, Fax: +1-214 648 9129, E-mail: steven.vernino@utsouthwestern.edu

Orthostatic symptoms are seen in more than 70% of cases, and most patients complain of lightheadedness or syncope on standing. An orthostatic drop in systolic blood pressure of 100 mmHg is not uncommon in patients with high gnAChR antibody levels (Gibbons and Freeman, 2009). Even with the dramatic drop in blood pressure on standing, there may be no significant heart rate increment in these patients. Patients may also notice decreased sweating, but might not volunteer this information unless specifically questioned about heat intolerance.

Parasympathetic dysfunction affects multiple organs. Patients may complain of sensitivity to bright light and trouble focusing because of impaired pupillary constriction to light. Dry mouth, dry eyes, and urinary retention are often present, and in some cases, these may be the presenting complaints. Men with AAG invariably report erectile failure. On examination, patients may have a resting tachycardia, and markedly reduced heart rate variability during deep breathing or Valsalva.

Characteristically, AAG affects autonomic function in isolation. Some patients (up to 25%) report sensory symptoms and paresthesias in limbs, but formal electrophysiological testing with nerve conduction studies are usually normal. Recently, there have been several reports of neurocognitive impairment in patients with AAG. Impaired memory seems to be common in AAG patients and may relate to recurrent episodes of cerebral hypoperfusion due to orthostatic hypotension (Hayashi and Ishii, 2009; Gibbons et al., 2010). In rare cases, severe cognitive and behavioral impairment appears to result from a coexisting autoimmune encephalopathy that is responsive to immunomodulatory treatment (Baker et al., 2009).

Patients with the acute and severe presentation of AAG tend to have a monophasic course with spontaneous but incomplete recovery over time. Chronic, insidious onset is also seen and sometimes may be misdiagnosed as MSA or PAF. Serological testing for gnAChR antibody and careful history and autonomic testing will help in differentiating these conditions (Goldstein et al., 2009). Patients who have AAG with gnAChR antibodies have more gastrointestinal dysmotility, sicca symptoms, and impaired pupillary function compared to seronegative AAG patients (Sandroni et al., 2004) or patients with degenerative dysautonomia.

Table 25.1

Diseases to be considered in differential diagnosis of autoimmune autonomic ganglionopathy (AAG)

Disease	Key differentiating features from gnAChR antibody-positive AAG
Guillain–Barré syndrome	Prominent weakness and areflexia associated with tachycardia, and blood pressure fluctuations
Neuromyotonia or neuromuscular hyperexcitability syndromes	Stiffness, fasciculations associated with increased sweating, tachycardia and change in sleep patterns
Diabetes	Sensorimotor somatic neuropathy
Amyloidosis	Painful small fiber neuropathy with autonomic impairment
Lambert–Eaton myasthenic syndrome	Weakness, areflexia, fatigue, dry mouth, impotence with predominantly cholinergic impairment on autonomic testing. Positive voltage-gated calcium channel antibodies
Paraneoplastic enteric or autonomic neuropathy	Severe gastrointestinal motility impairment, may be associated with other neurological symptoms, paraneoplastic antibodies, and small cell lung cancer
Sjögren's syndrome	Dry eyes and mouth, sensory neuropathy or neuronopathy and sometimes autonomic neuropathy
Pure autonomic failure	Chronic slowly progressive, predominantly adrenergic impairment with no central nervous system, gastrointestinal, or pupillary involvement
Multiple system atrophy	Chronic, relentlessly progressive parkinsonism or ataxia, neurogenic bladder and panautonomic impairment on formal testing

DIFFERENTIAL DIAGNOSIS

A number of diseases can mimic AAG by causing subacute and/or diffuse autonomic failure (see Table 25.1). In most instances, associated features or history helps to exclude these causes of autonomic dysfunction. Common causes for acute or subacute onset autonomic failure are toxic, metabolic, and autoimmune etiologies. Toxic and drug-induced autonomic failure can be differentiated by clear temporal relationship to exposure. Another important cause of acute autonomic dysfunction is Guillain–Barré syndrome (GBS). With GBS, patients have significant weakness and polyneuropathy along with blood pressure fluctuations, intermittent sweating and ileus. In neuromuscular hyperexcitabilitiy syndromes

(Isaac or Morvan syndrome), patients usually develop muscle stiffness and muscle hyperactivity along with hyperhidrosis and tachycardia, and may have cognitive impairment. A clear loss of autonomic function without sensory or motor neuropathy easily differentiates AAG from GBS or neuromuscular hyperexcitability syndromes where there is excessive and labile autonomic function associated with somatic neuropathy.

Sjögren syndrome (SS) is probably the most important differential diagnosis for AAG. The usual neurological manifestation of SS is sensory neuropathy or neuronopathy with autonomic dysfunction. The prominent sensory symptoms provides clear distinction from AAG. Parasympathetic secretomotor symptoms (dry eyes and mouth) and constipation are prominent in SS. Impaired pupil light reflex and urinary retention have also been reported. Sympathetic dysfunction, if present, is less severe than in AAG (Mandl et al., 2007). The autonomic features of Sjögren syndrome can have a subacute onset. Rare patients may have an overlap syndrome with low levels of serum gnAChR antibodies (Kondo et al., 2009). Diagnosis of SS can be facilitated by serological testing and salivary gland biopsy.

Other conditions associated with somatic neuropathy and prominent autonomic dysfunction include diabetes and amyloidosis. Autonomic failure with prominent gastrointestinal dysmotility and orthostatic hypotension may be seen. In particular, some cases of diabetic autonomic neuropathy have a rapid onset, which can mimic AAG. But these cases are easily distinguished by the presence of somatic neuropathy and evidence of poorly controlled hyperglycemia.

Autonomic failure can occur in association with various underlying malignancies as a paraneoplastic phenomenon. Paraneoplastic dysautonomia may be indistinguishable from AAG. Patients usually present with gastrointestinal dysmotility (especially gastroparesis), and this may be the only manifestation (paraneoplastic enteric neuropathy). In other cases, pandysautonomia occurs with prominent orthostatic hypotension and laboratory evidence of diffuse autonomic failure. Small cell lung cancer is the most commonly associated malignancy, but thymoma, breast cancer, and lymphoma have also been reported. As with many paraneoplastic disorders, an underlying malignancy may not be obvious at the time of initial presentation and extensive work-up (including PET scans) might be needed. Paraneoplastic antibody testing should be done. Anti-Hu antibody (also known as ANNA-1) and antiCRMP-5 (also known as anti-CV-2) are the most common serological markers (Vernino, 2009). Paraneoplastic dysautonomia is one of the common paraneoplastic neurological syndromes, seen in about 10% of all patients who are positive for anti-Hu antibody (Lucchinetti et al., 1998). Dysautonomia may occur as one manifestation of a multifocal paraneoplastic neurological syndrome, in association with limbic encephalitis, sensory neuronopathy, or other symptoms.

Lambert–Eaton myasthenic syndrome (LEMS) is another paraneoplastic syndrome with autonomic features. These patients usually have only mild cholinergic autonomic neuropathy (dry mouth, impotence, constipation) which is generally not confused with AAG. Most also complain of proximal weakness and fatigue. Electrodiagnostic testing and testing for voltage-gated calcium channel antibodies further differentiate LEMS from AAG (Vernino, 2009).

Chronic, slowly progressive AAG may mimic various degenerative causes of autonomic dysfunction, such as multiple system atrophy (MSA) or pure autonomic failure. In particular, AAG patients with severe orthostatic hypotension and bladder dysfunction may resemble multiple system atrophy. Prominent gastroenterological symptoms, pupillary involvement, positive gnAChR antibody, and the lack of extrapyramidal or cerebellar signs generally differentiate AAG from MSA (Table 25.1).

PATHOPHYSIOLOGY OF AUTOIMMUNE AUTONOMIC FAILURE

Ganglionic AChR antibodies

Approximately 50–60% of patients with AAG have antibodies to gnAChR and these specifically bind to α3 subunit. These receptors are present in all peripheral autonomic ganglia including sympathetic, parasympathetic, and enteric systems. Ganglionic nicotinic AChRs are antigenically different, but genetically and functionally similar to muscle nicotinic ACh receptors. The gnAChR consists of a pentamer with two α3 subunits associated with β2, β4 or α5 subunits. Antibodies against these receptors decrease the strength of synaptic transmission at the autonomic ganglia by reducing the number of functional gnAChR. Antibodies against the muscle AChR present in myasthenia gravis do not bind to the gnAChR, and gnAChR antibodies in AAG patients do not bind to the muscle AChR (Vernino et al., 2009).

Serum levels of gnAChR antibody correlate with the severity of symptoms and severity of autonomic dysfunction on objective testing. Recent studies revealed a strong positive correlation between the antibody titer and the magnitude of postural drop in blood pressure (Gibbons and Freeman, 2009). Importantly, a decrease in antibody titer generally correlates with improvement in autonomic function (Vernino et al., 2000), especially with plasma exchange (Gibbons et al., 2008). Antibody binding activity is determined quantitatively by radioimmunoprecipitation assay. Ganglionic AChR antibody levels greater than 0.05 nmol/L of serum is considered to be positive (Vernino et al., 2009). All symptoms and signs are more severe when the antibody level is higher than 1.00 nmol/L.

While high levels of gnAChR antibodies are quite specific for AAG, a low level of these antibodies is less specific (McKeon et al., 2009). Generally, patients with high antibody titer (>1.00 nmol/L) have pandysautonomia as seen in AAG. Patients with moderate antibody levels (0.10–0.99 nmol/L) tend to have milder, restricted, or more chronic autonomic dysfunction. Patients with low antibody levels (0.03–0.09 nmol/L) have a variety of non-specific neurological symptoms or non-neurological diseases with evidence of other autoantibodies. Low levels of gnAChR antibody may be found in about 3% of MG patients or in association with lung cancer with or without paraneoplastic dysautonomia. The clinical associations of gnAChR antibodies are summarized in Table 25.2.

Animal models

Experimental models of AAG (EAAG) have been described, and these studies have led to the confirmation that AAG is an antibody-mediated disease. Rabbits immunized with the α3 subunit of gnAChR develop severe autonomic dysfunction, with prominent pupillary, gastrointestinal, and bladder changes. Moreover, these autonomic features are associated with decreased cholinergic synaptic transmission in autonomic ganglia (Lennon et al., 2003).

When gnAChR antibodies were injected into healthy mice, these mice developed transient gastrointestinal dysmotility, urinary retention, dilated pupils, reduced heart rate variability, and impaired catecholamine response to stress, all features similar to human AAG (Vernino et al., 2004). In both rabbits with EAAG and humans with AAG, sympathetic failure with decreased plasma norepinephrine levels may be associated with evidence of intact cardiac sympathetic innervation on imaging studies (Goldstein et al., 2002; Vernino et al., 2009). These observations support the conclusion that the gnAChR antibodies produce a functional impairment of autonomic function through ganglionic blockade. Immunohistochemical studies of autonomic ganglia in rabbits with chronic EAAG revealed decrease in surface AChR (Vernino et al., 2003), but minimal reduction in neuronal density and no ultrastructural changes in synaptic organization (Tajzoy et al., 2011).

CLINICAL LABORATORY ASSESSMENT

The combination of subacute onset, diffuse autonomic impairment, and absence of sensory or motor involvement suggests AAG as the most likely diagnosis. Antibody testing and careful assessment in an autonomic laboratory provide additional information. Most autonomic laboratories can test for sudomotor, cardiovagal, and sympathetic functions and some have been able to assess the pupillary function quantitatively.

Patients with AAG will have decreased sweating and this can be assessed with thermoregulatory testing (TST) and quantitive sudomotor axon reflex testing (QSART). Typically there is diffusely decreased sweating on TST and distal more than proximal impairment on QSART (Kimpinski et al., 2009). These features suggest that perhaps there is both a ganglionic and postganglionic sudomotor dysfunction in AAG. Higher antibody levels correlate with more prominent postganglionic dysfunction.

Most patients have an orthostatic drop in blood pressure with a head-up tilt or stand test. Sympathetic impairment can also be identified by analysis of beat-to-beat blood pressure changes during Valsalva. Patients with AAG may have blunted to absent phase III and phase IV responses. Serum norepinephrine levels tend to be low and do not increase appropriately during orthostatic stress (Klein et al., 2003). A progressive fall in blood pressure during head-up tilt table test is most evident

Table 25.2

Diseases associated with GnAChR antibodies

Clinical disease	Antibody level (normal < 0.05 nmol/L)	Frequency of antibody positivity
Acute or subacute AAG	>1.0 nmol/L	50%
Chronic AAG	0.2–1.0 nmol/L	Unknown
Limited forms of AAG	0.1–0.5 nmol/L	Unknown
Postural orthostatic tachycardia syndrome (POTS)	<0.5 nmol/L	<10%
Gastrointestinal dysmotility	<0.5 nmol/L	<10%
Neuromuscular syndromes such as MG, LEMS or neuromyotonia	<0.2 nmol/L	<10%

AAG, autoimmune autonomic ganglionopathy; MG, myasthenia gravis; LEMS, Lambert–Eaton myasthenic syndrome.

in patients with gnAChR antibody levels greater than 1.00 nmol/L (Gibbons and Freeman, 2009). Parasympathetic cardiovagal dysfunction (assessed by heart rate deep breathing or Valsalva ratio) is also abnormal in patients with AAG; the heart rate may remain fixed at around 80 beats per minute. Pupillary function is also impaired in patients with AAG, and is most readily identified in patients with higher antibody levels (Gibbons and Freeman, 2009). Another unique feature of AAG is pupillary constriction fatigue, demonstrated by premature pupillary redilation during prolonged light stimulus (Muppidi et al., 2012). Ultimately, a diagnosis of AAG can usually be made on clinical grounds, complemented by gnAChR antibody and standard autonomic function tests. Autonomic tests and antibody levels are also useful in monitoring response to treatment.

Patients with AAG often have mild peripheral sensory symptoms, but in most instances nerve conduction studies are normal. Nerve conduction studies may be helpful in ruling out other conditions where both autonomic and somatic sensory nerves are involved, such as Sjögren syndrome, amyloidosis, or diabetes. Spinal fluid analysis can be performed as some AAG patients have elevated CSF protein level, but spinal fluid analysis is generally not necessary for patients with typical AAG presentation.

THERAPEUTIC APPROACHES

Immunomodulatory treatment

Most patients with severe and disabling autonomic failure will require both symptomatic and immunomodulatory treatment. There have been no controlled or randomized studies performed in patients with AAG, and the only available evidence is from single case reports or case series. Most treatment reports are from gnAChR antibody-positive patients, although a few antibody-negative patients were treated with IVIG.

Plasma exchange

Plasma exchange is considered effective because AAG is an antibody-mediated disease, and decreased antibody levels correlate with improved clinical status (Schroeder et al., 2005). The benefit from a course of plasma exchange treatment can be dramatic, but the duration of improvement tends to be short (a few weeks to months). If they relapse, patients need either repeated plasma exchanges or other immunomodulatory therapy (Gibbons et al., 2008; Iodice et al., 2009). Apart from the side-effects associated with catheter-related issues, one must be careful to avoid significant intravascular volume depletion as this might worsen orthostatic symptoms.

Intravenous immunoglobulin

Intravenous immunoglobulin (IVIG) is also commonly used as an initial therapy in AAG and is supported by multiple case series and case reports. As with plasma exchange, the benefit may be temporary. In several documented cases of seropositive AAG, IVIG was not effective, but patients responded to treatment with plasma exchange or a combined therapy with other immunomodulatory agents (Iodice et al., 2009). IVIG is generally well tolerated, and may provide an added benefit of reducing orthostatic symptoms by increasing plasma volume.

Corticosteroids

Corticosteroids, orally or in intravenous form, may be effective. Prednisone may be added to one of the above regimens, and in those patients who require long-term corticosteroid usage, other immunosuppressive (steroid-sparing) agents are used. Low-dose oral prednisone alone may be effective in patients with mild symptoms and low antibody level, but those with high antibody levels will usually require a combined regimen.

Other immunosuppressive agents

Various other immunosuppressive agents have also been shown to provide benefit in AAG patients. These include azathioprine, cyclophosphamide, and mycophenolate mofetil. A single case report demonstrated the benefit of rituximab after relapses with plasma exchange (Imrich et al., 2009). Because AAG is a relatively rare disease, there have been no controlled systematic studies to determine the efficacy of different treatments.

Other symptomatic treatments

Almost all patients with AAG require supportive treatments for orthostatic hypotension and other autonomic symptoms. However, symptomatic treatments alone often provide incomplete benefit and ultimately immunomodulatory treatments are needed. For orthostatic hypotension, patients may benefit from measures to increase intravascular volume. Vasoconstrictors such as midodrine or ephedrine may also be beneficial. Droxidopa (L-threo-dihydroxyphenylserine) has been shown to provide significant improvement in orthostatic hypotension in a single case even after the patient failed to respond to other treatments (Gibbons et al., 2005). Lower extremity compression stockings may also be tried for orthostatic symptoms.

Pyridostigmine can help relieve some cholinergic symptoms by increasing the synaptic transmission in autonomic ganglia and at parasympathetic and sudomotor target organs (Vernino et al., 2008). This increased cholinergic transmission can provide some modest benefit for orthostatic hypotension but may be more effective to improve salivation, lacrimation, or bowel motility. For urinary retention, intermittent catherizations or a direct cholinergic agonist (bethanechol) may be helpful. Constipation in AAG can be severe and should be managed aggressively to prevent impaction. Increasing fluid and fiber intake along with polyethylene glycol can be effective. Dry mouth can be relieved with oral glycerine spray.

SUMMARY

Autoimmune autonomic ganglionopathy is an antibody-mediated, potentially treatable autonomic disorder presenting with diffuse autonomic failure including prominent orthostatic intolerance and gastrointestinal dysmotility. Subacute onset of autonomic failure with pupillary and gastroenterological involvement in the absence of other known causes suggests AAG. High levels of ganglionic AChR antibodies are seen in approximately 50% of all AAG patients and confirm the diagnosis. These antibodies cause a similar phenotype of autonomic dysfunction in animal models and a functional impairment of cholinergic transmission in the autonomic ganglia. As this condition is uncommon, although increasingly being recognized by clinicians, the treatment options are based on case series. Patients seem to benefit from a combination of plasma exchange, steroids, or IVIG along with chronic immunosuppressive agents. Symptomatic therapy for orthostatic hypotension and constipation should be included. Clinical research to compare the effectiveness of different immunomodulatory therapies is still needed in patients with AAG.

References

Baker SK, Morillo C, Vernino S (2009). Autoimmune autonomic ganglionopathy with late-onset encephalopathy. Auton Neurosci 146: 29–32.

Gibbons CH, Freeman R (2009). Antibody titers predict clinical features of autoimmune autonomic ganglionopathy. Auton Neurosci 146: 8–12.

Gibbons CH, Vernino SA, Kaufmann H et al. (2005). L-DOPS therapy for refractory orthostatic hypotension in autoimmune autonomic neuropathy. Neurology 65: 1104–1106.

Gibbons CH, Vernino SA, Freeman R (2008). Combined immunomodulatory therapy in autoimmune autonomic ganglionopathy. Arch Neurol 65: 213–217.

Gibbons C, Centi J, Freeman R (2010). Autoimmune autonomic ganglionopathy: a reversible cortical dementia. Neurology 74 (Suppl 2): A493.

Goldstein DS, Holmes C, Dendi R et al. (2002). Pandysautonomia associated with impaired ganglionic neurotransmission and circulating antibody to the neuronal nicotinic receptor. Clin Auton Res 12: 281–285.

Goldstein DS, Holmes C, Imrich R (2009). Clinical laboratory evaluation of autoimmune autonomic ganglionopathy: preliminary observations. Auton Neurosci 146: 18–21.

Hayashi M, Ishii Y (2009). A Japanese case of autoimmune autonomic ganglionopathy (AAG) and a review of AAG cases in Japan. Auton Neurosci 146: 26–28.

Imrich R, Vernino S, Eldadah BA et al. (2009). Autoimmune autonomic ganglionopathy: treatment by plasma exchanges and rituximab. Clin Auton Res 19: 259–262.

Iodice V, Kimpinski K, Vernino S et al. (2009). Efficacy of immunotherapy in seropositive and seronegative putative autoimmune autonomic ganglionopathy. Neurology 72: 2002–2008.

Kimpinski K, Iodice V, Sandroni P et al. (2009). Sudomotor dysfunction in autoimmune autonomic ganglionopathy. Neurology 73: 1501–1506.

Klein CM, Vernino S, Lennon VA et al. (2003). The spectrum of autoimmune autonomic neuropathies. Ann Neurol 53: 752–758.

Kondo T, Inoue H, Usui T et al. (2009). Autoimmune autonomic ganglionopathy with Sjögren's syndrome: significance of ganglionic acetylcholine receptor antibody and therapeutic approach. Auton Neurosci 146: 33–35.

Lennon VA, Ermilov LG, Szurszewski JH et al. (2003). Immunization with neuronal nicotinic acetylcholine receptor induces neurological autoimmune disease. J Clin Invest 111: 907–913.

Lucchinetti CF, Kimmel DW, Lennon VA (1998). Paraneoplastic and oncologic profiles of patients seropositive for type 1 antineuronal nuclear autoantibodies. Neurology 50: 652–657.

Mandl T, Wollmer P, Manthorpe R et al. (2007). Autonomic and orthostatic dysfunction in primary Sjögren's syndrome. J Rheumatol 34: 1869–1874.

McKeon A, Lennon VA, Lachance DH et al. (2009). Ganglionic acetylcholine receptor autoantibody: oncological, neurological, and serological accompaniments. Arch Neurol 66: 735–741.

Muppidi S, Scribner M, Gibbons CH et al. (2012). A unique manifestation of pupillary fatigue in autoimmune autonomic ganglionopathy. Arch Neurol 69: 644–648.

Sandroni P, Vernino S, Klein CM et al. (2004). Idiopathic autonomic neuropathy: comparison of cases seropositive and seronegative for ganglionic acetylcholine receptor antibody. Arch Neurol 61: 44–48.

Schroeder C, Vernino S, Birkenfeld AL et al. (2005). Plasma exchange for primary autoimmune autonomic failure. N Engl J Med 353: 1585–1590.

Suarez GA, Fealey RD, Camilleri M et al. (1994). Idiopathic autonomic neuropathy: clinical, neurophysiologic, and follow-up studies on 27 patients. Neurology 44: 1675–1682.

Tajzoy E, Mukherjee S, Vernino S (2011). Autonomic ganglia neuronal density and synaptic structure in chronic experimental autoimmune autonomic ganglionopathy. Arch Neurol 68: 540–541.

Vernino S (2009). Antibody testing as a diagnostic tool in autonomic disorders. Clin Auton Res 19: 13–19.

Vernino S, Low PA, Fealey RD et al. (2000). Autoantibodies to ganglionic acetylcholine receptors in autoimmune autonomic neuropathies. N Engl J Med 343: 847–855.
Vernino S, Low PA, Lennon VA (2003). Experimental autoimmune autonomic neuropathy. J Neurophysiol 90: 2053–2059.
Vernino S, Ermilov LG, Sha L et al. (2004). Passive transfer of autoimmune autonomic neuropathy to mice. J Neurosci 24: 7037–7042.
Vernino S, Sandroni P, Singer W et al. (2008). Invited Article. Autonomic ganglia: target and novel therapeutic tool. Neurology 70: 1926–1932.
Vernino S, Hopkins S, Wang Z (2009). Autonomic ganglia, acetylcholine receptor antibodies, and autoimmune ganglionopathy. Auton Neurosci 146: 3–7.
Young RR, Asbury AK, Corbett JL et al. (1975). Pure pandysautonomia with recovery. Description and discussion of diagnostic criteria. Brain 98: 613–636.

Handbook of Clinical Neurology, Vol. 117 (3rd series)
Autonomic Nervous System
R.M. Buijs and D.F. Swaab, Editors

Chapter 26

The central sympathetic nervous system in hypertension

GIUSEPPE MANCIA[1]* AND GUIDO GRASSI[2]

[1]*Department of Prevention, Clinica Medica, Ospedale San Gerardo, Monza (Milan), Italy*

[2]*Department of Prevention and Clinical Medicine, University of Milano-Bicocca, Italy*

INTRODUCTION

There is overwhelming evidence that essential hypertension is accompanied by sympathetic activation, which participates not only in the development but also the progression of the hypertensive state. Many studies have shown in addition that a number of cardiovascular risk factors frequently accompanying hypertension are characterized by increased sympathetic influences, and that this is the case also for hypertension-related end organ damage. This chapter begins with a brief overview of the origin of sympathetic activation in essential hypertension, and then outlines the contribution of this activation to organ damage related to hypertension, together with the technical progress that has recently been made in the quantification of neural cardiovascular modulation, which may greatly help future studies in this area of research.

SYMPATHETIC ACTIVATION IN ESSENTIAL HYPERTENSION

The evidence that essential hypertension is accompanied by a sympathetic activation is multifold. First, using drugs that block cardiac sympathetic effects, vascular sympathetic effects and vagal cardiac effects, Julius (1996) showed that a noticeable fraction of borderline hypertensive subjects had an increased autonomic drive. Then Goldstein and his group (1983) reported that, even accounting for some negative results, an indirect marker of sympathetic tone such as plasma norepinephrine was significantly elevated in essential hypertensive compared to normotensive subjects. And finally, by subtracting plasma norepinephrine clearance from plasma norepinephrine values, Esler and coworkers (1997) showed that the rate of norepinephrine spillover from the neuroeffector junctions was increased in essential hypertension, particularly in young hypertensives. This was true particularly at the level of the kidney, the heart, and the central nervous system (Esler et al., 1989), i.e., in three cardiovascular areas of major importance in cardiovascular homeostasis regulation as well as in the pathogenesis of hypertension.

Interesting additional data on this issue have emerged through another approach, namely, the direct measurement of sympathetic nerve traffic to skeletal muscle circulation by a microelectrode inserted in the peroneal or brachial nerve. This approach has the limitation of offering only sectorial sympathetic information but has the advantage of being a direct, rather than an indirect measure of sympathetic activity. Further, the values taken in a given subject at a given time correlate to those taken an hour or few hours later more closely than plasma norepinephrine values, and this is true in both healthy and diseased subjects (Grassi et al., 1997, 2008b). Thus the muscle sympathetic nerve traffic measurement is more reproducible than other methods of quantifying sympathetic effects, which means that small intra- and intersubject increases in these effects have a greater chance of being detected.

This has indeed been the case in most studies that have quantified sympathetic nerve traffic in hypertensive subjects (Anderson et al., 1989; Floras et al., 1993; Grassi et al., 1998, 2007, 2008a). The results of these studies allow the following main conclusions to be drawn. First, the sympathetic activation occurring in hypertension is an early phenomenon in the clinical course of the disease. It also parallels the disease progression, becoming more and more evident from mild to moderate and more severe hypertensive state (Fig. 26.1) (Grassi et al., 1998). Interestingly, in addition some specific clinical forms of hypertension, e.g., those known as masked

*Correspondence to: Professor Giuseppe Mancia, Department of Prevention, Clinica Medica, Ospedale S. Gerardo dei Tintori, via Pergolesi 33, 20052 Monza (Milan), Italy. Tel: +39-039-233-357, Fax: +39-039-322-274, E-mail: giuseppe.mancia@unimib.it

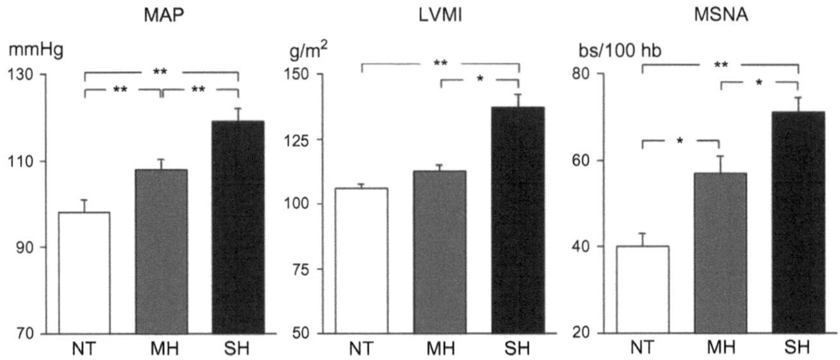

Fig. 26.1. Mean arterial pressure (MAP), left ventricular mass index (LVMI) and efferent postganglionic muscle sympathetic nerve traffic (MSNA) in normotensive subjects and in patients with mild (MH) and severe (SH) hypertension. Data are shown as means ± SE. Asterisks (*$p < 0.05$, **$p < 0.01$) refer to the statistical significance between groups. Note that there is a progressive sympathetic activation as the severity of the hypertensive state becomes greater. (Modified from Grassi et al., 1998.)

or white-coat hypertensive states or those defined under the term of dipping, nondipping or reverse dipping hypertension, are all characterized by a sustained adrenergic overdrive, whose magnitude, however, may be different according to the different types of high blood pressure (Grassi et al., 2007, 2008a).

SYMPATHETIC ACTIVATION AND CARDIOVASCULAR RISK FACTORS

Several cardiovascular risk factors frequently (and sometimes preferentially) associated with hypertension are characterized by a sympathetic activation which is triggered by a variety of mechanisms. For example, cigarette smoking causes a sharp and marked rise in blood pressure and heart rate which is so prolonged as to be still visible 20–30 minutes after its termination when the next cigarette is lit (Groppelli et al., 1992). This is caused by a nicotine-dependent stimulation of norepinephrine and epinephrine secretion from the sympathetic nerve terminals and adrenal medulla, the sympathetic nerve traffic being reflexly inhibited by the pressor response (Cryer et al., 1976; Grassi et al., 1994). Sympathetic nerve traffic is also markedly elevated in hypertensive patients with an early left ventricular diastolic dysfunction or in those in whom the blood pressure elevation is coupled with a less or more advanced congestive heart failure state (Grassi et al., 2003, 2009). An adrenergic overactivity also characterizes obese hypertensive or obese normotensive subjects (Andersson et al., 1991; Grassi et al., 1995c), possibly because of an impairment of the sympathetic restraint tonically exerted by the arterial baroreceptors (Grassi et al., 1995a). When obesity is associated with obstructive sleep apnea there is a further potentiation of the sympathetic activation (Grassi et al., 2005), as it has been described in the clinical condition known as metabolic syndrome, in which the obese state is associated with other metabolic or hemodynamic abnormalities (Grassi, 2009). Finally, sympathetic activation has also been documented in a condition commonly found not only in obese but also in lean hypertensive subjects, i.e., the insulin resistance state (Ferrannini et al., 1987), because the elevated insulin levels resulting from the insulin resistance state stimulate sympathetic cardiovascular activation, both centrally and peripherally (Rowe et al., 1981; Anderson et al., 1991; Berne et al., 1992).

Interestingly, in several instances the sympathetic nervous system reciprocates the stimulation it receives, thus forming a vicious circle in which the components reinforce each other. This is the case with insulin resistance, which is increased by sympathetic influences on trough vasoconstriction in the skeletal muscle regions (Jamerson et al., 1993; Lembo et al., 1994), leading to a further increase in insulin secretion and further sympathetic activation. It is also the case for the renin–angiotensin system, the activation of which has been regarded as an independent cardiovascular risk factor (Cambien et al., 1994). Animal studies have shown that sympathetic influences directly and indirectly stimulate renin secretion from the juxtaglomerular cells (Richardson et al., 1974) and that the resulting overproduction of angiotensin II activates the sympathetic nervous system both centrally and peripherally. This is now also clear in man (Mancia et al., 1996), because of the evidence that in normotensive man, increasing and reducing sympathetic activity by deactivating and stimulating cardiac receptors through nonhypotensive lower body negative suction and leg raising, respectively, was accompanied not only by marked increases and reductions in vascular resistance and plasma norepinephrine but also by marked changes in plasma renin activity, all effects being abolished in heart transplant patients in whom no reflex input from cardiac receptors was operative (Giannattasio et al., 1993). In patients with angiographic evidence of coronary atherosclerosis, the marked coronary vasoconstriction induced by reflexly activating the sympathetic nervous system through a diving maneuver was markedly reduced by blocking angiotensin II production through administration of an angiotensin-converting enzyme inhibitor. Conversely, in subjects with little angiographic evidence of coronary atherosclerosis, the small increase in coronary vascular resistance induced by diving was markedly potentiated by the intracoronary infusion of angiotensin Il at doses so small as to leave baseline coronary blood flow and vascular resistance unaffected (Perondi et al., 1992; Saino et al., 1997).

SYMPATHETIC ACTIVATION AND ORGAN DAMAGE

Conclusive evidence is now available that a sympathetic activation directly promotes cardiac and vascular alterations that lead to hypertension-related morbidity and mortality, independently of blood pressure elevation. In the experimental setting this has been shown to occur for left ventricular hypertrophy, which can be induced by subpressor doses of adrenergic agents (Patel et al., 1991) and prevented only if a blood pressure reduction is not accompanied by excessive reflex cardiac sympathetic stimulation (Sen, 1974). It has also been shown to occur for arteriolar wall hypertrophy or remodeling (Heagerty, 1997), a structural abnormality that can be found at an early stage of hypertension (Giannattasio et al., 1995) and helps maintain the blood pressure elevation (Folkow, 1978). The remodeling phenomenon, however, can be seen in other conditions not characterized by hypertension, such as the obese state (Grassi et al., 2010a) or when hypertension is complicated by other metabolic changes (Grassi et al., 2010b).

Animal and human evidence has been obtained, however, showing that sympathetic influences also stiffen the large artery wall. This has an unequivocal pathophysiological implication because in stiffer arteries the traumatic effect of intravascular pressure is greater and the progression of atherosclerosis probably faster. Recent data come, first, from a study in rats which showed that carotid artery distensibility, as measured by sonographic assessment of arterial diameter over the diastolosystolic pressure range, is markedly increased following sympathectomy (Mangoni et al., 1997). Second, radial artery distensibility as assessed by the above technique is markedly increased following anesthesia of the brachial plexus in humans undergoing surgery for Dupuytren disease (Grassi et al., 1995a). Third, similar findings have been obtained in the femoral artery of normal patients undergoing spinal anesthesia for arthroscopic surgery of the knee or lumbar sympathectomy for peripheral artery disease. Thus, the stiffening influence of the sympathetic nervous system (1) involves the wall of both elastic-type and muscle-type arteries and (2) is visible in both normal and diseased vessels. The mechanism involved is likely to be smooth muscle contraction because contracted muscle tissue is obviously less distensible than relaxed muscle. Sympathetic effects on vessel wall structure cannot, however, be excluded. Furthermore, sympathetic activation may also reduce arterial wall distensibility through indirect mechanisms. This is suggested by the evidence that in rats a progressive increase in the heart rate induced by cardiac pacing resulted in a progressive reduction of carotid artery distensibility (Mangoni et al., 1996). Because similar findings were obtained in sympathectomized animals, it may be concluded that tachycardia, a typical consequence of sympathetic activation, can reduce arterial distensibility, not only by directly affecting the arterial wall but probably also by indirectly reducing the time allowed for a viscoelastic structure such as the arterial wall to distend.

Sympathetic mechanisms are also involved in the process leading to hypertension-related cardiac structural alterations, i.e., left ventricular hypertrophy. Indeed, evidence has emerged that when hypertension is coupled with left ventricular hypertrophy, the degree of sympathetic activation is greater for magnitude than that seen in the uncomplicated hypertensive state (Greenwood et al., 2001; Burns et al., 2007).

ORIGIN OF SYMPATHETIC ACTIVATION IN ESSENTIAL HYPERTENSION

Despite years of investigation, the origin of the sympathetic activation in essential hypertension remains largely unknown. An attractive hypothesis is that this activation occurs because of an excessive number of and/or reactivity to stressful environmental stimuli which lead, through frequent transient blood pressure elevations, to a permanently hypertensive state. To date, however, this hypothesis has been confirmed only in animal studies, with controversial or only circumstantial support in man (Mancia et al., 1987; Grassi, 2009). First, the cardiovascular responses to laboratory stressors, the approach that has been mostly used, have not consistently been reported to be greater before or after development of hypertension than in normotension (Mancia et al., 1987), Second, human reactivity to stress has been difficult to quantify precisely, because the effect of any given stress has a limited reproducibility and because in any given subject an increased blood pressure response to a stressor can coexist with normal or even reduced responses to other stressors (Mancia et al., 1987). Finally, in man, blood pressure responses to laboratory or, even more, natural (e.g., the white-coat effect) stressful stimuli do not bear a close relationship to 24 hour blood pressure variability, presumably because behavioral influences are only one of the determinants of blood pressure changes in daily life (Mancia et al., 1997a).

Another attractive hypothesis (not exclusive of the previous one) is that the sympathetic activation originates from an impairment of the baroreflex, that is, of a major restraining mechanism of sympathetic tone (Mancia et al., 1997b; Grassi, 2009) (Fig. 26.2). This is supported by evidence that in congestive heart failure and other diseases, such as obesity, the muscle sympathetic nerve traffic increase is related to a reduced sympathetic modulation by the baroreflex (Fig. 26.3) (Grassi et al., 1995b). However, in hypertension, a baroreflex impairment has been demonstrated for the heart rate (Smyth et al., 1969; Mancia et al., 1997b) but not for the blood pressure component of the reflex (Mancia et al., 1978, 1982) (Fig. 26.4). Furthermore, studies of our group have shown that baroreceptor deactivation and stimulation by vasoactive drug-induced changes in blood pressure can influence sympathetic nerve traffic so that it remains unchanged in essential and secondary hypertension compared to

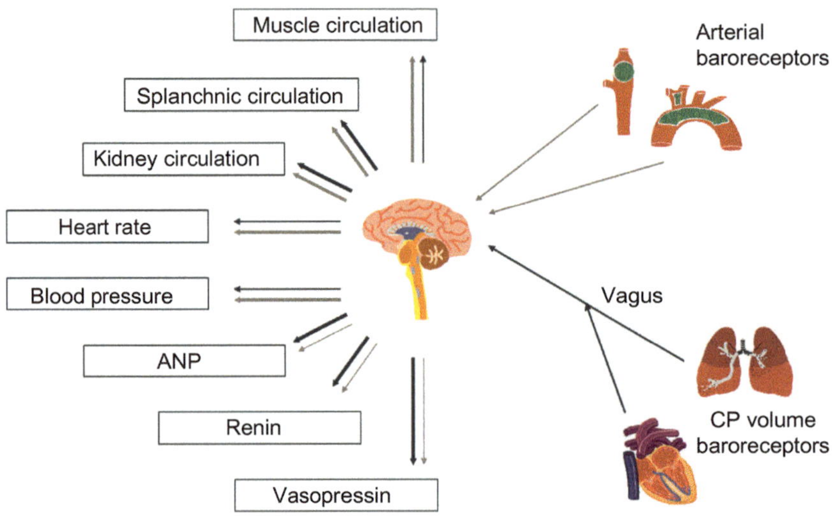

Fig. 26.2. Scheme illustrating the humoral and nonhumoral variables under control of arterial baroreceptors and cardiopulmonary volume receptors. CP, cardiopulmonary; ANP, atrial natriuretic peptides.

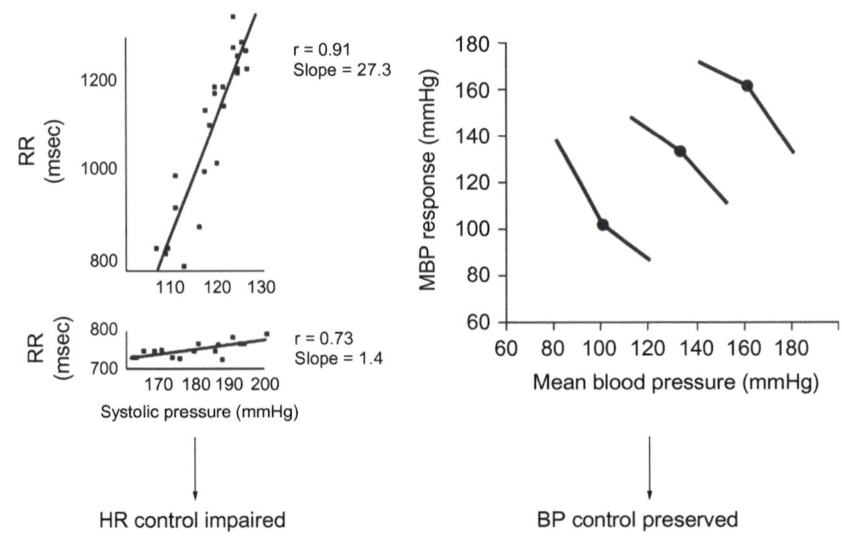

Fig. 26.3. Baroreceptor heart rate (HR, left panel) and blood pressure (BP, right panel) control at different normal and elevated mean blood pressure (MBP) values. Note that in hypertensive subjects baroreceptor modulation of heart rate, of vagal nature, is markedly flattened going from the normotensive state (left panel, upper box) to the hypertensive state (left panel, lower box). In contrast, baroreceptor modulation of blood pressure (right panel), of sympathetic nature, is unaltered in hypertension. (Modified from Mancia et al., 1978.)

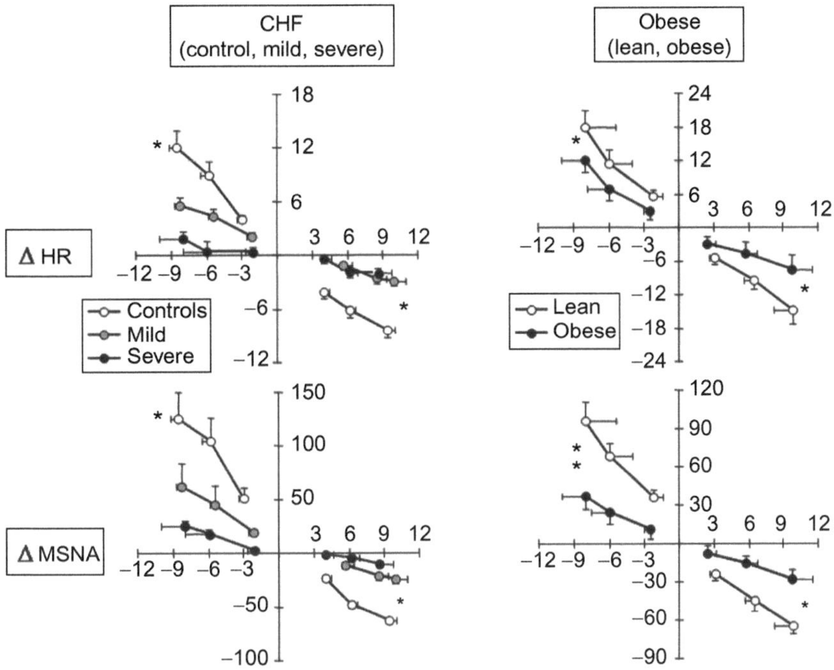

Fig. 26.4. Baroreceptor control of heart rate (HR) and muscle sympathetic nerve traffic (MSNA) in patients with congestive heart failure (CHF, left panels) or with obesity (obese, right panels). Data from healthy control subjects are also shown. Compared to healthy controls, patients with different degrees of heart failure and obesity show a marked impairment of baroreceptor modulation of heart rate and sympathetic drive. (Modified from Grassi et al., 1995b, c.)

normotension, at variance from the reduction in the heart rate alterations (Grassi et al., 1988, 2009). Thus the available evidence does not support the baroreflex as the origin of the sympathetic activation. The data do not rule out the possibility that reflex cardiovascular regulation may contribute to this phenomenon. First, although unimpaired, the baroreflex is reset toward elevated blood pressure values in hypertension, which means that its influence preserves rather than suppresses the increased sympathetic activity. Second, the sympathoinhibitory influence of cardiac receptors is impaired in essential hypertension, although only when accompanied by left ventricular hypertrophy (Grassi et al., 1988). Thus reflex mechanisms contribute to the sympathetic activation occurring in hypertension, although their effects appear to be late and nonspecific.

CONCLUSIONS

Multifold evidence is available that essential hypertension is characterized by a sympathetic activation. This activation can also be found in several cardiovascular risk factors commonly associated with hypertension, possibly because the sympathetic nervous system entertains positive feedback relationships with several metabolic and humoral mechanisms involved in cardiovascular control. Sympathetic activity promotes the organ damage accompanying hypertension, including atherosclerosis, which can be increased by its ability to reduce large-artery wall distensibility. The origin of the sympathetic activation of essential hypertension remains difficult to determine, but techniques that allow autonomic cardiovascular modulation to be assessed under daily life conditions may permit this aspect to be more properly investigated. Sympathetic deactivation is thus regarded as an important goal of antihypertensive treatment and of new therapeutic procedures recently employed in the treatment of resistant hypertension, such as the catheter-based renal sympathetic nerve ablation and the carotid baroreceptor stimulation (Grassi, 2009).

References

Anderson EA, Sinkey CA, Lawton WJ et al. (1989). Elevated sympathetic nerve activity in borderline hypertensive humans: evidence from direct intraneural recordings. Hypertension 14: 177–183.

Anderson EA, Hoffman RP, Balon TW et al. (1991). Hyperinsulinemia produces both sympathetic neural activation and vasodilation in normal humans. J Clin Invest 87: 2246–2252.

Andersson B, Elam M, Wallin BG et al. (1991). Effect of energy-restricted diet on sympathetic muscle activity in obese women. Hypertension 18: 783–789.

Berne C, Fagius J, Pollare T et al. (1992). The sympathetic response to euglycaemic hyperinsulinaemia: evidence from microelectrode nerve recordings in healthy subjects. Diabetologia 35: 873–879.

Burns J, Sivananthan MU, Ball SG et al. (2007). Relationship between central sympathetic drive and magnetic resonance imaging-determined left ventricular mass in essential hypertension. Circulation 115: 1999–2005.

Cambien F, Costerousse O, Tiret L et al. (1994). Plasma level and gene polymorphism of angiotensin converting enzyme in relation to myocardial infarction. Circulation 90: 669–676.

Cryer PE, Haymond MW, Santiago JV et al. (1976). Norepinephrine and epinephrine release and adrenergic mediation of smoking-associated hemodynamic and metabolic events. N Engl J Med 295: 573–577.

Esler M (1997). Sympathetic activity in experimental and human hypertension. In: A Zanchetti, G Mancia (Eds.), Handbook of Hypertension, Pathophysiology of Hypertension, Vol. 17. Elsevier Science, Amsterdam, pp. 628–673.

Esler M, Lambert G, Jennings G (1989). Regional norepinephrine turnover in human hypertension. Clin Exp Hypertens 11 (Suppl 1): 75–89.

Ferrannini E, Buzzigoli G, Bonadonna R et al. (1987). Insulin resistance in essential hypertension. N Engl J Med 317: 350–357.

Floras JS, Hara K (1993). Sympathoneural and haemodynamic characteristics of young subjects with mild essential hypertension. J Hypertens 11: 647–655.

Folkow B (1978). The Fourth Volhard Lecture: cardiovascular structural adaptation; its role in the initiation and maintenance of primary hypertension. Clin Sci Mol Med 4 (Suppl 3): 3S–22S.

Giannattasio C, Del Bo A, Cattaneo BM et al. (1993). Reflex vasopressin and renin modulation by cardiac receptors in humans. Hypertension 21: 461–469.

Giannattasio C, Cattaneo BM, Mangoni AA et al. (1995). Cardiac and vascular structural changes in normotensive subjects with parental hypertension. J Hypertens 13: 259–264.

Goldstein DS (1983). Plasma catecholamines and essential hypertension: an analytical review. Hypertension 5: 86–99.

Grassi G (2009). Assessment of sympathetic cardiovascular drive in human hypertension: achievements and perspectives. Hypertension 54: 690–697.

Grassi G, Giannattasio C, Cléroux J et al. (1988). Cardiopulmonary reflex before and after regression of left ventricular hypertrophy in essential hypertension. Hypertension 12: 227–237.

Grassi G, Seravalle G, Calhoun DA et al. (1994). Mechanisms responsible for sympathetic activation by cigarette smoking in humans. Circulation 90: 248–253.

Grassi G, Giannattasio C, Failla M et al. (1995a). Sympathetic modulation of radial artery compliance in congestive heart failure. Hypertension 26: 348–354.

Grassi G, Seravalle G, Cattaneo BM et al. (1995b). Sympathetic activation and loss of reflex sympathetic control in mild congestive heart failure. Circulation 92: 3206–3211.

Grassi G, Seravalle G, Cattaneo BM et al. (1995c). Sympathetic activation in obese normotensive subjects. Hypertension 25: 560–563.

Grassi G, Bolla GB, Seravalle G et al. (1997). Comparison between reproducibility and sensitivity of muscle sympathetic nerve traffic and plasma noradrenaline in men. Clin Sci 92: 285–289.

Grassi G, Cattaneo BM, Seravalle G et al. (1998). Baroreflex control of sympathetic nerve activity in essential and secondary hypertension. Hypertension 31: 68–72.

Grassi G, Seravalle G, Quarti-Trevano F et al. (2003). Effects of hypertension and obesity on the sympathetic activation of heart failure patients. Hypertension 42: 873–877.

Grassi G, Facchini A, Trevano FQ et al. (2005). Obstructive sleep apnea-dependent and -independent adrenergic activation in obesity. Hypertension 46: 321–325.

Grassi G, Seravalle G, Trevano FQ et al. (2007). Neurogenic abnormalities in masked hypertension. Hypertension 50: 537–542.

Grassi G, Seravalle G, Quarti-Trevano F et al. (2008a). Adrenergic, metabolic, and reflex abnormalities in reverse and extreme dipper hypertensives. Hypertension 52: 925–931.

Grassi G, Bolla G, Quarti-Trevano F et al. (2008b). Sympathetic activation in congestive heart failure: reproducibility of neuroadrenergic markers. Eur J Heart Fail 10: 1186–1191.

Grassi G, Seravalle G, Quarti-Trevano F et al. (2009). Sympathetic and baroreflex cardiovascular control in hypertension-related left ventricular dysfunction. Hypertension 53: 205–209.

Grassi G, Seravalle G, Scopelliti F et al. (2010a). Structural and functional alterations of subcutaneous small resistance arteries in severe human obesity. Obesity 18: 92–98.

Grassi G, Seravalle G, Brambilla G et al. (2010b). Impact of the metabolic syndrome on subcutaneous microcirculation in obese patients. J Hypertens 28: 1708–1714.

Greenwood JP, Scott EM, Stoker JB et al. (2001). Hypertensive left ventricular hypertrophy: relation to peripheral sympathetic drive. J Am Coll Cardiol 38: 1711–1717.

Groppelli A, Giorgi DMA, Omboni S et al. (1992). Persistent blood pressure increase induced by heavy smoking. J Hypertens 10: 495–499.

Heagerty AM (1997). Structural changes in resistance arteries in hypertension. In: A Zanchetti, G Mancia (Eds.), Handbook of Hypertension. Pathophysiology of Hypertension, Vol. 17. Elsevier Science, Amsterdam, pp. 426–437.

Jamerson KA, Julius S, Gudbrandsson T et al. (1993). Reflex sympathetic activation induces acute insulin resistance in the human forearm. Hypertension 21: 618–623.

Julius S (1996). The evidence for a pathophysiologic significance of the sympathetic overactivity in hypertension. Clin Exp Hypertens 18: 305–321.

Lembo G, Capaldo B, Rondina V et al. (1994). Acute noradrenergic activation induces insulin resistance in normal skeletal muscle. Am J Physiol 266: E242–E247.

Mancia G, Parati G (1987). Reactivity to physical and behavioural stress and blood pressure variability in hypertension. In: S Julius, DR Basset (Eds.), Handbook of

Hypertension, Behavioural Factors in Hypertension, Vol. 9. Elsevier, Amsterdam, pp. 104–122.

Mancia G, Ludbrook J, Ferrari A et al. (1978). Baroreceptor reflex in human hypertension. Circ Res 43: 170–177.

Mancia G, Ferrari AU, Leonetti G et al. (1982). Carotid sinus baroreceptor control of arterial pressure in renovascular hypertensive subjects. Hypertension 4: 47–50.

Mancia G, Saino A, Grassi G (1996). Interactions between the sympathetic nervous system and renin angiotensin system. In: JH Laragh, BM Brenner (Eds.), Hypertension: Pathophysiology, Diagnosis and Management, Raven Press, New York, pp. 399–408.

Mancia G, Parati G, Di Rienzo M et al. (1997a). Blood pressure variability. In: A Zanchetti, G Mancia (Eds.), Handbook of Hypertension, Vol. 17. Elsevier, Amsterdam, pp. 117–169.

Mancia G, Grassi G, Ferrari AU (1997b). Reflex control of the circulation in experimental and human hypertension. In: A Zanchetti, G Mancia (Eds.), Handbook of Hypertension, Vol. 17. Elsevier, Amsterdam, pp. 568–601.

Mangoni AA, Mircoli L, Giannattasio C et al. (1996). Heart rate dependence of arterial distensibility in vivo. J Hypertens 14: 897–901.

Mangoni AA, Mircoli L, Giannattasio C et al. (1997). Effect of sympathectomy on the mechanical properties of common carotid and femoral artery. Hypertension 30: 1085–1088.

Patel MB, Stewart JM, Loud AV et al. (1991). Altered function and structure of the heart in dogs with chronic elevation in plasma norepinephrine. Circulation 84: 2091–2100.

Perondi R, Saino A, Tio AR et al. (1992). ACE inhibition attenuates sympathetic coronary vasoconstriction patients with coronary artery disease. Circulation 85: 2004–2013.

Richardson DW, Stella A, Leonetti G et al. (1974). Mechanisms of renal release of renin by electrical stimulation of brain stem in the cat. Circ Res 34: 425–434.

Rowe JW, Young JB, Minaker KL et al. (1981). Effect of insulin and glucose infusion on sympathetic nervous system activity in normal man. Diabetes 30: 219–225.

Saino A, Pomidossi G, Perondi R et al. (1997). Intracoronary angiotensin II potentiates coronary sympathetic vasoconstriction in man. Circulation 96: 148–153.

Sen S, Tarazi RC, Khairallah P et al. (1974). Cardiac hypertrophy in spontaneously hypertensive rats. Circ Res 35: 775–781.

Smyth HS, Sleight P, Pickering GW (1969). Reflex regulation of arterial pressure during sleep in man: a quantitative method of assessing baroreflex sensitivity. Circ Res 24: 109–121.

Handbook of Clinical Neurology, Vol. 117 (3rd series)
Autonomic Nervous System
R.M. Buijs and D.F. Swaab, Editors

Chapter 27

Pathology of emesis: its autonomic basis

WILLIAM L. HASLER*

Division of Gastroenterology, Department of Internal Medicine, University of Michigan Health System, Ann Arbor, MI, USA

INTRODUCTION

The autonomic nervous system comprises parasympathetic and sympathetic subsystems that serve to regulate a range of internal functions including cardiorespiratory, vascular, and gastrointestinal activities. In concert with central structures in the brainstem and cerebrum, afferent and efferent components of the autonomic nervous system participate in the response to emetic stimulation as well as the motor acts involved in the act of vomiting itself. Autonomic dysfunction is prevalent in many clinical conditions complicated by nausea and vomiting; furthermore, nausea and vomiting are common complaints of patients with disorders of autonomic nervous system activity. Many antiemetic therapies act in part by action on autonomic pathways.

MECHANISMS OF EMESIS: RELATION TO AUTONOMIC NERVOUS SYSTEM

Induction of emesis involves a complex interplay of structures within the central nervous system, the peripheral nervous system including autonomic nerves, and the gastrointestinal tract. In contrast, the neural pathways involved in the sensation of nausea are poorly understood but may include the hypothalamus and inferior cerebral cortical gyrus (Yates et al., 1998). Several regions serve as receptor sites for emetic stimulation. Centrally, the area postrema at the base of the fourth cerebral ventricle has been termed the chemoreceptor trigger zone based on its responsiveness to neural and circulating chemical emetic stimuli. Peripheral receptor sites to emetic stimulation include vagal and nonvagal pathways including the vestibular apparatus.

Vagal pathways prominently participate in vomiting in response to several peripheral stimuli. Stimulation of C afferent fibers in the vagus is a potent stimulus of vomiting (Urayama et al., 2010). Abdominal vagal afferents terminate in the nucleus of the solitary tract. Solitary nucleus neurons then project to nuclei within the brainstem which serve to coordinate the act of emesis. Other neurons from the nucleus of the solitary tract project centrally to the hypothalamus and other structures. In shrews, abdominal vagotomy abolishes emesis evoked by gastric distention while carotid sinus denervation suppresses emesis during experimental motion sickness (Uchino et al., 2008). In another well-described model of peripherally induced emesis, intragastric administration of the irritant substance copper sulfate elicits prompt vomiting that is blocked by vagotomy with or without associated splanchnicectomy but is unaffected by area postrema ablation (Beleslin and Strbac, 1987; Fukui et al., 1994). Splanchnic denervation reduces prodromal responses before retching but does not prevent vomiting. Autonomic neural function is disrupted during experimental emesis from peripheral stimulation in animal models. In shrews, gastric mucosal stimulation elicits blood pressure increases and bradycardia prior to retching (Uchino et al., 2002).

Another peripheral model of emesis is experimental motion sickness, which is usually evoked by rotatory stimulation. Motion sickness is mediated by the vestibular complex, area postrema, vagal nuclei, parabrachial nuclei, and reticular formation (Pedigo and Brizzee, 1985). Muscarinic receptors are present in high densities in vagal nuclei and the area postrema, at intermediate levels in the reticular formation, and in low densities in the vestibular nuclei. Projections from vestibular nuclei to the subtrigeminal nucleus, lateral tegmental field, and nucleus ambiguus also are described (Balaban, 1996). Vagal cholinergic pathways are involved in gastric motor and myoelectric consequences of motion sickness. Atropine blocks gastric slow wave arrhythmias elicited during motion sickness in humans (Kim et al., 1997). In shrews, vagotomy blocks such

*Correspondence to: William L. Hasler, M.D., University of Michigan Hospital, 3912 Taubman Center, 5362, Ann Arbor, 48109, MI, USA. Tel: +1-734-936-4780, Fax: +1-734-936-7392, E-mail: whasler@umich.edu

dysrhythmias but does not prevent motion-induced emesis (Percie du Sert et al., 2010).

Studies of visceral motor events associated with retching and emesis confirm vagal and nonvagal components to the responses. Retching is preceded by increases in swallowing and relaxation of the lower esophageal sphincter and proximal stomach, followed by rhythmic relaxation and contraction of the pharynx and cervical esophagus (Lang et al., 1993). With emesis, the pharynx and cervical esophagus relax then contract in retrograde fashion and the geniohyoideus contracts. Distally, emesis in dogs is associated with retrograde peristaltic contractions (RPC) beginning in the jejunum, a peri-RPC inhibitory period, many post-RPC contractions, and a post-RPC inhibitory period on manometry (Lang et al., 1986). These motor responses are blocked by vagotomy, but only the RPC is mediated by cholinergic pathways. Antral mucosal stimulation in ferrets evokes reductions in gastric corpus pressure that are blocked by atropine but are not affected by splanchnicectomy or adrenalectomy (Andrews and Wood, 1988). Myoelectric correlates of these motor events include retrograde trains of spike potentials in the jejunum followed by slow wave tachyarrhythmias in the antrum during emesis elicited by dopamine and copper sulfate (Lee et al., 1985). Other phenomena associated with motion sickness include gastric bradyarrhythmias and aborally propagating intestinal giant migrating contractions that are associated with defecation (Lang et al., 1999).

Insights provided by fictive retching/emesis model

Studies of fictive retching/vomiting in decerebrate animals have provided detailed characterization of afferent and efferent neural pathways involved in the act of emesis. Fictive vomiting is defined by coordinated pre-emetic bursts of synchronous activity in the phrenic and abdominal nerves during retching that are followed by prolonged abdominal nervous activity with expulsion that occur in response to emetic stimuli (Grelot et al., 1992). Structures activated during fictive emesis include the afferent vagus, nucleus of the solitary tract, nucleus ambiguus, reticular formation, respiratory premotor neurons in the medulla, and motor neurons to the diaphragm and abdominal musculature (Fig. 27.1) (Onishi et al., 2007). Central participation in fictive retching is evidenced by suppression of the phenomenon by area postrema cooling during vagal stimulation (Koga and Fukuda, 1992). Studies involving directed section of different neural structures indicate that fictive vomiting during vagal stimulation is activated by vagal afferent fibers that enter the rostral bulb and descend through the solitary nucleus complex to the area subpostrema (Fukuda and Koga, 1991). Subpostrema neurons then project through the reticular column to the Botzinger complex. A central pattern generator for emesis is proposed to consist of nonrespiratory neurons in the reticular area dorsomedial to the retrofacial nucleus (Koga and Fukuda, 1997). Half of nonrespiratory neurons interacting with Botzinger complex neurons fire in synchrony with retching while one third fire between periods of retching (Fukuda and Koga, 1992). Fictive retching and vomiting are associated intimately with alterations in respiratory nerve activity. Respiratory inspiratory and expiratory group neurons exhibit altered activation patterns during these phenomena (Milano et al., 1992). Arterial carbon dioxide concentrations increase from retching to gastric expulsion due to cessation of pulmonary ventilation (Fukuda and Koga, 1993). Fictive retching elicited by vagal afferent activation is reduced by stimulation of the rostral part of the nucleus raphe magnus, indicating descending inhibition of the phenomenon (Urayama et al., 2010).

Insights provided by c-fos distribution studies

Studies of c-fos expression in response to emetic stimulation provide insight into afferent and efferent pathways that participate in vomiting. Emetic stimuli including electrical depolarization of the vagus and hypertonic saline increase c-fos protein in the nodose ganglia, area postrema, solitary nucleus, intermediate reticular zone of the lateral tegmental field, spinal nucleus of the trigeminal nerve, nucleus ambiguus, inspiratory propriospinal cell region at the level of C2, dorsal vagal and phrenic motor nuclei, ventrolateral medulla, locus coeruleus, subcoeruleus, lateral parabrachial nucleus, raphe magnus, subretrofacial nucleus, and spinal dorsal horn (Gieroba and Blessing, 1994; Miller and Ruggiero, 1994). More than 70% of c-fos staining is ipsilaterally localized.

Neurohumoral mediators of emesis: interaction with autonomic nervous system

Both afferent and efferent mechanisms of emesis involve activation of cholinergic and adrenergic pathways in the peripheral and central nervous system. Muscarinic receptors are present presynaptically on vagal afferent terminals in cats (Leslie et al., 1990b). Acetylcholine enhances emetic responses to motion stimuli (Uchino et al., 2001). Combined M_1/M_4 muscarinic agonists elicit emesis that is incompletely blocked by area postrema ablation (Beleslin et al., 1989). Atropine blocks the gastric motor effects but not necessarily the emetic actions of erythromycin (Qin et al., 1993). Intraventricular perfusion of nicotine and dimethylphenylpiperazinium elicit emesis as

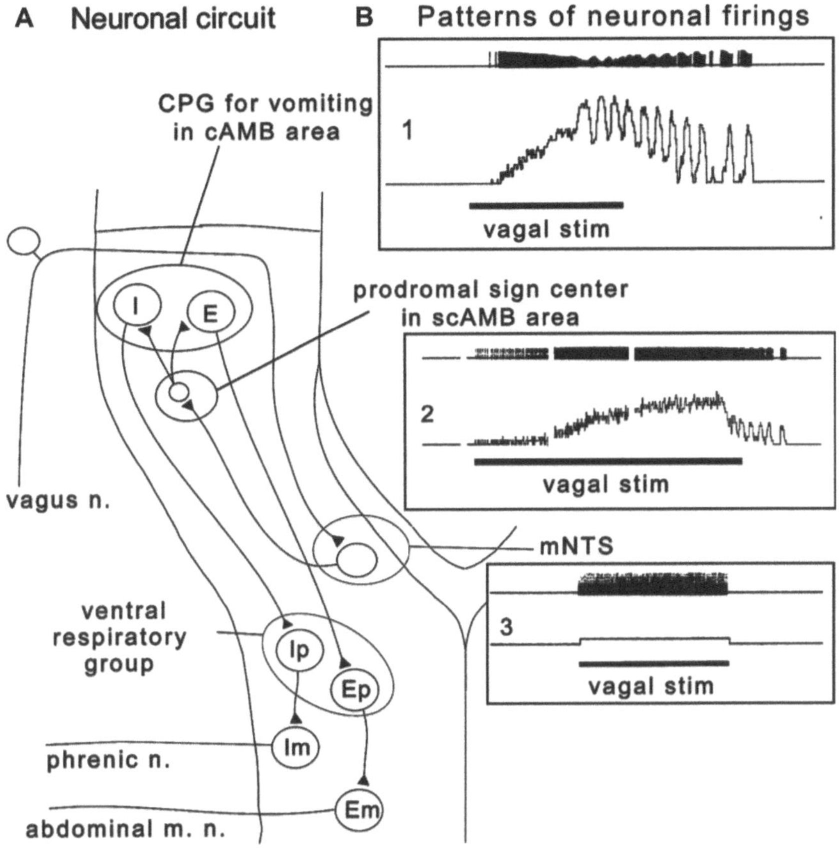

Fig. 27.1. A representation of the neuronal circuit that participates in the induction of emesis in dogs is shown in (**A**) (mNTS, nucleus of the solitary tract; CPG, central pattern generator for vomiting; cAMB, compact part of the nucleus ambiguus; scAMB, semicompact part of the nucleus ambiguus; Ep, premotor neurons of the abdominal muscles; Em, motor neurons of the abdominal muscles; Ip, premotor neurons of the diaphragm; Im, motor neurons of the diaphragm). Afferent vagal stimulation in decerebrated paralyzed dogs leads to emesis by progressive activation of (1) neurons of the mNTS, (2) neurons of the prodromal sign center near the scAMB, (3) neurons of the central pattern generator near the cAMB, (4) respiratory (Ip) and abdominal muscle (Ep) premotor neurons in the caudal medulla, and (5) motor neurons of the diaphragm (Im) and abdominal muscles (Em). Patterns of neuronal firings in response to vagal stimulation in dogs are shown in (**B**) including a firing pattern of a typical CPG neuron (**B-1**), a firing pattern of a neuron of the scAMB (**B-2**), and a firing pattern of a typical mNTS neuron (**B-3**). For each panel, the upper tracing shows rectangular pulses representing neuronal firings while the lower tracing shows a frequency histogram representing neuronal firings with 0.1 s bins. (Reproduced from Onishi et al., 2007.)

well, phenomena blocked by area postrema ablation indicating central nicotinic activation (Beleslin and Krstic, 1986). During apomorphine-induced nausea in humans, plasma epinephrine increases threefold (Grant et al., 1986). C-fos expression in the solitary nucleus in response to electrical vagal stimulation is localized to catecholamine neurons, as shown by simultaneous tyrosine hydroxylase labeling studies (Gieroba and Blessing, 1994). Likewise, activity of noradrenergic neurons in the locus coeruleus is increased by emetic stimulation in cats (Rasmussen et al., 1986). Area postrema ablation prevents emesis in response to central administration of epinephrine, norepinephrine, and dopamine (Jovanovic-Micic et al., 1989). α_2-Adrenoceptor agonists (xylazine, clonidine, naphazoline) activate emesis which can be blocked by α_2 antagonists such as yohimbine and tolazoline and by area postrema ablation (Japundzic-Zigon et al., 1997). The ability of scopolamine, chlorpromazine, the nonselective serotonin antagonist methysergide, and both H_1 and H_2 histamine antagonists to reduce vomiting in response to α agonists indicates interaction of several nerve pathways in this emetic response. Similarly, the α_2 antagonists yohimbine and idazoxan block the emetic actions of dopamine showing additional interplay (Jovanovic-Micic et al., 1995). Isoproterenol prolongs the latency to induction of emesis during motion sickness in shrews, indicating β-adrenoceptor contributions to emesis in some models (Uchino et al., 2001).

An extensive body of literature has confirmed the roles of serotonin (5-HT, 5-hydroxytryptamine) and neurokinin pathways in the induction of emesis. 5-HT_3 receptor binding is greatest in the dorsomedial nucleus of the solitary nucleus (Leslie et al., 1990a). 5-HT_3 receptor binding in the dorsal vagal complex is attenuated by nodose ganglion excision and abolished by vagotomy. 5-HT_3 agonists elicit emesis in a phenomenon blunted by vagotomy and splanchnicectomy but not area postrema ablation (Miller and Nonaka, 1992). Antagonists at 5-HT_3 receptor exert antiemetic effects by acting on vagal afferent nerves with lesser additional actions on the area postrema, solitary nucleus, and medulla (Costall and Naylor, 1992). 5-HT_3 antagonists prevent emesis and c-fos expression in the solitary nucleus in response to many stimuli including 5-HT_3 agonists, dopamine agonists, copper sulfate, and selected cancer chemotherapeutic agents (Reynolds et al., 1991). Other serotonin receptor subtypes participate in emesis as well. Emesis is induced by 5-HT_4 agonist administration, a response blocked by vagotomy (Fukui et al., 1994). 5-HT_4 antagonists block emesis in response to copper sulfate more effectively than 5-HT_3 antagonists in some studies. 5-HT_{1A} agonists block vomiting after the capsaicin analog resiniferatoxin in shrews (Andrews et al., 1996). Neurokinin NK_1 immunoreactivity is prominent in the reticular area near the nucleus ambiguus (Onishi et al., 2007). NK_1 receptors are present in the abdominal vagus, solitary nucleus, and area postrema (Navari, 2009). NK_1 antagonists abolish emesis evoked by vagal or solitary nucleus stimulation (Fukuda et al., 1998). Similarly, NK_1 antagonist injection into the reticular formation abolishes fictive vomiting in response to emetic stimulation (Onishi et al., 2007). The proposed site of antiemetic action of NK_1 antagonists is the dorsal vagal complex (Hornby, 2001). However, an additional abdominal vagal site of action of this drug class also has been proposed (Minami et al., 2001).

Miscellaneous mediators participate in emesis. High densities of cannabinoid CB_1 receptors are present in the dorsal motor nucleus of the vagus and in the medial subnucleus of the solitary nucleus, while lower densities are observed in the area postrema and dorsal subnucleus of the solitary nucleus (Van Sickle et al., 2003). CB_1 agonists such as δ-9-tetrahydrocannabinol bind in the dorsal vagal complex (Hornby, 2001). Administration of tetrahydrocannabinol rostral to the obex reduces c-fos expression in the solitary nucleus and dorsal motor nucleus of the vagus while THC delivered to the obex decreases c-fos protein in the area postrema and dorsal subnucleus of the solitary nucleus, providing potential pathways for the antiemetic effects of cannabinoid compounds (Van Sickle et al., 2003). Transient receptor potential vanilloid 1 (TRPV1) immunoreactivity is also prominent in the solitary nucleus and dorsal motor nucleus of the vagus and is less dense in the area postrema (Sharkey et al., 2007). TRPV1 activators such as capsaicin, resiniferatoxin, and olvanil elicit vomiting in some models but also inhibit emesis in response to several stimuli including copper sulfate, apomorphine, motion, nicotine, and selected chemotherapeutic agents (Andrews et al., 2000; Chu et al., 2010). In fictive vomiting studies, capsaicin and resiniferatoxin injections into the fourth ventricle abolish emesis to vagal stimulation but do not block retching in response to direct stimulation of the solitary nucleus confirming that emetic vagal afferents are capsaicin-sensitive C fibers (Shiroshita et al., 1997). Vasopressin is released by several different emetic stimuli including experimental motion sickness and electrical vagal stimulation (Kim et al., 1997). Exogenous vasopressin at supraphysiological doses is a potent emetogenic stimulus. However, no definitive studies blocking vomiting in response to vasopressin antagonists have been reported. Finally, inflammatory mechanisms can participate in emesis in some clinical conditions. Tumor necrosis factor-α (TNF_α) injection in the dorsal vagal complex elicits nausea and vomiting, causes gastric stasis, and suppresses gastric contractility elicited by thyrotropin-releasing hormone (Hermann et al., 2003). Conversely, perfusion of the fourth cerebral ventricle with TNFR:Fc reverses the gastric inhibition evoked by lipopolysaccharide (Hermann et al., 2002).

ROLE OF THE AUTONOMIC NERVOUS SYSTEM IN CLINICAL CONDITIONS WITH NAUSEA AND VOMITING

Autonomic function in gastrointestinal disease

Disordered autonomic dysfunction is described in many diseases of the gastrointestinal tract. The role of autonomic dysfunction in symptom genesis in these conditions is unproved; however, therapies relating to modulation of autonomic function may be effective in some cases. Cyclic vomiting syndrome (CVS) is characterized by recurrent episodes of relentless nausea and vomiting often in association with abdominal pain and diarrhea with intervening asymptomatic periods. In the most recent consensus document, CVS was defined as the report of stereotypical episodes of vomiting regarding onset (acute) and duration (<1 week) with ≥ 3 discrete episodes in the past year, a symptom onset at least 6 months previously, and an absence of nausea and vomiting between episodes (Tack et al., 2006). CVS affects both pediatric and adult populations and has been associated with personal or family histories of migraine headaches and (in adults) cannabis abuse (Abell et al., 2008). Some studies of autonomic function in CVS show reductions in cholinergic function; however, prominent decreases in the postural adjustment ratio (a sympathetic adrenergic measure) are also noted (Rashed et al., 1999). Pediatric CVS patients also exhibit increased sympathetic modulation of the sinus node as shown by decreases in the ratio of the low to high frequency components of R-R interval variability of the electrocardiogram (To et al., 1999). In an assessment of autonomic function in CVS in children, cardiovascular responses to deep breathing and Valsalva maneuvers were preserved while sudomotor findings were abnormal in all cases. Some children exhibit increased heart rates with tilt table testing or vasodepressor tendencies (Table 27.1) (Chelimsky and Chelimsky, 2007). These findings suggest the presence of sympathetic impairments involving both vasomotor and sudomotor pathways.

Gastroparesis is defined by symptoms of gastric retention in the absence of mechanical obstruction (Parkman et al., 2010). The condition presents with nausea, vomiting, bloating, discomfort, fullness, and an inability to complete a normal sized meal, and is commonly diagnosed by demonstrating delayed gastric emptying on scintigraphy. Most often, gastroparesis is idiopathic in nature or occurs as a complication of diabetes or after intentional or inadvertent vagotomy from surgery for ulcer disease or acid reflux, although other conditions cause gastroparesis in a third of cases. In an examination of vagal tissue from a diabetic with gastroparesis who underwent gastroenterostomy, electron microscopy revealed reduced density of unmyelinated axons and decreased caliber of those neurons that survived (Guy et al., 1984). In a study of gastroparesis in cirrhotic patients, composite autonomic scores were threefold higher than a group of hepatitis C patients without cirrhosis (Verne et al., 2004). Detailed characterization of autonomic function and structure was provided by study of a 31-year-old woman who developed severe gastroparesis in association with autonomic failure after a bout of acute gastroenteritis (Lobrano et al., 2006). Autonomic evaluation showed abnormally low heart rate variability and altered responses to Valsalva maneuvers. Full-thickness intestinal biopsies showed normal enteric neuronal and smooth muscle morphology. Her course was complicated by pulmonary embolus and death. On autopsy, analysis of her sympathetic chain revealed reductions in myelinated axons with vacuolar degeneration and inflammation which was interpreted as being similar to Guillain–Barré syndrome.

Autonomic dysfunction may be profound in motion sickness and other disorders of vestibular activity. Indeed, pallor, diaphoresis, and changes in heart rate and respirations are common with motion sickness, vertigo, and experimental caloric ear irrigation. In subjects watching an irregularly oscillating video that produced a sense of movement, nausea related to increases in heart rate, perspiration, and respiratory cycle activity (Fig. 27.2) (Himi et al., 2004). During motion sickness in parabolic flight, there is intolerance of upright tilt with associated falls in peripheral resistance (Schlegel et al., 2001). With vomiting, there is increased R-R variability on electrocardiography, elevated carotid-cardiac baroreflex responsiveness, and upright hypocapnia. Likewise, the high frequency component of the R-R interval increases during caloric irrigation of the external ear (Yasumura et al., 2005). Increases in circulating epinephrine are observed during motion sickness; however,

Table 27.1

Autonomic findings in patients with cyclic vomiting syndrome (from Chelimsky and Chelimsky, 2007)

Patient	Deep breathing	Valsalva maneuver	Tilt test	QSART
1	Normal	Normal	POTS with vasodepressor syndrome; abdominal pain at lowest blood pressure of 52/40	Absent in 3/5 sites
2	Normal	Normal	POTS only	Exaggerated responses in 2/4 sites; persistent sweat activity in all sites
3	Normal	Normal	POTS with vasodepressor syndrome; abdominal pain and headache at lowest blood pressure of 61/40	Absent in 2/3 sites
4	Normal	Normal	POTS only	Absent in 3/5 sites
5	Normal	Normal	POTS only; abdominal pain while upright	Absent in 1/5 sites; exaggerated responses in 2/5 sites
6	Normal	Normal	POTS with vasodepressor syndrome; no gastrointestinal symptoms	Absent in 4/4 sites

POTS, postural orthostatic tachycardia syndrome; QSART, quantitative sudomotor axon reflex test.

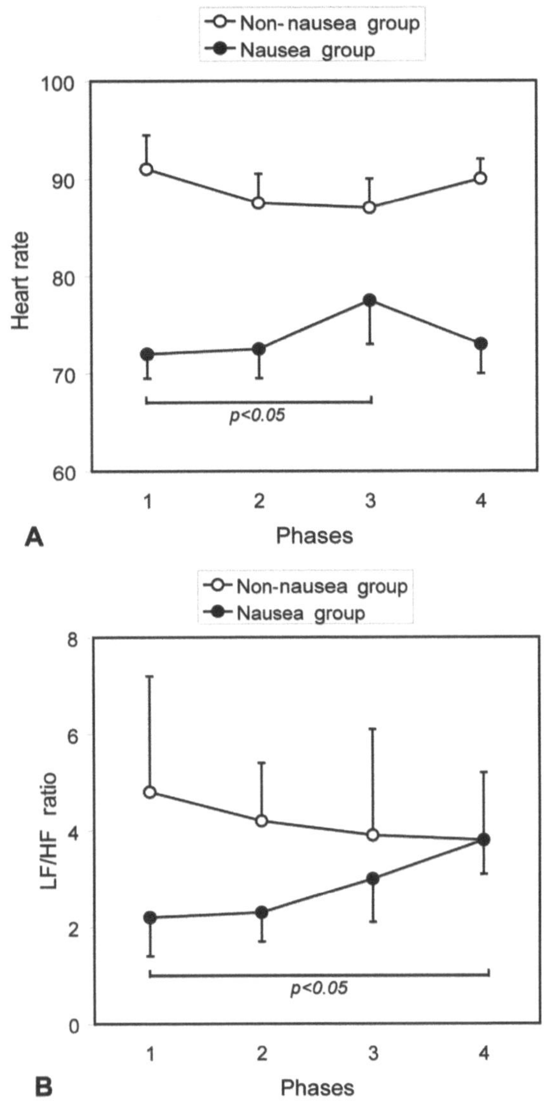

Fig. 27.2. Changes in cardiac reactions are shown for healthy subjects watching an irregularly oscillating video to produce sensations of motion sickness. Heart rate increases are greater in those who develop nausea vs. those without nausea (**A**) ($p < 0.05$). Additionally, subjects reporting nausea exhibit increases in the low to high frequency components of R-R ratios on electrocardiography (**B**) ($p < 0.05$). (Adapted from Himi et al., 2004.)

patients with the highest epinephrine levels show resistance to development of motion-induced nausea (Kohl, 1993).

Autonomic function in other conditions

Several nongastrointestinal conditions with associated nausea and vomiting may have autonomic function abnormalities. During migraines, patients commonly report a range of autonomic disturbances including flushing, pallor, and diaphoresis (Melek et al., 2007). Findings in migraine sufferers have been reported to be similar to those of CVS patients who also show low cholinergic function measures (Rashed et al., 1999). However, others have observed variable and transient increases in sympathetic activity resulting in sinus tachycardia as well as increases in parasympathetic activity that can lead to sinus bradycardia, atrioventricular block, and ST- and T-wave changes on electrocardiography (Melek et al., 2007). Cardiovascular conditions may relate to autonomic dysfunction. Hypotension leads to activation of the cardiodepressor reflex with decreased sympathetic and increased parasympathetic activity (Barnas et al., 1999). The carotid sinus syndrome can be complicated by severe nausea and vomiting (Toorop et al., 2007). Decreases in resting diastolic blood pressure and skin conductance and blunted responses to cold pressor tests are observed in anorexia nervosa (Abell et al., 1987). Sinus tachycardia, hypertension, and reduced R-R interval variability are seen during attacks of acute intermittent porphyria (Goto et al., 1989). Loss of sympathetic activity relates to delays in orocecal transit after spinal cord injury (Chen et al., 2004).

Nausea and vomiting with dysautonomia syndromes and related conditions

Familial dysautonomia presents in children with vomiting crises that can result in significant complications, including recurrent pulmonary aspiration, impaired nutrition, and repeated hospitalizations (Udassin et al., 1992). Associated abnormalities include pharyngoesophageal dyskinesia, gastric retention, and gastroesophageal reflux (Vinograd et al., 1985). Because of the pronounced morbidity and mortality of these complications, affected children with familial dysautonomia often undergo fundoplication to prevent vomiting episodes. In small case series, performance of fundoplication does reduce vomiting; however, dysautonomic manifestations of disease are unaffected and children may continue to retch violently (Vinograd et al., 1985; Udassin et al., 1992). Furthermore, some studies note that half of patients continue to experience aspiration pneumonia and 15% develop failure of the fundoplication wrap from recurrent retching (Udassin et al., 1992). Other complications of fundoplication include postoperative dysphagia, gastric dilation, and dumping syndrome (Vinograd et al., 1985).

Orthostatic intolerance includes the postural orthostatic tachycardia syndrome (POTS) as well as other conditions of neurally mediated hypotension (Fig. 27.3) (Sullivan et al., 2005). The most common gastrointestinal symptoms in affected patients are nausea, vomiting, and abdominal pain. Other symptoms include lightheadedness/syncope, blurred vision, fatigue, palpitations, diaphoresis,

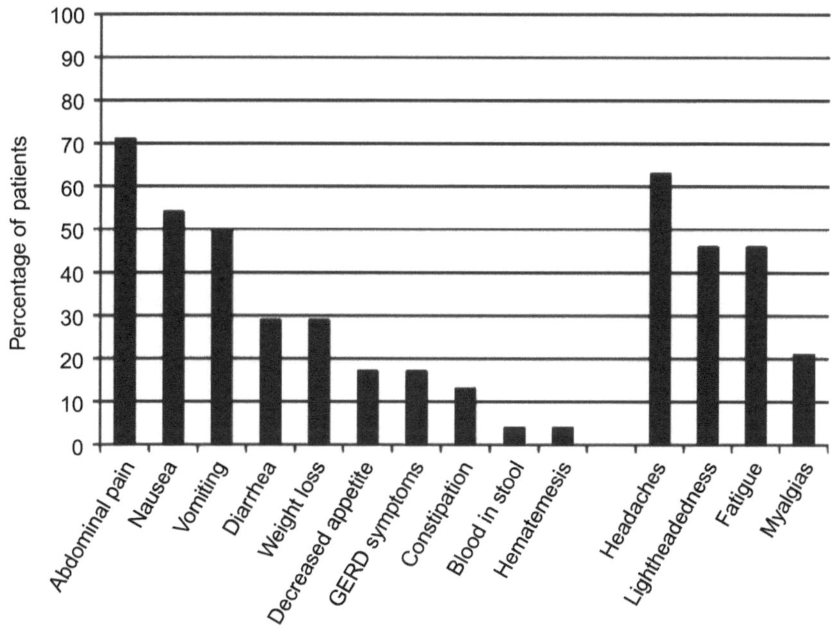

Fig. 27.3. Percentages of patients with orthostatic intolerance syndrome who report different gastrointestinal and nongastrointestinal symptoms are shown. Reports of abdominal pain, nausea, and vomiting are prominent. (From Sullivan et al., 2005.)

and tremors (Mack et al., 2010). The median number of gastrointestinal symptoms per patient is three. Nearly half of patients have intermittent nausea and vomiting for > 3 years; however, these symptoms commonly resolve with treatment of the orthostatic intolerance.

Other neurological conditions with altered autonomic function may have gastrointestinal consequences. In a case report, a 14-year-old girl with Friedreich's ataxia presented with frontal headaches, rash, abdominal pain, and recurrent nausea and vomiting (Margalith et al., 1984). She exhibited increases in plasma catecholamines that were suppressed with clonidine consistent with a dysautonomic response. A second case involved a 21-year-old man who developed acute ataxia, orthostatic hypotension, and associated nausea and vomiting after Epstein–Barr infection (Itoh et al., 1993). Autonomic testing showed dysfunction of sympathetic postganglionic nerves.

Nausea and vomiting complicating medication therapy or poisoning

Selected medication classes with activity on autonomic nervous pathways may cause nausea and vomiting as treatment side-effects. Acetylcholinesterase inhibitors are employed in a variety of settings. Physostigmine elicits a motion sickness-like presentation of nausea, vomiting, malaise, and increases in adrenocorticotropin hormone, β-endorphin, cortisol, and prolactin (Janowsky et al., 1984). Use of neostigmine or physostigmine during a range of surgeries (including gynecological, obstetric, urological, and vascular) frequently leads to increased postoperative nausea and vomiting (PONV), often in association with other manifestations of autonomic dysregulation including bradycardia (Ho et al., 2005). Cholinesterase inhibitors are increasingly used as therapies to improve cognitive function in mild to moderate Alzheimer's disease. In older studies, physostigmine use led to increases in nausea, vomiting, and abdominal discomfort in Alzheimer's patients (Thal et al., 1999). Likewise, newer agents with cholinesterase inhibitor activity, such as donepezil and galantamine, and drugs with dual acetyl- and butyrlcholinesterase inhibition, such as rivastigmine, elicit nausea, vomiting, anorexia, and diarrhea in dementia (Birks, 2006; Kavirajan and Schneider, 2007). In one report to poison centers in the US, exposure to such centrally active cholinesterase inhibitors was associated with vomiting in 34% and nausea in 28% of cases (McCain et al., 2007). In systematic reviews and meta-analyses, more patients with dementia have been reported to drop out of clinical trials on active acetylcholinesterase inhibitors compared to placebo (Birks, 2006; Kavirajan and Schneider, 2007).

Nicotine is used therapeutically for a variety of indications; its prescription is commonly complicated by nausea and vomiting. In healthy control nonsmokers receiving drug for depression, nicotine patches elicit nausea and anorexia as well as dysautonomic symptoms including lightheadedness and postural imbalance (Salin-Pascual et al., 1996; Zingler et al., 2007). Likewise, when employed in clinical trials in patients with Parkinson's disease, nicotine use is complicated by high rates of treatment-induced nausea and vomiting (Villafane et al., 2007). Conversely, both nicotine and tobacco smoking have been associated with reduced risks of developing PONV in patients undergoing laparoscopic cholecystectomy (Ionescu et al., 2007). Nicotine administration in standard clinical doses causes little nausea in smokers, reflecting development of tolerance to this effect (Srivastava et al., 1991).

Prescription of other drugs that act on the autonomic nervous system can have gastrointestinal consequences. Nausea, vomiting, and abdominal pain are reported after ingestion of dietary supplements containing ephedrine and caffeine (Schier et al., 2003). Intrathecal epinephrine increases nausea in women undergoing labor analgesia (Goodman et al., 2002). Conversely, one report observed less PONV when phenylephrine was administered for blood pressure control during Cesarean section (Ngan Kee et al., 2004).

Accidental or intentional poisoning with agents active against autonomic pathways can elicit prominent nausea and vomiting. Mushrooms in the genus *Inocybe* contain muscarine. When ingested, patients present with severe nausea, vomiting, diarrhea, and abdominal pain, with other dysautonomic manifestations including hypersalivation, lacrimation, diaphoresis, tachy- and bradycardia, and hypotension (Lurie et al., 2009). Likewise, consumption of wild onion bulbs (*Zigadenus*) leads to nausea, vomiting, bradycardia, and hypotension which resolve with atropine administration (West and Horowitz, 2009). In a misguided therapeutic experiment, one Alzheimer's patient developed nausea, diaphoresis, and hypotension after taking carbamylcholine in an effort to relieve his neurological symptoms (Schulz et al., 2006). Some 14% of those working in illicit methamphetamine laboratories report nausea and vomiting due to drug exposure (Thrasher et al., 2009). Exposure to one insecticide (amitraz) with clonidine-like properties can cause nausea, vomiting, hypotension, bradycardia, respiratory impairment, and coma (Elinav et al., 2005).

THERAPY OF NAUSEA AND VOMITING IN RELATION TO THE AUTONOMIC NERVOUS SYSTEM

Medication and nonmedication therapies with action on autonomic neural pathways show significant benefits in a range of clinical conditions.

Anticholinergics

Oral, transdermal, and parenteral anticholinergic medications which bind to muscarinic receptors on autonomic nervous and non-neural sites reduce symptoms with vestibular stimulation. Several studies in motion sickness induced experimentally in the laboratory by rotatory stimuli or Coriolis maneuvers, in sailors, during parabolic flight, and during immersion in virtual reality report significant reductions in nausea and/or vomiting with prophylactic transdermal or buccal scopolamine (Pyykko et al., 1985; Hasler et al., 1995; Dornhoffer et al., 2004). In experimental motion sickness in healthy volunteers, oral and intravenous anticholinergic treatment prolongs the time needed to experience nausea by 40% and reduces gastric myoelectrical rhythm disturbances (tachygastria) which uniformly develop during motion-induced illness (Hasler et al., 1995; Dornhoffer et al., 2004). Hyoscine treatment of motion sickness is associated with reduced pulse rate and skin conductance changes, emphasizing actions on autonomic sites (Golding and Scott, 1997). One study observed no benefits of the peripherally active antimuscarinic agent methscopolamine, suggesting that this drug class acts centrally rather than peripherally to prevent motion-induced nausea (Hasler et al., 1995). In systematic reviews, scopolamine was more effective than placebo and methscopolamine and equipotent as antihistamines for prophylaxis of motion sickness (Spinks et al., 2004). Abrupt withdrawal of transdermal scopolamine is reported to produce rebound nausea and vomiting in patients with motion sickness, necessitating a taper with oral meclizine (Patel and Ezzo, 2009). In related studies, scopolamine also reduces symptoms in response to cold caloric stimulation in healthy subjects.

Anticholinegic agents also have documented benefits in preventing PONV. Hyoscine reduces PONV in patients undergoing orthopedic or plastic surgery (Wilkinson et al., 1989). Parenteral atropine is effective in many settings including tonsillectomy and adenoidectomy and in children receiving ketamine (Chhibber et al., 1999). Transdermal scopolamine reduces PONV in those undergoing Cesarean section, laparoscopy, and other gynecological, abdominal, orthopedic, otologic, and cosmetic surgeries (Kranke et al., 2002). In patients at high risk for PONV, scopolamine given in combination with the 5-HT_3 antagonist ondansetron decreases the need for rescue antiemetics and increases rates of reports of no nausea or vomiting after surgery (Gan et al., 2009).

Agents with action on adrenoceptor pathways

Agents with action on adrenoceptor pathways in the peripheral and central nervous systems reduce nausea and vomiting in selected conditions. The sympathomimetic amine ephedrine is effective in prevention of PONV after gynecological surgery and can prevent some of the secondary central neurological effects on electroencephalographic activity during motion sickness

(Wood et al., 1990). However, others have reported no benefits of ephedrine in motion sickness and greater efficacy of the α_1 agonist phenylephrine in women undergoing spinal anesthesia prior to labor (Cooper et al., 2002). Counter to these findings is the observation that the α antagonist phentolamine blunts nausea and gastric slow wave rhythm disturbances in healthy subjects during experimental motion sickness (Hasler et al., 1995). Although the centrally acting α_2 agonist clonidine exhibits emetogenic effects in some models, the drug reduces rates of PONV after orthopedic, spinal, and otologic surgery (Farmery and Wilson-MacDonald, 2009). Clonidine was also reported to offer superior PONV prophylaxis versus diazepam in a meta-analysis of children undergoing surgery (Dahmani et al., 2010). Another α_2 agonist dexmedetomidine reduced antiemetic drugs needs after laparoscopic bariatric surgery (Tufanogullari et al., 2008). Clonidine has also shown antiemetic benefits in clinical conditions with prominent autonomic disturbances. The drug reduced symptoms in a teenager with refractory CVS and in two patients with a syndrome of catecholamine and prostaglandin E_2 overproduction (Palmer and Cameron, 2005). In one study, clonidine reduced symptoms in a group of patients with diabetic gastroparesis (Rosa-e-Silva et al., 1995). The mechanism of this benefit is uncertain as different investigations have shown variable effects on gastric emptying.

Nonmedication options

Some nonmedication therapeutic options for conditions with prominent nausea and vomiting act via modulation of autonomic function. Gastric electrical stimulation delivered via an implanted device to the corporoantral junction of the stomach reduces nausea and vomiting in uncontrolled case series of patients with refractory gastroparesis (Hasler, 2009). Depolarizing stimuli with this method do not entrain intrinsic gastric electrical activity, reverse gastric myoelectric rhythm disturbances, or accelerate gastric emptying and thus do not truly pace the stomach. In a physiological investigation, one group noted that gastric electrical stimulation increased the mechanical threshold needed to elicit discomfort during balloon distention of the proximal stomach and decreased the sympathovagal balance with observed increases in vagal activity during autonomic function testing (Table 27.2) (McCallum et al., 2010). On positron emission tomography, gastric stimulation promoted increased activation in the thalamus and caudate nuclei demonstrating central projections of its actions. In experimental canine studies examining potential novel stimulation protocols for next generation human devices, use of a dual pulse morphology with five short pulses and one intervening high-intensity long pulse reduced sympathetic and increased vagal parasympathetic activity (Lei and Chen, 2009). In another study in diabetics with gastroparesis and other associated diabetic complications, combined pancreas–kidney transplantation reduced overall gastrointestinal symptoms with improvement in the postural adjustment ratio and Valsalva ratio on autonomic testing (Hathaway et al., 1994). In contrast, no similar benefits were observed with kidney transplantation alone. Finally, alternative therapies of nausea and vomiting such as acupuncture are reported to affect autonomic function including increasing the high-frequency power of R-R variability on electrocardiographic testing (Huang et al., 2005).

Table 27.2

Heart rate and parameters of heart rate variability before and after gastric electrical stimulation in patients with gastroparesis (from McCallum et al., 2010)

Parameter	Before GES	During GES	p Value
Mean heart rate (beats/min)	98.7 ± 6.2	88.2 ± 5.4	NS
P_L/P_H ratio	2.2 ± 0.6	0.5 ± 0.1	0.04

GES, gastric electrical stimulation; P_L, percentage of power of the R-R variability in the low frequency band; P_H, percentage of power of the R-R variability in the high frequency band; p value determined by Wilcoxon signed-rank test; NS not significant.

CHEMOTHERAPY AND RADIOTHERAPY INDUCED NAUSEA AND VOMITING: PARTICIPATION OF THE AUTONOMIC NERVOUS SYSTEM

A large volume of investigation has been devoted to mechanisms of emesis after cancer chemotherapy and after irradiation for intrabdominal malignancy. Additionally, more studies have been conducted examining efficacy of treatments for chemotherapy-induced nausea and vomiting (CINV) than all other causes of nausea and vomiting.

Mechanisms

Much of the knowledge of mechanisms of CINV has come from study of highly emetogenic agents such as cisplatin. Cisplatin increases c-fos expression in the area postrema, solitary nucleus, dorsal motor nucleus of the vagus, central amygdala, dorsal raphe nucleus, parabrachial nucleus, and cells in the ependymal lining of the fourth ventricle (Reynolds et al., 1991; Horn, 2009). Unilateral vagotomy blocks c-fos protein in the ipsilateral solitary nucleus but not contralateral solitary nucleus, area postrema, or spinal cord. Similarly, c-fos increases in the solitary nucleus elicited by abdominal irradiation are prevented by subdiaphragmatic vagotomy (Yamada et al., 2000).

Highly emetogenic stimuli such as cisplatin exert emetic and other effects by complex actions on peripheral and central neural pathways. 5-HT_3 pathways are critical mediators of acute CINV. Cisplatin evokes 5-HT release from gut enterochromaffin cells that activate vagal afferent 5-HT_3 receptors (Minami et al., 2003). Cisplatin-induced emesis is blocked by 5-HT_3 antagonists and abdominal vagotomy plus splanchnicectomy but not splanchnicectomy alone, showing involvement of peripheral neural sites (Fukui et al., 1993). Likewise, c-fos expression patterns in the solitary nucleus in response to abdominal radiation treatments are blocked by 5-HT_3 antagonists (Yamada et al., 2000). However, cisplatin also increases 5-HT turnover in the brainstem, a region rich in 5-HT_3 receptors, reflecting central actions as well (Darmani et al., 2009). In contrast to acute emesis, delayed CINV occurs when substance P is released in the brainstem. Centrally but not peripherally active NK_1 antagonists block both acute and delayed emesis elicited by cisplatin in experimental animals (Rudd et al., 1996). Delayed emesis is abolished by area postrema ablation, but is unmodified by vagotomy plus splanchnicectomy (Percie du Sert et al., 2009). A possible role of adrenoceptor mechanisms in the delayed nausea observed one or more days after cisplatin is suggested by increases in norepinephrine levels during this time period (Fredrikson et al., 1994). Stimulatory efferent motor effects of cisplatin on the gastric corpus and antrum are reduced by acute vagotomy and are blocked by atropine but not ondansetron, emphasizing additional participation of peripheral muscarinic cholinergic pathways (Davidson and Andrews, 1993).

Prominent autonomic disruption accompanies CINV. Skin pallor increases from baseline to nausea to vomiting after chemotherapy (Fig. 27.4) (Morrow et al., 1992). Heart rates decrease with chemotherapy-induced nausea, reflecting increased cardiac parasympathetic activity, but increase again with emesis. The presence of autonomic dysfunction can predict CINV severity (Zachariae et al., 2007). In one study, the total number of abnormal autonomic tests correlated with nausea intensity after chemotherapy (Belig et al., 1995). In another report, patients who developed severe nausea had more abnormal autonomic function tests before symptom development (Morrow et al., 1999).

Implications for therapy

Between 70% and 80% of patients receiving chemotherapy develop nausea and vomiting. Certain agents (cisplatin, dacarbazine, melphalan, nitrogen mustard) are especially emetogenic (Table 27.3) (Grunberg et al., 2005; Navari, 2009). Risk factors for CINV include female gender, age < 50 years, and prior histories of chemotherapy-induced symptoms. Current prevention and treatment of CINV can include medications in a number of classes including dopamine antagonists, 5-HT_3 antagonists, NK_1 antagonists, corticosteroids, cannabinoids, and other miscellaneous agents (Navari, 2009). As a reflection of the importance of 5-HT_3 and NK_1 receptor pathways in the mediation of chemotherapy-induced nausea and vomiting, antagonists at both receptor subtypes have assumed prominent roles in prophylaxis of this treatment complication. First-generation 5-HT_3 antagonists such as ondansetron and granisetron are effective in preventing acute emesis from moderately or highly emetogenic cancer chemotherapy, but exhibit little efficacy against delayed emesis occurring 1–5 days after chemotherapy (Geling and Eichler, 2005). A newer agent, palonsetron, reduces vomiting during this delayed period by virtue of its increased receptor-binding affinity and longer half-life (Jordan et al., 2007). NK_1 antagonists such as oral aprepitant and intravenous fosaprepitant provide additive reductions in acute and delayed emesis to 5-HT_3 antagonist-based regimens (Van Belle et al., 2002). By themselves, NK_1 antagonists provide superior control of acute and delayed CINV (Hesketh et al., 2003). For highly emetogenic chemotherapy, current consensus guidelines for prophylaxis of nausea and vomiting include aprepitant plus a 5-HT_3 antagonist and dexamethasone with additional use of aprepitant and days 2 and 3 and dexamethasone on days 2 through 4 to prevent delayed symptoms (Navari, 2007). For moderately emetogenic regimens, prophylaxis of acute emesis includes a 5-HT_3 antagonist plus dexamethasone with use of either drug during the delayed phase.

CONCLUSIONS

The autonomic nervous system is critically involved in the initiation and expression of emesis in response to a number of stimuli. Several gastrointestinal disorders with prominent nausea and vomiting, such as cyclic vomiting syndrome, gastroparesis, and motion sickness, are associated with clinical manifestations of autonomic dysfunction and objective abnormalities on autonomic function testing. Conversely, dysautonomia syndromes commonly present with significant nausea and vomiting. Therapies acting in part on autonomic pathways relieve vomiting by action on diverse receptor subtypes. The most extensively studied etiology of severe emesis,

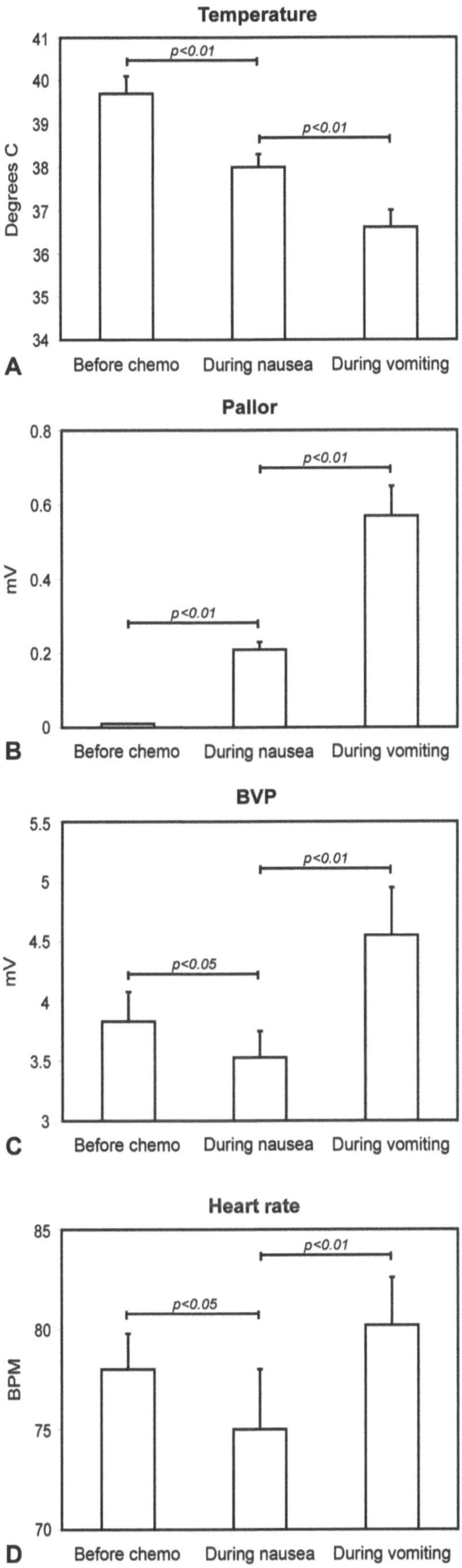

chemotherapy-induced nausea and vomiting, is preventable in many instances by agents acting on the peripheral and central nervous system.

ACKNOWLEDGMENTS

Dr. Hasler receives funding from grant 1 U01 DK073985-01 from the National Institutes of Health, Bethesda, MA, USA.

Table 27.3

Emetic potential of different chemotherapeutic agents (adapted from Grunberg et al., 2005; Navari, 2009)

Emetogenic potential	Representative chemotherapeutic agents
High (emesis in nearly all patients)	Cisplatin
	Dacarbazine
	High dose melphalan
	Nitrogen mustard
Moderate (emesis in > 70% of patients)	Anthracyclines
	Carboplatin
	High dose carmustine
	Cyclophosphamide
	Ifosfamide
	Irinotecan
	High dose methotrexate
	Oxaliplatin
	Topotecan
Low (emesis in 10–70% of patients)	Etoposide
	Fluorouracil
	Gemcitabine
	Mitoxantrone
	Taxanes
	Vinblastine
	Vinorelbine
Minimal (emesis in < 10% of patients)	Bortezomib
	Hormonal therapies
	Bleomycin

Fig. 27.4. Effects of chemotherapy on clinical measures of autonomic function are shown before chemotherapy, during nausea, and during vomiting. There are progressive decreases in skin temperature (**A**) and increases in pallor (**B**) with development of nausea followed by emesis after chemotherapy. With chemotherapy-induced nausea, blood volume pulse (BVP) (**C**) and heart rate (BPM, beats per minute) (**D**) decrease. As patients develop vomiting, these responses reverse and increase significantly. (Adapted from Morrow et al., 1992.)

References

Abell TL, Malagelada JR, Lucas et al. (1987). Gastric electromechanical and neurohumoral function in anorexia nervosa. Gastroenterology 93: 958–965.

Abell TL, Adams KA, Boles RG et al. (2008). Cyclic vomiting syndrome in adults. Neurogastroenterol Motil 20: 269–284.

Andrews PL, Wood KL (1988). Vagally mediated gastric motor and emetic reflexes evoked by stimulation of the antral mucosa in anaesthetized ferrets. J Physiol 395: 1–16.

Andrews P, Torii Y, Saito H et al. (1996). The pharmacology of the emetic response to upper gastrointestinal tract stimulation in *Suncus murinus*. Eur J Pharmacol 307: 305–313.

Andrews PL, Okada F, Woods AJ et al. (2000). The emetic and anti-emetic effects of the capsaicin analogue resiniferatoxin in *Suncus murinus*, the house musk shrew. Br J Pharmacol 130: 1247–1254.

Balaban CD (1996). Vestibular nucleus projections to the parabrachial nucleus in rabbits: implications for vestibular influences on the autonomic nervous system. Exp Brain Res 108: 367–381.

Barnas MG, Boer WH, Koomans HA (1999). Hemodynamic patterns and spectral analysis of heart rate variability during dialysis hypotension. J Am Soc Nephrol 10: 2577–2584.

Beleslin DB, Krstic SK (1986). Dimethylphenylpiperazinium-induced vomiting: nicotinic mediation in area postrema. Brain Res Bull 16: 5–10.

Beleslin DB, Strbac M (1987). Noradrenaline-induced emesis. Alpha-2 adrenoceptor mediation in the area postrema. Neuropharmacology 26: 1157–1165.

Beleslin DB, Strbac M, Jovanovic-Micic D et al. (1989). Area postrema: cholinergic and noradrenergic regulation of emesis. A new concept. Arch Int Physiol Biochim 97: 107–115.

Belig AJ, Morrow GR, Barry M et al. (1995). Autonomic measures associated with chemotherapy-related nausea: techniques and issues. Cancer Invest 13: 313–323.

Birks J (2006). Cholinesterase inhibitors for Alzheimer's disease. Cochrane Database Syst Rev 1: CD005593.

Chelimsky TC, Chelimsky GG (2007). Autonomic abnormalities in cyclic vomiting syndrome. J Pediatr Gastroenterol Nutr 44: 326–330.

Chen CY, Chuang TY, Tsai YA et al. (2004). Loss of sympathetic coordination appears to delay gastrointestinal transit in patients with spinal cord injury. Dig Dis Sci 49: 738–743.

Chhibber AK, Lustik SJ, Thakur R et al. (1999). Effects of anticholinergics on postoperative vomiting, recovery, and hospital stay in children undergoing tonsillectomy with or without adenoidectomy. Anesthesiology 90: 697–700.

Chu KM, Ngan MP, Wai MK et al. (2010). Olvanil: a non-pungent TRPV1 activator has anti-emetic properties in the ferret. Neuropharmacology 58: 383–391.

Cooper DW, Carpenter M, Mowbra P et al. (2002). Fetal and maternal effects of phenylephrine and ephedrine during spinal anesthesia for cesarean delivery. Anesthesiology 97: 1582–1590.

Costall B, Naylor RJ (1992). Neuropharmacology of emesis in relation to clinical response. Br J Cancer Suppl 19: S2–S7.

Dahmani S, Brasher C, Stany I et al. (2010). Premedication with clonidine is superior to benzodiazepines. A meta analysis of published studies. Acta Anaesthesiol Scand 54: 397–402.

Darmani NA, Crim JL, Janoyan JJ et al. (2009). A re-evaluation of the neurotransmitter basis of chemotherapy-induced immediate and delayed vomiting: evidence from the least shrew. Brain Res 1248: 40–58.

Davidson HI, Andrews PL (1993). Stimulation of gastrointestinal motility by cisplatin in the ferret: activation of an intrinsic cholinergic mechanism dissociated from emesis. Naunyn Schmiedebergs Arch Pharmacol 347: 506–513.

Dornhoffer J, Chelonis JJ, Blake D (2004). Stimulation of the semicircular canals via the rotary chair as a means to test pharmacologic countermeasures for space motion sickness. Otol Neurotol 25: 740–745.

Elinav E, Shapira Y, Ofran Y et al. (2005). Near-fatal amitraz intoxication: the overlooked pesticide. Basic Clin Pharmacol Toxicol 97: 185–187.

Farmery AD, Wilson-MacDonald J (2009). The analgesic effect of epidural clonidine after spinal surgery: a randomized placebo-controlled trial. Anesth Analg 108: 631–634.

Fredrikson M, Hursti TJ, Steineck G et al. (1994). Delayed chemotherapy-induced nausea is augmented by high levels of endogenous noradrenaline. Br J Cancer 70: 642–645.

Fukuda H, Koga T (1991). The Botzinger complex as the pattern generator for retching and vomiting in the dog. Neurosci Res 12: 471–485.

Fukuda H, Koga T (1992). Non-respiratory neurons in the Botzinger complex exhibiting appropriate firing patterns to generate the emetic act in dogs. Neurosci Res 14: 180–194.

Fukuda H, Koga T (1993). Hypercapnia and hypoxia which develop during retching participate in the transition from retching to expulsion in dogs. Neurosci Res 17: 205–215.

Fukuda H, Koga T, Furakawa N et al. (1998). The tachykinin NK_1 receptor antagonist GR205171 prevents vagal stimulation-induced retching but not neuronal transmission from emetic vagal afferents to solitary nucleus neurons in dogs. Brain Res 802: 221–231.

Fukui H, Yamamoto M, Sasaki S et al. (1993). Involvement of 5-HT_3 receptors and vagal afferents in copper sulfate- and cisplatin-induced emesis in monkeys. Eur J Pharmacol 249: 13–18.

Fukui H, Yamamoto M, Sasaki S et al. (1994). Possible involvement of peripheral 5-HT_4 receptors in copper sulfate-induced vomiting in dogs. Eur J Pharmacol 257: 47–52.

Gan TJ, Shiba AC, Kovac AL et al. (2009). A randomized, double-blind, multicenter trial comparing transdermal scopolamine plus ondansetron to ondansetron alone for the prevention of postoperative nausea and vomiting in the outpatient setting. Anesth Analg 108: 1498–1504.

Geling O, Eichler H (2005). Should 5-hydroxytryptamine-3 receptor antagonists be administered beyond 24 hours after chemotherapy to prevent delayed emesis? Systematic reevaluation of clinical evidence and drug cost implications. J Clin Oncol 23: 1289–1294.

Gieroba ZJ, Blessing WW (1994). Fos-containing neurons in medulla and pons after unilateral stimulation of the afferent abdominal vagus in conscious rabbits. Neuroscience 59: 851–858.

Golding JF, Scott JR (1997). Comparison of the effects of a selective muscarinic receptor antagonist and hyoscine (scopolamine) on motion sickness, skin conductance and heart rate. Br J Clin Pharmacol 43: 633–637.

Goodman SR, Kim-Lo SH, Ciliberto CF et al. (2002). Epinephrine is not a useful addition to intrathecal fentanyl or fentanyl-bupivacaine for labor analgesia. Reg Anesth Pain Med 27: 374–379.

Goto J, Inoue K, Mannen T (1989). Autonomic dysfunctions in acute intermittent porphyria. Rinsho Shinkeigaku 29: 774–777.

Grant PJ, Hughes JR, Dean HG et al. (1986). Vasopressin and catecholamine secretion during apomorphine-induced nausea mediate acute changes in haemostatic function in man. Clin Sci 71: 621–624.

Grelot L, Milano S, Portillo F et al. (1992). Membrane potential changes of phrenic motoneurons during fictive vomiting, coughing, and swallowing in the decerebrate cat. J Neurophysiol 68: 2110–2119.

Grunberg SM, Osoba D, Hesketh PJ et al. (2005). Evaluation of new antiemetic agents and definition of antineoplastic agent emetogenicity: an update. Support Care Cancer 13: 80–84.

Guy RJ, Dawson JL, Garrett JR et al. (1984). Diabetic gastroparesis from autonomic neuropathy: surgical considerations and changes in vagus nerve morphology. J Neurol 47: 686–691.

Hasler WL (2009). Methods of gastric electrical stimulation and pacing: a review of their benefits and mechanisms of action in gastroparesis and obesity. Neurogastroenterol Motil 21: 229–243.

Hasler WL, Kim MS, Chey WD et al. (1995). Central cholinergic and alpha-adrenergic mediation of gastric slow wave dysrhythmias evoked during motion sickness. Am J Physiol 268: G539–G547.

Hathaway DK, Abell T, Cardoso S et al. (1994). Improvement in autonomic and gastric function following pancreas-kidney versus kidney-alone transplantation and the correlation with quality of life. Transplantation 57: 816–822.

Hermann GE, Tovar CA, Rogers RC (2002). LPS-induced suppression of gastric motility relieved by TNFR:Fc construct in dorsal vagal complex. Am J Physiol Gastrointest Liver Physiol 283: G634–G639.

Hermann GE, Tovar CA, Rogers RC (2003). TNF_{α}-stimulation of cFos-activation of neurons in the solitary nucleus is suppressed by TNFR:Fc adsorbant construct in the dorsal vagal complex. Brain Res 976: 69–74.

Hesketh PJ, Grunberg SM, Gralla RJ et al. (2003). The oral neurokinin-1 antagonist aprepitant for the prevention of chemotherapy-induced nausea and vomiting: a multinational, randomized, double-blind placebo-controlled trial in patients receiving high-dose cisplatin: the Aprepitant Protocol 052 Study Group. J Clin Oncol 21: 4112–4119.

Himi N, Koga T, Nakamura E et al. (2004). Differences in autonomic responses between subjects with and without nausea while watching an irregularly oscillating video. Auton Neurosci 116: 46–53.

Ho KM, Ismail H, Lee KC et al. (2005). Use of intrathecal neostigmine as an adjunct to other spinal medications in perioperative and peripartum analgesia: a meta-analysis. Anaesth Intensive Care 33: 41–53.

Horn CC (2009). Brain Fos expression induced by the chemotherapy agent cisplatin in the rat is partially dependent on an intact abdominal vagus. Auton Neurosci 148: 76–82.

Hornby PJ (2001). Central neurocircuitry associated with emesis. Am J Med 111 (Suppl 8A): 106S–112S.

Huang ST, Chen GY, Lo HM et al. (2005). Increase in the vagal modulation by acupuncture at neiguan point in the healthy subjects. Am J Chin Med 33: 157–164.

Ionescu D, Badescu C, Acalovschi I (2007). Nicotine patch for the prevention of postoperative nausea and vomiting: a prospective randomised trial. Clin Drug Investig 27: 559–564.

Itoh Y, Oishi T, Ohnishi A et al. (1993). Acute cerebellar ataxia with sympathotonic orthostatic hypotension following Epstein–Barr virus infection: a case report. Rinsho Shinkeigaku 33: 503–506.

Janowsky DS, Risch SC, Ziegler M et al. (1984). A cholinomimetic model of motion sickness and space adaptation syndrome. Aviat Space Environ Med 55: 692–696.

Japundzic-Zigon N, Samardzic R, Beleslin DB (1997). Clonidine-induced emesis: a multitransmitter pathway concept. Pharmacol Res 35: 287–297.

Jordan K, Schmoll HJ, Aapro MS (2007). Comparative activity of antiemetic drugs. Crit Rev Oncol Hematol 61: 162–175.

Jovanovic-Micic D, Strbac M, Krstic SK et al. (1989). Ablation of the area postrema and emesis. Metab Brain Dis 4: 55–60.

Jovanovic-Micic D, Samardzic R, Beleslin DB (1995). The role of alpha-adrenergic mechanisms within the area postrema in dopamine-induced emesis. Eur J Pharmacol 272: 21–30.

Kavirajan H, Schneider LS (2007). Efficacy and adverse effects of cholinesterase inhibitors and memantine in vascular dementia: a meta-analysis of randomized controlled trials. Lancet Neurol 6: 782–792.

Kim MS, Chey WD, Owyang C et al. (1997). Role of plasma vasopressin as a mediator of nausea and gastric slow wave dysrhythmias in motion sickness. Am J Physiol 272: G853–G862.

Koga T, Fukuda H (1992). Neurons in the nucleus of the solitary tract mediating inputs from emetic vagal afferents and the area postrema to the pattern generator for the emetic act in dogs. Neurosci Res 14: 166–179.

Koga T, Fukuda H (1997). Descending pathway from the central pattern generator of vomiting. Neuroreport 8: 2587–2590.

Kohl RL (1993). Autonomic function and plasma catecholamines following stressful sensory stimuli. Aviat Space Environ Med 64: 921–927.

Kranke P, Morin AM, Roewer N et al. (2002). The efficacy and safety of transdermal scopolamine for the prevention of postoperative nausea and vomiting: a quantitative systematic review. Anesth Analg 95: 133–143.

Lang IM, Sarna SK, Condon RE (1986). Gastrointestinal motor correlates of vomiting in the dog: quantification and characterization as an independent phenomenon. Gastroenterology 90: 40–47.

Lang IM, Sarna SK, Dodds WJ (1993). Pharyngeal, esophageal, and proximal gastric responses associated with vomiting. Am J Physiol 265: G963–G972.

Lang IM, Sarna SK, Shaker R (1999). Gastrointestinal motor and myoelectric correlates of motion sickness. Am J Physiol 277: G642–G652.

Lee KY, Park HJ, Chey WY (1985). Studies on mechanism of retching and vomiting in dogs. Effect of peripheral dopamine blocker on myoelectric changes in antrum and upper small intestine. Dig Dis Sci 30: 22–28.

Lei Y, Chen JD (2009). Effects of dual pulse gastric electrical stimulation on gastric tone and compliance in dogs. Dig Liver Dis 41: 277–282.

Leslie RA, Reynolds DJ, Andrews PL et al. (1990a). Evidence for presynaptic 5-hydroxytryptamine$_3$ recognition sites on vagal afferent terminals in the brainstem of the ferret. Neuroscience 38: 667–673.

Leslie RA, Shah Y, Thejomayen M et al. (1990b). The neuropharmacology of emesis: the role of receptors in neuromodulation of nausea and vomiting. Can J Physiol Pharmacol 68: 279–288.

Lobrano A, Blanchard K, Abell TL et al. (2006). Postinfectious gastroparesis related to autonomic failure: a case report. Neurogastroenterol Motil 18: 162–167.

Lurie Y, Wasser SP, Taha M et al. (2009). Mushroom poisoning from species of genus *Inocybe* (fiber head mushroom): a case series with exact species identification. Clin Toxicol 47: 562–565.

Mack KJ, Johnson JN, Rowe PC (2010). Orthostatic intolerance and the headache patient. Semin Pediatr Neurol 17: 109–116.

Margalith D, Dunn HG, Carter JE et al. (1984). Friedrich's ataxia with dysautonomia and labile hypertension. Can J Neurol Sci 11: 73–77.

McCain KR, Sawyer TS, Spiller HA (2007). Evaluation of centrally acting cholinesterase inhibitor exposures in adults. Ann Pharmacother 41: 1632–1637.

McCallum RW, Dusing RW, Sarosiek I et al. (2010). Mechanisms of symptomatic improvement after gastric electrical stimulation in gastroparetic patients. Neurogastroenterol Motil 22: 161–167.

Melek IM, Seyfeli E, Duru M et al. (2007). Autonomic dysfunction and cardiac repolarization abnormalities in patients with migraine attacks. Med Sci Monit 13: RA47–RA49.

Milano S, Grelot L, Bianchi AL et al. (1992). Discharge patterns of phrenic motoneurons during fictive coughing and vomiting in decerebrate cats. J Appl Physiol 73: 1626–1636.

Miller AD, Nonaka S (1992). Mechanisms of vomiting induced by serotonin-3 receptor agonists in the cat: effect of vagotomy, splanchnicectomy or area postrema lesion. J Pharmacol Exp Ther 260: 509–517.

Miller AD, Ruggiero DA (1994). Emetic reflex arc revealed by expression of the immediate-early gene c-fos in the cat. J Neurosci 14: 871–888.

Minami M, Endo T, Yokoda H et al. (2001). Effects of CP-99,994, a tachykinin NK-1 receptor antagonist, on abdominal afferent vagal activity in ferrets: evidence for involvement of NK-1 and 5-HT$_3$ receptors. Eur J Pharmacol 428: 215–220.

Minami M, Endo T, Hirafuji M et al. (2003). Pharmacological aspects of anticancer drug-induced emesis with emphasis on serotonin release and vagal nerve activity. Pharmacol Ther 99: 149–165.

Morrow GR, Angel C, Dubeshter B (1992). Autonomic changes during cancer chemotherapy induced nausea and vomiting. Br J Cancer Suppl 19: S42–S45.

Morrow GR, Hickok JT, Dubeshter B et al. (1999). Changes in clinical measures of autonomic nervous system function related to cancer chemotherapy-induced nausea. J Auton Nerv Syst 78: 57–63.

Navari RM (2007). Review of updated antiemetic guidelines for chemotherapy-induced nausea and vomiting. Commun Oncol 4 (1 Suppl.): 3S–11S.

Navari RM (2009). Pharmacological management of chemotherapy-induced nausea and vomiting: focus on recent developments. Drugs 69: 515–533.

Ngan Kee WD, Khaw KS, Ng FF (2004). Comparison of phenylephrine infusion regimens for maintaining maternal blood pressure during spinal anaesthesia for Caesarean section. Br J Anaesth 92: 469–474.

Onishi T, Mori T, Yanagihara M et al. (2007). Similarities of the neuronal circuit for the induction of fictive vomiting between ferrets and dogs. Auton Neurosci 136: 20–30.

Palmer GM, Cameron DJ (2005). Use of intravenous midazolam and clonidine in cyclical vomiting syndrome: a case report. Paediatr Anaesth 15: 68–72.

Parkman HP, Camilleri M, Farrugia G et al. (2010). Gastroparesis and functional dyspepsia: excerpts from the AGA/ANMS meeting. Neurogastroenterol Motil 22: 113–133.

Patel PN, Ezzo DC (2009). Withdrawal symptoms after discontinuation of transdermal scopolamine therapy: treatment with meclizine. Am J Health Syst Pharm 66: 2024–2026.

Pedigo NW, Brizzee KR (1985). Muscarinic cholinergic receptors in area postrema and brainstem areas regulating emesis. Brain Res Bull 14: 169–177.

Percie du Sert N, Rudd JA, Moss R et al. (2009). The delayed phase of cisplatin-induced emesis is mediated by the area postrema and not the abdominal visceral innervations in the ferret. Neurosci Lett 465: 16–20.

Percie du Sert N, Chu KM, Wai MK et al. (2010). Telemetry in a motion-sickness model implicated the abdominal vagus in motion-induced gastric dysrhythmia. Exp Physiol 95: 768–773.

Pyykko I, Schalen L, Jantti V et al. (1985). Transdermally administered scopolamine vs. dimenhydrinate. I. Effect on nausea and vertigo in experimentally induced motion sickness. Acta Otolaryngol 99: 588–596.

Qin XY, Pilot MA, Thompson H et al. (1993). Effects of cholinoceptor and 5-hydroxytryptamine$_3$ receptor antagonism on erythromycin-induced canine intestinal motility disruption and emesis. Br J Pharmacol 108: 44–49.

Rashed H, Abell TL, Familoni BO et al. (1999). Autonomic function in cyclic vomiting syndrome and classic migraine. Dig Dis Sci 44: 74S–78S.

Rasmussen K, Morilak DA, Jacobs BL (1986). Single unit activity of locus coeruleus neurons in the freely moving cat. I. During naturalistic behaviors and in response to simple and complex stimuli. Brain Res 37: 324–334.

Reynolds DJ, Barber NA, Grahame-Smith DG et al. (1991). Cisplatin-evoked induction of c-fos protein in the brainstem of the ferret: the effect of cervical vagotomy and the anti-emetic 5-HT_3 receptor antagonist granisetron (BRL 43694). Brain Res 565: 231–236.

Rosa-e-Silva L, Troncon LE, Oliviera RB et al. (1995). Treatment of diabetic gastroparesis with oral clonidine. Aliment Pharmacol Ther 9: 179–183.

Rudd JA, Jordan CC, Naylor RJ (1996). The action of the NK_1 tachykinin receptor antagonist, CP 99,994, in antagonizing the acute and delayed emesis induced by cisplatin in the ferret. Br J Pharmacol 119: 931–936.

Salin-Pascual RJ, Rosas M, Jimenez-Genchi A et al. (1996). Antidepressant effect of transdermal nicotine patches in nonsmoking patients with major depression. J Clin Psychiatry 57: 387–389.

Schier JG, Traub SJ, Hoffman RS et al. (2003). Ephedrine-induced cardiac ischemia: exposure confirmed with a serum level. J Toxicol Clin Toxicol 41: 849–853.

Schlegel TT, Brown TE, Wood SJ et al. (2001). Orthostatic intolerance and motion sickness after parabolic flight. J App Physiol 90: 67–82.

Schulz M, Graefe T, Stuby K et al. (2006). Case report: acute unintentional carbachol intoxication. Crit Care 10: R84.

Sharkey KA, Cristino L, Oland LD et al. (2007). Arvanil, anandamide and N-arachidonoyl-dopamine (NADA) inhibit emesis through cannabinoid CB1 and vanilloid TRPV1 receptors in the ferret. Eur J Neurosci 25: 2773–2882.

Shiroshita Y, Koga T, Fukuda H (1997). Capsaicin in the 4th ventricle abolishes retching and transmission of emetic vagal afferents to solitary nucleus neurons. Eur J Pharmacol 339: 183–192.

Spinks AB, Wasiak J, Villanueva EV et al. (2004). Scopolamine for preventing and treating motion sickness. Cochrane Database Syst Rev 3: CD002851.

Srivastava ED, Russell MA, Feyerabend C et al. (1991). Sensitivity and tolerance to nicotine in smokers and nonsmokers. Psychopharmacology (Berl) 105: 63–68.

Sullivan SD, Hanauer J, Rowe PC et al. (2005). Gastrointestinal symptoms associated with orthostatic intolerance. J Pediatr Gastroenterol Nutr 40: 425–428.

Tack J, Talley NJ, Camilleri M et al. (2006). Functional gastroduodenal disorders. Gastroenterology 130: 1466–1479.

Thal LJ, Ferguson JM, Mintzer J et al. (1999). A 24-week randomized trial of controlled-release physostigmine in patients with Alzheimer's disease. Neurology 52: 1146–1152.

Thrasher DL, Von Derau K, Burgess J (2009). Health effects from reported exposure to methamphetamine labs: a poison center-based study. J Med Toxicol 5: 200–204.

To J, Issenman RM, Kamath MV (1999). Evaluation of neurocardiac signals in pediatric patients with cyclic vomiting syndrome through power spectral analysis of heart rate variability. J Pediatr 135: 363–366.

Toorop RJ, Scheltinga MR, Huige MC et al. (2007). Excessive vomiting abolished by carotid denervation. Auton Neurosci 133: 175–177.

Tufanogullari B, White PF, Peixoto MP et al. (2008). Dexmedetomidine infusion during laparoscopic bariatric surgery: the effect on recovery outcome variables. Anesth Analg 106: 1741–1748.

Uchino M, Ishii K, Kuwahara M et al. (2001). Role of autonomic nervous system for development and suppression of motion sickness in *Suncus murinus*. Auton Neurosci 94: 46–51.

Uchino M, Ishii K, Kuwahara M et al. (2002). Role of the autonomic nervous system in emetic and cardiovascular responses in *Suncus murinus*. Auton Neurosci 100: 32–40.

Uchino M, Ito K, Kuwahara M et al. (2008). Interactions of carotid sinus or aortic input with emetic signals from gastric afferents and vestibular system. Auton Neurosci 144: 36–42.

Udassin R, Seror D, Vinograd I et al. (1992). Nissen fundoplication in the treatment of children with familial dysautonomia. Am J Surg 164: 332–336.

Urayama Y, Yamada Y, Nakamura E et al. (2010). Electrical and chemical stimulation of the nucleus raphe magnus inhibits induction of retching by afferent vagal fibers. Auton Neurosci 152: 35–40.

Van Belle S, Liechinitser M, Navari RM et al. (2002). Prevention of cisplatin-induced acute and delayed emesis by the selective neurokinin-1 antagonists, L-758,298 and MK-869. Cancer 94: 3032–3041.

Van Sickle MD, Oland LD, Mackie K et al. (2003). $Delta_9$-tetrahydrocannabinol selectively acts on CB_1 receptors in specific regions of dorsal vagal complex to inhibit emesis in ferrets. Am J Physiol Gastrointest Liver Physiol 285: G566–G576.

Verne GN, Soldevia-Pico C, Robinson ME et al. (2004). Autonomic dysfunction and gastroparesis in cirrhosis. J Clin Gastroenterol 38: 72–76.

Villafane G, Cesaro P, Rialland A et al. (2007). Chronic high dose transdermal nicotine in Parkinson's disease: an open trial. Eur J Neurol 14: 1313–1316.

Vinograd I, Udassin R, Beilin B et al. (1985). The surgical management of children with familial dysautonomia. J Pediatr Surg 20: 632–636.

West P, Horowitz BZ (2009). Zigadenus poisoning treated with atropine and dopamine. J Med Toxicol 5: 214–217.

Wilkinson AR, Frampton CM, Glover PW et al. (1989). Preoperative transdermal hyoscine for the prevention of postoperative nausea and vomiting. Anaesth Intensive Care 17: 285–289.

Wood CD, Stewart JJ, Wood MJ et al. (1990). Therapeutic effects of antimotion sickness medications on the secondary symptoms of motion sickness. Aviat Space Environ Med 61: 157–161.

Yamada Y, Tsukamoto G, Kobashi M et al. (2000). Abdominal vagi mediate c-Fos expression induced by X-ray irradiation in the nucleus tractus solitarii of the rat. Auton Neurosci 83: 29–36.

Yasumura S, Shojaku H, Watanabe Y (2005). Prediction of vestibule-autonomic symptoms during the caloric test: evaluation of autonomic activity by spectral analysis of the electrocardiographic R-R interval. Acta Otolaryngol 125: 1265–1271.

Yates BJ, Miller AD, Lucot JB (1998). Physiological basis and pharmacology of motion sickness: an update. Brain Res Bull 47: 395–406.

Zachariae R, Paulsen K, Mehlsen M et al. (2007). Chemotherapy-induced nausea, vomiting, fatigue: the role of individual differences related to sensory perception and autonomic reactivity. Psychother Psychosom 76: 376–384.

Zingler VC, Denecke K, Jahn K et al. (2007). The effect of nicotine on perceptual, ocular motor, postural, and vegetative functions at rest and in motion. J Neurol 254: 1689–1697.

Handbook of Clinical Neurology, Vol. 117 (3rd series)
Autonomic Nervous System
R.M. Buijs and D.F. Swaab, Editors

Chapter 28

Sympathetic microneurography

VAUGHAN G. MACEFIELD*
School of Medicine, University of Western Sydney, Sydney, and Neuroscience Research Australia, Sydney, Australia

INTRODUCTION

The sympathetic nervous system, which, together with the parasympathetic and enteric nervous systems, constitute the autonomic nervous system, plays a key role in homeostasis – maintaining the internal state of the body and correcting for any variation through appropriate increases or decreases in activity of its many effector organs. The sympathetic nervous system plays important roles in the ongoing regulation of arterial pressure, the control of blood flow through various organs, and the maintenance of core temperature through thermoregulatory processes. Given that the brain depends on an adequate supply of oxygenated blood and glucose, at an optimal perfusion pressure, it is not difficult to see that disturbances in sympathetic control of blood flow can have neurological implications. However, it is its status as an important component of the nervous system that will be considered here. While much has been learned about the operation of the sympathetic nervous system in experimental animals, until the mid 1960s only indirect measures of sympathetic outflow were available in humans.

Microneurography, in which a tungsten microelectrode is inserted percutaneously into an accessible peripheral (or cranial) nerve, to record action potentials from groups of axons, or individual axons, was developed by Karl-Erik Hagbarth and Åke Vallbo in the Department of Clinical Neurophysiology at the University Hospital in Uppsala, Sweden. While microneurography is a skilled and time-consuming invasive procedure that cannot be applied routinely for clinical diagnostic purposes, it has nevertheless been applied in an investigative manner to increase our understanding of neurological disorders. Indeed, the original aim in developing the technique was to record from muscle spindles (stretch receptors located in skeletal muscles which, because they encode changes in muscle length, serve as our primary proprioceptors) in order to address the prevailing idea of motor control at the time – the servo control theory. Although initially satisfied with the information that multiunit recordings could provide (Hagbarth and Vallbo, 1968a), to their surprise Hagbarth and Vallbo managed to impale the sensory axons of individual muscle spindle endings, and to record action potentials from single afferents during muscle stretch and during contractions (Hagbarth and Vallbo, 1969). The servo control theory of muscle contraction, established on the basis of experiments conducted in the anaesthetized cat, was demolished by data obtained from recordings of muscle spindles in human subjects performing voluntary contractions, which showed that muscle spindles are not activated before α motor neurons (Vallbo, 1971). Recordings from human muscle spindles have contributed significantly to clinical neurology, but its greatest application has been to increase our understanding of the sympathetic nervous system, disorders of which feature in many diseases. When Hagbarth and Vallbo first heard over the loudspeakers the sound of "waves approaching the distant shore," they thought they were artifacts; the identity of these waves as bursts of sympathetic nerve activity has generated thousands of studies that have taken advantage of the capacity to record action potentials directly from postganglionic sympathetic axons in awake human subjects. Indeed, the technique of microneurography is primarily used for recording sympathetic outflow to muscle and, to a smaller extent, skin. And while many of these diseases involve the cardiovascular or respiratory systems, there are specific neurological conditions – Guillain–Barré syndrome, for instance – in which the primary disturbance is at the axonal level, yet the systems affected are widespread. An early review on the application of microneurography to studies of the somatosensory and sympathetic nervous systems (Vallbo et al., 1979), and a more recent review that focuses on the sympathetic nervous system

*Correspondence to: Professor Vaughan G. Macefield, School of Medicine, University of Western Sydney, Locked Bag 1797, Penrith, NSW 2751, Australia. Tel: +61-2-4620-3779, Fax: +61-2-4620-3890, E-mail: v.macefield@uws.edu.au

(Vallbo et al., 2004), should be consulted by those readers interested in the technical development of microneurography. Here I shall commence by providing a brief overview of the methodological approach used to record sympathetic nerve activity directly in awake human subjects.

MICROELECTRODE RECORDINGS OF SYMPATHETIC NERVE ACTIVITY IN HUMANS

All current microneurographic recordings employ tungsten microelectrodes, typically with a shaft diameter of 200 μm and a rounded tip with a diameter of ~5 μm, electrolytically sharpened and insulated; an area at the tip from which the insulation has been removed exposes a small surface area of tungsten and thereby provides a very focal recording tip. The quality of a microelectrode is measured by its impedance to a high-frequency (typically 1 kHz) current passed through the microelectrode: the higher the impedance, the smaller the recording surface and hence the more focal the recording. An accessible peripheral nerve – typically the fibular nerve at the fibular head or the tibial nerve in the popliteal fossa – is located by brief pulses of electrical stimuli delivered through a small-diameter (1–2 mm) surface probe. The optimal site for inserting the microelectrode is defined as that with the lowest electrical threshold for evoking twitches in the innervated muscles and/or radiating paresethesiae in the cutaneous innervation territory of the nerve. A subdermal microelectrode, in which ~1 mm of insulation has been removed from the tip, is inserted ~1 cm away from the recording microelectrode and serves as the reference.

While some investigators do not use intraneural stimulation to guide the insertion of the microelectrode tip into a muscle or cutaneous fascicle of the nerve, I prefer the systematic approach afforded by intraneural stimulation, in which electrically evoked responses are used to define the location of a particular nerve fascicle and the electrode depth and angle adjusted while reducing the current as the fascicle is approached. We know we have entered a fascicle because either twitches or paresthesiae are generated at currents of 0.02 mA or lower. Once in a muscle fascicle, the investigator applies muscle stretch or tendon percussion to identify muscle spindle afferents and uses this to guide the microelectrode tip into an area of the fascicle in which spontaneous bursts of muscle sympathetic nerve activity (MSNA) can be found, facilitated by asking the subject to perform an end-expiratory apnea, an inspiratory-capacity apnea, or Valsalva maneuver – each of which increases bursts of MSNA. Conversely, these maneuvers are ineffective when recording from a cutaneous fascicle: rather one defines the innervation territory of the fascicle by light stroking of the skin and evoking bursts of skin sympathetic nerve activity (SSNA) by unexpected stimuli (such as a loud shout) or brisk sniffs that evoke solitary bursts of SSNA. Both the levels of MSNA and SSNA differ across individuals.

Because the nerve signals are small, they need to be amplified ~20 000 times, via a headstage located close to the recording site, so as to avoid picking up electrical noise from long leads connected to a high-impedance microelectrode. Most headstages, which of course need to be electrically isolated, preamplify the signal 10–100 times, with further amplification and filtering being undertaken in the main amplifier. Sympathetic nerve activity is typically recorded with a bandpass of 300 Hz to 3 kHz, or 700 Hz to 2 kHz. To facilitate identification of the bursts, the filtered neural signal is also full-wave rectified and displayed as a mean voltage neurogram, produced as an integrated signal (time constant 100 ms) or as an RMS-processed (root mean square) moving average (time constant 100 or 200 ms); the advantage of the RMS-processed signal is that it does not introduce the time lag associated with integration. Figure 28.1 shows a schematic representation of a peripheral nerve and a recording of muscle sympathetic nerve activity.

Analysis of multiunit sympathetic nerve activity

It has been known since the first direct recordings of sympathetic neural traffic were made in awake human subjects that sympathetic outflow occurs in bursts (Delius et al., 1972a, b). The standard approach to recording these bursts – *multiunit* recordings from many postganglionic axons – has been and continues to be very useful, but there is much information in the patterning of sympathetic bursts that has not been extracted from multiunit recordings, though recent analytical approaches have begun to address this (Steinback et al., 2010; Salmanpour et al., 2011). The traditional analysis of sympathetic nerve activity involves counting the number of visible bursts occurring per minute (burst frequency) or per 100 heart beats (burst incidence), or measuring the total neural activity (the cumulative sum of burst amplitudes in 1 minute). However, burst area and amplitude are determined not just by the intensity of a sympathetic burst, but are also dependent on the proximity of the electrode tip to the group of active axons from which the recording is made. For this reason, one cannot compare burst amplitudes across subjects. There have been some attempts to quantify burst amplitude by normalizing burst area to the largest burst encountered in a recording and calculating the distribution of burst amplitudes (Sverrisdottir et al., 1998, 2000).

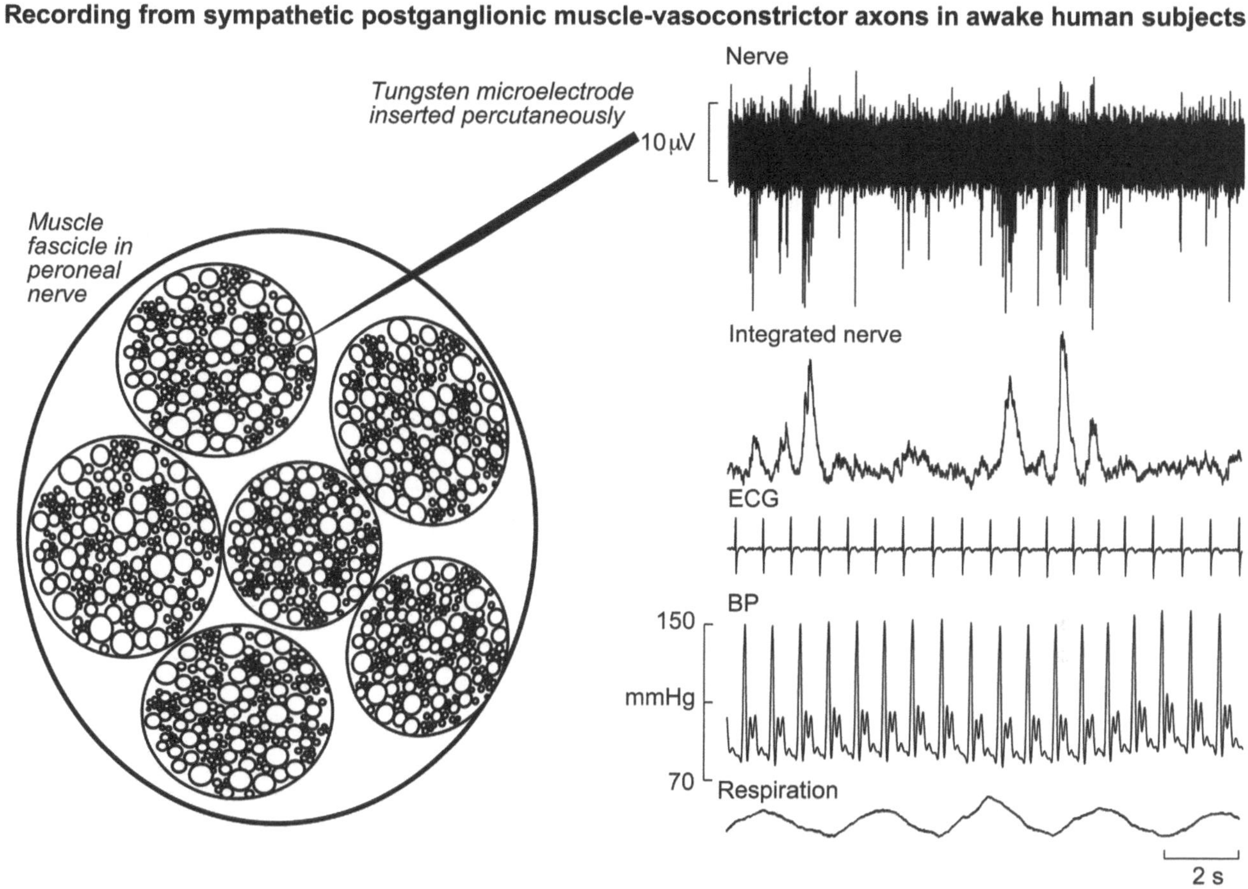

Fig. 28.1. Spontaneous bursts of muscle sympathetic nerve activity recorded via a tungsten microelectrode inserted into a muscle fascicle of the fibular nerve.

While these approaches have had some success in identifying differences in overall burst intensity across subjects, the information content derived from a multiunit sympathetic burst is still relatively coarse. In multiunit recordings, the absolute strength (intensity) of the recorded activity cannot be determined because of the dependence on the proximity of the electrode tip to the active nerve fibers.

Analysis of single-unit sympathetic nerve activity

Because postganglionic sympathetic axons are unmyelinated (C fibers) and travel within the nerve as a group, it is technically very difficult to record from individual sympathetic axons in human subjects. Nevertheless, using highly selective microelectrodes it is possible to record from individual postganglionic axons. The first study detailing the firing properties of single muscle vasoconstrictor neurons in human subjects was undertaken nearly 20 years ago (Macefield et al., 1994). Since then, the single-unit approach has been extended to characterizing the firing properties of cutaneous vasoconstrictor neurons (Macefield and Wallin, 1999a) and sudomotor neurons (Macefield and Wallin, 1999b), as well as studying muscle vasoconstrictor neurons in different disease states (see below). The approach used in recording from single sympathetic neurons in humans is the same as that used to record multiunit activity: a tungsten microelectrode is manually inserted percutaneously into an accessible peripheral nerve, the only difference being that the microelectrode has to have a much higher impedance, so as to limit the area over which neural activity is picked up. When an intraneural site is located in which spontaneous or evoked oligounitary sympathetic bursts are present, small electrode adjustments are made until a position is found in which individual spikes from a putative single fiber stand out from the noise. Spike superimposition is used to determine whether a recording is likely to be from a single nerve fiber: because these are extracellular recordings from C fibers the spikes are typically triphasic with a dominant negativity.

Sympathetic nerve fibers travelling in muscle fascicles supply blood vessels in the skeletal muscles: their identification as *muscle vasoconstrictor neurons* is based purely on behavioral criteria – i.e., the known properties of muscle vasoconstrictor neurons as recorded in experimental animals. Because muscle vasoconstrictor outflow is tightly coupled to the arterial baroreceptors, the discharge of individual neurons possess a strong cardiac rhythmicity: firing probability increases with maneuvers that unload the baroreceptors and decreases with maneuvers that load the baroreceptors; they also exhibit respiratory modulation (Macefield et al., 1994, 2002). However, sympathetic axons travelling in cutaneous fascicles may supply blood vessels, sweat glands, hairs and/or adipose tissue, so it is important to identify the target tissue of a given cutaneous sympathetic neuron. To record from individual *cutaneous vasoconstrictor neurons* we have used whole-body cooling to bias the thermoregulatory sympathetic outflow toward an increase in cutaneous vasoconstrictor discharge, in which skin blood flow is negligible and sweat release is abolished (Macefield and Wallin, 1999a). Conversely, whole-body warming was used to increase the spontaneous activity of *sudomotor neurons* and decrease the activity of cutaneous vasoconstrictor neurons, as evidenced by the increases in sweating and skin blood flow (Macefield and Wallin, 1996). External disturbances need to be minimized in these studies to avoid coactivation of sudomotor and cutaneous vasoconstrictor neurons, as cutaneous sympathetic outflow in the awake state is very labile and critically dependent on the level of arousal or anxiety.

Regardless of the type of postganglionic sympathetic axon from which activity is recorded – muscle vasoconstrictor, cutaneous vasoconstrictor or sudomotor – the firing properties are remarkably similar. The spontaneous activity of each unit consists of a low-level intermittent discharge, typically around 0.5 Hz, that is dominated by solitary spikes. Because each type of neuron possesses some degree of cardiac rhythmicity, the cardiac interval (the time between two successive R-waves) provides a means of dividing a recording into defined epochs and determining whether an individual unit fires or remains quiescent. An example of a unitary recording from a sudomotor neuron is shown in Figure 28.2. The proportion of cardiac intervals in which a neuron fires one or more spikes is referred to as the "firing probability" and is the unitary equivalent of burst incidence – the percentage of cardiac intervals in which multiunit sympathetic bursts occur. While the firing probability varies across units, mean firing probability is below 40% in healthy subjects, and does not differ significantly between different types of neurons.

As noted above, individual neurons tend to fire only once in a sympathetic burst. This is illustrated graphically in Figure 28.3 for each of the three classes of neurons studied so far. The open columns represent those

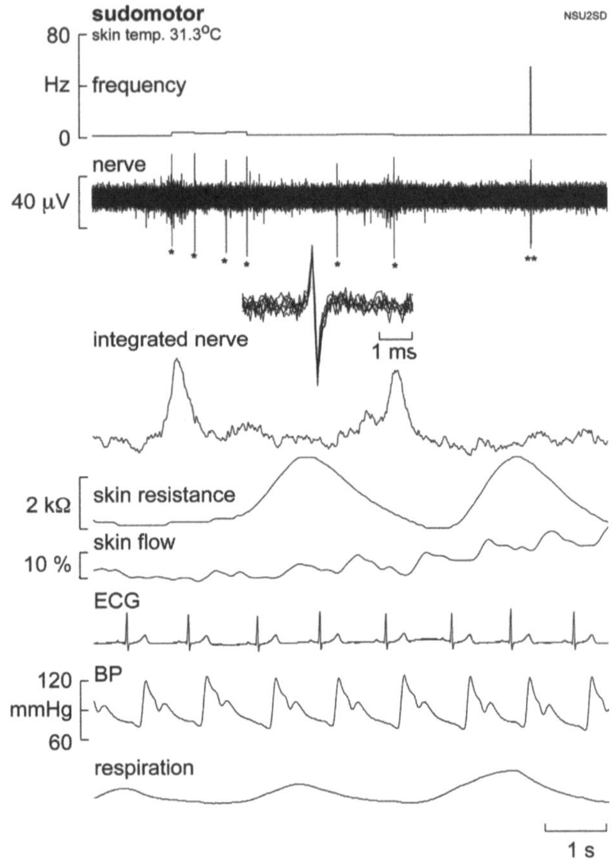

Fig. 28.2. Unitary recording from a single sudomotor neuron recorded in a healthy subject during thermally induced sweating. This unit generally fired only one spike (indicated by asterisks) per cardiac interval, but occasionally fired multiple spikes within a burst of MSNA. Superimposed spikes show a uniform spike morphology, indicating that the action potentials originated from the same single sympathetic axon. The integrated nerve signal primarily reflects far-field activity of many sympathetic neurons firing. Decreases in skin resistance, indicating sweat release, are shown as upward deflections. (Reproduced with permission from Macefield et al., 2002.)

cardiac intervals in which the neurons were quiescent; these intervals have been excluded in the lower panels to illustrate the distribution of solitary and multiple firing when the neurons were active. The distributions are remarkably similar for all three unit types, each type generating solitary spikes in some 70% of all cardiac intervals in which it was active. Moreover, while individual units could fire up to seven times within a cardiac interval (i.e., within a sympathetic burst) such behavior was rare; the majority of neurons fired at most two to four spikes per cardiac interval. Thus, a sympathetic multiunit burst is brought about predominantly by many axons firing a single action potential, rather than a few neurons discharging many spikes. Indeed,

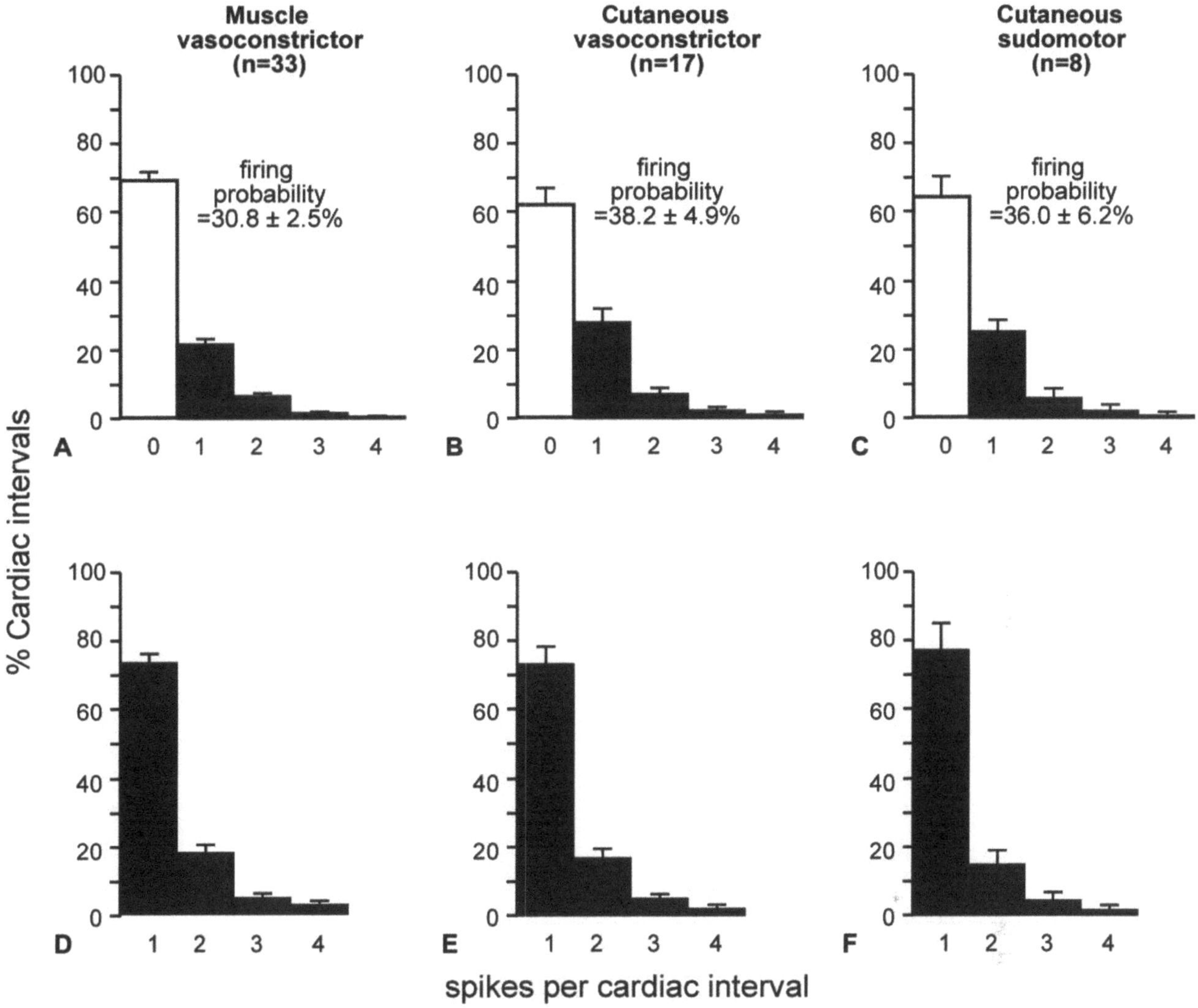

Fig. 28.3. Upper rows: percentage of cardiac intervals in which populations of single muscle vasoconstrictor neurons (**A**), cutaneous vasoconstrictor neurons (**B**), and sudomotor neurons (**C**) are quiescent (white columns) or fire 1, 2, 3 or 4 spikes (black columns) per cardiac interval. Lower rows: same data after excluding those cardiac intervals in which a unit was silent. (Data reanalyzed from Macefield et al., 1994; Macefield and Wallin, 1996, 1999b.)

modeling supports the empirical data: an individual postganglionic sympathetic neuron receives excitatory drive from, on average, two preganglionic neurons (Macefield, 2011).

In multiunit recordings the absolute strength of the activity cannot be determined because of the dependence on the proximity of the electrode tip to the active axons. This is not a problem with single-unit recordings where the information content is quantal in nature: the discrete events (i.e., the action potentials) can be represented as digital bits. In single-unit terms, differences in the strength of multiunit sympathetic activity (i.e., the number and the intensity of multiunit bursts) may be brought about in two ways: by differences in (1) the number of active neurons or (2) the mean firing frequency of the neurons (or a combination of both). A difference in mean firing frequency may, in turn, be due to (1) a difference in firing probability of active neurons, i.e., the percentage of heart beats in which the neurons are active or (2) a difference in the number of spikes the neurons discharge within the bursts (or a combination of both). For example, the sustained increase in MSNA brought about by a maximal inspiratory breath-hold is associated with an increase in mean firing frequency of individual neurons, from 0.23 ± 0.04 Hz (n = 9) at rest to 1.04 ± 0.14 Hz during the apnea (Macefield and Wallin, 1999a); due both to an increase in firing probability (from 32.5 ± 6.2% at rest to 56.3 ± 3.1% during the apnea) and an increased probability of multiple firing (from 14.9 ± 4.5% to 38.7 ± 5.6%).

MUSCLE SYMPATHETIC NERVE ACTIVITY IN HEALTH AND DISEASE

Muscle sympathetic nerve activity (MSNA) is vasoconstrictor in function, its primary role in health being to buffer acute falls in blood pressure, via the baroreflex, and thereby maintain a constant perfusion pressure. MSNA is also modulated by respiration, decreasing during inspiration and increasing during expiration (Hagbarth and Vallbo, 1968b; Eckberg et al., 1985; Seals et al., 1990; Macefield and Wallin, 1995). MSNA remains constant in a given individual, day to day, year to year (Fagius and Wallin, 1993). Some subjects have high levels of MSNA at rest while some have low levels; identical twins have similar levels, whereas the burst incidence in fraternal twins differs widely (Wallin et al., 1993). However, it is also known that, after the age of 60, MSNA increases with age (Sverrisdottir et al., 2001; Hart et al., 2009a). Body weight also influences MSNA: it is known that a body mass index above 30 (obese) increases the resting level of MSNA, but individual muscle vasoconstrictor neurons do not show an increased incidence of multiple firing in obesity (Lambert et al., 2007). Given that there is a poor relationship between the level of resting MSNA and resting blood pressure, one cannot predict the level of MSNA in a given healthy individual by recording blood pressure (or heart rate) alone (Joyner et al., 2008). However, it is known that bursts of MSNA occur when diastolic pressure is low (and hence baroreceptor input is low) during the spontaneous fluctuations in arterial pressure that occur at rest (Sundlöf and Wallin, 1978), the strongest correlation being to cardiac output in men (Charkoudian et al., 2005) but not in women (Hart et al., 2009b).

Orthostatic hypotension and syncope

Changes in body posture induce hydrostatic challenges to the cardiovascular system, which must respond to these challenges in order to provide an adequate supply of blood to the brain. Rapid adjustments to the vascular system are mediated primarily by the autonomic nervous system and normally meet this challenge; failure of the cardiovascular system to respond to the effects of gravity on the vascular system can produce dizziness, lightheadedness, and fainting and can be debilitating in clinical postural hypotension (Mathias, 1995) and the postural orthostatic tachycardia syndrome (POTS; Low et al., 1995). Given that muscle vasoconstrictor drive serves to buffer acute falls in blood pressure, it might be expected that MSNA would be elevated in cases of hypotension. Certainly, MSNA to leg muscles increases linearly from the horizontal to vertical position (Burke et al., 1977; Iwase et al., 1987; Fu et al., 2001). It has been shown that whether a given individual can adequately compensate for this gravitational challenge depends on how much "vasoconstrictor reserve" they have; those who show an appropriate increase in MSNA on head-up tilt, and show a further increase in MSNA during a superimposed maneuver that also increases MSNA (immersion of the hand in ice-cold water), tolerated passive tilt for longer than those with a lower reserve (Fu et al., 2004). During prolonged head-up tilt, in most people blood pressure tends to remain stable, but in some individuals with orthostatic intolerance blood pressure is not maintained. In these people blood pressure and heart rate start to fall (presyncope), and unless cerebral perfusion can be maintained, fainting (syncope) occurs. A comparison of MSNA, hemodynamic responses, and baroreflex sensitivity during early tilting in people susceptible to presyncope found no differences between controls, but cardiac output fell in presyncopal subjects and in the majority total peripheral resistance also fell (Fu et al., 2012). However, in some subjects total peripheral resistance was maintained even at presyncope, though in all presyncopal subjects MSNA fell rapidly at presyncope *after* the onset of hypotension. Another study followed subjects through to syncope, and found that – despite the rapid fall in blood pressure – MSNA was maintained, arguing against inhibition of MSNA as being responsible for the fall in blood pressure in syncope (Vaddadi et al., 2010).

Lower body negative pressure (LBNP) has been used to emulate the gravitational shifts on standing, and has the advantage that even more pronounced fluid shifts can be induced than those produced by gravity. Ichinose and colleagues (2006) showed that the relationship between diastolic pressure and MSNA burst incidence was preserved at levels of LBNP extending from – 10 mmHg to –60 mmHg, but were progressively shifted upwards until presyncope occurred, at which point the slope of the relationship declined. This suggests that baroreflex control over MSNA decreases with increasing orthostatic stress, particularly during the period immediately preceding the severe hypotension associated with orthostatic syncope. Cooke and colleagues (2009) also showed that baroreflex control of MSNA is reduced before presyncope, resulting in fused bursts of MSNA, and that despite the fall in blood pressure during presyncope, withdrawal of MSNA does not necessarily occur. Interestingly, the prolonged bursts of MSNA during syncope lose their tight cardiac rhythmicity and bursts of MSNA can be evoked by arousal stimuli (i.e., MSNA takes on the character of skin sympathetic nerve activity) – evidently because of the withdrawal of baroreceptor input (Iwase et al., 2002).

Cardiovascular and respiratory disease

Standard multiunit recordings have shown that elevated levels of muscle sympathetic nerve activity feature in many diseases. In addition to congestive heart failure (Leimbach et al., 1986; Rundqvist et al., 1997; Macefield et al., 1999), obstructive sleep apnea (Carlsson et al., 1993; Narkiewicz et al., 1999; Elam et al., 2002), chronic obstructive pulmonary disease (Heindl et al., 2001; Ashley et al., 2010), and bronchiectasis (Ashley et al., 2010), MSNA is increased in many different forms of hypertension: essential (Grassi et al., 1998; Schlaich et al., 2004), pregnancy-induced (Schobel et al., 1996; Fischer et al., 2004), renovascular (Johansson et al., 1999; Miyajima et al., 2001), and that associated with chronic kidney disease (Hausberg et al., 2002; Tuncel et al., 2002; Schlaich et al., 2009). However, it should be emphasized that a high level of MSNA does not, on its own, mean there is any underlying pathology: many healthy young individuals have high MSNA at rest yet normal blood pressures and no evidence of cardiovascular disease (Macefield and Wallin, 1999a). This is one of the advantages of single-unit recordings: because of the quantal nature of unitary recordings, the firing properties of individual neurons can be compared across subjects and diseases. This means that by providing evidence of increases in firing probability, mean firing rate, or multiple firing one can see whether the elevated level of MSNA reflects an increase in central sympathetic drive that may be being driven by some underlying pathology. Individual muscle vasoconstrictor neurons in healthy subjects with elevated resting levels of MSNA have *low* firing probabilities, *low* firing rates, and a *low* incidence of multiple firing (Macefield and Wallin, 1999b). Conversely, pathophysiological increases in MSNA are associated with at least one of the following features: *high* firing probabilities, *high* firing rates, and a *high* incidence of multiple firing of individual neurons. Moreover, the patterns of increase differ according to the type of pathology.

For example, comparison of the firing properties of single muscle vasoconstrictor neurons in chronic obstructive pulmonary disease (COPD), congestive heart failure (CHF), bronchiectasis (BE), and obstructive sleep apnea (OSA) reveals differences that cannot be seen in multiunit recordings. A unitary recording is illustrated in Figure 28.4. The overall burst incidence in COPD (86%; Ashley et al., 2010) is comparable to that seen in CHF (88%; Macefield et al., 1999), which for both conditions is greater than that seen in either the patients with bronchiectasis (78%; Ashley et al., 2010) or those with obstructive sleep apnea (77%; Elam et al., 2002). Indeed, COPD and CHF represent two of the highest pathophysiological causes of augmented spontaneous MSNA ever recorded. The firing probability of muscle vasoconstrictor neurons was also as high in COPD (52%; Ashley et al., 2010) as in CHF (55%; Macefield et al., 1999), as was also the case for OSA (51%). It is interesting that the firing probability of single neurons in BE (38%) was no different from that of the healthy subjects with high levels of MSNA (35%). We know that hypoxemia and hypercapnia are not clinical features of bronchiectasis, so perhaps this accounts for the low firing probabilities seen in this disease. Nevertheless, given that the overall burst incidence were comparable in the two groups, one can

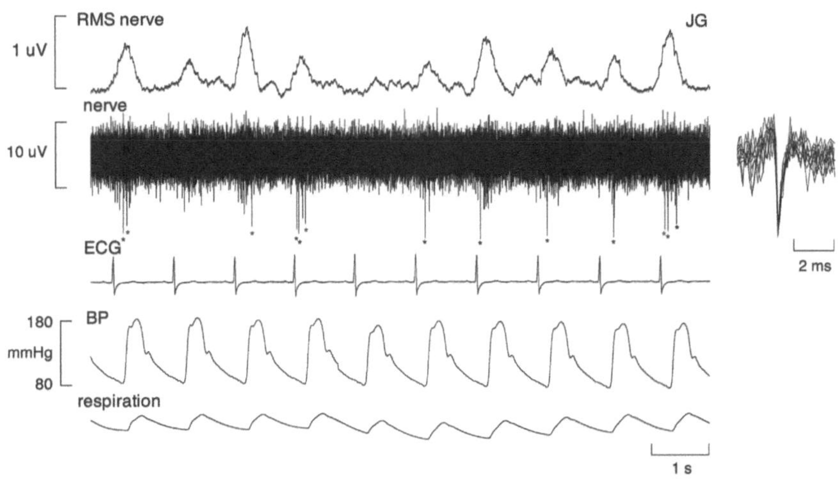

Fig. 28.4. Unitary recording from a single muscle vasoconstrictor neuron in an awake patient with chronic obstructive pulmonary disease (COPD). Typical of muscle vasoconstrictor neurons, this unit generally fired only one spike (indicated by asterisks) per cardiac interval, but occasionally fired multiple spikes within a burst of MSNA. Superimposed spikes show a uniform spike morphology, indicating that the action potentials originated from the same single sympathetic axon. The RMS-processed nerve signal primarily reflects far-field activity of many sympathetic neurons firing. (Unpublished data from Macefield.)

conclude that, unless one of the firing properties is elevated, a high level of MSNA does not equate to an increase in central sympathetic drive.

Mean firing rates in the COPD patients (0.92 Hz) were similar to those found in OSA (0.96 Hz) and CHF patients (0.98 Hz), but higher than those in the bronchiectasis patients (0.72 Hz) and the healthy subjects (0.33 Hz). However, unlike the CHF patients, who showed no decrease in the proportion of solitary spikes generated within a burst, the patients with COPD, OSA, and BE each showed a significant trend towards multiple firing. This is illustrated in Figure 28.5. We had already demonstrated that a shift towards multiple firing can occur in CHF, but only during the prolonged cardiac intervals following ventricular ectopic beats (Elam and Macefield, 2001). While the shift towards multiple firing can be explained by the elevated chemical drive in COPD and OSA, it is difficult to account for the increase seen in BE; perhaps it is related to the increased work of breathing in this disease. Regardless of the underlying mechanisms by which multiple within-burst firing occurs, it is not unreasonable to posit that a shift towards multiple firing will increase the liberation of noradrenaline/norepinephrine (and colocalized neurotransmitters) from the vasoconstrictor nerve terminals – thereby increasing the degree of neurally mediated vasoconstriction. Indeed, Lambert and colleagues recently showed that cardiac noradrenaline spillover is higher when muscle vasoconstrictor neurons (and presumably, in parallel, cardiac sympathetic neurons) exhibit an increase in multiple firing (Lambert et al., 2011), arguing that multiple firing does cause an increase in neurotransmitter release.

Neurological and psychiatric disorders

Although most disturbances in muscle sympathetic nerve activity are related to cardiovascular, cardiorespiratory and renal diseases, MSNA is affected in certain neurological conditions. In *Parkinson's disease*, there is an inverse relationship between MSNA and age or disease progression, whereas there was a positive relationship in control subjects, which indicates a progressive deterioration of sympathetic function in Parkinson's disease (Shindo et al., 2003). In *multiple system atrophy* (MSA) there is an absence of spontaneous MSNA, resulting in postural hypotension (Donadio et al., 2010). As its name suggests, *pure autonomic failure* (PAF) involves all branches of the autonomic nervous system: because muscle (and splanchnic) vasoconstrictor drive is interrupted, patients suffer from severe orthostatic hypotension;

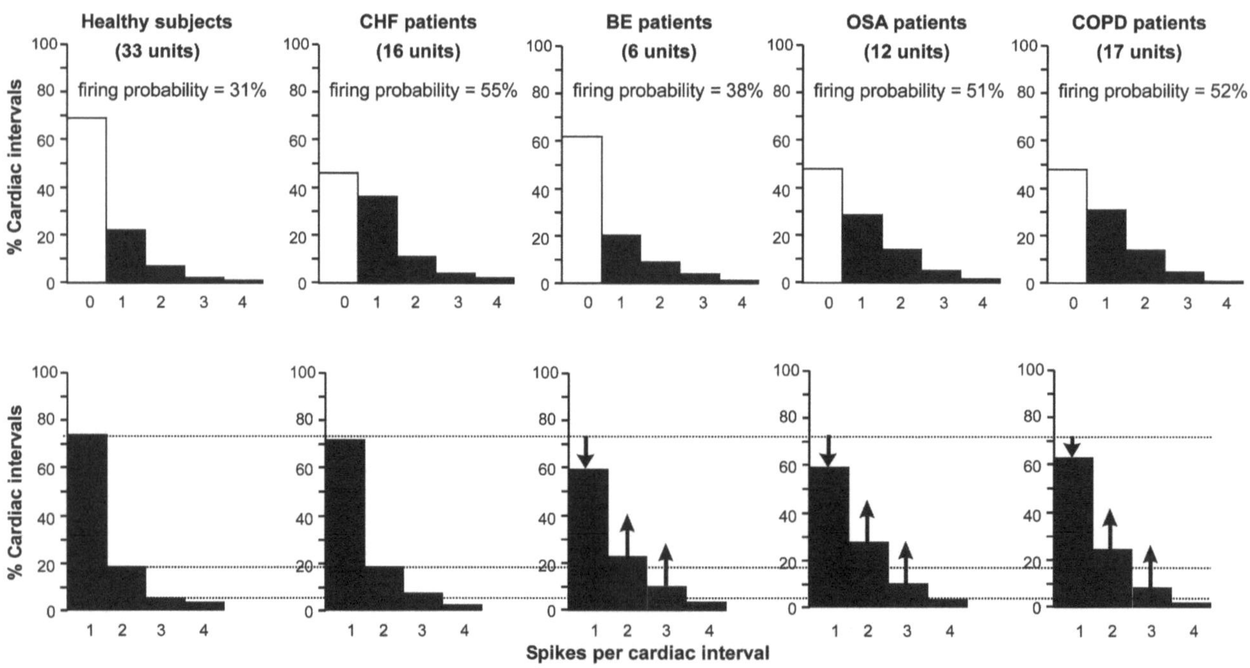

Fig. 28.5. Firing distributions of single muscle vasoconstrictor neurons in healthy subjects. Data combined from subjects with low (Macefield et al., 1994) or high (Macefield and Wallin, 1999b) resting levels of MSNA, and in patients with congestive heart failure (CHF; Macefield et al., 1999), bronchiectasis (BE; Ashley et al., 2010), obstructive sleep apnoea (OSA; Elam et al., 2002), or chronic obstructive pulmonary disease (COPD; Ashley et al., 2010). The histograms showing pooled data on the percentage of cardiac intervals in which units were quiescent, fired a single spike or 2, 3 or 4 spikes. In the top panel, all cardiac intervals have been included: the open columns represent those cardiac intervals in which the neurons were silent. The lower panels show data in which only those cardiac intervals were included in which a unit was active.

MSNA is absent in PAF (Thompson et al., 1995; Donadio et al., 2010). Conversely, MSNA is greatly increased in *Guillain–Barré syndrome*, an autoimmune disease that targets myelinated axons, presumably due to involvement of baroreceptor afferents (Fagius and Wallin, 1983); bilateral experimental blockade of the glossopharyngeal and vagus nerves, which also interrupts baroreceptor afferent transmission, likewise causes a pronounced increase in MSNA, coupled with a marked increase in blood pressure (Fagius et al., 1985). Augmented MSNA and sustained hypertension is also found in cases of *baroreflex failure*, typically brought about by surgery and radiotherapy for cancers of the neck (Heusser et al., 2005).

Hereditary sensory and autonomic neuropathy type III (HSAN III), also known as familial dysautonomia or Riley–Day syndrome, is a rare congenital disorder associated with postural hypotension yet marked increases in blood pressure during emotional excitement; these patients have a loss of functional baroreceptors – noradrenaline is not released during upright tilt but is released during emotional excitement or anxiety (Norcliffe-Kaufmann et al., 2010). MSNA – though greatly reduced – appears as tonic activity without any cardiac modulation, increasing during anxiety (Macefield et al., 2013). A more common neurological condition affecting sympathetic outflow is *spinal cord injury*: a complete cervical or high thoracic spinal lesion interrupts the descending drive to the preganglionic sympathetic neurons in the intermediolateral column of the thoracolumbar spinal cord, abolishing spontaneous bursts of MSNA and thereby interfering with the ongoing control of blood flow and blood pressure (Stjernberg et al., 1986). Importantly, because the reflex circuitry within the isolated spinal cord is still intact, visceral or somatic sensory inputs originating below the lesion can reflexly excite vasoconstrictor neurons, resulting in a dramatic and uncontrolled increase in blood pressure (*autonomic dysreflexia*) that can lead to stroke, cardiac arrest, and death (Karlsson, 1999; Macefield et al., 2012).

Muscle sympathetic nerve activity is increased in patients with panic disorder (Lambert et al., 2006, 2010; Barton et al., 2007), with single muscle vasoconstrictor neurons showing a shift towards more multiple firing (Lambert et al., 2006). MSNA is also increased in major depressive disorder, and alleviated following treatment with selective serotonin reuptake inhibitors (Barton et al., 2007).

SKIN SYMPATHETIC NERVE ACTIVITY IN HEALTH AND DISEASE

The primary role of sympathetic outflow to the skin is to control heat loss from the body, through its actions on sweat glands, cutaneous blood vessels, and, phylogenetically, the hairs. In addition to its role in thermoregulation, it has also been commandeered as a means of emotional expression, its activity increasing during brief arousal stimuli (Hagbarth et al., 1972) or by states of elevated emotional engagement (Brown et al., 2012). Skin sympathetic nerve activity (SSNA) supplying hairy skin is primarily composed of cutaneous vasoconstrictor and sudomotor activity (though pilomotor and lipomotor axons also travel within cutaneous fascicles of peripheral nerves). In addition to a reduction in cutaneous vasoconstrictor drive and increase in sudomotor drive during whole-body heating, active vasodilatation occurs (Kamijo et al., 2005). Unlike MSNA, sympathetic outflow to skin is not influenced by the arterial baroreceptors (Wilson et al., 2001). Skin sympathetic nerve activity is not really affected by most of the diseases that affect MSNA, but there are exceptions: pure autonomic failure and multiple system atrophy also involve skin sympathetic outflow (Donadio et al., 2010). However, most of the conditions that affect SSNA are related to disturbances in thermoregulation.

Disorders of thermoregulation

Palmar plantar hyperhidrosis is associated with an increase in sudomotor outflow, as shown by recordings from the tibial and median nerves (Iwase et al., 1997; Macefield et al., 2008); individual sudomotor neurons show an increase in firing probability and multiple firing (Macefield et al., 2008). Hyperhidrosis is a purely sudomotor disturbance, without any involvement of other components of the sympathetic or parasympathetic nervous systems. Less common neurological conditions include *pure autonomic failure*, referred to above: because sudomotor drive is interrupted, anhidrosis also features. Autoimmune autonomic ganglionopathy targets nicotinic cholinergic receptors at autonomic ganglia, resulting in loss of function; one feature of this is anhidrosis and compromised thermoregulation (Kimpinski et al., 2009). *Congenital insensitivity to pain with anhidrosis* (CIPA), one of the hereditary sensory and autonomic neuropathies (HSAN type IV), is due to a genetic mutation that results in a selective loss of unmyelinated afferent axons, resulting in a loss of nociception, and a loss of sympathetic axons to the skin (Axelrod, 2002); because these individuals cannot sweat, thermoregulation is greatly compromised, resulting in dangerous febrile episodes that can lead to death in infancy. Conversely, SSNA is preserved in patients with HSAN III (Macefield et al., 2013). An even rarer condition affecting the autonomic nervous system is Ross syndrome, which features anhidrosis as well as a tonic (Adie's) pupil and loss of deep tendon reflexes (Ross, 1958). Spontaneous and evoked SSNA is absent in this condition, as are sudomotor and cutaneous vasoconstrictor responses, but spontaneous MSNA is preserved, which fits the clinical picture of there being little disturbance to blood pressure regulation in this condition. So, while *pure autonomic failure* affects both SSNA and MSNA (Donadio et al., 2010), microneurography has revealed that Ross syndrome affects SSNA but not MSNA (Donadio et al., 2012). In another rare condition associated with anhidrosis, *idiopathic pure sudomotor failure*, microneurography revealed that SSNA is intact; evidently the failure occurs at the cholinergic synapse with the sweat glands (Nakazato et al., 2004). Of course, given the involvement of sympathetic and parasympathetic innervation of the pupil, the autonomic signs of Ross syndrome are not limited to the skin, but at least the study by Donadio and colleagues (2012) provided further differentiation in the expression of autonomic disturbances across neurological conditions.

As noted above, in complete high thoracic or cervical spinal lesions there are *no spontaneous bursts of MSNA or SSNA below the lesion* because the descending supraspinal excitatory drive to the preganglionic sympathetic neurons – located in the intermediolateral column of the thoracolumbar spinal cord – is interrupted; nor are there spontaneous fluctuations in skin blood flow or sweat release below the lesion (Wallin and Stjernberg, 1984).

CONCLUSIONS

Sympathetic microneurography has contributed significant information to our understanding of the operation of the cardiovascular system. While it is too difficult to use routinely in the clinic, it has proven invaluable in investigative studies in diverse clinical conditions, and there is no doubt it will continue to do so.

References

Ashley C, Burton D, Sverrisdottir YB et al. (2010). Firing probability and mean firing rates of human muscle vasoconstrictor neurones are elevated during chronic asphyxia. J Physiol 588: 701–711.

Axelrod FB (2002). Hereditary sensory and autonomic neuropathies. Familial dysautonomia and other HSANs. Clin Auton Res 12 (Suppl 1): I2–I14.

Barton DA, Dawood T, Lambert EA et al. (2007). Sympathetic activity in major depressive disorder: identifying those at increased cardiac risk? J Hypertens 25: 2117–2124.

Brown R, James C, Henderson LA et al. (2012). Autonomic markers of emotional processing: skin sympathetic nerve activity in humans during exposure to emotionally-charged images. Front Physiol 3: 394.

Burke D, Sundlöf G, Wallin BG (1977). Postural effects on muscle nerve sympathetic activity in man. J Physiol 272: 399–414.

Carlsson JT, Hedner J, Elam M et al. (1993). Augmented resting sympathetic activity in awake patients with obstructive sleep apnea. Chest 103: 1763–1768.

Charkoudian N, Joyner MJ, Johnson CP et al. (2005). Balance between cardiac output and sympathetic nerve activity in resting humans: role in arterial pressure regulation. J Physiol 568: 315–321.

Cooke WH, Rickards CA, Ryan KL et al. (2009). Muscle sympathetic nerve activity during intense lower body negative pressure to presyncope in humans. J Physiol 587: 4987–4999.

Delius W, Hagbarth KE, Hongell A et al. (1972a). Manoeuvres affecting sympathetic outflow in human muscle nerves. Acta Physiol Scand 84: 82–94.

Delius W, Hagbarth KE, Hongell A et al. (1972b). Manoeuvres affecting sympathetic outflow in human skin nerves. Acta Physiol Scand 84: 177–186.

Donadio V, Cortelli P, Elam M et al. (2010). Autonomic innervation in multiple system atrophy and pure autonomic failure. J Neurol Neurosurg Psychiatry 81: 1327–1335.

Donadio V, Cortelli P, Giannoccaro MP et al. (2012). Muscle and skin sympathetic activities in Ross syndrome. Clin Neurophysiol 123: 1639–1643.

Eckberg DL, Nerhed C, Wallin BG (1985). Respiratory modulation of muscle sympathetic and vagal cardiac outflow in man. J Physiol 365: 181–196.

Elam M, Macefield VG (2001). Multiple firing of single muscle vasoconstrictor neurons during cardiac dysrhythmias in human heart failure. J Appl Physiol 91: 717–724.

Elam M, Macefield VG, McKenzie D (2002). The firing pattern of single muscle vasoconstrictor neurons in awake patients with the obstructive sleep apnoea syndrome. J Appl Physiol 93: 297–303.

Fagius J, Wallin BG (1983). Microneurographic evidence of excessive sympathetic outflow in the Guillain–Barré syndrome. Brain 106: 589–600.

Fagius J, Wallin BG (1993). Long-term variability and reproducibility of resting human muscle nerve sympathetic activity at rest, as reassessed after a decade. Clin Auton Res 3: 201–205.

Fagius J, Wallin BG, Sundlöf G et al. (1985). Sympathetic outflow in man after anaesthesia of the glossopharyngeal and vagus nerves. Brain 108: 423–438.

Fischer T, Schobel HP, Frank H et al. (2004). Pregnancy-induced sympathetic hyperactivity: a precursor of preeclampsia. Eur J Clin Invest 34: 443–448.

Fu Q, Iwase S, Niimi Y et al. (2001). Effects of lower body positive pressure on muscle sympathetic nerve activity response to head-up tilt. Am J Physiol 281: R778–R785.

Fu Q, Witkowski S, Levine BD (2004). Vasoconstrictor reserve and sympathetic neural control of orthostasis. Circulation 110: 2931–2937.

Fu Q, Verheyden B, Wieling W et al. (2012). Cardiac output and sympathetic vasoconstrictor responses during upright tilt to presyncope in healthy humans. J Physiol 590: 1839–1848.

Grassi G, Colombo M, Seravalle G et al. (1998). Dissociation between muscle and skin sympathetic nerve activity in essential hypertension, obesity, and congestive heart failure. Hypertension 31: 64–67.

Hagbarth KE, Vallbo ÅB (1968a). Discharge characteristics of human muscle afferents during muscle stretch and contraction. Exp Neurol 22: 674–694.

Hagbarth KE, Vallbo ÅB (1968b). Pulse and respiratory grouping of sympathetic impulses in human muscle-nerves. Acta Physiol Scand 74: 96–108.

Hagbarth KE, Vallbo ÅB (1969). Single unit recordings from muscle nerves in human subjects. Acta Physiol Scand 76: 321–334.

Hagbarth KE, Hallin RG, Hongell A et al. (1972). General characteristics of sympathetic activity in human skin nerves. Acta Physiol Scand 84: 164–176.

Hart EC, Charkoudian N, Wallin BG et al. (2009a). Sex differences in sympathetic neural-hemodynamic balance: implications for human blood pressure regulation. Hypertension 53: 571–576.

Hart EC, Joyner MJ, Wallin BG et al. (2009b). Age-related differences in the sympathetic-hemodynamic balance in men. Hypertension 54: 127–133.

Hausberg M, Kosch M, Harmelink P et al. (2002). Sympathetic nerve activity in end-stage renal disease. Circulation 106: 1974–1979.

Heindl S, Lehnert M, Criée et al. (2001). Marked sympathetic activation in patients with chronic respiratory failure. Am J Respir Crit Care Med 164: 597–601.

Heusser K, Tank J, Luft FC et al. (2005). Baroreflex failure. Hypertension 45: 834–839.

Ichinose M, Saito M, Fujii N et al. (2006). Modulation of the control of muscle sympathetic nerve activity during severe orthostatic stress. J Physiol 576: 947–958.

Iwase S, Mano T, Saito T (1987). Effects of graded head-up tilting on muscle sympathetic activities in man. Physiologist 30 (Suppl. 1): S62–S63.

Iwase S, Ikeda T, Kitazawa H et al. (1997). Altered response in cutaneous sympathetic outflow to mental and thermal stimuli in primary palmoplantar hyperhidrosis. J Auton Nerv Syst 64: 65–73.

Iwase S, Mano T, Kamiya A et al. (2002). Syncopal attack alters the burst properties of muscle sympathetic nerve activity in humans. Auton Neurosci 95: 141–145.

Johansson M, Elam M, Rundqvist B et al. (1999). Increased sympathetic nerve activity in renovascular hypertension. Circulation 18: 2537–2542.

Joyner MJ, Charkoudian N, Wallin BG (2008). A sympathetic view of the sympathetic nervous system and human blood pressure regulation. Exp Physiol 93: 715–724.

Kamijo Y, Lee K, Mack GW (2005). Active cutaneous vasodilation in resting humans during mild heat stress. J Appl Physiol 98: 829–837.

Kimpinski K, Iodice V, Sandroni P et al. (2009). Sudomotor dysfunction in autoimmune autonomic ganglionopathy. Neurology 73: 1501–1506.

Lambert E, Hotchkin E, Alvarenga M et al. (2006). Single-unit analysis of sympathetic nervous discharges in patients with panic disorder. J Physiol 570: 637–643.

Lambert E, Straznicky N, Schlaich M et al. (2007). Differing pattern of sympathoexcitation in normal-weight and obesity-related hypertension. Hypertension 50: 862–868.

Lambert E, Dawood T, Straznicky N et al. (2010). Association between the sympathetic firing pattern and anxiety level in patients with the metabolic syndrome and elevated blood pressure. J Hypertens 28: 543–550.

Lambert EA, Schlaich MP, Dawood T et al. (2011). Single-unit muscle sympathetic nervous activity and its relation to cardiac noradrenaline spillover. J Physiol 589: 2597–2605.

Leimbach WN, Wallin BG, Victor RG et al. (1986). Direct evidence from intraneural recordings for increased central sympathetic outflow in patients with heart failure. Circulation 73: 913–919.

Low PA, Opfer-Gehrking TL, Textor SC et al. (1995). Postural tachycardia syndrome (POTS). Neurology 45: S19–S25.

Macefield VG (2011). On the number of preganglionic neurones driving human postganglionic sympathetic neurones: a comparison of modelling and empirical data. Front Neurosci 5: 132.

Macefield VG, Wallin BG (1995). Modulation of muscle sympathetic activity during spontaneous and artificial ventilation and apnoea in humans. J Auton Nerv Syst 53: 137–147.

Macefield VG, Wallin BG (1996). The discharge behaviour of single sympathetic neurones supplying human sweat glands. J Auton Nerv Syst 61: 277–286.

Macefield VG, Wallin BG (1999a). Firing properties of single vasoconstrictor neurones in human subjects with high levels of muscle sympathetic drive. J Physiol 516: 293–301.

Macefield VG, Wallin BG (1999b). Respiratory and cardiac modulation of single vasoconstrictor and sudomotor neurones to human skin. J Physiol 516: 303–314.

Macefield VG, Vallbo ÅB, Wallin BG (1994). The discharge behaviour of single vasoconstrictor motoneurones in human muscle nerves. J Physiol 481: 799–809.

Macefield VG, Rundqvist B, Sverrisdottir YB et al. (1999). Firing properties of single muscle vasoconstrictor neurones in the sympathoexcitation associated with heart failure. Circulation 100: 1708–1713.

Macefield VG, Elam M, Wallin BG (2002). Firing properties of single postganglionic sympathetic neurones recorded in awake human subjects. Auton Neurosci 95: 146–159.

Macefield VG, Sverrisdottir YB, Elam E et al. (2008). Firing properties of sudomotor neurones in hyperhidrosis and thermal sweating. Clin Auton Res 18: 325–330.

Macefield VG, Burton AR, Brown R (2012). Somatosympathetic vasoconstrictor reflexes in human spinal cord injury: responses to innocuous and noxious sensory stimulation below lesion. Front Physiol 3: 215.

Macefield VG, Norcliffe-Kaufmann LJ, Axelrod F et al. (2013). Cardiac-locked bursts of muscle sympathetic nerve activity are absent in familial dysautonomia. J Physiol 591: 689–700.

Mathias CJ (1995). Orthostatic hypotension: causes, mechanisms, and influencing factors. Neurology 45: S6–S11.

Miyajima E, Yamada Y, Yoshida Y et al. (2001). Muscle sympathetic nerve activity in renovascular hypertension and primary aldosteronism. Hypertension 17: 1057–1062.

Nakazato Y, Tamura N, Ohkuma A et al. (2004). Idiopathic pure sudomotor failure: anhidrosis due to deficits in cholinergic transmission. Neurology 63: 1476–1480.

Narkiewicz K, Kato M, Phillips BG et al. (1999). Nocturnal continuous positive airway pressure decreases daytime sympathetic traffic in obstructive sleep apnea. Circulation 100: 2332–2335.

Norcliffe-Kaufmann L, Axelrod F, Kaufmann H (2010). Afferent baroreflex failure in familial dysautonomia. Neurology 75: 1904–1911.

Ross AT (1958). Progressive selective sudomotor denervation: a case with coexisting Adie's syndrome. Neurology 8: 809–817.

Rundqvist B, Casale R, Bergmann-Sverrisdottir Y et al. (1997). Rapid fall in sympathetic nerve hyperactivity in patients with heart failure after cardiac transplantation. J Card Fail 3: 21–26.

Salmanpour A, Brown LJ, Steinback CD et al. (2011). Relationship between size and latency of action potentials in human muscle sympathetic nerve activity. J Neurophysiol 105: 2830–2842.

Schlaich MP, Lambert E, Kaye DM et al. (2004). Sympathetic augmentation in hypertension: role of nerve firing, norepinephrine reuptake, and angiotensin neuromodulation. Hypertension 43: 169–175.

Schlaich MP, Socratous F, Hennebry S et al. (2009). Sympathetic activation in chronic renal failure. J Am Soc Nephrol 20: 933–939.

Schobel HP, Fischer T, Heuszer K et al. (1996). Preeclampsia – a state of sympathetic hyperactivity. N Engl J Med 335: 1480–1485.

Seals DR, Suwarno NO, Dempsey JA (1990). Influence of lung volume on sympathetic nerve discharge in normal humans. Circ Res 67: 130–141.

Shindo K, Watanabe H, Tanaka H et al. (2003). Age and duration related changes in muscle sympathetic nerve activity in Parkinson's disease. J Neurol Neurosurg Psychiatry 74: 1407–1411.

Steinback CD, Salmanpour A, Breskovic T et al. (2010). Sympathetic neural activation: an ordered affair. J Physiol 588: 4825–4836.

Stjernberg L, Blumberg H, Wallin BG (1986). Sympathetic activity in man after spinal cord injury. Outflow to muscle below the lesion. Brain 109: 695–715.

Sundlöf G, Wallin BG (1978). Human muscle nerve sympathetic activity at rest. Relationship to blood pressure and age. J Physiol 274: 621–637.

Sverrisdottir YB, Rundqvist B, Elam M (1998). Relative burst amplitude in human muscle sympathetic nerve activity: a sensitive indicator of altered sympathetic traffic. Clin Auton Res 8: 95–100.

Sverrisdottir YB, Rundqvist B, Johannsson G et al. (2000). Sympathetic neural burst amplitude distribution: a more specific indicator of sympatho-excitation in human heart failure. Circulation 102: 2076–2081.

Sverrisdottir YB, Johannsson G, Jungersten L et al. (2001). Is the somatotropic axis related to sympathetic nerve activity in healthy ageing men? Hypertension 19: 2019–2024.

Thompson JM, O'Callaghan CJ, Kingwell BA et al. (1995). Total norepinephrine spillover, muscle sympathetic nerve activity and heart-rate spectral analysis in a patient with dopamine beta-hydroxylase deficiency. J Auton Nerv Syst 55: 198–206.

Tuncel M, Augustyniak R, Zhang W et al. (2002). Sympathetic nervous system function in renal hypertension. Curr Hypertens Rep 4: 229–236.

Vaddadi G, Esler MD, Dawood T et al. (2010). Persistence of muscle sympathetic nerve activity during vasovagal syncope. Eur Heart J 31: 2027–2033.

Vallbo ÅB (1971). Muscle spindle response at the onset of isometric voluntary contractions in man. Time difference between fusimotor and skeletomotor effects. J Physiol 218: 405–431.

Vallbo ÅB, Hagbarth KE, Torebjörk HE et al. (1979). Somatosensory, proprioceptive, and sympathetic activity in human peripheral nerves. Physiol Rev 59: 919–957.

Vallbo ÅB, Hagbarth KE, Wallin BG (2004). Microneurography: how the technique developed and its role in the investigation of the sympathetic nervous system. J Appl Physiol 96: 1262–1269.

Wallin BG, Stjernberg L (1984). Sympathetic activity in man after spinal cord injury. Outflow to skin below the lesion. Brain 107: 183–198.

Wallin BG, Kunimoto MM, Sellgren J (1993). Possible genetic influence on the strength of human muscle nerve sympathetic activity at rest. Hypertension 22: 282–284.

Wilson TE, Cui J, Crandall CG (2001). Absence of arterial baroreflex modulation of skin sympathetic activity and sweat rate during whole-body heating in humans. J Physiol 536: 615–623.

Handbook of Clinical Neurology, Vol. 117 (3rd series)
Autonomic Nervous System
R.M. Buijs and D.F. Swaab, Editors
2013 Published by Elsevier B.V.

Chapter 29

Sympathetic neuroimaging

DAVID S. GOLDSTEIN*
Clinical Neurocardiology Section, Clinical Neurosciences Program, Division of Intramural Research, National Institute of Neurological Disorders and Stroke, National Institutes of Health, Bethesda, MD, USA

INTRODUCTION

Until recently, clinical laboratory means to test patients with known or suspected abnormalities of the autonomic nervous system consisted of physiological independent manipulations (e.g., heat exposure, the Valsalva maneuver, tilt table testing) and physiological dependent measures (e.g., sweating, blood pressure, heart rate and heart rate variability, skin and core temperature, forearm vascular resistance, skin electrical conductance, cutaneous humidity, microcirculatory blood flow, skeletal muscle sympathetic microneurography). Since the late 1970s, clinical research on autonomic disorders has also included neurochemical measures (e.g., plasma norepinephrine during supine rest and orthostasis, norepinephrine spillover, dihydroxyphenylglycol, neuronal nicotinic receptor antibodies) and physiological or neurochemical effects of neuropharmacological agents (e.g., tyramine, clonidine, yohimbine, desipramine, trimethaphan, isoprenaline (isoproterenol), glucagon, reboxetine, and acetylcholine in the quantitative sudomotor axon reflex test).

Sympathetic neuroimaging provides an important supplement to physiological, neurochemical, and neuropharmacological approaches in the evaluation of patients with clinical autonomic disorders. Sympathetic neuroimaging using planar scintigraphy became available in the 1980s (Wieland et al., 1981), followed by imaging using single-photon emission computed tomography (SPECT) and positron emission tomography (PET).

This chapter discusses concepts underlying sympathetic neuroimaging, summarizes methods and agents used, highlights some applications of clinical sympathetic neuroimaging, and emphasizes the utility of sympathetic neuroimaging to detect cardiac sympathetic denervation in Lewy body diseases.

Neuroimaging assessments can be categorized in terms of anatomic, functional nonspecific, and functional specific. Examples of anatomic neuroimaging are computed tomography and magnetic resonance imaging of the brain. Functional nonspecific approaches include functional magnetic resonance imaging and assessments of regional perfusion or metabolism by PET after administration of ^{18}F-fluorodeoxyglucose, ^{13}N-ammonia, or ^{15}O-water. None of these specifically visualizes sympathetic innervation or function. Functional specific methods examine transporters, receptors, or uptake and storage of neurotransmitters. For instance, measurement of putamen ^{18}F-DOPA-derived radioactivity enables assessment of dopaminergic innervation.

SYMPATHETIC NEUROIMAGING METHODS AND AGENTS

Almost all sympathetic neuroimaging to date has involved visualization of noradrenergic innervation in the left ventricular myocardium. This is because the heart is a relatively large, solid, homogeneous organ that possesses a high concentration of sympathetic nerve fibers, which surround the myocardial cells. Cardiac sympathetic images therefore resemble myocardial perfusion scans. Other thoracic organs and tissues possess much lower concentrations of sympathetic nerves.

SPECT scanning after injection of the sympathomimetic amine ^{123}I-metaiodobenzylguanidine (^{123}I-MIBG) constitutes by far the most commonly used means worldwide to assess cardiac sympathetic innervation. The principle underlying ^{123}I-MIBG scanning (as well as sympathetic neuroimaging by PET imaging agents) is not binding of the tracer-labeled compound to receptors on target cells but physical translocation into the

*Correspondence to: Dr. David S. Goldstein, NINDS, NIH, 10 Center Drive MSC-1620, 9000 Rockville Pike, Building 10 Room 5N220, Bethesda, MD 20892-1620, USA. iPhone: +1-301-675-1110, Tel: +1-301-496-2103, Fax: +1-301-402-0180, E-mail: goldsteind@ninds.nih.gov

sympathetic nerves via the cell membrane norepinephrine transporter and then subsequent active transport of the cytosolic compound into vesicles via the vesicular monoamine transporter. Thus, sympathetic neuroimaging depends on radiolabeling of vesicles in sympathetic nerves.

Cardiac sympathetic neuroimaging by ^{123}I-MIBG scanning has been used extensively in Japan, to a lesser extent in Europe, and hardly at all in the US. The dearth of American experience seems to be the result of a vicious cycle involving lack of insurance coverage because of lack of clinical trials because of lack of availability because of lack of insurance coverage. ^{123}I-MIBG scanning is available at many centers in the US for diagnostic evaluation of pheochromocytoma, since third party payers generally cover this application.

^{123}I-MIBG, an analog of the sympathomimetic amine guanethidine, differs structurally from norepinephrine, the neurotransmitter that mediates sympathetic regulation of the cardiovascular system. Injected MIBG undergoes a different fate from catecholamines such as norepinephrine. These differences must be considered when interpreting changes in ^{123}I-MIBG-derived radioactivity in terms of sympathetic neuronal function. First, MIBG is not as avidly or selectively removed by sympathetic nerves as are catecholamines. Second, MIBG that is taken up into the sympathoneural axoplasm is not a substrate for monoamine oxidase, the main enzyme metabolizing cytosolic catecholamines in sympathetic nerves. Third, MIBG is not a substrate for catechol-*O*-methyltransferase, a major enzyme metabolizing catecholamines in non-neuronal cells. The metabolic fate of MIBG remains poorly understood.

Since ^{123}I is not a positron-generating isotope, the usual method for detecting ^{123}I-MIBG-derived radioactivity is SPECT, not PET scanning. After intravenous injection of ^{123}I-MIBG, a relatively long time is required for obtaining interpretable images. ^{123}I-MIBG "uptake" is measured about 15–30 minutes after injection of the tracer. Since by then myocardial uptake of tracer-labeled catecholamines is already complete, ^{123}I-MIBG-derived radioactivity in the "uptake" phase actually probably reflects a complex combination of transmembrane neuronal uptake, intraneuronal vesicular uptake, loss of radioactivity via exocytosis, and loss of radioactivity from vesicular leakage back into the cytosol and reverse transport from the cytosol into the extracellular fluid. ^{123}I-MIBG "washout" is measured 3–4 hours after tracer injection. Radioactivity concentrations in SPECT scans are difficult to calibrate, and myocardial concentrations of ^{123}I-MIBG-derived radioactivity are usually expressed in terms of heart:mediastinum (H:M) ratios. In patients with coronary arterial or microvascular narrowing, as in atherosclerotic cardiovascular disease and diabetes mellitus, low H:M ratios of ^{123}I-MIBG-derived radioactivity might reflect decreased delivery of the imaging agent by blood perfusion; ^{123}I-MIBG SPECT studies rarely take myocardial perfusion into account.

Based on H:M ratios of ^{123}I-MIBG-derived radioactivity, decreased uptake, increased washout, or both have been reported in many disorders, including cardiovascular disorders (congestive heart failure, myocardial infarction, ventricular arrhythmias, dilated cardiomyopathy, Chagas disease, essential hypertension with left ventricular hypertrophy, Brugada syndrome), neurodegeneration (Parkinson disease, pure autonomic failure, Lewy body dementia), emotional distress, liver or kidney failure, and hypothyroidism. In general, attenuated uptake and accelerated washout are associated with relatively poor prognosis (Fanciulli et al., 2013; Kinbara et al., 2013). This pattern may reflect decreased vascular delivery of ^{123}I-MIBG, increased cardiac sympathetic nerve traffic, decreased vesicular sequestration of cytoplasmic amines, or decreased activity of the cell membrane norepinephrine transporter abnormalities that may coexist.

PET scanning offers potential advantages over SPECT scanning. Spatial resolution is better, the amount of injected radioactivity is smaller, radioactivity concentrations in tissues can be measured in absolute terms, and analyses of curves relating tissue radioactivity detected by PET scanning over time (time–activity curves) can yield information about not only innervation but also function. Several agents have been developed for sympathetic neuroimaging by PET scanning. These include ^{11}C-epinephrine, ^{18}F-dopamine, ^{11}C-hydroxyephedrine, and ^{18}F-metaraminol. ^{11}C-Epinephrine and ^{18}F-dopamine are catecholamines, whereas ^{11}C-hydroxyephedrine and ^{18}F-metaraminol are noncatecholamine sympathomimetic amines. All are investigational agents, and none has been approved yet for clinical diagnostic purposes.

Although ^{18}F-DOPA, which is used to assess striatal dopaminergic innervation in neurodegenerative disorders such as Parkinson disease, is converted to ^{18}F-dopamine in sympathetic nerves, ^{18}F-DOPA is not a sympathetic neuroimaging agent. Structurally, ^{18}F-DOPA is a neutral amino acid, not a catecholamine, and so ^{18}F-DOPA is taken up by all cells via the neutral amino acid transporter. ^{18}F-DOPA in non-neuronal cells may be converted to ^{18}F-dopamine by L-aromatic-amino-acid decarboxylase, but this is probably a minor fate compared to incorporation into peptides or proteins and enzymatic breakdown independent of catecholamine formation. After intravenous injection of ^{18}F-DOPA, plasma concentrations of ^{18}F-dopamine are too low to enable visualization of cardiac sympathetic innervation.

SYMPATHETIC NEUROIMAGING IN DYSAUTONOMIAS

The term, "dysautonomia," has been used to refer to conditions in which altered activity of one or more components of the autonomic nervous system adversely affects health. The following discussion focuses on dysautonomias that involve functional abnormalities or loss of sympathetic noradrenergic nerves.

Diabetic neuropathy

One of the first forms of dysautonomia to be evaluated by clinical sympathetic neuroimaging was diabetic autonomic neuropathy. In general, patients with diabetic autonomic neuropathy have decreased H:M ratios of ^{123}I-MIBG-derived radioactivity, especially in delayed images. A factor complicating interpretation of the results in terms of sympathetic denervation is that patients with diabetic autonomic neuropathy have a high frequency of coronary arterial and arteriolar narrowing, so that low H:M ratios ^{123}I-MIBG-derived radioactivity may reflect a combination of denervation and decreased delivery of the tracer to sympathetic nerves via coronary hypoperfusion. Analyses of ^{123}I-MIBG SPECT scans rarely adjust ^{123}I-MIBG-derived radioactivity for perfusion in the same regions of interest. ^{11}C-Hydroxyephedrine PET scanning, with perfusion assessed by $^{13}NH_3$ scanning in the same subjects, has revealed imaging evidence for heterogeneously decreased sympathetic innervation in relatively distal (apical, inferior, lateral) cardiac segments (Allman et al., 1993). A study of painful diabetic neuropathy reported decreased local concentrations of perfusion-adjusted ^{18}F-dopamine-derived radioactivity, along with decreased entry of norepinephrine into the venous drainage, indicating local sympathetic denervation (Tack et al., 2002).

Heart failure with risk of sudden death

Many studies have reported abnormalities of H:M ratios ^{123}I-MIBG-derived radioactivity in patients with heart failure. Treatments that successfully alleviate the failure generally improve the sympathetic neuroimaging abnormalities. Accelerated loss of ^{123}I-MIBG-derived radioactivity is associated with a worse prognosis (Kasama et al., 2008).

Patients with heart failure and low cardiac ejection fractions have an increased risk of sudden cardiac death. In such patients, a low H:M ratio of ^{123}I-MIBG-derived radioactivity at 4 hours after tracer injection is an independent, adverse prognostic factor (Jacobson et al., 2010). It is possible that ^{123}I-MIBG cardiac sympathetic neuroimaging might help identify a subgroup of heart failure patients at high risk for sudden death from ventricular fibrillation, in whom implantation of an cardioverter-defibrillator would be warranted.

Bilateral thoracic sympathectomies

Upper thoracic sympathectomy is used to treat several disorders – especially hyperhidrosis; however, sympathetic nerve fibers emanating from thoracic ganglia also innervate the heart. ^{18}F-Dopamine PET scanning was used to test whether unilateral or bilateral upper thoracic sympathectomy affects cardiac sympathetic innervation in humans (Moak et al., 2005). Low rates of entry of norepinephrine into the arm venous drainage verified upper limb sympathectomy. Patients with bilateral sympathectomy had low septal myocardial concentrations of ^{18}F-dopamine-derived radioactivity compared with normal volunteers but greater radioactivity than in patients with pure autonomic failure. Patients with unilateral sympathectomy had normal radioactivity. Bilateral upper thoracic sympathectomy therefore partly decreases cardiac sympathetic innervation density.

Complex regional pain syndrome

Patients with complex regional pain syndrome (formerly called reflex sympathetic dystrophy) have posttraumatic pain disproportionate to the injury and spreading beyond the distribution of any single peripheral nerve. Sympathetic neurocirculatory function and the role of sympathetic postganglionic nerve traffic in maintaining the pain was examined in patients with unilateral complex regional pain syndrome (Goldstein et al., 2000b). Most patients had had the condition for more than 1 year. Perfusion-adjusted ^{18}F-dopamine-derived radioactivity was if anything higher on the affected side. Therefore, chronic complex regional pain syndrome is not attended by evidence of local sympathetic denervation.

Chronic orthostatic intolerance

In postural tachycardia syndrome (often referred to as POTS) and repeated neurocardiogenic presyncope, chronic orthostatic intolerance occurs without persistent sympathetic neurocirculatory failure. Patients with postural tachycardia syndrome or neurocardiogenic presyncope underwent measurements of neurochemical indices of cardiac release, reuptake, and synthesis of norepinephrine based on entry of norepinephrine into the cardiac venous drainage (cardiac norepinephrine spillover) and measurement of left ventricular myocardial sympathetic innervation density using ^{18}F-dopamine PET scanning. Mean cardiac norepinephrine spillover was increased in POTS and decreased in neurocardiogenic presyncope, compared to healthy control subjects (Goldstein et al., 2002). Both patient groups had normal myocardial ^{18}F-dopamine-derived radioactivity. POTS and neurocardiogenic presyncope therefore differ in tonic cardiac sympathetic function, with increased cardiac norepinephrine release in the former and decreased release in the latter; however, both conditions involve normal density of myocardial sympathetic innervation. Because POTS and neurocardiogenic presyncope include specific abnormalities of cardiac sympathetic function even while at supine rest, both can be considered forms of dysautonomia.

Hypertrophic cardiomyopathy

Patients with hypertrophic cardiomyopathy often have clinical findings consistent with increased cardiac sympathetic outflow. The status of sympathetic innervation specifically in hypertrophic regions was examined by PET scanning using the perfusion imaging agent ^{13}N-ammonia and the sympathoneural imaging agent ^{18}F-dopamine in patients with hypertrophic cardiomyopathy and in healthy control subjects (Li et al., 2000). At all time points, the ^{18}F:^{13}N ratio was lower in hypertrophied than in nonhypertrophied regions, indicating decreased neuronal uptake of catecholamines in the hypertrophied myocardium. Decreased uptake could reflect local relative hypoinnervation, decreased numbers of neuronal uptake sites, or metabolic limitations on cell membrane transport. By enhancing norepinephrine delivery to adrenoceptors for a given amount of sympathetic nerve traffic, decreased neuronal uptake might help explain major clinical features of hypertrophic cardiomyopathy, including increased risk of sudden cardiac death.

Familial dysautonomia

Familial dysautonomia involves incomplete development of the sympathetic nervous system. Adult familial dysautonomia patients underwent thoracic ^{18}F-dopamine PET scanning, to determine the status of sympathetic innervation of the heart (Goldstein et al., 2008). The patients had decreased uptake and accelerated loss of ^{18}F-dopamine-derived radioactivity in the interventricular myocardial septum and ventricular free wall, compared to untreated controls. Desipramine-treated controls had decreased uptake but normal loss of radioactivity. Familial dysautonomia therefore seems to involve cardiac noradrenergic hypoinnervation. Since accelerated loss of ^{18}F-dopamine-derived radioactivity cannot be explained by decreased neuronal uptake alone, the disease may also involve augmented norepinephrine loss from extant terminals, either from increased exocytosis or decreased vesicular uptake.

ASSOCIATION OF CARDIAC NORADRENERGIC DENERVATION WITH LEWY BODY DISEASES

Lewy body diseases such as Parkinson disease, dementia with Lewy bodies, and pure autonomic failure are characterized by intraneuronal precipitates of the protein α-synuclein and are now therefore subsumed under the heading of synucleinopathies. Multiple system atrophy is also considered to be a form of synucleinopathy; however, in multiple system atrophy α-synuclein deposits are found in glial cells rather than neurons (Wakabayashi et al., 1998).

Lewy body synucleinopathies are associated with imaging evidence of substantial cardiac sympathetic denervation, both by ^{123}I-MIBG SPECT (Satoh et al., 1999) and by ^{11}C-hydroxyephedrine (Raffel et al., 2006) and ^{18}F-dopamine PET (Goldstein et al., 1997, 2000a) scanning. In these diseases, cardiac noradrenergic denervation has been confirmed by profoundly decreased tyrosine hydroxylase immunoreactivity in epicardial nerves (Orimo et al., 2002, 2006). In contrast, most (but not all) patients with multiple system atrophy have neuroimaging evidence for normal cardiac sympathetic innervation (Druschky et al., 2000; Braune, 2001; Orimo et al., 2001, 2007). Although imaging evidence for cardiac sympathetic denervation does not exclude multiple system atrophy, the finding of normal cardiac sympathetic innervation probably does exclude Parkinson disease with orthostatic hypotension.

Nonmotor findings associated with Parkinson disease, such as dementia, loss of sense of smell (anosmia), REM behavior disorder, baroreflex failure, and orthostatic hypotension, have all been reported to be associated with cardiac noradrenergic denervation (Orimo et al., 2005; Lee et al., 2006; Miyamoto et al., 2006; Yoshita et al., 2006; Goldstein et al., 2009; Kashihara et al., 2010).

When in the course of Parkinson disease does cardiac sympathetic denervation occur? Based on the concept proposed by Braak for the pathogenetic sequence (Braak et al., 2004), there is early deposition of α-synuclein in the olfactory bulb and autonomic nerves in the stomach, with subsequent ascending pathology in the autonomic ganglia, dorsal motor nucleus of the vagus in the caudal medulla, rostral ventrolateral medulla, pontine locus coeruleus, midbrain substantia nigra, and finally diffuse lesions in the cortex. This concept predicts that imaging evidence of cardiac sympathetic denervation may be a premotor biomarker of Parkinson disease. Although such evidence can precede the movement disorder by several years (Goldstein et al., 2007) and is apparent in at least some patients with *de novo* Parkinson disease (Oka et al., 2006, 2011) or incidental Lewy body disease (Orimo et al., 2008b), the frequency and consistency of this abnormality as an antecedent of parkinsonism have not yet been determined (Goldstein et al., 2011a).

Cardiac sympathetic neuroimaging and postmortem neuropathological findings have linked α-synucleinopathy with noradrenergic denervation in Lewy body diseases (Orimo et al., 2008b). Thus, patients with familial Parkinson disease from abnormalities of the gene encoding α-synuclein have cardiac sympathetic denervation (Goldstein et al., 2001; Singleton et al., 2004; Orimo et al., 2008a). The pathological changes seem to progress in a retrograde, centripetal manner.

Bases for the association of α-synucleinopathy with catecholaminergic denervation remain obscure. According to the "catecholaldehyde hypothesis," catecholaldehydes produced from enzymatic deamination of cytosolic catecholamines exert cytotoxic effects because of oxidative stress and oligomerization of α-synuclein (Burke et al., 2008; Panneton et al., 2010), resulting in deleterious positive feedback loops. Consistent with this view, postmortem putamen tissue from patients with Parkinson disease contains increased levels of dihydroxyphenylacetaldehyde, relative to dopamine (Goldstein et al., 2011b).

Especially because of the utility of cardiac sympathetic neuroimaging in distinguishing Parkinson disease from multiple system atrophy in patients with clinical evidence of central neurodegeneration and orthostatic hypotension, sympathetic neuroimaging seems a valuable addition to physiological, neuropharmacological, and neurochemical approaches in the diagnostic evaluation of selected patients with autonomic and neurodegenerative disorders (Goldstein and Sharabi, 2009).

References

Allman KC, Stevens MJ, Wieland DM et al. (1993). Noninvasive assessment of cardiac diabetic neuropathy by carbon-11 hydroxyephedrine and positron emission tomography. J Am Coll Cardiol 22: 1425–1432.

Braak H, Ghebremedhin E, Rub U et al. (2004). Stages in the development of Parkinson's disease-related pathology. Cell Tissue Res 318: 121–134.

Braune S (2001). The role of cardiac metaiodobenzylguanidine uptake in the differential diagnosis of parkinsonian syndromes. Clin Auton Res 11: 351–355.

Burke WJ, Kumar VB, Pandey N et al. (2008). Aggregation of alpha-synuclein by DOPAL, the monoamine oxidase metabolite of dopamine. Acta Neuropathol 115: 193–203.

Druschky A, Hilz MJ, Platsch G et al. (2000). Differentiation of Parkinson's disease and multiple system atrophy in early disease stages by means of I-123-MIBG-SPECT. J Neurol Sci 175: 3–12.

Fanciulli A, Strano S, Colosimo C et al. (2013). The potential prognostic role of cardiovascular autonomic failure in alpha-synucleinopathies. Eur J Neurol 20: 231–235.

Goldstein DS, Sharabi Y (2009). Neurogenic orthostatic hypotension: a pathophysiological approach. Circulation 119: 139–146.

Goldstein DS, Holmes C, Cannon RO 3rd et al. (1997). Sympathetic cardioneuropathy in dysautonomias. N Engl J Med 336: 696–702.

Goldstein DS, Holmes C, Li ST et al. (2000a). Cardiac sympathetic denervation in Parkinson disease. Ann Intern Med 133: 338–347.

Goldstein DS, Tack C, Li ST (2000b). Sympathetic innervation and function in reflex sympathetic dystrophy. Ann Neurol 48: 49–59.

Goldstein DS, Li ST, Kopin IJ (2001). Sympathetic neurocirculatory failure in Parkinson disease: evidence for an etiologic role of alpha-synuclein. Ann Intern Med 135: 1010–1011.

Goldstein DS, Holmes C, Frank SM et al. (2002). Cardiac sympathetic dysautonomia in chronic orthostatic intolerance syndromes. Circulation 106: 2358–2365.

Goldstein DS, Sharabi Y, Karp BI et al. (2007). Cardiac sympathetic denervation preceding motor signs in Parkinson disease. Clin Auton Res 17: 118–121.

Goldstein DS, Eldadah B, Sharabi Y et al. (2008). Cardiac sympathetic hypo-innervation in familial dysautonomia. Clin Auton Res 18: 115–119.

Goldstein DS, Sewell L, Holmes C (2009). Association of anosmia with autonomic failure in Parkinson disease. Neurology 74: 245–251.

Goldstein DS, Sewell L, Sharabi Y (2011a). Autonomic dysfunction in PD: a window to early detection? J Neurol Sci 310: 118–122.

Goldstein DS, Sullivan P, Holmes C et al. (2011b). Catechols in post-mortem brain of patients with Parkinson disease. Eur J Neurol 18: 703–710.

Jacobson AF, Senior R, Cerqueira MD et al. (2010). Myocardial iodine-123 meta-iodobenzylguanidine imaging and cardiac events in heart failure. Results of the prospective ADMIRE-HF (AdreView Myocardial Imaging for Risk Evaluation in Heart Failure) study. J Am Coll Cardiol 55: 2212–2221.

Kasama S, Toyama T, Sumino H et al. (2008). Prognostic value of serial cardiac ^{123}I-MIBG imaging in patients with stabilized chronic heart failure and reduced left ventricular ejection fraction. J Nucl Med 49: 907–914.

Kashihara K, Imamura T, Shinya T (2010). Cardiac ^{123}I-MIBG uptake is reduced more markedly in patients with REM sleep behavior disorder than in those with early stage Parkinson's disease. Parkinsonism Relat Disord 16: 252–255.

Kinbara T, Hayano T, Otani N et al. (2013). Iodine-123 metaiodobenzylguanidine imaging can predict future cardiac events in Japanese patients with Parkinson's disease. Ann Nucl Med 27: 123–131.

Lee PH, Yeo SH, Kim HJ et al. (2006). Correlation between cardiac ^{123}I-MIBG and odor identification in patients with Parkinson's disease and multiple system atrophy. Mov Disord 21: 1975–1977.

Li ST, Tack CJ, Fananapazir L et al. (2000). Myocardial perfusion and sympathetic innervation in patients with hypertrophic cardiomyopathy. J Am Coll Cardiol 35: 1867–1873.

Miyamoto T, Miyamoto M, Inoue Y et al. (2006). Reduced cardiac ^{123}I-MIBG scintigraphy in idiopathic REM sleep behavior disorder. Neurology 67: 2236–2238.

Moak JP, Eldadah B, Holmes C et al. (2005). Partial cardiac sympathetic denervation after bilateral thoracic sympathectomy in humans. Heart Rhythm 2: 602–609.

Oka H, Mochio S, Onouchi K et al. (2006). Cardiovascular dysautonomia in de novo Parkinson's disease. J Neurol Sci 241: 59–65.

Oka H, Toyoda C, Yogo M et al. (2011). Cardiovascular dysautonomia in de novo Parkinson's disease without orthostatic hypotension. Eur J Neurol 18: 286–292.

Orimo S, Ozawa E, Oka T et al. (2001). Different histopathology accounting for a decrease in myocardial MIBG uptake in PD and MSA. Neurology 57: 1140–1141.

Orimo S, Oka T, Miura H et al. (2002). Sympathetic cardiac denervation in Parkinson's disease and pure autonomic failure but not in multiple system atrophy. J Neurol Neurosurg Psychiatry 73: 776–777.

Orimo S, Amino T, Itoh Y et al. (2005). Cardiac sympathetic denervation precedes neuronal loss in the sympathetic ganglia in Lewy body disease. Acta Neuropathol (Berl) 109: 583–588.

Orimo S, Amino T, Takahashi A et al. (2006). Cardiac sympathetic denervation in Lewy body disease. Parkinsonism Relat Disord 12 (Suppl 2): S99–S105.

Orimo S, Kanazawa T, Nakamura A et al. (2007). Degeneration of cardiac sympathetic nerve can occur in multiple system atrophy. Acta Neuropathol (Berl) 113: 81–86.

Orimo S, Uchihara T, Nakamura A et al. (2008a). Cardiac sympathetic denervation in Parkinson's disease linked to SNCA duplication. Acta Neuropathol (Berl) 116: 575–577.

Orimo S, Uchihara T, Nakamura A et al. (2008b). Axonal alpha-synuclein aggregates herald centripetal degeneration of cardiac sympathetic nerve in Parkinson's disease. Brain 131: 642–650.

Panneton WM, Kumar VB, Gan Q et al. (2010). The neurotoxicity of DOPAL: behavioral and stereological evidence for its role in Parkinson disease pathogenesis. PLoS One 5: e15251.

Raffel DM, Koeppe RA, Little R et al. (2006). PET measurement of cardiac and nigrostriatal denervation in parkinsonian syndromes. J Nucl Med 47: 1769–1777.

Satoh A, Serita T, Seto M et al. (1999). Loss of ^{123}I-MIBG uptake by the heart in Parkinson's disease: assessment of cardiac sympathetic denervation and diagnostic value. J Nucl Med 40: 371–375.

Singleton A, Gwinn-Hardy K, Sharabi Y et al. (2004). Association between cardiac denervation and parkinsonism caused by alpha-synuclein gene triplication. Brain 127: 768–772.

Tack CJ, van Gurp PJ, Holmes C et al. (2002). Local sympathetic denervation in painful diabetic neuropathy. Diabetes 51: 3545–3553.

Wakabayashi K, Yoshimoto M, Tsuji S et al. (1998). Alpha-synuclein immunoreactivity in glial cytoplasmic inclusions in multiple system atrophy. Neurosci Lett 249: 180–182.

Wieland DM, Brown LE, Rogers WL et al. (1981). Myocardial imaging with a radioiodinated norepinephrine storage analog. J Nucl Med 22: 22–31.

Yoshita M, Taki J, Yokoyama K et al. (2006). Value of ^{123}I-MIBG radioactivity in the differential diagnosis of DLB from AD. Neurology 66: 1850–1854.

Handbook of Clinical Neurology, Vol. 117 (3rd series)
Autonomic Nervous System
R.M. Buijs and D.F. Swaab, Editors

Chapter 30

Skin biopsies in the assessment of the autonomic nervous system

NINGSHAN WANG AND CHRISTOPHER H. GIBBONS*
Center for Autonomic and Peripheral Nerve Disorders, Beth Israel Deaconess Medical Center, Boston, MA, USA

INTRODUCTION

A host of neurophysiological techniques have been developed over the last century to measure autonomic nervous system function (Low, 2003). These methodologies were developed, in part, because of the general inaccessibility of the autonomic nervous system to direct measurement. Autonomic testing, with the exception of microneurography, indirectly measures functional responses to pharmacological or physiological provocation (Hilz and Dutsch, 2006). Similar challenges have been faced in the evaluation of the peripheral nervous system, where functional testing has predominated in the clinical investigation of disease. Structural assessments of peripheral nerve fibers are typically obtained through sural nerve biopsies, a relatively invasive technique (Bosboom et al., 2001; Jann et al., 2003; Malik et al., 2005; Bennett et al., 2008). Sural nerve biopsies are limited both by their location (they are constrained to a single anatomical site) and their repeatability (only a single biopsy from each side is traditionally performed).

In the 1990s, a technique to measure cutaneous peripheral nerve fibers was developed. This technique, utilizing a standard dermatological punch skin biopsy, was combined with immunoflorescent staining of nerve fibers using the panaxonal marker protein gene product 9.5 (PGP 9.5) (McCarthy et al., 1995; Kennedy and Wendelschafer-Crabb, 1996). Epidermal nerve fibers can be easily identified and quantified (Lauria et al., 2010b). The direct clinical applicability of this technique was quickly recognized and several laboratories began to publish reports of skin biopsy findings across various disease states. Since then, the rapid dissemination of this technique has resulted in the widespread utilization of skin biopsy in the evaluation of small fiber neuropathy.

During this period of time there was interest in the study of autonomic innervation of the skin, largely pioneered by Dr. William Kennedy. Kennedy developed several confocal immunoflorescent staining protocols to visualize sweat glands, blood vessels, hair follicles, and other dermal structures (Kennedy et al., 1994). Over the past decade, the use of skin biopsies has expanded from the original study of small unmyelinated sensory nerve fibers to widespread investigation of cutaneous autonomic nerve fibers in a variety of disease states (Dabby et al., 2006; Nolano et al., 2006, 2010; Gibbons et al., 2009, 2010a).

The skin provides a unique window into the autonomic nervous system. Peripheral adrenergic and cholinergic fibers innervate a number of cutaneous structures and can easily be sampled through punch skin biopsies. Skin biopsies allow for both regional sampling, in diseases with patchy distribution, and the opportunity for repeated sampling in progressive disorders.

ANATOMY OF THE SKIN

A major demarcation in the anatomy of the skin is the dermal–epidermal junction. Above the basement membrane, within the epidermal layer, unmyelinated nociceptive C fibers extend (Fig. 30.1). Autonomic nerve fibers are not found within the epidermal layer. Just below the basement membrane run bundles of nerve fibers, described as the subdermal plexus. Deeper within the dermal tissue reside a number of adnexal structures including hair follicles, arrector pili muscles, sweat glands, and blood vessels (Fig. 30.1).

*Correspondence to: Christopher H. Gibbons, M.D., M.M.Sc., Center for Autonomic and Peripheral Nerve Disorders, Beth Israel Deaconess Medical Center, 1 Deaconess Road, Palmer 111, Boston, MA 02215, USA. Tel: +1-617-632-8454, Fax: +1-617-632-0852, E-mail: cgibbons@bidmc.harvard.edu

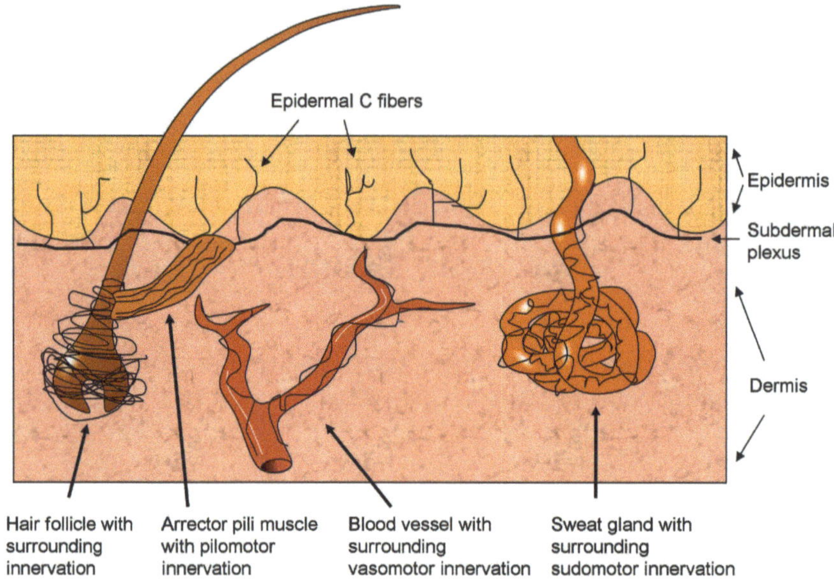

Fig. 30.1. Anatomy of the skin. Representative dermal structures are shown with their accompanying innervation. Structures are not drawn to scale.

Hair follicles

Throughout life, hair follicles undergo continuous cycles of growth and repair. This has resulted in recent interest in hair follicles as a primary source of stem cells (Cotsarelis et al., 1999; Webb et al., 2004; Kaur, 2006; Webb and Kaur, 2006; Abbas and Mahalingam, 2009). The stem cells ensure epidermal growth, but also facilitate epidermal repair. Regions of epidermis between hair follicles, known as the intrafollicular epidermis, continuously shed terminally differentiated cells that are replaced throughout life (Webb and Kaur, 2006; Abbas and Mahalingam, 2009). Specific niches of stem cells are sequestered in regions of the hair follicle; the bulge region is the best characterized, although additional stem cells have been identified in the interfollicular epidermis and even within the sebaceous gland (Kaur, 2006; Tiede et al., 2007). Stem cells have provided fruitful discussion for treatment of a number of diseases, but to date, no practical applications have successfully targeted the autonomic nervous system.

Hair follicles extend from the deep dermal tissue through the epidermal layer before extending beyond the epithelial layer. Hair follicles are anchored to the dermal tissue by arrector pili muscles and sebaceous glands. Hair follicles also have large numbers of sensory fibers that surround the base of the follicle and extend up the shaft of the hair to the epidermal surface, providing feedback for local stimulation, as shown in Figure 30.2. Autonomic nerve fibers, primarily sympathetic cholinergic with some adrenergic as well, also provide innervation to the base of the hair follicle (Botchkarev et al., 1999). Sensory nerve fibers outnumber autonomic nerve fibers in the total innervation of the hair follicle.

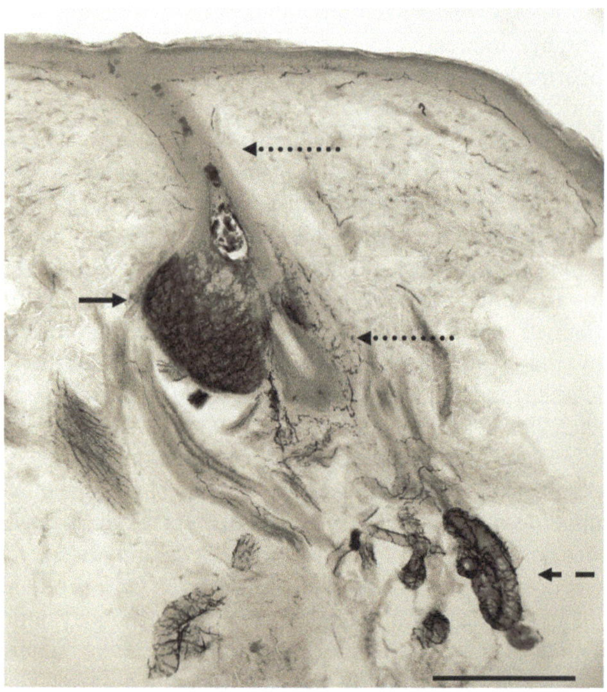

Fig. 30.2. Hair follicle. A light microscope image of a hair follicle is shown (dotted arrows), with an attached sebaceous gland (solid arrow). Note the network of sensory nerve fibers shown surrounding the base of the hair follicle. A sweat gland is seen in the deeper dermal tissue (dashed arrow). The unbroken line at the bottom is the scale bar, and is 500 μm.

Sweat glands

Sweat glands are located 3–5 mm below the epidermal layer, and are made up of sweat gland tubules that are intertwined with capillaries and nerve fibers (Fig. 30.3A–C).

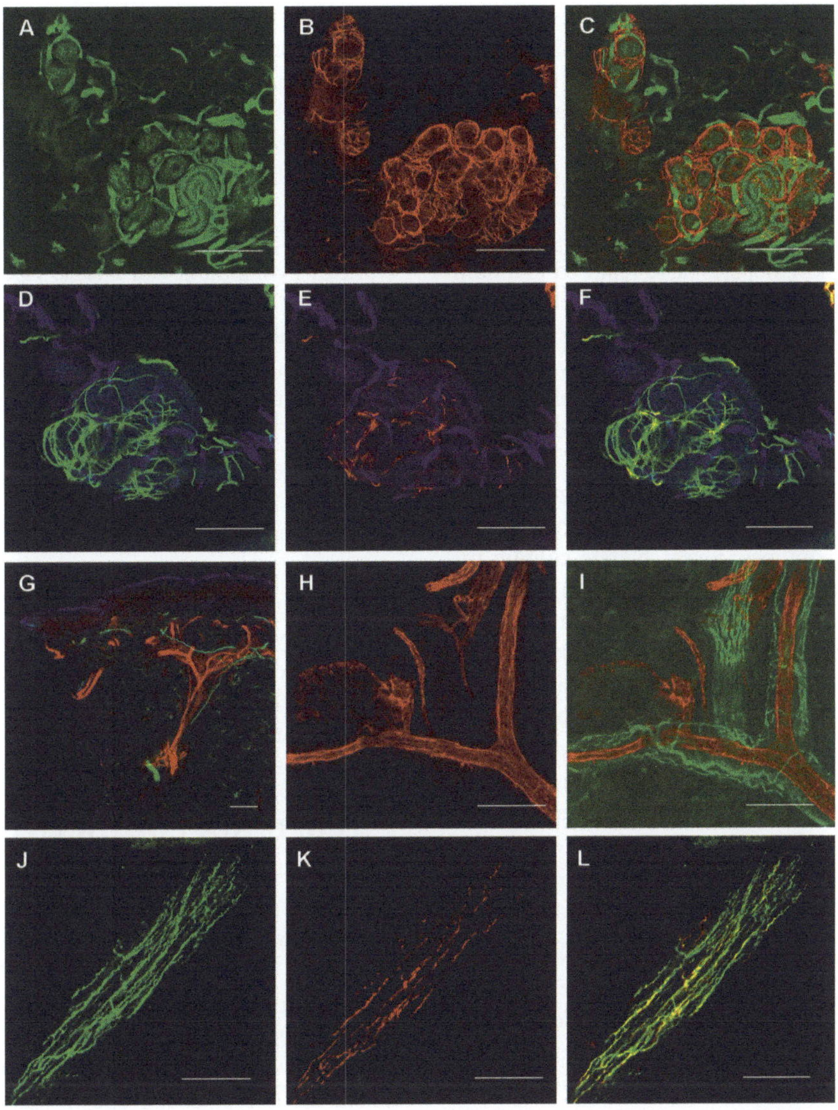

Fig. 30.3. Cutaneous innervation. Various dermal structures are shown, with their specific autonomic innervation highlighted. (**A–C**) A sweat gland. In (**A**), the green stain is CD31, an endothelial marker that is highlighting the blood vessels; there is also some background stain picked up by the sweat glands tubules. In (**B**), the red stain is from PGP 9.5, a panaxonal marker, showing the innervation around the sweat gland. In (**C**), the merged picture highlighting both the innervation and the vasculature is shown. (**D–F**) A sweat gland. In (**D**), the blue stain is CD31 (blood vessels) while the green stain is vasoactive intestinal peptide (VIP, stains for sympathetic cholinergic fibers). In (**E**), the blue stain is CD31 (blood vessels) while the red stain is tyrosine hydroxylase (TH, stains for sympathetic adrenergic fibers). (**F**) Shows the merged picture with sympathetic adrenergic and cholinergic fibers seen surrounding a sweat gland and accompanying blood vessels. (**G–I**) Cutaneous blood vessels. In (**G**), the red stain is CD31 (blood vessels), the green is PGP 9.5 (all nerve fibers). In (**H**), the red stain is CD31 highlighting the dermal capillaries at a magnified view. In (**I**), the innervation (PGP 9.5) is shown with the accompanying blood vessels. (**J–L**) Arrector pilorum muscle. In (**J**), the sympathetic adrenergic fibers (stained with TH) are shown in an arrector pili muscle. In (**K**), the sympathetic cholinergic innervation (stained with VIP) is shown, and is much less than the sympathetic adrenergic innervation. In (**L**), the merged image showing both cholinergic and adrenergic innervation within an arrector pili muscle is shown. White scale bars at the bottom of each image indicate 100 μm.

(Gibbons et al., 2009, 2010a). Many tools used to assess the integrity of the peripheral autonomic nervous system investigate the sympathetic cholinergic, or sudomotor, response. The quantitative sudomotor axon reflex test (QSART) has become a widely used and reliable marker of peripheral autonomic and peripheral nerve dysfunction (Low et al., 1983). A number of other measures of sudomotor function have been developed, such as quantitative direct and indirect sweat testing (Gibbons et al., 2008), the dynamic sweat test (Provitera et al., 2010), and silicone impressions (Vilches and Navarro, 2002). Such widespread investigation of sudomotor function is, in part, secondary to reports that peripheral sudomotor dysfunction may be the most sensitive finding in patients with distal small fiber neuropathy (Low et al., 2006).

Although structural investigation of sudomotor fibers has existed for a number of years (Kennedy et al., 1994), recent advances in methodology and the widespread use of skin biopsies have resulted in a number of recent publications on the subject (Gibbons et al., 2009). Clinical skin biopsies stained with the panaxonal marker PGP 9.5, imaged by light microscopy, can highlight the network of innervation surrounding sweat glands, and may be reduced in diseases affecting the peripheral nervous system (Gibbons et al., 2009, 2010a). Despite the clinical utility of a panaxonal marker, such as PGP 9.5, it cannot differentiate sensory and autonomic nerve fiber subtypes. Sweat glands contain primarily sympathetic cholinergic innervation, although some sympathetic adrenergic fibers are noted as well (Fig. 30.3D–F). There are also dense networks of capillaries within sweat glands that are accompanied by their own innervation (Fig. 30.3A–C). The use of multiple immunoflorescent stains that can identify nerve fiber subtypes with confocal imaging may play an increasing role in diagnosing specific diseases and defining disease severity.

Blood vessels

Blood vessels of various sizes course throughout the dermis layer, but they are not located within the epidermal layer (Fig. 30.3G–I). Due to the relative complexity of the vasomotor plexus, few data exist about quantitation of vasomotor nerve fibers (Rayman et al., 1995). A recent report describes a method to trace the length of nerve fibers associated with a given length of capillary and reports a result as a vasomotor density (Gibbons et al., 2010b). The phenotypic expression of vasomotor nerve fibers has not been fully elucidated. Pathological evidence suggests that sensory, sympathetic, and parasympathetic fibers are expressed within cutaneous blood vessels (Ruocco et al., 2002; Ramien et al., 2004). However, physiological evidence suggests that cholinergic transmission is responsible for cutaneous vasodilatation (Crandall et al., 2002; Shibasaki et al., 2002). There is little information about changes to vasomotor innervation in various disease states.

Arrector pili muscles

The arrector pili muscles extend from deeper dermal tissue to the intradermal shaft of a hair follicle (Fig. 30.3J–L). Upon contraction, the arrector pili muscle causes piloerection and formation of cutis anseri, or goose bumps. Compared to other dermal neural networks, arrector pili muscles contain relatively simple patterns of innervation. Nerve fibers run parallel to the muscle and can be easily identified through either light or confocal microscopy with the use of a panaxonal marker, such as PGP 9.5 (Nolano et al., 2010). Selective immunohistochemical markers, such as the sympathetic adrenergic markers dopamine β-hydroxylase or tyrosine hydroxylase, reveal a dense network of sympathetic adrenergic fibers running through the muscle (Gibbons et al., 2010b; Nolano et al., 2010). Arrector pili muscle innervation has not been well studied and only a single recent publication has attempted to quantify the density of arrector pili muscles in healthy and disease states (Nolano et al., 2010).

Limitations

Skin biopsies have substantial evidence supporting their use in measurement of intraepidermal nerve fiber density, but they do not yet have the same degree of evidence in the analysis of autonomic nerve fibers. Skin biopsy is unable to differentiate the various etiologies that may cause the specific sensory or autonomic neuropathy. In addition, skin biopsy would not replace the use of nerve biopsy in cases of suspected vasculitis.

IMMUNOFLORESCENT STAINING OF CUTANEOUS TISSUE

A number of different methodologies have been developed to immunohistochemically stain skin biopsies for evaluation of nerve fiber density. Typically, a 3 mm punch skin biopsy is obtained from the area of interest. The most extensive reported data exist for the distal leg (10 cm above the lateral maleolus), the distal thigh (10 cm above the lateral knee), and the proximal thigh (10 cm below the hip) (Polydefkis et al., 2001; Gibbons et al., 2006; Lauria et al., 2010a). In cases where symptoms are asymmetric, samples may be taken from the affected region and the contralateral (unaffected) side for comparison. Tissue studied by light microscopy is typically stained with the panaxonal marker PGP 9.5. For confocal images, tissue is often stained with a number of different cutaneous markers to highlight the various dermal structures (Perretti et al., 2003; Donadio et al., 2008; Di et al., 2010). Skin blisters are occasionally used to study epidermal nerve fiber densities, but this method separates the epidermal layer from the dermal layer, so does not allow the study of autonomic nerve fibers (Panoutsopoulou et al., 2009).

Most laboratories fix their tissue in a paraformaldehyde-lysine-periodate or Zamboni (paraformaldehyde, picric acid) solution (McCarthy et al., 1995; Lauria et al., 2010b). After fixation, tissue is stored in a cryoprotectant. Thick sections ranging from 50 μm to 100 μm are cut perpendicular to the epidermis to visualize both dermis and epidermis, using a freezing sliding microtome. For epidermal nerve fibers, staining of only a few tissue sections is typically necessary. For adequate sampling of autonomic nerve fibers, which may be sequestered within sweat glands and not evenly distributed throughout a tissue sample, greater numbers of sections are required (Gibbons et al., 2009, 2010a).

QUANTIFICATION OF AUTONOMIC NERVE FIBERS

Sweat glands contain some of the most complex innervation of any of the dermal structures (Kennedy et al., 1994). Unfortunately, this convoluted wrapping of nerve fibers around sweat gland tubules prohibits use of standard counting techniques. Although descriptive semiquantitative analysis of sudomotor density has been reported, our own analysis of this method has not suggested a high degree of reliability or reproducibility, although others have reported more consistent findings (Dabby et al., 2006, 2007; Gibbons et al., 2010a). We have recently reported both a manual unbiased stereological approach and a rapid automated technique for quantitation of sudomotor fibers (Gibbons et al., 2009, 2010a). Both techniques are described using light microscope images and the panaxonal marker PGP 9.5 (although they can be easily adapted to use of confocal images). The manual stereological approach utilizes a grid of circles placed over the image of a sweat gland to count the number of nerve fibers that intersect the grid of circles. Results are reported as the intercept density (number of circles containing nerve fibers/total number of circles). This method has high inter- and intraexaminer correlation coefficients, is reduced in patients with diabetic peripheral neuropathy, and correlates with quantified physical examination scores (Gibbons et al., 2009). The drawback to this approach is that it is very time consuming and labor-intensive. The automated sudomotor nerve fiber density assessment utilizes the same image as the manual method but traces the nerve fibers using computer software to quantify the density of nerve fibers within the area of the sweat gland (Gibbons et al., 2010a). Both methods are able to differentiate groups of patients with diabetic neuropathy from healthy control subjects. The automated method is much faster, but has wider confidence intervals, thereby reducing the utility in clinical practice, but providing a valuable tool in large studies where time constraints may play a factor. To date, quantitation of sudomotor nerve fibers has been limited to healthy control subjects and patients with diabetic neuropathy. Ongoing investigation into other disease states will expand the utility of these techniques in the future.

Pilimotor nerve fibers, expressed within arrector pili muscles, have recently been reported using a novel method of quantitation (Nolano et al., 2010). The number of nerve fibers running in parallel with the length of the muscle is counted in cross-section. The number of nerve fibers/mm is reported as a result. The density of nerve fibers is reduced in diabetic subjects compared to controls using the panaxonal marker PGP 9.5, a sympathetic adrenergic marker dopamine β-hydroxylase, and the sympathetic cholinergic marker vasoactive intestinal peptide. The authors do report some selective loss of cholinergic or adrenergic nerve fiber innervation in patients with diabetes. Further study confirming these findings in patients with diabetes and other diseases is necessary before this technique achieves more widespread utilization.

At this time, no rigorous study of vasomotor innervation has yet been performed. A recent report of healthy subjects treated with capsaicin cream describes a reduction in the vasomotor density (Gibbons et al., 2010b). These findings have not been reproduced in patients with disease. There are no reports at this time of quantitation of innervation around hair follicles in human subjects.

SKIN BIOPSY EVALUATION OF SPECIFIC AUTONOMIC DISORDERS

One of the earliest descriptions of autonomic innervation highlights the age-related changes that occur in healthy elderly compared to young, healthy subjects (Abdel-Rahman et al., 1992). This study reported an age-related decline in sudomotor nerve fiber density and sudomotor function. The decline in innervation was associated with target organ (sweat gland) atrophy and a decrease in functional sweat production. These findings highlight the initial foray into the structural–functional aspects of cutaneous autonomic innervation in human subjects.

Kennedy expanded the utility of cutaneous skin biopsies by using a number of different immunohistochemical stains in combination with confocal microscopy to capture the complexity of cutaneous autonomic innervation. He described the first laser scanning confocal microscopy images of sweat glands and examined the complex innervation in healthy subjects (Kennedy et al., 1994).

One of the first publications to utilize immunoflorescent imaging of autonomic nerve fibers highlighted the clinical utility of skin biopsy by differentiating subjects with Ross syndrome and Holmes–Adie syndrome (Nolano et al., 2006). In patients with Ross syndrome there was a complete absence of sympathetic cholinergic innervation in anhidrotic skin (as measured by vasoactive intestinal peptide), with only rare sympathetic cholinergic fibers seen in hyperhidrotic skin. There was also an absence of sympathetic cholinergic innervation within blood vessels, hair follicles, and arrector pilorum muscles in both anhidrotic and hyperhidrotic skin in the patients with Ross syndrome. In contrast, patients with Holmes–Adie syndrome had numerous structural abnormalities to sympathetic cholinergic nerves, but had relatively preserved sympathetic cholinergic nerve fiber density (Nolano et al., 2006). This study reported very similar clinical results between Ross and Holmes–Adie syndromes when assessed by thermoregulatory sweat testing, suggesting the potential for clinical overlap and confusion. Thus the potential value of skin biopsies was highlighted as a diagnostic tool in diseases affecting the autonomic nervous system.

Additional studies have evaluated sudomotor nerve fiber density in patients with multiple system atrophy and Parkinson's disease (Dabby et al., 2006; Donadio et al., 2006, 2008; Nolano et al., 2008). These neurodegenerative diseases have significant clinical overlap, but different clinical courses and outcomes. Both disorders may display prominent autonomic dysfunction, although it is presumed that subjects with multiple system atrophy experience autonomic dysfunction from preganglionic dysfunction, while patients with Parkinson's disease experience autonomic dysfunction from peripheral nerve damage (Goldstein, 2003, 2004). Reports of skin biopsies in these two populations support this theory; patients with Parkinson's disease appear to have peripheral autonomic nerve fiber density reduction, while those with multiple system atrophy exhibit normal peripheral sudomotor innervation (Dabby et al., 2006; Donadio et al., 2008).

Cold-induced sweating type 1 (CISS1) is a very rare disorder where profuse sweating is seen in cold environments and is associated with mutations to cytokine receptor-like factor 1 (Di et al., 2010). A recent case report describes generalized anhidrosis by thermoregulatory sweat testing with absent sudomotor cholinergic innervation and increased sudomotor adrenergic innervation, suggesting a failure of the embryological phenotypic transition of sudomotor fibers from adrenergic to cholinergic. This simple report underscores the potential utility of skin biopsy in defining the structural underpinnings of disease.

In patients with familial dysautonomia, there is normal sweat production as measured by quantitative sudomotor axon reflex testing (Bickel et al., 2004). Skin biopsy evaluation of the same individuals revealed normal numbers of sweat glands, although many were atrophic in appearance (Hilz et al., 2004). The density of innervation around the sweat glands was reduced by more than 50% compared to control subjects, despite the normal sweat response to acetylcholine iontophoresis. The reason for the discrepancy between measures of sudomotor structure and function in familial dysautonomia is not known.

The use of skin biopsies in the evaluation of peripheral autonomic function has been most widely examined in patients with diabetes. Attempts to quantifying both sudomotor and pilomotor nerve fiber density have occurred in healthy subjects and in patients with diabetes (Gibbons et al., 2009, 2010a; Nolano et al., 2010). These studies show a length-dependent pattern to autonomic nerve fiber damage, with the distal legs exhibiting the earliest and most severely reduced nerve fiber densities. These studies have consistently revealed a decrease in sudomotor and pilomotor nerve fiber density in patients with diabetes, and the density decline parallels neuropathy progression. These techniques supplement existing analysis of intraepidermal nerve fiber density by identifying specific individuals with isolated or predominant damage to a particular nerve fiber subtype, be that sensory or autonomic.

ADDITIONAL TECHNIQUES

The use of catecholamine histofluorescence has been reported in a number of animal studies (Crivellato et al., 1990; Rao et al., 1994; Landis, 1999; Tian et al., 2000). Tissue is typically homogenized in order to quantitate cutaneous catecholamines, and may provide interesting avenues of investigation in diseases of disordered catecholamine expression and production. This method, although highly sensitive to catecholamine expression, does not allow localization of catecholamine expression to specific adnexal structures. Catecholamine histofluorescence can also be applied to sectioned tissue, and when combined with specific staining of nerve fibers in adjacent tissue sections, may provide insight into mechanisms of human disease. Further study in both humans and animals is necessary to determine the potential utility of catecholamine histofluorescence.

SUMMARY

Skin biopsies provide a unique window into the autonomic nervous system. Easily accessible nerve fibers can be sampled in different regions and over time to directly investigate structural changes to autonomic nerve fibers in health and disease. Skin biopsies are easy to perform, although they may be technically challenging to process. The number of academic centers performing skin biopsies is increasing, and some now investigate autonomic innervation as well as sensory nerve fiber density. Despite these advances, there are still few published data in many diseases affecting autonomic nervous system function, including those that are acquired, hereditary, neurodegenerative, or autoimmune. In addition, further standardization of techniques is necessary for results to be comparable across centers and a greater quantity of normative data for control subjects is necessary to adequately describe the range of expression in healthy individuals. Ongoing investigation into the use of stem cells from cutaneous tissue will continue to expand the utility of skin biopsies, and may some day play a role in developing novel treatments for patients with autonomic disorders.

References

Abbas O, Mahalingam M (2009). Epidermal stem cells: practical perspectives and potential uses. Br J Dermatol 161: 228–236.

Abdel-Rahman TA, Collins KJ, Cowen T et al. (1992). Immunohistochemical, morphological and functional changes in the peripheral sudomotor neuro-effector system in elderly people. J Auton Nerv Syst 37: 187–197.

Bennett DL, Groves M, Blake J et al. (2008). The use of nerve and muscle biopsy in the diagnosis of vasculitis: a 5 year retrospective study. J Neurol Neurosurg Psychiatry 79: 1376–1381.

Bickel A, Axelrod FB, Marthol H et al. (2004). Sudomotor function in familial dysautonomia. J Neurol Neurosurg Psychiatry 75: 275–279.

Bosboom WM, van den Berg LH, Franssen H et al. (2001). Diagnostic value of sural nerve demyelination in chronic inflammatory demyelinating polyneuropathy. Brain 124: 2427–2438.

Botchkarev VA, Peters EM, Botchkareva NV et al. (1999). Hair cycle-dependent changes in adrenergic skin innervation, and hair growth modulation by adrenergic drugs. J Invest Dermatol 113: 878–887.

Cotsarelis G, Kaur P, Dhouailly D et al. (1999). Epithelial stem cells in the skin: definition, markers, localization and functions. Exp Dermatol 8: 80–88.

Crandall CG, Shibasaki M, Yen TC (2002). Evidence that the human cutaneous venoarteriolar response is not mediated by adrenergic mechanisms. J Physiol 538: 599–605.

Crivellato E, Damiani D, Mallardi F (1990). Comparison between the L-DOPA histofluorescence procedure and the indirect immunofluorescence with anti-T6 and -HLA-DR monoclonal antibodies in visualizing Langerhans cells of human epidermis. Acta Histochem 88: 59–64.

Dabby R, Djaldetti R, Shahmurov M et al. (2006). Skin biopsy for assessment of autonomic denervation in Parkinson's disease. J Neural Transm 113: 1169–1176.

Dabby R, Vaknine H, Gilad R et al. (2007). Evaluation of cutaneous autonomic innervation in idiopathic sensory small-fiber neuropathy. J Peripher Nerv Syst 12: 98–101.

Di LR, Nolano M, Boman H et al. (2010). Central and peripheral autonomic failure in cold-induced sweating syndrome type 1. Neurology 75: 1567–1569.

Donadio V, Nolano M, Provitera V et al. (2006). Skin sympathetic adrenergic innervation: an immunofluorescence confocal study. Ann Neurol 59: 376–381.

Donadio V, Nolano M, Elam M et al. (2008). Anhidrosis in multiple system atrophy: a preganglionic sudomotor dysfunction? Mov Disord 23: 885–888.

Gibbons CH, Griffin JW, Polydefkis M et al. (2006). The utility of skin biopsy for prediction of progression in suspected small fiber neuropathy. Neurology 66: 256–258.

Gibbons CH, Illigens BM, Centi J et al. (2008). QDIRT: quantitative direct and indirect test of sudomotor function. Neurology 70: 2299–2304.

Gibbons CH, Illigens BM, Wang N et al. (2009). Quantification of sweat gland innervation: a clinical-pathologic correlation. Neurology 72: 1479–1486.

Gibbons CH, Illigens BM, Wang N et al. (2010a). Quantification of sudomotor innervation: a comparison of three methods. Muscle Nerve 42: 112–119.

Gibbons CH, Wang N, Freeman R (2010b). Capsaicin induces degeneration of cutaneous autonomic nerve fibers. Ann Neurol 68: 888–898.

Goldstein DS (2003). Imaging of the autonomic nervous system: focus on cardiac sympathetic innervation. Semin Neurol 23: 423–433.

Goldstein DS (2004). Functional neuroimaging of sympathetic innervation of the heart. Ann N Y Acad Sci 1018: 231–243.

Hilz MJ, Dutsch M (2006). Quantitative studies of autonomic function. Muscle Nerve 33: 6–20.

Hilz MJ, Axelrod FB, Bickel A et al. (2004). Assessing function and pathology in familial dysautonomia: assessment of temperature perception, sweating and cutaneous innervation. Brain 127: 2090–2098.

Jann S, Bramerio MA, Beretta S et al. (2003). Diagnostic value of sural nerve matrix metalloproteinase-9 in diabetic patients with CIDP. Neurology 61: 1607–1610.

Kaur P (2006). Interfollicular epidermal stem cells: identification, challenges, potential. J Invest Dermatol 126: 1450–1458.

Kennedy WR, Wendelschafer-Crabb G (1996). Utility of skin biopsy in diabetic neuropathy. (Review; 40 refs.). Semin Neurol 16: 163–171.

Kennedy WR, Wendelschafer-Crabb G, Brelje TC (1994). Innervation and vasculature of human sweat glands: an immunohistochemistry-laser scanning confocal fluorescence microscopy study. J Neurosci 14: 6825–6833.

Landis SC (1999). Development of muscarinic receptors and regulation of secretory responsiveness in rodent sweat glands. Life Sci 64: 381–385.

Lauria G, Bakkers M, Schmitz C et al. (2010a). Intraepidermal nerve fiber density at the distal leg: a worldwide normative reference study. J Peripher Nerv Syst 15: 202–207.

Lauria G, Hsieh ST, Johansson O et al. (2010b). European Federation of Neurological Societies/Peripheral Nerve Society Guideline on the use of skin biopsy in the diagnosis of small fiber neuropathy. Report of a joint task force of the European Federation of Neurological Societies and the Peripheral Nerve Society. Eur J Neurol 17: 903–909.

Low PA (2003). Testing the autonomic nervous system. Semin Neurol 23: 407–421.

Low PA, Caskey PE, Tuck RR et al. (1983). Quantitative sudomotor axon reflex test in normal and neuropathic subjects. Ann Neurol 14: 573–580.

Low VA, Sandroni P, Fealey RD et al. (2006). Detection of small-fiber neuropathy by sudomotor testing. Muscle Nerve 34: 57–61.

Malik RA, Tesfaye S, Newrick PG et al. (2005). Sural nerve pathology in diabetic patients with minimal but progressive neuropathy. Diabetologia 48: 578–585.

McCarthy BG, Hsieh ST, Stocks A et al. (1995). Cutaneous innervation in sensory neuropathies: evaluation by skin biopsy. Neurology 45: 1848–1855.

Nolano M, Provitera V, Perretti A et al. (2006). Ross syndrome: a rare or a misknown disorder of thermoregulation? A skin innervation study on 12 subjects. Brain 129: 2119–2131.

Nolano M, Provitera V, Estraneo A et al. (2008). Sensory deficit in Parkinson's disease: evidence of a cutaneous denervation. Brain 131: 1903–1911.

Nolano M, Provitera V, Caporaso G et al. (2010). Quantification of pilomotor nerves: a new tool to evaluate autonomic involvement in diabetes. Neurology 75: 1089–1097.

Panoutsopoulou IG, Wendelschafer-Crabb G, Hodges JS et al. (2009). Skin blister and skin biopsy to quantify epidermal nerves: a comparative study. Neurology 72: 1205–1210.

Perretti A, Nolano M, De Joanna G et al. (2003). Is Ross syndrome a dysautonomic disorder only? An electrophysiologic and histologic study. Clin Neurophysiol 114: 7–16.

Polydefkis M, Hauer P, Griffin JW et al. (2001). Skin biopsy as a tool to assess distal small fiber innervation in diabetic neuropathy. Diabetes Technol Ther 3: 23–28.

Provitera V, Nolano M, Caporaso G et al. (2010). Evaluation of sudomotor function in diabetes using the dynamic sweat test. Neurology 74: 50–56.

Ramien M, Ruocco I, Cuello AC et al. (2004). Parasympathetic nerve fibers invade the upper dermis following sensory denervation of the rat lower lip skin. J Comp Neurol 469: 83–95.

Rao MS, Jaszczak E, Landis SC (1994). Innervation of footpads of normal and mutant mice lacking sweat glands. J Comp Neurol 346: 613–625.

Rayman G, Malik RA, Sharma AK et al. (1995). Microvascular response to tissue injury and capillary ultrastructure in the foot skin of type I diabetic patients. Clin Sci (Colch) 89: 467–474.

Ruocco I, Cuello AC, Parent A et al. (2002). Skin blood vessels are simultaneously innervated by sensory, sympathetic, and parasympathetic fibers. J Comp Neurol 448: 323–336.

Shibasaki M, Wilson TE, Cui J et al. (2002). Acetylcholine released from cholinergic nerves contributes to cutaneous vasodilation during heat stress. J Appl Physiol 93: 1947–1951.

Tian H, Habecker B, Guidry G et al. (2000). Catecholamines are required for the acquisition of secretory responsiveness by sweat glands. J Neurosci 20: 7362–7369.

Tiede S, Kloepper JE, Bodo E et al. (2007). Hair follicle stem cells: walking the maze. Eur J Cell Biol 86: 355–376.

Vilches JJ, Navarro X (2002). New silicones for the evaluation of sudomotor function with the impression mold technique. Clin Auton Res 12: 20–23.

Webb A, Kaur P (2006). Epidermal stem cells. Front Biosci 11: 1031–1041.

Webb A, Li A, Kaur P (2004). Location and phenotype of human adult keratinocyte stem cells of the skin. Differentiation 72: 387–395.

Handbook of Clinical Neurology, Vol. 117 (3rd series)
Autonomic Nervous System
R.M. Buijs and D.F. Swaab, Editors

Chapter 31

Heart rate variability

IWONA CYGANKIEWICZ[1]* AND WOJCIECH ZAREBA[2]
[1]*Department of Electrocardiology, Medical University of Lodz, Lodz, Poland*
[2]*Heart Research Follow-up Program, University of Rochester Medical Center, Rochester, NY, USA*

INTRODUCTION

Heart rate variability (HRV) reflects beat-to-beat changes in RR intervals, which are related to the ongoing interplay between two arms of the autonomic nervous system (Task Force, 1996; Kleiger et al., 2005; Bilchick and Berger, 2006; Lahiri et al., 2008). Sinus node, the major heart's pacemaker, presents its own intrinsic activity; nevertheless, a variety of internal and external stimuli that change the balance between sympathetic and vagal tone influence the final basic heart rate. Heart rate changes may occur as a response to mental or physical stress, cardiac or noncardiac diseases, or pharmacological or invasive treatment. Autonomic nervous system imbalance, with a shift toward increased sympathetic and decreased vagal tone, has been proven to be associated with higher risk of cardiac mortality. Therefore HRV has become an important and well-recognized tool in identifying patients at risk of cardiovascular death (Task Force, 1996; Perkiomaki et al., 2000; Lombardi, 2001; Malik, 2004; Kleiger et al., 2005; Bilchick and Berger, 2006; Lahiri et al., 2008). Not surprisingly, taking into consideration strict heart–brain relationships, HRV assessment has become an important diagnostic tool in the detection of autonomic impairment and prediction of prognosis in several neurological disorders.

HEART RATE VARIABILITY MEASUREMENT

Heart rate variability measurement may be based on a battery of simple bedside reflex tests or may implement more advanced computer-based algorithms to reflect RR interval changes. Electrocardiogram (ECG)-based evaluation of HRV may be performed by several methods, including time- and frequency-domain analyses as well as nonlinear techniques (Task Force, 1996; Perkiomaki et al., 2000; Lombardi, 2001; Malik, 2004; Kleiger et al., 2005; Bilchick and Berger, 2006; Lahiri et al., 2008). The analysis is usually based on long-term (at least 18 hours) Holter ECG recordings or short-term (usually 5 minutes) ECG recordings obtained under controlled standardized conditions to avoid any influence from external stimuli that may affect autonomic nervous tone.

Bedside heart rate variability analysis

Cardiac dysautonomia may be assessed by means of a series of simple bedside reflex tests, known for several decades as the Ewing battery of tests (Ewing et al., 1985). This includes three tests based on the assessment of heart rate response to stimuli, and two tests to evaluate blood pressure changes. Heart rate changes are evaluated as a response to (1) Valsava maneuver, (2) orthostatic stress, and (3) deep breathing. The first test assesses the ratio between the longest RR interval following the release phase to the shortest RRR interval during the strain phase. The second one is computed as the ratio between the longest RR interval at the 30th beat to the 15th beat after standing from supine position. The last one reflects the index as the mean of the differences between the longest and the shortest RR interval during three consecutive breathing cycles. These tests were found to reflect cardiac autonomic neuropathy in diabetic patients and were implemented into basic screening clinical tests to be performed (Boulton et al., 2005; Tesfaye et al., 2010). One should not forget that a simple marker such as resting tachycardia may reflect an increased sympathetic tone and was found to be related to unfavorable outcome in several cohorts of patients (Kannel et al., 1987; Palatini et al., 1999).

*Correspondence to: Iwona Cygankiewicz, M.D., Ph.D., Department of Electrocardiology, Medical University of Lodz, Ul.Sterlinga 1/3, 91-425 Lodz, Poland. Tel: +48-426-644-269/304, Fax: +48-426-644-269/304, E-mail: cygankiewicz@interia.pl

Time-domain heart rate variability analysis

The most common HRV evaluation in clinical practice is based on time-domain analysis. In this analysis HRV parameters are derived from direct measurements of the NN (normal-to-normal) RR intervals, or from the differences between NN intervals (Task Force, 1996; Malik, 2004; Kleiger et al., 2005; Bilchick and Berger, 2006; Lahiri et al., 2008). The simplest and most widely used HRV parameter is the standard deviation of all NN intervals (SDNN) calculated over a long-term period (Fig. 31.1). The other parameters of the time domain analysis are displayed in Table 31.1. While SDNN is believed to reflect overall variability, the rMSSD and pNN50 are considered as measurements predominantly reflecting parasympathetic modulation of the heart (Task Force, 1996; Kleiger et al., 2005). Among time-domain HRV parameters, there is no single parameter that could be considered as representing mostly sympathetic modulation of the heart. Time-domain-based HRV parameters are frequently decreased in patients with cardiac and noncardiac conditions, and may reflect higher risk of mortality during a follow-up (Kleiger et al., 1987, 2005; Task Force, 1996; La Rovere et al., 1998; Malik, 2004; Bilchick and Berger, 2006; Lahiri et al., 2008).

Geometric methods convert series of RR intervals into geometric patterns such as sample density distribution of NN intervals or Lorenz plot of NN intervals (Kleiger et al., 1987; Malik, 2004). These methods based on histograms of various interval frequencies are believed to be less affected by noise and artifacts during computerized processing of ECG recordings. The HRV triangular index is the most frequently studied parameter; it is determined by dividing the total number of NN intervals by the maximum of the density distribution (Ewing et al., 1985). Lower values of HRV triangular index were associated with increased mortality (Malik et al., 1989). When HRV is evaluated by means of Lorenz or Poincaré plots, a fan-shaped plot characterizes patients with preserved HRV, while narrow plots correspond to those with depressed HRV and unfavorable outcome during a follow-up (Task Force, 1996; Kleiger et al., 2005; Bilchick and Berger, 2006).

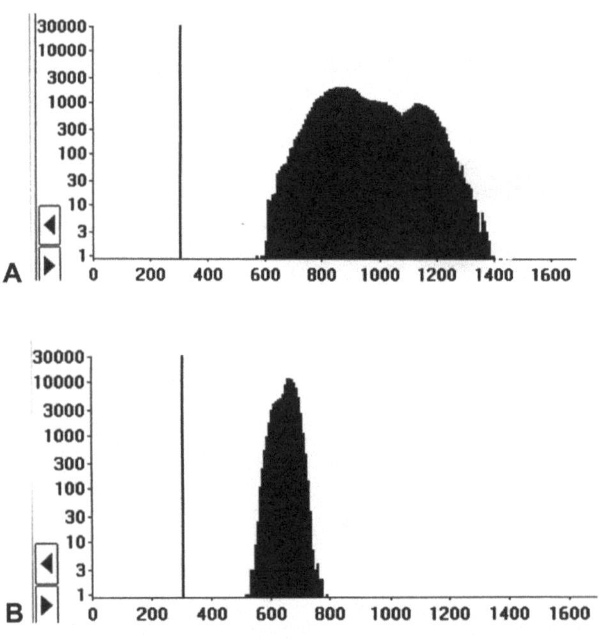

Fig. 31.1. Exemplary heart rate variability measures in (**A**) a healthy subject, (**B**) a patient with diabetes and advanced autonomic neuropathy.

For time-domain analysis it is recommended that long-term, at least 18 hour, ECG recordings encompassing morning and night hours should be used. The length of recording significantly influences the overall variability values so it is emphasized that time-domain HRV parameters obtained from periods of different durations should not be compared (Task Force, 1996; Crawford et al., 1999; Kleiger et al., 2005). The following measures for time domain assessment are recommended by the NASPE/ESC Task Force on HRV analysis: SDNN, HRV triangular index, SDANN, and rMSSD (Task Force, 1996).

Frequency-domain heart rate variability analysis

Spectral analysis provides information on how the power of HRV is distributed as a function of frequency. This analysis is usually performed on the basis of short-term (5 minute) recordings under controlled conditions. Spectral indices are assessed at baseline as well as in a reflex to a series of maneuvers. Nevertheless, long-term recordings are also frequently used (Task Force, 1996; Crawford et al., 1999; Lombardi, 2001). Spectral analysis aims to separate different frequency components of an entire RR intervals modulation (Fig. 31.2). The most frequently used approach to compute spectral indices is based on fast Fourier transformation (FFT). The total power of RR interval variability is the total variance, and corresponds to the sum of the four spectral bands, LF (low frequency), HF (high frequency), VLF (very low frequency) and ULF (Table 31.1). Spectral analysis obtained from short-term recordings is characterized by three major components: the HF; LF, and VLF components. While vagal tone is considered to be a major contributor to the HF component, the exact pathogenesis of LF has not been fully elucidated. It is believed to reflect both sympathetic and vagal influence and has been correlated with baroreflex sensitivity (La Rovere et al., 1998). Recently, Moak et al. (2007) documented that LF power reflects baroreflex function more than cardiac

Table 31.1

Heart rate variability parameters

Time-domain analysis		
Variable	Units	Description
SDNN	ms	Standard deviation of all normal-to-normal (NN) intervals
SDANN	ms	Standard deviation of the averages of NN intervals in all 5 minute segments of the entire recording
RMSSD	ms	Square root of the mean of the sum of the squares of differences between adjacent NN interval
pnn50	%	Percentage difference between adjacent NN intervals that are greater than 50 ms
Frequency-domain analysis		
Total power	ms^2	Variance of all NN intervals < 0.4 Hz
ULF	ms^2	Ultra low frequency < 0.003 Hz
VLF	ms^2	Very low frequency < 0.003–0.04 Hz
LF	ms^2	Low frequency power 0.04–0.15 Hz
HF	ms^2	High frequency power 0.15–0.4 Hz
LF/HF	Ratio	Ratio of low–high frequency power

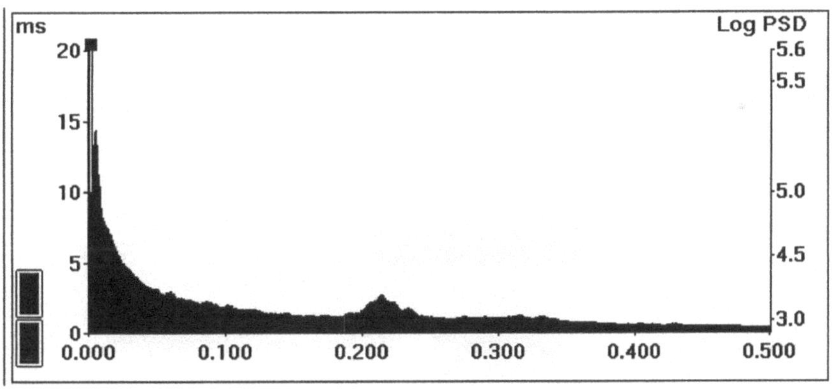

Fig. 31.2. A sample spectral analysis of a healthy subject (5 minute recording).

sympathetic innervation. In this study, LF power was unrelated to cardiac norepinephrine spillover, while it showed significant correlation with baroreflex sensitivity. Patients with decreased BRS displayed lower LF values independently on cardiac innervation. The lack of correlation between LF and cardiac norepinephrine spillover was further confirmed by Baumert et al. (2009) in a study of patients with depression and panic disorder. Sympathovagal balance is frequently described by LF/HF ratio. During resting controlled conditions, healthy subjects show a slight predominance of LF over HF, therefore the LF/HF ratio is usually between 1 and 2. However, in light of above mentioned associations between baroreflex sensitivity and LF, the LF/HF ratio represents a mutual relationship between baroreflex sensitivity and parasympathetic modulation rather than so-called sympathovagal balance.

The VLF power represents several factors influencing the heart such as thermoregulation, the renin–angiotensin system, and endothelial factors. It is also considered as a measure of sympathetic activity. The ULF spectral component represents very low oscillations and it might reflect circadian and neuroendocrine rhythms. Power of spectral components may be expressed in absolute (ms^2) and normalized units (nu). Normalized units are obtained as follows:

$$LF\ or\ HF\ norm\ (nu) = (LF\ or\ HF\ (ms^2)) \\ *100/(total\ power\ (ms^2) - VLF\ (ms^2))$$

Five minute recordings under controlled conditions are recommended for frequency domain analysis. It is assumed that at least a 1 minute recording is needed to assess the HF component while 2 minutes are required for LF analysis (Task Force, 1996; Crawford et al., 1999). Spectral analysis could be also performed on the entire 24 hour period from 5 minute segments averaged over the entire 24 hour period giving averages of the modulations attributable to the LF and HF components.

It has been documented that time and frequency parameters are strongly correlated with each other. While time-domain methods are preferred for long-term recording analysis, spectral methods are preferred to assess reflex responses under controlled conditions. The majority of commercial Holter devices provide an automatic measurement of HRV parameters; however, it should be emphasized that independently on the method applied the careful editing of RR intervals is required in each case. The methodology of HRV measurement was standardized in a special report of the Task Force of ESC/NASPE in 1996 and in ACC/AHA guidelines for ambulatory electrocardiography (Task Force, 1996; Crawford et al., 1999).

Nonlinear heart rate variability analysis

Nonlinear analysis is considered to be less dependent on the preprocessing of a recording and to better express the complex nature of heart rate variability. The exact pathophysiological background of nonlinear indices has not been fully clarified yet; however, there is some evidence that this more sophisticated mathematical approach toward RR variability may be superior to conventional HRV parameters in risk stratification. Among methods that have been proven to provide prognostic information are detrended fluctuation analysis (DFA), power law relationship analysis, approximate entropy, and PD2i (Guzzetti et al., 2000; Mäkikallio et al., 2001; Maestri et al., 2007; Skinner, 2007).

CLINICAL COVARIATES OF HEART RATE VARIABILITY

Various physiological and pathological factors may influence sympathovagal status, and consequently heart rate and its variability. Age and gender are considered as a major determinant of HRV. A constant decline in time- and frequency-domain parameters is observed through the life course (Umetani et al., 1998). It was reported that for every 10 years, an approximately 15% decline in LF and HF power is observed (Moolgard et al., 1994). Women are known to have lower SDNN and LF, but higher HF values, than men. Gender-related differences weaken with age, and diminish around the time of menopause, which may suggest a potential hormonal influence on the autonomic nervous system (Stein et al., 1997).

A decrease in HRV has been consistently reported in patients after myocardial infarction, and contributed to both structural changes of the left ventricle and decreases in vagal activity or blunted responses of the sinus node to autonomic regulation (Lombardi et al., 1987; Schwartz et al., 1988; Casolo et al., 1992). Such a shift toward increased sympathetic activity and loss of vagal protection may contribute to enhanced arrhythmogenesis and subsequent sudden death. Decreased HRV parameters are consequently reported also in patients with nonischemic cardiomyopathy, and in those with heart failure, independently of its origin. The extent of HRV reduction correlates with the advancement of heart failure expressed by measurements of ejection fraction, NYHA class, or brain natriuretic peptide (BNP) levels (Casolo et al., 1989; Nolan et al., 1992; Guzetti et al., 1995). Regarding noncardiac conditions, depressed HRV is frequently observed in patients with endocrinopathies (mostly in diabetes), and various neurological and psychiatric diseases. Heart rate variability is known to be modified by a variety of drugs, as well as by numerous invasive procedures.

CLINICAL APPLICATION OF HEART RATE VARIABILITY

According to the Task Force on HRV evaluation (1996), there are two clear clinical situations where HRV analysis should be performed: to assess mortality risk in postinfarction patients, and to detect early evidence of cardiac autonomic neuropathy in diabetic patients. Nevertheless, since the publication of the Task Force in 1996, depressed HRV has been consistently linked to higher mortality in other patient populations; thus there is increasing evidence that such evaluation may be useful in the early detection of autonomic dysfunction and in the selection of a subgroup of patients at risk of unfavorable outcome during follow-up. The association between abnormal HRV, impaired autonomic nervous tone, and cardiovascular mortality is well documented. The clinical relevance of HRV as a prognostic tool was first appreciated in 1965 by Hon and Lee (Hon and Lee, 1965); however, its history in cardiac risk stratification emerged in the late 1980s, when the first publications on HRV as a risk stratifier in postinfarction patients appeared (Kleiger et al., 1987; Malik et al., 1989). In 1987, Kleiger and coworkers from the Multicenter Post-Infarction Project (MPIP) reported than SDNN < 50 ms was associated with over fivefold higher risk of mortality as compared to those with

SDNN > 100 ms in a cohort of over 800 myocardial infarction survivors (Kleiger et al., 1987). Multiple later publications consistently confirmed the predictive value of HRV parameters in postinfarction patients (Bigger et al., 1992; Zuanetti et al., 1996; Huikuri et al., 2000, 2009; Camm et al., 2004; Mäkikallio et al., 2005; Exner et al., 2007). Decreased HRV is also strongly linked with the progression of heart failure and all-cause mortality in patients with cardiomyopathies and congestive heart failure (CHF) (Ponikowski et al., 1997; Nolan et al., 1998; Boveda et al., 2001; La Rovere et al., 2003). Even though experimental data indicated an association between impaired HRV and a preponderance toward ventricular arrhythmias, data on the association between decreased HRV and sudden death is conflicting. Depressed HRV mostly identifies patients at risk of all-cause mortality, with less evidence for its association with arrhythmic sudden death. It cannot be excluded that the paucity of clear evidence for the association between depressed HRV parameters and sudden death might be associated with difficulty in categorizing the sudden or arrhythmic nature of death. The autonomic nervous system operates differently in various patients depending not only on the disease but also on the advancement of the disease process. HRV parameters successfully predict CHF worsening and total mortality in CHF patients, indicating that autonomic dysfunction is a part of the overall clinical picture in such patients, but these parameters seem to have little or no prognostic significance for predicting arrhythmic events in these patients.

Diabetes

Autonomic neuropathy may involve every system in the human body; nevertheless, it is mainly cardiovascular impairment that is related to increased mortality. Cardiac autonomic neuropathy is one of the major complications of diabetes mellitus, contributing significantly to increased morbidity and mortality in this group (Bernardi et al., 1992; Vinik et al., 2003; Boulton et al., 2005; Schonauer et al., 2008; Lefrandt et al., 2010; Tesfaye et al., 2010). Autonomic dysfunction in diabetes has been recognized since the 1970s, initially based on a series of simple bedside maneuvers such as the Valsalva maneuver, deep breathing, or standing up (Ewing et al., 1985). In recent years, HRV assessment has been recommended as an easy, noninvasive, and reliable and easily applicable method to quantify the sympathetic and parasympathetic components of the ANS. The American Diabetes Association recommends HRV analysis as a part of the diagnosis of autonomic neuropathy (Boulton et al., 2005).

Several studies have demonstrated that HRV is diminished in diabetic patients, and that impairment of HRV parameters is correlated with the progression of disease as well as with unfavorable prognosis (Schonauer et al., 2008). It is postulated that HRV reduction may precede clinical manifestations of autonomic neuropathy. Possible mechanisms contributing to altered HRV in diabetic patients include endothelium dysfunction, increased intima-media thickness at bar receptor sites, and reduced vascular reactiveness (Bernardi et al., 1992; Pagani, 2000; Gerritsen et al., 2001; Vinik et al., 2003; Schonauer et al., 2008; Lefrandt et al., 2010).

Autonomic dysfunction may be observed in a significant percentage of diabetic patients (Ewing et al., 1985) and is related to unfavorable prognosis, being an independent predictor of renal or vascular dysfunction, microretinopathy. The prevalence of cardiac autonomic dysfunction (CAN) in patients with diabetes varies according to advancement of the disease, being reported as 7.7% in patients with newly diagnosed diabetes and up to 90% in patients scheduled for pancreas transplant (Ziegler et al., 1992; Kennedy et al., 1995; Coppini et al., 2000). Compared with nondiabetic patients, the risk of cardiovascular disease is three times higher, and is associated with complications of diabetes. Therefore, one may assume that the increased cardiovascular risk is closely related to the presence of autonomic neuropathy. CAN is also related to an increased risk of sudden death, often contributing to silent myocardial infarctions.

Diabetes-associated cardiac autonomic dysfunction is characterized by resting sinus tachycardia, loss of circadian variation in heart rate, abnormal chronotropism, or even fixed heart rate at advanced stages of the disease, silent angina, and orthostatic hypotension. Cardiac autonomic dysfunction is accompanied by other dysfunctions such as gastroparesis, sudomotor dysfunction, erectile dysfunction, urinary bladder dysfunction, and hypoglycemic autonomic failure (Bernardi et al., 1992; Vinik et al., 2003; Lefrandt et al., 2010).

Assessment of autonomic dysfunction in diabetes consists of a series of bedside tests of cardiovascular reflexes (Ewing's tests), HRV evaluation and assessment of baroreceptor sensitivity. Simple Ewing's tests, recognized for more than 40 years, aim to evaluate the heart rate response to Valsalva maneuvers, orthostatic stress, and heart rate changes during deep breathing. These reflexes are believed to reflect mostly parasympathetic activity. The American Diabetes Association (ADA) recommends the following tests to assess CAN (Boulton et al., 2005):

- resting heart rate (> 100 bpm abnormal)
- beat-to-beat HRV with a patient at rest and supine breathing according to metronome at a pace of 6 breaths per minute (a difference in heart rate > 15 bpm is normal, < 10 bpm is abnormal) (expiration/inspiration ratios are dependent on age)

- heart rate response to standing (RR interval measured at beats 15 and 30 after standing – 30:15 ratio should be > 1.03)
- heart rate response to Valsalva maneuver (the ratio of longest/shortest RR should be > 1.2)
- systolic blood pressure response to standing
- diastolic blood pressure response to isometric exercise
- ECG QT/QTc interval measurement
- spectral HRV analysis
- neurovascular flow.

According to an ADA statement, patients with CAN are characterized by decreased VLF, LF, HF power and diminished LF/HF ratio. Patients should be screened for CAN at the time of diabetes type 2 diagnosis and within 5 years after diagnosis of type 1 diabetes. If screening is negative, tests should be repeated annually. The ESC/NASPE Task Force (1996) recommends three types of HRV methods in the valuation of diabetic patients: long-term time-domain analysis, long-term frequent domain analysis and short-term spectral analysis. Recently published conclusions from the Toronto Consensus Panel on Diabetic Neuropathy stated that the most specific and sensitive approach toward CAN evaluation in clinical research should be based on: (1) heart rate variability; (2) baroreflex sensitivity; (3) muscle nerve activity; (4) plasma catecholamines; (5) heart sympathetic imaging (Bernardi et al., 2011).

It is still not clear which of the tests should be preferred: Ewing's functional test based on time-domain analysis or short-term 5 minute spectral HRV analysis based on commercial algorithms for HRV. Obviously, conventional HRV tests are easier to perform in an ambulatory setting in the absence of computerized ECG analysis. On the other hand, spectral analysis seems to be faster, patient-independent, low-cost and has good intra-individual reproducibility. Even though bedside tests are considered as a gold standard to evaluate autonomic dysfunction in a clinical practice, it has been demonstrated that 33% of patients with normal conventional tests present abnormal HRV parameters (Moolgard et al., 1992). Similarly, Malpas and Maling (1990) demonstrated that SDNN was reduced in patients with "normal" bedside tests. Therefore, it seems that HRV analysis has higher sensitivity in the detection of early autonomic impairment.

Decreased time-domain HRV parameters have been consistently demonstrated in diabetic patients. Similarly to what was observed in respect to time-domain HRV parameters, spectral indices are also able to select patients with dysfunction among those with normal bedside tests. Patients with diabetes, especially those in advanced neuropathic stages, are characterized by reduced power in all spectral bands, as well as failure to increase LF on standing. Nonlinear methods such as approximate entropy or PD2i are also reduced in diabetic patients and are believed to provide some incremental value to the aforementioned time-, and frequency-domain HRV tests (Bellavere et al., 1992; Spallone et al., 1993; Takase et al., 2002; Skinner et al., 2011).

The relationship between CAN and mortality has only been studied in small groups of patients. The mortality ranges from 9% to 53% (Ewing et al., 1980; Orchard et al., 1996). Maser et al. (2003), in a meta-analysis, included 15 studies published between 1966 and 2001 with a follow-up ranging from 0.5 to 16 years; they documented that mortality rates were higher in patients with CAN as compared to those with normal baseline assessment. The relative risks observed in the studies included in the meta-analysis ranged from 0.91 to 9.20. Nevertheless, it should be emphasized that CAN in the evaluated studies was defined based mostly on traditional bedside Ewing tests.

The rationale for performing diagnostic tests is based on the assumption that early diagnosis may stop further progression of autonomic dysfunction. As symptomatic CAN occurs late in the course of a disease, and is associated with a higher mortality risk, identification of CAN may result in prophylaxis of macro- and microangiopathy. It should be emphasized that the presence of autonomic dysfunction should alert clinicians to the possibility of coexisting cardiovascular risk factors. Early detection of the preclinical phase of CAN may lead to implementation of more aggressive treatment and control of cardiovascular risk factors. Whether such a treatment may resolve autonomic dysfunction unfortunately remains unclear. In a study by Athyros et al. (1998), 1 year of quinapril treatment restored HF and LF/HF ratios in patients with previously recognized autonomic dysfunction. The influence of more aggressive treatment on the prognosis among diabetes patients was evaluated in the ACCORD Study. The subanalysis of this study showed that CAN associated with decreased SDNN values based on 10 second ECG recordings carried twofold higher risk of mortality during 3.5 years of follow-up. Interestingly, the increased mortality in patients with CAN was independent of the allocation of patients to intensive versus standard glycemia control (Pop-Bosui et al., 2010).

Stroke

Not only cardiovascular, but also cerebrovascular diseases are known to be associated with disturbances of the autonomic nervous system. Whether this dysautonomia contributes to high mortality in stroke survivors has been intensively studied. The mortality in stroke survivors remains high and is mainly due to cardiac causes. It is well recognized that stroke predisposes to cardiac arrhythmias, and consequently to sudden cardiac death. The incidence of sudden cardiac death in stroke patients is estimated at 6% (Silver et al., 1984). Autonomic imbalance, which is observed in stroke victims, is most evident in the acute phase; nevertheless, these changes may persist and be related to subsequent increased risk of mortality. Heart rate variability changes accompanying ischemic stroke have been observed since the early 1990s. Even though the exact pathomechanism has not been fully elucidated, there are data indicating that such an autonomic imbalance was related to an increased risk of arrhythmias, and even sudden death, in the future (Korpelainen et al., 1997, 1999; He et al., 2010; Orlandi et al., 2000).

Cardiovascular dysfunction is most evident in the acute phase and is believed to shift autonomic tone toward increased sympathetic activity, and withdrawal of the vagal one. It was in 1994 that Barron et al. (Barron et al., 1994) first observed that spectral HRV indices are reduced in patients evaluated 4–11 days after stroke, as compared to healthy controls.

Heart rate variability is disturbed immediately after stroke and may remain altered for at least several months after the event (Lakusic et al., 2005). Data regarding HRV alteration in the latter phase are, however, conflicting. As documented by Korpelainen et al. (1997), such an autonomic dysregulation may be reversible within 6 months following an ischemic stroke. Similarly to what is observed in myocardial infarction, this may indicate a higher risk for arrhythmias and sudden death in the early period poststroke. A transient profile of autonomic dysfunction and arrhythmias was also observed by Orlandi et al. (2000), who documented a higher percentage of arrhythmias in patients with right hemisphere stroke as compared to left-sided stroke. Spectral indices were decreased when examined within 10 hours from the onset of symptoms and were significantly decreased after 3 days from admission in patients who presented with arrhythmias. Lakusic et al. (2005) also observed a gradual improvement of HRV parameters between the 2nd and the first 6 months after stroke. However, as compared to healthy subjects, poststroke patients still presented with lower time-domain and spectral HRV indices. On the other hand, in elderly patients, HRV depression was found to persist over 9 months after stroke. McLaren et al. (2005), in a case-control study of 76 stroke patients compared with 70 matched controls, evaluated autonomic impairment after stroke recovery in elderly patients. Lower TP and LF spectrum were observed 9 months after stroke. However, HRV in this study was assessed based on short-term 5 minute recordings.

Several papers addressed the problem of infarct localization and its influence on HRV parameters. Most studies showed that HRV dysregulation was more pronounced in patients with right-sided infarcts. In a study by Tokgozoglu et al. (1999), the most pronounced decrease in spectral HRV indices was observed in patients with stroke located in right insular cortex region. More importantly, right-sided infarct location accompanied by more reduced HRV was related to higher frequency of in-hospital sudden deaths. These data were confirmed by Colivicchi et al. (2004), who found that HRV depression was more frequently observed in patients with right hemispheric or insular stroke. Additionally, the authors documented that lower values of SDNN, rMSSD, and a higher LF/HF ratio observed in right insular infarcts were associated with more frequent and complex arrhythmias. The same group observed that decreased SDNN, together with right stroke location and nonsustained tachycardias, were predictive for mortality during a follow-up (Colivicchi et al., 2005). Therefore, it seems that, similarly to what is observed for other cardiovascular diseases, abnormal HRV in stroke victims is associated with increased mortality.

A number of clinical studies have documented that HRV reduction may be considered as a marker of poststroke morbidity and mortality. Mäkikallio et al. (2004), in a series of 84 stroke victims followed for up to 7 years, showed that power law slope β ($\beta < -1.5$) was the only multivariate predictor of mortality, while conventional HRV parameters showed no prognostic value. In a study by Gujjar et al. (2004), in 25 patients with acute stroke, LF and VLF power significantly correlated with mortality. After adjustment for significant clinical covariates, LF power remained an independent significant risk marker. In, to date, the largest cohort of 327 patients with an acute first stroke, fractal dimension assessed on the second day after admission was significantly associated with mortality during 1 month of follow-up (He et al., 2010). This study also confirmed different behavior of HRV in right- and left-sided strokes. Bassi et al. (2007) documented that lower SDNN values along age, stroke severity, and other scores were independently related with unfavorable outcome in poststroke patients undergoing a rehabilitation program. Therefore, the authors suggested that HRV analysis may provide additional information on the probability of functional recovery in stroke victims. Interestingly the same group a few years later demonstrated that SDNN was predictive for unfavorable outcome only in males and was not related to functional outcome in females undergoing a rehabilitation program (Bassi et al., 2010).

The early recognition of coexisting autonomic dysfunction may provide information on potentially unfavorable outcome related mostly to cardiac events.

Interestingly, depressed HRV assessed at nighttime was significantly associated with a risk of stroke in apparently healthy subjects. The Copenhagen Holter Study including 678 patients with no history of cardiovascular diseases documented that the risk of stroke during a median 76 months follow-up was independently associated with SDNN (HR 0.67, CI $= 0.51$–0.89, $p = 0.005$ per 10 ms increments of SDNN). In total, 81% of strokes occurred in patients presenting with low SDNN ($< = 38$ ms) at nighttime (Bicini et al., 2011).

Multiple sclerosis

Autonomic impairment, including cardiovascular autonomic dysfunction, is not uncommon in patients with multiple sclerosis (MS) (McDougall and McLeod, 2003; Merkelbach et al., 2006; Mahovic and Lakusic, 2007). Cardiovascular involvement may be expressed as poor physical fitness, fatigue, orthostatic hypotonia, or cardiac arrhythmias.

Association between multiple sclerosis and arrhythmias, or even sudden death, has been reported. The autonomic dysfunction observed in MS is believed to be associated with cardiac innervations in the medulla oblongata. A few case reports have documented paroxysmal atrial fibrillation associated with an exacerbation of disease (Chagnac et al., 1986; Schroth et al., 1992). Even though the exact relationship between MS and arrhythmias has not been fully clarified, one cannot exclude that cardiovascular autonomic dysfunction may play an important role. Several reports have documented abnormal HRV in patients with MS. All the reports are based on small groups of patients; however, they show a similar pattern of decreased HRV in MS as compared to healthy subjects. Most of the studies showed a significant shift toward loss of vagal and increased sympathetic tone. In a study by Frontoni et al. (1996), 16 MS patients underwent classic cardiovascular reflex tests and short-term HRV evaluation. Spectral analysis showed lower VLF and LF values in MS patients as compared to a control group, while classic reflex tests were abnormal in only 25% of studied patients. In a study by Brezinova et al. (2004), all spectral indices were lower in MS patients. On the other hand, Monge-Argiles et al. (2000) found higher LH power, LH/HF ratio as well as VLF and HF power in MS patients as compared to normal subjects in spectral HRV analysis based on 24 hour Holter recordings. Whether this difference is based on the length of a recording remains unclear. Not only spectral but also time-domain indices assessed from long-term ambulatory Holter monitoring show significant impairment of the autonomic nervous system in MS as compared to healthy subjects (Tombul et al., 2011).

Autonomic dysfunction in MS patients is likely to be associated with plaques located in the brainstem as well as spinal cord which affect autonomic areas and their connections.

Some authors contributed cardiovascular dysfunction to brain MS-related lesions but the results are contradictory. Saari et al. (2004) observed that autonomic dysfunction correlates with lesions observed in MRI while Frontoni et al. (1996) found no correlations between HRV indices and lesion areas detected on brain MRI. Monge-Argiles et al. (1998) compared spectral indices between MS patients and those with isolated brainstem lesions and found that heart rate LF, VLF and LF/HF ratio were significantly higher in MS patients.

Similarly to what was observed in other diseases, the degree of autonomic impairment depends on the advancement of the disease. As documented by Mahovic and Lakusic (2007), patients with MS lasting over 5 years present with significantly lower overall HRV than those with a shorter duration of disease (SDNN 88 vs 94, $p = 0.008$, TP $= 1683$ vs 2028, $p = 0.006$).

Muscular dystrophies

Cardiovascular involvement has been frequently observed in patients with muscular dystrophies. Atrial and ventricular arrhythmias, as well as conduction disturbances, are not uncommon in these patients. Even though the exact link between ventricular arrhythmias and sudden death has not been found, the risk of sudden death is increased in patients with Duchenne (DMD) and Becker muscular dystrophies. Similarly, the risk of sudden death in myotonic dystrophies is high and contributed to cardiac arrhythmias (Schoser et al., 2004; Hermans et al., 2010).

Impairment of HRV in patients with muscular dystrophies has been described since the early 1990s. Yotsukura et al. (1995) were the first to describe autonomic imbalance in patients with Duchenne dystrophy. Based on 24 hour Holter monitoring, they described significantly lower pNN50 and HF power and higher LF and LF/HF ratio in dystrophy patients as compared to controls. Furthermore, alteration in HRV was related to the progression of disease. The same group evaluated changes in HRV over the course of a disease during a 9 year follow-up, and found consequently lower HRV indices, especially HF ratio, in patients with dystrophy as compared to healthy controls (Yotsukura et al., 1998). The same pattern was observed for time-domain indices. These findings were later confirmed by other authors. Lanza et al. (2001) described lower HRV values in 60 patients with Duchenne muscular dystrophy as compared to controls. The most significant differences between studied patients and a control group were observed for pNN50 and HF power. Interestingly, no significant correlation except for HF power was observed between HRV indices and left ventricular ejection fraction. Inoue et al. (2009) evaluated HRV indices in 57 DMD patients and related them to BNP levels and left ventricle shortening fraction. They found depressed HRV measures even in patients with normal BNP and LV shortening fraction, which may suggest that HRV impairment may be considered as a preclinical marker of cardiovascular involvement in this group.

However, the question arose whether HRV may be used for risk stratification as it is recommended in post-infarction patients (Politano et al., 2008). Two groups have addressed this problem. Ducceschi et al. (1997) compared HRV parameters, repolarization (QTc), and arrhythmias between 20 patients with Becker muscular dystrophy and 14 healthy controls, and found lower SDNN and higher number of VPBs in patients with dystrophy. In 2006, Ammendola et al. (2006) found significantly lower SDNN (120 vs 135 ms, $p < 0.05$) and higher LF/HF ratio (2.4 vs. 1.1, $p < 0.03$) in a group of 30 men with Becker muscular dystrophy as compared to 30 healthy controls. During 8 years of follow-up, eight patients died suddenly. Sudden death victims were characterized by significantly lower SDNN (109 vs. 126 ms) and higher LF/HF ratio (Fig. 31.3).

Di Leo et al. (2004) confirmed HRV abnormalities in patients with myotonic dystrophy type 1; reduction in LF ratio was inversely correlated with duration of disease. Nevertheless, these findings were based on conventional bedside autonomic tests.

Parkinson's disease

Various autonomic nervous system dysfunctions accompany the classic symptoms of akinesia, rigidity, and tremor observed in Parkinson's disease. Similarly to other neurological diseases, HRV is consistently reported as decreased in patients with Parkinson's disease (Turkka et al., 1987; Ludin et al., 1987; Van Dijk et al., 1993). Autonomic dysfunction in PD has been

Fig. 31.3. Comparison of SDNN values in patients with Becker muscular dystrophy who died and those who survived. (Reproduced with permission from Ammendola et al., 2006.)

demonstrated by a variety of measures from bedside autonomic reflex test to long-term HRV analysis. As demonstrated by Kallio et al. (2000), HRV alteration is present already in untreated patients with mild clinical symptoms indicating early involvement of autonomic nervous system in this group. Furthermore, classic autonomic reflex tests showed that autonomic dysfunction is more attenuated in patients presenting with hypokinesia/rigidity as the initial symptom as compared to a subgroup with tremor at the beginning of a disease (Kallio et al., 2000). The same group later compared traditional HRV parameters, as well as selected nonlinear parameters analyzed from 24 hour recordings, between untreated patients with Parkinson's disease and a healthy control group (Haapaniemi et al., 2001). Lower SDNN, all spectral indices, and slope of the power–law relation were found in a Parkinson group (Fig. 31.4). When HRV was correlated with total and motor rating score, a significant inverse relationship with VLF, and LF power spectra was observed.

It is not clear whether HRV impairment in parkinsonian patients should be considered as a consequence of abnormal autonomic nervous system functioning or is merely related to reduced physical activity. Devos et al. (2003) evaluated HRV parameters at different stages of a disease comparing three groups: (1) less than 2 years of disease duration without L-dopa treatment; (2) mildly impaired with L-dopa; (3) patients with advanced disease and motor complications. The first two groups were found to present lower daytime LF and LF/HF ratio values, while the third group was characterized by a decrease in HF and pNN50. Autonomic dysfunction was found to be associated with motor dysfunction such as brady- and hypokinesia. This association, even though significant, was weak but related to disease severity.

The effect of deep brain stimulation (DBS) on HRV in Parkinson's disease is controversial. The majority of papers reported no changes in HRV after deep brain stimulation (Erola et al., 2006; Azevedo et al., 2010). Interesting application of HRV analysis was presented by Chen et al. (2011), who correlated HRV changes with response to deep brain stimulation. This study demonstrated that only good clinical responders to DBS showed a significant increase in LF during a follow-up of 9–32 months. Compared with preoperative values, the good response group showed a significant improvement in HRV indices as compared to a group with only fair response to therapy.

Epilepsy

Data on HRV in untreated patients with early stages of epilepsy are contradictory. Persson et al. (2007) compared 22 untreated patients with epilepsy with a healthy

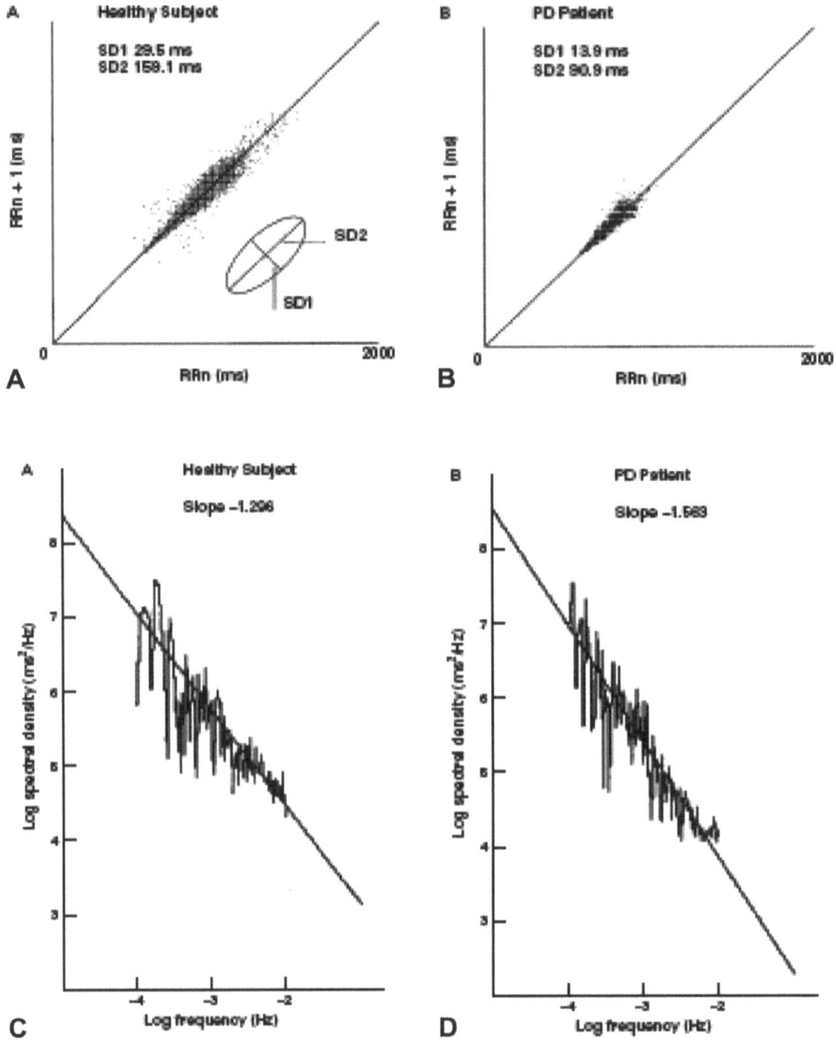

Fig. 31.4. Comparison of heart rate variability parameters between a healthy subject and a patient with Parkinson's disease. (**A**) and (**B**) Poincaré plots, (**C**) and (**D**) power law slopes. (Reproduced with permission from Haapaniemi et al., 2001.)

control group and found no differences in HRV measures assessed from 24 hour Holter monitoring. On the other hand, Evrengül et al. (2005) described higher SDNN, lower HF and higher LF values assessed from short-term recordings in untreated patients with epilepsy, suggesting increased sympathetic tone. Nevertheless, it is difficult to compare these results as they were obtained from recordings of long and short duration.

In more advanced stages of a disease, especially in refractory epilepsy, significant impairment of HRV has been consistently reported over decades (Drake et al., 1998; Tomson et al., 1998). Not only classic HRV parameters, but also nonlinear measures were found abnormal in patients with epilepsy. In a study by Ansakorpi et al. (2002), nonlinear HRV parameters, power law slope and ApEn were significantly lower in patients with temporal epilepsy as compared to healthy subjects. Further deterioration of HRV was observed in refractory patients as compared to those with well-controlled disease. These changes were concomitant with a significant decrease in conventional HRV measures. Autonomic nervous system impairment may be found in the period immediately following seizures as well as in the postictal period. Toth et al. (2010) recorded EEG-ECG during seizures and in the postictal period in patients undergoing preoperative evaluation and found decreased time-domain HRV parameters immediately after seizures as well as during the subsequent 5–6 h of observation. There was a decrease in LF power spectrum shortly after seizures while HF decreased in the late postseizure period. Furthermore, HRV changes were more pronounced in patients with generalized tonic-clonic seizures as compared to those with partial ones.

A recently published meta-analysis by Lotufo et al. (2012) aimed to evaluate the impact of epilepsy as well as its treatment on HRV measures. Thirty out of 366 initially reviewed papers were included in the meta-analysis. Primary analysis focused on LF and HF power spectra and revealed that patients with epilepsy present with lower values of HF, while no significant differences were observed for LF power spectrum. HRV parameters were also influenced by treatment. Patients receiving treatment (either pharmacological or vagal nerve stimulation) showed lower values of HF and a trend toward higher LF values. The authors also compared well-controlled patients versus those with refractory disease and found no significant differences, although, as emphasized by the authors, this lack of difference might be attributed to the limited number of studies included in the analysis. Secondary analysis included evaluation of time-domain parameters that confirmed lower HRV values (SDNN and rMSSD) in epileptic patients. Lower values of HF power and rMSSD might suggest autonomic imbalance toward decreased vagal activity which is known to be associated with increased risk of arrhythmias. Deterioration in HRV was reported to precede sudden death in patients with epilepsy (Rauscher et al., 2011).

The increased risk of sudden cardiac death in patients with seizures has been recognized for more than a century. Sudden unexpected death in epilepsy (SUDEP) is defined as a sudden, unexpected, witnessed or unwitnessed nontraumatic death in a patient with epilepsy excluding status epilepticus. It is the most frequent cause of death related to epilepsy with its incidence ranging from 0.1 to 9 per 1000 person-years with the highest incidence observed in patients with an advanced stage of disease scheduled for surgical treatment. The most significant risk factors associated with SUDEP include long duration of a disease, polytherapy with antiepileptic drugs, and poorly controlled seizures. The mechanism of SUDEP is probably multifactorial, and cardiac dysautonomia may be related to occurrence of life-threatening cardiac arrhythmias. Whether HRV analysis may help to stratify patients at risk of SUDEP has been a subject of numerous studies. The results, however, have been conflicting (Surges et al., 2009a, b).

Not only epilepsy itself, but what is even more important, pharmacotherapy also may influence the autonomic nervous system, promoting the occurrence of arrhythmias. An association between carbamazepine treatment and depressed HRV has been described since the early 1990s (Tomson and Kenneback, 1997; Persson et al., 2003). On the other hand, an abrupt withdrawal of antiepileptic drugs was also described to be associated with a significant decrease in SDNN, TP, VLF, and LF power spectra within the first 5 days after carbamazepine and phenytoin withdrawal. In a double-blinded randomized study, Lossius et al. (2007) analyzed 24 hour HRV parameters before and after withdrawal of antiepileptic treatment and found SDNN, pNN50, LF and LF in recordings after treatment discontinuation (Kenneback et al., 1997).

SUMMARY

Heart rate variability analysis provides opportunities to assess changes in the autonomic nervous system at rest, in response to physiological conditions, and in response to a disease process. The time-domain HRV parameters reflect overall autonomic modulation with parasympathetic components well represented by the rMSSD and pNN50 parameters. In frequency-domain HRV analysis, HF is useful in assessing parasympathetic modulation, but LF is now believed to represent baroreflex sensitivity instead of sympathetic modulation. There is no HRV parameter reflecting directly sympathetic modulation and the assessment of sympathetic activation requires more complex testing of muscle sympathetic activity response. HRV parameters are altered (mainly parasympathetic withdrawal) in several neurological conditions, including stroke, Parkinson disease, multiple sclerosis, and epilepsy. Nonlinear analysis of heart rate was found to be useful in predicting outcome in stroke patients. In other conditions, there is evidence for depressed HRV, mostly parasympathetic withdrawal, but the clinical and prognostic significance of these findings is as yet unknown.

References

Ammendola E, Russo V, Politano L et al. (2006). Is heart rate variability a valid parameter to predict sudden death in patients with Becker's muscular dystrophy? Heart 92: 1686–1687.

Ansakorpi H, Korpelainen JT, Huikuri HV et al. (2002). Heart rate dynamics in refractory and well controlled temporal lobe epilepsy. J Neurol Neurosurg Psychiatry 72: 26–30.

Athyros VG, Didangelos TP, Karamitsos DT et al. (1998). Long-term effect of converting enzyme inhibition on circadian sympathetic and parasympathetic modulation in patients with diabetic autonomic neuropathy. Acta Cardiol 53: 201–209.

Azevedo E, Santos R, Freitas J et al. (2010). Deep brain stimulation does not change neurovascular coupling in non-motor visual cortex: an autonomic and visual evoked blood flow velocity response study. Parkinsonism Relat Disord 16: 600–603.

Barron SA, Rogovski Z, Hemli J (1994). Autonomic consequences of cerebral hemisphere infarction. Stroke 25: 113–116.

Bassi A, Colivicchi F, Santini M et al. (2007). Cardiac autonomic dysfunction and functional outcome after ischemic stroke. Eur J Neurol 14: 917–922.

Bassi A, Colvicchi F, Santini M et al. (2010). Gender-specific predictors of functional outcome after stroke rehabilitation:

potential role of the autonomic nervous system. Eur Neurol 63: 279–284.

Baumert M, Lambert GW, Dawood T et al. (2009). Short-term heart rate variability and cardiac norepinephrine spillover in patients with depression and panic disorder. Am J Physiol Heart Circ Physiol 297: H674–H679.

Bellavere F, Belzani I, De Masi G et al. (1992). Power spectral analysis of heart rate variations improves assessment of diabetic autonomic neuropathy. Diabetes 41: 633–640.

Bernardi L, Ricordi L, Lazzari P et al. (1992). Impaired circadian modulation of sympathovagal activity in diabetes. A possible explanation for altered temporal onset of cardiovascular disease. Circulation 86: 1443–1452.

Bernardi L, Spallone V, Stevens M et al. on behalf of the Toronto Consensus Panel on Diabetic Neuropathy (2011). Investigation methods for cardiac autonomic function in human research studies. Diabetes Metab Res Rev. Jun 21. http://dx.doi.org/10.1002/dmrr.1224. [Epub ahead of print].

Bicini Z, Mouridsen MR, Kober L et al. (2011). Decreased nighttime heart rate variability is associated with increased stroke risk. Stroke 42: 3196–3201.

Bigger JT Jr, Fleiss JL, Steinman RC et al. (1992). Frequency domain measures of heart period variability and mortality after myocardial infarction. Circulation 85: 164–171.

Bilchick K, Berger RD (2006). Heart rate variability. J Cardiovasc Electrophysiol 17: 691–694.

Boulton AJM, Vinik AI, Arezzo JC et al. (2005). Diabetic neuropathies. A statement by the American Diabetes Association. Diabetes Care 28: 956.

Boveda S, Galinier M, Pathak A et al. (2001). Prognostic value of heart rate variability in time domain analysis in congestive heart failure patients. J Interv Card Electrophysiol 5: 181–187.

Brezinova M, Goldenberg Z, Kucera P (2004). Autonomic nervous system dysfunction in multiple sclerosis patients. Bratisl Lek Listy 105: 404–407.

Camm AJ, Pratt CJ, Schwartz PJ et al. (2004). Azimilide post Infarct surVival Evaluation (ALIVE) investigators. A randomized, placebo-controlled trial of azimilide using heart rate variability for risk stratification. Circulation 109: 990–996.

Casolo G, Balli E, Taddei T et al. (1989). Decreased spontaneous heart rate variability in congestive heart failure. Am J Cardiol 64: 1162–1167.

Casolo GC, Stroder P, Signorini C et al. (1992). Heart rate variability during the acute phase of myocardial infarction. Circulation 85: 2073–2079.

Chagnac Y, Martinovits G, Tadmor R et al. (1986). Paroxysmal atrial fibrillation associated with an attack of multiple sclerosis. Postgrad Med 62: 385–387.

Chen SY, Yang CC, Kuo TB et al. (2011). Association of heart rate variability with clinical outcome in parkinsonian patients after subthalamic deep brain stimulation: a retrospective cohort study. J Formos Med Assoc 110: 593–599.

Colivicchi F, Bassi A, Santini M et al. (2004). Cardiac autonomic derangement and arrhythmias in right-sided stroke with insular involvement. Stroke 35: 2094–2098.

Colivicchi F, Bassi A, Santini M et al. (2005). Prognostic implications of right-sided insular damage, cardiac autonomic derangement, and arrhythmias after acute ischemic stroke. Stroke 36: 1710–1715.

Coppini DV, Bowtell PA, Weng C et al. (2000). Showing neuropathy is related to increased mortality in diabetic patients – a survival analysis using an accelerated failure time model. J Clin Epidemiol 53: 519–523.

Crawford MH, Bernstein SJ, Deedwania PC et al. (1999). ACC/AHA guidelines for ambulatory electrocardiography: a report of the American College of Cardiology/American Heart Association Task Force on Practice Guidelines. J Am Coll Cardiol 34: 912–948.

Devos D, Kroumova M, Bordet R et al. (2003). Heart rate variability and Parkinson's disease severity. J Neural Transm 110: 997–1011.

Di Leo R, Rodolico C, De Gregorio C et al. (2004). Cardiovascular autonomic control in myotonic dystrophy type 1: a correlative study with clinical and genetic data. Neuromuscul Disord 14: 136–141.

Drake ME Jr, Andrews JM, Castleberry CM (1998). Electrophysiologic assessment of autonomic function in epilepsy. Seizure 7: 91–96.

Ducceschi V, Nigro G, Sarubbi B et al. (1997). Autonomic nervous system imbalance and left ventricular systolic dysfunction as potential candidates for arrhythmogenesis in Becker muscular dystrophy. Int J Cardiol 59: 275–279.

Erola T, Haapaniemi T, Heikkinen E et al. (2006). Subthalamic nucleus deep brain stimulation does not alter long-term heart rate variability in Parkinson's disease. Clin Auton Res 16: 286–288.

Evrengül H, Tanriverdi H, Dursunoglu D et al. (2005). Time and frequency domain analyses of heart rate variability in patients with epilepsy. Epilepsy Res 63: 131–139.

Ewing DJ, Campbell IW, Clarke BF (1980). The natural history of diabetic autonomic neuropathy. Q J Med 49: 95–108.

Ewing DJ, Martyn CN, Young RJ et al. (1985). The value of cardiovascular autonomic tests: 10 years experience in diabetes. Diabetes Care 8: 491–498.

Exner DV, Kavanagh KM, Slawnych MP et al. (2007). Noninvasive risk assessment early after a myocardial infarction the REFINE study. J Am Coll Cardiol 50: 2275–2284.

Frontoni M, Fiorini M, Strano S et al. (1996). Power spectrum analysis contribution to the detection of cardiovascular dysautonomia in multiple sclerosis. Acta Neurol Scand 93: 241–245.

Gerritsen J, Dekker JM, TenVoorde BJ et al. (2001). Impaired autonomic function is associated with increased mortality, especially in subjects with diabetes, hypertension, or a history of cardiovascular disease: the Hoorn Study. Diabetes Care 24: 1793–1798.

Gujjar AR, Sathyaprabha TN, Nagaraja D et al. (2004). Heart rate variability and outcome in acute severe stroke: role of power spectral analysis. Neurocrit Care 1: 347–353.

Guzetti S, Cogliati C, Turiel M et al. (1995). Sympathetic predominance followed by functional denervation in the progression of chronic heart failure. Eur Heart J 16: 1100–1107.

Guzzetti S, Mezzetti S, Magatelli R et al. (2000). Linear and non-linear 24 h heart rate variability in chronic heart failure. Auton Neurosci 86: 114–119.

Haapaniemi TH, Pursiainen V, Korpelainen JT et al. (2001). Ambulatory ECG and analysis of heart rate variability in

Parkinson's disease. J Neurol Neurosurg Psychiatry 70: 305–310.

He L, Li C, Luo Y et al. (2010). Clinical prognostic significance of heart abnormality and heart rate variability in patients with stroke. Neurol Res 32: 530–534.

Hermans MC, Pinto YM, Merkies IS et al. (2010). Hereditary muscular dystrophies and the heart. Neuromuscul Disord 20: 479–492.

Hon EH, Lee ST (1965). Electronic evaluations of the fetal heart rate patterns preceding fetal death: further observations. Am J Obstet Gynecol 87: 814–826.

Huikuri HV, Mäkikallio TH, Peng CK et al. for the DIAMOND Study Group (2000). Fractal correlation properties of RR interval dynamics and mortality in patients with depressed left ventricular function after an acute myocrdial infarction. Circulation 101: 47.

Huikuri HV, Raatikainen MJ, Moerch-Joergensen R et al. Cardiac Arrhythmias and Risk Stratification after Acute Myocardial Infarction Study Group (2009). Prediction of fatal or near-fatal cardiac arrhythmia events in patients with depressed left ventricular function after an acute myocardial infarction. Eur Heart J 30: 689–698.

Inoue M, Mori K, Hayabuchi Y et al. (2009). Autonomic function in patients with Duchenne muscular dystrophy. Pediatr Int 51: 33–40.

Kallio M, Haapaniemi T, Turkka J et al. (2000). Heart rate variability in patients with untreated Parkinson's disease. Eur J Neurol 7: 667–672.

Kannel WB, Kannel C, Paffenburger RS Jr (1987). Heart rate and cardiovascular mortality: the Framingham Study. Am Heart J 113: 1489–1494.

Kenneback G, Erickson M, Tomson T et al. (1997). Changes in arrhythmia profile and heart rate variability during abrupt withdrawal of antiepileptic drugs. Implications for sudden death. Seizure 6: 369–375.

Kennedy WR, Navarro X, Sutherland DE (1995). Neuropathy profile of diabetic patients in a pancreas transplantation program. Neurology 45: 73–80.

Kleiger RE, Miller JP, Bigger JT et al. the Multicenter Post-Infarction Research Group (1987). Decreased heart rate variability and its association with increased mortality after acute myocardial infarction. Am J Cardiol 59: 256–262.

Kleiger RE, Stein PK, Bigger TJ (2005). Heart rate variability: measurement and clinical utility. Ann Noninvasive Electrocardiol 10: 88–101.

Korpelainen JT, Sotaniemi KA, Huikuri HV et al. (1997). Circadian rhythm of heart rate variability is reversibly abolished in ischemic stroke. Stroke 28: 2150–2154.

Korpelainen JT, Sotaniemi KA, Mäkikallio A et al. (1999). Dynamic behavior of heart rate in stroke. Stroke 30: 1008–1013.

La Rovere MT, Bigger JT Jr, Marcus FI et al.for the ATRAMI Investigators (1998). Baroreflex sensitivity and heart rate variability in prediction of total cardiac mortality after myocardial infarction. Lancet 351: 478–484.

La Rovere MT, Pinna GD, Maestri R et al. (2003). Short-term heart rate variability strongly predicts sudden death in chronic heart failure. Circulation 107: 565–570.

Lahiri MK, Kannankeril PJ, Goldberger JJ (2008). Assessment of autonomic function in cardiovascular disease. Physiological basis and prognostic implications. J Am Coll Cardiol 51: 1725–1733.

Lakusic N, Mahovic D, Babic T (2005). Gradual recovery of impaired cardiac autonomic balance within first six months after cerebral stroke. Acta Neurol Belg 105: 39–42.

Lanza GA, Dello Russo A, Giglio V et al. (2001). Impairment of cardiac autonomic function in patients with Duchenne muscular dystrophy: relationship to myocardial and respiratory function. Am Heart J 141: 808–812.

Lefrandt JD, Smit AJ, Zeebregts CJ et al. (2010). Autonomic dysfunction in diabetes: a consequence of cardiovascular damage. Curr Diabetes Rev 6: 348–358.

Lombardi F (2001). Frequency domain analysis of heart rate variability. In: W Zareba, P Maison-Blanche, EH Locati (Eds.), Noninvasive Electrocardiology in Clinical Practice. Futura Publishing, Armonk, NY, pp. 163–180.

Lombardi F, Sandrone G, Pernpruner S et al. (1987). Heart rate variability as an index of sympathovagal interaction after myocardial infarction. Am J Cardiol 60: 1239–1245.

Lossius MI, Erikssen JE, Mowinckel P et al. (2007). Changes in autonomic cardiac control in patients with epilepsy after discontinuation of antiepileptic drugs: a randomized controlled withdrawal study. Eur J Neurol 14: 1022–1028.

Lotufo PA, Valiengo L, Benseñor IM et al. (2012). A systematic review and meta-analysis of heart rate variability in epilepsy and antiepileptic drugs. Epilepsia 53: 272–282.

Ludin SM, Steiger UH, Ludin HP (1987). Autonomic disturbances and cardiovascular reflexes in idiopathic Parkinson's disease. J Neurol 235: 10–15.

Maestri R, Pinna GD, Accardo A et al. (2007). Nonlinear indices of heart rate variability in chronic heart failure patients: redundancy and comparative clinical value. J Cardiovasc Electrophysiol 18: 425–433.

Mahovic D, Lakusic N (2007). Progressive impairment of autonomic control of heart rate in patients with multiple sclerosis. Arch Med Res 38: 322–325.

Mäkikallio TH, Huikuri HV, Hintze U et al. the DIAMOND Study Group (Danish Investigations of Arrhythmia and Mortality ON Dofetilide) (2001). Fractal analysis and time- and frequency-domain measures of heart rate variability as predictors of mortality in patients with heart failure. Am J Cardiol 87: 178–182.

Mäkikallio AM, Mäkikallio TH, Korpelainen JT et al. (2004). Heart rate dynamics predict poststroke mortality. Neurology 62: 1822–1826.

Mäkikallio TH, Barthel P, Schneider R et al. (2005). Prediction of sudden cardiac death after acute myocardial infarction: role of Holter monitoring in the modern treatment era. Eur Heart J 26: 762–769.

Malik M (2004). Standard measurement of heart rate variability. In: M Malik, AJ Camm (Eds.), Dynamic Electrocardiography. Blackwell Futura, New York, US, pp. 13–21.

Malik M, Farrell T, Cripps T et al. (1989). Heart rate variability in relation to prognosis after myocardial infarction:

selection of optimal processing techniques. Eur Heart J 10: 1060–1074.

Malpas SC, Maling TJB (1990). Heart rate variability and cardiac autonomic function in diabetes. Diabetes 39: 1177–1181.

Maser RE, Mitchell BD, Vinik AI et al. (2003). The association between cardiovascular autonomic neuropathy and mortality in individuals with diabetes. A meta-analysis. Diabetes Care 26: 1895–1901.

McDougall AJ, McLeod JG (2003). Autonomic nervous system function in multiple sclerosis. J Neurol Sci 215: 79–85.

McLaren A, Kerr S, Allan L et al. (2005). Autonomic function is impaired in elderly stroke survivors. Stroke 36: 1026–1030.

Merkelbach S, Heansch CA, Hemmer B et al. (2006). Multiple sclerosis and the autonomic nervous system. J Neurol 253 (Suppl 1): I21–I25.

Moak JP, Goldstein DS, Eldadah BA et al. (2007). Supine low-frequency power of heart rate variability reflects baroreflex function, not cardiac sympathetic innervation. Heart Rhythm 4: 1523–1529.

Monge-Argiles JA, Palacio Ortega F, Vila Sobrino JA et al. (2000). Brainstorm lesions decrease heart rate variability. Neurologia 15: 158–163.

Moolgard H, Christensen PD, Sorensen KE et al. (1992). Association of 24-h cardiac parasympathetic activity and degree of neuropathy in IDDM patients. Diabetes 41: 812–817.

Moolgard H, Hermansen K, Bjerregaard P (1994). Spectral components of short-term RR interval variability in healthy subjects and effects of risk factors. Eur Heart J 15: 1174–1183.

Nolan J, Flapan AD, Capewell S et al. (1992). Decreased cardiac parasympathetic activity in chronic heart failure and its relation to left ventricular function. Br Heart J 69: 761–767.

Nolan J, Batin PD, Andrews R et al. (1998). Prospective study of heart rate variability and mortality in chronic heart failure: results of the United Kingdom heart failure evaluation and assessment of risk trial (UK-heart). Circulation 98: 1510–1516.

Orchard TJ, Lloyd CE, Maser RE et al. (1996). Why does diabetic autonomic neuropathy predict IDDM mortality? An analysis from the Pittsburgh Epidemiology of Diabetes Complications Study. Diabetes Res Clin Pract 34 (Suppl): S165–S171.

Orlandi G, Fanucchi S, Strata G et al. (2000). Transient autonomic nervous system dysfunction dyring hyperacute stroke. Acta Neurol Scand 102: 317–321.

Pagani M (2000). Heart rate variability and autonomic diabetic neuropathy. Diabetes Nutr Metab 13: 341–346.

Palatini P, Casiglia E, Julius S et al. (1999). High heart rate – a risk factor for cardiovascular death in elderly men. Arch Intern Med 159: 585–592.

Perkiomaki JS, Mäkikallio TH, Huikuri HV (2000). Nonlinear analysis of heart rate variability: fractal and complexity measures of heart rate behaviour. Ann Noninvasive Electrocardiol 5: 179–187.

Persson H, Ericson M, Tomson T (2003). Carbamazepine affects autonomic cardiac control in patients with newly diagnosed epilepsy. Epilepsy Res 57: 69–73.

Persson H, Ericson M, Tomson T (2007). Heart rate variability in patients with untreated epilepsy. Seizure 16: 504–508.

Politano L, Palladino A, Nigro G et al. (2008). Usefulness of heart rate variability as a predictor of sudden cardiac death in patients with muscular dystrophies. Acta Myol 27: 114–122.

Ponikowski P, Anker SD, Chua TP et al. (1997). Depressed heart rate variability as an independent predictor of death in chronic congestive heart failure secondary to ischemic or idiopathic dilated cardiomyopathy. Am J Cardiol 79: 1645–1650.

Pop-Bosui R, Evans GW, Gernstein HC et al. the ACCORD Study Group (2010). Effects of Cardiac Autonomic Dysfunction On Mortality Risk In Diabetes (ACCORD) Trial. Diabetes Care 33: 1578–1584.

Rauscher G, De Giorgio AC, Miler PR et al. (2011). Sudden unexpected death in epilepsy associated with progressive deterioration in heart rate variability. Epilepsy Behav 21: 103–105.

Saari A, Tolonen U, Paakko E et al. (2004). Cardiovascular autonomic dysfunction correlates with brain MRI leasion load in MS. Clin Neurophysiol 115: 1473–1478.

Schonauer M, Thomas A, Morbach S et al. (2008). Cardiac autonomic diabetic neuropathy. Diab Vasc Dis Res 5: 336–344.

Schoser BG, Ricker K, Schneider-Gold C et al. (2004). Sudden cardiac death in myotonic dystrophy type 2. Neurology 63: 2402–2404.

Schroth WS, Tenner SM, Rappaport BA et al. (1992). Multiple sclerosis as a cause of atrial fibrillation and electrocardiographic changes. Arch Neurol 49: 422–424.

Schwartz PJ, Vanoli E, Stramba-Badiale M et al. (1988). Autonomic mechanisms and sudden death: new insights from the analysis of baroreceptor reflexes in conscious dogs with and without a myocardial infarction. Circulation 78: 969–979.

Silver FI, Norris JW, Lewis AJ et al. (1984). Early mortality following stroke: a prospective review. Stroke 15: 492–496.

Skinner JE (2007). New paradigms in heart–brain medicine: nonlinear physiology, state-dependent proteomics. Cleve Clin J Med 74 (Suppl 1): S79–S85.

Skinner JE, Weiss DN, Anchin JM et al. (2011). Nonlinear PD2i heart rate complexity algorithm detects autonomic neuropathy in patients with type 1 diabtes mellitus. Clin Neurophysiol 122: 1457–1462.

Spallone V, Bernardi L, Ricordi L et al. (1993). Relationship between the circadian rhythms of blood pressure and sympathovagal balance in diabetic autonomic neuropathy. Diabetes 42: 1745–1752.

Stein PK, Kleiger RE, Rottman JN (1997). Differing effect of age on heart rate variability in men and women. Am J Cardiol 80: 302–305.

Surges R, Thijs RD, Tan HL et al. (2009a). Sudden unexpected death in epilepsy: risk factors and potential pathomechanisms. Nat Rev Neurol 5: 492–504.

Surges R, Henneberger C, Adlei P et al. (2009b). Do alterations in inter-ictal heart rate variability predict sudden unexpected death in epilepsy? Epilepsy Res 87: 277–280.

Takase B, Kitamura H, Noritake M et al. (2002). Assessment of diabetic autonomic neuropathy using 24 h spectral analysis of heart rate variability: a comparison with the findings of the Ewing battery. Jpn Heart J 43: 127–135.

Task Force (1996). Heart Rate Variability. Standards of measurement, physiological interpretation and clinical use. Task Force of the European Society of Cardiology and the North American Society of Pacing and Electrophysiology. Circulation 93: 1043–1065.

Tesfaye S, Boulton AJM, Dyck PJ et al. on behalf of the Toronto Diabetic Neuropathy Expert Group (2010). Diabetic neuropathies: update on definitions, diagnostic criteria, estimation of severity, and treatments. Diabetes Care 33: 2285–2293.

Tokgozoglu SL, Batur MK, Topcuoglu MA et al. (1999). Effect of stroke localization on cardiac autonomic balance and sudden death. Stroke 30: 1307–1311.

Tombul T, Anlar O, Tuncer M et al. (2011). Impaired heart rate variability as a marker of cardiovascular dysfunction in multiple sclerosis. Acta Neurol Belg 111: 116–120.

Tomson T, Kenneback G (1997). Arrhythmia, heart rate variability, and epileptic drugs. Epilepsia 38 (11 Suppl): S48–S51.

Tomson T, Ericson M, Ihrman C et al. (1998). Heart rate variability in patients with epilepsy. Epilepsy Res 30: 77–83.

Toth V, Hejjel L, Fogarasi A et al. (2010). Periictal heart rate variability analysis suggest long-term postictal autonomic disturbances in epilepsy. Eur J Neurol 17: 780–787.

Turkka JT, Tolonen U, Myllylä VV (1987). Cardiovascular reflexes in Parkinson's disease. Eur Neurol 26: 104–112.

Umetani K, Singer DH, McCraty R et al. (1998). Twenty-four hour time domain heart rate variability and heart rate: relations to age and gender over nine decades. J Am Coll Cardiol 31: 593–601.

Van Dijk JG, Haan J, Zwinderman K et al. (1993). Autonomic nervous system dysfunction in Parkinson's disease: relationships with age, medication, duration, and severity. J Neurol Neurosurg Psychiatry 56: 1090–1095.

Vinik AL, Maser RE, Mitchell BD et al. (2003). Diabetic autonomic neuropathy. Diabetes Care 26: 1553–1579.

Yotsukura M, Sasaki K, Kachi E et al. (1995). Circadian rhythm and variability of heart rate in Duchenne-type progressive muscular dystrophy. Am J Cardiol 76: 947–951.

Yotsukura MM, Fujii K, Katayama A et al. (1998). Nine-year follow up study of heart rate variability in patients with Duchenne-type progressive muscular dystrophy. Am Heart J 136: 289–296.

Ziegler D, Gries FA, Spuler M et al. (1992). The epidemiology of diabetic neuropathy: Diabetes Cardiovascular Autonomic Neuropathy Multicenter Study Group. J Diabetes Complications 6: 49–57.

Zuanetti G, Neilson JMM, Latini R et al. on behalf of GISSI2 investigators (1996). Prognostic significance of heart rate variability in post-myocardial infarction patients in the fibrinolytic era. The GISSI-2 results. Circulation 94: 432–436.

Handbook of Clinical Neurology, Vol. 117 (3rd series)
Autonomic Nervous System
R.M. Buijs and D.F. Swaab, Editors

Chapter 32

Chronic activation of the baroreflex and the promise for hypertension therapy

THOMAS E. LOHMEIER[1]* AND RADU ILIESCU[2]

[1]*Department of Physiology, University of Mississippi Medical Center, Jackson, MS, USA*

[2]*Cardiovascular Research Department, Gr. T. Popa Center for Biomedical Research and the University of Medicine and Pharmacology, Gr. T. Popa, Iasi, Romania*

THE BAROREFLEX AS A LONG-TERM DETERMINANT OF SYMPATHETIC ACTIVITY AND ARTERIAL PRESSURE

Despite the overwhelming evidence that the sympathetic nervous system plays a key role in the pathogenesis of primary hypertension, the mechanisms that account for sympathetic activation are largely unclear (Esler, 2000; Esler et al., 2006). Nonetheless, one common abnormality in primary hypertension, especially when associated with obesity, is baroreflex dysfunction, which involves both the parasympathetic and sympathetic limbs of the baroreflex (Grassi et al., 1998, 2000). Given the importance of the baroreflex in the acute regulation of arterial pressure, it is often stated that impaired ability of the baroreflex to suppress sympathetic outflow contributes to the sympathoexcitation in primary hypertension. However, extrapolation of acute reflex responses to the chronic state is not warranted, particularly in light of the fact that baroreceptors reset in the direction of a sustained change in arterial pressure (McCubbin et al., 1956; Chapleau et al., 1991; Cowley, 1992). Indeed, it is well established that baroreceptor resetting results in a higher threshold pressure for baroreceptor activation in hypertension. While this preserves the fundamental role of the baroreflex in the acute regulation of arterial pressure, it also diminishes its ability chronically to suppress sympathetic activity, and thereby attenuate the severity of hypertension. Thus, on one hand, if resetting of the baroreflex is complete and there is no continuous influence of the baroreflex on sympathetic activity, then it would be unlikely that baroreflex dysfunction could contribute to the sustained sympathoexcitation that plays a causal role in the pathogenesis of primary hypertension. On the other hand, if the baroreflex does not entirely adapt to long-term changes in arterial pressure, baroreflex dysfunction could play a role in the pathogenesis of primary hypertension. Unfortunately, due to the paucity of techniques available to assess the chronic effects of the baroreflex on sympathetic activity and arterial pressure, the time course of baroreflex resetting has not been established with certainty and, therefore, its role in long-term control of arterial pressure is unclear. However, as discussed below, recent chronic studies using novel approaches to assess sympathetic activity and cardiovascular function indicate that resetting is incomplete during the natural activation of the baroreflex in experimental models of hypertension (Lohmeier et al., 2000, 2005c; Barrett et al., 2005; Lohmeier and Drummond, 2007; Malpas, 2010).

Although resetting at the level of the baroreceptors is usually considered the major cause of resetting of the baroreflex in chronic hypertension, central adaptations may also play a role in this process (Chapleau et al., 1991). Central resetting in hypertension refers to alterations within the CNS that diminish the magnitude of sympathoinhibition induced by a given level of baroreceptor input. These alterations may result from changes in neural activity within the CNS or changes in circulating hormones that act centrally to influence sympathetic outflow. One approach to demonstrate central resetting has been to compare sympathetic or blood pressure responses to acute electrical stimulation of baroreceptor afferents in hypertensive and control animals (Gonzalez et al., 1983; Chapleau et al., 1991; Yin and Sved, 1996; Zhang and Mifflin, 2000). When compared to normotensive

*Correspondence to: Thomas E. Lohmeier, Ph.D., Department of Physiology, University of Mississippi Medical Center, 2500 North State Street, Jackson, MS 39216-4505, USA. Tel: +1-601-984-1822, Fax: +1-601-984-1822, E-mail: tlohmeier@umc.edu

controls, blunted reductions in sympathetic activity and arterial pressure in response to acute electrical stimulation of baroreceptor afferents have been reported in several forms of hypertension, supporting the concept of central resetting. However, because of the acute nature of these studies and technical limitations precluding chronic stimulation of baroreceptor afferents, the importance of central resetting in attenuating baroreflex function in chronic hypertension has been unclear.

RATIONALE FOR ELECTRICAL STIMULATION OF THE CAROTID BAROREFLEX FOR HYPERTENSION THERAPY

Clinical studies conducted in the 1960s challenged the concept that central resetting is an important mechanism that diminishes suppression of sympathetic activity during chronic increases in arterial pressure. During this decade, various medical devices were designed to activate the carotid baroreflex by electrical stimulation of baroreceptor afferents with the intent of lowering arterial pressure in patients with severe hypertension whose blood pressure was inadequately controlled by medication (Schwartz et al., 1967; Tuckman et al., 1968; Lohmeier et al., 2005a). While the findings from these early studies lacked the requisite rigor of current clinical trials, they did suggest that long-term electrical stimulation of the carotid baroreflex might chronically attenuate the severity of hypertension. However, despite these encouraging findings, this approach for the treatment of hypertension was abandoned because the technology was too crude to achieve reliable sustained baroreflex activation and chronic antihypertensive responses, particularly in the absence of adverse effects, including extraneous muscle and nerve stimulation, pain, and dysphonia. Furthermore, this was also a point in time when several new antihypertensive drugs were becoming available to clinicians for the treatment of hypertension.

Since these early studies more than 50 years ago, a multitude of drug classes with diverse actions have been developed to treat hypertension. Given the proven success of pharmacological therapy in lowering blood pressure and reducing the associated adverse cardiovascular effects of hypertension, it would appear incongruous that there has been a resurgence of interest in carotid baroreflex activation for the treatment of hypertension. In this regard, three factors have provided the impetus to move forward with the current clinical trials using carotid baroreflex activation for hypertension therapy. First, despite the availability of pharmacological therapies, the percentage of patients achieving adequate blood pressure control remains unacceptably low (Chobanian et al., 2003; Calhoun et al., 2008). While the failure to achieve adequate blood pressure control can largely be accounted for by underuse of adequate drug therapy and patient noncompliance, in some patients, blood pressure is still uncontrollable despite adherence to optimal drug regimens. These patients have resistant hypertension (Chobanian et al., 2003; Kaplan, 2005; Calhoun et al., 2008). Although different definitions have been used, a recent statement from the American Heart Association defined resistant hypertension as follows: "blood pressure that remains above goal despite the concurrent use of 3 antihypertensive agents of different drug classes." "Ideally, one of the 3 agents should be a diuretic and all agents should be prescribed at optimal dose amounts (Calhoun et al., 2008)." Although the true prevalence of resistant hypertension is unknown, it is not uncommon. In a recent study from a large cohort (68 045 subjects) of treated hypertensive patients from a Spanish Ambulatory Blood Pressure Monitoring Registry, ~8% of the patients were classified as having true resistant hypertension after eliminating those subjects with the "white-coat effect" (De la Sierra et al., 2011). This compares to the current best estimate of a 9% prevalence of resistant hypertension in the US (Persell, 2011). Thus, there is a need in these patients for a new treatment option for lowering blood pressure, such as baroreflex activation.

Second, current clinical trials evaluating the efficacy and safety of baroreflex activation therapy (BAT) for the treatment of resistant hypertension would not have been possible without the development of modern device technology for baroreflex activation. As previously reviewed in greater detail (Lohmeier et al., 2005a), this modern technology has overcome the technical limitations encountered in early clinical trials using electrical stimulation of the carotid baroreflex. With the present approach for activation of the baroreflex by the Rheos system (CVRx, Inc.), the stimulating electrodes are implanted in the perivascular space around each carotid sinus and not around the carotid sinus nerve, as in some of the early studies. This essentially eliminates the possibility of damage to the carotid sinus nerve and the likelihood of activating carotid body chemoreceptor afferents which, along with the baroreceptor fibers, are present in the carotid sinus nerve. By increasing sympathetic activity and respiratory rate, activation of chemoreceptor afferents would be expected to limit the therapeutic efficacy of baroreflex activation and confound interpretation of baroreflex responses. Additionally, the problems reported in early studies related to extraneous nerve and muscle stimulation have been solved by technological advances in electrode design that permit localized stimulation of carotid baroreceptors. Another limitation in early studies was the inability to achieve consistent individualized control of the level and characteristics of baroreflex activation because the electrical current delivery to the electrodes was either fixed internally or external control was unreliable. With the current Rheos system, the internally implantable, miniature pulse generator is completely programmable externally by radiofrequency control. This allows for controlled current delivery throughout the day, including the possibility of providing customized day–night patterns of baroreflex activation. An additional desirable feature of the Rheos system is that blood pressure responses are dose-dependent and readily controllable. Finally, by the very nature of the implant, this device-based approach for antihypertensive therapy guarantees patient compliance.

Third, recent long-term experimental studies conducted in chronically instrumented animals have provided a strong physiological rationale for pursuing the current clinical trials to evaluate the efficacy of baroreflex activation for hypertension therapy (Lohmeier et al., 2000, 2004, 2005b, 2007a, b, 2009, 2010). Up until the end of the 20th century, the dogma, based primarily on acute studies, was that the baroreflex is unimportant in long-term control of arterial pressure because it totally resets to the prevailing level of arterial pressure (Cowley, 1992). However, around the turn of the century, a number of experimental observations were made in chronically instrumented animals that challenged this paradigm. Most significantly, these studies demonstrated sustained baroreflex-mediated suppression of renal sympathetic nerve activity (RSNA) and attendant increases in renal excretory function in experimental models of hypertension (Carroll et al., 1984; Lohmeier et al., 2000, 2005c; Barrett et al., 2005; Malpas, 2010). As discussed below, such sustained neurally mediated increases in renal excretory function are consistent with the hypothesis that the natural activation of the baroreflex is a long-term compensatory mechanism that attenuates the severity of hypertension. Finally, and of greater relevance to the issue of BAT, several preclinical studies using the Rheos system have demonstrated impressive long-term reductions in arterial pressure during prolonged activation of the baroreflex (Lohmeier et al., 2004, 2005b, 2007a, b, 2009, 2010). Results from recent clinical trials and the experimental studies that preceded and were expanded during the clinical trials are discussed below. The experimental studies are especially important because they provide an understanding of the fundamental mechanisms that not only mediate the chronic blood pressure-lowering effects of baroreflex activation, but the cardiovascular responses to suppression of central sympathetic outflow in general. Further, by elucidating the conditions wherein optimal blood pressure responses to BAT may be achieved, these studies are valuable for identifying those patients with resistant hypertension that have the greatest likelihood of benefiting from BAT.

CHRONIC ELECTRICAL STIMULATION OF THE CAROTID BAROREFLEX HAS SUSTAINED EFFECTS TO SUPPRESS SYMPATHETIC ACTIVITY AND ARTERIAL PRESSURE

A recurrent argument (and one of the few based on long-term cardiovascular responses) supporting the concept of central resetting, and discounting the role of baroreflexes in long-term control of arterial pressure, is that the rise in sympathetic activity and arterial pressure following complete surgical resection of arterial baroreceptor afferent fibers is only transient, and that little or no hypertension occurs chronically (Cowley, 1992). The absence of sustained increases in arterial pressure following sinoaortic denervation (SAD) has been attributed to central resetting. It must be appreciated, however, that there are striking adaptations that occur in the central baroreflex circuitry after elimination of baroreceptor afferents to their central projections. Indeed, the nucleus tractus solitarius, the initial site of termination of baroreceptor afferents, loses its ability to tonically regulate arterial pressure (Schreihofer et al., 1992). Furthermore, if such stark adaptations were to occur under physiological conditions and, therefore, if central resetting were a dominant process that counteracts changes in baroreceptor input into the brain, then chronic electrical activation of the carotid sinus would not be expected to produce appreciable, sustained reductions in sympathetic activity and arterial pressure. Clearly this is not the case. Because chronic electrical stimulation of the carotid sinus bypasses mechanotransduction at the level of the baroreceptors, it allows for precise control of baroreceptor input into the brain. This permits a quantitative, time-dependent evaluation of the role of central resetting in progressively attenuating the suppression of sympathetic activity and arterial pressure in response to controlled increases in baroreceptor input.

The differential acute and long-term effects of carotid baroreflex activation and SAD on mean arterial pressure (MAP) and heart rate have been carefully documented in a recent experimental study conducted in chronically instrumented dogs (Lohmeier et al., 2010) (Fig. 32.1). In confirmation of previous findings in dogs (Thrasher, 2005), both MAP and heart rate increased substantially after SAD; however, both returned to control levels within 1–2 weeks after baroreceptor deafferentation. A sustained twofold increase in the SD of MAP throughout the 3 week period following SAD confirmed the completeness of arterial baroreceptor denervation. Although indices of sympathetic activity were not

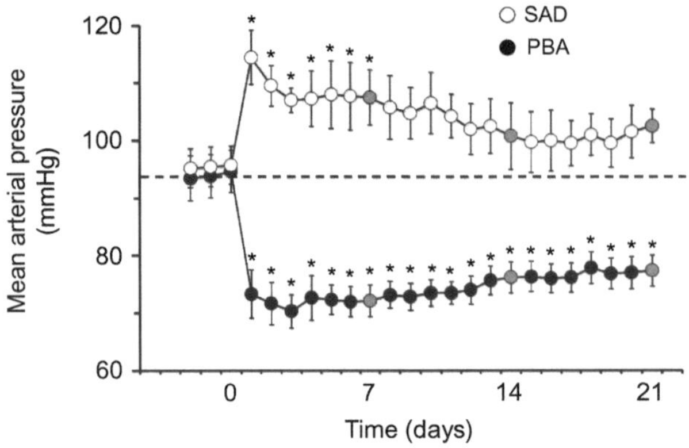

Fig. 32.1. Changes in mean arterial pressure after sinoaortic denervation (SAD; n = 4) and during prolonged baroreflex activation (PBA; n = 6). Values are mean ± SE. *$p < 0.05$ vs. control. Shaded circles indicate end-of-week responses. (Reproduced from Lohmeier et al., 2010.)

measured in this study, others have shown that the initial sympathoexcitation after SAD wanes in parallel with the return of MAP to control levels (Cowley, 1992; Thrasher, 2005). In striking contrast to the transient cardiovascular responses to SAD, substantial reductions in MAP and heart rate persisted during continuous electrical stimulation of the carotid sinus. Furthermore, these reductions in MAP and heart rate were associated with marked and sustained suppression of sympathetic activity, as reflected by substantial reductions in whole body plasma norepinephrine (NE) spillover (an index of centrally generated sympathetic outflow) and plasma NE concentration (Fig. 32.2). After discontinuation of baroreflex activation, both MAP and heart rate increased abruptly toward control levels. Further, when measured 1 week after baroreflex activation, the above measures of sympathetic activity had returned to baseline values. Additional observations indicated that there were no increases in the reactivity of peripheral α_1-adrenergic receptors during chronic baroreflex activation, despite chronic suppression of sympathetic activity. Thus, adaptations within the brain and in the peripheral circulation do not appreciably offset the long-term sympathoinhibitory blood pressure-lowering effects of electrical stimulation of the carotid baroreflex. These experimental findings during 3 weeks of baroreflex activation are consistent with the chronic antihypertensive effects of BAT in patients with resistant hypertension (see below).

INTERACTION OF THE CAROTID BAROREFLEX WITH THE RENIN–ANGIOTENSIN SYSTEM IN LONG-TERM CONTROL OF ARTERIAL PRESSURE

Activation of the renin–angiotensin system counteracts the lowering of arterial pressure in response to some

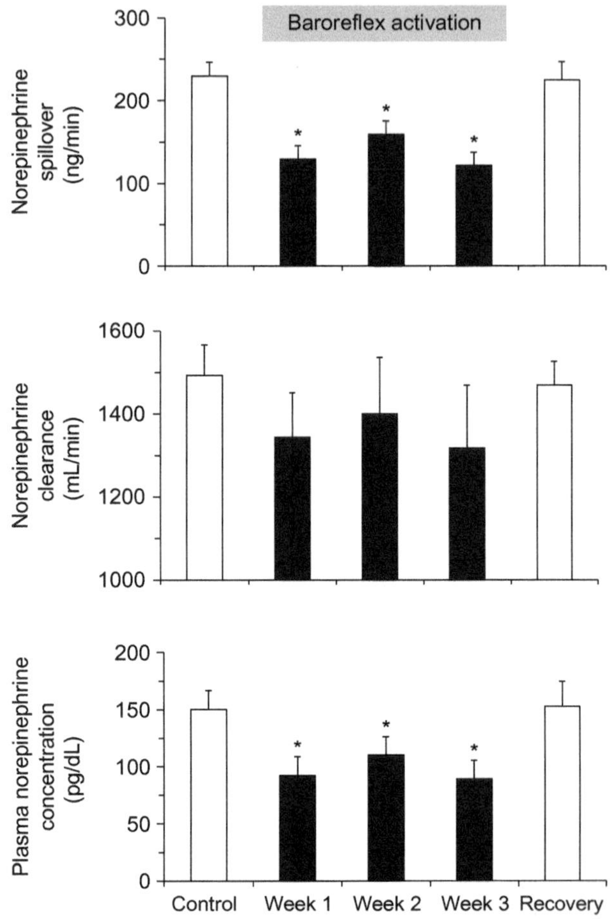

Fig. 32.2. Changes in whole body plasma norepinephrine spillover and clearance, and plasma norepinephrine concentration in response to 3 weeks of baroreflex activation. Values are mean ± SE (n = 6). $p < 0.05$ vs. control. (Reproduced from Lohmeier et al., 2010.)

antihypertensive agents, such as diuretics and calcium channel blockers. In marked contrast, in normotensive dogs in which Rheos stimulation parameters are programmed to achieve 20–25 mmHg reductions in MAP, a particularly striking and important finding is that there is no activation of the renin–angiotensin–aldosterone system (Lohmeier et al., 2004, 2005b, 2007a, b, 2009, 2010). The absence of an increase in plasma renin activity (PRA) in response to pressure reductions of this magnitude is particularly impressive in light of quantitative studies in chronically instrumented dogs demonstrating marked increases in renin secretion when renal perfusion pressure decreases more than 15 mmHg (Kirchheim et al., 1985; Yang et al., 1989). The blunting of the renin secretion in response to reductions in arterial pressure during chronic baroreflex activation suggests the following. First, baroreflex activation must have persistent inhibitory effects on renin secretion. Further, because increases in renal adrenergic activity stimulate renin secretion and suppression of renal adrenergic activity has the opposite effects (DiBona and Kopp, 1997), it is likely that chronic electrical stimulation of the carotid baroreflex has sustained effects to inhibit RSNA. Although this possibility has not yet been tested, it is consistent with the finding that the natural activation of the baroreflex in hypertensive animals, including dogs, has sustained renal sympathoinhibitory effects (Carroll et al., 1984; Lohmeier et al., 2000; Barrett et al., 2005). Second, if this powerful arterial pressure hormonal feedback control mechanism did respond in a normal manner to reduced arterial pressure and the renin response to hypotension was not blunted, it is probable that the ultimate lowering of arterial pressure in response to baroreflex activation would be greatly diminished.

These hypotheses were tested in dogs during equivalent degrees of carotid sinus stimulation before and during continuous infusion of angiotensin II (ANG II) (Lohmeier et al., 2005b). Chronic infusion of ANG II produced three- to fivefold increases in plasma ANG II and aldosterone concentration, and a concomitant increase in MAP of ~35 mmHg. Following induction of ANG II hypertension, acute reductions in MAP in response to baroreflex activation were pronounced and comparable to those achieved before induction of hypertension. However, the acute robust falls in MAP were not sustained, but progressively, markedly attenuated, compared to the responses during control conditions (Fig. 32.3). Thus, this study demonstrates that the long-term antihypertensive effects of baroreflex activation are distinctly diminished in the presence of modest increases in circulating ANG II. Further, this study emphasizes the importance of reflex-suppression of renin secretion in counteracting pressure-dependent renin secretion and permitting a substantial long-term lowering of arterial pressure during activation of the baroreflex. But what accounts for the disparate acute and long-term MAP responses to baroreflex activation in ANG II hypertension? Further, how can chronic baroreflex activation lower arterial pressure if it causes modest sodium retention (Fig. 32.3)? These are commonly raised questions.

INCREASES IN RENAL EXCRETORY FUNCTION ARE THE PRIMARY DETERMINANT OF THE CHRONIC BLOOD PRESSURE-LOWERING RESPONSE TO BAROREFLEX ACTIVATION

A fundamental law of the circulation is that arterial pressure is the product of cardiac output and total peripheral resistance. Because short-term regulation of arterial pressure is critically dependent upon the rapid changes in cardiac output and peripheral resistance evoked by the baroreflex, it is easy to overlook the potential importance of reflexes on renal excretory function and in long-term control of arterial pressure. Long-term regulation of arterial pressure is closely

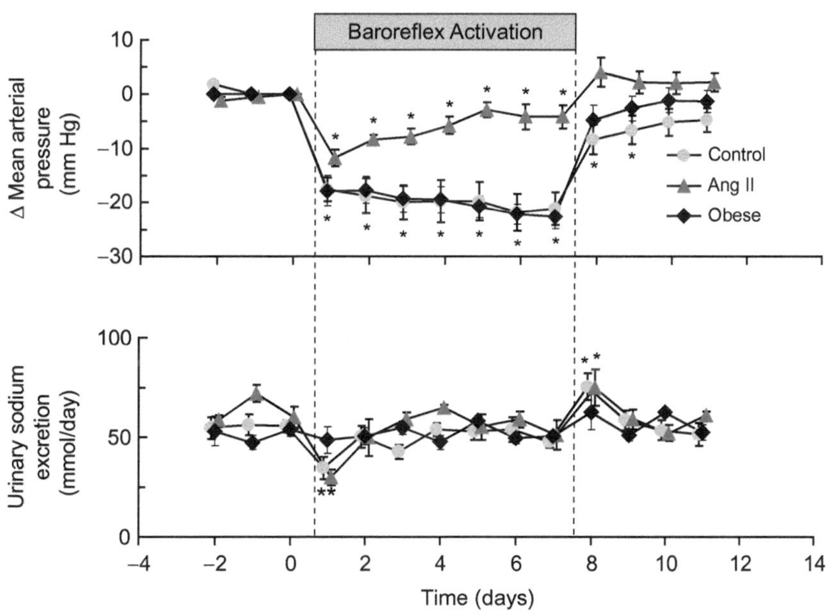

Fig. 32.3. Changes in mean arterial pressure and sodium excretion in response to baroreflex activation in dogs before (control) and after induction of ANG II hypertension. Responses to baroreflex activation in an additional group of dogs with obesity hypertension are also illustrated. Values are mean ± SE (n = 6). $p < 0.05$ vs. control. Values for mean arterial pressure before baroreflex activation: control = 93 ± 1 mmHg, ANG II hypertension = 129 ± 3 mmHg, and obese hypertension = 110 ± 3 mmHg. (Reproduced from Lohmeier et al., 2005b, 2007a.)

linked to the control of body fluid volume, which is determined by the balance between fluid intake and renal excretion (Guyton, 1980). According to the concept of long-term volume and arterial pressure control by the kidneys, for a given level of sodium intake, chronic changes in arterial pressure cannot occur without alterations in renal excretory capacity (the renal set point for pressure natriuresis). On one hand, if renal excretory capacity is not altered, perturbations that tend to raise arterial pressure by increasing cardiac output and peripheral resistance will increase fluid excretion by pressure natriuresis and diuresis. This leads to a progressive decrease in extracellular fluid volume, reducing venous return and cardiac output until arterial pressure returns to control. Only then is fluid balance achieved. Conversely, when blood pressure decreases, the kidneys retain salt and water until arterial pressure is restored to normal. On the other hand, if there is a shift in pressure natriuresis to a new set point, this will lead to long-term changes in arterial pressure. Conditions that impair pressure natriuresis will necessitate higher levels of arterial pressure to achieve fluid balance, whereas lower arterial pressures will occur concomitantly with increases in renal excretory capacity, such as during prolonged baroreflex activation.

The transient sodium retention commonly associated with the onset of baroreflex activation presumably occurs because the initial abrupt fall in arterial pressure, due to attendant changes in peripheral resistance, vascular capacity, and cardiac function, falls below the new renal set point for pressure natriuresis. Although the acute hemodynamic effects of baroreflex activation account for the initial fall in arterial pressure (and sodium excretion), the final level of arterial pressure is dominated by the more sluggish renal mechanism for control of body fluid volume. Because baroreflex activation has pronounced effects to chronically lower arterial pressure without activating the renin–angiotensin system (Fig. 32.3, Control), inhibition of renin secretion likely plays an important role in permitting increased renal excretory capacity. In direct support of this hypothesis is the observation that the final reduction in arterial pressure during baroreflex activation is greatly diminished in the presence of increased circulating plasma levels of ANG II (Fig. 32.3, ANG II). When plasma ANG II levels are elevated, the antinatriuretic effects of ANG II must substantially counteract the neurally induced mechanisms that promote sodium excretion and lower arterial pressure during baroreflex activation. A corollary of this logic is the following. To the extent that suppression of RSNA plays the primary role in increasing renal excretory capacity during chronic activation of the baroreflex, the preexisting level of RSNA may be an important determinant of the long-term arterial pressure response to carotid sinus stimulation. As RSNA is suppressed in ANG II hypertension as a result of the natural engagement of the baroreflex (Carroll et al., 1984; Lohmeier et al., 2000; Barrett et al., 2005), it would be reasonable to expect impaired blood pressure-lowering effects of carotid sinus stimulation due to the limited ability to evoke further reflex reductions in RSNA. If this rationale is correct, considerable renal sympathoinhibition and associated antihypertensive responses to baroreflex activation may occur in forms of hypertension associated with increased RSNA.

ELECTRICAL ACTIVATION OF THE CAROTID BAROREFLEX ABOLISHES OBESITY HYPERTENSION

Many patients with resistant hypertension are obese and obesity-related hypertension has an important neurogenic component (Hall et al., 2001; Esler et al., 2006; Shibao et al., 2007). Further, obesity-hypertension is associated with increased RSNA, and bilateral renal denervation markedly diminishes the severity of the hypertension (Kassab et al., 1995; Vaz et al., 1997; Rumantir et al., 1999). In marked contrast to the ANG II model of hypertension, electrical activation of the baroreflex has pronounced and sustained blood pressure-lowering effects in dogs with obesity hypertension, indicating a sustained increase in renal excretory capacity (Fig. 32.3). In fact, baroreflex activation totally abolishes obesity-induced increments in plasma NE concentration and the attendant sympathetically mediated hypertension (Lohmeier et al., 2007a). An additional benefit of baroreflex activation in the canine model of obesity hypertension is that it markedly diminishes the associated tachycardia. With weight gain, the early progression of the hypertension is also associated with sharp increases in PRA; however, despite sustained sympathetic activation, large increases in PRA do not persist, presumably because the sympathetic drive for renin secretion is offset by the rise in arterial pressure. Furthermore, and of particular relevance to the impressive antihypertensive response, there is no increase in PRA with baroreflex activation, despite the large fall in arterial pressure. Once again, this suggests that reductions in RSNA counteract pressure-dependent renin secretion during baroreflex activation, a response that permits the robust blood pressure-lowering effect of prolonged baroreflex activation. In summary, the collective observations from these studies suggest greater degrees of baroreflex-induced renal sympathoinhibition in obesity than in ANG II hypertension, with preexisting levels of RSNA being high in obesity and low in ANG II hypertension.

SURPRISINGLY, THE RENAL NERVES ARE NOT ESSENTIAL FOR LONG-TERM REDUCTIONS IN ARTERIAL PRESSURE DURING BAROREFLEX ACTIVATION

Based on the above discussion, it is reasonable to suggest that the renal nerves play a critical role in linking increased renal excretory function and lowering of blood pressure to carotid sinus stimulation and attendant reductions in central sympathetic outflow. In an experiment designed to evaluate whether the renal nerves are truly of paramount importance in shifting pressure natriuresis to lower pressures during chronic baroreflex activation, equivalent degrees of carotid sinus stimulation were applied for 7 days in the same dogs before and after bilateral renal denervation (Lohmeier et al., 2007b). By eliminating the influence of the renal nerves on renal excretory function, transient but not sustained reductions in arterial pressure were expected during baroreflex activation. That is, the temporal changes in MAP during baroreflex activation were projected to mimic the pattern that occurred in the ANG II model of hypertension (Fig. 32.3). Contrary to this hypothesis, the findings in this study were astonishing – renal denervation had no impact on the pattern of fall in arterial pressure during prolonged activation of the baroreflex. There was, however, a tendency for greater sodium retention and a smaller fall in arterial pressure after renal denervation when compared to control conditions. Nonetheless, this study clearly demonstrates that the presence of the renal nerves is not an obligate requirement for achieving long-term reductions in arterial pressure during prolonged baroreflex activation (PBA).

An established integrative mathematical model of human physiology (Quantitative Human Physiology, QHP2008) was used to gain insight into this unexpected finding (Iliescu and Lohmeier, 2009). QHP closely reproduced the actual hemodynamic, renal excretory, and neurohormonal responses to baroreflex activation, providing important validation of the mathematical model. The simulations indicated that prolonged activation of the baroreflex produces sustained suppression of RSNA and concomitant increases in renal excretory function. Furthermore, the simulations also showed that non-neural mechanisms also participate in promoting sodium excretion during baroreflex activation, and that the contribution of these mechanisms intensifies at the expense of excessive fluid retention when there is impaired suppression of renal adrenergic activity, such as after renal denervation. The model predicted that after renal denervation, increased secretion of atrial natriuretic peptide and increased renal interstitial fluid pressure are among the most important redundant non-neural natriuretic mechanisms that increase renal excretory function during PBA. While the predictions from these simulations have not been tested experimentally, they do provide a theoretical basis supporting the importance of renal mechanisms in chronically lowering arterial pressure during baroreflex activation.

BLOOD PRESSURE LOWERING BY ANTIHYPERTENSIVE DRUGS IS POTENTIATED BY CHRONIC BAROREFLEX ACTIVATION

Patients with resistant hypertension have inadequate blood pressure control despite taking multiple classes of drugs, including those that can lead to chronic activation of the sympathetic nervous system, such as diuretics and calcium channel blockers (De Champlain et al., 1998; Ligtenberg et al., 1999; Fu et al., 2005). The sustained sympathetic activation associated with the antihypertensive effects of these drugs is consistent with experimental evidence of incomplete resetting during chronic unloading of the baroreceptors (Thrasher, 2005). Thus, sympathetic activation may diminish the antihypertensive efficacy of some drugs and contribute to the intractable hypertension in patients with resistant hypertension. The significance of sympathetic activation in counteracting antihypertensive therapy would be expected to be diminished in those patients taking α_1- and β-adrenergic blocking agents, but these drugs are often not taken concurrently and in the doses necessary to achieve complete blockade of these adrenergic receptors. Because the prohypertensive effects of increased RSNA are mediated by increases in vascular resistance, sodium reabsorption, and renin secretion as a result of stimulating renal α_1- and β-adrenergic receptors (DiBona and Kopp, 1997), complete blockade of both receptors is necessary totally to counteract the neurally induced alterations in renal function that lead to increased arterial pressure. Additionally, by stimulating α_2-adrenergic receptors and suppressing RSNA, centrally acting sympatholytics provide more extensive sympathetic inhibition because they reduce the potential hypertensive effects of cotransmitters released along with NE from adrenergic nerve terminals. However, their use is limited because of the associated adverse side-effects.

A significant experimental observation of relevance to hypertension therapy is that suppression of central sympathetic outflow by baroreflex activation has substantial chronic effects to lower arterial pressure by additional mechanisms other than by decreasing activation of α_1- and β-adrenergic receptors (Lohmeier et al., 2009). In normotensive dogs, chronic administration of prazosin and propranolol decreased arterial pressure ~20 mmHg when administered in doses that completely blocked the normal agonist-induced responses to stimulation of α_1- and β-adrenergic receptors. However, this lowering of arterial pressure during chronic adrenergic blockade was associated with sustained activation of the sympathetic nervous system, as reflected by a threefold increase in plasma NE concentration. Subsequently, and most importantly, during continued adrenergic blockade, arterial pressure decreased an additional 10 mmHg concomitantly with a reduction of plasma NE concentration to control levels during 7 days of baroreflex activation. Acute reductions in arterial pressure in response to a peripherally acting α_2-adrenergic receptor antagonist were also quantified. These responses were greater during the sympathoexcitation associated with blockade of α_1- and β-adrenergic receptors alone than when plasma NE concentration was suppressed to control levels during blockade of α_1- and β-adrenergic receptors and simultaneous baroreflex activation. Taken together, these responses suggest that baroreflex activation may contribute to achieving optimal therapy for the control of hypertension by blocking the compensatory increases in sympathetic activity that naturally counteract the blood pressure-lowering effects of antihypertensive drugs. Further, the mechanism for this additional lowering of blood pressure with BAT may include diminished activation of postsynaptic α_2-adrenergic receptors as well as blockade of those receptors – α_1- and β-adrenergic receptors – believed to be most important in mediating the actions of NE.

Experimental observations indicate that baroreflex activation may also be efficacious in lowering arterial pressure when used in combination with another commonly used class of antihypertensive drugs, calcium channel blockers. Similar to the response discussed above with adrenergic blockade, the chronic blood pressure-lowering effects of amlodipine in normotensive dogs were found to be associated with sustained activation of the sympathetic nervous system (Iliescu et al., 2010), a response not uncommon during hypertension therapy with this and other calcium channel blockers (De Champlain et al., 1998; Ligtenberg et al., 1999). Along with a 15 mmHg fall in MAP during chronic administration of amlodipine, PRA also increased substantially in parallel with activation of the sympathetic nervous system. Moreover, analogous to the response found with adrenergic blockade, there were further substantial reductions in both plasma NE concentration and arterial pressure during continued amlodipine administration and concurrent chronic activation of the baroreflex. PRA also decreased in parallel with baroreflex suppression of central sympathetic outflow, indicating that the additional fall in arterial pressure with baroreflex activation was partially mediated by neurally induced suppression of renin secretion.

Thus, baroreceptor unloading and persistent baroreflex-mediated sympathetic activation during antihypertensive therapy may be a mechanism contributing to resistant hypertension. Furthermore, the ability of carotid sinus stimulation to counteract drug-induced sympathetic activation by suppressing central sympathetic outflow may at least partially account for the lowering of arterial pressure with BAT in patients with resistant hypertension.

BAROREFLEX ACTIVATION THERAPY IN PATIENTS WITH RESISTANT HYPERTENSION: BENEFITS BEYOND BLOOD PRESSURE REDUCTION

The Device Based Therapy in Hypertension Extension Trial (DEBuT-HET)

The DEBuT-HET (clinicaltrials.gov identifier NCT00710294) was a multicenter, prospective, nonrandomized feasibility trial evaluating the safety and efficacy of BAT in 45 patients with resistant hypertension (Scheffers et al., 2010). The trial was conducted in nine European centers. All subjects received full doses of antihypertensive medications, with the median number of drugs being five. No significant changes in medication occurred during the study. Office blood pressure and heart rate at entry were 179/105 mmHg and 80 bpm, respectively. Office blood pressures decreased by 21/13 (n = 37), 30/20 (n = 26), and 33/22 (n = 17) mmHg at 3 months, 1 year, and 2 years of BAT therapy, respectively (Fig. 32.4). Corresponding reductions in heart rate were 8, 8, and 11 bpm. In a subset of patients, ambulatory blood pressures dropped 6/4 mmHg at 3 months, 13/8 mmHg at 1 year, and 24/13 mmHg at 2 years. All patients were stimulated bilaterally and the stimulation parameters were individually titrated

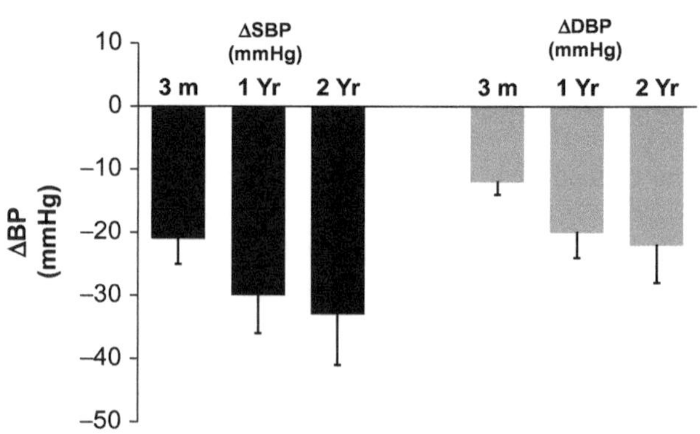

Fig. 32.4. Reductions in systolic blood pressure (SBP) and diastolic blood pressure (DBP) after 3 months (n = 37), 1 year (n = 26), and 2 years (n = 17) of baroreflex activation therapy in patients with resistant hypertension. (Data from Scheffers et al., 2010.)

for therapy. At each follow-up visit, blood pressure returned toward baseline values when the device was temporarily turned off before falling again when stimulation was resumed. Additionally, the procedure and device were well tolerated with favorable measures of safety during 2 years of therapy. Thus, although devoid of a control group, this study suggested that chronic baroreflex activation safely decreases arterial pressure in patients with resistant hypertension beyond what can be achieved with drug treatment.

Studies in subgroups of patients from the DEBuT-HET have provided mechanistic insight into the cardiovascular responses to baroreflex activation in patients with resistant hypertension. By employing microneurography to measure postganglionic sympathetic nerve activity in muscle, Heusser et al. (2010) provided direct evidence that electrical stimulation of the carotid baroreflex suppresses central sympathetic outflow. In this study, acute depressor responses to baroreflex activation were correlated with reductions in muscle sympathetic nerve activity. Another important acute observation in this study was that continuous electrical stimulation of the baroreflex activation did not impair spontaneous baroreflex control of either sympathetic activity or heart rate. Similarly, no impairment of baroreflex heart rate regulation was reported during simultaneous 24 hour recordings of arterial pressure and heart rate in a canine study (Lohmeier et al., 2010). In fact, despite the fact that continuous electrical stimulation of the carotid sinus is not linked to the cardiac cycle, the sensitivity of the spontaneous baroreflex control of heart rate actually increased in this experimental study. In another subgroup of patients from the DEBuT-HET, 24 hour Holter recordings were made before and after 3 months of baroreflex activation therapy (Wustmann et al., 2009). Along with sustained reductions in heart rate and blood pressure, the ECG recordings showed increased heart rate variability during prolonged baroreflex activation, consistent with continuous hemodynamic measurements in dogs, which indicated decreased blood pressure variability as well (Lohmeier et al., 2010). Taken together, these measures of autonomic and circulatory function are consistent with favorable changes in sympathovagal balance during prolonged activation of the baroreflex. Furthermore, from a mechanistic standpoint, the findings from the experimental studies suggest that BAT extends the operating pressure range of the baroreflex by diminishing the degree of resetting that normally occurs at the baroreceptors, thereby allowing not only a lower blood pressure set point during hypertension but also adequate buffering of short-term cardiovascular variability. By decreasing blood pressure variability and associated target organ damage (Parati, 2005) and diminishing the likelihood of adverse cardiac events, these dynamic alterations in baroreflex function may contribute to the clinical benefit of baroreflex activation beyond what is achieved by reduction of blood pressure alone. Finally, in a subgroup of patients with resistant hypertension, serial echocardiograms showed improvement in both cardiac structure and function after 3 and 12 months of BAT when compared to baseline values (Bisognano et al., 2011b). One of the most impressive changes was an 25% reduction in the elevated left ventricular mass index after 12 months of BAT.

The Rheos® Pivotal Trial

The Rheos® Pivotal Trial (clinicaltrials.gov identifier NCT00442286) was the first large-scale, double-blind, randomized placebo-controlled device trial in patients with resistant hypertension. The trial just concluded and the salient findings have been reported (Bisognano et al., 2011a). In this trial, 265 patients were implanted with the Rheos device at 37 U.S. centers and two centers in Europe. One month after implantation, a 2:1 randomization scheme was used – the pacemaker was turned on in 181 patients for the full 12 months, and was off for the first 6 months and then turned on for the final 6 months in the other 84 subjects. The trial had five predetermined coprimary end points, two related to efficacy and three related to safety. The end points for efficacy were based on prespecified reductions in systolic blood pressure at 6 and 12 months. The safety end points reflected procedural events within 30 days of device implantation and adverse responses to chronic activation of the device when evaluated after 6 and 12 months of therapy. The trial showed significant benefit for the end points of efficacy at 12 months and safety during activation of the device. However, it did not meet end points for efficacy at 6 months and procedural safety. The failure to meet the 6 month efficacy end point was related to a number of issues in trial design and to the challenges of clinical trial methodology and the interpretation of responses in patients with resistant hypertension (Bakris et al., 2010; Martin and Victor, 2010). Prespecified performance criteria were rather arbitrary and based on historical literature on implantable cardioverter defibrillators and pacemakers; however, the majority of the procedure-related adverse events completely resolved.

In spite of not meeting two of the five coprimary end points, the criteria for long-term efficacy and safety of the device were met. Furthermore, an additional analysis showed that the percentage of patients attaining a systolic blood pressure < 140 mmHg at 6 months was significantly greater when therapy was on (42%) than when it was off (24%). And at 12 months, more than 50% of the patients achieved a systolic blood pressure of <140 mmHg. It is also noteworthy that reductions in systolic blood pressure at 12 months were comparable to those reported in the DEBuT-HET discussed above (Fig. 32.4). Finally, although subjects had bilateral implants, the majority were programmed for unilateral activation. This suggests that the complexity of the implant procedure could be reduced by performing unilateral implants, thereby reducing the likelihood of procedure-related safety issues. Future clinical trials will address and likely resolve the limitations of this study. New technology which includes a less invasive implant procedure has been developed by CVRx® and is currently undergoing study in Europe.

CONCLUSIONS

The experimental studies reviewed above have provided novel insight into the fundamental mechanisms that account for the chronic lowering of blood pressure during sustained suppression of central sympathetic outflow and, more specifically, during prolonged activation of the baroreflex. Some of the mechanisms identified have been predicable; others have been surprising and will require further investigation. The mechanistic insight from these studies is particularly relevant to the current fascination with device-based therapy for the treatment of resistant hypertension because it helps to identify subsets of this heterogeneous hypertensive population who stand to benefit the most. Because obesity, sleep apnea, renal disease, and the use of some classes of antihypertensive drugs are commonly associated with both increased sympathetic activity and resistant hypertension (Chobanian et al., 2003; Kaplan, 2005; Calhoun et al., 2008), it has not been surprising that BAT has been particularly efficacious in lowering blood pressure in some patients with this intractable form of hypertension, given the ability of continuous electrical stimulation of the carotid sinus to chronically suppress sympathetic activity. However, considering that renal disease is not an uncommon feature of resistant hypertension and the fact that the kidneys play a critical role in long-term control of arterial pressure, it will be especially important in future studies to elucidate the impact of renal dysfunction on the subsequent cardiovascular, and renal hemodynamic and excretory responses to global suppression of sympathetic activity by BAT. Despite the encouraging findings from recent clinical trials, more basic research and additional clinical trials are needed to better define the benefit of baroreflex activation therapy in resistant hypertension and in other states of sympathetic activation such as heart failure and advanced renal disease.

REFERENCES

Bakris GL, Lindholm LH, Black HR et al. (2010). Divergent results using clinic and ambulatory blood pressures. Report of a Darusentan-resistant hypertension trial. Hypertension 56: 824–830.

Barrett CJ, Guild S-J, Ramchandra R et al. (2005). Baroreceptor denervation prevents sympathoinhibition during angiotensin II-induced hypertension. Hypertension 46: 1–5.

Bisognano JD, Bakris G, Nadim MK et al. (2011a). Baroreflex activation therapy lowers blood pressure in patients with resistant hypertension: results from the double-blind, randomized, placebo-controlled Rheos Pivotal trial. J Am Coll Cardiol 58: 765–773.

Bisognano JD, Kaufman CL, Bach DS (2011b). Improved cardiac structure and function with treatment using an implantable device in resistant hypertension: results from European and United States trials of the Rheos system. J Am Coll Cardiol 57: 1787–1788.

Calhoun DA, Jones D, Textor S et al. (2008). Resistant hypertension: diagnosis, evaluation, and treatment. A scientific statement from the American Heart Association Professional Education Committee of the Council for High Blood Pressure Research. Hypertension 117: e510–e526.

Carroll RG, Lohmeier TE, Brown AJ (1984). Chronic angiotensin II infusion decreases renal norepinephrine overflow in the conscious dog. Hypertension 6: 675–681.

Chapleau MW, Hajduczok G, Abboud FM (1991). In: IH Zucker, JP Gilmore (Eds.), Reflex control of the circulation. CRC Press, Boca Raton, FL, pp. 165–194.

Chobanian AV, Bakris GL, Black HR et al. (2003). Seventh report of the National Committee on Prevention, Detection, Evaluation, and Treatment of High Blood Pressure. Hypertension 42: 1206–1252.

Cowley AW Jr (1992). Long-term control of arterial pressure. Physiol Rev 72: 231–300.

De Champlain J, Karas M, Nguyen P et al. (1998). Different effects of nifedipine and amlodipine on circulating catecholamine levels in essential hypertensive patients. J Hypertens 16: 1357–1369.

De la Sierra A, Segura J, Banegas JR (2011). Clinical features of 8295 patients with resistant hypertension classified on the basis of ambulatory blood pressure monitoring. Hypertension 57: 898–902.

DiBona GF, Kopp UC (1997). Neural control of renal function. Physiol Rev 77: 75–197.

Esler M (2000). The sympathetic system and hypertension. Am J Hypertens 13: 99S–105S.

Esler M, Straznicky N, Eikelis N et al. (2006). Mechanisms of sympathetic activation in obesity-related hypertension. Hypertension 48: 1–10.

Fu Q, Zhang R, Witkowski S et al. (2005). Persistent sympathetic activation during chronic antihypertensive therapy. A potential mechanism for long-term morbidity? Hypertension 45: 513–521.

Gonzalez ER, Krieger AJ, Sapru HN (1983). Central resetting of baroreflex in spontaneously hypertensive rat. Hypertension 5: 346–352.

Grassi G, Cattaneo B, Seravalle G et al. (1998). Baroreflex control of sympathetic nerve activity in essential and secondary hypertension. Hypertension 31: 68–72.

Grassi G, Seravalle G, Dell'Oro R et al. (2000). Adrenergic and reflex abnormalities in obesity-related hypertension. Hypertension 36: 538–542.

Guyton AC (1980). Arterial Pressure and Hypertension. Saunders, Philadelphia, PA.

Hall JE, Hildebrandt DA, Kuo J (2001). Obesity hypertension: role of leptin and sympathetic nervous system. Am J Hypertens 14: 103s–115s.

Heusser K, Tank J, Engeli S et al. (2010). Carotid baroreceptor stimulation, sympathetic activity, baroreflex function, and blood pressure in hypertensive patients. Hypertension 55: 619–626.

Iliescu R, Lohmeier TE (2009). Lowering of blood pressure during chronic suppression of central sympathetic outflow: insight from computer simulations. Clin Exp Pharmacol Physiol 37: e24–e33.

Iliescu R, Dwyer TM, Irwin ED et al. (2010). Electrical activation of carotid baroreflex enhances the chronic blood pressure lowering effects of calcium channel blockade by abolishing sympathoexcitation (abstract). Hypertension 56: e71–e72.

Kaplan NM (2005). Resistant hypertension. J Hypertens 23: 1441–1444.

Kassab S, Kato T, Wilkins C et al. (1995). Renal denervation attenuates the sodium retention and hypertension associated with obesity. Hypertension 25: 893–897.

Kirchheim HR, Finke R, Hackenthal E et al. (1985). Baroreflex sympathetic activation increases threshold pressure for pressure-dependent renin release. Pflugers Arch 405: 127–135.

Ligtenberg G, Blankestijn PJ, Oey PL et al. (1999). Reduction of sympathetic hyperactivity by enalapril in patients with chronic renal failure. N Engl J Med 340: 1321–1328.

Lohmeier TE, Drummond HA (2007). The baroreflex in the pathogenesis of hypertension. In: GYH Lip, JE Hall (Eds.), Comprehensive Hypertension. Elsevier, Philadelphia, PA, pp. 265–279.

Lohmeier TE, Lohmeier JR, Haque A et al. (2000). Baroreflexes prevent neurally-induced sodium retention in angiotensin hypertension. Am J Physiol Regul Integr Comp Physiol 279: R1437–R1448.

Lohmeier TE, Irwin ED, Rossing MA et al. (2004). Prolonged activation of the baroreflex produces sustained hypotension. Hypertension 43: 306–311.

Lohmeier TE, Barrett AM, Irwin ED (2005a). Prolonged activation of the baroreflex: a viable approach for the treatment of resistant hypertension? Curr Hypertens Rep 7: 193–198.

Lohmeier TE, Dwyer TM, Hildebrandt DA et al. (2005b). Influence of prolonged baroreflex activation on arterial pressure in angiotensin hypertension. Hypertension 46: 1194–1200.

Lohmeier TE, Hildebrandt DA, Warren S et al. (2005c). Recent insights into the interactions between the baroreflex and the kidneys in hypertension. Am J Physiol Regul Integr Comp Physiol 288: R828–R836.

Lohmeier TE, Dwyer TM, Irwin ED et al. (2007a). Prolonged activation of the baroreflex abolishes obesity-induced hypertension. Hypertension 49: 1307–1314.

Lohmeier TE, Hildebrandt DA, Dwyer TM et al. (2007b). Renal denervation does not abolish sustained baroreflex-mediated reductions in arterial pressure. Hypertension 49: 373–379.

Lohmeier TE, Hildebrandt DA, Dwyer et al. (2009). Prolonged activation of the baroreflex decreases arterial pressure even during chronic adrenergic blockade. Hypertension 53: 833–838.

Lohmeier TE, Iliescu R, Dwyer TM et al. (2010). Sustained suppression of sympathetic activity and arterial pressure during chronic activation of the carotid baroreflex. Am J Physiol Heart Circ Physiol 299: H402–H409.

Malpas SC (2010). Sympathetic nervous system overactivity and its role in the development of cardiovascular diseases. Physiol Rev 90: 513–557.

Martin WA, Victor RG (2010). Premise, promise, and potential limitations of invasive devices to treat hypertension. Curr Cardiol Rep 13: 86–92.

McCubbin JW, Green JH, Page IH et al. (1956). Baroreceptor function in chronic renal hypertension. Circ Res 4: 205–210.

Parati G (2005). Blood pressure variability: its measurement and significance. J Hypertens 23: s19–s25.

Persell SD (2011). Prevalence of resistant hypertension in the United States, 2003–2008. Hypertension 57: 1076–1080.

Rumantir MS, Vaz M, Jennings GL et al. (1999). Neural mechanisms in human obesity-related hypertension. J Hypertens 17: 1125–1133.

Scheffers IJM, Kroon AA, Schmidli et al. (2010). Novel baroreflex activation therapy in resistant hypertension; results of a European multi-center feasibility study. J Am Coll Cardiol 56: 1254–1258.

Schreihofer AM, Ito S, Sved AF (1992). Nucleus tractus solitarius and control of blood pressure in chronic sinoaortic denervated rats. Am J Physiol Regul Integr Comp Physiol 262: R258–R266.

Schwartz SI, Griffith LSC, Neistadt A et al. (1967). Chronic carotid sinus nerve stimulation in the treatment of essential hypertension. Am J Surg 114: 5–15.

Shibao C, Gamboa A, Diedrich A et al. (2007). Autonomic contribution to blood pressure and metabolism in obesity. Hypertension 49: 17–33.

Thrasher TN (2005). Effects of chronic baroreceptor unloading on blood pressure in the dog. Am J Physiol Regul Integr Comp Physiol 288: R863–R871.

Tuckman J, Reich T, Lyon AF et al. (1968). Electrical stimulation of the sinus nerves in hypertensive patients. Clinical evaluation and physiological studies. In: JE Wood (Ed.), Neural Control of Arterial Pressure. Hypertension. Vol. 16. American Heart Association, New York, NY, pp. 23–26.

Vaz M, Jennings G, Turner A et al. (1997). Regional sympathetic nervous activity and oxygen consumption in obese normotensive human subjects. Circulation 96: 3423–3429.

Wustmann K, Kucera JP, Scheffers I et al. (2009). Effects of chronic baroreceptor stimulation on the autonomic cardiovascular regulation in patients with drug-resistant arterial hypertension. Hypertension 54: 530–536.

Yang HM, Lohmeier TE, Kivlighn SD et al. (1989). Sustained increases in plasma epinephrine concentration do not modulate renin release. Am J Physiol Endocrinol Metab 257: E57–E64.

Yin M, Sved AF (1996). Role of γ-aminobutyric acid-B receptors in baroreceptor reflexes in hypertensive rats. Hypertension 27: 1291–1298.

Zhang J, Mifflin SW (2000). Integration of aortic nerve inputs in hypertensive rats. Hypertension 35: 430–436.

Index

NB: Page numbers in *italics* refer to figures and tables.

B

C

I

J

K

L

Q

R

S

T

FSC
www.fsc.org
MIX
Papier | Fördert gute Waldnutzung
FSC® C083411

Kolibri 360 GmbH
Ferdinand-Jühlke-Straße 7
99095 Erfurt, Deutschland
produktsicherheit@kolibri360.de